OXFORD LOGIC GUIDES: 28

OXFORD LOGIC GUIDES

1. Jane Bridge: *Beginning model theory: the completeness theorem and some consequences*
2. Michael Dummett: *Elements of intuitionism*
3. A.S. Troelstra: *Choise sequences: a chapter of intuitionistic mathematics*
4. J.L. Bell: *Boolean-valued models and independence proofs in set theory* (1st edition)
5. Krister Segerberg: *Classical propositional operators: and exercise in the foundations of logic*
6. G.C. Smith: *The Boole-De Morgan correspondence 1842–1864*
7. Alec Fisher: *Formal number theory and computability: a work book*
8. Anand Pillay: *An introduction to stability theory*
9. H.E. Rose: *Subrecursion: functions and hierachies*
10. Michael Hallett: *Cantorian set theory and limitation of size*
11. R. Mansfield and G. Weitkamp: *Recursive aspects of descriptive set theory*
12. J.L. Bell: *Boolean-valued models and independence proofs in set theory* (2nd edition)
13. Melvin Fitting: *Computability theory: semantics and logic programming*
14. J.L. Bell: *Toposes and local set theories: an introduction*
15. Richard Kaye: *Models of Peano arithmetic*
16. J. Chapman and F. Rowbottom: *Relative category theory and geometric morphisms*
17. S. Shapiro: *Foundations without foundationalism*
18. J.P. Cleave: *A study of logics*
19. R.M. Smullyan: *Gödel's incompleteness theorems*
20. T.E. Forster: *Set theory with a universal set*
21. C. McLarty: *Elementary categories, elementary toposes*
22. R.M. Smullyan: *Recursion theory for metamathematics*
23. Peter Clote and Jan Krajíček: *Arithmetic, proof theory, and computational complexity*
24. Tarski: *Introduction to logic and to the methodology of deductive sciences*
25. Malinowski: *Many valued logics*
26. Alexandre Borovik and Ali Nesin: *Groups of finite morley rank*
27. Smullyan: *Diagonalization and self-reference*
28. Dov M. Gabbay, Ian Hodkinson, and Mark Reynolds: *Temporal logic: mathematical foundations and computational aspects* (Volume 1)

Temporal Logic

Mathematical Foundations and Computational Aspects

Volume 1

DOV M. GABBAY
IAN HODKINSON

and

MARK REYNOLDS

Department of Computing, Imperial College of Science, Technology and Medicine, London

CLARENDON PRESS • OXFORD
1994

Oxford University Press, Walton Street, Oxford OX2 6DP
Oxford New York Toronto
Delhi Bombay Calcutta Madras Karachi
Kuala Lumpur Singapore Hong Kong Tokyo
Nairobi Dar es Salaam Cape Town
Melbourne Auckland Madrid
and associated companies in
Berlin Ibadan

Oxford is a trade mark of Oxford University Press

Published in the United States
by Oxford University Press Inc., New York

A catalogue record for this book is available from the British Library

Library of Congress Cataloging in Publication Data
(Data available)

ISBN 0 19 853769 7

Typeset by the authors using TEX

Printed in Great Britain by
Bookcraft (Bath) Ltd
Midsomer Norton, Bath

Preface

We are happy to present the reader with volume 1 of our book on temporal logic. Work on this volume was begun in 1979 by Dov Gabbay,[1] who prepared a draft manuscript [Gabbay, 1981c] which covered earlier versions of chapters 1–3, 6–11, and 13. The 1981 manuscript was intended as a research monograph covering Gabbay's results on expressive completeness, H-dimension, the Irreflexivity Rule and separation theory. In 1983 Gabbay moved to the computing department at Imperial College and his research activities were enriched by the multitude of theoretical and practical problems confronting the computing community. Temporal logic research took a more practical turn (see volume 2). In 1985 Mark Reynolds,[2] and later Ian Hodkinson,[3] joined the temporal logic group. A serious theoretical research programme (with PhD and MSc students) into the mathematical foundations of temporal logic and its applications to practical problems was launched. Following the new research, further chapters were written by Gabbay, Hodkinson, Reynolds and Marcelo Finger. The old chapters, namely, chapters 1–3, 6–11, and 13, were edited, corrected, and expanded by Hodkinson and Reynolds. The following is a guide to the work done in 1983–92.

Chapters 1 and 2 are essentially the original versions by Gabbay with some editing and additions by Hodkinson and Reynolds.

Chapter 3 is based on an original version by Gabbay but has been thoroughly revised by Hodkinson.

Sections 4.1 and 4.2 are by Gabbay, the rest of chapter 4 is by Reynolds.

Chapter 5 is by Gabbay and also appears as [Gabbay, 1991a].

Chapter 6 is mostly by Gabbay. Section 6.9 is based on a joint paper by Hodkinson and Gabbay (see [Gabbay and Hodkinson, 1990]). Sections 6.10 and 6.11 were written by Hodkinson. The rest of chapter 6 was edited by Hodkinson.

Chapter 7 is based on the original in [Gabbay, 1981c] but has been corrected by Hodkinson and Reynolds.

Sections 8.1–8.3 are by Gabbay and Reynolds. The rest of chapter 8 is by Hodkinson and will appear as [Hodkinson, 1992a].

[1]Dov Gabbay is currently a senior research fellow supported by the UK Science and Engineering Research Council (B/91/SF/382).

[2]Mark Reynolds' work was partially supported by the UK Science and Engineering Research Council under the Metatem project (GR/F/28526) and the Syntactical Foundations of Non-monotonic Reasoning Project (GR/G/46671).

[3]Ian Hodkinson's work was partially supported by the UK Science and Engineering Research Council's Pure Logic Language project and Advanced Fellowship B/ITF/266.

Chapters 9 and 10 were originally by Gabbay, but have been edited and corrected by Reynolds.

Chapter 11 was so thoroughly revised that it can now be considered a joint paper between Gabbay and Reynolds.

Chapter 12 appears in [Gabbay *et al.*, 1993]. Section 12.1 is by Reynolds, section 12.2 is by Hodkinson, sections 12.3–12.5 are by Reynolds, and the rest of chapter 12 is by Hodkinson with stated obligation to [Gabbay *et al.*, 1980].

Sections 13.1–13.3 were edited by Hodkinson from Gabbay's manuscript. Section 13.4 is by Hodkinson. Section 13.5 is to be considered a joint paper between Gabbay and Hodkinson. Sections 13.6–13.7 are by Hodkinson.

Chapter 14 is a joint paper [Finger and Gabbay, 1991] by Gabbay with Marcelo Finger.

Chapter 15 is a survey by Hodkinson.

Parts of [Gabbay *et al.*, 1993] and [Gabbay, 1989a] appear by permission of Springer-Verlag and parts of [Gabbay, 1981a], [Gabbay, 1981b], [Hodkinson, 1992b], and [Finger and Gabbay, 1991] appear by permission of Kluwer Academic Publishers.

We would like to thank Rob Hubbard for help compiling the references, Hilfred Chau for help with the indexing, Hans Jürgen Ohlbach for indispensible help with the diagrams, Antoinette Ratnam for typing some earlier versions of chapters and Jane Spurr for help with the LaTeXproduction.

Thanks are also extended to Howard Barringer and our colleagues in Manchester, Marcelo Finger, Robin Hirsch, Ildiko Sain, Andras Simon, Edith Spaan, Andrzej Szałas, Yde Venema, and Alberto Zanardo.

Our research policy, and this is our advice to our readers, is to concentrate on problems and notions of general mathematical value, possibly abstracted from application areas. We have tried not to be drawn to transient short term local solutions, which although they may be of great temporary use, are out of date and out of interest in a very short time. We have concentrated on results and methodologies which we believe will be of continued interest both for logical theory and practical applications. Adopting this point of view our intention in this book is to give a systematic, thematic coverage of the mathematical foundations and computational aspects of temporal logic. We stopped to present volume 1 simply in order not to delay publication, and more research on the topics covered in volume 1 as well as on the computational topics will continue in volume 2.

London
December 1993

DMG
IMH
MAR

Contents

Tentative contents of volume 2

Algorithmic proof for temporal logic

- Basic ideas of theorem proving: natural deduction, tableaux, resolution
- Labelled modal system
- LDS approach

Temporal databases

- Introduction
- Temporal data representations
- Temporal query languages
- Historical versus rollback databases
- Two-dimensional databases

Temporal active databases

- Introduction
- Semantics
- Changing the past
- Effects of changing the past
- Algorithms

Intervals and planning

- Introduction
- Evaluation in intervals and points
- A simple interval logic
- Completeness questions
- Recognising interval logics
- Planning problem
- Tractability of relation algebras

Temporal logic and automata

- Introduction
- Automata

Branching time

- CTL
- Axiomatizations
- Implementability of reactive modules
- Probabilistic approach

Clocks

- Names
- Definability
- IRR and clock rules
- Axiomatizability
- Generalising Sahlqvist

Chapter 1

The handling of time: introduction and survey

1.1 Introduction

This book is about the handling of time, both in pure logic and in computing. It attempts to present a unified point of view, where computational and mathematical properties of the logic of time are interwoven in an intrinsic way.

Time is a very exciting subject. It is one of the few subjects on which everyone is an expert. We move through time continuously, and in order to survive and manage ourselves sensibly we constantly have to make temporal decisions. Philosophy, since the days of Aristotle, has been trying to analyse the way in which we make these decisions. With the rise of computer science, where, ideally, one wants the machine to do the job for, or of, the human, there is a new urgency in the precise logical analysis of human temporal activity.

Human (and hence computer) time-related activity can be divided into several main areas, all very familiar to us. One of the simplest, and the most important area, is our handling of time-dependent data. In computing this is the area of databases. To us ordinary people, it is just time-dependent information, involving such questions as when to go to the dentist, when to pick up the child from school, how long one can delay in filling in one's tax return, and so on. There is another temporal dimension involved in the area of time-dependent data besides direct dependency on time. This is the dimension of when a data item is presented to us. For example, if we get a bill to pay our tax on 1 January 1990, it is important when the bill was sent or received, e.g. received September 1989 for January 1990. In database terms two times are involved: the time dependency of the data and the time when it was introduced into the database. Surprisingly, computing is only now beginning to cope with such things.

Another important area in both human activity and computing is planning. If I have to do the shopping, take my child to visit a friend, and cook supper, I must organize the sequence properly.

The above is a simple planning problem, but to organize an airport is a more complex planning problem. To be able to let the computer solve it for us we need to develop a logical theory and analyse the steps involved correctly.

Everybody knows the term 'time sharing'; what it means in practice is that if neither of us can afford something (e.g. a car or a flat in Spain) then we buy it together and time share. (Computers are more humble; they share such things as memory or a printer in order to be more efficient.) We can formulate some intuitive principles on how to share (in computing this is called the *specification*) but there is always the question of exactly how we are going to manage it (on what dates am I going to be in the flat and on what dates are you? who is going to do the garden and collect the garbage?, etc.). This is the *implementation* of the principles. Given such an implementation, we have the problem of how to show that it satisfies the specification. One way of doing this is to formulate the procedures in 'temporal' logic and then formally prove that it satisfies the specification. In computing, the official name is program *specification and verification.*

We hope that by now you, the reader, are convinced that you are an expert on time handling. Please apply your intuition in judging whether the views that follow are reasonable or not.

1.1.1 Short survey of uses

Before we proceed with the next section, let us describe the main areas of temporal activity in computer science.

Since time is involved in every aspect of human activity, and as computer science is being applied in wider and wider areas of our lives, it has become essential to develop logical systems which can describe time-dependent phenomena.

There are at least eight main areas of computer science where time is involved in an essential way and where temporal and modal logics are used. These are as follows:

1. Databases: their management, their updating, their use of knowledge representation involving time, and logical deduction involving time-dependent data.

2. Program specification, development, and verification for concurrent programs and processes. Control of program execution.

3. Hardware: VLSI, reasoning about circuits.

4. Natural language processing.

5. Distributed systems. Protocols for sharing knowledge.

6. Temporal logic as an imperative/declarative programming language.

7. Planning.

8. Real-time systems.

Let us describe what is involved in each of these topics in a little more detail. Of course we will be able to say more once we are more familiar with systems of temporal and modal logic.

Databases

Time is involved in four ways:

1. Temporal logic is needed to describe a theory of updating and maintaining databases. This requires an ability to talk about the logic of the updating and the evolution of databases in time. For example, we may have time-dependent updating rules or some global constraints involving time.

2. Putting information which is time-dependent in the databases requires the choice of a good temporal language. How to represent time-dependent knowledge, how to access it, and how to reason with it.

3. The real world is a combination of the two types of dependency given above. We need to develop a logic of actions in time. These notions are of key importance to new programming languages.

4. Time is involved in artificial intelligence via its connection with non-monotonic reasoning and deductive databases. We believe that there can be no satisfactory theory of non-monotonic reasoning without addressing the problems of temporally evolving knowledge bases.

Program specification and verification

(a) We need to develop a temporal logic which can describe the properties of programs and the way in which they change states with execution, and we need automated temporal reasoning capabilities to prove the properties of these programs.

(b) We can use temporal logic to control the execution of programs.

(c) In concurrent programs, the knowledge of the input/output relations of each participant does not yield full knowledge of the input/output relation of the parallel execution. We need a logic that we can use to prove the correct behaviour of the program from the initial description.

VLSI

One needs asynchronous timing. We need a temporal logic specifying the behaviour of pieces of hardware and possibly we need to prove theorems on how to put systems together and maintain certain properties.

Natural language processing

Time is involved in an indirect way. We need to analyse the uses of the tense and aspect of the language in order to prepare it (interface) for the computer. This is aside from the theoretical value of studying the logical structure of time use in language. It is a vast, complex, and challenging area.

The need for temporal logic in distributed systems

One describes the partial knowledge that different agents have as dependent on past, present, and future states of the world. One describes protocols for communication between stations and temporal conditions for turning stations on and off.

Executing temporal logic

Temporal logic can also be used as a programming language. The basic paradigm is *review the past* **and then** *take action in the future.* The reviewing of the past is temporal database querying, and the action in the future is in effect a logical imperative language. Several languages of this type have been proposed and implemented.

Planning

This is an important area of artificial intelligence which involves all aspects of temporal reasoning. Abstractly we have an initial state, and certain actions that can be performed in a given state if it satisfies certain conditions. Performing an action with respect to a state brings about a new state. The goal is to bring about a state satisfying the 'goal' conditions, and the task is to find a sequence of actions that achieves this end.

Planners should derive the actions required to reach a goal state. But we can treat planning as a form of reasoning about actions and plans in which time and modality play essential rôles. We can conduct this reasoning in an executable temporal logic, as in example 1.1.7. The temporal dimension concerns the order of actions in a plan and the duration of the resulting states. The modal dimension concerns what actions are possible in a state and how they can be ordered to achieve the goal state. We can bring 'possibility' and other modal notions into the object language where they can play a rôle in reasoning. We can also accommodate concurrent actions and extraneous events not initiated by the planner.

The so-called frame problem is typical of the difficulties encountered in this area. Suppose that we have a database Δ, which from an initial state Δ_0 undergoes a series of discrete changes leading to Δ_n. Δ satisfies some integrity constraints, and so these changes probably affect other parts of Δ. One instance of the frame problem is to find a systematic way of checking, for a given entry in Δ_0, whether it is affected in Δ_n.

Real-time systems

Doing things in real time is not always just a matter of speed. The problem is logical. We survey the application area and our temporal handling of it and propose a logical course of action. It may take an hour to do all that is necessary. On a faster computer it may take only one second. That does not make our solution real time. What we must do is organize and layer out the solution logically into several layers of priorities. We perform the full operation if there is enough time, otherwise there are several options of approximations. We must allow for interrupts, milestones, checking points, etc. The problem is therefore logical.

We shall give more detail about these areas later in this chapter. Let us now turn to temporal logic.

Summary

The main present-day research areas of temporal logic are:

1. Philosophical applications. Temporal logic is used in philosophy to clarify various concepts which have been studied since the time of Aristotle. Some examples are causality, historical necessity, identity through time, the notions of events and actions, etc.

2. Applications in computer science as described above.

3. Natural language. Logical analysis of the use of tense and aspect in natural languages. Logical time models for natural language.

4. Pure logical study of temporal logic within the framework of logic itself. Special topics here include:

   (a) axiom systems, theorem proving and proof theory; decidability; model theory;

   (b) expressive power of temporal languages;

   (c) applications of temporal logic to the pure theory of other logics (e.g. the notion of provability as a modal logic, etc.);

   (d) deductive reasoning involving time.

   All the above four aspects of pure logical theory are of great importance to computer science.

Temporal logics can be presented in several different ways:

1. Use predicate logic with an additional parameter for time.

2. Use special temporal logics to express temporal phenomena. There are two methods of presentation here:

   (a) semantic presentation;

   (b) presentation using axiomatic or other deductive systems for the connectives.

3. The final method is via direct reference to events.

1.1.2 The declarative view

Our purpose is to present our perception of the place of temporal logic within the framework of logic in general, and that of logic and computation in particular. Temporal logic has changed and developed incredibly since its conception as a discipline by Arthur Prior 30 years ago. It is studied by many researchers of numerous and different backgrounds. Different research groups have different conceptions of what temporal logic is and of what exactly they do themselves. On many occasions we have heard such comments as 'that is not logic' referring to a system presented by a colleague. The subject is certainly in a state of accelerated dynamic growth, and a new orientation and point of view is currently needed, as well as a good coverage of its mathematical and computational aspects. A good understanding will enable the subject and the community of researchers to face the challenges of the future.

Our purpose in this chapter is to present, step by step, a comprehensive mathematical and computational point of view of temporal logic and to indicate the scope and place of the remaining chapters of this book.

The initial step is to clarify the basic concepts involved in the reasoning about time and change through time. The first issue is how to represent our perception of time and change. Is the primary concept the flow of time, and is change a secondary concept measured against it, or is the primary concept that of change, and is time perceived to flow only because of change?

In the philosophical literature there are supporters for both views. One can also adopt a mixed view, where both time and change appear as primary, with postulates connecting them. For the purpose of this introduction, we will only discuss and illustrate the issues involved briefly.

If we adopt the concept of time as the primary one, we must decide whether we are dealing with points or intervals. The choice means whether the basic propositions are evaluated at time points (written $\|A\|_t = \textit{truth value}$, and meaning '$A$ is true at time t') or at intervals (written $\|A\|_i = \textit{truth value}$, and reading '$A$ is true at an interval i'). The point approach may be one or many dimensional, depending on whether we use parameters or not, that is, whether we evaluate a formula at a single point (i.e. $\|A\|_t = \textit{truth value}$) or at several points (i.e. $\|A\|_{t,t_1,\ldots,t_n} = \textit{truth value}$ where the points $t_1, \ldots, t_n$ are parameters). In many cases, an interval system can be reduced to a point system by taking certain auxiliary points as parameters. For example, we may be able to reduce the interval evaluation $\|A\|_i$ to a point evaluation $\|A\|_{i^+,i^-}$, where i^+, i^- are the interval ends. Dealing with pure intervals would require the introduction of several relations on intervals, such as earlier, later, subset, meet, next, intersect, and disjoint.

The temporal systems involving time points or intervals can be presented either semantically or axiomatically. Change in events can be defined in terms of time in the most intuitive way. Propositions have truth values over time and when these values hold for a while, we can perceive ourselves to be in a state. When they change, then there is a change of state. Events can be taken as changes in state.

The other approach is to take events and change as primary concepts; we may or may not define a flow of time using change and events. One can define moments of time as equivalence classes of states in some cases, namely, if the system remains in the same state for a while then the whole interval can be taken as one point. The other way is to take maximal families of nested intervals (states) and regard them as points, which can be used to redefine the states. This is more like Dedekind completion. Examples of this approach occur naturally in databases, which may stay in the same state until they are updated into a new state. Relevant updating actions depend on states. If the database shows, for example, that stocks are low, it can issue an order for new stock. This action depends on the state of the data and not on the time. To compare, if we make actions time dependent (e.g. pay cheque every first of the month), then we need a mixed approach, time and state.

Typical application areas where the flow of time is the primary concept are control problems or protocols for communication or the use of resources. In these examples 'time' may not necessarily be real time but only a sequence of 'snapshots' of the system with which we are dealing.

The above discussion described a few of the conceptual problems that are associated with time. Once a position is taken on that issue, we are faced with the problem of formal time representation and reasoning. We must adopt a syntactical and logical approach to time. This approach involves constructing representations and logical systems for handling time. One needs to decide how to represent time and events. The main purpose of this book is a syntactical study of temporal logic where a flow of time is the primary semantic concept. The approach taken in this book is to start at points and move on to events. We present logical systems declaratively representing temporal behaviour. To achieve this we need to put forward our view of what a logical system is. This view is developed and motivated in the first author's book on labelled deductive systems.

We assume that we are working in the context of a language $\mathbf{L}$ which is just a set of *well-formed formulae* or *wffs*. A logical system to us is a pair, $(\vdash, \mathbf{S}_\vdash)$, where $\vdash$ is a consequence relation and $\mathbf{S}_\vdash$ is an algorithmic system (recursively enumerable[1]) for generating the pairs (Δ, Q) such that $\Delta \vdash Q$ holds, where Δ is a set of wffs and Q is a wff. We refer to Δ as the database, Q as the query, and $\mathbf{S}_\vdash$ as the algorithmic system for $\vdash$ (sometimes also referred to as the answering mechanism). Thus, for example, according to our view, the classical logic consequence relation $\vdash$, with $\mathbf{S}_1$ being a resolution-based automated system and $\mathbf{S}_2$ being a Gentzen or a Hilbert

[1]Most systems we consider are recursively enumerable, i.e. partially decidable. However, there may be cases in which one might want to consider relativizations with respect to some oracles which do not necessarily have to be partially decidable (see [Hinman, 1987] for a discussion) or to consider infinitary proof systems (cf. [Goldblatt, 1987b]). It was not really necessary for us to rule these cases out here.

system, gives rise to two different logics, $(\vdash, \mathbf{S}_1)$ and $(\vdash, \mathbf{S}_2)$. This is not the place to justify this view, but certainly some of the results of this book contribute to its justification.

$\vdash$ can be presented mathematically in some manner, either through a semantical interpretation or through some inductive definition or in any other mathematically proper way. It is just a set of pairs $\{(\Delta, Q)\}$ satisfying the axioms of the monotonic consequence relation, namely:

1. **Reflexivity:**
$$\Delta \cup \{A\} \vdash A.$$

2. **Monotonicity:**
$$\frac{\Delta \vdash A}{\Delta \cup \Delta' \vdash A}.$$

3. **Cut:**
$$\frac{\Delta \vdash A; \quad \Delta' \cup \{A\} \vdash B}{\Delta \cup \Delta' \vdash B}.$$

We regard the problem of whether $\Delta \vdash Q$ (in the logic $(\vdash, \mathbf{S}_\vdash)$), denoted by $\Delta \vdash ?Q$, as a problem of querying a set of data. Suppose we are given a set of assumptions Δ written in some formal language $\mathbf{L}$. This set represents our data concerning some aspect of an application. We may have queries Q from the data, represented in (probably the same) language $\mathbf{L}'$. The basic (non-temporal) logical problem associated with this situation is: Does the query follow from the data, or in symbols, do we have $\Delta \vdash Q$ $(\Delta \vdash ?q)$? A logical system $(\vdash, \mathbf{S}_\vdash)$ gives us the pairs Δ, Q such that the answer is yes. There are two mainstream types of logics involved: the monotonic ones and the non-monotonic ones. The difference has to do with the properties of the consequence relation $\vdash$. A monotonic logic would satisfy the three rules of consequence relation given above. The non-monotonic logics do not necessarily satisfy (2) or (3).

The conceptual difference between the two is that in the monotonic case $\Delta \vdash A$ means that A gets an answer yes from a part of Δ and it does not matter what the rest of Δ is. Thus $\Delta \vdash A$ iff for some minimal $\Delta_0 \subseteq \Delta, \Delta_0 \vdash A$; and so we also have $\Delta \cup \Delta' \vdash A$, because the answer really depends only on Δ_0. In the non-monotonic case the answer depends on the entire set Δ. Equivalently, in the non-monotonic case the answer may be obtained from some $\Delta_0 \subseteq \Delta$, but it *does* matter what else is in Δ besides Δ_0. Thus if more data is added so that we have a new set, $\Delta \cup \Delta'$, the answer may not continue to be yes.

Example 1.1.1

1. Consider the database Δ in the language with implication and negation:

$$(a \to b) \to a$$
$$c$$

   and the query $?a$.
   Whether or not $\Delta \vdash a$ depends on the underlying logic. If it is classical logic the answer is yes. For intuitionistic logic the answer is no.

2. Consider the same database with the query $?{\sim}b$. In classical and intuitionistic logic we do not get the answer yes. If the logic is non-monotonic with the closed world assumption the answer would be yes! (The closed world assumption includes the understanding that any atom which is not a head of a clause is negated by the database.)

Note that, so far, no temporal dimension is involved. We have only a database Δ and a logic (a system for obtaining answers from Δ). The logic may be one of a great variety available in the literature. There is classical logic, intuitionistic logic, relevance logic, inheritance systems, circumscription, Horn clause logic, etc. We have listed a mixed variety of systems because any one of them will do for obtaining answers from data.

Each logic has two features: the first is the representation and query language, and the second is the deductive system (possibly non-monotonic) for obtaining answers from data. In the above list, classical, intuitionistic, and relevance logics are all based on the same language but contain slightly different answering mechanisms $\mathbf{S}_{\vdash}$. Circumscription is still based on the same language but yields answers in a totally different way, and an inheritance system is completely different even in its representation.

We now proceed to widen our horizon and consider several related databases. Denote them by $\{(\Delta_t, \vdash_t) \mid t \in T\}$. Each database Δ_t has its own answering logical system $\vdash_t$. For simplicity let us assume that all the Δ_t are based on the same language and employ the same logic $\vdash_t = \vdash$. We can thus present the system of databases as $\{\Delta_t \mid t \in T\}$, the $\vdash$ being implicit. The area of indexed (e.g. modal and temporal) logics deals with logical connections between such databases. The queries we want to ask are of the form: *Which databases give answer yes to Q_i and how are they related?*

Thus the next step is pluralistic. We move from one single set of assumptions and logic to a whole related system of them. The added dimension is the study of connections between the databases. Let us consider, for example, the simplest connection, namely: If the answer to Q is yes from *one* database, the answer is yes from *all* of them.

We are not able to express such a relationship without a richer, possibly external, language. The language and logic of Δ_t involves Δ_t alone and is not able to deal with the above relation, which is a metarelation to $\{\Delta_t \mid t \in T\}$. The problem can be further complicated when T, the index set, has its own structure (e.g. it may be partially ordered by $<$ or it may have a binary function $+$ on it, so that whenever t, s are indices so is $t+s$).

Thus the research area of indexed (modal and temporal) logics deals with systems of databases $\{\Delta_t \mid t \in T\}$ where the index set T has a structure of its own. (We express this fact by writing $(T, <, +)$ as a typical example with one relation symbol and one function symbol.)

The queries either relate to any particular databases $\Delta_t, t \in T$ or to some patterns of relationships between them.

Example 1.1.2 Figure 1.1 is an example of a system of databases.

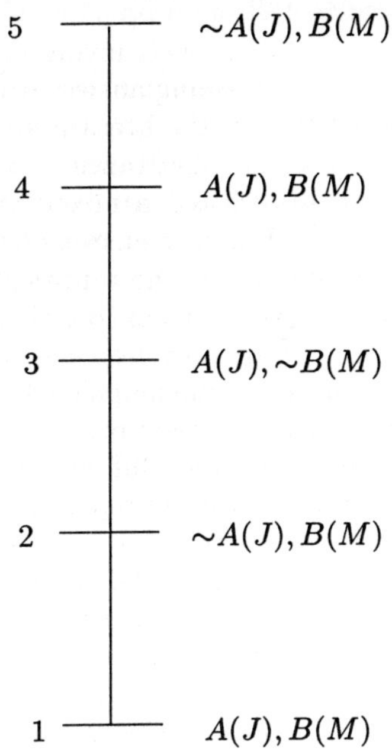

Figure 1.1

We see a pattern here that whenever $A(J)$ and $B(M)$ are true, $\sim A(J)$ is true at the next time point. We cannot express this without special additional language capability.

The reader may think that our presentation and point of view so far is traditional. However, we have already departed from the traditional view and have laid the ground for further generalizations, which can address the variety of systems available in the area of logic and computation.

First, under the present conceptual framework, the structure (e.g. $(T, <, +)$) of indices receives more prominence and is allowed to have a (logical) life of its own.

Second, we are under no obligation to disallow a system $(\Delta_t, \vdash_t)$ where $\vdash_t$ are different logics (though it makes some sense to ask that they be based on the same language).

Third, the logics $\vdash_t$ need not be monotonic.

Fourth, in the system $(\Delta_t, \vdash_t)$ we may have relationships not only among queries but also among $\vdash_t$. We may have that the logics relate to each other in conformity with the index structure $(T, <, +)$. For example, if t represents time, $\vdash_t$ may be a logic evolving with time.

We shall see that in practical computer science problems which use logic we have occasion and need to call upon many combinations of these features of an indexed system.

We defined the notion of a logical system to be a pair $(\vdash, \mathbf{S}_\vdash)$. We must therefore specify how we see the system $\{\Delta_t, \vdash_t\}$ as a logical system. There are many ways of defining the composite system. One simple way is to take the formulae of the composite system to be of the form $t : A$, where $A \in \Delta_t$. The relation $\{t_i : A_i\} \vdash s : B$ is defined somehow using $\{\vdash_t\}$ and the algorithmic system $\mathbf{S}_\vdash$ for the composite consequence relation $\vdash$ is also defined in some way using $\{\mathbf{S}_{\vdash_t}\}$. Thus we do obtain a composite $(\vdash, \mathbf{S}_\vdash)$.

Having described the scope of possibilities with systems of databases $(\Delta_t, \vdash_t)$, we have yet to ask what exactly are our options with regard to a language describing the relationships between the Δ_ts. Here there are two extreme possibilities: the external and the internal ones, and, of course, many combinations of them.

The external view is to have a language **M** which can describe the notion of $\Delta_t \vdash_t A$. **M** is a metalanguage, allowing explicit reference to t and to $(T, <, +)$, having names for the elements of Δ_t and, by using its own (i.e. **M**s) connectives and logical means, being able to describe relationships among the Δ_ts. The metalanguage **M** can be either some especially rich language or classical logic playing the rôle of metalanguage.

The second approach is to enrich the language of each Δ_t with special additional connectives and operators allowing it to relate to its 'neighbours' in the structure $(T, <, +)$. This involves a slight change of point of view. Δ_t is no longer a database of data isolated on its own but it will now contain data (in the enriched language) of how it is related to neighbouring

$\Delta_s, s \neq t$. So, for example, $\Box A \in \Delta_t$ may put in the database at t the information that all other neighbouring databases answer the query $?A$ as yes. A side effect of this change in view is that we can no longer take any old family of databases and put them together to form a system, as the databases also contain information about their neighbours. They must fit together.

The experienced modal logician will notice that the system $(T, <, +, \Delta_t, \vdash_t, t \in T)$ is a mixture of syntax and semantics. $(T, <)$ can be regarded as a possible world structure, where at each $t \in T$ we do not have the traditional classical model but a theory Δ_t in the logic $\vdash_t$. This is in itself an innovation, in the context of possible world semantics. It is especially suited for a computational point of view. To find the value of an atom q at a possible world t, we do not consult a mathematical assignment $h(t, q)$ as in the traditional Kripke semantics, but, rather, we submit the query $?q$ to an algorithmic proof procedure $\Delta_t \vdash ?q$.

Example 1.1.3 Using the case of example 1.1.2 and figure 1.1 above, let us illustrate the pure metalanguage approach. We can name a term or a predicate t by 't'. The metalanguage **M** has a **Derive** predicate and we can write

Derive (1, 'A', 'J')
Derive (1, 'B', 'M')
Derive (2, '$\sim A$', 'J')

⋮

Derive (5, '$\sim A$', 'J')
Derive (5, 'B', 'M').

The aforementioned temporal connection can be expressed by

$$\forall t[\textbf{Derive}\ (t, `A', `J') \wedge \textbf{Derive}\ (t, `B', `M') \\ \rightarrow \textbf{Derive}\ (t+1, `{\sim}A', `J')].$$

Example 1.1.4 To illustrate the internal approach enrich the language of example 1.1.2 with an additional internal connective **Next** A. Its meaning at database Δ_n is that A obtains answer yes from the neighbouring Δ_{n+1}. To express the connection of example 1.1.2 we can add the extra data $A(J) \wedge B(M) \rightarrow$ **Next** $\sim A(J)$ at each point.

Example 1.1.5 In example 1.1.3, instead of writing **Derive** (t, 'A',x) we can write in classical logic $A^*(t, x)$, where A^* is a two-sorted predicate. This presentation is an object language presentation. It has the advantage of drawing directly on the resources of first-order classical logic. Its disadvantages are its limited expressive power (as compared with modal and temporal logics with connectives which can express second-order properties) and the unintuitive mode of presentation of classical logic.

The next step in the discussion is the question when is such a system to be considered temporal? The structure $(T, <, +)$ need not, in the general case, be associated with time at all. It could represent, for example, a network of underground stations and the way in which they are related (with $+$ meaning the next section in line or nothing at all). Even if we do impose a *direction* on $(T, <, +)$, it still does not necessarily mean that we are dealing with time.

Practical experience in handling temporal logics and their applications leads us to adopt the following intuitive characterization of a temporal system.

The system $(T, <, +)$ is considered temporal if it has an (irreflexive transitive) ordering $<$. We regard $<$ as a flow of 'time' (where there are several possible futures) and the data at Δ_t as a 'snapshot' of what is happening at t. Thus $\{\Delta_t, t \in T\}$ is a family of collections of data with a direction $<$ which is in most cases irreflexive and transitive. T need not be time itself but could just represent an index set of 'snapshots'. We use $+$ to give an example of the use of additional function symbols which may be necessary for some applications.

Note the following. The logic of all the $\vdash_t$ can be the same logic $\mathbf{L}$, while the logic of $(T, <, +)$ can be a different logic $\mathbf{L}'$. $(T, <, +)$ can, for example, be a model of an intuitionistic theory of linear ordering, and to use properties of $<$ in our reasoning we need to prove them in intuitionistic logic. Note that the logics $\mathbf{L}$ and $\mathbf{L}'$ need not be the same. To have, for example, intuitionistic temporal logic we may have (1) $\mathbf{L}$ intuitionistic and $\mathbf{L}'$ classical, or (2) $\mathbf{L}$ classical and $\mathbf{L}'$ intuitionistic. (1) and (2) are not the same. Take, for example, linear time and the modal axiom

$$\Diamond A \wedge \Diamond B \to \Diamond(A \wedge B) \vee \Diamond(A \wedge \Diamond B) \vee \Diamond(B \wedge \Diamond A).$$

To show its validity, we assume that $\Diamond A \wedge \Diamond B$ holds at t and show the consequent. From the assumption, we infer the existence of s_1 and s_2 such that $t < s_1$, $t < s_2$, A holds at s_1 and B holds at t_2. We now use the theory of linear order to show that either $s_1 = s_2$, $s_1 < s_2$, or $s_2 < s_1$. If the theory of linear order is classical this can be done. If it is intuitionistic, then this depends on the version of the intuitionistic theory. We may or may not show the above. Thus the modal axiom may or may not hold. For each t, the reasoning locally could be classical; in this case, $\Box(A \vee {\sim}A)$ is valid, since $A \vee {\sim}A$ classically holds at each t.

Having given a temporal direction to the databases Δ_t, via the $<$ in $(T, <, +)$, we now have several options as to the type of relationship the Δ_t

and $\vdash_t$ can have. The relationship can be an extensional patterns relationship, describing how the answers to various queries at different times from the Δ_ts are related and how they are distributed, or it could be a more complex relationship about the logical evolution of the rules in the logics at t with respect to the time t.

This new type of change and evolution in the logic $\vdash_t$ itself is of special interest in the case where the logics involved are non-monotonic. The essential feature in non-monotonic reasoning is that new information coming in may affect the answer. Thus if our current database is Δ_t and $\Delta_t \vdash_t A$ non-monotonically, any new information B can cause $\Delta_t, B \not\vdash_t A$. Let $\Delta_{t+1} = \Delta_t \cup \{B\}$. We thus obtain an evolving set of theories which may or may not predict correctly. After a series of wrong predictions we may change the non-monotonic mechanism itself. Consider a database Δ_1 and a non-monotonic logic $\vdash$. Let $B_1, B_2, B_3, \ldots$ be a sequence of wffs in the language of Δ. Let $\Delta_{n+1} = \Delta_n \cup \{B_n\}$. If we consider B_n as a stream of updates (inputs) we may have that $\Delta_n \vdash \sim B_n$. Thus the nth input is contrary to the prediction of Δ_n (via the logic $\vdash$). We may, after a while, revise $\vdash$ to $\vdash'$ to obtain better predictions. The connection between non-monotonicity and time is not well studied and it is our hope to study it (in research papers).

Another possibility of evolution and change is to be more specific on how each Δ_t can change into another $\Delta_{t'}$. In the metalanguage **M** we can allow specific operations (actions) which connect the theories Δ_t, Δ_s for $t < s$. A practical example of this is database update systems. We start with a single or several databases and we have a stock of operations such as *delete* and *add*, which transform one database to another. We are now bordering on the area of planning. Consider, for example, the *blocks world,* where movements of blocks prompt the change from one database to another. In this area of research the problems considered are planning problems. Given a Δ_{t_1} and a set of actions $\{a_i\}$ which can act on databases to give other databases, we can pose the exercise: *For a given goal database Δ_{t_2}, find a sequence of Δ_{s_i} that would lead from Δ_{t_1} to Δ_{t_2} via the applications of $\{a_i\}$.*

The variety of applications of modal and temporal logic and of indexed systems of databases and logics prompted the first author to develop a new theory of inference systems called Labelled Deductive Systems. We have already seen that the basic question of $\Delta \vdash ?Q$ is transformed to the basic question of

$$\{\Delta_t, \vdash_t\}?\{t_i : Q_i\},$$

where $\{t_i : Q_i\}$ is a finite set. That is, given a system of databases and given the way they are connected, does it follow that Q_i holds at t_i (i.e.

$t_i : Q_i$)? Does the system force the answer yes on the query $\Delta_t?Q$?

We are thus led to a new concept, the concept of a labelled database. The new database is considered as a (finite) amount of information on various t_i taken from various Δ_{t_i}. Formally, it is a set of labelled formulae $\{t_i : A_i\}$. A query is a labelled formula $t : A$ and the basic logic can have rules of the form $t_i : A_i \vdash t : A$. The rules may depend on $(T, <, +)$.

Of course in the different aspects of temporal systems outlined above, the research emphasis is different and they might be perceived in the mind of the reader as separate research areas altogether. This is not the case. The connections exist and, especially in application areas, the more unifying point of view is of value.

Sometimes in the metalanguage $\mathbf{M}$ it is possible to code the information of several of the Δ_{t_i} into one Δ_{t_0}. This is especially possible in the framework where each Δ_t can talk about neighbouring Δ_ss. For example, in modal logic the labelled database $\{t_1 : A_1,\ t_2 : A_2,\ t_3 : A_3\}$ where $t_1 < t_2 < t_3$ can be essentially represented as

$$\{t_1 : A_1 \wedge \Diamond(A_2 \wedge \Diamond A_3)\},$$

where $\Diamond$ denotes possibility.

We now address the issue of computational tractability. The logical set-up that we have now is that of asking a query $t : Q$ from a database system of the form $\{\Delta_t \mid t \in T\}$ for some particular $(T, <, +)$. We can represent the data as $\{t_i : A_i\}$ with the understanding that $t : A$ means $A \in \Delta_t$ and t_i are from $(T, <, +)$. The reasoning for obtaining the answer may be in the metalogic $\mathbf{M}$ or in the object logic involved. If we want a practical tractable system for answering queries, we find that we have to simplify the representation of the database system. To give an example of such a simplification we assume that all theories $\Delta_t, t \in T$ contain *atomic* statements only. We further assume that $(T, <)$ is linear time, and that in the intended interpretation of the theories, the atoms are true persistently, namely that the set of points where each atom is true can be represented as a (finite) union of maximal intervals or halflines.

In such a case we discover that the best representation of the database system (from the computational point of view) is to indicate the extension of each atom q via maximal intervals of the form $[t_1, t_2] : q$, with the understanding that $q \in \Delta_t$ for each $t \in [t_1, t_2]$ and $\sim q \in \Delta_t$ if t is not in any such interval. We further discover that it is sufficient to list atomic formulae of the form $t_1 : q^+, t_2 : q^-$, where q^+, q^- are new atoms associated with q indicating when q begins to be true and terminates being true. The interval $[t_1, t_2] : q$ can be read from the above data.

The reader should note that we are now using a way of *coding* the representation $\{\Delta_t \mid t \in T\}$ via a new representation $\{t_i : q_i^{\pm}\}$ and in general we need a program to obtain one from the other.

Example 1.1.6 The data has the form

1. $\{[1,5] : q,\ [7,11] : q\}$, which we replace by the data
2. $\{1 : q^+,\ 5 : q^-,\ 7 : q^+,\ 11 : q^-\}$.

To obtain (1) from (2) we need a small program.

1.1.3 The procedural and the imperative views

Our view of modal and temporal logic so far has been essentially declarative. The basic idea has been that we have a database in some form denoted by Δ and a query Q and we want to know whether $\Delta \vdash Q$ in the appropriate logic. Δ can be a sophisticated database, perhaps a system $\{\Delta_t \mid t \in T\}$ of databases, or a labelled database. Whatever Δ and Q are, the basic question we ask is always declarative: Do Δ and Q share the relation $\Delta \vdash Q$?

Example 1.1.7 $\Delta = \{(a \to b) \to a, a \to \Diamond(b \to c), \Box b\}$, $Q = \Diamond c$.

The question is, does $\Delta \vdash Q$ in an underlying logic which is an intuitionistic modal logic?

The next mental step is to read the database *procedurally* and use the form of the data clauses to guide our search for an answer. This view is the same as the logic programming view for classical logic. It is not an imperative view; there are no actions to be taken, just a procedural reading of the syntactic structure of the database to obtain an answer. This area in modal and temporal logic is identified as modal and temporal logic programming. The reader should note that what we mean here is a procedural reading of the natural modal and temporal logic formalism, and not the use of classical logic programming to simulate temporal phenomena. Thus when one uses Horn clause logic programming (e.g. Prolog) or indeed just classical logic to talk about (simulate) temporal phenomena, one is not doing modal and temporal logic programming in our sense.

The procedural view of logic, then, is just a refinement of the declarative view, which uses the database in a prescribed procedural way to find answers. Both views are based on the perception that the data is laid out in front of us and that we are querying it. We are external to the flow of time. Thus in the declarative view, the entire flow of time is given to us.

Now consider the ancient Aristotelean question: 'Will there be a sea battle tomorrow?'

This sentence is either true, in which case there will be a battle, or is false, in which case there will not be a battle. It may seem paradoxical because we think that tomorrow has not happened yet!

But according to the declarative view, tomorrow has happened. The entire history has happened and we are outside it, viewing it. If we adopt the reference point, which we call today, and ask about the next point, which we call tomorrow, either the sea battle took place there or not.

There is, of course, a new point of view, that of being inside time: the point of view where the future has not happened yet. The past has happened and can be queried in the usual way and the future has not happened. This point of view is very familiar to us. We go through time every day looking at the future from this point of view. The problem is: what does this point of view mean logically? How can we understand a future sentence (e.g. Tq, q true tomorrow)? The imperative future point of view reads the sentence imperatively: 'create a future in which the statement is true'. Thus Tq is read imperatively as: 'make q true in the next state'.

Once the future has been (successfully) done, the same statement Tq can be read declaratively.

The imperative future can serve as an imperative command language for control and action. It is interesting to compare the imperative view with the control features in Prolog. The clause $print \wedge p \to q$ is not purely declarative. In querying q one asks for p, succeeds, and then 'asks' for 'print' and succeeds by printing (so 'print' is always imperative, never declarative in Prolog). This may look parallel to 'T *print*' in which we make the wff true by actually printing. Time is hidden in the Prolog interpretation. It is the stepwise movement of the pointer. Therefore, no clearly defined model is available after execution in which the formulae can be interpreted declaratively 'as is'. In the Prolog case one needs to move to a metalanguage describing the pointer movements (the interpreter) and, in this metalanguage, statements like '*print*' have a declarative meaning. However, going to a metalanguage is not really making the object language declarative. To take an extreme intuitive example, one can describe the most violent murder case in a metalanguage, making it a declarative relation between successive moments in time, which is a very unintuitive and uninvolving way of looking at the action.

Let $a_1, \ldots, a_n$ be the successive atomic actions involved in the murder case. Let **Happen** (t, x) be a metapredicate saying that x happened at moment t. t could be proper time or just a step-by-step counter. Consider

$$\varphi = \exists x_1, \ldots x_n \bigwedge_{i=1}^{n} (\mathbf{Happen}(i, x_i) \wedge x_i = a_i).$$

The above describes declaratively, in the metalanguage, the sequence of events of the object language.

1.1.4 Mathematical studies

Once temporal languages and logics are introduced, they can be studied for their own sake as mathematical systems. Traditional mathematical questions usually studied in logic can be asked and answered in the temporal case. Among them are the following:

1. Hilbert axiomatizations and natural deduction proof theory for temporal logics;
2. correspondence between axioms and conditions on the flow of time;
3. functional completeness (expressive completeness) of temporal connectives;
4. predicate temporal logics; propositional quantifiers;
5. decision procedures for various temporal systems;
6. translations and interpretations between various temporal systems and other systems;
7. connection of temporal logic with automata theory;
8. applications in computing, philosophy, and linguistics.

The rest of this chapter is a survey, giving a flavour of what is included in this book.

1.2 How to present a temporal logic in the predicate calculus

The first method of accounting for temporal phenomena is to do it within the framework of the classical predicate calculus. The exact nature of this method will be studied in chapter 5. The idea is to add an extra variable for time to the language of predicate logic and make all predicates time dependent.

Thus if $L(x, y)$ represents x loves y, we can express in predicate logic such statements as

John loves all those who do not love themselves

by:

$$\forall x[\sim L(t,x,x) \to L(t,J,x)].$$

The parameter t appears everywhere.

One can now quantify over t and express time-dependent properties such as

$$\forall t > now\ (L(t,M,J)) : \quad \textit{Mary will always love John.}$$

This approach is favoured by many because of our direct control over time points. We can quantify over t, s, compare them, even use Skolem functions involving them. Here is another example:

John is taking a pill today and will take one every two days!

$$\begin{gathered}\text{Take}(now, J, \text{pill}) \\ \wedge\ \forall t, s \geq now\ (\text{Take}(t, J, \text{pill}) \wedge (t+2=s) \to \text{Take}(s, J, \text{pill})).\end{gathered}$$

There are some problems about identity through time, etc., but the computer scientists leave these problems to the philosophers. The problems crop up in automated reasoning and some formal approach to identity and constant terms is needed.

Theorem-proving machinery for classical logic can sometimes be adapted for these problems, and this is an advantage. However, even for simple sentences like the above, this approach is not ideal. The first 'problem' is that the 'logic' of the time is hidden. The temporal logic as described above is not predicate logic. The parameter t ranges over a special flow of time, e.g. $t = 0, 1, 2, \ldots$, and this makes the logic special. For example, we are committed to all the axioms involving the time parameter t dictated by the properties of the chosen flow of time.

The second problem with using predicate logic as a temporal logic is more fundamental and has to do with the representation of time-dependent data. Consider a simple example:

Since M became Prime Minister the pound has steadily been going down, and in fact it will continue to go down for as long as he or she remains Prime Minister.

This sentence is crystal clear to us and is understood immediately without any mental effort. Let us write it in predicate logic with time parameters.

Let $PM(t,x)$ be read as:

x is Prime Minister, at time t

Let $VP(t,x)$ be read as:

the value of the pound at t is x

Then the sentence becomes:

$$\begin{aligned}
&\exists t < now\ [\mathrm{PM}(t, M) \\
&\qquad \wedge \exists s < t\, \forall u (s < u < t \rightarrow {\sim}\mathrm{PM}(u, M)) \\
&\qquad \wedge \forall u (t < u < now \rightarrow \mathrm{PM}(u, M)) \\
&\qquad \wedge \forall u, v, x, y (t < u < v \leq now \wedge \mathrm{VP}(u, x) \wedge \mathrm{VP}(v, y) \rightarrow x > y)] \\
&\wedge \forall s > now\ [\forall u (now < u < s \rightarrow \mathrm{PM}(u, M)) \\
&\qquad \rightarrow \forall u, v, x, y (now \leq u < v < s \wedge \mathrm{VP}(u, x) \wedge \mathrm{VP}(v, y) \rightarrow x > y)]
\end{aligned}$$

The above sentence of predicate logic is not readable.

A point always claimed in favour of predicate logic is its rich, highly developed theorem-proving techniques. Take, for example, resolution methods; to present the above as a 'ready for resolution' disjunctive clause we have to skolemize and then write it in disjunction forms. The 'meaning' will be lost forever. In any case, as we explain below, available theorem-proving methods for classical logic often cannot be used directly.

Another awkward example is the following. To take a mortgage on a house one needs an insurance policy stating that if the owner dies, the mortgage will be paid. Thus the conditions required by the bank are:

x obtains a loan at time t if house of x is of suitable value and in good condition at time t and x's financial position is acceptable at time t and x takes an insurance policy at time t.

How do we express the insurance policy clause in predicate logic? Our first attempt is to use a time parameter; after all, the insurance is a committment to pay in the future. Thus we write, x takes an insurance policy (of the above kind) at time t with company y iff for all $s > t$ (if x dies at s then y pays at s). However, this is not sufficient, because we can say in English:

John took an insurance policy but when he died, the company did not pay.

To express this property we need to outline the planned future and the real future separately, for each moment of time t. It becomes awkward to do so in predicate logic.

The above objections are qualitative and stylistic but not mathematical. It is like saying that PASCAL is nicer to write in than BASIC but that both languages are equivalent in mathematical power. Can pure temporal logics, such as, for example, axiomatic systems, say things which the predicate calculus *cannot say at all*? Here is a well-known example:

Last year, every time I went to London I came back with a headache.

What we need for the correct representation of this example is to be able to express a functional relationship between visits to London and headaches. If after each visit to London there was a headache before the next visit, we can then express the relationship, but if the headache developed after the next visit, we cannot express the functional dependence. However, note that, from the point of view of computer science, the notion of expressiveness is different from the logical notion. For example, the finiteness of time is not a first-order property but computationally it is very convenient.

So far we have outlined how reference to time within predicate logic can be done by isolating a special variable t for time. Let us write predicates involving time as $A^*(t, x)$, reading, for example, 'x attends the lecture at time t', where x ranges over students and t over time. This only looks like first-order predicate logic but may not be exactly so. Let us elaborate.

Classical model theory is related to classical provability or theorem proving via the completeness theorem. So, for example, to know whether

$$\text{Data} \vDash B \text{ (in first-order semantics)}$$

we use a proof system or a theorem prover to prove B from the data. In our special case the quantifiers in the Data are restricted: t ranges over time and x ranges over students.

We cannot use an ordinary proof system or a well-known theorem prover for first-order classical logic to check whether B follows from the data. The completeness theorem tells us that the Data $\nvDash B$ iff there exists a first-order model of the data in which B is false. Here 'a model' means *any model, including* those in which $\forall t$ ranges over elements of the domain, i.e. students, and $\exists x$ ranges over times.

Since we restrict the type of model allowed, we may obtain a new notion of $\vDash$ (semantic consequence) and we may need a new proof system or theorem prover. In fact, the new $\vDash$ may not even be arithmetical.

It can be proved that the monadic predicate temporal logic with time represented by the natural numbers is not arithmetical. If we are committed to quantifying over the natural numbers as time, this gives us the power to define the natural model of arithmetic. We will not go into details here, but give one example to illustrate. Let

$$\begin{aligned} A &= \forall x \exists t \forall s (t < s \to A^*(s, x)). \\ B &= \exists t \forall x (t < s \to A^*(s, x)). \end{aligned}$$

A is the data. A says that for every student (every x, as x ranges over students) there is a time after which the student always attends the lecture.

B is the query. B says that there is a time after which all students attend the lecture.

If we want to check whether B follows from A we cannot just take a theorem prover for classical logic, put A in as data, B as query, push the button, and let it compute. We will not obtain the correct answer. What we have to do is supplement the data A with axioms about the flow of time. For example, in predicate logic we can have the predicates $\mathbf{T}(u)$ for 'u is a moment of time', and $\mathbf{D}(u)$ for 'u is an element of the domain', and we can write the following set of axioms (in which $\leq$ is the usual abbreviation):

1a. $\forall u(\mathbf{D}(u) \vee \mathbf{T}(u))$.

1b. $\sim\exists u(\mathbf{D}(u) \wedge \mathbf{T}(u))$.

2. $\forall u, v, w(\mathbf{T}(u) \wedge \mathbf{T}(v) \wedge \mathbf{T}(w) \wedge u < v \wedge v < w \rightarrow u < w)$
   (transitivity of $<$).

3. $\forall u{\sim}[u < u \wedge \mathbf{T}(u)]$
   (irreflexivity of $<$).

4. $\exists u(\mathbf{T}(u) \wedge \forall v(\mathbf{T}(v) \rightarrow u \leq v))$
   (existence of a first point of time).

5. $\forall u(\mathbf{T}(u) \rightarrow \exists v(\mathbf{T}(v) \wedge u < v \wedge {\sim}\exists w(\mathbf{T}(w) \wedge u < w \wedge w < v)))$
   (every point has an immediate successor).

6. $\forall u(\mathbf{T}(u) \wedge \exists v(\mathbf{T}(v) \wedge v < u)$
   $\rightarrow \exists v(\mathbf{T}(v) \wedge v < u \wedge {\sim}\exists w(\mathbf{T}(w) \wedge v < w \wedge w < u)))$
   (any point which is not the first point has an immediate predecessor).

7. $\forall u, v(\mathbf{T}(u) \wedge \mathbf{T}(v) \rightarrow u < v \vee u = v \vee v < u)$
   (linearity).

Let Δ be the above set of axioms along with axioms for equality. It characterizes 'natural-number-like' time (i.e. discrete with a first element). There is no axiom system Δ' that has natural number time as its only model.

Let $A\# = \forall x(\mathbf{D}(x) \rightarrow \exists t(\mathbf{T}(t) \wedge \forall s(\mathbf{T}(s) \wedge t < s \rightarrow A^*(s, x))))$.
$B\# = \exists t(\mathbf{T}(t) \wedge \forall x \forall s(\mathbf{D}(x) \wedge \mathbf{T}(s) \wedge t < s \rightarrow A^*(s, x)))$.

Then to ask whether B follows from A in the 'integer-like, discrete flow of time with a first moment' one can ask instead whether

$$\Delta \cup \{A\#\} \vdash B\#$$

in first-order classical logic. One can therefore use a classical theorem prover to check the answer.

It seems, therefore, that we have found a way of overcoming the restrictions on the special ranges of the variables x and t, and that we can reduce the query $A?B$ to the query $\Delta, A\#?B\#$. Note, however, that this can only be done because an axiomatization Δ exists for the class of 'integer-like' flows of time. In the case of natural numbers time $\{1, 2, 3, \ldots\}$ *no such* Δ *exists*!

The above explains how ordinary 'many-sorted' databases can manage to be first-order. For example, consider a database with predicates like $S(John, 1000)$ (the salary of John is £1000). The range of employees and possible salaries is not necessarily potentially infinite but very large and finite and hence can be characterized by some Δ. In the case of natural numbers time we cannot say that time is a large but finite set of moments. It is by nature infinite.

1.3 Semantic presentation of temporal connectives

The second way of presenting a temporal logic is via connectives. We imagine a transitive flow of time $(T, <, =)$ and the various predicates, e.g.

John loves Mary and M is PM,

have values *true* or *false* at each moment of time. So far, it is equivalent to writing $L(t, J, M)$ and $PM(t, m)$, except that the t is suppressed.

The difference comes when we want to talk about the distribution of truth values through time. We allocate *connectives* to the logic which describe behaviour through time using the suppressed *now* as a reference point. This is very similar to the way in which we use connectives in English.

Example 1.3.1 Consider the following, very basic, connectives:

GA	=	A will always be true
HA	=	A was always true
FA	=	A will sometimes be true
PA	=	A was sometimes true
$S(A, B)$	=	B was continuously true since A was true
$U(A, B)$	=	B will continuously be true until A is true
YA	=	A was true yesterday
TA	=	A will be true tomorrow

Thus, for example, *John will love Mary* becomes:

$$FL(J, M),$$

meaning, in predicate logic:

$$\exists t > now\ L(t, J, M).$$

John has loved Maggie since Maggie was PM is:

$$S(PM(m), L(J, m)),$$

whereas in predicate logic we have to write:

$$\exists t < now\ [PM(t, m) \wedge \forall s(t < s < now\ \rightarrow L(s, J, m))].$$

Let us translate the sentence about the pound. Time is the integers:

$$S({\sim}PM(m), PM(m) \wedge \textit{Pound going down}) \wedge PM(m)$$
$$\wedge\, U({\sim}PM(m), PM(m) \wedge \textit{Pound going down}).$$

We have to translate '*Pound going down*':

$$\forall y \forall x[VP(x) \wedge YVP(y) \rightarrow y > x].$$

That is, '*If the value of the pound now is x, and if its value was y the day before* (i.e. if $YVP(y)$), *then $y > x$.*'

The above translations are good no matter whether the temporal connectives are introduced via axioms or via a semantic interpretation.

We now describe the components needed to specify a temporal logic in a semantic way. These are:

(1a) the flow of time of the logic;

(1b) the units of time needed to give truth values;

(1c) possible restrictions on the assignments to the atoms;

(2a) the temporal connectives used;

(2b) the truth conditions for the connectives.

The meaning of (1a) is clear. The flow of time can be the integers, the real numbers, partially ordered sets, and so on. (1b) chooses the units of time at which atomic sentences have truth values. These can be integers, intervals of integers, pairs of integers, intervals, infinite paths of moments in time, etc. (1c) describes possible restrictions on the truth values given to the atoms. A typical example could be that each atom can be true at only a finite number of points (units of time). (2) chooses the connectives and determines their meaning.

We give five temporal logics, all based on the integer flow of time. They differ either in (1b), the unit of time for truth values, in (1c), the assignment to the atoms, or in (2), the choice of connectives.

Example 1.3.2

(1a) Time is the integers.

(1b) The unit of time for truth values is an integer. An atom has a truth value at each moment of time.

(1c) The assignment to the atoms is arbitrary.

(2a) The connectives are (besides the classical $\sim, \wedge, \vee, \rightarrow$): $S(A, B)$ and $U(A, B)$.

(2b) The truth conditions for the connectives are defined as follows. $S(A, B)$ is true at n ('now') if for some $t < n$, A is true at t and for all points between t and n, B is true. $U(D, C)$ is true at n if for some $t' > n$, D is true at t' and for all points between n and t', C is true.

Diagrammatically, see figure 1.2.

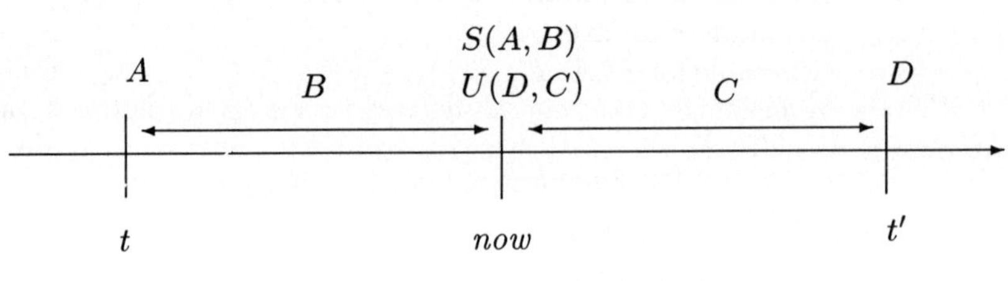

Figure 1.2

FA and PA of the preceding example can be defined as $U(A, \top)$ and $S(A, \top)$ respectively. We also have $YA = S(A, \bot)$ and $TA = U(A, \bot)$.

Example 1.3.3

(1a) Time is the integers.

(1b) The unit of time for truth values is a closed interval $[m, n], m \leq n$.

(1c) There is no restriction on the assignment to the atoms. An atom is true or false at an interval. Thus we may have q true at [1, 3], q false at [1, 2], and q true at [2, 3]. We have no persistence criteria or any principles for subintervals.

(2a) The connectives are $Next(A)$ and $(A + B)$.

(2b) The truth conditions for the connectives are: $Next\, A$ is true at $[m, n]$ iff $m + 1 \leq n$ and A is true at $[m + 1, n]$. $(A + B)$ is true at $[m, n]$ iff for some k, $m \leq k \leq n$ and we have: A is true at $[m, k]$ and B is true at $[k, n]$.

Example 1.3.4

(1a) Time is the integers.

(1b) The unit of time for truth values is a pair (m, n) of integers.

(1c) There is no restriction on the assignments to atoms.

(2a) The connectives are F^* and P^*.

(2b) In this logic an atom is given truth values at points. The pair (m, n) is considered as an evaluation point m and a reference point n. Thus the truth value of an atom q at (m, n) depends on m only, though the truth value of an arbitrary A may depend on both (m, n).
Here is the table for F^*A. F^*A is true at (m, n) iff one of the following cases holds:

(a) $m = n$ and for some $k > n, A$ is true at (k, n);

(b) $m > n$ and A is true at (m, m);

(c) $m < n$ and for all k such that $m < k < n, A$ is true at (k, k).

P^*A has the symmetrical condition. P^*A is true at (m, n) iff

(a) $m = n$ and for some $k < n, A$ is true at (k, n);

(b) $m < n$ and A is true at (m, m);

(c) $m > n$ and for all k such that $m > k > n$, A is true at (k, k).

Example 1.3.5 The fourth logic is exactly like the logic in 1.3.2 with the added restriction on the assignment to atoms that the set of points in which an atom is true is either finite itself or its complement is finite.

Example 1.3.6 The fifth logic is the logic of execution sequences of a program.

1. Let the set of units of time be all infinitely proceeding sequences of integers, of the form $s = (x_0, x_1, x_2, \ldots)$.
   Let $s^n = (x_n, x_{n+1}, \ldots)$ be the tail of s. Atomic sentences q are given a truth value at each s.

2. We define the temporal connectives $\Box q, \Diamond q, U^*(p,q)$ and $\bigcirc q$:
   $\Box q$ is true at s iff $\forall n \geq 0$(q is true at s^n);
   $\Diamond q$ is true at s iff for some $n \geq 0, q$ is true at s^n;
   $\bigcirc q$ is true at s iff q is true at s^1;
   $U^*(p,q)$ is true at s iff for some n, p is true at s^n and for all $m \leq n$, q is true at s^m.

   $\Box$ corresponds to our familiar G, $\Diamond$ corresponds to F, and $\bigcirc$ corresponds to *Next*. Note, however, that the future in this system includes *now* as well.

1.4 Axiomatic presentation of temporal logic

The axiomatic presentation follows the tradition of defining the meaning of the logical connectives using proof rules and deductive axioms.

Our first aim in this section is to give some examples and try to show that axioms on temporal connectives are very expressive. In fact, it seems that even the simplest set of connectives, namely $\{F, P\}$, which are really nothing more than the existential 'quantifiers', enable us to write axiom systems which can distinguish between different properties of the flow of time.

The weakest possible propositional temporal logic is the logic K_t of Arthur Prior. All it describes is the fact that if t is in the future of s, then s is in the past of t. It is presented as a Hilbert system with the following axiom schemas.

- Axioms for classical logic (take all tautologies as axioms);
- $G(A \wedge B) \leftrightarrow GA \wedge GB$;
- $H(A \wedge B) \leftrightarrow HA \wedge HB$;

- $A \to GPA$;
- $A \to HFA$.

We also add the proof rules, called generalizations:

$$\frac{A}{GA} \quad \text{and} \quad \frac{A}{HA}$$

to the classical modus ponens.

We need to clarify the notion of a model. A temporal structure has the form $(T, <, t_0, h)$, where $(T, <)$ is a flow of time, $t_0 \in T$ and h is an assignment. Usually $(T, <)$ is a flow of time only if it is transitive and irreflexive but here, for greater generality, we will let $<$ be any binary relation on T. $(T, <, t_0, h)$ is a model of a formula A iff $\|A\|^h_{t_0} = 1$. We say A is valid in $(T, <, h)$ if for all $t \in T$, $\|A\|^h_t = 1$. We say A is valid in $(T, <)$ iff for all h, A is valid at $(T, <, h)$. Given a class $\mathbf{K}$ of flows of time, we say A is valid at $\mathbf{K}$ if A is valid at all $(T, <) \in \mathbf{K}$. The class $\mathbf{K}$ may be defined by a condition, i.e. $\mathbf{K} = \{(T, <) \mid (T, <) \vDash \varphi\}$, φ a sentence of the first-order theory of $(T, <)$.

A temporal logic $\mathbf{L}$ is said to be (*sound and*) *complete* for a class $\mathbf{K}$, if the theorems of the logic are exactly the formulae which are valid in $\mathbf{K}$. A temporal logic $\mathbf{L}$ is said to *characterize* the class $\mathbf{K}$ if the theorems of the logic are all valid at $\mathbf{K}$ and for any $(T, <) \notin \mathbf{K}$ there exists a theorem A of $\mathbf{L}$ and an h and $x \in T$ such that $\|A\|^h_x = 0$. The difference between these two notions is important. We will see in chapter 3 that the class of irreflexive flows of time cannot be characterized by any logic which uses only the usual inference rules. Nevertheless, the system K_t is sound and complete for the class of irreflexive flows of time. Another example is linearity. Although this cannot be characterized using the usual rules the system of K_t, plus axioms for transitivity and linearity, is sound and complete for the linear flows.

A logic is said to characterize a condition φ if it characterizes the class of $(T, <)$ satisfying φ.

A schema A of temporal logic characterizes a class $\mathbf{K}$ (or a condition φ defining the class $\mathbf{K}$) within a larger class $\mathbf{K}_0 \supseteq \mathbf{K}$ iff A is valid in all structures of $\mathbf{K}$ and for any $(T, <) \in \mathbf{K}_0$ not in $\mathbf{K}$, A is not valid at $(T, <)$.

We also say that A 'expresses' φ just in case A characterizes the class $\mathbf{K} = \{(T, <) \mid (T, <) \vDash \varphi\}$.

In general, there is a good correspondence between semantic conditions and axioms, with few exceptions. Different conditions on $(T, <)$ can make it possible to choose an assignment h which can validate or invalidate wffs. Temporal logic wffs can be made false in one type of flow of time and hold true in others. There are wffs that are always true in dense time, discrete time, or even complete time. Some are always true in branching time, and so on.

Consider Hf (where f is falsity, e.g. $g \wedge {\sim}g$ for an atom g). We have:

$$\|Hf\|_t^h = 1 \text{ iff } {\sim}\exists s[s < t].$$

Thus in a flow of time satisfying $\forall t \exists s[s < t]$ (there is no first moment of time), $PHf \vee Hf \vee FHf$ is always false. Otherwise it is true. This shows that the truth of wffs depends on the flow of time. In fact, some properties of the flow of time may be characterized by wffs. Take, for example, $U({\sim}f, f) \wedge S({\sim}f, f)$. This is true at a point t iff t is *discrete*. That is, there are predecessor and successor moments relative to t; formally, t satisfies $\exists x \exists y[y < t < x \wedge \forall z(z > t \to z \geq x \wedge z < t \to z \leq y)]$. Discreteness can also be expressed using P, F only.

Let us summarize the listing and table below, 'correspondence' between conditions and wffs 'expressing' these conditions. We will try to use P, F, only, and not U, S. We write GA for ${\sim}F{\sim}A$ for any A, and HA for ${\sim}P{\sim}A$. We assume the standard (first-order) truth conditions for the connectives, as in example 1.3.1. Here, T is non-empty and $<$ is a binary relation on T. The only assumption that we make on $<$ is that T is connected, in the sense that if $t, t' \in T$, then for some integer n there are $u_1, u_2, \ldots, u_n \in T$ such that $tRu_1Ru_2R \ldots Ru_nRt'$, where xRy means that x and y are comparable using $\leq$ (i.e. xRy abbreviates $x = y \vee x < y \vee y < x$).

(a) The condition that the present moment is in the past of any future moment corresponds to $A \to GPA$, or to $A \to HFA$. This condition is already incorporated in the notion of flow of time since when we write $t < s$ we say that t is in the past of s but also that s is in the future of t. The fact that we are using a binary relation as semantics for F, P, with their usual truth conditions, forces $A \to GPA$ and $A \to HFA$ to hold. Symbolically, $t < s$ is the same as $s > t$.

(b) The transitivity of $<$ corresponds to $GA \to GGA$ or, equivalently, to $HA \to HHA$.

(c) The existence of an endpoint above any moment corresponds to wff $FG\bot$.

(d) The condition that there exists one future doomsday common to all futures corresponds to $FG\bot \wedge [F(A \wedge G\bot) \wedge F(B \wedge G\bot) \to F(A \wedge B \wedge G\bot)]$.

(e) The condition of weak past linearity corresponds to the axiom $PA \wedge PB \to P(A \wedge B) \vee P(A \wedge PB) \vee P(PA \wedge B)$.

(f) The condition of existence of a tomorrow point corresponds to

$$F(GA \wedge A \wedge {\sim}B) \wedge F(GB \wedge {\sim}A)$$
$$\rightarrow F(F(GA \wedge A \wedge {\sim}B) \wedge F(GB \wedge {\sim}A)).$$

We saw before that $U({\sim}\bot, \bot)$ corresponds to this condition. However, U is not definable using F, P but F is definable using U : $FA = U(A, A \rightarrow A)$.

(g) The condition $\forall x \forall y(x < y \rightarrow \exists u(x < u < y))$ is a condition of total density. We can take the opposite of discreteness for a point, namely $\forall s > t \exists u(t < u < s)$. The corresponding wff is $FA \rightarrow FFA$ (within linear flows).

(h) The condition of well foundedness of the past (i.e. no infinite regress into the past) corresponds to $P({\sim}A) \rightarrow P({\sim}A \wedge HA)$.

(k) The condition of completeness (no gaps) corresponds to the wff

$$FGA \wedge F{\sim}A \wedge G({\sim}A \rightarrow F{\sim}A) \rightarrow F(GA \wedge {\sim}PGA).$$

It is doubtful whether the condition can be expressed using F only. It can be expressed using U only as

$$FGA \wedge F{\sim}A \wedge G({\sim}A \rightarrow F{\sim}A) \rightarrow U(A \wedge GA, F{\sim}A).$$

Recall that $FA = U(A, \top)$ and $GA = {\sim}F{\sim}A$.

Let us summarize this surprising correspondence between additional axioms and properties of flows of time in Table 1.1.

For example, the axioms for K_t, together with the axioms for linearity, density, and no end points comprise an axiom system for rational time flow.

We see that axioms can correspond to properties of time which first-order logic cannot express. This is partly because our notion of correspondence quantifies over all temporal models with a given flow of time, and so is itself not first-order. Thus the presentation of a temporal logic via axioms and rules can sometimes be more powerful than a first-order semantic presentation. On the other hand, there are semantic presentations which have no known corresponding axiom systems, such as some logics of Historical Necessity (discussed in later chapters).

We stated in the table that irreflexivity cannot be characterized by an axiom. It can be characterized, however, by a rule of inference, namely:

$$\frac{{\sim}q \wedge Gq \wedge Hq \rightarrow A}{A} \quad : \text{ atom } q \text{ not in } A.$$

Table 1.1

Properties of time	**Axiom**
Irreflexivity ($\sim t < t$)	No axiom; cannot be characterized by axioms, only by a rule (to be explained later)
Reflexivity ($t < t$)	$GA \rightarrow A$ or, alternatively, $HA \rightarrow A$
Transitivity of $<$	$GA \rightarrow GGA$ or, alternatively, $HA \rightarrow HHA$
Linearity (within transitive flows)	$G(A \wedge GA \rightarrow B) \vee G(B \wedge GB \rightarrow A)$ and similarly for H
density (within linear irreflexive time)	$FA \rightarrow FFA$ and $PA \rightarrow PPA$ (with axioms for linear irreflexive time)
Time is Dedekind complete (like the real numbers) (for linear time)	$(L(GA \rightarrow PGA) \wedge GA) \rightarrow HA$ (where $LX \equiv X \wedge GX \wedge HX$), and its mirror image swapping P with F, G with H (with axioms for linear time)
Time is irreflexive and well ordered in both directions	$FA \rightarrow F(A \wedge G{\sim}A)$ and $PA \rightarrow P(A \wedge H{\sim}A)$
Time is irreflexive and has an infinite chain both ways	$GF({\sim}\bot)$ and $HP({\sim}\bot)$
Time is finite	$FA \rightarrow F(A \wedge G{\sim}A)$ $PA \rightarrow P(A \wedge H{\sim}A)$ (note that this means that axioms can correspond to second-order conditions)
No end points	$F\top \wedge P\top$

The rule says that if A is true whenever time is irreflexive ($\sim q \wedge Gq \wedge Hq$ can hold at t only when $\sim t < t$), then A is true anyway ($\vdash A$), because for any counter model of A in a flow of time with reflexive t (i.e. $t < t$), we can always duplicate t and make the time irreflexive. We can always change the assignment on the atom q in the new model to make $\sim q \wedge Gq \wedge Hq$ true without affecting A, since q is not in A.

Given a certain property of flows of time (for example reflexivity, $\forall x(x < x)$), and given an axiom schema (such as $GA \to A$ (all A)), we say that the schema strongly characterizes the property if the following holds:

1. For any flow of time $(T, <)$ and any h, if $(T, <)$ satisfies the property, then h validates the axiom schema: i.e. every instance of the axiom is true in $(T, <, h)$ at every $t \in T$.

2. If $(T, <)$ does not satisfy the property, then there exists an h and $t \in T$ and an instance of the schema that is false at t under h.

According to this definition, the schema $GA \to A$ strongly characterizes reflexivity. Irreflexivity ($\forall x \sim (x < x)$) cannot be strongly characterized by a schema, because it can be shown that for any model $(T, <, h)$ there exists a model $(T', <', h')$ that is irreflexive and *essentially* validates the same formulae in the following sense. There exists a homomorphism α from $(T', <', h')$ on to $(T, <, h)$ such that

- $\forall t, s \in T'(t \leq s \to \alpha t \leq \alpha s)$;
- $\forall t, s \in T(t < s \to \exists u, v \in T'(u < v \wedge u \neq v \wedge \alpha u = t \wedge \alpha v = s))$;
- $(\forall t \in T')(\forall \text{ atoms } q)(h'(t, q) = h(\alpha t, q))$.

One can prove that for all A and all $t \in T'$, $\|A\|_t^{h'} = \|A\|_{\alpha t}^h$.

We can now describe how a condition on the flow of time (e.g. irreflexivity) can be characterized by a rule. We say that a rule *weakly characterizes* a condition on T if the following hold:

1. The rule is valid in the class of all models satisfying the condition.

2. Whenever a theory D cannot prove a formula A using modus ponens and the rule as its rules of inference, then $\sim A$ is true at some point of a model of D that satisfies the condition.

This says that the class of models satisfying the condition provides a sound and complete semantics for the logic associated with the rule.

Using the above irreflexivity rule, we can prove the following wholesale axiomatization lemma, Lemma 1.4.2, below.

First some definitions.

Definition 1.4.1 Let $\mathbf{K}$ be a set of ordered sets of the form $(T,<)$, and let $\sharp_i(q_j)$ be symbols for connectives. Let $\psi_i(t, Q_j)$ be second-order formulae in the monadic theory with Q_j, t free. Given any formula $A(q_j)$ in the propositional language with $\sharp_i$, we can translate it to $\psi_A(t, Q_j)$ in the monadic theory.

We write $(T,<) \vDash A$ iff $(T,<) \vDash \forall t \forall Q \psi_A$. We write $\mathbf{K} \vDash A$ iff for all $(T,<) \in \mathbf{K}, (T,<) \vDash A$.

Now let $\mathrm{Logic}(\mathbf{K}) = \{A \mid \mathbf{K} \vDash A\}$.

1. Let Δ be a theory and $\mathbf{K}$ a set of ordered sets. We say that Δ characterizes $\mathbf{K}$ if $\Delta = \mathrm{Logic}(\mathbf{K})$.

2. Let α be a condition on $(T,<)$. We say that Δ characterizes α if $\Delta = \mathrm{Logic}\{(T,<) \mid (T,<) \vDash \alpha\})$.

3. Let $\mathbf{K}$ be a set of ordered sets and $\Delta = \mathrm{Logic}(\mathbf{K})$, where Δ is in a language with connectives $\sharp_i$. We say that Δ can be axiomatized if we can write a deductive system for Δ.

4. Let $\{\sharp_i\}$ be a set of connectives. We say that $\{\sharp_i\}$ can be axiomatized over $(T,<)$ if we can write a deductive system for the theory $\Delta = \mathrm{Logic}(\{(T,<)\})$.

Similar definitions can be made if by a *model* we mean $(T,<,h)$. $\mathbf{K}$ would be the set of such $(T,<,h)$ and, as above,

$$(T,<,h) \vDash \psi_A(t, Q_j) \text{ if } (T,<) \vDash \forall t \psi_A(t, h(Q_j)).$$

Lemma 1.4.2 *Assume a given set of connectives $\sharp_1, \ldots, \sharp_m$ for a temporal logic over a flow of time $(T,<,=)$, such that these connectives have first-order truth tables (see section 1.6). (The semantics is that there are no restrictions over h, and evaluation is at time points $t \in T$.) Then more connectives $\sharp_{m+1}, \ldots, \sharp_{m+k}$ can be* effectively *found and added such that the logic with the entire set of connectives $\sharp_1, \ldots, \sharp_{(m+k)}$ can be effectively axiomatized. Of course, the irreflexivity rule is used in the axiomatization. The entire procedure of finding the new connectives and the axioms is effective and computable.*

In fact, for any given axiom system using P, F only for a given class of flows of time, there is an extension to the connectives $\sharp_1, \ldots, \sharp_m$ as described above.

Example 1.4.3 We can now axiomatize the logic with Since and Until for any flow of time. Take the corresponding axioms for the same flow of time using P, F, G, H and the rule for irreflexivity, and add the axioms

$$Hq \wedge {\sim}q \wedge Gq \to [U(A,B) \leftrightarrow F(A \wedge H(B \vee Hq))].$$
$$Hq \wedge {\sim}q \wedge Gq \to [S(A,B) \leftrightarrow P(A \wedge G(B \vee Gq))].$$

What we are really doing in these axioms is stating the truth conditions of U and S.

Example 1.4.4 To give another example take the connective Nq, reading: *q is true around now*, namely:

$$Nq \text{ is true at } t \text{ iff } \exists x, y(x < t < y \wedge \forall z(x < z < y \to q \text{ is true at } z))$$

The properties of N together with F and P can be characterized, without irreflexivity, by the following axioms, say for linear rational time:

1. Axioms for F and P for linear rational time (see Table 1.1).

2. $q \wedge Hq \wedge Gq \to Nq$
   $Nq \to NNq$
   $Nq \to q$
   $N(p \wedge q) \leftrightarrow Np \wedge Nq$
   $G{\sim}q \to {\sim}Nq$
   $H{\sim}q \to {\sim}Nq$.

Using irreflexivity we can write the following axioms for N.

(1*) Axioms for P, F for linear rational time, including the irreflexivity rule.

(2*) $r \wedge H{\sim}r \wedge G{\sim}r \to [Nq \leftrightarrow (q \wedge FH(Pr \to q) \wedge PG(Fr \to q))]$.

Notice that the wff involving q in (2*) uses r as a parameter name for the point t and states the table of N. In the same manner we can axiomatize any connective. We will give more detail in the main body of the book.

We mentioned that sometimes we cannot axiomatize a connective directly but only with the help of other connectives, called intermediate connectives. The table of a connective $\sharp$ may be so complex that some intermediate connectives $\sharp_i$ are needed and an axiom system is written for $\{\sharp, \sharp_i\}$ together.

Using irreflexivity we can also easily axiomatize various further properties of the flow of time. For example, let us write an axiom stating that each point has an immediate successor:

$$q \wedge H{\sim}q \wedge G{\sim}q \to FHH{\sim}q.$$

There will be more discussion, of course, in the chapters dealing with axiomatizations.

Example 1.4.5 Here is an axiom system for the temporal logic of Manna and Pnueli [Manna and Pnueli, 1981] for program specification and verification for a shared variable concurrent system. This system was represented as example 1.3.6.

Time is the integers. We deal with future only. The connectives are:

$\Box A$	=	A will always be true (beginning now)
$\Diamond A$	=	A will sometimes be true including now
$\bigcirc A$	=	A is true tomorrow
$U^*(A,B)$	=	B begins now to be true until A is true.

Axioms:

$$\sim\Diamond A \leftrightarrow \Box\sim A$$
$$\Box(A \to B) \to (\Box A \to \Box B)$$
$$\Box A \to A$$
$$\bigcirc\sim A \leftrightarrow \sim\bigcirc A$$
$$\bigcirc(A \to B) \to (\bigcirc A \to \bigcirc B)$$
$$\Box A \to \bigcirc A$$
$$\Box A \to \bigcirc\Box A$$
$$\Box(A \to \bigcirc A) \wedge A \to \Box A$$
$$U^*(A,B) \leftrightarrow A \vee (B \wedge \bigcirc U^*(A,B))$$
$$U^*(A,B) \to \Diamond A$$

The inference rules are modus ponens and $\Box$-generalization, $\frac{A}{\Box A}$, and of course, all substitution instances of classical propositional tautologies are axioms.

When we apply the above system to program verification we add the following groups of axioms:

1. axioms about the program application domain;

2. axioms about the concurrent programming principles.

We obtain a logic which can be applied in program verification. We can then, for a given program, describe what it is and state the correctness lemma for it. We can then try to prove the correctness statement using the above axioms.

Unfortunately, as we will see in chapter 4, we need an infinitary axiom system in order to handle full, first-order temporal logic with quantifiers.

Let us look at another example.

Example 1.4.6 Consider a propositional language with the connectives H and P (P being $\sim H\sim$) and the classical connectives $\wedge, \vee, \sim, \to$. The intended interpretation of HA is 'A has always been true', and the flow of time we have in mind is *finite past linear chain*:

$$-n, \ldots, -3, -2, -1, 0.$$

We define the system $Y1$:

1. The connectives are $H, \wedge, \vee, \sim, \rightarrow$ (P is defined as $\sim H \sim$).
2. The language is propositional. Thus any atom q is a wff, and if A and B are wffs so are $A \wedge B, A \vee B, A \rightarrow B, \sim A, HA$ and PA.
3. A flow of time for $Y1$ is any finite chain of the form:

   $$-n < -(n-1) < \ldots < -1 < 0,$$

   where $n > 0$, 0 is the present and $-1, -2, \ldots, -n$ are the past points.
4. An assignment of truth values is any function h giving values 1 or 0 to any atom q at every point of time, i.e. $h(-m, q) \in \{1, 0\}$.
5. The interpretation of HA is: HA is true at $-m$ iff A is true at all points $-n < -m$.
6. A wff A is said to be valid if it has value 1 at any point in any flow of time for any h.
7. The following is an axiom/rule of inference system for $Y1$:
   (a) A, for any truth functional tautology A;
   (b) $\frac{A}{HA}$;
   (c) $H(A \wedge B) \leftrightarrow HA \wedge HB$;
   (d) $PA \wedge PB \rightarrow P(A \wedge B) \vee P(A \wedge PB) \vee P(B \wedge PA)$
   (axiom for linearity);
   (e) $PA \rightarrow P(A \wedge H \sim A)$ (axiom for finiteness).

Theorem 1.4.7 (the completeness theorem for Y1) $Y1 \vdash A$ *iff* A *is valid* (*in every finite linear time model*).

There are several methods for proving completeness and we mention two here. The first is a semantic tableau method. We notice that axiom (e) says that if B was true then there is a first time in which B was true, and axiom (d) says that all past truths are linearly ordered. Thus given a wff A for which we want to find a model, we look at all subformulae B_i of A and if they were true in the past, we can order linearly the first times at which they were true. This gives us the tableau model. The second method is via complete theories and is more general. It is given in chapter 3.

1.5 Algorithmic proof presentation of temporal logic

The usual automated reasoning systems for classical logic cannot be used directly for temporal logics. In this section we give you an idea of the type of problems involved.

Classical resolution methods, the most widespread automated reasoning method for classical logic, cannot be applied directly to temporal logic even in the propositional case. The system is based on normal form where the data clauses are disjunctions of atoms or their negations. Given a set Δ in this form, consider any two such clauses. These may be of the form:

$$A = X_1 \vee \ldots \vee X_n \vee y \text{ and } B = Z_1 \vee \ldots \vee Z_m \vee {\sim}y.$$

In this case, A and B can be resolved to form $C = X_1 \vee \ldots \vee X_n \vee Z_1 \vee \ldots \vee Z_m$ which can be added to Δ. If the empty clause is obtained in this manner, then the original set of data is inconsistent.

The above is classical resolution.

Resolution steps are taken by identifying the two disjuncts y in A and ${\sim}y$ in B. Crucial to this is the rewriting of data clauses in normal form. In temporal logic the above normal form cannot be achieved. The units Gy, Fz, for example, cannot be reduced further. We can thus achieve only a partial normal form, where the 'atomic' units have the form either q atomic, a GA, an FA, an HA, or a PA.

It is quite possible in temporal logic for two units FA and $G{\sim}A$ to do the job of y, ${\sim}y$, because FA and $G{\sim}A$ are contradictory. Furthermore, the contradiction may be embedded inside several nestings of G and F as, for example, is the case with GFA and $FG{\sim}A$. Furthermore, what can also happen is that the contradiction is hidden in several clauses and so several resolution steps have to be taken. For example, $F(A \vee B), G{\sim}A, G{\sim}B$ are contradictory. Such complications do not exist in the classical case, where the basic units for contradiction are atoms or their negations. This requires an inductive definition which will generalize the basic situation of resolving $y, {\sim}y$.

In the sequel we assume for simplicity that the language contains G and F only.

Given two clauses in the new normal form, say $A = X_1 \vee \ldots \vee X_n \vee y$, $B = Z_1 \vee \ldots \vee Z_m \vee y'$, we can have several possibilities:

1. if $y' = {\sim}y$ (or $y = {\sim}y'$) we can add

$$C = X_1 \vee \ldots \vee X_n \vee Z_1 \vee \ldots \vee Z_m.$$

2. $y = GA_1$ and $y' = FB_1$ (or symmetrically, $y = FA_1$ and $y' = GB_1$), in which case let

$$R(y, y') = F(A_1 \wedge B_1 \wedge GA_1).$$

If A_1, GA_1 and B_1 *contradict each other,* then we can add C as in (1). How can we check if GA_1, A_1, B_1 contradict each other? We bring $\Delta_1 = \{A_1, GA_1, B_1\}$ into conjunctive normal form and apply the above resolution process to obtain the empty clause. Note that the complexity of Δ_1 is less than Δ, where the measure of complexity is the maximal number of nested temporal connectives. After a finite number of steps there will be no temporal connectives left.

3. It may be that y, y' do not contradict each other, in which case other resolutions are necessary.

Different temporal logics will have more rules on how $R(y, y')$ can be constructed. The above rules take into account the transitivity axiom $GA \rightarrow GGA$.

Example 1.5.1 Consider the data

$$FFA, GG(A \rightarrow B)$$

and the query

$$FFB.$$

Form the set

$$\Delta = \{FFA, GG(\sim A \vee B), GG\sim B\}$$

we resolve the first and last clause and obtain $F(FA \wedge G{\sim}B)$. This item is not contradictory. We resolve it with the middle clause and obtain

$$F(FA \wedge G{\sim}B \wedge G(\sim A \vee B)).$$

We should really include $GG{\sim}B$ and $GG(\sim A \vee B)$ as conjuncts under the F, but they follow from $G{\sim}B$ and $G(\sim A \vee B)$. Now we have a contradictory clause, as we shall see. Resolve (within the scope of the outer F) $FA \wedge G{\sim}B$ to obtain $F(A \wedge {\sim}B \wedge G{\sim}B)$ and resolve that with $G(\sim A \vee B)$ to obtain $F(A \wedge G{\sim}B \wedge G(\sim A \vee B) \wedge {\sim}B \wedge (\sim A \vee B))$, all within the scope of the outer F, i.e. we have

$$FF(A \wedge {\sim}B \wedge G{\sim}B \wedge G(\sim A \vee B) \wedge (\sim A \vee B)),$$

which classically resolves to

$$FF(\textit{ false }) = \textit{false}.$$

The above method does generalize resolution, but it is not a simple one. It does not apply so successfully in temporal logic, where both P and F are present. The reason is that when we resolve FA with GB to obtain $F(A \wedge B \wedge GB)$ we must include under the F the present as past, i.e. if the database is

$$q \wedge FA \wedge GB$$

after resolution we need

$$F(A \wedge B \wedge GB \wedge P(q \wedge FA \wedge GB))$$

and not, as in the pure future fragment, only $F(A \wedge B \wedge GB)$. In pure future temporal logic, or in modal logic, such complications never arise. In temporal logic they most certainly and most frequently do. Take, for example, $B = (Pq \rightarrow \sim A)$; without the Pq in the resolvent, we never obtain a contradiction.

Obviously in the case of temporal logic it is better to separate the search for contradictions, i.e. movement in reasoning from one moment of time to another, out from the temporal formulae themselves. The second approach we take is to label the formulae with time labels. Thus a database is a set of labelled formulae $t_i : A_i$ together with a graphical representation of how t_i are temporally related. Rules allow us to deduce formulae from other formulae at each label and across labels.

The following are sample rules:

1. **G-Rule:**

$$\frac{t : GA, \ \ t < s}{s : A}.$$

2. **F-Rules:**

$$\frac{t : FA}{\text{create an } s \text{ with } t < s \text{ and } s : A}.$$

3.

$$\frac{t < s; s : A}{t : FA}.$$

4. For K_t one assumes that $<$ is transitive.

We can work through example 1.5.1 using this notation.

Example 1.5.2 Show that

$$t : FFA; \ t : GG(A \rightarrow B) \vdash t : FFB$$

The following is a proof:

1. $t : FFA$ assumption.

2. Create $s, t < s$ with $s : FA$ from F-Rule.

3. Create $s', s < s'$ with $s' : A$ from F-Rule.

4. $t : GG(A \to B)$ assumption.

5. $s : G(A \to B)$ from (4) and G-Rule.

6. $s' : A \to B$ from (5).

7. $s' : B$ modus ponens.

8. $s : FB$ from (7).

9. $t : FFB$ from (8).

If we examine the resolution steps in example 1.5.1, the first resolution created the set of wff parallel to the point s and the second created point s'. With the point notation we have the advantage of being able to run up and down time in any direction.

1.6 Expressive power of temporal connectives

We have seen how to specify a temporal logic semantically. We need a flow of time, a language with connectives, and an evaluation function for atoms and connectives relative to the relevant unit of time. The evaluation function need not be expressible in the first-order theory of the flow of time. So, conceivably, we can have a connective saying: 'I am true at t if and only if the future is a well-ordered set.' As is well known, being well ordered is not a first-order property.

It is interesting to look at temporal logics where the truth tables of the connectives are first-order definable. Certainly in such cases we can hope to axiomatize the temporal logic in some way, and if this logic is used for specification or correctness, then we can hope for an automatic theorem prover to do the checking for us. We now proceed to define more precisely what we mean by a first-order temporal connective of dimension n.

Let $(T, <, =)$ be a partially ordered set, called the flow of time. Consider the monadic first-order theory of this flow of time. This means that we allow, besides $<$ and $=$, unary predicates over time such as $P(t), Q(s)$, and quantifiers over points. Thus we can write

$$\Psi_1(t, P, Q) = \exists t' < t[P(t') \wedge \forall s(t' < s < t \to Q(s))].$$

This is a formula with the free variable t involving P, Q. Another formula could be

$$\Psi_2(t, P, Q) = \exists t' > t[P(t') \wedge \forall s(t < s < t' \to Q(s))].$$

A one-dimensional temporal logic with connectives $\sharp_1, \ldots, \sharp_m$, where the connective $\sharp_i$ has $k(i)$ places, is defined by formulae $\mathbf{C}_1, \ldots, \mathbf{C}_m$ of the monadic predicate logic of time. $\mathbf{C}_i$ has $k(i)$ atomic monadic unary predicates and one free variable t. Thus $\sharp_i(q_1, \ldots, q_{k(i)})$ holds at t if $\mathbf{C}_i(t, Q_1, \ldots, Q_{k(i)})$ holds, where

$$Q_j = \{s \mid q_j \text{ true at } s\}.$$

For example, Ψ_1 above is the table for *Since*(p, q) and Ψ_2 is the table for *Until*(p, q). The above is just a formal definition of a first-order truth table.

An n-dimensional truth table will be a formula $\mathbf{C}(R_1, \ldots, R_k, t_1 \ldots, t_n)$ where $R_1, \ldots, R_k$ are all n-place predicates and $t_1, \ldots, t_n$ are n free variables. Here we allow the use of full predicate logic with $T, <, =$, and require $R_1, \ldots, R_k$ to be n-place and not only monadic. Thus the table for a connective $\sharp(A_1, \ldots, A_k)$ is defined by:

$\sharp(q_1, \ldots, q_k)$ is true at points $t_1, \ldots, t_n$ iff $\mathbf{C}(R_1, \ldots, R_k, t_1, \ldots, t_n)$ holds,

where $R_i = \{(s_1, \ldots, s_n) \mid q_i \text{ is true at } (s_1, \ldots, s_n)\}$.

So, for example, the table for *Next* q is:

$$\mathbf{C}_1(R, m, n) = m + 1 \leq n \wedge R(m + 1, n).$$

The table for $(q_1 + q_2)$ is

$$\mathbf{C}_2(R_1, R_2, m, n) = \exists k(m \leq k \leq n \wedge R_1(m, k) \wedge R_2(k, n)).$$

Since we are using temporal logic to specify and prove properties in some application area, it is of vital importance to us to know the expressive power of temporal logics and to be able to axiomatize the logic. In this section we shall deal with expressive power. We must know how to find expressive connectives and how to use them. We first explain what kind of problem we face.

Take the connectives Since and Until, for integer time. The wff which is the table of Since is Ψ_1. Thus '$S(A, B)$ is true at t' means:

$$\exists s < t \forall u[A(s) \wedge (s < u < t \rightarrow B(u))].$$

The expressions of the logic are built up from S, U by substitution. Thus we can write

$$U(S(S(A,B), {\sim}S(B,B)), U(B,A)).$$

This is a complex expression. Its truth at a point t corresponds to a first-order formula

$$\Psi(A,B,t).$$

Ψ is built up by the quantifiers of S and U which come in *blocks* of two, namely $(\exists\forall)$.

Problem 1.6.1 Can we express all wffs $\Psi(t,A,B)$ of the predicate logic of the time by substituting S, U within each other?

How about

$$\forall s \exists t' \forall u (t < s \vee s > u \vee A(s) \vee B(u) \vee (A(t') \rightarrow A(u)))?$$

Is there a $C(A,B)$, built up from A, B, Since and Until, that is true at t iff the above wff holds? We have three quantifiers here. Can we obtain an equivalent formula using S, U (which have blocks of two?)

The problem of expressive power of a temporal logic has several aspects:

1. How expressive are (i.e. how many Ψs can one generate from) the temporal connectives of our logic?
2. Given a limited number of connectives, which we need for our area of application, how do we find what to add in order to make them more expressive?
3. Can we find enough (but finitely many) connectives which give us the expressive power of all (full) quantification.
4. How does this depend on the flow of time and the unit of time for evaluation?

Clearly for computer science applications we would like some nice, natural connectives which are useful and intuitive, very strong in expressive power and can easily be axiomatized. For example, for integer time, Since and Until are *fully expressive*. This means that for any formula Ψ of predicate logic of the flow of time there is a propositional formula E with Ψ equivalent to (in the flow of time) its truth table, i.e.

$$E \text{ is true at } t \text{ iff } \Psi(t) \text{ holds.}$$

Since and Until are not sufficient for rational number time. For example, we cannot say using Since and Until that we have the situation of Figure 1.3 which is when we have a sort of $(A + {\sim}A)$ connective with a 'gap' in the middle. So there is no last point satisfying A and no first point satisfying ${\sim}A$. We might have A true at all $t < \pi$ and false at all $t > \pi$ (π is an irrational number).

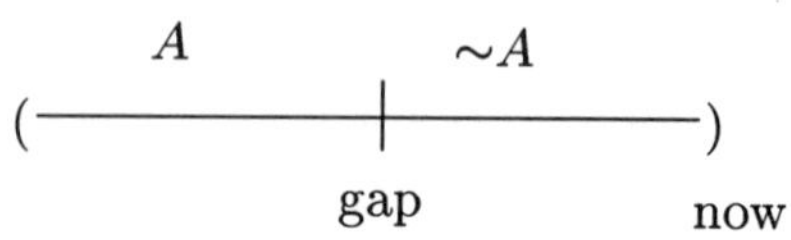

Figure 1.3

We need more connectives for rational time, namely, the connective $U'(A, B)$: 'B is true until a gap with A true on the other side of the gap and ${\sim}B$ true on the other side of the gap at points arbitrarily near the gap' (see figure 1.4), and similarly for $S'(A, B)$: 'B is true since a gap with A true on the other side of the gap and ${\sim}B$ true on the other side of the gap at points arbitrarily near the gap.'

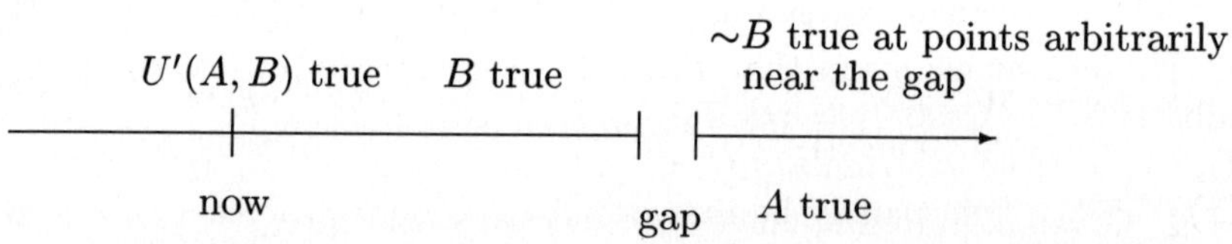

Figure 1.4. $U'(A, B)$.

The following concepts are related to expressive completeness:

1. separation;

2. H-dimension.

We explain separation first. Suppose we have a flow of time and some connectives on it. Suppose we can, for any t, decompose the flow of time into a finite number of regions. Consider, for example, figure 1.5. The obvious regions around t are:

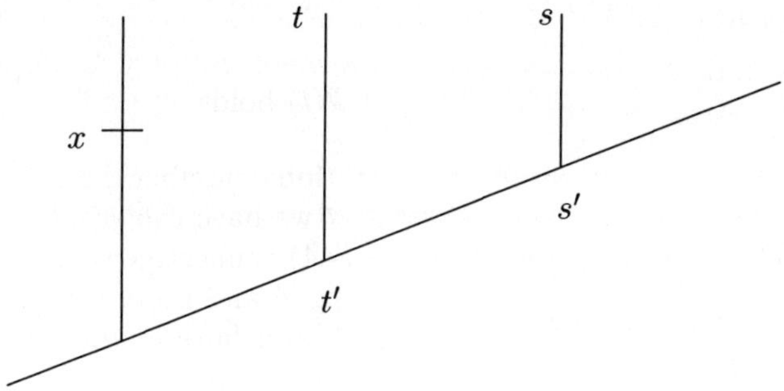

Figure 1.5

1. points above t;

2. t itself;

3. points between t and t';

4. t' itself;

5. points below t' and above to the side (including x);

6. points above t'.

By regions we mean that T can be decomposed into pairwise disjoint subsets of the form $T_1 = \{t\}, T_2, \ldots, T_n$, such that $T = \bigcup_i T_i$ and, for each T_i, there exists a formula $\varphi_i(t, <, =, x)$ such that $x \in T_i$ iff $T \vDash \varphi_i(t, <, =, x)$. The φ_i will usually be first-order but do not have to be. If they are not, then the connectives (which are to be functionally complete) will not have first-order truth tables. This is so because some of the connectives that we are going to use will have the form (roughly) $\|F_i A\| = 1$ iff $\exists s \in T_i(\|A\|_s = 1)$.

Suppose that the connectives we have allow us to distinguish between the regions and the intersection of regions of t and regions of any $s \neq t$ (in our case we need 6×6 intersections). Suppose, further, that the following holds:

> *Any wff of the logic can be rewritten in an equivalent way as a boolean combination of wffs, each wff depending only on a single region (i.e. a pure region wff).*

Then we say that the original set of connectives has the separation property over the flow of time.

Here is the general idea of a theorem which will play an important role in this book:

Theorem 1.6.2 (the separation theorem) *A set of connectives is expressively complete iff it has the separation property (plus some minor conditions which are detailed in theorem 9.4.2).*

Let us see what this means in terms of Since and Until and integer time (see figure 1.6).

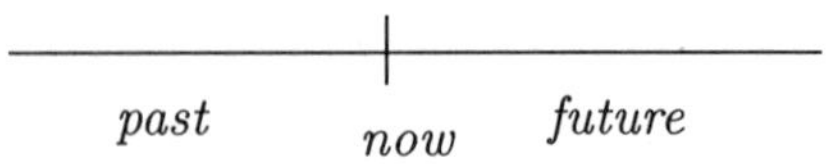

Figure 1.6

The natural regions are Past, Now, and Future. Separation means that every sentence involving Since and Until can be rewritten as a combination of pure past, present, and pure future sentences.

Here is an example. Consider $D = S(a \wedge U(A,B), q)$. This is not a pure sentence. $U(A,B)$ is embedded inside an S. Let us draw a diagram. See figure 1.7.

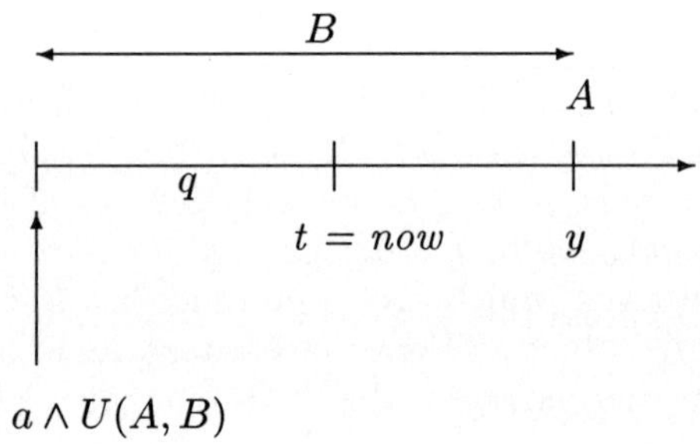

Figure 1.7

There are three cases depending on where y is:

1. If $y < t$ then $D \equiv D_1 = S(A \wedge q \wedge S(a, B \wedge q), q)$.
2. If $y = t$ then $D \equiv D_2 = S(a, B \wedge q) \wedge A$.
3. If $y > t$ then $D \equiv D_3 = S(a, B \wedge q) \wedge B \wedge U(A, B)$.

These are the only cases; thus

$$D \equiv D_1 \vee D_2 \vee D_3.$$

Thus D is equivalent to a boolean combination of pure sentences.

We examine all possible cases of impure nesting of U within S, i.e.
$S(a \wedge U, q)$
$S(a \wedge {\sim}U, q)$
$S(a, q \vee U)$
$S(a, q \vee {\sim}U)$
$S(a \wedge U, q \vee U)$
$S(a \wedge {\sim}U, q \vee {\sim}U)$
$S(a \wedge {\sim}U, q \vee U)$
$S(a \wedge U, q \vee {\sim}U)$
(U appearing twice in an abbreviation above indicates two occurrences of the *same* formula beginning with U). We rewrite each case as a combination of pure wffs. This, together with a careful induction argument, will prove separation of S, U over the integers and show that our language has the same power of expression as quantifying over integer time itself.

If we have a set of connectives and we do not know whether it is expressively complete, we can find what is needed by actually trying to separate and see how we get stuck. Assume we have F, P, G, H and integer time. Take $F(a \wedge Hq)$ (see figure 1.8). At t, q is true, Hq is true, but q is true

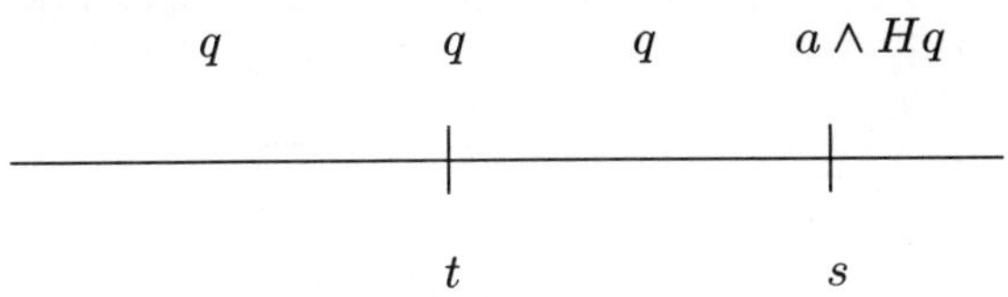

Figure 1.8

until s. We cannot express this and find that we need Until. Thus:

$$F(a \wedge Hq) = q \wedge Hq \wedge U(a, q).$$

In fact *Until cannot be defined using* F, P, H, G.

1.7 Fixed point extensions of the U, S logic

In this section the flow of time is the natural numbers $\mathbb{N} = \{0, 1, 2, \ldots\}$. Atoms are true at time points, without restrictions.

We saw earlier that adding U and S to the logic of F, P, H, G increases the expressive power so that it matches that of the monadic first-order predicate logic of $(\mathbb{N}, <)$. However, we may still not be satisfied. Suppose we want to define the days of the week. We want a formula **Sunday** that is true in any $(\mathbb{N}, <, h)$ at 0,7,14,... only. So the truth values of **Sunday** must be independent of h. In fact, in the U, S logic there is no such formula. Any formula A that is independent of h is true at finitely many points only, or, alternatively, false at finitely many points only, whereas **Sunday** is true at infinitely many points $(0, 7, 14, \ldots)$ and also false at infinitely many $(1, 2, 3, 4, 5, 6, 8, 9, 10, \ldots)$. The fact that we cannot define $\{0, 7, 14, \ldots\}$ in monadic first-order logic either is no consolation: we want to be able to do so.

So we want to extend our logic again. There are various ways of doing this. Three well-known ones are to use ω-regular expressions, to quantify over atomic propositions (see chapter 8), and to allow fixed point expressions. These three all have the same expressive power. Just as the U, S logic 'US' was equivalent to monadic first-order logic, so the first two of these new extended logics are equivalent to monadic second-order logic on the structure $(\mathbb{N}, <)$. This logic is known as $S1S$.

Let us look at the fixed point extension in more detail. It is called *USF*, the 'F' standing for fixed points. *USF* is also equivalent to S1S, at least in a weak sense. See chapter 8. Let A be a pure past formula of US, involving just one atom q. Take $(\mathbb{N}, <, h)$. Then of course the value of h on $q, h(q)$, is a subset of $\mathbb{N}$. We examine what happens to A when we vary this value.

Since A is pure past, the truth value of A at 0 does not depend on h. For example, recall that Yq ('q was true yesterday') abbreviates $S(q, \perp)$. Yq is true at n iff $n - 1 \geq 0$ and q is true at $n - 1$. So if A is $\sim Yq$, then A is true at 0 whatever h is.

Let us now change $h(q)$ so that the truth value of q at 0 is the same as A's. In our example, we make q true at 0, since $\sim Yq$ is. Then as A is pure past, its value at 1 only depends on previous values of q, i.e. the value of q at 0, which we have just set up. So A now has a well-defined value at 1: false, in our example.

We can continue. Define q at 1 to be the new value of A at 1 (still keeping its old value at 0). Now A has a well-defined value at 2. In the example, $\sim Yq$ is now true at 2. We see that we can build up a complete value for q in this way. For $\sim Yq$, we obtain $\{0, 2, 4, \ldots\}$, the even numbers. The extension *USF* of US has a new formula $\varphi q A$, whose interpretation is just this value of q. So here, $\varphi q \sim Yq$ is true at $0, 2, 4, \ldots$ only, in any $(\mathbb{N}, <, h)$.

We can do this formally.

1. If A is a pure past formula of *USF* with free atoms $q, r_1, \ldots, r_k$, then $\varphi q A$ is also a well-formed formula. It is a pure past formula and q is

a bound atom within it: q is no longer free, as the truth value of φqA does not depend on it. Thus the free atoms of φqA are $r_1, \ldots, r_k$.

NOTE: since φqA is also pure past, nested φs are possible.

2. If $(\mathbb{N}, <, h)$ is a flow of time and A is as in (1), define h_i $(i = 0, 1, 2, \ldots)$ by

   - $h_0 = h$
   - $h_{n+1}(x) = h(x)$ if $x \neq q$
   - $h_{n+1}(q) = \{t \in \mathbb{N} \mid \|A\|_t^{h_n} = 1\}$.

   Then we define $\|\varphi qA\|_t^h = \|A\|_t^{h_t}$ for all $t \in \mathbb{N}$.

We defined the value of $\varphi q{\sim}Yq$ at $n+1$ using the values we had already set up for q at $0, 1, \ldots, n$. But, by definition, these values for q are also the values of $\varphi q{\sim}Yq$. So we had: $\varphi q{\sim}Yq$ is true at n iff ${\sim}Y(\varphi q{\sim}Yq)$ is true at n. We see that the value of $\varphi q{\sim}Yq$ is defined *recursively*, in terms of its own previous values.

Equally, we can see the definition of the value of φqA as a *fixed point definition*. In our example, suppose we start with $h(q) = \{0, 2, 4, \ldots\}$. Then it is easy to see that ${\sim}Yq$ is true at $0, 2, 4, \ldots$ also, exactly as q is. Viewing '$q \to {\sim}Yq$' as a set function $\mathbf{f}$, we see that the set $\{0, 2, 4, \ldots\}$ is a *fixed point* of $\mathbf{f}$. So $\varphi q{\sim}Yq$ solves the equation $q \leftrightarrow {\sim}Yq$. In fact it is the only solution.

This is generally true. If A as above is pure past, then fix $(\mathbb{N}, <, h)$. For subset $S \subseteq \mathbb{N}$, define a function $\mathbf{f}$ by $\mathbf{f}(S) = \{t \in \mathbb{N} \mid \|A\|_t^{h_s} = 1\}$, where $h_S(x) = h(x)$ if $x \neq q$ and $h_S(q) = S$. Then $\mathbf{f}$ has a unique fixed point: there is always a unique subset S of $\mathbb{N}$ such that $\mathbf{f}(S) = S$. And we have $S = \{t \in N \mid \|\varphi qA\|_t^h = 1\}$. Again, φqA provides the unique solution of the equation $q \leftrightarrow A(q)$ (in any model $(\mathbb{N}, <, h)$, as we vary $h(q)$).

Example 1.7.1 We define $S(p, q)$. First, define $A = \varphi xY(p \vee (q \wedge x))$. In table 1.2 we tabulate the truth values of A for a few points in a model. We construct the value of x recursively as the value of $Y(p \vee (q \wedge x))$ at each point. The resulting values of x are, by definition, the values of $\varphi xY(p \vee (q \wedge x)) = A$.

We observe that A and $S(p, q)$ agree everywhere. In fact it is easily seen that, for any t, h,

$$\|S(p, q)\|_t^h = \|Y(p \vee (q \wedge S(p, q))\|_t^h.$$

Hence $\{t \in \mathbb{N} \mid \|S(p, q)\|_t^h = 1\}$ is a fixed point of the function $\mathbf{f}$ corresponding to $Y(p \vee (q \wedge x))$ as above.

Table 1.2

t	p	q	x	$Yp \vee (q \wedge x))$	$S(p,q)$
0	0	1	0	0	0
1	1	0	0	0	0
2	0	0	1	1	1
3	0	0	1	1	1
4	1	0	0	0	0
5	0	1	1	1	1
6	1	1	1	1	1
7	0	0	1	1	1
8	0	1	0	0	0

So by the fixed point semantics for φ, A and $S(p,q)$ are actually equivalent formulae.

So if we have Y and φ we can dispense with S.

We have seen fixed point and recursive characterizations for unnested φs. What about nested applications of φs? The fixed point characterization of the semantics of nested φs is still valid. What about the recursive description?

In fact, nested applications of the φ operator correspond to 'multiple recursion', where we define several atoms at t in terms of their previous values. See chapter 8.4. We can ask whether nested applications of φ increase the expressive power of *USF*. The answer is 'not much'. We will prove that every *USF* formula is equivalent to one with at most one nesting of φs. (In fact, it is equivalent to one with at most one level of nesting of φ, and using only Y, U and the boolean connectives. Any formula not involving U is equivalent to one with no nested φs at all.)

So recursion offers an alternative definition of the semantics of φqA to fixed points. Which is better? Since the two forms are equivalent, the differences are minor. The fixed point characterization often allows very short proofs, but for implementation and many more difficult proofs the recursive description is better. In practice one uses both.

We stated above that *USF* and other similar logics are equivalent in expressive power to the monadic second-order logic $S1S$. The proof uses the equivalence of $S1S$ to Muller automata.

Since $S1S$ is known to be decidable, a consequence is that *USF* is also decidable. There is an algorithm that, given a formula A of *USF*, decides whether or not there is h such that A is true at all points in $(\mathbb{N}, <, h)$. This is another way of showing that US is decidable: US is a sublogic of *USF*.

We may not even be satisfied with the expressive power of *USF*. A logic called *USF2* extends the equation–solution idea above. Recall that $\varphi x {\sim} Yx$ was the 'solution' of the 'equation' $x \leftrightarrow {\sim}Yx$. We can extend *USF* to allow formulae providing solutions of more complicated equations where the 'unknown' is not an atom but a connective. For example, we can define a binary connective $M(p,q)$ in this way:

$$\begin{aligned} M(p,q) \quad &\leftrightarrow Y(p \wedge M(p,q)) \vee Y({\sim}p \wedge {\sim}q \wedge M(p,q)) \\ &\vee Y({\sim}p \wedge M([p \wedge Pp], q)). \end{aligned}$$

Notice that $M(p \wedge Pp, q)$ appears on the right-hand side, and $M(p,q)$ on the left. This is not allowed in ordinary *USF*. In *USF*, the arguments on both sides must be the same.

There is a unique solution as the formula on the right is pure past. *USF2* allows M as a valid connective. It is not difficult to see that given any $(\mathbb{N}, <, h)$, $M(p,q)$ is true at t iff p is true at more points in the set $\{0, 1, \ldots, t-1\}$ than q is.

USF2 is strictly more expressive than *USF*.

1.8 H-Dimension

The property of whether or not one can possibly find a *finite* set of fully expressive connectives on a flow of time is a property of the flow of time itself.

Can we characterize this property by some other means, hopefully very simple and intuitive and then use it to show, for an application-oriented flow of time, that full expressiveness is not available? The answer is 'yes' and the property is H-dimension.

Let us begin by writing down that there exist at least two different elements of time. How do we write that? It is very simple. We write

$$\exists xy(x \neq y).$$

How about three different elements?

$$(\exists xyz)(x \neq y \wedge y \neq z \wedge x \neq z).$$

Now try to say the same thing *without* using three different variable letters. Use only two. Is it possible to say that there exist three different elements using two letters only? The answer is, generally not. Sometimes the answer is 'yes', depending on the flow of time. If time is linear, for example, we can write

$$\exists x[\exists y(y < x) \wedge \exists y(y > x)].$$

We used only x here, and y was reused. We know that the second y $(y > x)$ is different from the first y $(y < x)$. If we did not have a total linear order we could not guarantee the existence of three elements in this way, because the above formula would have said *more.*

A flow of time has an H-dimension $< m$ if any sentence about the flow of time involving $<$, unary predicates $P(t), Q(t)$, free variables, and quantifiers, can be written equivalently using at most m different bound variable letters. Linear order can manage with three.

The relationship between H-dimension and expressive completeness is not as straightforward or as useful as one would like but, given the existence of many open problems in the area, there is plenty of room left for new stronger theorems. Because of the technical complications with the current results we will have to await chapter 13 to state them properly but, essentially, we have one positive relationship and one negative result depending on the notion of expressive completeness employed.

If one considers multi-dimensional connectives but only asks that they express all the monadic formulae in one free variable, then we do have the following, which appears later as theorem 13.2.3 after its notation has been defined:

Theorem 1.8.1 (H-dimension theorem) *Let $(T, <)$ be a flow of time. Then the following are equivalent:*

1. *There exists a finite-dimensional temporal logic with a finite number of connectives which is expressively complete over $(T, <)$ in one dimension.*
2. *$(T, <)$ has finite H-dimension.*

A corollary of this theorem is that a partially ordered set which allows for arbitrary large co-chains cannot have a finite set of fully expressive connectives. For example, the tree in figure 1.9 cannot have a set of fully expressive connectives.

Unfortunately, as we will also see in chapter 13, a long-standing conjecture that expressive completeness for *one-dimensional* connectives is also equivalent to finite H-dimension has turned out to be false. Expressive completeness of a finite set of one-dimensional connectives does imply finite H-dimension, but we will see examples with finite H-dimension but without a finite set of expressively complete one-dimensional connectives.

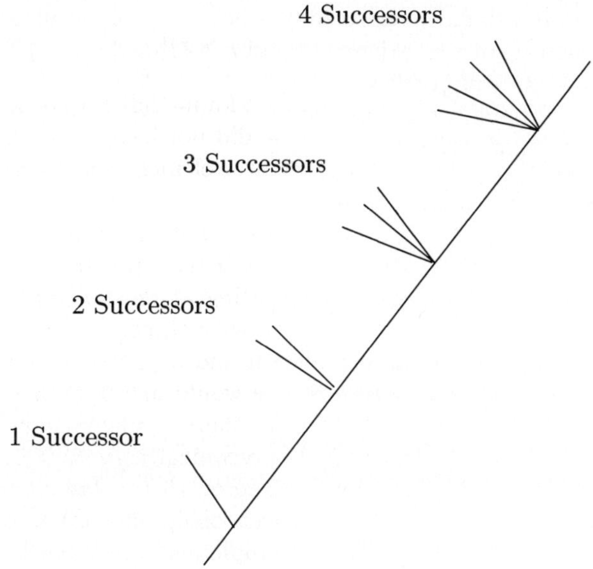

Figure 1.9

Perhaps there exist reasonably workable extra conditions which, along with finite H-dimension, are enough to ensure one-dimensional expressive completeness.

1.9 Decidability

We say that a logic is decidable — or has a decision procedure — if there is an algorithm for determining the validity or otherwise of formulae. Throughout this book we see many applications of various temporal logics. Obviously these logics are much more useful if, besides expressing problems in them, we can effectively reason within them. It becomes very important to find decision procedures or just answer the question whether a logic is decidable or not.

Most propositional modal and temporal systems are decidable, although it is possible to cook up examples of undecidable ones. In chapter 8 we will see that if we add propositional quantifiers then we tend to obtain undecidable logics unless we restrict the semantics (as we do to the natural numbers in that chapter). This is because we may be able to express arithmetic. We will see in chapter 4 that predicate temporal logics are even more complex and, for the same reason, are often undecidable.

To prove that a logic is undecidable one could code into it a known undecidable theory such as arithmetic, all of first-order predicate logic, or the first-order theory of graphs.

There are four main sorts of methods of proving decidability in temporal logic: there are proofs relying on

- automata;
- the finite model property;
- Rabin's theorem [Rabin, 1969]; and
- Ramsey's theorem or similar phenomena.

One key observation in many of these proofs is that the decidability of a temporal logic based on connectives with first-order tables follows from the decidability of the universal monadic second-order theory over the appropriate class of structures. This is because we can ask whether all subsets of the structure make a first-order table true instead of asking whether all valuations for the atoms make the temporal formula true. (Of course, we first have to translate the temporal formula into a monadic formula but, as seen in the proof of lemma 9.1.5, the translation is effective.) Thus it also becomes sufficient to prove the decidability of the full monadic second-order theory of the class of structures.

We will study decidability in detail in chapter 15 but let us now quickly look at these four methods.

Automata are usually only useful on discrete structures with countable height; but they often show the decidability of the full monadic second-order theory. The idea is to define a species of finite state automata and what it means for such an automaton to accept or reject a temporal structure, e.g. $(\mathbb{N}, <, h)$, with an assignment to only finitely many atoms. Then one shows that to each monadic second-order sentence σ there corresponds an automaton M such that

$$M \text{ accepts } (\mathbb{N}, <, h) \text{ iff } (\mathbb{N}, <, h) \models \sigma.$$

For certain classes of structures it is then easy to decide by examining M whether M will accept a structure because its runs will become cyclic. We use such a proof in considering the decidabilty of temporal logics with fixed point operators.

In certain circumstances, if we are dealing with the language with F and P only, then there is no need to use monadic logics to prove decidability: we can use the finite model property. Indeed, sometimes we must use the finite model property, because, as in the example of the class of all flows of time, Prior's logic may be decidable while the universal monadic second-order logic is not. Suppose that for such a logic over the class $\mathcal{K}$ of Kripke frames we have

- a sound and complete axiomatization; and
- the finite model property, i.e. that if A has a model in $\mathcal{K}$ then A has a finite model in $\mathcal{K}$.

Then the logic is decidable. To see this we simply notice that one of the two algorithms (a) going through the possible proofs of $\sim A$ and (b) going through the possible finite models of A in $\mathcal{K}$ will succeed. An important technique for proving the finite model property is that of filtrations, which we will see in a later chapter.

In [Rabin, 1969] the decidability of the full monadic second-order theory $S2S$ of the binary tree is proved. To show that another theory is decidable one may be able to code all its models into the binary tree, as was done with the full monadic theory of the rationals by Rabin. Other examples abound in [Gabbay, 1973]. Rabin's result, which incidently uses automata in its proof, has been generalized by Shelah and Stupp. Gurevich and Harrington used games to give a simple proof (see chapter 15).

Ramsey's theorem [Ramsey, 1930] concerns the surprising behaviour of maps from k-sized subsets of natural numbers into a finite set. It turns out that there must be an infinite subset of the natural numbers on whose k-sized subsets such a function is constant. In [Läuchli and Leonard, 1966] Läuchli and Leonard used this to show that the monadic first-order theory of $(\mathbb{N}, <)$ is decidable. The same method works on structures such as the rationals and integers on which automata cannot be used. A similar method was used by Gurevich in [Burgess and Gurevich, 1985] to prove the decidability of the universal second-order theory over the reals: a case in which the full monadic second-order theory is not decidable. The method was considerably extended by Shelah [Shelah, 1975].

1.10 Sample application to computing, 1

Below, time is the natural numbers $\{0, 1, 2, \ldots\}$. Atoms are true or false at time points, and there is no restriction on their values.

There are two competing approaches to programming: the declarative one and the imperative one. There are advantages to each approach and on first impressions there seems to be a genuine conflict between the two. The executable temporal logic described here shows that these two approaches can truly complement each other in a natural way.

To summarize the idea, take the example of a temporal sentence of the form $Ya \to Tb$. One can view this sentence as a formula of temporal logic which can be either true or false in a temporal model. One can use a temporal axiom system to check whether it is a temporal theorem, etc. In other words, we treat it as a formula of logic.

There is another way of looking at it. Suppose we are at time n. In a real ticking-forward temporal system, time $n+1$ has not happened yet. We can find the truth value of Ya by checking the past. We do not yet know the value of Tb because tomorrow has not yet come. Checking the truth value of Ya is a declarative reading of Ya. However, we need not read Tb declaratively. We do not need to wait and see what happens to b tomorrow. *Since tomorrow has not come yet, we can make b true tomorrow if we want to, and are able to.* We are thus reading Tb imperatively: 'Make b true tomorrow.'

If we are committed to maintaining the truth of the specification $Ya \to Tb$ throughout time, we can read it as 'at any time t, if Ya holds then execute Tb', or schematically, 'if declarative past then imperative future'.

Imperative languages have a little of this feature. For example, if $x < 5$ let $x = x+1$ else $x = 0$. The if $x < 5$ is a very restricted form of declarative test, and the *let* $x = x + 1$ *else* $x = 0$ is imperative. On the other hand, even supposedly declarative languages such as Prolog cheat and allow for imperative statements. Prolog clauses can have the form Print 'B' $\to b$, and the goal Print 'B' is satisfied by Printing. In fact, one can accomplish a string of imperative commands in Prolog just by stringing goals together in a clever way.

The reader may object that above we are considering a very special temporal formula, of the form $Ya \to Tb$, or, schematically, *declarative past* $\to$ *imperative future.* Most formulae are not of this form. Our response is that given any formula A of a temporal logic for which the separation theorem holds, we can effectively find a formula B, a boolean combination of formulae of the form *past* $\to$ *present, future,* that is globally equivalent to A in the sense that in any model, A holds at all times if and only if B does. So our technique is quite general.

Another objection is that we may be asked to execute things over which we have no control. In $Ya \to Tb$ above, maybe b is 'there is a read request from outside', whilst a is 'print the screen'. Executing such a formula makes little sense. The answer is that presumably in such cases it is the original specification that made little sense. In the event, we 'execute' b by waiting and hoping for the best.

We thus see our temporal language as a pointer in the direction of unifying the declarative and imperative approaches in a logical way. Let us now discuss this in more detail.

The conceptual framework follows the paradigm that time is dynamic, and involves both actions and querying databases. These two are connected via logic: the logic of time behaviour. The schematic diagram

REVIEW THE PAST ... AND THEN TAKE APPROPRIATE ACTION

describes intuitively what is going on on the conceptual level.

Imagine a lecturer giving a course. The lecturer maintains, via a research assistant, an attendance chart recording which students attended which lecture, and a similar chart for homework. After the first three lectures the lecturer decides to review the past — to check the attendance and homework sheets — and to take appropriate action: maybe to reduce the homework, terminate the attendance of certain students, etc.

Another example is process control. A group of police are detailed to merge two queues (e.g. traffic queues). They may decide to allow (more or less) equal numbers in from each queue. Every few minutes they review the situation. They may find that they have inadvertently favoured one queue, and may then give precedence to the other queue to compensate.

Conceptually, then, we have:

DECLARATIVE QUERYING OF PAST gives rise to FUTURE ACTION.

The above involves no logic, only basic intuitions. Even babies follow the above rule.

We need a formal logical framework to describe the conceptual framework. It involves a system of temporal logic. Note that the conceptual framework already demands a temporal logic richer than the ones in the current literature. Events are not sufficient. We are dealing with future actions. Time is dynamic; we are within time, acting upon it, changing it, trying to influence events to suit our purpose (specification). Currently available temporal systems are static, in that they only describe.

What kind of logic should we have? First an example. Imagine a trading company. Saying that it is good for business, the boss wants the accounting department to pay invoices within a week: 'There will be no time delays and it will improve our image.'

This specification can be written as

$$\text{ALWAYS}(\text{I} \rightarrow \text{C IN 7 DAYS}).$$

Here, 'I' means 'invoice received' and 'C' means 'cheque sent off'. We must be careful about exactly what we mean by I and C. 'I' can mean 'a letter was opened by the post room and inside was an invoice; this was then stamped received 17 January'. In this case, I was true on 17 January. 'C' can mean that a cheque was printed by the computer on a certain date, and was automatically posted. At this stage we do not need to go into more details of how it is done. So 'C true on 24 January' means that the computer printed the cheque on 24 January.

Using the temporal connectives $\square$, meaning 'always', and T, meaning 'tomorrow', we write the specification as:

$$\square(I \to TC \vee TTC \vee \ldots \vee T^7C).$$

We can also write the same specification as

$$\square(Y^3I \to YYC \vee YC \vee C \vee TC \vee \ldots \vee T^4C).$$

Y means 'yesterday'.

Assume that we are dealing with only one invoice and cheque: the propositional case. Assume that the board of directors want to review whether the accounts department have indeed complied with their instructions. The boss asks the accounts people for the history of invoices and cheques. Here is what they come up with:

1	2	3	4	5	......	19	20
$\sim I$	$\sim I$	I	$\sim I$	$\sim I$	......	$\sim I$	$\sim I$
$\sim C$	$\sim C$	$\sim C$	$\sim C$	$\sim C$	......	$\sim C$	C

This history does not satisfy the specification. The invoice came on the 3rd of the month, and was only paid on the 20th.

What the boss sees is a temporal model. It is static. However, when the accounts department are conducting their day-to-day operations, things are still going on. They want to make sure that outstanding invoices are paid on time. Suppose that today is 20 January. There may be an invoice I' which came in on the 14th, and has not yet been paid. In order to meet the specification, the chief accountant may say, 'please pay this one immediately' (i.e. make C' true at time 21). Here we do have an instance of REVIEW THE PAST, THEN ACT ON THE FUTURE.

The deputy accountant may say that it is too late to pay on the 21st, 'but I'll leave a note for tomorrow (the 21st) to effect payment on the 22nd'. Here we have message passing, which transfers an obligation to act.

Having paid on the 22nd, this event has to be recorded, so that when the boss next conducts an audit, C' will be recorded as true on 22 January.

The reader will see that we are developing a logic here that on the one hand can be used for specification (of what we want the program to do) and, on the other, for execution. (How to pass from the specification to the executable part requires some mathematical theorems, essentially the separation theorem discussed above). Since logically the two formulations are equivalent we will be able to use logic and proof theory to prove correctness.

We start with a declarative specification A, which is a formula of temporal logic. A is transformed into an executable form EF, which is a conjunction of expressions of the form

$$\textbf{Hold } C \text{ in the past} \rightarrow \textbf{Execute } B \text{ now.}$$

At any moment in time, the past is given to us as a database. Thus **Hold** C can be evaluated as a goal from this database in a Prolog program. **Execute** B can be performed imperatively. This creates more data for the database, as the present becomes the past. We have to prove logically that:

1. If the program follows the instruction of EF, then the execution sequences satisfy A, i.e. the temporal model created by the program execution satisfies the temporal formula A.

2. Any execution sequence satisfying A is non-deterministically realizable by the program using EF.

The temporal language can be used for planning. If we want to achieve B we try to execute B. The temporal logic will give several ways of satisfying B while *Hold C* remains true. Any such successful way of executing B is a plan. In logical terms we are looking for a model for our atoms but since we are dealing with a temporal model and the atoms can have an imperative meaning, we obtain a plan.

1.11 Another application

We consider two examples here, two temporal systems which arose from application to computing. These are the systems described in section 1.3: Manna and Pnueli's system, 1.3.6 [Manna and Pnueli, 1981], and Ben Moszkowski's system, 1.3.3 [Moszkowski, 1986].

Example 1.11.1 The Manna–Pnueli system with Until, $\bigcirc, \Diamond$ and $\Box$. This temporal logic can be used to talk about the execution of a program; by formalizing the states and the transition rules of a program we can prove properties of the program.

We base our use of temporal logic on the view of programs as generators of execution sequences. If we allow non-deterministic programs, then each input to program generates a set of possible execution sequences of the form given above. By stating properties which hold for all the execution sequences of a program, we are stating properties of the generating program.

To illustrate this point, let a program be labelled by $l_o, \ldots, l_e$, l_o being the entry point and l_e being the exit point. Let $at\, l_i$ be a predicate true at stage n if at that stage the program is about to execute l_i. *Partial correctness* can thus be stated ((**A**,**B**) is the input–output specification):

$$at\ l_o \wedge (y = x) \wedge \mathbf{A}(x) \to \Box(at\ l_e \to \mathbf{B}(x, y)).$$

This reads: If at any time the variable y equals input x and $\mathbf{A}(x)$ holds, then whenever we reach termination point $\mathbf{B}(x, y)$ holds, where y is the current values of the program variable.

Total correctness is expressed as:

$$at\ l_o \wedge y = x \wedge \mathbf{A}(x) \to \Diamond(at\ l_e \wedge \mathbf{B}(x, y)).$$

$\Diamond$ says we will terminate with the correct values of the program variable y.

We can classify properties according to the form of the temporal formula.

Form $q \to \Box q$ is used for invariance properties such as safety, partial correctness, mutual exclusion, clear performance, deadlock freedom.

$q \to \Diamond q$ expresses eventualities, e.g. liveness, total correctness, accessibility, livelock.

Here are some more examples of what this language can express.

(a) **Response to insistence**

The weakest form of responsiveness states that a permanent holding of a condition or a request p will eventually force a response q. This can be written as: $\Box p \to \Diamond q$, or, if stated over all future behaviours, $\Box(\Box p \to \Diamond q)$.

Sometimes, the response q frees the requester from being frozen in the requesting state. In this case once q becomes true, p ceases to hold, apparently falsifying the hypothesis $\Box p$. This difficulty is only interpretational and we can write instead the logically equivalent condition $\sim\Box(p \wedge \sim q)$, namely, it is impossible for p to be frozen indefinitely and q never to realize.

To illustrate the utilization of such a statement, consider a component of a process which is busy waiting for a condition (say, $x > 0$) which presumably some other process is expected to generate:

$$m: \ \text{if } x = 0 \ \textit{then go to } m.$$

We use the proposition '*at m*' which is true at each time instant in which the execution of our process is at m, namely about to execute m. Then, the only statement we can make about this situation and its resolution is:

$$\sim\Box(at\ m \wedge x > 0),$$

i.e. it is impossible for the process to be stuck at m while x is continuously positive. It implies that the scheduler will eventually schedule this process, which will find x positive and proceed beyond m.

This is also the weakest definition of a semaphore's behaviour, as well as the minimal fairness requirement from any scheduler. It requires that any process will eventually be scheduled for execution. Note that for semaphores this allows infinite overtaking.

(b) **Response to persistence**

A stronger requirement, and the one which most suits the behaviour of a semaphore, is that the infinitely repeating occurrence of the condition p will eventually cause q. We do not require p to hold continuously, but only to be true infinitely often. The statement of this eventuality is:

$$\Box\Diamond p \rightarrow \Diamond q,$$

or, if required continuously:

$$\Box(\Box\Diamond p \rightarrow \Diamond q).$$

Similarly to the case above, the first form is logically equivalent to $\sim\Box(\Diamond p \wedge \sim q)$ and to $\Diamond(\Diamond p \rightarrow q)$, which we will sometimes prefer.

Consider a semaphore instruction:

$$m : p(x).$$

The proper statement of its behaviour is:

$$\sim\Box(\Diamond(x > 0) \wedge at\, m).$$

This states that no process can be waiting indefinitely at m for a semaphore entry while the semaphore's variable turns true (positive) infinitely often. Note that to ensure this it is not sufficient to guarantee that this process will be scheduled infinitely often, because by some highly improbable run of bad luck all of these scheduling instances could coincide exactly with the instances in which $x = 0$, and the process will never proceed. The same criterion should also apply to the fairness of conditional critical section instructions:

$$m : \ \textit{with } r \textit{ when } B \textit{ do } S.$$

Here we should also require:

$$\sim\Box(\Diamond(r > 0 \wedge B) \wedge at\, m),$$

i.e. it is impossible to remain stuck forever at m while states in which both B hold and the resource r is free repeat infinitely often. This can be implemented by means of semaphores and a queue, or by using an unbounded counter. It cannot be implemented with semaphores alone.

A weaker interpretation of the conditional critical section is also possible:

$$\sim\Box(\Diamond(r > 0) \wedge at\, m \wedge B).$$

This one guarantees admission only if B is permanently held true from a certain point on, while r becomes free (positive) infinitely often. The implementation of this weaker construct by semaphores is easy.

(c) **Response to an impulse**

The strongest type of responsiveness is the one in which a single occurrence of p guarantees q. This is written as: $p \rightarrow \Diamond q$ (more generally $\Box(p \rightarrow \Diamond q)$). It is a natural and useful expression for the discussion of total correctness, 'sometimes' reasonings, and other temporal causalities of *internal* events. It is too strong, on the other hand, for the expression of responses to external signals and conditions. This is so because it is seldom the case that a system is so attentive that it could always detect a single occurence of an incoming signal which does not repeat.

Having formulated the different types of responsiveness, we can augment them with requirements for fairness. While managing quite well up to now with just the $\Box, \Diamond$ operators, it is time now to introduce the Until operator U.

(d) **Absence of unsolicited responses**

We may wish to complement the statement that p will eventually cause q, $(p \rightarrow \Diamond q)$, by saying that on the other hand, q will never happen unless preceded by a p. We may wish to state, for example, that a resource allocation system will not grant a resource to somebody who did not request it. Ignoring boundary effects (i.e. p and q happening simultaneously), this can be expressed as:

$$\Diamond q \rightarrow U(p, \sim q),$$

that is, if q is going to happen at all, it cannot happen until p happened first (or concurrently).

(e) **Strict fairness**

Suppose there are two requests p_1 and p_2 and two corresponding responses q_1 and q_2. We may wish to impose strong FIFO (first in first out) discipline on the responding agent and state that if p_1 preceded p_2 then q_1 will precede q_2.

For this it is convenient to express the fact that starting from the present, the first occurrence of p_2 must be preceded by an occurrence of

p_1. In fact this is exactly the expression used in (d) above and we define it generally:

$$Pr(p_1, p_2) = \Diamond p_2 \to U(p_1, {\sim}p_2).$$

The strict responsiveness discipline (FIFO) can now be written as:

$$Pr(p_1, p_2) \to Pr(q_1, q_2).$$

Consider, for example, two competing semaphore instructions:

$$\begin{array}{ll} m : P(x) & n : P(x) \\ m' : & n' : \end{array}$$

A FIFO implementation of semaphores would require:

$$Pr(at\, m, at\, n) \to Pr(at\, m', at\, n').$$

One could use this as a basic axiom for the behaviour of semaphores under a given FIFO implementation and then deduce from it a global FIFO behaviour of larger parts of the program. (d) and (e) are really only a first approximation to the properties we had in mind. For example, $Pr(p, q)$ strictly ensures only that the *first* occurrence of q is preceded by an occurrence of p. What exactly we would like to have is: 'Every q is preceded by a p which happened after the last q, if any'. Thus we cannot be satisfied with a single p succeeded by many qs, even though this situation formally satisfies the requirement: 'every q is preceded by a p'. The property described above, i.e. interleaving of at least one p between consecutive qs, can be described as:

$$Pr(p, q) \wedge \Box(q \to Pr(p, q)),$$

which states that picking as a reference point either the initial state, or any instant in which q holds, ensures that the first instance of q *after* the reference point is preceded by an instance of p occurring between the reference point and the instance of q.

(f) **Mutual exclusion**

P_1 and P_2 are two processes wanting to enter a critical area. P_1 controls the variable y_1 exclusively. The variable t is shared. P_1 makes y_1 true when interested in entering the critical section and makes $t = 1$. P_1 can enter if either y_2 is false or $t = 1$. It makes y_1 = false when exiting. P_2 performs the symmetrical actions when it wants to enter the area. We want to prove in logic that there is no deadlock, i.e.

$$P_1 \text{ is waiting (i.e. } y_1 = t) \to \Diamond(P1 \text{ is in critical section}).$$

Here is the diagram of the sequence of events!

	start:	$y1 =$ false $y2 =$ false $t = 1$		
l_o	non critical		m_0	
l_1	$y_1 :=$ true		m_1	$y_2 :=$ true
l_2	t := 1		m_2	t := 2
l_3	if y_2 is false go to l_5		m_3	if y_1 is false go to m_5
l_4	if t := 1 go to l_3		m_4	if t := 2 go to m_3
l_5	critical section		m_5	critical section
			m_6	$y_2 :=$ false
l_6	$y_1 :=$ false			
l_7	go to l_0		m_7	go to m_0

Example 1.11.2 Let us now consider the interval time logic of example 1.3.3 and see what we can express using its connectives. Recall that evaluation is at intervals $[m, n]$. The connectives are *Next* A and $(A + B)$.

1. $\Box A$ is true at $[m, n]$ if for all $m \leq m' \leq n' \leq n$, A is true at $[m', n']$. This can be expressed as: $\sim(\top + \sim A + \top)$.

2. Let *empty* $= \sim$ *Next* $\top$. Then it holds at $[m, n]$ iff $m = n$.

3. Let *skip* $=$ *Next empty*. It says that the interval is one unit long. Thus, *skip* + *skip* + ... + *skip* (n times) says the interval has length n.
   Thus we can say:

4. (*skip* + *skip* + q) $= q$ holds after two units of time.

5. (*skip* $\wedge q$) $= q$ is true at some unit subinterval.

6. The yield operator $A \rightarrow\!\!\rightarrow B$ holds at $[m, n]$ iff for any k with $m \leq k \leq n$, if A holds at $[m, k]$ then B holds at $[k, n]$. $A \rightarrow\!\!\rightarrow B$ can be defined as $\sim(A + \sim B)$.

7. $(\top \rightarrow\!\!\rightarrow A)$ says A is true at some final subinterval.

8. $\Diamond_i q = (q \rightarrow\!\!\rightarrow \top)$: q true at some initial subinterval.

9. Let *beg* $q =$ ((*empty* $\wedge q) \rightarrow\!\!\rightarrow \top$). It means that q begins.

10. Let *fin* $q = (\top \rightarrow\!\!\rightarrow$ (*empty* $\wedge q$)). It means that q is the final state.

11. Let $\Diamond_t q = (\top \rightarrow\!\!\rightarrow q)$. This reads: q is true at some terminal subinterval. Let $\Box_t q$ read: at all terminal subintervals q is true. Then:

12. halt $q = \Box_t(q \leftrightarrow \mathit{empty})$ reads: the interval finishes only when q is true.

13. Let $\mathit{keep}\ q = \Box_t(\sim \mathit{empty} \rightarrow q)$.
    This reads that q is true in all non-empty terminal intervals.

    This example is based on the logic of true intervals, called Tempora by B. Moszkowski in [Moszkowski, 1986]; it allows for propositional quantification, which will be discussed in detail in a later chapter. Meanwhile, assuming we can quantify over propositions, we can define:

14. $q\ \mathit{until}\ p = \exists x[\mathit{beg}\, x \wedge \Box_t(\mathit{beg}\ x \rightarrow p \vee (q \wedge\ \mathit{Next}\ \mathit{beg}\ x))]$. x is a new propositional variable, x is initially true and inductively remains so until q is true.

15. We can iterate:
    $2q = q + q$
    $nq = q + (n-1)q$
    $\star q = \Sigma_1^\infty q = q + q + q + \cdots$.
    $* q = \exists x(\mathit{beg}\ x \wedge \Box_t\ \mathit{beg}\ x \rightarrow \mathit{empty} \vee \Diamond_i(q \wedge \mathit{Next}\ \mathit{halt}\ \mathit{beg}\ x))$
    This says that the interval is of infinite length.

16. $\mathit{While}\ p\ \mathit{do}\ q = [*((\mathit{beg}\, p) \wedge q) \wedge \mathit{fin}(\sim p)]$.

Chapter 2

Semantical presentation of temporal connectives

In this chapter we discuss in more detail some of the aspects involved in presenting a temporal logic semantically. We deal with propositional temporal logics but the same applies to predicate temporal logics.

2.1 The flow of time

The choice of flow of time depends, of course, on the area of application that one has in mind. Generally, a flow of time has the form of a partially ordered set $(T, <, =, 0)$, where $0 \in T$ is a distinguished moment and $<$ is an irreflexive and transitive relation, called the earlier–later relation. 0 is taken to be the present moment or some special moment of time and is optional. Some flows of time are just $(T, <, =)$.

Instead of one flow of time we may have several flows linked in some manner, or a flow and certain of its subflows. In such a case we write the flow as a system $(T_a, <_a, =), a \in A$.

Below are some examples of choices of flows.

Example 2.1.1 In logics arising in program verification one is interested in linear discrete time. Thus one may take as a flow of time the natural numbers $\mathbb{N}$. Each discrete unit corresponds to one execution of a deterministic program. 0 is the first moment.

Example 2.1.2 Suppose we are interested in the analysis of future sentences in English. We look at some examples of future sentences that have similar but yet different shades of meaning, for example:

1. The train will leave tomorrow.
2. The train leaves tomorrow.
3. The train is leaving tomorrow.
4. The train is going to leave tomorrow.
5. The train will be leaving tomorrow.
6. The train was leaving tomorrow.

The precise, correct analysis of these sentences is not so important at the moment, as we are not analysing the future, but only giving another example of the choice of flow of time. We may therefore decide approximately that the differences between the sentences lie in different degrees of planning and expected certainty. We will therefore choose branching future, allowing for several future possibilities, together with one linear path for the planned future. The above may be realized in a flow of time of the following form: $(T, <, =, T_a), a \in T$, where T is the set of all moments of time, and $<$ is an irreflexive and transitive relation on T. $<$ is branching into the future and linear into the past. Thus for any $t \in T$ we have the type of diagram shown in figure 2.1. T_t is the planned maximal linear future passing through t.

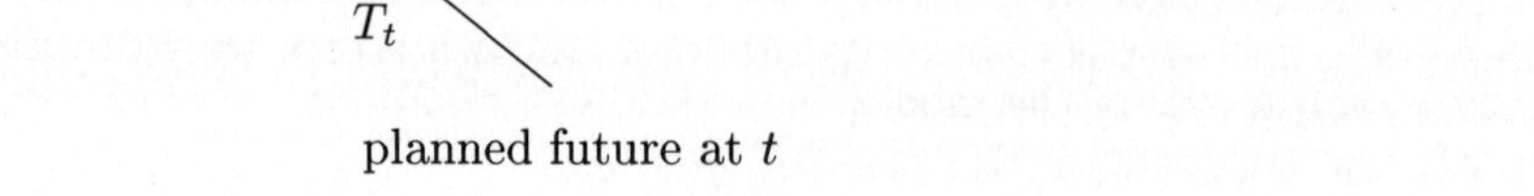

Figure 2.1

Axiomatically $(T, <, =, T_a), a \in T$ has the following properties:

(a) $<$ is irreflexive and transitive, i.e.

(a1) $\forall x {\sim} (x < x)$,

(a2) $(\forall x, y, z)(x < y \wedge y < z \rightarrow x < z)$.

(b) The past is linear:
$(\forall x, y, z)(x < z \wedge y < z \ \rightarrow \ x < y \vee y < x \vee x = y)$.

(c) For each $a \in T, T_a$ is a maximal linear path through a, i.e.

(c1) $T_a \subseteq T$,

(c2) $a \in T_a$,

(c3) $(T_a, <)$ is a linear ordering,

(c4) $(\forall x, y, z)(x \in T_a \wedge x < y \wedge y < z \wedge z \in T_a \rightarrow y \in T_a)$,

(c5) $\sim\exists x(\forall y \in T_a(x < y) \vee \forall y \in T_a(y < x))$.

(d) Note that for any $a \in T$ the past of a is the actual history but the future may be branching and unknown.

Example 2.1.3 A flow of time may be philosophically motivated. We may want to study the notion of historical necessity and thus may want a model of the form $(T, <, =, T_i), i \in J$, where $(T, <, =)$ is as in the preceding example and T_i are some downward closed linear intervals indexed by an arbitary set J. T_i are some possible histories. We may demand that $T = \bigcup T_i$, i.e. through each point of time t there passes a possible history. There may be more than one possible history through a given $t \in T$.

The notion of historical necessity is formalized by saying that at a point $t \in T$, A is historically necessary if it is true in all histories T_i in which $t \in T_i$.

The above same model also arose in the analysis of events. Someone may be 'building a house' at a moment $t \in T_i$: T_i being the path of actual history. The house may never be finished. How do we know she was really building a house at t and not building a restaurant? One way of dealing with the problem (though not without its own difficulties) is to say that in some other possible histories, T_j with $j \neq i$ and $t \in T_j$, the house was indeed built.

The exact details of the model depend on the data, but certainly such flows of time are useful in this case.

Example 2.1.4 The most common flows of time are of the form $(T, <, =)$ where $<$ is assumed, in most cases, to be irreflexive and transitive, and where both the future and past may be branching. These flows are studied in connection with the simplest temporal language with connectives 'A will be true' and 'A was true'.

Various restrictions are put on, and their effect on the valid sentences of the temporal language is studied. Here are some of these natural restrictions:

(a) $\forall t \exists s (t \leq s \wedge \sim\exists u(s < u))$.
In words: above each point (i.e. t) there exists an endpoint (s). See figure 2.2 for an example.

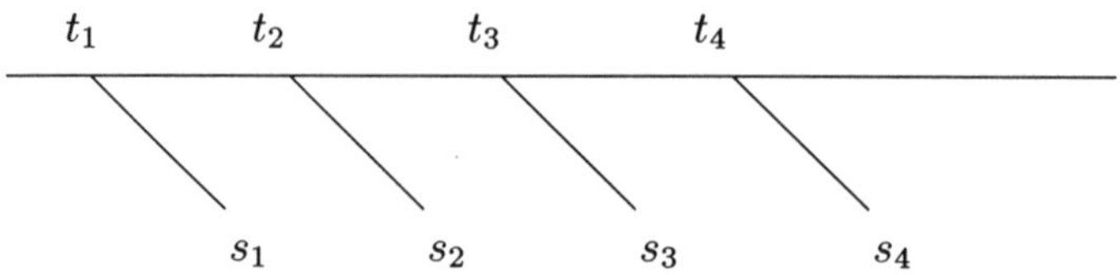

Figure 2.2

(b) We can strengthen our demand for a last (doomsday) point: $\exists t \forall s (s \leq t)$.
See figure 2.3 for an example.

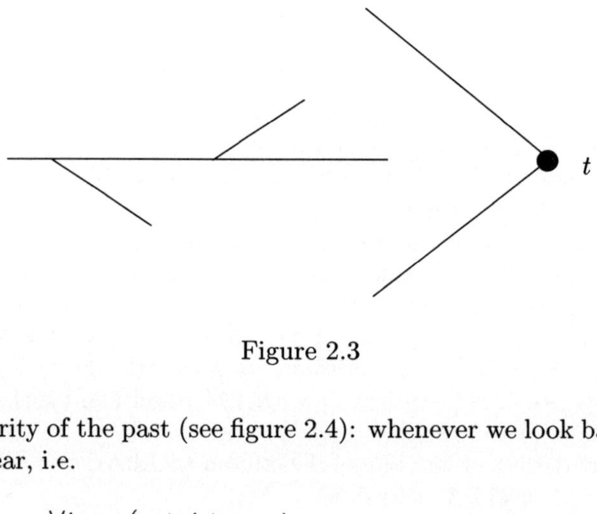

Figure 2.3

(c) Linearity of the past (see figure 2.4): whenever we look back the past is linear, i.e.

$$\forall t, x, y(x < t \wedge y < t \rightarrow x = y \vee x < y \vee y < x).$$

We can, similarly, require future linearity. If we require both we obtain full linearity.

(d) Property of discreteness of a point t:

(d1) Existence of tomorrow point: $\exists s(t < s \wedge \forall u(t < u \rightarrow s \leq u))$.

(d2) Existence of yesterday point: $\exists s'(t > s' \wedge \forall u(t > u \rightarrow s' \geq u))$.

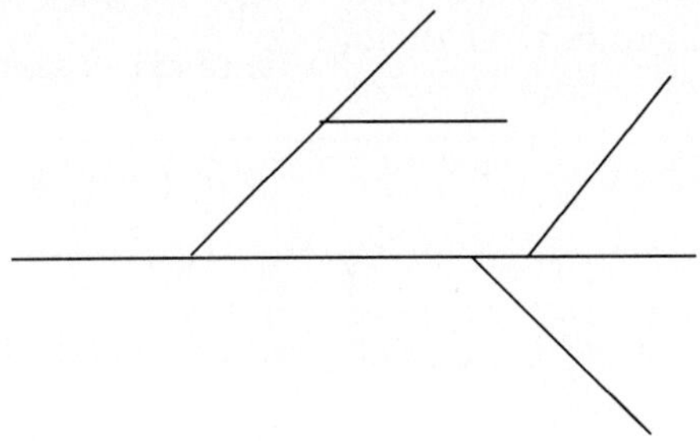

Figure 2.4

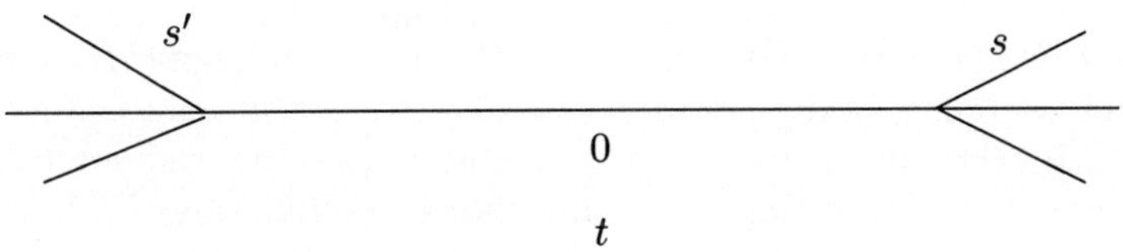

Figure 2.5

(d) = (d1) and (d2) as in figure 2.5.

(e) A discrete flow of time:
The flow of time of the form shown in figure 2.6 is both total linear and discrete. It is obtained from T = integers.

Figure 2.6

(f) Density: $\forall xy(x < y \rightarrow \exists u(x < u < y))$.

All the above conditions on the flow of time are first order. We can ask for conditions that are higher order. For example:

(g) The flow of time is well ordered, i.e. any descending (into the past) set of moments has a first element, or equivalently there does not exist an infinite sequence of points with $t_1 > t_2 > t_3 \ldots > t_n > \ldots$.

(h) The flow of time is complete, i.e. it has no gaps. Figure 2.7 gives an example of a gap in the rationals (at π).

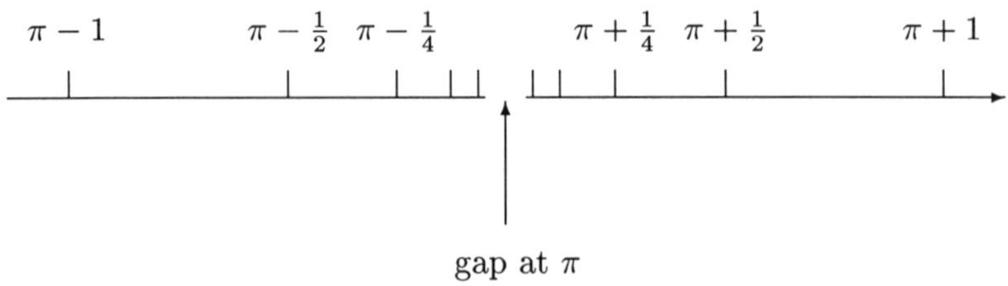

Figure 2.7

The notion of event is relevant here. When we talk about a flow of time, we also have points in time or moments of time. The atomic sentences may have truth values at points of time or at intervals. One common feature to all such models is that time ticks on although nothing may be happening. Thus the temporal model shown in figure 2.8 is mathematically equivalent

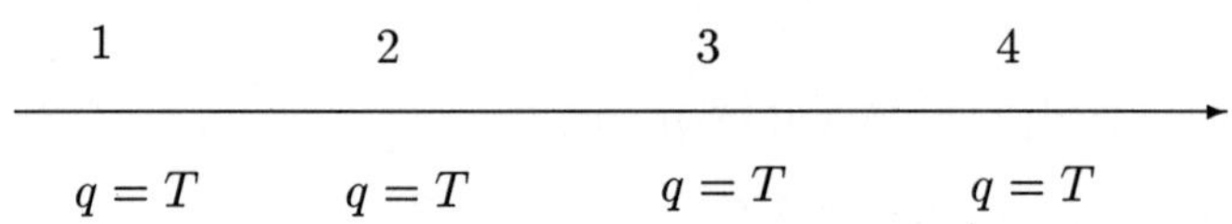

Figure 2.8

to that shown in figure 2.9, a reflexive model with one point, because only

Figure 2.9

one event has occurred, namely q.

If we adopt the view that there is no actual metric time but, rather, moments of time exist only because some *change* has occurred, then we are dealing with a very primitive form of a theory or calculus of events. We say 'primitive' because the notion of an event is very complicated. We will discuss this in a later chapter. Here let us only note that for computer science applications, philosophical theories of events are less likely to be useful. We all go around with our wristwatches and are used to global time counting, hours, minutes, seconds, etc. There are event driven operations, but they can easily be dealt with within a metric (time flow) temporal logic and do not justify the adoption of an event-only framework. We do not want to have to introduce a new event every time another hour or second goes by.

2.2 The assignment to the atoms

Let us now examine the possible assignments we give to the atoms. We confine ourselves to propositional logics. An atom q is given different values at different moments of time. So we can take a function h such that at each moment t of time, $h(t, q) = 0$ or 1 (or T or F, depending on one's notation). 1 is true, 0 is false. We can impose restrictions on h. For example, for the predicate 'is alive' we can require that 'once dead always dead'; the type of requirement we will have is:

1. $h(t, q) = 0 \wedge t < s \rightarrow h(s, q) = 0.$

Another possible requirement is continuity:

2. $h(t, q) = 1 \wedge h(t', q) = 1 \wedge t < x < t' \rightarrow h(x, q) = 1.$

We sometimes use the notation of $h(q)$, given by: $h(q) = \{t \mid h(t, q) = 1\}$, i.e. $h(q)$ is the set where q is true. We shall use this notation whenever it is more convenient.

Note that the kind of requirement on h can make an important difference to the type and strength of the logical system obtained. To give an extreme example, the condition $\forall x, y\{h(x, q) = h(y, q)\}$ yields classical propositional logic.

Another decision related to atoms is whether the value depends on moments or on paths. Why should h be only of the form $h(t, q), t \in T$? It could be $h(T_1, q)$ where T_1 is a history or a path, or possibly also $h(T_1, t, q)$? What we choose depends on the applications, as we have seen in the preceding section.

Another component in the definition of the unit of time for truth values is the choice of dimension, which means the number of indices for evaluation of formulae. There are many examples illustrating the need for indices. The simplest, perhaps, is the analysis of the future perfect. We need a future evaluation point and we must record the speech point as a reference point.

Thus in such cases the value of a wff A should be computed at two indices. We use the notation $\|A\|^h_{t,r}$ where h is the assignment to atoms, r is the reference point, and t is the evaluation point. In this case we say the dimension is 2. The evaluation point is important because atoms are evaluated there. Atoms disregard reference points. Thus we have

1. $\|q\|^h_{t,r} = h(t,q)$, for q atomic;

or equivalently using our other notation we can write

1*. $\|q\|^h_{t,r} = 1$ iff $t \in h(q)$, q atomic.

For general A, $\|A\|^h_{t,r}$ will depend on r as well as on t. A may say something like 'q is true between r and t and $r < t$' using some special connectives.

This can also happen for implementation reasons, when the points of time at which q is true are stored via maximal intervals on which q is true.

Of course it is also quite possible to make the atoms q dependent on more than one index, i.e. make h a function $h(t,r,q)$. An example is in databases. When we update a value in a database we may want to enter the time of the update. Thus, $h(s,t,q)$ would mean that the item of data 'q is true at time t' was entered in the database at time s. We can thus have

$h(1,2,q) = 1$	—	q put in the database at time 1, to hold at time 2, i.e. q was planned at time 1.
$h(2,2,q) = 0$	—	q recorded at time 2, to be false at time 2.
$h(4,2,q) = 1$	—	q corrected at time 4, to have been true at time 2.

Such logics rarely arise in other applications. What does arise is evaluation in intervals or paths but that is a special topic to be treated independently. We call such semantics 'interval semantics'. What we are dealing with now is 'point semantics', as opposed to interval semantics.

To summarize the notion formally, we say an interpretation is m-dimensional if evaluation is performed at m index points comprised of one evaluation point and $m-1$ reference points. Any formula A is therefore evaluated as follows:

1. $\|A\|^h_{t,r_1,\ldots r_{m-1}}$ = value 0 or 1.

2. $\|q\|^h_{t,r_1,\ldots r_{m-1}} = h(t,q)$, for q atomic.

We mentioned before that for the future perfect one needs at least one evaluation point and one reference point. Are there examples in English that need more? Reichenbach [Reichenbach, 1957] suggested three indices.

There may be a need for even more; we believe perhaps three or four are sufficient. At any rate, we certainly believe that a finite fixed number is sufficient, though some authors claim to have examples which require more and more indices, i.e. an unbounded number of indices.

Note that one can write axiom systems for simple temporal connectives which force the semantical interpretation to have more than one index. Similarly, there are systems (Hilbert axioms and modus ponens) which force restrictions on the assignment h. It can be proved for these systems that without restrictions on h or without more indices we cannot obtain a completeness theorem.

Example 2.2.1 Consider the following perfectly natural description:

> *When the police inspector arrested the burglar, he knew that by the time the criminal is brought before the judge, he will have realized that his best policy would be to deny that he bought the gun.*

To represent the temporal sequence here correctly, we must record the following indices:

(a) a reference point of 'arrest' because the point of 'buying the gun' should come before the arrest;

(b) a reference point of 'facing the judge' because the realization point comes before that;

(c) a point for evaluation;

(d) any more points?

The question of how many indices are required for English is important technically. If the number needed is not fixed and bounded, i.e. we need more and more indices, then we obtain a propositional temporal logic which is equivalent to the predicate logic on $(T, <, =)$. We must have certain connectives in the propositional temporal logic, to be able to realize the equivalence. If we have these connectives in the temporal logic then essentially we obtain a variable free predicate logic on $(T, <, =)$. The connectives we need are of the form: 'There is a reference point that ....' [Recall that 'John will come' has the form 'There is an evaluation point ....'] So these connectives are simple and natural.

We are not saying that we cannot obtain a variable free predicate logic of $(T, <, =)$ without unbounded dimension. In fact, for many 'nice' flows of time and 'well-chosen' connectives, even in dimension 1, the temporal propositional logic with these connectives is essentially the predicate logic on the flow of time. More about the subject will be said in the chapters on expressive completeness.

2.3 The connectives

We now turn to the fourth component in the semantic definition of a temporal logic, namely the choice of connectives. Each well-formed formula A has a set of indices $h(A)$ in which it is true. In symbols:

$$h(A) = \{(t, r_1, \ldots, r_{n-1}) \mid \|A\|^h_{t,r_1,\ldots,r_{n-1}} = 1\},$$

where h is the original assignment to the atoms. We assumed above that we are dealing with an n-dimensional logic with one evaluation point and $n-1$ reference points. A connective is characterized by its number of places, i.e. unary, binary, etc., and by its table. The table tells us how to compute the truth values for the connectives.

So if $\sharp(q_1, \ldots, q_m)$ is an m-place connective, its table $\psi_\sharp$ will tell us what is the set $h(B)$ in terms of the sets $h(A_i), i = 1, \ldots, m$, where $B = \sharp(A_1, \ldots, A_m)$. In simple words, $\psi_\sharp$ tells us where B is true, if we know where each A_i is true. So the choice of connectives is really a choice of tables. Depending on what we want to represent in what area of application, we choose different connectives.

Example 2.3.1 Consider a one-dimensional logic with evaluation points only. Thus for a flow of time $(T, <)$, we evaluate wffs at $t \in T$. In symbols: $\|A\|^h_t$. The most natural connectives are the existential ones. They are unary, denoted by Fq and Pq. They represent the English:

$$Fq = \text{`It will be the case that } q\text{'};$$
$$Pq = \text{`It was the case that } q\text{'}.$$

Their truth conditions are:

$$\|Fq\|^h_t = 1 \text{ iff } \exists s > t \|q\|^h_s = 1;$$
$$\|Pq\|^h_t = 1 \text{ iff } \exists s < t \|q\|^h_s = 1.$$

There are two other extremely important binary connectives for one dimension. These are Since and Until. Their presence in a logic give it, in many cases, the expressive power of first-order predicate logic. Their truth conditions are

$$\begin{aligned} \|U(p,q)\|^h_t = 1 \quad &\text{iff} \quad \exists s > t(\|p\|^h_s = 1 \\ &\qquad \wedge \forall x(t < x < s \to \|q\|^h_x = 1)); \\ \|S(p,q)\|^h_t = 1 \quad &\text{iff} \quad \exists s < t(\|p\|^h_s = 1 \\ &\qquad \wedge \forall x(s < x < t \to \|q\|^h_x = 1)). \end{aligned}$$

$U(p,q)$ reads 'q will be continuously true until a point where p is true' and $S(p,q)$ reads 'q has been continuously true since a point where p was true'.

Formally, using the notation ψ for tables, we can write:

$$\begin{aligned}
\psi_F(T_0) = &\quad \{s \in T \mid \exists x \in T_0(s < x)\}. \\
\psi_P(T_0) = &\quad \{s \in T \mid \exists x \in T_0(x < s)\}. \\
\psi_U(T_0, T_1) = &\quad \{t \in T \mid \exists s \in T(s \in T_0 \wedge \forall x(t < x < s \to x \in T_1))\}. \\
\psi_S(T_0, T_1) = &\quad \{t \in T \mid \exists s \in T(s \in T_0 \wedge \forall x(s < x < t \to x \in T_1))\}.
\end{aligned}$$

We will not insist in the future on formal definitions for ψ. It is quite sufficient to explain how to obtain the new truth sets from the old ones.

Similarly, we can define the universal connectives G and H:

$$\begin{aligned}
\|Gq\|_t^h = 1 \text{ iff } \forall s > t \|q\|_s^h = 1. \\
\|Hq\|_t^h = 1 \text{ iff } \forall s < t \|q\|_s^h = 1
\end{aligned}$$

It is easy to see that

$$\begin{aligned}
Fq = {\sim}G{\sim}q = U(q, q \to q). \\
Pq = {\sim}H{\sim}q = S(q, q \to q).
\end{aligned}$$

The connectives F, P are really the basic quantifiers and we therefore expect any logic to contain them.

Example 2.3.2 Again, given a flow of time $(T,<)$ and evaluation at points $t \in T$, we can consider the monadic language $(T,<,=,Q_i)$ where $Q_i \subseteq T$ are subset variables and we allow point quantification $\forall x \in T, \exists x \in T$. Any wff $\varphi(t,<,=,Q_i)$ containing t free only, and $Q_1,\ldots,Q_m \subseteq T$, can serve as a *table* for a one-dimensional m-place connective $\sharp$. We can use φ as follows:

If $B = \sharp(q_1,\ldots,q_m)$ we let

$$h(B) = \{t \mid (T,<,=) \vDash \varphi(t,<,=,h(q_1),\ldots,h(q_m))\}.$$

In other words, φ tells us where B is true. In fact, in the preceding examples, we can take for F, P, U and S the following formulae:

- $\varphi_F(t,Q) = \exists x > tQ(x)$;
- $\varphi_U(t,Q_0,Q_1) = \exists s > t(Q_0(s) \wedge \forall x(t < x < s \to Q_1(x)))$;
- $\varphi_P(t,q) = \exists x < tQ(x)$;
- $\varphi_S(t,Q_0,Q_1) = \exists s < t(Q_0(s) \wedge \forall x(s < x < t \to Q_1(x)))$.

Of course these tables φ are first order. There may be other tables for connectives which are not first order. These tables can be expressed in some second-order theory such as $S1S$, the monadic second-order theory of one successor function.

Example 2.3.3 We said above that F and P are basic connectives for one-dimensional temporal logics because they are the existential quantifiers. Let us examine what kind of connectives we can use for the existential quantifiers in two-dimensional logic. In other words, we evaluate at two points, a reference point r and an evaluation point t: in symbols: $\|A\|_{r,t} = 0$ or 1. Of course, we have three possibilities here, $r = t$, $r > t$, and $r < t$. Let us take the case of $r < t$. We must have existential connectives saying that there exists an evaluation point x which is:

(a) before r;

(b) equal to r;

(c) between r and t;

(d) equal to t; or

(e) after t.

The above is not enough because we must also say what the reference point is, for each of the cases (a)–(e). The reference point can be taken (1) as x; or (2) as r; or (3) as t. We have the following 13 possibilities for the new existence connectives E:

$\|Eq\|_{r,t} = 1$ iff one of the conditions below holds:

Case (a): $x < r$.

(a1) $\exists x < r \|q\|_{x,x} = 1$

(a2) $\exists x < r \|q\|_{r,x} = 1$

(a3) $\exists x < r \|q\|_{t,x} = 1$

Case (b): $x = r$.

(b2) $\|q\|_{r,r} = 1$

(b3) $\|q\|_{t,r} = 1$

Case (c): $r < x < t$.

(c1) $\exists x[(r < x < t) \wedge \|q\|_{x,x} = 1]$

(c2) $\exists x[(r < x < t) \wedge \|q\|_{r,x} = 1]$

(c3) $\exists x[(r < x < t) \wedge \|q\|_{t,x} = 1]$

Case (d): $x = t$.

(d2) $\|q\|_{r,t} = 1$

(d3) $\|q\|_{t,t} = 1$

Case (e): $x > t$.

(e1) $\exists x > t \|q\|_{x,x} = 1$

(e2) $\exists x > t \|q\|_{r,x} = 1$

(e3) $\exists x > t \|q\|_{t,x} = 1$

So we can obtain different connectives depending on the condition we choose, e.g. E_{a1} or E_{e2}, etc.

In all the above cases we quantified over the evaluation point. We can also quantify over the reference point. For example, we can rewrite (a1)–(e3) with x as reference point and x, r or t as evaluation point. Not all possibilities are interesting or arise naturally in applications.

Another useful set of connectives should be mentioned here. These connectives allow one to define all the existential connectives of the preceding example.

Example 2.3.4 Consider a temporal evaluation system with two indices. Evaluation has the form $\|A\|_{r,t}$, where r is the reference point and t is the evaluation point. h is omitted from our notation here. We define the following connectives by the following tables:

(a) $\|Fq\|_{r,t} = 1$ iff $\exists x > t \|q\|_{r,x} = 1$;

(b) $\|Pq\|_{r,t} = 1$ iff $\exists x < t \|q\|_{r,x} = 1$;

(c) $\|J^+q\|_{r,t} = 1$ iff $\|q\|_{r,r} = 1$
$\|J^-q\|_{r,t} = 1$ iff $\|q\|_{t,t} = 1$;

(d) $\|Jq\|_{r,t} = \|q\|_{t,r}$;

(e) $\|C^<\|_{r,t} = 1$ iff $t < r$;

(f) $\|C^>\|_{r,t} = 1$ iff $r < t$
$\|C^=\|_{r,t} = 1$ iff $t = r$.

The connectives F, P are the usual existential ones and ignore the reference point completely. The constants $C^<, C^>, C^=$ record the relation between the reference and the evaluation points, and the connectives $J^\pm$ and J equate or switch the reference and evaluation points.

Using the above, all possible quantification connectives can be defined. For example, for the case E_{a3} for the case $r < t$ we have

$$\|E_{a3}q\|_{r,t} = 1 \text{ iff } \exists x < r(\|q\|_{t,x} = 1).$$

The case $r < t$ can be expressed by saying $C^{>}$ holds. In this case we have

$$C^{>} \to (E_{a3}q = JPq).$$

For the case E_{c2} we obtain

$$C^{>} \to (E_{c2}q = P(C^{>} \wedge q)).$$

For the case of E_{c3} we obtain

$$C^{>} \to (E_{c3}q = JF(q \wedge C^{<})).$$

Example 2.3.5 Evaluation at one point can be regarded as a special case of evaluation at two points, where the reference point is always the same as the evaluation point. Thus the one-dimensional connectives Fq and $U(p,q)$ in example 2.3.1 can be presented as two-dimensional F^{+}, U^{+} as follows:

$$\begin{array}{rcl} \|F^{+}q\|_{t,t} = 1 & \text{iff} & \exists s > t(\|q\|_{s,s} = 1), \\ \|U^{+}(p,q)\|_{t,t} = 1 & \text{iff} & \exists s > t\{(\|p\|_{s,s} = 1) \text{ and} \\ & & \forall u(t < u < s \to \|q\|_{u,u} = 1)\}. \end{array}$$

Another possibility for presenting a one-dimensional connective as two-dimensional is to ignore the reference point completely.

Formally, the first order notion of an n-dimensional table for a connective with m places $\sharp(A_1, \ldots, A_m)$ can be taken in the predicate language of $(T, <, =)$. For each A_i the set

$$h(A_i) = \{(t, r_1, \ldots, r_{n-1}) \mid \|A_i\|^h_{t,r_1,\ldots,r_{n-1}} = 1\} \subseteq T^n$$

is the set of indices at which A_i is true. It depends on the assignment h. Thus for different h, we obtain different sets.

We need a wff

$$\varphi(t', s_1, \ldots, s_{n-1}, R_1, \ldots, R_m, <, =)$$

to tell us how to obtain the truth set of $\sharp(A_1, \ldots, A_m)$ from the truth sets of the A_i. Thus we can define:

Definition 2.3.6 φ is a table for $\sharp$ iff for all h, t', $s_1, \ldots, s_{n-1}$, and $A_1, \ldots, A_m$, we have:

$$\|\sharp(A_1, \ldots, A_m)\|^h_{t', s_1, \ldots, s_{n-1}} = 1$$
$$\text{iff } (T, <, h) \vDash \varphi(t', s_1, \ldots, s_{n-1}, h(A_1), \ldots, h(A_m)).$$

Thus consider the connective E_{a2} of example 2.3.4:

$$\|E_{a2}\|_{r,t} = 1 \text{ iff } \exists x < r \|A\|_{r,t} = 1.$$

Here, φ is:

$$\varphi_{a2}(r, t, R) = [r < t] \to (\exists x < r) R(r, x),$$

where the antecedent $[r < t]$ is put in because in case (a2) we assumed $r < t$. The way that φ is written makes $E_{a2} =$ *truth* if $\sim(r < t)$ holds.

2.4 In search of a good temporal logic

We would like to find a good temporal system to be formulated either within predicate logic or as a separate temporal logic.

Before we continue let us ask if there is any specially useful and particularly interesting flow of time. We are looking for a flow that arises in many applications and is worthy of detailed study.

At first consideration, linear time, or possibly a special case of it (integer time, for example), may come to mind. It is true that linear time is used in many model applications and deserves, without doubt, further detailed study. But even more useful, we think, are the many variations of future branching and linear past flows of time: in other words, the flow of example 2.1.2 and variations of it.

There seems to be unanimous agreement that the past should be linear, though there are some special cases, e.g. temporal logics for special relativity, where the past is not linear. These cases are few. There are many cases, on the other hand, which require branching future. Some of these have been mentioned before.

First we have a lot of evidence from attempts at the logical analysis of English. We saw in section 2.1 that to give a proper analysis of the future tenses in English one needs branching future temporal logic. This need is but a reflection of our inability to predict the future. There are always several possibilities. These manifest themselves in our use of language and should therefore be carried over to the temporal systems suggested for the analysis of language.

Another linguistic manifestation can be seen in the properties of the tensed conditional. A statement such as:

If John had said another word Mary would have slapped him

can be understood as saying that on the basis of whatever had happened up to a point t, including Mary's mood and John's 'suggestions', John faced a fork at t as shown in figure 2.10:

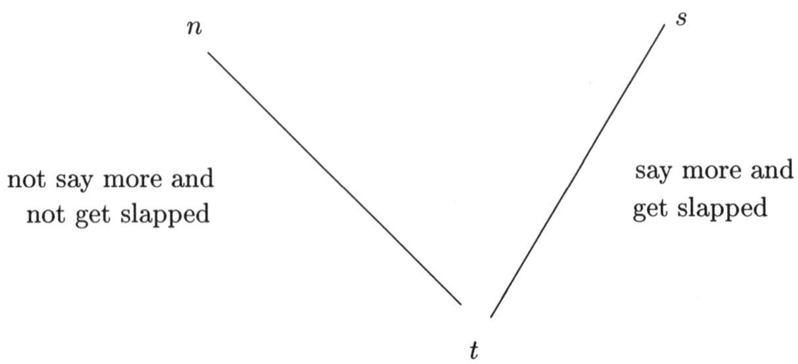

Figure 2.10

The point n is 'now', the point t is in the past of 'now', and s is an alternative to 'now'.

Of course, we gave one special example suitable for 'tense-logical' analysis. There are many problems and many suggested systems for tensed conditionals where not a single one of them could manage without branching future.

A third major indication from language for branching future is the analysis of such events as 'building a house'. We mentioned this example earlier. We can say

She started building a house but never finished.

How do we indicate that the sequence of her doings is 'house-building'? Again, there could be many theories of events but all will have to allow for some sort of possible future branching (figure 2.11). Again, n is 'now', s is the alternative.

From philosophical considerations we also arrive at branching future. Firstly, one may hold the view that the future is 'open'. In other words, it is not the case that we don't know the future and hence must have branching models for dealing with it, but that it is actually branching. Secondly, philosophical systems dealing with causality of historical necessity would most certainly call for a branching future. The exact nature of the system

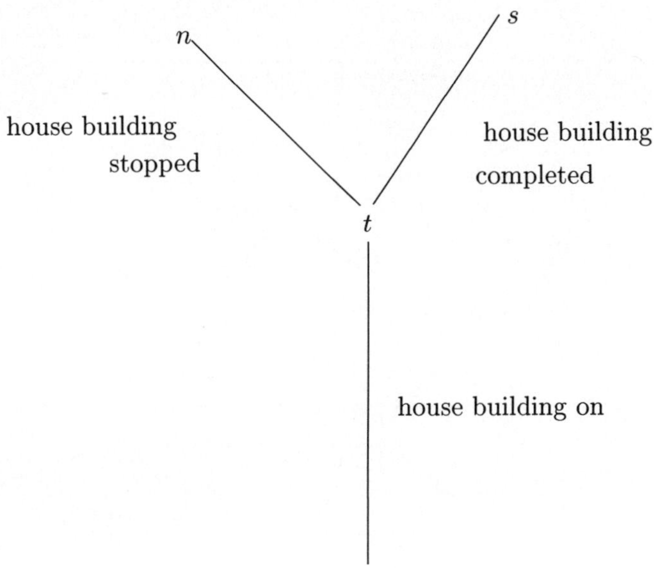

Figure 2.11

involved will depend on the nature and needs of the application but the future will have to be branching.

These systems arising in the philosophy of science also have branching future (in the case of relativity, they may also have non-linear past).

In computer science, branching future arises in the analysis of non-deterministic programs. Suppose we are now in the state of affairs t and we are about to execute the next line in the program a. Being non-deterministic there will be several possibilities which can put us in the states $x_1, \ldots, x_n$. Thus the flow of 'states' should take the form shown in figure 2.12, i.e., branching 'future'.

Planning is another area where branching future is essential. The reason that we took certain actions in the past may have been because of certain possible futures, which unfortunately did not come true.

Another problem we have to settle, having agreed on branching future and linear past, is what are the other properties of the flow of time? There are still many possibilities, but we think that in the general case nothing special can be said. One property which may be true, however, is that whenever we have a branching point, it is discrete, i.e. it has the form of figure 2.13.

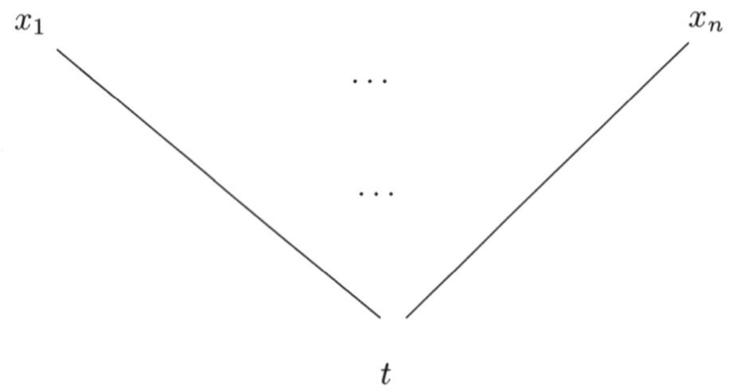

Figure 2.12

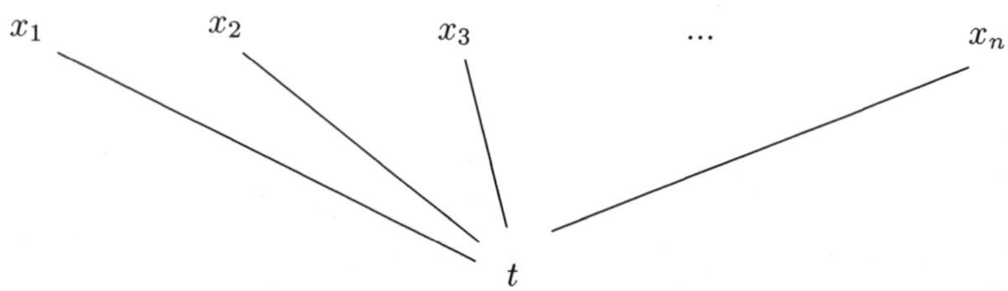

$$t < x_i, i = 1, 2, \ldots, n, \quad \text{and } \forall y (t < y \to \bigvee_{i=1}^{n} x_i \leq y)$$

Figure 2.13

In computer science applications it is reasonable to assume that all points are discrete. For comparison, an example of branching in non-discrete form is given in figure 2.14.

The simplest branching flow of time is the completely discrete one. An example is a tree, with (say) a fixed number of successors for each point. Figure 2.15 shows the first four levels of the infinite binary tree. Each point in the tree has two different immediate successors and one immediate predecessor.

The discrete binary flow is technically sufficient to describe many other flows. For example, we can take the flow with branching past and branching future as a special case of the flow with branching future only. Each point has two futures in the binary flow. If we regard the left future as 'past'

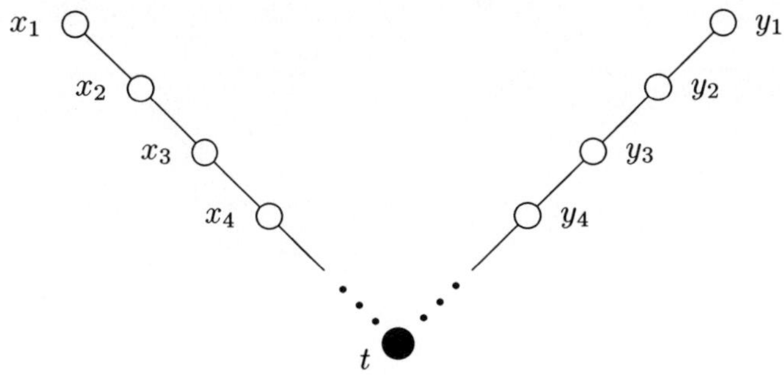

$$t < \ldots x_n < \ldots < x_2 < x_1$$
$$t < \ldots y_n < \ldots < y_2 < y_1.$$

Figure 2.14

and the right future as 'future' we obtain branching future and past.

This is best explained in a diagram (figure 2.16). t has two future points y_1 and y_2 and two past points x_1 and x_2. Figure 2.17 represents this situation in pure future branching logic. Here, p ('past') and f ('future') are intermediate points and are not counted in the 'flow of time'. Thus one can partition the future branching time into two subsets, T_0 and T_1. T_1 is used to represent a 'new flow of time with branching past and future' and the other subset T_0 helps indicate how this 'new flow' 'runs'. So, for example, let $(T, <)$ be the infinite binary tree, as in figure 2.15. $(T, <)$ has linear past and branching future. Let $T_0, T_1, T_p, T_f \subseteq T$ be such that the following hold.

- For all $x, y \in T$, if y is an immediate successor of x then exactly one of x, y is in T_0. Thus, T_0 consists of alternate levels of the tree. For definiteness, we could add that the root of the tree is in T_0. Then T_0 will be the levels of even height: $0, 2, \ldots$.
- $T_1 = T - T_0$. So T_1 consists of all the levels of T that are not in T_0.
- $T_0 = T_p \cup T_f$, and $T_p \cap T_f = \varnothing$.
- For all $x \in T - T_0$, one of the immediate successors of x is in T_f, and the other is in T_p.

T_1 is the set of points in the new flow of time, and T_0 is the set of intermediate points such as p, f in figure 2.17. T_p can be taken as the set of points p, and T_f the set of points f.

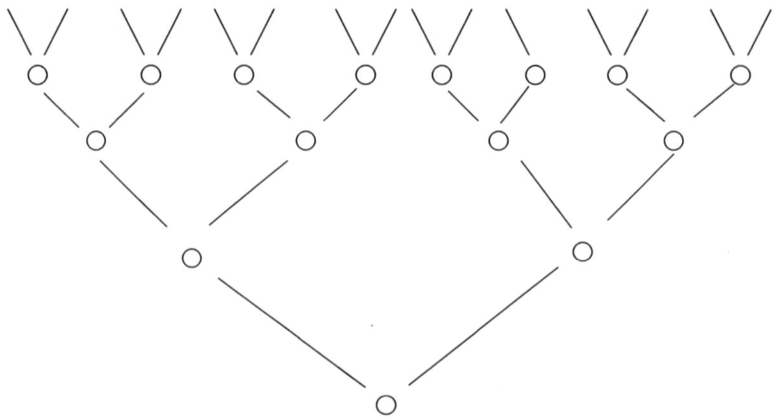

Figure 2.15

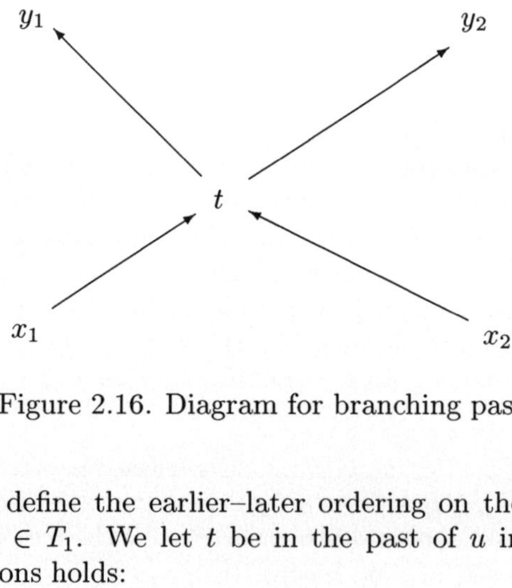

Figure 2.16. Diagram for branching past.

We can now define the earlier–later ordering on the new flow T_1, as follows. Let $t, u \in T_1$. We let t be in the past of u in T_1 if one of the following conditions holds:

1. in T we have $t < u \wedge \forall v \in T_0(t < v < u \to v \in T_f)$; or

2. in T we have $t > u \wedge \forall v \in T_0(u < v < t \to v \in T_p)$.

We should, therefore, pay special attention to discrete future-branching past-linear flows of time. Any results about such flows can have many applications in logical systems arising in computing, natural language, and philosophy.

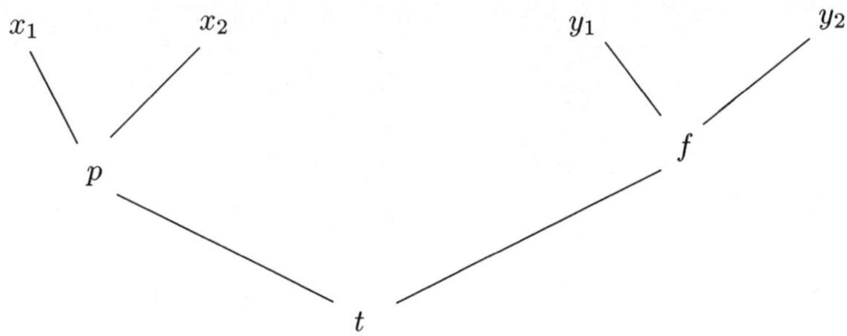

Figure 2.17. Diagram for representing past as left future.

Having settled on a flow of time, we have the problem of assignments h. Here it seems that there are no special requirements. For each atom q the value could be $h(t, q) = 0$ or 1. If we have several histories T_i we can have values $h(t, T_i, q)$, with $t \in T_i$. Thus atoms may have different values at t depending on which history is envisaged as containing t. We think atoms are so simple that they depend only on t.

More difficult is the choice of indices. In how many points do we evaluate, i.e. if we evaluate at

$$\|A\|^h_{t,x_1,\ldots,x_{n-1}},$$

what is n? The first problem is: is n bounded, or are there more and more examples requiring n to be bigger and bigger? We think n is bounded and probably $n = 4$ will be sufficient for most purposes. In principle, there may be technical differences between the cases of $n = 1$ and $n = 2$ but we think the cases $n = 2$ or any $n > 3$ are technically similar.

Coming now to the choice of connectives, it seems that all the quantifier connectives are of the form connective$_i$, where

$$\|\text{connective}_i q\|_{(\text{indices})} = 1 \text{ iff } \exists x \in T_i(\|q\|_{(x \text{ and indices})} = 1).$$

Let us explain what the sets T_i are in the above equation. There are natural regions into which the indices divide the flow of time. For example, for linear time and two indices, say $t < s$, the regions are as in figure 2.18. In the case of branching future, linear past time, and one index, t, the regions are shown in figure 2.19. They are $T_1 = \{t\}$, $T_2 =$ future of t, and $T_3 =$ linear past of t and any branching from any past of t. In each application, the T_i in the equation above run over all the natural regions.

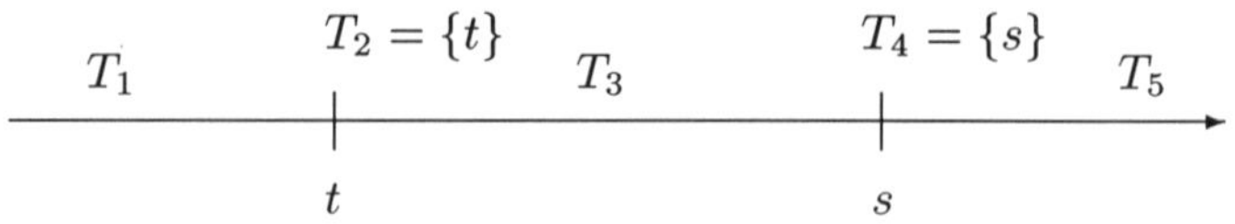

Figure 2.18

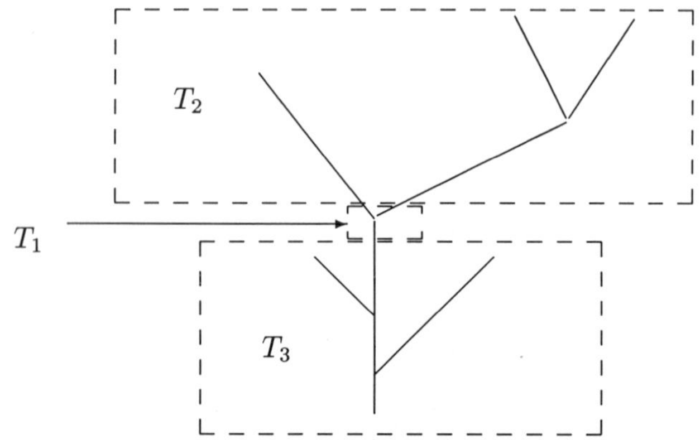

Figure 2.19

Experience shows that the Since and Until connectives are also required in each region. The above list of connectives is a minimal one.

Let us sum up by giving an example of a very useful one-dimensional temporal logic with branching future and linear past.

Example 2.4.1

(a) The flow of time has the form $(T, <)$ where $<$ is a discrete ordering where each point has two different immediate successors and one immediate predecessor. See figure 2.20. The binary tree of figure 2.15 is an example, but there are others, such as those with infinite sequences $t_0 > t_1 > \ldots$ going into the past.

(b) The connectives are the following:

$$\begin{array}{rll} \|U(p,q)\|_t = 1 & \text{iff} & \exists x > t(\|p\|_x = 1 \wedge \\ & & \forall y(t < y < x \to \|q\|_y = 1)) \\ \|S(p,q)\|_t = 1 & \text{iff} & \exists x < t(\|p\|_x = 1 \wedge \\ & & \forall y(x < y < t \to \|q\|_y = 1)) \\ \|Z^{\pm}q\|_x = 1 & \text{iff} & \|q\|_{x^{\pm}} = 1, \end{array}$$

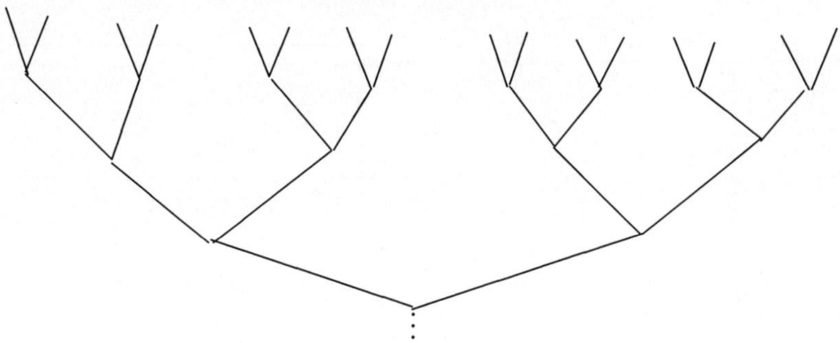

Figure 2.20

where x^- is the predecessor of x and x^+ is obtained from x by going to the predecessor x^- and taking its other successor. See figure 2.21.

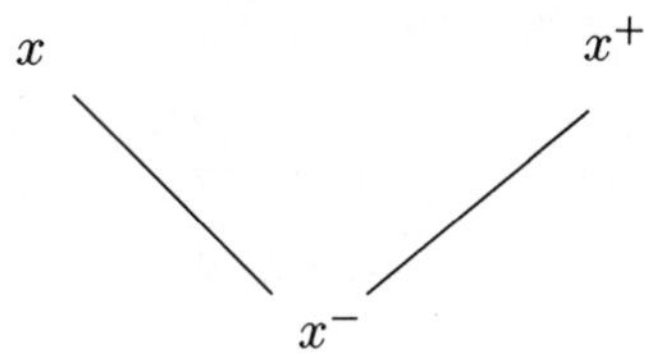

Figure 2.21

We believe the above system is expressively complete for the binary tree.

Exercise 2.4.2 Try to perform separation (see chapters 1 and 9–11) for the above set of connectives. This will give you a feel for them.

Chapter 3

Axiomatic presentation of propositional temporal connectives

3.1 Introduction

As with many logics, one of the crucially important properties of the sorts of temporal logics that we met semantically in the last chapter is that there are also syntactic ways of characterizing validity. We will look at decidability questions in a later chapter but here we will concentrate on the axiomatic presentation of temporal systems.

The traditional Hilbert-system approach to temporal systems began with unpublished work on the language with F and P by Lemmon and Scott in the late 1960s. Below we see that to the usual rules and axioms for classical propositional logic we add temporal versions of Kripke's modal axiom (axiom (2a) below), an axiom (2b) to say that F and P are opposites, and the generalization rules (3) which allow one to infer that any valid formula is always valid.

Gabbay's IRR-rule [Gabbay, 1981b] is an intuitive inference rule which makes many of the completeness proofs much easier. We will introduce this rule here, although we will not study it in detail until chapter 6.

3.2 The Hilbert system K_t

K_t is the weakest temporal logic. It is formulated in the language with the classical connectives $\sim, \wedge, \vee, \rightarrow$, and the two basic 'quantifiers' F and P. G and H are definable as $G = {\sim}F{\sim}$ and $H = {\sim}P{\sim}$. The axioms and rules for K_t are as follows.

Axioms

1. All substitution instances of truth functional tautologies.
2. All substitution instances of

   (a) $G(p \rightarrow q) \rightarrow (Gp \rightarrow Gq)$
   $H(p \rightarrow q) \rightarrow (Hp \rightarrow Hq)$,

   (b) $q \rightarrow GPq$
   $q \rightarrow HFq$,

   (c) and, although they are not usually included in K_t, all substitution instances of
   $Gp \rightarrow GGp$
   $Hp \rightarrow HHp$,
   because, as we will see, they can be used to guarantee transitive models and these are the only models we are interested in here.

Inference rules

3. Generalization $\dfrac{A}{GA}, \dfrac{A}{HA}$.

4. Modus ponens $\dfrac{A,\ A \rightarrow B}{B}$.

The following is the irreflexivity rule, which is a valid rule in all our systems. At this stage we are not adopting it as part of K_t.

5. IRR-rule

$$\frac{{\sim}q \wedge Hq \rightarrow A}{A}, \quad \text{or, alternatively,} \quad \frac{{\sim}q \wedge Hq \wedge Gq \rightarrow A}{A},$$

   where q is an atom that does not appear in A.

Definition 3.2.1 We let Z be any logical system of axioms and rules that contains those of K_t. In this chapter, if Z is not specified it will be given by the context; usually it will be $Z = K_t$.

(a) Let $Z \vdash A$ iff for some sequence of wffs $B_1, \ldots, B_m = A$, we have that for each $i = 1, \ldots, m$, B_i is either an axiom of Z or is obtained from some formula or formulae in $\{B_1, \ldots, B_{i-1}\}$ by the application of an inference rule of Z.

(b) A set of formulae Δ proves A, written $\Delta \vdash A$, iff there are formulae $B_1, \ldots, B_m \in \Delta$ such that $Z \vdash (B_1 \wedge \ldots \wedge B_m) \to A$.

(c) A set Δ is consistent iff $\Delta \nvdash \bot$ ($\bot$ is *falsity*).

Exercise 3.2.2

$$\text{Show}\quad \begin{array}{l} \vdash G(A \wedge B) \leftrightarrow GA \wedge GB, \\ \vdash H(A \wedge B) \leftrightarrow HA \wedge HB. \end{array}$$

Theorem 3.2.4 below is useful in connection with the use of the IRR-rule. We will use it in chapter 6.

Definition 3.2.3 Let $\mathbf{L}$ be either G or H and let $\mathbf{M} = {\sim}\mathbf{L}{\sim}$, i.e. if $\mathbf{L} = G$ then $\mathbf{M} = F$ and if $\mathbf{L} = H$ then $\mathbf{M} = P$. Let $\mathbf{L}^* = G$ if $\mathbf{L} = H$ and $\mathbf{L}^* = H$ if $\mathbf{L} = G$. Similarly, let $\mathbf{M}^* = {\sim}\mathbf{L}^*{\sim}$.

Axiom (2b) can be written as $q \to \mathbf{L}\mathbf{M}^*q$. For $\mathbf{L} = G$ we obtain $q \to GPq$ and for $\mathbf{L} = H$ we obtain $q \to HFq$. Since $\mathbf{L}$ can be either G or H we can write $\mathbf{L}_1, \ldots, \mathbf{L}_m$ for different choices of $\mathbf{L}_i \in \{G, H\}$.

Theorem 3.2.4 *If for wffs* $x, y, B_1, \ldots, B_m$,

$$(*) \qquad \vdash x \to \mathbf{L}_0(B_1 \to \mathbf{L}_1(B_2 \to \cdots \mathbf{L}_{m-1}(B_m \to \mathbf{L}_m{\sim}y))\cdots),$$

then:

$$(**) \qquad \vdash y \to \mathbf{L}^*_m(B_m \to \mathbf{L}^*_{m-1}(B_{m-1} \to \cdots \mathbf{L}^*_1(B_1 \to \mathbf{L}^*_0{\sim}x))\cdots).$$

For example, we have

$$\begin{array}{rcll} \vdash x \to G{\sim}y & \Rightarrow & \vdash y \to H{\sim}x & (m = 0); \\ \vdash x \to G(B_1 \to H{\sim}y) & \Rightarrow & \vdash y \to G(B_1 \to H{\sim}x) & (m = 1). \end{array}$$

Proof. We prove, by induction on $j = 1, \ldots, m$, that:

$$\begin{array}{lll} (\dagger) & \vdash & \mathbf{M}_{j-1}(B_j \wedge \mathbf{M}_j(B_{j+1} \wedge \ldots \wedge \mathbf{M}_{m-1}(B_m \wedge \mathbf{M}_m y))\cdots) \\ & & \to (B_{j-1} \to \mathbf{L}^*_{j-2}(B_{j-2} \to \cdots \to \mathbf{L}^*_1(B_1 \to \mathbf{L}^*_0{\sim}x))\cdots). \end{array}$$

First from (*) we obtain, by changing sides,

$$\vdash \mathbf{M}_0(B_1 \wedge \mathbf{M}_1(B_2 \wedge \ldots \wedge \mathbf{M}_{m-1}(B_m \wedge \mathbf{M}_m y)) \cdots) \to \sim x.$$

This is (†) for $j = 1$.

Assume (†) for j. Then by rule (3) and axiom (2a) we obtain:

$$\begin{aligned}\vdash \quad & \mathbf{L}^*_{j-1}\mathbf{M}_{j-1}(B_j \wedge \mathbf{M}_j(B_{j+1} \wedge \ldots \wedge \mathbf{M}_{m-1}(B_m \wedge \mathbf{M}_m y)) \cdots) \\ & \to \mathbf{L}^*_{j-1}(B_{j-1} \to \mathbf{L}^*_{j-2}(B_{j-2} \to \cdots \to \mathbf{L}^*_1(B_1 \to \mathbf{L}^*_0 \sim x)) \cdots).\end{aligned}$$

By axiom (2b) and some propositional work we obtain

$$\begin{aligned}\vdash \quad & B_j \wedge \mathbf{M}_j(B_{j+1} \wedge \ldots \wedge \mathbf{M}_{m-1}(B_m \wedge \mathbf{M}_m y)) \cdots) \\ & \to \mathbf{L}^*_{j-1}(B_{j-1} \to \mathbf{L}^*_{j-2}(B_{j-2} \to \cdots \to \mathbf{L}^*_1(B_1 \to \mathbf{L}^*_0 \sim x)) \cdots),\end{aligned}$$

and so:

$$\begin{aligned}\vdash \quad & \mathbf{M}_j(B_{j+1} \wedge \ldots \wedge \mathbf{M}_{m-1}(B_m \wedge \mathbf{M}_m y)) \cdots) \\ & \to (B_j \to \mathbf{L}^*_{j-1}(B_{j-1} \to \mathbf{L}^*_{j-2}(B_{j-2} \to \cdots \to \mathbf{L}^*_1(B_1 \to \mathbf{L}^*_0 \sim x)) \cdots),\end{aligned}$$

which is (†) for $j + 1$.

For the case $j = m$ we obtain

$$\vdash \mathbf{M}_m y \to (B_m \to \mathbf{L}^*_{m-1} \cdots \mathbf{L}^*_1(B_1 \to \mathbf{L}^*_0 \sim x)) \cdots),$$

and then, by rule (3) and axiom (2a) again,

$$\vdash \mathbf{L}^*_m \mathbf{M}_m y \to \mathbf{L}^*_m(B_m \to \cdots \mathbf{L}^*_0 \sim x)) \cdots),$$

and hence by axiom (2b) we obtain (**). ■

Remark 3.2.5 Note that the properties we used to prove theorem 3.2.4 were the following: there are connectives $\mathbf{L}$ and $\mathbf{L}^*$ such that

(a) $\vdash \mathbf{L}(A \wedge B) \leftrightarrow (\mathbf{L}A \wedge \mathbf{L}B)$
$\vdash \mathbf{L}^*(A \wedge B) \leftrightarrow (\mathbf{L}^*A \wedge \mathbf{L}^*B)$,

(b) $\vdash A \to \mathbf{L}\mathbf{M}^*A$
$\vdash A \to \mathbf{L}^*\mathbf{M}A$,

(c) $\dfrac{A}{\mathbf{L}A,\ \mathbf{L}^*A}$.

A completeness theorem for K_t without the irreflexivity rule

Consider a flow $(T, <, =, h)$ with $<$ irreflexive and transitive and h an assignment.

Definition 3.2.6

(a) For any wff in the language of K_t, i.e. with $\{\sim, \vee, \rightarrow P, F\}$, we can compute the value $\|A\|_t^h$ by induction as follows:

- $\|A\|_t^h = 1$ iff $h(t, A) = 1$, if A is atomic
- $\|{\sim}A\|_t^h = 1$ iff $\|A\|_t^h = 0$
- $\|A \wedge B\|_t^h = 1$ iff $\|A\|_t^h = \|B\|_t^h = 1$
- $\|A \vee B\|_t^h = 1$ iff $\|A\|_t^h = 1$ or $\|B\|_t^h = 1$
- $\|A \rightarrow B\|_t^h = 1$ iff $\|A\|_t^h = 0$ or $\|B\|_t^h = 1$
- $\|PA\|_t^h = 1$ iff $\exists s < t(\|A\|_s^h = 1)$
- $\|FA\|_t^h = 1$ iff $\exists s > t(\|A\|_s^h = 1)$.

(b) A formula A is true at t in h iff $\|A\|_t^h = 1$. We also say in this case — if there is such a t — that $(T, <, h)$ is a model for A.

(c) A formula A is h-valid iff $\|A\|_t^h = 1$ for all $t \in T$.

(d) A wff A is said to be valid in $(T, <)$ iff $\|A\|_t^h = 1$ for all h and all $t \in T$.

(e) Let **K** be a class of flows of time. A formula A is said to be valid in the class iff for any $(T, <, h)$ with $(T, <) \in \mathbf{K}$, and for any $t \in T$, $\|A\|_t^h = 1$.

Theorem 3.2.7 (completeness theorem for K_t) *Let Δ_0 be any theory (i.e. a set of formulae) in the language with P, F, and the classical connectives. Then Δ_0 is consistent iff for some $(T, <, h)$ with $<$ irreflexive and transitive, and some $t \in T$, $\|A\|_t^h = 1$ for all $A \in \Delta_0$.*

Proof. We begin with the notion of a *complete K_t theory.* Δ is a complete K_t theory iff (i) Δ is consistent, and (ii) for any B, $B \in \Delta$ or ${\sim}B \in \Delta$.

The same process as known from classical logic allows for a consistent theory Δ_0 to be extended to a maximally consistent (i.e. complete) theory $\Delta \supseteq \Delta_0$.

Let S be the set of all complete K_t theories. Define a relation R on S by:

$$\Delta R \Gamma \text{ iff for all wff } B \text{ we have } GB \in \Delta \Rightarrow B \in \Gamma.$$

Lemma 3.2.8

(*a*) *R is transitive.*

(*b*) *$\Delta R\Gamma$ iff for all wff B, if $HB \in \Gamma$ then $B \in \Delta$.*

Proof.

(a) Let $\Gamma, \Delta, \Xi \in S$ and suppose that $\Gamma R\Delta$ and $\Delta R\Xi$. We show that $\Gamma R\Xi$. So let $GB \in \Gamma$. By axiom (2c), $GGB \in \Gamma$. Hence $GB \in \Delta$, and $B \in \Xi$. So $\Gamma R\Xi$, as required.

(b) Assume $\Delta R\Gamma$ and $HB \in \Gamma$. Then $FHB \in \Delta$, and so $B \in \Delta$ (by axiom (2b)). Conversely, assume that $\Delta R\Gamma$ fails. Then for some B, $GB \in \Delta$ and $\sim B \in \Gamma$. But $\vdash \sim B \to H{\sim}GB$, and so, for $B' = {\sim}GB$, we have $HB' \in \Gamma$ and $\sim B' \in \Delta$.

■

Lemma 3.2.9 *For any $\Delta \in S$, the following hold.*

(*a*) *Suppose that $FA \in \Delta$. Then for some Γ, $A \in \Gamma$ and $\Delta R\Gamma$.*

(*b*) *Suppose that $PA \in \Delta$. Then for some Γ, $A \in \Gamma$ and $\Gamma R\Delta$.*

Proof. The method of this proof will occur again and again throughout temporal logic. The set S of theories is the set of moments of time, and the wffs in the theories themselves are the wffs true at each moment of time. (The wffs true *at* Δ are the wffs *in* Δ.) R is the earlier–later relation. Thus condition (a) says that if FA is supposed to be true at Δ, then indeed there exists a future moment where A is true.

To show (a) let $FA \in \Delta$ be given. Let $\Gamma_0 = \{A\} \cup \{B \mid GB \in \Delta\}$. If A is true at a future moment, all these Bs must be there.

Γ_0 is consistent. For, otherwise, there would be $B_1, \ldots, B_k$ such that $GB_1, \ldots, GB_k \in \Delta$ and $K_t \vdash (\bigwedge_{i\leq k} B_i) \to {\sim}A$. Hence, by generalization and axiom (2a),

$$K_t \vdash G(\textstyle\bigwedge_i B_i \to {\sim}A),$$
$$K_t \vdash (\textstyle\bigwedge_i GB_i) \to G{\sim}A,$$

and since $GB_i \in \Delta$ for each i, we obtain $G{\sim}A \in \Delta$, a contradiction.

Thus Γ_0 is consistent and is contained in a complete theory $\Gamma \in S$. We have $\Delta R\Gamma$, because $GB \in \Delta$ implies $B \in \Gamma_0 \subseteq \Gamma$. Also, $A \in \Gamma_0 \subseteq \Gamma$. This proves ($a$).

To prove (b), for $PA \in \Delta$, we let $\Gamma_0 = \{A\} \cup \{B \mid HB \in \Delta\}$. Γ_0 is consistent and is contained in a $\Gamma \in S$. $\Gamma R\Delta$ holds by the preceding lemma. ■

We would like to take (S, R, g) as our model, where $g(\Gamma, q) = 1$ iff $q \in \Gamma$. However, we have one problem. R is not necessarily irreflexive. It is quite possible that $\Delta R \Delta$ holds.

To eliminate this possibility we make copies of Δ. This method is sometimes called *bulldozing*.

Let $\mathbb{Z}$ be the set of integers and let $T = S \times \mathbb{Z}$, i.e. T is the set of all pairs (Δ, m), where $\Delta \in S$ and m is an integer. Define $<$ by:

$$(\Delta, m) < (\Gamma, n) \quad \text{iff} \quad \Delta R \Gamma \text{ and } m < n.$$

Then $<$ is irreflexive and transitive. Let h be defined, for atomic q and $t = (\Delta, n) \in T$, by:

$$h(t, q) = 1 \text{ iff } q \in \Delta.$$

We have thus obtained a structure $(T, <, h)$. This structure, and sometimes the structure (S, R, g) also, shall be referred to as the *canonical model for* K_t.

Lemma 3.2.10 *For any B, Δ, n, we have $\|B\|^h_{(\Delta,n)} = 1$ iff $B \in \Delta$.*

Proof. By induction on B. The case of atomic q is true by definition. The cases of the classical connectives can be seen to hold immediately.

If $FB \in \Delta$, then by the preceding lemma, for some $\Gamma \in S$ we have $\Delta R \Gamma$ and $B \in \Gamma$. By the induction hypothesis for $(\Gamma, n+1)$, $\|B\|^h_{(\Gamma,n+1)} = 1$, and because $(\Delta, n) < (\Gamma, n+1)$, $\|FB\|^h_{(\Delta,n)} = 1$.

Conversely, if $\|FB\|^h_{(\Delta,n)} = 1$ then there must be $(\Gamma, m) > (\Delta, n)$ with $\|B\|^h_{(\Gamma,m)} = 1$. Inductively, $B \in \Gamma$, so since $\Delta R \Gamma$, we obtain $FB \in \Delta$, as required.

The case of P is treated similarly. ■

We are now ready to prove the completeness theorem.

First note that if $K_t \vdash B$ then B is valid. Hence if a theory Δ_0 has a model, it is consistent (because *falsity* is not valid).

Conversely, assume that Δ_0 is consistent. Hence $\Delta_0 \subseteq \Delta$ for some $\Delta \in S$. Take the canonical model $(T, <, h)$ and let $t = (\Delta, 0)$. By the preceding lemma, $\|B\|^h_t = 1$ for all $B \in \Delta$. ■

Irreflexivity

We saw in the course of the proof that we could not exclude the possibility that $\Delta R \Delta$, and so we had to do some special construction to make our $<$ irreflexive. On the other hand, we had no difficulty in showing R to be transitive. We have a special axiom for that, namely $GA \to GGA$. Perhaps we can add a special axiom for irreflexivity?

We can see from the construction of the canonical model that this is not possible. Suppose for contradiction that α is such an axiom. So α is valid in every irreflexive transitive flow; and the flow (S, R) as built above is irreflexive, where S, this time, is the set of all $(K_t + \alpha)$-consistent theories.

Take any non-irreflexive flow $(T, <)$, so that there is $t \in T$ with $t < t$. Let h be an arbitrary assignment into T, and let $\Gamma = \{B \mid \|B\|_t^h = 1\}$. Since $t < t$, we have $\|GB\|_t^h = 1 \Rightarrow \|B\|_t^h = 1$. That is, $\Gamma R \Gamma$. If α is valid in $(T, <)$, then Γ is $(K_t + \alpha)$-consistent, and so $\Gamma \in S$, which is impossible since (S, R) is irreflexive. Hence α is not valid in $(T, <)$.

We now see that the following holds.

(**) $(T, <)$ is irreflexive iff for all h and all $t \in T$, $\|\alpha\|_t^h = 1$.

That this is impossible is shown by the following theorem.

Theorem 3.2.11 *There is no α such that (**) holds for any $(T, <)$.*

Proof. Assume such an α exists and obtain a contradiction. Take a one-point flow $\{a\}$ with $a < a$. Clearly, $(\{a\}, <)$ is reflexive and hence for some g, $\|\alpha\|_a^g = 0$.

Let $(\mathbb{Z}, <)$ be the integers with *smaller-than* and let h be defined by $h(m, q) = g(a, q)$ for all atomic q. This makes a copy of the model $\{a\}$ at each point m.

Lemma 3.2.12 *For each wff A and each t, $\|A\|_t^h = \|A\|_a^g$.*

Proof. By induction. For atomic q this is the definition. For $\sim, \wedge, \vee, \to$ the lemma follows easily from the induction hypothesis. We now check the case of F.

$\|FA\|_t^h = 1$ iff $\exists s > t \|A\|_s^h = 1$ iff (by the induction hypothesis) $\|A\|_a^g = 1$ iff $\|FA\|_a^g = 1$. Similarly, we obtain the clause for P. ■

In particular, we deduce that $\|\alpha\|_t^h = 0$. But this is a contradiction because according to (**) since $<$ is irreflexive α must hold for any h. This proves theorem 3.2.11. ■

We thus see that the condition of irreflexivity cannot be characterized by an axiom α. Yet, this condition is a basic requirement for any flow of time. Is there anything that can be done?

What we can do is add a rule which is valid in all irreflexive flows of time and use it to obtain irreflexivity.

Consider the rule

$$\text{IRR:} \quad \frac{\sim q \wedge Hq \to A}{A},$$

where q is atomic and A does not contain q.

This rule is valid in K_t. For assume $\nvdash A$. Then by the completeness theorem for K_t there exists a $(T, <, h)$, *with* $<$ *irreflexive*, and $t \in T$, such that $\|A\|_t^h = 0$. Let h' be like h except that $h'(x, q) = 0$ iff $x = t$. Clearly, since q does not appear in A,

$$\|A\|_t^{h'} = \|A\|_t^h = 0.$$

However, $\|\sim q \wedge Hq\|_t^{h'} = 1$. Thus $\|\sim q \wedge Hq \to A\|_t^{h'} = 0$, and hence $\nvdash \sim q \wedge Hq \to A$.

Thus this rule, IRR, is a valid rule for K_t. If we add it to K_t we obtain no new theorems. What have we gained?

If we extend a logic, i.e. include more theorems, then any axioms in an axiomatization of the original logic will be kept as theorems of the extended logic. This is not the case for rules that happen to be valid. For example, we can add to K_t the axioms $Gq \to q$, $Hq \to q$, and obtain a proper extension '$K_t^{\leq}$' complete for flows of time $(T, \leq)$ where $\leq$ is reflexive. The rule IRR will no longer be valid, since $K_t^{\leq} \vdash Hq \to q$ and so $K_t^{\leq} \vdash \sim q \wedge Hq \to \bot$.

However, if IRR is officially adopted as a rule of K_t (and not only seen to be valid), then IRR will have to appear in any extension of K_t. Thus we can no longer add the axioms $Gq \to q$, $Hq \to q$, because we obtain an inconsistent extension.

Since in all our temporal logics $<$ is irreflexive, we are justified in adopting this rule. We shall see later how this rule characterizes irreflexivity. In all the temporal systems, including K_t, we will have IRR as a rule.

3.3 Filtration for K_t

We saw that K_t is complete for the class of all flows of time of the form $(T, <, =)$ where $<$ is irreflexive and transitive. T may be infinite. We want to see what happens if the requirement that T is finite is made.

Theorem 3.3.1 *For all A, $K_t \vdash A$ iff for all $(T, <, =, h)$, with T finite, and $<$ not necessarily irreflexive but certainly transitive, and for all $t \in T$, we have $\|A\|_t^h = 1$.*

Proof. We shall use a method called *filtration*. Given $K_t \nvdash A$ we know from the preceding section that some irreflexive and transitive $(S, <, =, h)$

and $t \in S$ exist such that $\|A\|_t^h = 0$. Since A has only a finite number of subformulae and $\|A\|$ depends on the values $\|B\|$, for all subformulae B of A, we are interested only in a finite amount of information from S. We shall therefore obtain a finite T such that in T, $\|A\|_t^h = 0$ for some $t \in T$.

Let Δ be the closure of the set of all subformulae of A under the boolean operations (up to equivalence). Then Δ is finite and $A \in \Delta$.

Definition 3.3.2 For $x, y \in S$ let $x \sim y$ be defined by: $x \sim y$ iff for all $B \in \Delta$, $\|B\|_x^h = \|B\|_y^h$. This means that as far as Δ is concerned x and y are the same. $\sim$ is an equivalence relation. It divides S into a finite set of equivalence classes.

Let T be this set of $\sim$-classes. We have $|T| \leq 2^{|\Delta|}$. Define $<_T$ as follows, where x^* denotes the $\sim$-equivalence class of x.

(a) $x^* <'_T y^*$ iff for some $x' \in x^*$ and $y' \in y^*$ we have $x' < y'$.

(b) Let $<_T$ be the transitive closure of $<'_T$.

(c) Let $h^*(x^*, q) = 1$ iff $h(x, q) = 1$. This definition of h^* is good for any $q \in \Delta$.

Lemma 3.3.3 *For any $GB \in \Delta$ or $HB \in \Delta$ we have*

(a) $\|GB\|_x^h = 1$ *and* $x^* <_T y^*$ *imply* $\|B\|_y^h = 1$.

(b) $\|HB\|_y^h = 1$ *and* $x^* <_T y^*$ *imply* $\|B\|_x^h = 1$.

Proof. We show that

(c) $\|GB\|_x^h = 1$ and $x^* <'_T z^*$ imply $\|B \wedge GB\|_z^h = 1$.

Assume $\|GB\|_x^h = 1$. Since $x^* <'_T z^*$, it follows that for some $x' \in x^*$, $z' \in z^*$, we have $x' < z'$. Since $x' \in x^*$ and $GB \in \Delta$, we obtain $\|GB\|_{x'}^h = 1$ and so also $\|GGB\|_{x'}^h = 1$ and hence $\|B \wedge GB\|_{z'}^h = 1$. But $B, GB \in \Delta$, so $\|B \wedge GB\|_z^h = 1$. This proves (c).

From (c) we obtain (a) by induction on the 'distance' from x^* to y^* in $<'_T$. We can similarly prove (b). ∎

Lemma 3.3.4 *For any $B \in \Delta$ and any $x^* \in T$, $\|B\|_{x^*}^{h^*} = \|B\|_x^h$.*

Proof. By induction on B. The case of atomic q holds by definition. The connectives $\sim, \wedge, \vee, \rightarrow$ present no difficulties. Let $FB \in \Delta$. Assume $\|FB\|_x^h = 1$. Then $\exists y(x < y \wedge \|B\|_y^h = 1)$ and so $\exists y^*(x^* <_T y^* \wedge \|B\|_{y^*}^{h^*} = 1)$, by the induction hypothesis. This means that $\|FB\|_{x^*}^{h^*} = 1$ by definition.

Assume $\|FB\|_x^h = 0$. Then $\|G{\sim}B\|_x^h = 1$ and, because Δ is closed under negation, $G{\sim}B \in \Delta$. Hence by lemma 3.3.3 and the induction hypothesis, $\|B\|_{y^*}^{h^*} = 0$ for all y^* such that $x^* <_T y^*$, and so $\|FB\|_{x^*}^{h^*} = 0$.

A similar argument will establish the case of PB. Thus the induction is complete and lemma 3.3.4 is proved. ■

The preceding lemma shows that $\|A\|_{t^*}^{h^*} = 0$. Thus we saw that if $K_t \nvdash A$, then for some finite $(T, <_T, h)$ with $<_T$ transitive (but maybe not irreflexive) and some $t \in T$, $\|A\|_t^h = 0$. Clearly, if $K_t \vdash A$, then A holds in all such models. Thus theorem 3.3.1 is proved. ■

Theorem 3.3.5 *K_t is decidable.*

Proof. We have proof procedures for K_t via its axioms and rules. We also have disproof procedures by checking all finite $(T, <, h)$. Thus, together, we obtain decision procedures for K_t. ■

Remark 3.3.6 K_t is not complete for finite flows $(T, <, =)$ with $<$ irreflexive as well as transitive. The following axioms are valid in all finite irreflexive flows. But they are not valid in all irreflexive flows, and so are not provable from K_t.

(a) $Fq \to F(q \wedge {\sim}Fq)$.

(b) $Pq \to P(q \wedge {\sim}Pq)$.

Theorem 3.3.7 *Consider K_i, the extension of K_t with the axiom schema of remark 3.3.6. Then for any A, $K_i \vdash A$ iff for all finite irreflexive flows $(T, <, =, h)$ and all $t \in T$, $\|A\|_t^h = 1$.*

Proof. First show that if the axiom is false at some $t \in T$ under some assignment h, then T is infinite. Assume $\|Fq \wedge G(q \to Fq)\|_{t_0}^h = 1$, then for some t_1, $t_0 < t_1$ and $\|q \wedge (q \to Fq)\|_{t_1}^h = 1$. We also have $\|G(q \to Fq)\|_{t_1}^h = 1$. We must therefore have a t_2 with $t_1 < t_2$; and again $\|Fq \wedge G(q \to Fq)\|_{t_2}^h = 1$, and so on. Thus we obtain the sequence $t_0 < t_1 < t_2 < \ldots$, which is infinite because of irreflexivity.

It follows that if $K_i \vdash A$ then A is valid in any finite irreflexive flow.

For completeness, assume that A is K_i-consistent. Let S be the set of all maximal K_i-consistent sets. Let A^* be the closure under the boolean operations (up to logical equivalence) of the set

$$\{B, PB, FB, PPB, FFB \mid B \text{ a subformula of } A\}.$$

Note that A^* is finite and is closed under taking subformulae. Finally, let

$$S^* = \{\Gamma \cap A^* \mid \Gamma \in S,\ B \wedge ({\sim}FB \vee {\sim}PB) \in \Gamma \cap A^* \text{ for some } B\}.$$

Notice that S^* is finite (indeed, its size is at most $|A^*|$). Also, if $\Sigma \in S^*$, $X \in A^*$ and $\Sigma \vdash X$, then $X \in \Sigma$.

There is $\Sigma_A \in S^*$ with $A \in \Sigma_A$. This can be seen as follows. Choose $\Gamma \in S$ containing A. If ${\sim}PA \in \Gamma$, then we may take Σ_A to be $\Gamma \cap A^* \in S^*$. If $PA \in \Gamma$, then by axiom (b) of remark 3.3.6, $P(A \wedge {\sim}PA) \in \Gamma$. So by the analogue of lemma 3.2.9 for K_i, there is $\Delta \in S$ with $A \wedge {\sim}PA \in \Delta$, and we can take Σ_A to be $\Delta \cap A^* \in S^*$.

Define an ordering $<$ on S^*, by:

$$\Sigma < \Phi \text{ iff } \quad \begin{array}{l}\text{for all } X \in \Sigma, \text{ if } PX \in A^* \text{ then } PX \in \Phi, \text{ and} \\ \text{for all } X \in \Phi, \text{ if } FX \in A^* \text{ then } FX \in \Sigma.\end{array}$$

Then $<$ is irreflexive and transitive.

For if $\Sigma \in S^*$ then there is B with (say) $B \wedge {\sim}FB \in \Sigma$. If $\Sigma < \Sigma$ then $FB \in \Sigma$, so Σ is inconsistent and cannot be in S^*. Hence $<$ is irreflexive.

For transitivity, assume that $\Sigma < \Phi < \Psi$ in S^*. Let X be a formula such that $X \in \Sigma$ and $PX \in A^*$. Certainly, $PX \in \Phi$. There are now two cases.

If $PPX \in A^*$, then $PPX \in \Psi$, so by the transitivity axiom, $PX \in \Psi$.

If not, then as $PX \in A^*$, we must have $X = PY$ for some $Y \in A^*$. Thus, $PPY \in \Phi$, so the transitivity axiom, $PY = X \in \Phi$. Hence $PX \in \Psi$, as required.

The case of $X \in \Psi$ is similar. Transitivity is proved.

Now define an assignment h by $h(\Sigma, q) = 1$ iff $q \in \Sigma$, for all $\Sigma \in S^*$ and all atoms q.

Lemma 3.3.8 *For any subformula B of A, and any $\Sigma \in S^*$, we have $\|B\|^h_\Sigma = 1$ iff $B \in \Sigma$.*

Proof. By induction on B. The only difficult cases are FB and PB. So assume that FB is a subformula of A, and that the lemma holds for B. Choose $\Gamma \in S$ extending Σ.

First, if $\|FB\|^h_\Sigma = 1$ then there is $\Phi > \Sigma$ in S^* with $\|B\|^h_\Phi = 1$. By the inductive hypothesis, $B \in \Phi$, so, as $FB \in A^*$, we have $FB \in \Sigma$.

Conversely, if $FB \in \Sigma$, then $FB \in \Gamma$, so by axiom (a) of the remark, $F(B \wedge {\sim}FB) \in \Gamma$. By lemma 3.3.9 (for K_i), there is $\Delta \in S$ with $B \wedge {\sim}FB \in \Delta$, $X \in \Gamma \Rightarrow PX \in \Delta$, and $X \in \Delta \Rightarrow FX \in \Gamma$. Let $\Phi = \Delta \cap A^*$. Then $\Phi \in S^*$ and $B \in \Phi$. By the inductive hypothesis, $\|B\|^h_\Phi = 1$, and, clearly, $\Sigma < \Phi$. So $\|FB\|^h_\Sigma = 1$.

The case of PB is similar. ■

Hence $\|A\|^h_{\Sigma_A} = 1$, and we have found a finite, irreflexive, transitive model of A. This proves completeness. ■

3.4 Löb's axiom

Recall the logic $Y1$ of example 1.4.6. It was a logic with the connectives H and P, and the axioms and rules were:

(a) $\vdash A$, for any truth functional tautology.

(b) modus ponens and $\vdash A \Rightarrow\ \vdash HA$.

(c) $\vdash H(A \wedge B) \leftrightarrow HA \wedge HB$
$\vdash HA \to HHA$.

(d) Linearity axiom
$PA \wedge PB \to P(A \wedge B) \vee P(A \wedge PB) \vee P(B \wedge PA)$.

(e) Finiteness axiom (also called Löb's axiom)
$PA \to P(A \wedge H{\sim}A)$.

Theorem 3.4.1 *$Y1$ is complete for the class of flows of time consisting of of all finite chains.*

Proof. Given a consistent A, we want to find a temporal model whose flow of time is a finite chain and at some point of which A is true. Let S be the set of all maximal $Y1$-consistent sets, and define $<$ on S as usual, by $\Gamma < \Delta$ iff for all formulae C, $GC \in \Gamma \Rightarrow C \in \Delta$.

Take $\Delta \in S$ containing A. We can assume that Δ is irreflexive. For if $\Delta < \Delta$, then $PA \in \Delta$, so by Löb's axiom, $P(A \wedge H{\sim}A) \in \Delta$. Hence, by the analogue of lemma 3.2.9 for $Y1$, there is $\Delta' < \Delta$ containing $A \wedge H{\sim}A$. Clearly, we cannot have $\Delta' < \Delta'$, so we can replace Δ by Δ'.

Claim: $\{\Gamma \in T \mid \Gamma < \Delta\}$ is linearly ordered.

Proof of claim: Suppose for contradiction that there are $\Gamma, \Xi < \Delta$ that are not equal and are not related by $<$. So there are formulae B, C, D such that $B \in \Gamma, {\sim}B \in \Xi$, $HC \in \Gamma, {\sim}C \in \Xi$, and ${\sim}D \in \Gamma, HD \in \Xi$. Let

$$\begin{aligned} X &= B \wedge HC \wedge {\sim}D, \\ Y &= {\sim}B \wedge {\sim}C \wedge HD. \end{aligned}$$

Then $PX \wedge PY \in \Delta$. By the linearity axiom, we have $P(X \wedge Y) \in \Delta$, $P(X \wedge PY) \in \Delta$, or $P(Y \wedge PX) \in \Delta$, so that one of $X \wedge Y$, $X \wedge PY$, $Y \wedge PX$ must be consistent. This is impossible, and proves the claim.

Now let $T \subseteq S$ be the set consisting of Δ together with all $\Gamma \in S$ such that (i) $\Gamma < \Delta$, and (ii) for some subformula B of A, $B \wedge H{\sim}B \in \Gamma$. Observe that T is finite. For, by linearity, for any B there is at most one $\Gamma < \Delta$ with $B \wedge H{\sim}B \in \Gamma$; and there are finitely many subformulae B of A.

Define $<$ on T as on S: for $\Gamma, \Xi \in T$, $\Gamma < \Xi$ in T iff $\Gamma < \Xi$ in S. Define h on T as on S: $h(\Gamma, q) = 1$ iff $q \in \Gamma$. Then $(T, <, h)$ is a finite, transitive, irreflexive, linear model.

Lemma 3.4.2 *For all $\Gamma \in T$ and all subformulae B of A, we have $\|B\|^h_\Gamma = 1$ (in $(T, <, h)$) iff $B \in \Gamma$.*

Proof. As in lemma 3.3.8. ∎

Since $\|A\|^h_\Delta = 1$, we obtain a finite temporal linear model for A. This finishes the proof of the completeness theorem (theorem 3.4.1). ∎

3.5 A system $Y2$ corresponding to a higher-order semantic condition on the flow of time

We modify our example $Y1$ of the preceding section and try to show that the new example cannot be translated into predicate logic.

Consider a random number generator program. Consider a temporal logic describing the steps of the program execution.

At each stage n, we may want to say something about what happened in the sequence $0, 1, \ldots, n$. We also want to say something about the entire execution sequence.

Definition 3.5.1 Thus our temporal logic has the form:

(a) The flow of time is as indicated in figure 3.1.

(b) Our temporal language for this logic is the same as the system $Y1$, i.e. containing H and P.

(c) The time table is different from $Y1$:

 (c1) HA is considered true at any n iff A is true at all $m < n$;

 (c2) HA is considered true at ω iff A is true on a set of measure 1, i.e. $\{n \mid A \text{ true at } n\}$ is *large*.

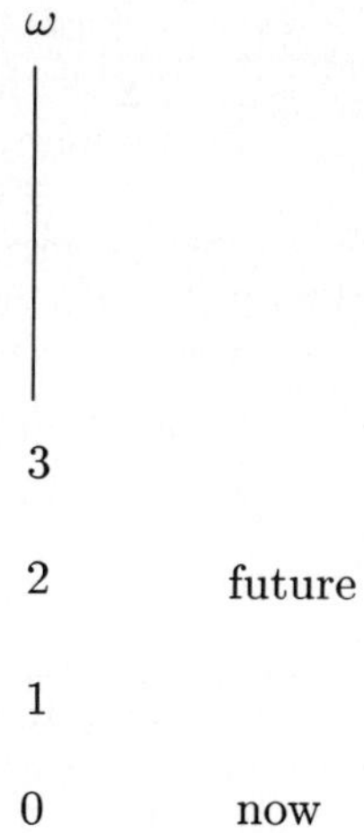

Figure 3.1:

(d) The notion of *large* is given by a non-principal ultrafilter F over the flow of time $T = \{0, 1, \ldots\}$: i.e. a set of subsets of T with the following properties:

(d1) $T \in F$, $\emptyset \notin F$;

(d2) $X \in F$ and $Y \in F \Rightarrow X \cap Y \in F$;

(d3) $X \in F$ and $X \subseteq Y \Rightarrow Y \in F$;

(d4) For all X, $X \in F$ or $T - X \in F$;

(d5) If $X \in F$ then X is infinite.

This is a reasonable definition because we are interested in the 'global' behaviour of the process.

(e) We define the logic $Y2$ as the set of all wffs valid at ω, under *any* assignment and *any* non-principal ultrafilter F.

The following are properties of $Y2$:

(a) $Y2 \vdash$ all truth functional tautologies.

(b) If $Y1 \vdash A$ then $Y2 \vdash HA$.

(c) $Y2 \vdash HHA \rightarrow HA$.

(d) $Y2 \vdash H(PHA \wedge H(HA \to A) \to HA)$.

(e) $HP(\sim q \wedge Hq)$ is consistent (i.e. $Y2$ does not prove its negation).

(f) $Y2 \vdash HA \leftrightarrow \sim H \sim A$.

We do not know at the moment any exact complete axiomatization of $Y2$. However, $Y2^*$, which consists of (a)–(f) above, cannot be characterized by any flow of time without using F. There cannot be a temporal logic model with a fixed flow of time $(S, >, \omega)$ in which always under any assignment h the axioms of $Y2^*$ are valid and also $HP(\sim q \wedge Hq)$ is true at ω under some assignment.

We *must* use an ultrafilter.

Let us show this fact.

Theorem 3.5.2 *The semantics of $Y2^*$ cannot be characterized by a family of flows of time.*

Proof. Suppose for contradiction that a flow of time $(S, <)$ exists — without an ultrafilter — in which the axioms (a), (c) and (f) are valid, and that h is an assignment into S such that

$$\|HP(\sim q \wedge Hq)\|^h_\omega = 1 \text{ for some } \omega \in S.$$

The connectives H and P have their usual meanings: i.e. $\|Hp\|^h_t = 1$ iff $\|p\|^h_u = 1$ for all $u < t$ in S, etc.; notice that axiom (f) is just $HA \leftrightarrow PA$, so *the connectives P and H are equivalent in S*. Thus, $\|PP(\sim q \wedge Hq)\|^h_\omega = 1$, so there exists $x < \omega$ in S with $\|P(\sim q \wedge Hq)\|^h_x = 1$. ($x = \omega$ is possible!)

We are going to prove that x is the unique point $< x$.

Let r be any atom and g any assignment such that $g(r) = \{x\}$. Hence $\|Pr\|^g_\omega = 1$. Now by axioms (a) and (c), $Y2 \vdash Pr \to PPr$, so $\|PPr\|^g_\omega = 1$. By axiom (f), we obtain $\|HPr\|^g_\omega = 1$, so that $\|Pr\|^g_x = 1$. But $g(r) = \{x\}$, so we see that $x < x$. Moreover, by (f) again, $\|Hr\|^g_x = 1$, so that $\{y \in S \mid y < x\} = \{x\}$.

But now, because $\|P(\sim q \wedge Hq)\|^h_x = 1$, we must have $\|\sim q \wedge Hq\|^h_x = 1$. This is impossible, as $x < x$.

This completes the impossibility proof. ■

Thus (a)–(f) cannot be modelled on a first-order flow of time without the use of some ultrafilter.

The proof hinges on our ability to find an r which is true at exactly x. If we impose limitations on the possible assignment g, then we may not have an allowable r.

Can we characterize $Y2^*$ using some condition on the assignment h? In other words, is $Y2^*$ complete for a class of models $(S, >, \omega, h)$ such that h satisfies certain restrictions?

Open problem

Find an axiomatic system $Y3$ (similar to $Y2$) which is complete for a linear flow with a filter, as defined above.

Chapter 4

Predicate temporal logics

4.1 Introduction

The move from the propositional to the predicate version of a logic is simple in principle, though one has several options to consider. Temporal logic is a language for talking about several databases and the connections between them. Predicate logic talks about domains of individuals. Put together we have to deal with several domains of individuals and the question is what are the individuals of the combined domains? This is the source of our options.

Technically, the move to predicate logic is undertaken by adding individual variables to the language and possibly individual constants, and adding atomic predicates and quantifiers $\forall$ and $\exists$. Thus predicate temporal logic is syntactically like classical predicate logic with the addition of the temporal connectives. The problem is what do the variables range over? Let us be more specific.

Consider a flow of time $(T,<)$. In the propositional case, a proposition can be true at different moments of time, thus we can view the temporal structure as a family of classical structures $\mathcal{A}_t, (T,<,\mathcal{A}_t), t \in T$. Each $\mathcal{A}_t$ is a function from the set of atoms $\{q_i\}$ into $\{0,1\}$.

This structure can also be given by the assignment h, where

$$h(t,q) = \mathcal{A}_t(q).$$

In the predicate case, $\mathcal{A}_t$ should be a classical model with domain D_t and an interpretation I_t. Thus $\mathcal{A}_t = (D_t, I_t)$ and $D_t \neq \varnothing$ and for all n place predicates Q, $I_t(Q) \subseteq D_t^n$.

Thus a temporal predicate structure has the form $(T,<,D_t,I_t), t \in T$.

For $t \neq s$ D_t and D_s are not necessarily related. We can have $D_t \subseteq D_s$, $D_t \supseteq D_s$, $D_t \cap D_s = \varnothing$, etc.

An element a may be in D_t, it may not be in D_s, and again it may be in D_r with $t < s < r$. No continuity is assumed. Formulae are evaluated at time points and thus $\|A(x)\|_t^A$ must be given a value. If x is a variable, we need an assignment g to the variables. The question is what is $g(x)$? Is it independent of t or do we make g dependent on t?

Also another decision to make, in $\|\forall x A(x)\|_t$, does x range over all the elements of D_t or over all elements of $\cup_s D_s$ or maybe only over elements of $\cup_{s \leq t} D_s$?

Let us take the simplest course of action and let $g(x) \in D = \cup_t D_t$ and let $\forall x$ range over D_t. Let us check what to expect, even in this simple case.

Consider $B = \exists x A(x)$. B can hold or not hold in different times. Different elements $a \in D_t$ can make $A(a)$ true and hence make B hold.

Let c_t be an element at time t such that $A(c_t)$ holds at t.

We may wish to be able to skolemize in temporal logic, just like we do in classical logic, and write $A(c)$ and expect a skolem rule

$$\frac{\exists x A(x)}{A(c)}, c \text{ a skolem constant.}$$

Obviously, then, we need to interpret a constant c as having different values in different times, thus

$$g(c)(t) = c_t \in D_t, \text{ for all } t$$

(i.e. $g(c) = \lambda t \,.\, c_t$). Such a view may be satisfactory for $\exists$ but is problematic for $\forall$. The question is can we write the universal rule

$$\frac{\forall x A(x)}{A(y)}$$

for all cases y?

How about $y = c$?

Consider

$$\forall x \Diamond A(x).$$

At time t, x ranges over the elements $a \in D_t$ of the domain D_t. Assume $t < s$ and assume that for all $a \in D_t$, $A(a)$ holds at s. Assume that for some $b \in D_s$, $A(b)$ is false and that $b \notin D_t$.

If we allow for constants c as mentioned above, we may have a constant c with $c_t = a, c_s = b$. Thus we may have that $\forall x \Diamond A(x)$ is true at t but $\Diamond A(c)$ can be false because $\|A(c)\|_s$ is $\|A(b)\|_s$ and its value is 0. Thus the new rule

$$\frac{\forall x A(x)}{A(y)}$$

cannot be used without care, in the presence of constants c as above.

On the other hand, if we do not allow for constants c like the above, we will not be able to skolemize, at least not naïvely as we do in classical logic.

Suppose for the moment that we do allow for skolemization. Thus we want to allow, whenever $\exists x A(x)$ holds at t, that for some constant c possibly dependent on t we have that $A(c)$ holds at t, namely $\|A(c_t)\|_t = 1$. If we allow a c in the syntax, we can equally evaluate $\|A(c_s)\|_s$.

If $\|A(c_s)\|_s = 1$, we obtain $\|\exists x A(x)\|_s = 1$, and so we must be careful in the choice of c. We must have that $\|A(c_t)\|_t = 1$ if $\|\exists x A(x)\|_t = 1$ but the value we give to c_s must be one for which $\|A(c_s)\| = 1$, if possible.

Suppose $\|\forall x{\sim}A(x)\|_s = 1$, then we obtain $\|{\sim}A(c_s)\|_s = 1$, by the above. Thus c skolemizes both the universal and existential quantifiers.

This is not so simple to do simultaneously, because we know in classical logic that the universal and existential quantifiers are skolemized differently.

The next step is to say that perhaps we let x range over all (constant) 'elements' c as well as all real elements $a \in D_t$. Thus

$$\|\forall x A(x)\|_t = 1$$

iff for all

$$a \in D_t, \|A(a)\|_t = 1$$

and for all c

$$\|A(c_t)\|_t = 1.$$

This would effectively mean that our 'elements' are all elements of $D = \cup_t D_t$, because we may always have a c such that $c_s = b$, for any $b \in D_s$. We may not wish to have this commitment. It is thus by now obvious that we are going to have some technical difficulties in sorting out our domain and constants policy.

There is, however, another problem, an intuitive one. Our n-place relations Q are, so far, assumed to be defined on each D_t only, i.e. $\mathcal{I}_t(Q) \subseteq D_t^n$. We thus cannot write relations such as $R(a, b)$, where $a \in D_t$ and $b \in D_s$, and where $b \notin D_t$ and $a \notin D_s$. For example, a sentence like 'John loves his grandmother', is meaningful only if they co-exist in some time, and John certainly cannot continue to love his grandmother after she is dead. However, intuitively we agree that there should be a truth value to the statement in such cases.

In view of the above, we may wish to try and adopt a different strategy. We may wish to assume that the predicates at world t can talk about all elements in all the other worlds, namely

$$\mathcal{I}_t(q) \subseteq D^n.$$

where

$$D = \cup_s D_s.$$

Alternatively we may try the more modest assumption that

$$\mathcal{I}_t(Q) \subseteq D^n_{\leq t} = (\cup_{s \leq t} D_s)^n.$$

In this case, John can love his grandmother. It may still be reasonable to assume, though, that if $a_1, \ldots, a_n \in \mathcal{I}_t(Q)$, then for some i, $a_i \in D_t$.

The next question is, what does $\forall\exists$ range over? We have two options. Either over D_t or over all of D.

If $\forall$ and $\exists$ range over all of D, then we will have the problem of how can we tell whether $a \in D_t$ or not? i.e. whether a exists at world t? In this case we would need an existence predicate $E(x)$ with the rule

$$\|E(x)\|_t = 1 \text{ iff } x \in D_t.$$

On the other hand, if $\forall$ and $\exists$ range over D_t only, then we will have the problem of how can we tell that John loves everybody (alive) but not his grandmother (who is dead). We will have that $\forall x A(x) \wedge \sim A(a)$ may be consistent, for $a \notin D_t$ at world t. So what do we do; do we accept the value or not?

If we accept the value then the axiom

$$A(x) \to HFA(x)$$

is acceptable as holding in the semantics. If we do not accept the value then some restrictions are necessary, possibly

$$A(x) \to H(\text{`} x \text{ exists'} \to FA(x)).$$

Even if we assume that all relations $R(x, y)$ are meaningful only for $x, y \in D_t$, we still have another option. Consider $\|\Diamond A(b)\|_t$, for $b \notin D_t$ and $b \in D_s$ and $t < s$. We can adopt the view that the above cannot have a truth value because b is not at world t. However, since the $\Diamond$ operator requires $A(b)$ to be evaluated at world s, we can obtain a value for the formula.

There is yet another intuitive difficulty. Our relations Q are between elements from different worlds but only between elements. We cannot write, for example, $\|Q(a, c)\|_t$ to mean a relation between a and c. It can only be a relation between a and c_t. Thus a sentence such as 'I respect you as a prime minister (i.e. as a 'c') but not as a person (i.e. as a 'c_t')' are contradictory because

$$\|R(a, c)\|_t = \|R(a, c_t)\|_t.$$

If we want our semantics to capture that intuition we must allow the predicates to relate to c as a function, namely to $\lambda t c_t$.

An entirely new (but related) problem is how to define $=$ in temporal logic.

The problem is how to evaluate $\|c = d\|_t$. One possibility is that it holds iff $c_t = d_t$. Another is that it holds iff $\forall t[c_t = d_t]$. Similarly, the value of $\|x = c\|_t$ should be decided. Obviously, we use the assignment g. $g(x)$ is a fixed element. Do we say, $\|x = c\|_t$ holds at t if $g(x) \in D_t$ and $g(x) = c_t$ or do we require that wherever $g(x) \in D_s$ then $g(x) = c_s$?

Different axioms hold for different choices, e.g.

$$\begin{array}{l} c = d \to G(c = d) \\ c \neq d \to G(c \neq d) \\ x = c \to G(x = c) \\ x \neq c \to G(x \neq c) \\ x = y \to G(x = y) \\ x \neq y \to G(x \neq y). \end{array}$$

4.2 Some basic quantificational systems

We start with two very basic systems, the quantifications counterpart of K_t, with varying domains, constants, and function symbols.

Definition 4.2.1 The predicate language Q_1K_t contains atomic predicates $\{Q_1^{n_1}, Q_1^{n_2}, \ldots\}$, where $Q_1^{n_i}$ is n_i place; function symbols $\{F_1^{m_1}, F_1^{m_2}, \ldots\}$, where $F_1^{m_i}$ is m_i place function; the usual connectives, the quantifiers $\forall$ and $\exists$, and the temporal connectives F, G, P, and H. (Different versions of these connectives will be denoted by $G_1, G_2, F_1, F_2, H_1, H_2, P_1, P_2$, etc.)

The notions of a well-formed formula and of a term are defined in the traditional manner.

Definition 4.2.2 (the system Q_1K_t) A Q_1K_t temporal structure has the form $(T, <, 0, D_t, I_t)$, where $(T, <)$ is an irreflexive ordering $0 \in T$, and $D_t \neq \varnothing$ for $t \in T$ is the domain in time t. We have no requirement on the relationship between the domains D_t and D_s for $t \neq s$.

I_t is the assignment, giving values to the atomic predicates and function symbols. We have:

1. $I_t(F_i^{n_i})$ is a function such that for each $a_1, \ldots, a_{n_i} \in D_t$,

$$I_t(F_i^{n_i})(a_1, \ldots, a_{n_i}) \in D_t,$$

namely, $I_t(F_i^{n_i}) \in D_t^{D_t^{n_i}}$.

2. $I_t(Q_i^{n_i}) \subseteq D_t^{n_i}$, i.e. $I_t(q_i^{n_i})$ is an n_i-place predicate.

3. Let g be an assignment, $g(x) \in \cup_t D_t$, to the variables. g can be extended to all function symbols whenever defined, by letting

$$g_t(F_i^{n_i}(t_1, \ldots, t_n)) = I_t(F_i^{n_i})(g_t(t_1), \ldots, g_t(t_n)).$$

We define the value $\|A\|_t^g$ in the structure $(T, <, 0, D_t, I_t)$ as follows:

(a) $\|Q_i^{n_i}(x_1, \ldots, x_{n_I})\|_t^g = 1$ iff $(g(x_1), \ldots, g(x_{n_i})) \in I_t(Q_i^{n_i})$ and $g(x_j) \in D_t$ for $j = 1, \ldots, n_i$.
$\|Q_i^{n_i}(x_1, \ldots, x_{n_i})\|_t^g = 0$ iff $(g(x_1), \ldots, g(x_{n_i})) \notin I_t(Q_i^{n_i})$ and $g(x_j) \in D_t$, for $j = 1, \ldots, n_i$.
$\|Q_i^{n_i}(x_1, \ldots, x_{n_i})\|_t^g =$ undefined iff for some $j, g(x_j) \notin D_t$.

(b) $\|A \wedge B\|_t^g = 1$ iff $\|A\|_t^g = 1$ and $\|B\|_t^g = 1$.
$\|A \wedge B\|_t^g = 0$ if $\|A\|_t^g = 0$ or $\|B\|_t^g = 0$.

(c) $\|{\sim}A\|_t^g = 1$ iff $\|A\|_t^g = 0$.
$\|{\sim}A\|_t^g = 0$ iff $\|A\|_t^g = 1$.

(d) $\|\forall x A(x)\|_t^g = 1$ iff for all $g' \equiv_x g$ such that $g'(x) \in D_t$.
$\|A\|_t^{g'} = 1$, where $g' \equiv_x g$ means $\forall y \neq x[g(x) = g'(x)]$.
$\|\forall x A(x)\|_t^g = 0$ iff for some $g' \equiv_x g, g'(x) \in D_t$ and $\|A\|_t^{g'} = 0$.

(e) $\|\exists x A(x)\|_t^g = 1$ iff for some $g' \equiv_x g$, $g'(x) \in D_t$ and $\|A(x)\|_t^{g'} = 1$.
$\|\exists x A(x)\|_t^g = 0$ iff or all $g' \equiv_x g$, such that $g'(x) \in D_t$ we have $\|A\|_t^{g'} = 0$.

(f) $\|F_1 A\|_t^g = 1$ iff for some s such that $t < s, \|A\|_s^g = 1$.
$\|F_1 A\|_t^g = 0$ iff for all s such that $t < s$, we have $\|A\|_s^g = 0$.

(g) $\|P_1 A\|_t^g = 1$ iff for some $s < t, \|A\|_s^g = 1$.
$\|P_1 A\|_t^g = 0$ iff for all $s < t$ we have $\|A\|_s^g = 0$.

(h) $\|G_1 A\|_t^g = 1$ iff for all s such that $t < s$, we have $\|A\|_s^g = 1$.
$\|G_1 A\|_t^g = 0$ iff for some s such that $t < s$, we have $\|A\|_s^g = 0$.

(i) $\|H_1 A\|_t^g = 1$ iff for all $s < t$ we have $\|A\|_s^g = 1$.
$\|H_1 A\|_t^g = 0$ iff for some $s < t$ we have $\|A\|_s^g = 0$.

4. A formula A is undefined at t under g iff it can have neither value 1 nor value 0 at t under g.

We say that A holds in $(T, <, 0, D_t, I_t)$ under g iff $\|A\|_0^g = 1$.

We say that $\models_1 A$ iff for all structures and all $g, \|A\|_0^g = 1$.

Examples 4.2.3

1. Let $(T, <, 0, D_t, I_t)$ be a structure. Consider $a \in D_t$ such that $a \notin D_s$, $t < s$. Consider an atom Q and assume $\|Q(a)\|_t = 1$. Consider $\|P_1Q(x)\|^g_s$ for a g such that $g(x) = a$. Since in the inductive definition of $\|P_1A(x)\|_s$, we did not require that $g(x) \in D_s$, we are allowed to evaluate $\|P_1Q(x)\|^g_s = 1$ iff for some $t < s$, $\|Q(x)\|^g_t = 1$. Indeed, such a t exists. $Q(x)$ is atomic and to evaluate $\|Q(x)\|^g_t$ we need to check whether $g(x) \in D_t$. In our case it does hold.

   We thus saw that
   $$\vDash A(x) \rightarrow G_1P_1A(x)$$
   holds. In fact it holds for any A, not only for A atomic.

2. Consider $\|G_1A(x)\|^g_t$.
   $\|G_1A(x)\|^g_t = 1$ iff for all s such that $t < s$, $\|A(x)\|^g_s = 1$.

   For the case that $A(x)$ is atomic the latter holds iff $g(x) \in D_s$ and $g(x) \in I_s(A)$. Consider $\|F_1A(x)\|^g_t = 1$. This holds iff for some $s, t < s$ and $\|A\|^g_s = 1$. For the case A is atomic the latter holds iff $g(x) \in D_s$ and $g(x) \in I_s(A)$.

   $\|{\sim}F_1{\sim}A\|^g_t = 1$ iff $\|F_1{\sim}A\|^g_t = 0$ iff for all s such that $t < s$, $\|{\sim}A\|^g_s = 0$ iff for all s such that $t < s$, $\|A\|^g_s = 1$
   Thus $G_1 = {\sim}F_1{\sim}$.
   Similarly, $H_1 = {\sim}P_1{\sim}$.

3. Let $Q(x)$ be atomic.
   Consider $\|{\sim}(Q(x) \wedge {\sim}Q(x))\|^g_t$. This has value 1 iff $\|(Q(x) \wedge {\sim}Q(x))\|^g_0 = 0$ iff either $\|Q(x)\|^g_t = 0$ or $\|Q(x)\|^g_t = 1$ iff $g(x) \in D_t$.

   Thus $E_1(x) = {\sim}(Q(x) \wedge {\sim}Q(x))$ is an existence predicate for x.

4. We can also define a definability connective $\delta_1(A)$ for the language, with the following table:
   $\|\delta_1(A)\|^g_t = 1$ iff A is definable at t under g iff $\|A\|^g_t = 0$ or $\|A\|^g_t = 1$.

   (a) For a formula $A(x_1, \ldots, x_n)$ without temporal conectives and with free variables $x_1, \ldots, x_n$ let $\delta(A(x_1, \ldots, x_n))$ be $E_1(x_1) \wedge \ldots \wedge E_1(x_n)$.

   (b) $\delta_1({\sim}A) = \delta_1(A)$.

   (c) $\delta_1(A \wedge B) = (\delta_1(A) \wedge {\sim}A) \vee (\delta_1(B) \wedge {\sim}B) \vee (\delta_1(A) \wedge \delta_1(B))$.

   (d) $\delta_1(\forall x A(x) = \forall x \delta_1(A(x)) \vee (\exists x {\sim}A(x) \wedge \delta_1(A(x))$.

   (e) $\delta_1(\exists x A(x)) = \exists x[A(x) \wedge \delta_1(A(x))] \vee \forall x[{\sim}A(x) \wedge \delta_1(A(x))]$.

(f) $\delta_1(F_1 A) = F_1(A \wedge \delta_1(A)) \vee G_1(\delta_1(A) \wedge {\sim}A)$.

(g) $\delta_1(P_1 A) = P_1(A \wedge \delta_1(A)) \vee H_1(\delta_1(A) \wedge {\sim}A)$.

(h) $\delta_1(G_1 A) = G_1(A \wedge \delta_1(A)) \vee F_1(\delta_1(A) \wedge {\sim}A)$.

(i) $\delta_1(H_1 A) = H_1(A \wedge \delta_1(A)) \vee P_1(\delta_1(A) \wedge {\sim}A)$.

Definition 4.2.4 (the system Q_2K_t) A Q_2K_t temporal structure has the same form as a Q_1K_t temporal structure. The assignments g are also the same. The inductive definition of the truth table, however, is different, and is as follows:

1. $\|Q_i^{n_i}(x_1, \ldots, x_{n_i})\|_t^g = 1$ iff whenever $g(x_j) \in D_t$ for $i = 1, \ldots, n_i$ then also $(g(x_1), \ldots, g(x_{n_i})) \in I_t(Q_i^{n_i})$.
   Otherwise the value is 0.

2. $\|A \wedge B\|_t^g = 1$ iff $\|A\|_t^g = 1$ and $\|B\|_t^g = 1$.

3. $\|{\sim}A\|_t^g = 1$ iff $\|A\|_t^g = 0$.

4. $\|\forall x A(x)\|_t^g = 1$ iff for all $g' \equiv_x g$ such that $g'(x) \in D_t$, we have $\|A(x)\|_t^{g'} = 1$.

5. $\|\exists x A(x)\|_t^g = 1$ iff for some $g' \equiv_x g$ such that $g'(x) \in D_t$ we have $\|A\|_t^{g'} = 1$.

6. $\|F_2 A(x)\|_t^g = 1$ iff for some $s > t$, $g(x) \in D_s$ and $\|A\|_s^g = 1$.

7. $\|G_2 A(x)\|_t^g = 1$ iff for all $s > t$, if $g(x) \in D_s$ then $\|A\|_s^g = 1$.

8. $\|P_2 A(x)\|_t^g = 1$ iff for some $s < t$, $(x) \in D_s$ and $\|A\|_s^g = 1$.

9. $\|H_2 A(x)\|_t^g = 1$ iff for all $s < t$ if $g(x) \in D_s$ then $\|A\|_s^g = 1$.

10. We say A holds in the structure under g iff $\|A\|_0^g = 1$.

11. We say $\vDash_2 A$ iff for all structures and all g, $\|A\|_0^g = 1$.

Example 4.2.5 The connectives of Q_2K_t are different from those of Q_1K_t. First, for atomic Q, there are only two values $0, 1$, not three. To obtain value 0 an atom needs to be defined and false.

Second, the meaning of $G_2A(x)$ is that $A(x)$ holds whenever x is defined. Thus suppose $A(a)$ holds at t and $a \in D_t$, and that $a \notin D_s$ for any $s > t$. We would also have $\|G_2P_2A(a)\|_t = 1$.

The reason is that we do not obtain a value for $\|P_2A(a)\|_s$ for $s > t$ because $a \notin D_s$, even though $A(a)$ holds at t.

Example 4.2.6 Consider the connectives G_{11} and H_{11} defined in Q_1K_t: $\|G_{11}A\|_t^g = 1$ iff for all s such that $t < s$ we have that either $\|A\|_s^g$ is undefined or $\|A\|_s^g = 1$.

These can be written as $\|G_{11}A\|_t^g = 1$ iff for all $s > t, \|A\|_s^g \neq 0$. $\|H_{11}A\|_t^g = 1$ for all $s < t, \|A\|_s^g \neq 0$.

G_{11} and H_{11} are definable from G_1 and H_1 as follows:

$$G_{11}A = G_1(\delta(A) \to A),$$
$$H_{11}A = H_1(\delta(A) \to A).$$

4.3 An axiomatization

As we have seen there is a bewildering variety of formalizations of first-order temporal logic. There are very good surveys in [Garson, 1984] and [van Benthem, 1992]. In order to investigate the axiomatizability of such logics we will need to concentrate here on one particular formalization. Below we will define the first-order temporal logic of F and P over various classes of flows but with fixed domains, rigid terms, and rigid variable assignments. However, it is not too difficult to see that many other formalizations can be comfortably coded into ours by using such tricks as coding functions as relations and using a special 'exists' predicate.

Many of the results were first achieved by Per Lindström and Dana Scott in the 1960s but unfortunately do not exist in published form. Some appear in [Hughes and Cresswell, 1968] and [Gabbay, 1975b]. We follow the proofs in [Reynolds, 1992b].

Although we confine ourselves to investigations of languages with just F and P it will be noted that the general axiomatization methods of chapter 6 are able to be used here to find axiomatizations with other connectives. There are also other methods of finding axiom systems without using the IRR-rule. For example, in [Reynolds, 1992a] we see an axiomatization for the first-order temporal logic of U and S over linear time and over the rationals.

Now we present our formalism. Let us introduce the language first.

Fix a first-order signature $\mathcal{S}$: so $\mathcal{S}$ contains function symbols of various arities (0-ary function symbols sometimes being called constants) and predicate symbols of various arities.

The predicate temporal language with F and P has the following constituents:

- a countable collection $x_1, x_2, \ldots$ of *variable symbols*;

- a set of *terms* including the variable symbols and being the smallest such set which also includes $f(t_1, ..., t_n)$ for any terms $t_1, ..., t_n$ and any n-ary function symbol f;
- the set of *closed terms*, being those terms in which variable symbols do not appear;
- *atomic formulae*, being of the form either $t_1 = t_2$ for terms t_1 and t_2, $\top$, or $p(t_1, ..., t_n)$ for terms $t_1, ..., t_n$ and n-ary predicate symbol p;
- the set of *formulae*, being the smallest set which includes all the atomic formulae and also $\sim\alpha$, $\alpha \wedge \beta$, $\forall x\alpha$, $F\alpha$, and $P\alpha$ for any formulae α and β;
- the usual concepts of free variables, bound variables, substitution, and sentences;
- abbreviations $\perp$, $\neq$, $\vee$, $\rightarrow$, $\leftrightarrow$, $\exists x$, G and H.

We use this language to talk about temporal structures with a fixed domain and rigid function symbols. There will be rigid assignments to variable symbols. More formally, a temporal structure is a 4-tuple $\mathcal{T} = (T, <, \mathcal{D}, h)$ where

- T is a set of *points of time*;
- $<$ is a binary relation on T, and $(T, <)$ is called the *flow of time*;
- $\mathcal{D}$ is a first-order structure in the language with just the function symbols of $\mathcal{S}$: so it has a domain, called the *object domain*, and interpretations for the function symbols as functions of the appropriate arity;
- h is a map such that for any n-ary predicate symbol p, and for any $t \in T$, $h(p, t)$ is a subset of $\mathcal{D}^n$.

Applying assignments ν to terms in the usual way, we can also define the value (0 or 1) of ${}^{\mathcal{T}}\|\alpha\|_t^\nu$ by induction on the construction of α as usual:

- ${}^{\mathcal{T}}\|t_1 = t_2\|_t^\nu = 1$ iff $\nu(t_1) = \nu(t_2)$;
- ${}^{\mathcal{T}}\|\top\|_t^\nu = 1$;
- ${}^{\mathcal{T}}\|p(t_1, ..., t_n)\|_t^\nu$ iff $(\nu(t_1), ..., \nu(t_n)) \in h(p, t)$;
- ${}^{\mathcal{T}}\|\sim\alpha\|_t^\nu = 1 - {}^{\mathcal{T}}\|\alpha\|_t^\nu$;

- ${}^{\mathcal{T}}\|\alpha \wedge \beta\|_t^\nu = ({}^{\mathcal{T}}\|\alpha\|_t^\nu)({}^{\mathcal{T}}\|\beta\|_t^\nu)$;
- ${}^{\mathcal{T}}\|\forall x\alpha\|_t^\nu = 1$ iff for all $d \in \mathcal{D}$, ${}^{\mathcal{T}}\|\alpha\|_t^\mu = 1$ where μ is the assignment which is just like ν except that x is mapped to d;
- ${}^{\mathcal{T}}\|F\alpha\|_t^\nu = 1$ iff there is some $s > t$ such that ${}^{\mathcal{T}}\|\alpha\|_s^\nu = 1$;
- ${}^{\mathcal{T}}\|P\alpha\|_t^\nu$ iff there is some $s < t$ such that ${}^{\mathcal{T}}\|\alpha\|_s^\nu = 1$.

Write ${}^{\mathcal{T}}\|\alpha\|_t$ for ${}^{\mathcal{T}}\|\alpha\|_t^\nu$ when α is a sentence and so ν might as well be the empty map.

Using this semantics we can define many useful logics. Suppose that $\mathcal{K}$ is a class of flows of time. For a set of sentences Γ and a sentence α write $\Gamma \models_{\mathcal{K}} \alpha$ iff for all temporal structures $\mathcal{T} = (T, <, \mathcal{D}, h)$ such that $(T, <) \in \mathcal{K}$, for all $t \in T$, if for all $\beta \in \Gamma$, ${}^{\mathcal{T}}\|\beta\|_t = 1$, then ${}^{\mathcal{T}}\|\alpha\|_t$. Call the resulting logic qFP/$\mathcal{K}$ or qFP/T if $\mathcal{K}$ consists of a single flow T. Later we will examine the logics qFP/L, the class of all linear flows, qFP/$\mathbb{Q}$, qFP/$\mathbb{R}$, qFP/$\mathbb{Z}$, and qFP/$\mathbb{N}$, but now let us try to find an axiomatization for the logic qFP/$\mathcal{K}$ when $\mathcal{K}$ is the set of all flows of time (transitive and not).

As rules we have modus ponens, universal generalization and the two temporal generalizations.

The axioms of QFP/$\mathcal{K}$ are all formulae of the following forms where α and β are formulae, t and u are terms and x is a variable symbol:

1. any tautology of classical propositional logic with formulae of our language substituted for atoms;
2. $\forall x\alpha \to \alpha[t\backslash x]$;
3. $t = t$;
4. $x = t \to (\alpha \to \alpha[t\backslash x])$;
5. $G(\alpha \to \beta) \to (G\alpha \to G\beta)$ and dual;
6. $\alpha \to GP\alpha$ and dual;
7. $(t = u) \to G(t = u)$ and dual;
8. $\forall xG\alpha \to G\forall x\alpha$ and dual.

Schema (8) above is a version of the Barcan formula.

For a set of sentences Γ and a sentence α, we write $\Gamma \vdash \alpha$ iff there is a proof of α from Γ using the rules and axioms above.

Soundness is clear, so once we have proved completeness below we have

Theorem 4.3.1 *The system QFP/$\mathcal{K}$ above is sound and complete for the predicate temporal logic qFP/$\mathcal{K}$ of F and P (as defined above) over the class $\mathcal{K}$ of all flows of time.*

With a certain predicate signature fixed in context we have the following definitions. We say that a set Γ of sentences is *consistent* iff we do *not* have $\Gamma \vdash \bot$. Γ is *maximal consistent* iff it is not a proper subset of any other consistent set (in the given language). Γ is *omega complete* iff whenever α is a formula with at most one free variable x and $\Gamma \vdash \alpha[t\backslash x]$ for all closed terms t, then $\Gamma \vdash \forall x\alpha$. Γ is *saturated* iff it is maximal consistent and omega complete.

Our first step is to show that any consistent set Σ_0 can be extended to a saturated set Γ_0. To do this we need to expand our language by a countable set $\{c_0, c_1, ...\}$ of new constant symbols which do not appear in Σ_0. Now enumerate the sentences of this expanded language as $\alpha_0, \alpha_1, \ldots$. Define consistent sets Σ_i recursively as follows. If $\Sigma_i \cup \{\alpha_i\}$ is inconsistent, then put $\Sigma_{i+1} = \Sigma_i$. If it is consistent and α_i is not of the form $\sim\forall x\beta$, then put $\Sigma_{i+1} = \Sigma_i \cup \{\alpha_i\}$. Now suppose that $\Sigma_i \cup \{\alpha_i\}$ is consistent and $\alpha_i = \sim\forall x\beta$. We choose a new constant c_j not appearing in α_i nor in Σ_i. Now note that $\Sigma_{i+1} = \Sigma_i \cup \{\sim\forall x\beta, \beta[c\backslash x]\}$ is consistent. Finally, put $\Gamma_0 = \bigcup_{i<\omega} \Sigma_i$. It is not difficult to prove that it is saturated.

Let T contain all the saturated sets Γ such that for any closed terms t_1 and t_2, $(t_1 = t_2) \in \Gamma$ iff $(t_1 = t_2) \in \Gamma_0$.

Order T by $<$ in the usual way: $\Gamma < \Delta$ iff for all sentences α, $G\alpha \in \Gamma$ implies $\alpha \in \Delta$.

Note that using axiom (6) and its dual we can easily prove that $\Gamma < \Delta$ iff for all β, $H\beta \in \Delta$ implies $\beta \in \Gamma$.

$(T, <)$ will be the flow of time of our structure.

For each closed term t, let $[t] = \{t' \mid (t = t') \in \Gamma_0\}$. Take the domain of $\mathcal{D}$ to be $\{[t] \mid t \text{ a closed term }\}$. This is non-empty because there is at least one constant symbol (for example c_0). It is easy to show that for any n-ary function symbol $f \in \mathcal{S}$, the following n-ary function $f^{\mathcal{D}}$ is well-defined on $\mathcal{D}$: $f^{\mathcal{D}}([t_1], ..., [t_n]) = [f(t_1, ..., t_n)]$. This will be the rigid extension of f.

For each time Γ, and each n-ary predicate symbol p we define $h(p, \Gamma) = \{([t_1], ..., [t_n]) \mid p(t_1, ..., t_n) \in \Gamma\}$. It is easily proved that $([t_1], ..., [t_n])$ is in $h(p, \Gamma)$ iff $p(t_1, ..., t_n) \in \Gamma$.

Put $\mathcal{T} = (T, <, \mathcal{D}, h)$.

Theorem 4.3.2 *For all $\Gamma \in \mathcal{T}$, for all sentences α,*

$$^{\mathcal{T}}\|\alpha\|_{\Gamma} = 1 \textit{ iff } \alpha \in \Gamma.$$

Proof. Proceed by induction on the complexity of formulae.

The difficult cases are:

$\forall x\alpha$, $\Rightarrow$: Suppose ${}^{\mathcal{T}}\|\forall x\alpha\|_\Gamma = 1$. Then for every closed term t,

$$ {}^{\mathcal{T}}\|\alpha[t\backslash x]\|_\Gamma = 1 $$

and by the inductive hypothesis, $\alpha[t\backslash x] \in \Gamma$. Since Γ is omega complete, $\forall x\alpha \in \Gamma$ as required.

$\forall x\alpha$, $\Leftarrow$: Suppose $\forall x\alpha \in \Gamma$. Then for any d in $\mathcal{D}$, say $d = [t]$, we have $\alpha[t\backslash x]$ in Γ by axiom (1), so by the inductive hypothesis, ${}^{\mathcal{T}}\|\alpha[t\backslash x]\|_\Gamma = 1$. Let ν be an assignment mapping x to the interpretation d, of t in $\mathcal{T}$. Then ${}^{\mathcal{T}}\|\alpha\|_\Gamma^\nu = 1$. This is true for any d, so ${}^{\mathcal{T}}\|\forall x\alpha\|_\Gamma = 1$.

$F\alpha$, $\Rightarrow$: Suppose that

$$ {}^{\mathcal{T}}\|F\alpha\|_\Gamma = 1 $$

so there is $\Delta > \Gamma$ such that, by the inductive hypothesis, $\alpha \in \Delta$. We cannot have $G{\sim}\alpha \in \Gamma$ for then, by definition of $<$, ${\sim}\alpha \in \Delta$. So we must have $F\alpha \in \Gamma$ instead.

$F\alpha$, $\Leftarrow$: Suppose that $F\alpha \in \Gamma$. Now we claim that $\Delta_0 = \{\beta \mid G\beta \in \Gamma\} \cup \{\alpha\}$ is consistent. If it is not consistent then $\{\beta \mid G\beta \in \Gamma\} \vdash {\sim}\alpha$, so by lemma 4.3.3 below (which can be proved by a straightforward induction), $\{G\beta \mid G\beta \in \Gamma\} \vdash G{\sim}\alpha$. This cannot be as Γ is consistent.

Lemma 4.3.3 *Let Σ be any set of sentences. If $\Sigma \vdash \gamma$ then $\{G\delta \mid \delta \in \Sigma\} \vdash G\gamma$.*

Next we show that $\{\beta \mid G\beta \in \Gamma\}$ is omega complete. So assume that $\{\beta \mid G\beta \in \Gamma\} \vdash \gamma(t)$ for every closed term t. By lemma 4.3.3 above, $\Gamma \vdash G\gamma[t\backslash x]$ and, since Γ is omega complete, $\Gamma \vdash \forall x G\gamma$. By schema (8), $\Gamma \vdash G\forall x\gamma$ and thus $\forall x\gamma \in \Delta_0$, as required.

Now the following lemma proves that Δ_0 is omega complete.

Lemma 4.3.4 *If Δ is omega complete, then so is $\Delta \cup \delta$ for any sentence δ.*

Proof. Suppose that β is a formula with a single free variable x and that $\Delta \cup \{\delta\} \vdash \beta[t\backslash x]$ for all closed terms t. Thus $\Delta \vdash (\delta \to \beta)[t\backslash x]$ for all t and by the omega completeness of Δ we have $\Delta \vdash \forall x(\delta \to \beta)$. Since x is not free in the sentence δ we have $\Delta \cup \{\delta\} \vdash \forall x\beta$, as required. ■

Lemma 4.3.5 *Next we show how to extend the consistent omega complete set* Δ_0 *to a saturated set* Δ *in the same language.*

Proof. First enumerate the sentences of the language as $\alpha_0, \alpha_1, \ldots$. We are going to construct recursively a consistent omega complete set Δ_i for each $i = 0, 1, \ldots$. We already have Δ_0 so suppose now that we have Δ_i. If $\Delta_i \cup \{\alpha_i\}$ is inconsistent, then put $\Delta_{i+1} = \Delta_i$. If it is consistent and α is not of the form $\sim\forall x\beta$, then put $\Delta_{i+1} = \Delta_i \cup \{\alpha\}$. This will be omega complete by lemma 4.3.4 above.

Now consider the case when $\alpha_i = \sim\forall x\beta$ and $\Delta_i \cup \{\sim\forall x\beta\}$ is consistent. We cannot have $\Delta_i \vdash \beta[t\backslash x]$ for all closed terms t for then, by its omega completeness, $\Delta_i \vdash \forall x\beta$. Thus there is t such that $\Delta_i \cup \{\sim\beta[t\backslash x]\}$ is consistent and we can form a consistent set $\Delta_{i+1} = \Delta_i \cup \{\sim\beta[t\backslash x], \sim\forall x\beta\}$. This is omega complete by lemma 4.3.4.

Putting $\Delta = \bigcup_{i<\omega} \Delta_i$ it is easy to prove that Δ is consistent and omega complete. ∎

Now we claim that $\Delta \in T$. Suppose that t_1 and t_2 are closed terms. If $(t_1 = t_2) \in \Gamma$, then the consistency of Γ with schema (7) shows that $G(t_1 = t_2) \in \Gamma$ so $(t_1 = t_2) \in \Delta_0 \subset \Delta$. While if $\sim(t_1 = t_2) \in \Gamma$, then $GP\sim(t_1 = t_2) \in \Gamma$ so $P\sim(t_1 = t_2) \in \Delta$. Thus we cannot have $(t_1 = t_2) \in \Delta$ by the dual of (7).

So $\Delta \in T$ and by the induction hypothesis ${}^{\mathcal{T}}\|\alpha\|_\Delta = 1$. Since $\Gamma < \Delta$ by construction we have ${}^{\mathcal{T}}\|F\alpha\|_\Gamma = 1$, as required.

∎

To finish the completeness proof just note that for all $\alpha \in \Sigma_0$,

$$^{\mathcal{T}}\|\alpha\|_{\Gamma_0} = 1$$

so we have our model.

4.4 Linear time: strongly complete

To axiomatize F and P over the class of linear flows of time we simply add four more schemas:

trans $G\alpha \to GG\alpha$ and dual;

lin $F\alpha \wedge F\beta \to F(\alpha \wedge \beta) \vee F(\alpha \wedge F\beta) \vee F(\beta \wedge F\alpha)$ plus dual.

Since these are clearly sound we have the following theorem once we prove completeness:

Theorem 4.4.1 *The system QFP/L above is sound and complete for the predicate temporal logic qFP/L of F and P over the class of linear flows of time.*

Suppose that we have a QFP/L-consistent set Σ_0 of sentences. Σ_0 is also QFP/K-consistent so we find the canonical model $(\mathcal{T}, \Gamma_0)$, as above (so $\mathcal{T} = (T, <, \mathcal{D}, h)$).

First of all note that $(T, <)$ is transitive. For suppose that $\Delta_1 < \Delta_2 < \Delta_3$. Then $G\alpha \in \Delta_1$ implies, since Δ_1 is maximal consistent with the schema trans, that $GG\alpha \in \Delta_1$. Thus $G\alpha \in \Delta_2$ and $\alpha \in \Delta_3$. So $\Delta_1 < \Delta_3$ as claimed.

Say that Ξ_1 and Ξ_2 from T are *comparable* iff either $\Xi_1 = \Xi_2$, $\Xi_1 < \Xi_2$, or $\Xi_2 < \Xi_1$. Being comparable is clearly a reflexive and symmetric relation while the next lemma shows that it is transitive.

Lemma 4.4.2 *Suppose that Δ_1, Δ_2, and Δ_3 are from T, that Δ_1 and Δ_2 are comparable, and that Δ_1 and Δ_3 are comparable.*

Then Δ_2 and Δ_3 are comparable.

Proof. Of the nine cases (depending on the nature of the comparisons assumed), the only interesting case is when that $\Delta_1 < \Delta_2$ and $\Delta_1 < \Delta_3$. Suppose for contradiction that Δ_2 and Δ_3 are not comparable so that

- since $\Delta_2 \neq \Delta_3$, there is $\alpha_0 \in \Delta_2$ such that $\sim\alpha_0 \in \Delta_3$;
- since $\Delta_2 \not< \Delta_3$, there is α_2 such that $G\alpha_2 \in \Delta_2$ and $\sim\alpha_2 \in \Delta_3$; and
- since $\Delta_3 \not< \Delta_2$, there is α_3 such that $G\alpha_3 \in \Delta_3$ and $\sim\alpha_3 \in \Delta_2$.

Let $\beta_2 = G\alpha_2 \wedge \sim\alpha_3 \wedge \alpha_0$ and $\beta_3 = G\alpha_3 \wedge \sim\alpha_2 \wedge \sim\alpha_0$. Then the orderings give us $F(\beta_2) \wedge F(\beta_3) \in \Delta_1$. We have our contradiction as axiom **lin** implies one of the following hold:

- $F(\beta_2 \wedge \beta_3) \in \Delta_1$. But this cannot be as $\beta_2 \wedge \beta_3$ is inconsistent.
- $F(\beta_2 \wedge F(\beta_3)) \in \Delta_1$. But this cannot be as it implies $F(G\alpha_2 \wedge F(\sim\alpha_2)) \in \Delta_1$.
- $F(\beta_3 \wedge F(\beta_2)) \in \Delta_1$. But this too cannot be as it implies $F(G\alpha_3 \wedge F(\sim\alpha_3)) \in \Delta_1$.

■

Let

$$C = \{\Delta \in T \mid \Gamma_0 \text{ and } \Delta \text{ are comparable }\}.$$

Let $(C, <)$ inherit $<$ from $(T, <)$. It is easy to see that

- $(C, <)$ is transitive;
- $(C, <)$ is a *total* order in the sense that for every Γ and Δ from C either $\Gamma < \Delta$, $\Gamma = \Delta$ or $\Delta < \Gamma$; and
- $(C, <)$ inherits the fundamental lemma from $(T, <)$ so that for every $\Gamma \in C$, for every sentence α, $\alpha \in \Gamma$ iff ${}^{(C,<,\mathcal{D},h)}\|\alpha\|_\Gamma = 1$.

$(C, <)$ may not be anti-symmetric or irreflexive. Let $C^i = \{\Gamma \in C \mid \Gamma \not< \Gamma\}$ be the set of irreflexive points. By transitivity of $(C, <)$, the following relation $\sim$ on $C^r = C - C^i = \{\Gamma \in C \mid \Gamma < \Gamma\}$ is an equivalence relation: $\Gamma \sim \Delta$ iff both $\Gamma < \Delta$ and $\Delta < \Gamma$. Call the $\sim$-classes *clusters*.

Choose a map $*_c$ from $\mathbb{R}$ onto each cluster c such that for each $\Delta \in c$, for each $r \in \mathbb{R}$, there are s and t in $\mathbb{R}$

- $s < r < t$, and
- $*_c(s) = \Delta = *_c(t)$.

We can do this because any cluster contains at most $2^{\aleph_0}$ maximal consistent sets. This is true because there are $\aleph_0$ formulae in the language with F and P. Thus there are $2^{\aleph_0}$ sets of formulae and so at most $2^{\aleph_0}$ maximal consistent sets altogether.

For the sake of simplified notation also define $*_{\{\Gamma\}}(0) = \Gamma$ for any $\Gamma \in C^i$.

Define

$$C^\flat = \{(\{\Gamma\}, 0) \mid \Gamma \in C^i\} \cup \{(c, s) \mid c \text{ a cluster and } s \in \mathbb{R}\}.$$

Order $C^\flat$ by $\prec$ by putting $(a, s) \prec (b, r)$ iff

- $a \neq b$ and there are $\Gamma_a \in a$, $\Gamma_b \in b$ such that $\Gamma_a < \Gamma_b$; or
- $a = b$ and $s < r$.

To check that $(C^\flat, \prec)$ is indeed linear, the only difficult part is establishing anti-symmetry. So suppose for contradiction that $(a, s) \prec (b, r) \prec (a, s)$. Because of the linearity of $\mathbb{R}$ we must have $a \neq b$ so that there are Γ_a and Δ_a in a and Γ_b and Δ_b in b such that $\Gamma_a < \Gamma_b$ and $\Delta_b < \Delta_a$. Since they are in the same clusters $\Gamma_b < \Delta_b$ and $\Delta_a < \Gamma_a$, so by the transitivity of $(C, <)$, $\Gamma_b < \Gamma_a$. Thus Γ_b and Γ_a must be in the same cluster after all, which is the contradiction.

Now define a temporal structure on $(C^\flat, \prec)$ with domain $\mathcal{D}$ by the following predicate valuation g. For n-ary p, $(a, s) \in C^\flat$ and $\overline{d} \in \mathcal{D}^n$,

$$\overline{d} \in g(p, (a, s)) \text{ iff } p(\overline{d}) \in *_a(s).$$

So $g(p, (a, s)) = h(p, *_a(s))$.

The fundamental lemma for $(C^\flat, \prec, \mathcal{D}, g)$ is:

$$^{(C^\flat, \prec, \mathcal{D}, g)}\|\alpha\|_{(a,s)} = 1 \text{ iff } \alpha \in *_a(s).$$

Proof. As usual, by induction on the construction of α. The difficult cases are:

$u_1 = u_2$**:** as in lemma 4.3.2.

$p(u_1, ..., u_n)$**:** holds because $g(p, (a, s)) = h(p, *_a(s))$.

$\forall x\alpha$**:** as in lemma 4.3.2.

$F\alpha, \Rightarrow$**:** Under the assumption the inductive hypothesis implies that there is (b, r) such that $(a, s) \prec (b, r)$ and $\alpha \in *_b(r)$. Then it follows that $*_a(s) < *_b(r)$, so by the fundamental lemma for $(C, <, \mathcal{D}, h)$, $F\alpha \in *_a(s)$.

$F\alpha, \Leftarrow$**:** Assume $F\alpha \in *_a(s)$. So by the fundamental lemma for $(C, <, \mathcal{D}, h)$, there is some Γ in C such that $*_a(s) < \Gamma$ and $\alpha \in \Gamma$. Either Γ is in the class a when there is some $s > r$ such that $*_a(s) = \Gamma$ or Γ is $*_b(r)$ for some later class b. In either case $(a, s) < (b, r)$ and the inductive hypothesis implies that $^{(C^\flat, \prec, \mathcal{D}, g)}\|\alpha\|_{(b,r)}$. Thus $^{(C^\flat, \prec, \mathcal{D}, g)}\|F\alpha\|_{(a,s)} = 1$, as required.

■

Now to finish our completeness proof we just let a be either the cluster containing Γ_0 or just $\{\Gamma_0\}$ if Γ_0 is irreflexive in C. In this latter case put $s = 0$ so that $*_a(s) = \Gamma_0$, while in the former case we know from the construction of $*_a$ that there is also such an $s \in \mathbb{R}$. By the preceding lemma we then have

$$^{(C^\flat, \prec, \mathcal{D}, g)}\|\alpha\|_{(a,s)} = 1$$

for all α in Σ_0.

4.5 Rationals

To axiomatize predicate F and P over the rationals we simply add four more schemas to our system QFP/L:

no-ends $F\top$, $P\top$

density $GG\alpha \to G\alpha$ and dual.

It is clear that this system QFP/Q is sound for the logic.

To show strong completeness supppose that we have a consistent set Σ_0 of sentences. Let $(C^\flat, \prec, \mathcal{D}, h)$ be the linear model obtained as above. Say that all the formulae of Σ_0 hold at $t_0 \in C^\flat$.

We now show that $(C^\flat, \prec)$ is dense. Suppose that $(a,s) \prec (b,r)$ in $C^\flat$. If $a = b$ then we are done as $(a,s) \prec (a, \frac{r+s}{2}) \prec (b,r)$. So suppose that $a \neq b$. If a is a cluster then we are also done because $(a,s) \prec (a, s+1) \prec (b,r)$. Similarly if b is a cluster.

It remains to consider the case when we have $a = \{\Gamma_a\}$, $b = \{\Gamma_b\}$, $\Gamma_a < \Gamma_b$, and both Γ_a and Γ_b are irreflexive points. Note that since $\Gamma_b \not< \Gamma_a$ (as they are not in the same cluster), there must be a sentence ε, say, such that $G\varepsilon \in \Gamma_b$ and $\sim\varepsilon \in \Gamma_a$. Also, since $\Gamma_a \neq \Gamma_b$ there is δ such that $\delta \in \Gamma_b$ and $\sim\delta \in \Gamma_a$.

We will find a set $\Delta \in C$ such that $\Gamma_a < \Delta < \Gamma_b$. Suppose for contradiction that there is no such Δ in C. Thus (by the totality of the ordering on C), for all $\Delta < \Gamma_b$, either $\Delta = \Gamma_a$ or $\Delta < \Gamma_a$. So, by the fundamental lemma for C, $\sim\delta \vee F{\sim}\varepsilon \in \Delta$. Using the dual of the definition of $<$ on C or T, we have $H(\sim\delta \vee F{\sim}\varepsilon) \in \Gamma_b$. That is,

$$(*)\ \ P(\delta \wedge G\varepsilon) \notin \Gamma_b.$$

It is clear that $\{\beta \mid G\beta \in \Gamma_a\}$is consistent. In fact we have already seen that such sets are omega complete. Thus by lemma 4.3.4:

$$\Delta_0 = \{\beta \mid G\beta \in \Gamma_a\} \cup \{FH(\sim\delta \vee F{\sim}\varepsilon)\}$$

is also omega complete.

Now we will show that Δ_0 is consistent. Suppose it is not, so that there are $\beta_1, ..., \beta_n$ such that each $G\beta_i \in \Gamma_a$ and $\bigwedge_{i=1}^n \beta_i \vdash \sim FH(\sim\delta \vee F{\sim}\epsilon)$. So by lemma 4.3.3 $\Gamma_a \vdash G(\sim FH(\sim\delta \vee F{\sim}\epsilon))$. This is just $\Gamma_a \vdash GGP(\delta \wedge G\epsilon)$, which, by the axiom for density, implies that $GP(\delta \wedge G\epsilon) \in \Gamma_a$. Thus $P(\delta \wedge G\epsilon) \in \Gamma_b$, which contradicts $(*)$.

By lemma 4.3.5 Δ_0 can be extended to a saturated set Δ. It is clear that $\Gamma_a < \Delta$ in T so Δ will be in C. Since $FH(\sim\delta \vee F{\sim}\varepsilon) \in \Delta$, the fundamental lemma for C implies that there is $\Xi \in C$ such that $\Delta < \Xi$ and $H(\sim\delta \vee F{\sim}\varepsilon) \in \Xi$.

Now we show that $\Delta < \Gamma_b$. Suppose not. Then the transitivity of $<$ on C implies that $\Gamma_b < \Xi$. This leads to the contradiction that $\sim\delta \vee F{\sim}\varepsilon \in \Gamma_b$.

By construction of $C^\flat$ there is $(c,t) \in C^\flat$ such that $*_c(t) = \Delta$. We've seen that $\Gamma_a < \Delta < \Gamma_b$ in C so since neither Γ_a nor Γ_b is reflexive we know that Δ is not equal to either. Thus $(a,s) \prec (c,t) \prec (b,r)$, as required.

It is clear that $(C^\flat, \prec)$ also has no end points so we have a model for Σ_0 which is linear, dense, and without end points. Theorem 4.5.1 below then finds us a model which has countable, linear, dense time without end points, i.e. a model over rational time.

Theorem 4.5.1 *Suppose that $\mathcal{T} = (T, <, \mathcal{D}, h)$ is a temporal $\mathcal{S}$-structure and that $t_0 \in T$.*

Then there is an $\mathcal{S}$-structure $\mathcal{T}' = (T', <', \mathcal{D}', h')$ and $t_0' \in T'$ such that

- *T' is countable;*
- *for all temporal sentences α in either the predicate language QFP (over $\mathcal{S}$), ${}^{\mathcal{T}}\|\alpha\|_{t_0}$ iff ${}^{\mathcal{T}'}\|\alpha\|_{t_0'}$;*
- *and $(T, <)$ is elementarily equivalent to $(T', <')$ in the first-order language of $<$.*

Proof. First we construct a new first-order language based on a signature $\mathcal{S}_\circ$ defined as follows. For each n-ary predicate p in $\mathcal{S}$ we include an $n{+}1$-ary predicate symbol $p_\circ$ in $\mathcal{S}_\circ$. The function symbols are the same. Also include the 2-ary predicate symbol $<$, constants Λ and c_0, and 1-ary predicates τ and δ.

For convenience assume that the language has all the variable symbols $x_0, x_1, \ldots$ of the predicate temporal language as well as two extra ones called temporal variable symbols. Now we are going to define a translation $\circ_s$ from the temporal predicate language into the predicate one for each s of the two temporal variable symbols. We will simultaneously define the maps $\circ_s$ between the languages by recursion on the construction of formulae in the temporal predicate language. For each temporal variable symbol s,

- $\circ_s(u = v) = (u = v)$,
- $\circ_s(\top) = \top$,
- $\circ_s(p(\overline{u})) = p_\circ(s, \overline{u})$,
- $\circ_s(\sim\alpha) = \sim \circ_s(\alpha)$,

- $\circ_s(\alpha \wedge \beta) = \circ_s(\alpha) \wedge \circ_s(\beta)$,
- $\circ_s(\forall x\alpha) = \forall x(\delta(x) \to \circ_s(\alpha))$,
- $\circ_s(F\alpha) = \exists t(\tau(t) \wedge (s < t) \wedge \circ_t(\alpha))$,
- $\circ_s(P\alpha) = \exists t(\tau(t) \wedge (t < s) \wedge \circ_t(\alpha))$,

where t is the other temporal variable symbol.

For any temporal structure $\mathcal{T} = (T, <, \mathcal{D}, h)$ and $t_0 \in T$, define a first-order structure $\mathcal{T}_\circ$ in the first-order language as follows:

- The domain is the disjoint union of T, the domain of $\mathcal{D}$ and a new element called Λ.
- Λ is interpreted as Λ.
- δ holds on the elements of $\mathcal{D}$ while τ holds on those of T.
- For any n-ary function symbol f from $\mathcal{S}$, for any n-tuple $\overline{e}$ from the domain, $f^{\mathcal{T}_\circ}(\overline{e})$ is $f^{\mathcal{D}}(\overline{e})$ if each e_i is from D and is Λ otherwise.
- For any n-ary predicate symbol p from $\mathcal{S}$, $(t, \overline{d}) \in p_\circ{}^{\mathcal{T}_\circ}$ iff $t \in T$ and $\overline{d} \in h(p, t)$.
- $(s, t) \in <^{\mathcal{T}_\circ}$ iff s and t are from T and $s < t$, and
- c_0 is interpreted as t_0.

Suppose that α is a formula in our temporal language, ν is an assignment of its free variables to elements of $\mathcal{D}$, $t \in T$, and s is a temporal variable symbol. Then, clearly, $\circ_s(\alpha)$ has free variables, those of α as well as s. Define μ to assign the former variables in the same way as ν does and to assign s to t. Now I claim that

$$^{\mathcal{T}}\|\alpha\|_t^\nu = 1 \text{ iff } (\mathcal{T}_\circ, \mu) \models \circ_s(\alpha).$$

This is shown by a straightforward induction on the construction of α.

Now $\mathcal{T}_\circ$ is a normal first-order structure described by our normal first-order language so we can use the Löwenheim–Skolem theorem to find a countable elementarily equivalent structure $\mathcal{T}'_\circ$.

Now we construct a temporal structure $\mathcal{T}' = (T', <', \mathcal{D}', h')$ and an element $t'_0 \in T'$. Let T' consist of exactly those elements of $\mathcal{T}'_\circ$ which satisfy τ and note that $<'$ will only be defined on them. Let $\mathcal{D}'$ be those elements which satisfy δ. These will be all the rest except Λ. Similarly, extract the functions on $\mathcal{D}'$. For each n-place predicate p from $\mathcal{S}$, each

$t \in T'$, and each $\overline{d} \in \mathcal{D}^n$, put $\overline{d} \in h'(p,t)$ iff $(t, \overline{d}) \in p_\circ^{\mathcal{T}'_\circ}$. Finally, let t'_0 be the interpretation of c_0.

Because we can say everything we need to in a first-order way about those elements of T and T' which satisfy τ, $(T', <')$ and $(T, <)$ are elementarily equivalent as required.

Also note that putting $(\mathcal{T}', t'_0)$ through the same construction $(\mathcal{T}, t_0) \mapsto \mathcal{T}_\circ$ gives us back $\mathcal{T}'_\circ$. Thus the corresponding result holds: i.e.

$$^{\mathcal{T}'}\|\alpha\|^{\nu}_{t'_0} = 1 \text{ iff } (\mathcal{T}'_\circ, \mu) \models \circ_s(\alpha).$$

where μ interprets variables as ν but also assigns s to t'_0.

Thus, given a sentence α in our temporal language, we have

$$\begin{array}{rll} ^{\mathcal{T}}\|\alpha\|_{t_0} = 1 & \text{iff} & (\mathcal{T}_\circ, [s \mapsto c_0]) \models \circ_s(\alpha) \\ & \text{iff} & \mathcal{T}_\circ \models \circ_s(\alpha)(c_0) \\ & \text{iff} & \mathcal{T}'_\circ \models \circ_s(\alpha)(c_0) \\ & \text{iff} & (\mathcal{T}'_\circ, [s \mapsto c_0]) \models \circ_s(\alpha) \\ & \text{iff} & ^{\mathcal{T}'}\|\alpha\|_{t'_0} = 1. \end{array}$$

■

4.6 Incompleteness via completeness

In the preceding few sections we found a complete axiomatization for the predicate logic of F and P over the class of all frames of time and over some subclasses. However, when we turn to the integers, the natural numbers or the reals—flows useful for computer science—we find that the techniques used in the propositional case do not carry across to the predicate case. In fact, as shown by Scott and Lindström in the 1960s, it is impossible to find a recursive axiom scheme for these logics. As we will see below, this is because it is possible to code arithmetic into these logics and so they fall foul of Gödel's incompleteness result. Proofs of incompleteness results appear in [Andréka *et al.*, 1979] (using Matiasievits' theorem), [Szalas and Holenderski, 1988], [Szalas, 1986], [Abadi, 1987a] (with corrections in [Abadi, 1990] and [Andréka *et al.*, 1991]) and [Garson, 1984]. There is a very general result in [Reynolds, 1992b] which we present below.

Because it is such a natural logic for dealing with the behaviour of programs, much work has been done on finding useful (although not both complete and recursive) axiomatizations for predicate temporal logic over the natural numbers time. Systems in [Sain, 1991b] and [Andréka *et al.*, 1992] are complete for classes of flows which include the natural numbers and only other 'similar' flows. In [Szalas, 1987b] we find a complete but infinitary system.

There is still much work to be done in this area. The interested reader is referred to [Garson, 1984] or the more recent [van Benthem, 1992] and [Andréka *et al.*, 1992] for excellent surveys.

So let us present a fairly general incompleteness result. As in chapter 12, we identify gaps in a flow of time with supremum-less, non-empty, proper initial segments of the order and insert the gap in the appropriate place in the order. Dedekind complete orders, then, are those without gaps. The completion $(T*, <)$ of an order $(T, <)$ is another order consisting of T and all the gaps in the right places and is Dedekind complete. The gap flow of an order is just the suborder $(T * -T, <)$ of $(T*, <)$. A flow of time is *riddled* iff its gap flow has a suborder isomorphic to the irrationals.

Theorem 4.6.1 *Suppose that $\mathcal{K}$ is a class of linear irreflexive flows which includes at least one infinite flow but no riddled flows.*

Then the predicate temporal logic of F and P over $\mathcal{K}$ is not recursively axiomatizable.

Proof. We work with the predicate temporal language of F and P with the following first-order symbols: 1-ary predicate symbols e and q, 2-ary predicate symbol $\prec$, a constant 0, a 1-ary function symbol $'$ (with $'(x)$ written as x'), and 2-ary function symbols $+$ and $\times$ (written infix). Recall that constant and function symbols are rigid. We will use the usual useful abbreviations $\Box\alpha$ for $H\alpha \wedge \alpha \wedge G\alpha$ and $\Diamond\alpha$ for $P\alpha \vee \alpha \vee F\alpha$.

Let α_0 be the conjunction of the following sentences:

1. $\forall x \Diamond(e(x) \wedge G({\sim}e(x)) \wedge H({\sim}e(x)))$.
2. $\Box\exists x(e(x) \wedge \forall y((x \neq y) \to {\sim}e(y)))$.
3. $\Box\forall x\forall y((x \prec y) \leftrightarrow \Diamond(e(x) \wedge Fe(y)))$.
4. $\Box\forall x(q(x) \to \Box q(x))$.
5. $\Box[q(0) \wedge \forall y((y \prec 0) \to {\sim}q(y))]$.
6. $\Box\forall x[q(x) \to ((x \prec x') \wedge q(x') \wedge \forall z((x \prec z) \wedge (z \prec x') \to {\sim}q(z)))]$.
7. $\Box\forall x\forall y[(q(x) \wedge q(y)) \to (q(x+y) \wedge q(x \times y))]$.
8. $\Box\forall x(q(x) \to (x + 0 = x))$.
9. $\Box\forall x\forall y(q(x) \wedge q(y) \to (x + y' = (x + y)'))$.
10. $\Box\forall x(q(x) \to (x \times 0 = 0))$.

11. $\Box\forall x\forall y(q(x) \wedge q(y) \rightarrow (x \times y' = (x \times y) + x))$.

12.

$$\begin{array}{ll} \Box\forall x\forall y(& q(x) \wedge q(y) \wedge (x \prec y) \\ \rightarrow & \exists z((x \prec z) \wedge (z \prec y) \\ & \wedge(z + z = x + y) \vee ((z + z)' = x + y))). \end{array}$$

13. $\Box\forall x\forall y\forall z(q(x) \wedge q(y) \wedge q(z) \rightarrow ((x + y) + z = x + (y + z)))$.

14. $\Box\forall x\forall y(q(x) \wedge q(y) \rightarrow (x + y = y + x))$.

15. $\Box\forall x\forall y\forall z(q(x) \wedge q(y) \wedge q(z) \wedge (x + y = x + z) \rightarrow (y = z))$.

16.

$$\begin{array}{ll} \Box\forall x[& (0 \prec x) \wedge \exists z(q(z) \wedge (x \prec z)) \\ \rightarrow & \exists y[(y \prec x) \wedge q(y) \wedge \forall z((y \prec z) \wedge (z \prec x) \rightarrow {\sim}q(z))]]. \end{array}$$

It is convenient to introduce a little notation for the case when, in a structure $(T, <, \mathcal{D}, h)$ of our temporal language, we find that the unary predicate q has a rigid interpretation on the domain D of $\mathcal{D}$ according to h. Then we write $\mathcal{D}\mid_q$ for the set of individuals from D which always satisfy q.

We will also need to work with the first-order language of arithmetic in order to make use of Gödel's result about the non-axiomatizability of arithmetic. The language of (Peano) arithmetic is the first-order language with $=$, a constant 0, a 1-ary function symbol $'$, and 2-ary function symbols $+$ and $\times$. A structure in this language is *a standard model of arithmetic* iff it is isomorphic to the model

- with domain the natural numbers;
- 0 interpreted as zero;
- $'$ as the successor function;
- $+$ as addition; and
- $\times$ as multiplication.

This particular structure is usually referred to as *the* standard model of arithmetic but we will find it convenient not to distinguish between it and other isomorphic structures.

Lemma 4.6.2 *Suppose that* $\mathcal{T} = (T, <, \mathcal{D}, h)$ *is a structure of our temporal language such that*

- *$(T, <)$ is not riddled; and*
- *for some $t \in T$, ${}^{\mathcal{T}}\|\alpha_0\|_t = 1$.*

Then $(\mathcal{D}\mid_q, 0, ', +, \times)$ exists and is a standard model of arithmetic.

Proof. Suppose that $\mathcal{T} = (T, <, \mathcal{D}, h)$ and $t \in T$ are as given. It is clear from sentence (3) that $\prec$ has a rigid interpretation on the domain D of $\mathcal{D}$. We will call this relation $\prec$ as well. Then we know from sentences (1),2 and (3) that $(D, \prec)$ is isomorphic to $(T, <)$.

From sentence (4) it is clear that the interpretation of q is also rigid. Thus, as defined above, $\mathcal{D}\mid_q \subseteq D$ is this set of individuals which always satisfy q.

Now consider $(\mathcal{D}\mid_q, \prec)$. From sentence (5) we know that 0 is its first element and from sentence (6) that every point d has a successor d'. Sentence (7) then confirms that $(\mathcal{D}\mid_q, 0, ', +, \times)$ is indeed a structure in the language of arithmetic.

We can define a sequence $\{s_n \mid n \in \mathbb{N}\}$ in $\mathcal{D}\mid_q$ recursively by $s_0 = 0$ and $s_{i+1} = s_i'$. If this sequence exhausts $\mathcal{D}\mid_q$, then it is straightforward to use sentences (8)–(11) to show that $(\mathcal{D}\mid_q, 0, ', +, \times)$ is a standard model of arithmetic.

We claim that the sequence will always exhaust $\mathcal{D}\mid_q$. Suppose for contradiction that there is some element of $\mathcal{D}\mid_q$ not in the sequence. Now define a sequence of binary relations $\sim_n$ on $\mathcal{D}\mid_q$ as follows:

- $b \sim_0 c$ iff $b = c$; and
- $b \sim_{n+1} c$ iff $d \sim_n c$ and either $b = d'$ or $d = b'$.

Let $\sim$ be the union of all the $\sim_n$ for $0 \leq n < \omega$. This is clearly an equivalence relation on $\mathcal{D}\mid_q$.

From sentences (8) and (9) we can see that $a \sim b$ iff there is some $m \in \mathbb{N}$ such that either $a + s_m = b$ or $b + s_m = a$.

Note that by sentence (6) the $\sim$-classes are each contained within non-overlapping intervals of $(D, \prec)$: i.e. if a, b, and c are from $\mathcal{D}\mid_q$ and $a \prec b \prec c$ then if $a \sim c$ then $a \sim b$.

Note also that since 0 has no predecessors in $\mathcal{D}\mid_q$ (by sentences (5) and (6)) the sequence s_n covers exactly the $\sim$-class of 0. Thus we have assumed (for contradiction) that there is more than one class.

Now I claim that the $\sim$-classes are densely ordered. Specifically we show that $\forall a, b \in \mathcal{D}\mid_q$, if $a \prec b$ and $a \not\sim b$, then there is $c \in \mathcal{D}\mid_q$ such that $a \prec c \prec b$ and $a \not\sim c$ and $c \not\sim b$.

Suppose that a and b are as in the claim. Since sentence (12) holds, there must be a c such that $a \prec c \prec b$ and either

- $c + c = a + b$; or
- $(c + c)' = a + b$.

The two cases are much the same to deal with so assume that we have c such that $a \prec c \prec b$ and $c + c = a + b$.

If $a \sim c$, then there is some m such that $a + s_m = c$. Thus $(a + s_m) + c = a + b$. By sentences (13), (14) and (15) we have $c + s_m = b$ and so $c \sim b$. By transitivity of $\sim$, $a \sim b$ which is a contradiction.

If $c \sim b$, then there is some m such that $c + s_m = b$. Thus $c + c = a + c + s_m$ and sentences (13), (14) and (15) tell us that $c = a + s_m$ so that $a \sim c$ and $a \sim b$. This also is a contradiction.

Thus we have proved the claim and the $\sim$-classes are ordered densely. Since there are at least two classes, then there are an infinite number. We can thus map the rationals $(\mathbb{Q}, <)$ into the $\sim$-classes of $\mathcal{D}\mid_q$ in an order preserving way. Say that for each rational r we choose $d_r \in \mathcal{D}\mid_q$ such that if $r < s$ in $\mathbb{Q}$, then $d_r \prec d_s$ and $d_r \not\sim d_s$.

Now I show how to find a gap d_γ in D for each irrational γ. Given such a γ let

$$D_\gamma = \{d \in D \mid \text{there is some rational } r < \gamma \text{ and some } e \in \mathcal{D}\mid_q \text{ such that } e \sim d_r \text{ and } d \prec e\}.$$

Choosing any rational $r > \gamma$ it is clear that d_r bounds D_γ above so that D_γ either has a supremum in D or ends on the right at a gap.

Suppose that D_γ does have a supremum s in D. Choosing any two rationals $r_1 < r_2 < \gamma$, we know that $d_{r_1} \prec d_{r_2}$ are in $\mathcal{D}\mid_q$ and in D_γ. So the supremum s of D_γ is strictly greater than the first element 0 (see sentence (5)) of $\mathcal{D}\mid_q$.

Since we know that for any rational $r^+ > \gamma$ we have d_{r^+} bounding D_γ above it is clear that there are elements of $\mathcal{D}\mid_q$ greater than s.

Thus, by (16), s has an immediate predecessor y in $\mathcal{D}\mid_q$. Since $y \prec s$ and D_γ is closed downwards, $y \in D_\gamma$. Thus there is a rational $r < \gamma$ and $e \in \mathcal{D}\mid_q$ such that $e \sim d_r$ and $y \prec e$.

Since y is the immediate predecessor of s in $\mathcal{D}\mid_q$ we must have either $s \prec e$ or $s = e$, both of which imply $s \prec e'$. But now since $e'' \in \mathcal{D}\mid_q$, $e'' \sim d_r$, and $e' \prec D_\gamma$ we have $e' \in D_\gamma$. This contradicts s being the supremum.

Thus D_γ ends at a gap d_γ on the right. Furthermore, it is clear that the maps $r \to d_r$ and $\gamma \to d_\gamma$ preserve the combined orderings of the rationals and irrationals so we have an order preserving map of the irrationals into the gaps of $(D, \prec)$. But since $(D, \prec)$ is isomorphic to $(T, <)$, we know that both these orders are riddled.

This contradicts our assumptions and implies that there cannot have been any extra elements of $\mathcal{D}\mid_q$ besides those in the sequence $\{s_n \mid n \in \mathbb{N}\}$. Thus we have our result. ∎

Now we define a translation taking a sentence A of arithmetic to the sentence ϕ_A of our predicate temporal logic which is just A with all quantifiers relativized to q: i.e. replace recursively $\forall x B$ by $\forall x(q(x) \to \phi_B)$ and $\exists x B$ by $\exists x(q(x) \wedge \phi_B)$ for any formula B and then restrict the map to sentences. Then it is straightforward to prove the following.

Lemma 4.6.3 *Suppose that $\mathcal{T} = (T, <, \mathcal{D}, h)$ is a structure in our temporal language, that h interprets q rigidly as $\mathcal{D}\mid_q$, that $(\mathcal{D}\mid_q, 0, ', +, \times)$ is a substructure of $(D, 0, ', +, \times)$, that $t \in T$, and that A is a sentence of the language of arithmetic.*

Then $(\mathcal{D}\mid_q, 0, ', +, \times) \models A$ iff ${}^{\mathcal{T}}\|\phi_A\|_t = 1$.

We know that $\mathcal{K}$ contains some infinite flow. A simple induction establishes that such a $(T, <)$ has either a suborder which is isomorphic to the natural numbers or a suborder which is isomorphic to their reverse order. If we only have infinite orders in $\mathcal{K}$ with reverse natural number suborders, then it is easier to proceed with the proof below considering the predicate temporal logic of F and P over the class of reverses of orders in $\mathcal{K}$ instead of over $\mathcal{K}$ itself. Of course, either of these logics are recursively axiomatizable iff the other is. Thus assume now that $\mathcal{K}$ contains an order $(T, <)$ with a suborder isomorphic to the natural numbers.

A sentence A of the language of arithmetic is *true* iff it holds in any (all) standard models of arithmetic. We know from Gödel that the true sentences are not recursively enumerable.

Lemma 4.6.4 *For any sentence A of the language of arithmetic, A is true iff $\models_{\mathcal{K}} \alpha_0 \to \phi_A$.*

Proof.

($\Rightarrow$) Suppose that A is true and for some $\mathcal{T} = (T, <, \mathcal{D}, h)$ we have $(T, <) \in \mathcal{K}$ and for some $t \in T$, ${}^{\mathcal{T}}\|\alpha_0\|_t = 1$.

By lemma 4.6.2, $(D\mid_q, 0, ', +, \times)$ is a standard model of arithmetic. Thus $(D\mid_q, 0, ', +, \times) \models A$ and lemma 4.6.3 gives us the required ${}^{\mathcal{T}}\|\phi_A\|_t = 1$.

($\Leftarrow$) Suppose that $\models_{\mathcal{K}} \alpha_0 \to \phi_A$. Choose some infinite $(T, <) \in \mathcal{K}$ and some $S \subseteq T$ such that $(S, <)$ is isomorphic to $(\mathbb{N}, <)$. So we can think of S as the natural numbers.

Now define a structure $\mathcal{D}$ in the language of arithmetic in any way so that

- the domain is T;
- 0 is (interpreted as) $0 \in S$;
- $'$ is the successor function on $(S, <)$;
- $+$ is addition on S; and
- $\times$ is multiplication on S.

The value of the functions can be anything whenever any of their arguments are not from S.

Next we define interpretations for e by $h(e,t) = \{t\}$ and q by $h(q,t) = S$ for any $t \in T$. Finally, interpret $\prec$ rigidly by $h(\prec, t) = \{(t_1, t_2) \in T \times T \mid t_1 < t_2\}$. Choosing any $t_0 \in T$, it is clear that $(T, <, \mathcal{D}, h, t_0) \models \alpha_0$ and that $(D\mid_q, 0, ', +, \times)$ is a standard model of arithmetic.

Thus ${}^{(T,<,\mathcal{D},h)}\|\phi_A\|_{t_0} = 1$ and lemma 4.6.3 implies that $(D\mid_q, 0, ', +, \times) \models A$.

Thus A is true.

■

Now suppose that we do have a weakly complete recursive axiomatization of $\models_{\mathcal{K}}$. Then, given any sentence A of the language of arithmetic, we can effectively translate it into the 'equivalent' formula ϕ_A in our predicate temporal langauge. By lemma 4.6.4 above A is true iff $\models_{\mathcal{K}} \alpha_0 \to \phi_A$ so we have a semi-decision procedure for arithmetic. This contradicts Gödel's theorem.

■

Corollary 4.6.5 *The predicate temporal logics of F and P over $\mathbb{R}$, $\mathbb{Z}$, and $\mathbb{N}$ are not recursively axiomatizable.*

From this it follows that the first-order temporal logic with U and S is incomplete over such classes of flows.

Chapter 5

Temporal logics presented in the predicate calculus: language–metalanguage

5.1 Introduction

Given a system of databases $(\Delta_t, \vdash_t), t \in T$, we say in chapter 1 that there are two ways of describing the logical connections between the theories $(\Delta_t, \vdash_t)$. One is to talk about the system declaratively from the outside, using a metalanguage **M**, and the other is to enrich the language of the theories Δ_t, add connectives and formulae to theory Δ_t, enabling it to express the way in which it is related to other theories.

If $(T, <)$ is a flow of time and $\Delta_t, t \in T$ is the theory which holds at t, then the second method allows for temporal connectives such as

$$Fq, Pq, U(p,q), S(p,q)$$

to be part of the languages of Δ_t, and talk about the values of q around t. We have not yet examined the first approach in depth, that of using a rich metalanguage **M** for talking about time and about the formulae which hold at each moment of time. Such a metalanguage must be able to name the moments of time and the earlier–later relation $<$, and also act as a metalanguage for the theory and logic $(\Delta_t, \vdash_t)$. In other words, we also need names and proof predicates for formulae of the logic $\vdash_t$. We can add the simplifying assumption that all $\vdash_t$ are the same logic $\vdash$; then we need one proof predicate for the logic $\vdash$. Let us examine our options in a general setting.

Our conceptual framework will revolve around two pure notions, the notions of a *proper metalanguage* and that of an object language $\mathbf{L}_1$ *implementing metalanguage features* for a language $\mathbf{L}_2$.

Given a language $\mathbf{L}$, we understand by a proper metalanguage $\mathbf{M}$ to the language $\mathbf{L}$, any language which is able to name every well-formed symbol of $\mathbf{L}$ and to describe all the logical operations of $\mathbf{L}$. Thus formulae φ of $\mathbf{L}$ become terms t_φ of $\mathbf{M}$ and logical relations of $\mathbf{L}$ become predicates in $\mathbf{M}$. $\mathbf{M}$ must have axioms to ensure that every model of $\mathbf{M}$ gives rise, through the embedding of $\mathbf{L}$ in $\mathbf{M}$, to a model of $\mathbf{L}$. For a given $\mathbf{L}$, $\mathbf{M}$ is not unique. Many languages can serve as metalanguages for $\mathbf{L}$, provided they are strong enough. In computer language terms, any sufficiently strong language, e.g. Basic, can be used to write an interpreter for any other language. In such a case one *simulates* one language in another. The other pure notion of one language, $\mathbf{L}_1$, serving as an object language, $\mathbf{L}_2$, is more difficult to explain. Intuitively, suppose we have two interpreters I_1 and I_2 which are linked in some way. Thus each interpteter has its allowable options of how to run. However, since they are linked, not all of these individual options can be realized. I_2 can choose to run in certain ways which will *limit* the options of I_1 because they are linked. Suppose we use a proper metalanguage $\mathbf{M}_1$ to talk and describe the options of I_1, and suppose further that we express, through a wff φ_1 of $\mathbf{M}_1$, our wish to have I_1 run only in certain ways. So φ_1 (if true) limits the runs (or options) of I_1. There is another way of limiting the runs of I_1, through its linkage with I_2.

It may be that I_2 can be modified to I_2' so that I_1 (when linked with I_2') can run exactly in a way which satisfies φ_1. If this is the case, then we say that I_2, through its object language modification to I_2' and its linkage with I_1, can *implement* φ_1, i.e. implement the metastatement φ_1. It may be that for each φ_n of $\mathbf{M}_1$, there exists a way of modifying I_2 into I_2^n which, through the link with I_1, allows I_1 to run only in a way which satisfies φ_n. In that case we have object language implementation of the set $\{\varphi_n\}$ of the language $\mathbf{M}_1$.

The reader should note that the concept of I_2 acting as an object level implementation of some metalevel statements $\{\varphi_n\}$ of I_1 is a relative one. Because I_1 and I_2 are linked, we can equally say that I_1 is an object level implementation of some metalevel statements $\{\varphi_n'\}$ of I_2. Thus the concept is relative to a metalevel set of sentences Δ.

Of course, the above pure concepts have to be formally defined. This may not be easy. We can now take the above language $\mathbf{L}$ and examine in depth the possibilities of using a metalanguage $\mathbf{M}$ for talking about time and about the formulae of **L** which hold at each moment of time. Such a metalanguage must be able to name the moments of time and the earlier–later relation $<$, and we also need names and proof predicates for formulae of the temporal logic $\vdash$. Let us assume, for example, that $\vdash$ is based on intuitionistic logic.

Let 't' name the moment of time t and '$<$' name $<$. Let 'φ' be the name of φ. Let **Hold**('$<$', 't', 's') be the metapredicate saying $t < s$. Let **Data**('t', 'φ') be the metapredicate saying $\varphi \in \Delta_t$. Let **Prove**('t', 'Ψ_1', 'ψ_2') be the predicate indicating that $\Psi_1 = \bigwedge_{s,\varphi}$ **Data** ('s', 'φ') $\vdash$ **Data** ('t', 'ψ_2'). Note that Ψ_1 is a conjunction of sentences of the form **Data**('s', 'φ') and therefore **Prove** is meta to **Data**.

The following is a metalanguage system defining the temporal relationship.

Definition 5.1.1 (the metalanguage M)

1. Let **M** be a predicate metalanguage with predicates **Hold**(x, y, z), **Data**(x, y) and **Prove**(x, y, z). The language is sorted. The sorts are as follows:

   (a) x in **Hold** is a relation name ($<$).

   (b) y, z in **Hold** and x in **Prove** and x in **Data** are time points names.

   (c) y in **Data** is an object language relation name.

   (d) y and z in **Prove** are **Data** relation names. Thus **Prove** is a metapredicate to **Data**.

2. The language **M** also has function symbols $\mathbf{f_{Prove}}$, $\mathbf{f_{Hold}}$, $\mathbf{f_{Data}}$, $\mathbf{f}_\wedge, \mathbf{f}_\vee, \mathbf{f}_\sim, \mathbf{f}_\rightarrow, \mathbf{f}_F$, and $\mathbf{f}_P$ satisfying the following:
   $\mathbf{f}_\wedge$ ('φ', 'ψ') = '$\varphi \wedge \psi$'
   $\mathbf{f}_\vee$ ('φ', 'ψ') = '$\varphi \vee \psi$'
   $\mathbf{f}_\rightarrow$ ('φ', 'ψ') = '$\varphi \rightarrow \psi$'
   $\mathbf{f}_\sim$ ('φ') = '$\sim \varphi$'
   $\mathbf{f}_F$ ('φ') = '$F\varphi$'
   $\mathbf{f}_P$ ('φ') = '$P\varphi$'
   $\mathbf{f_{Data}}$ ('t', 'A') = '**Data** ('t', 'A')'

   $\mathbf{f_{Hold}}$ ('$<$', 't', 's') = '$t < s$'
   $\mathbf{f_{Prove}}$ ('t', 'Ψ', 'φ') = '**Prove** ('t', 'Ψ', 'φ')'
   The following axioms hold in **M**.

   (a) **Prove**('t', 'φ', '$A \wedge B$') $\leftrightarrow$ (**Prove**('t', 'φ', 'A') $\wedge$ **Prove**('t', 'φ', 'B')).

(b) **Prove** ('t', 'φ', '$A \vee B$') $\leftrightarrow$ (**Prove**('t', 'φ', 'A') $\vee$ **Prove**('t', 'φ', 'B')).

(c) **Prove** ('t', 'φ', '$A \to B$') $\leftrightarrow$ **Prove**('t', $\mathbf{f}_\wedge$ ('φ' $\mathbf{f}_{\mathbf{Data}}$('$t$', '$A$')), '$B$').

(d) **Prove** ('t', 'φ', '$\bot$') $\to \forall x$ **Prove** ('t', 'φ', x).

(e) **Prove** ('t', 'φ', 'FA')
$\leftrightarrow \exists x$[**Hold** ('<', '$t$', x) $\wedge$ **Prove** (x, 'φ', 'A').

(f) **Prove** ('t', '**Data**' ('t', 'A'), 'A').

(g) **Prove** $(x, y, z) \to \forall y'$ **Prove** $(x, \mathbf{f}_\wedge(y, y'), z)$.

(h) **Prove** $(x, y, z) \wedge$ **Prove** $(x, y, \mathbf{f}_\to(z, z')) \to$ **Prove** (x, y, z').

Example 5.1.2 Note that the underlying logic is intuitionistic both in the object language **L** and the metalanguage **M**. Consider the system shown in figure 5.1. At time 1 A holds, at time 2 $P(A \wedge FB) \to C$ holds. We

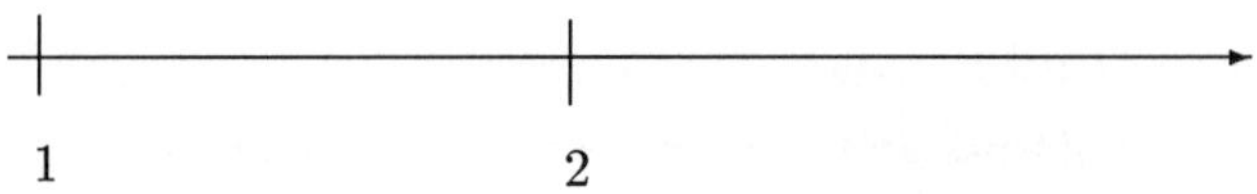

Figure 5.1

query at time 1 the formula $Q = FB \to FC$. We will argue informally. The actual proof is the formalization, in the metalanguage, of our informal proof:

1. Assume FB at time 1. We will show FC at time 1.

2. If FB holds at time 1, then $P(A \wedge FB)$ holds at time 2 and hence C holds at time 2.

3. If C holds at time 2, then FC holds at time 1.

4. From (1)–(3) it follows that $FB \to FC$ holds at time 1.

Note that the above reasoning is intuitionistic. It is proof theoretic and not model theoretic. In other words, we assume we know that certain formulae hold at a certain configuration of points and we reason intuitionistically using the proof theoretic properties of the connectives that more formulae hold at some other points. This is done in **L**. The metalanguage **M** can represent or formalize this reasoning.

Example 5.1.3 Use the same database as example 5.1.2. Try to show $FB \vee \sim FB$.

This will fail, because we have to show either FB or $\sim FB$ (i.e. $FB \rightarrow \bot$) and neither can be shown, even if the metalanguage is classical logic, because it does not follow in the object language.

The next question to ask ourselves is, can intuitonistic logic itself, which is used as the underlying logic of **M**, be used as the metalanguage **M** itself? This means that we do not use naming functions but, represent the system directly in intuitionistic logic. Actually, one needs exact definitions of the concept 'using intuitionistic logic directly', which is not easy to give. Instead of describing what it is we want to do, we will just do it.

We are going to translate the metalanguage predicates just defined into intuitionistic logic. We associate for each n-place atomic predicate formula $A(x_1, \ldots, x_n)$ of the temporal logic, a two-sorted formula $A^*(t, x_1, \ldots, x_n)$ of intuitionistic logic, where t ranges over a new sort for time. We also allow for the predicate $<$ to be binary over the time sort. Other formulae of this two-sorted logic are built as usual from atoms of the form $t < s, A^*(t, x_1, \ldots, x_n)$, the connectives, and the quantifiers on each sort.

Definition 5.1.4 We now translate the metapredicates **Data**('t', 'A'), **Hold**('<', 't', 's') and **Prove** as follows:

1. **Data**('t', '$A(x_2, \ldots, x_n)$')* $= A^*(t, x_1, \ldots, x_n)$ where $\bot^* = \bot$.

2. **Data**('t', '$A \sharp B$')* $=$ **Data**('t', 'A')$^* \sharp$ **Data** ('t', 'B')* where $\sharp$ is any of the connectives $\wedge, \vee, \rightarrow$.

3. **Data**('t', 'GA')* $= \forall s(t < s \rightarrow$ **Data** ('s', 'A')*).

4. **Data**('t', 'HA')* $= \forall s(s < t \rightarrow$ **Data** ('s', 'A')*).

5. **Data**('t', 'FA')* $= \exists s(t < s \wedge$ **Data** ('s', 'A')*).

6. **Data**('t', 'PA')* $= \exists s(s < t \wedge$ **Data** ('s', 'A')*).

7. **Hold** ('<', 't', 's')* $= (t < s)$.

8. **Prove** ('t','Ψ', 'φ')* $= (\Psi^* \rightarrow \varphi^*(t))$.

Note that in (8) $\varphi^*(t)$ is really **Data**('t', 'φ')*.

The importance of (8) is that the translation of '**Prove**' is the intuitionistic implication '$\rightarrow$'.

It is easy to show that the translation '$\rightarrow$' of '**Prove**' satisfies the properties listed under definition 5.1.1(c).

The above discussion deliberately chose intuitonistic logic in order to separate ourselves and the reader from any prejudices for and against classical logic. The process of 'implementing' metalanguage features in the object language is conceptually complicated and needs extensive study in its own right. Gabbay [Gabbay, 1992a] is currently involved in developing this area.

We have seen above two ways of representing our temporal logic **L** in another language **M** (functioning as a metalangauge). One was pure meta, through the predicates **Hold**, **Data**, and **Prove**, and one was an object level 'meta', through (intuitionistic or classical) predicate logic. We need to systematize and formally develop machinery for both approaches. For the first approach, the proper metalanguage approach, we offer the language **HFP**, to be studied later in this paper. For the predicate logic approach, we must give the proper definitions of how predicate logic can be used. We begin with predicate logic in the next section.

5.2 Classical predicate calculus as a metalanguage

We saw in preceding chapters that the majority of non-classical logics and certainly temporal logics are usually presented as extensions of the classical predicate calculus. The language of classical logic is extended and enriched syntactically by additional (temporal) connectives, and the logical properties of these new connectives is then determined either syntactically through special additional axioms and rules or semantically through some set theoretical interpretation.

Our task in this chapter is to examine whether predicate logic itself, without any additional connectives (for time) can handle adequately all temporal phenomena within itself. This may be accomplished either through explicit reference in predicate logic to time points, intervals, and events by special time variables and predicates, as we saw in the preceding section, or through some other more sophisticated means such as using classical logic as an object language implementation of the metalanguage, as hinted in the preceding section. There is the view that predicate logic is quite sufficient to handle time phenomena. This chapter will develop the machinery to examine this view. We shall see that although there is some merit in presenting temporal logics within the predicate calculus, this method is by no means comprehensive and strongly expressive, nor is classical logic the only logic that can be used.

Thus, our first problem is the following: What is the temporal expressive power of first-order predicate logic, and how can it be used effectively and correctly?

This problem has two aspects. First is the mathematical aspect, namely, are there temporal logics defined using new special connectives and axioms, which cannot be presented mathematically in first-order predicate logic? Are there such logics which are intuitive and arise in applications? One such candidate was the logic $Y2$ of section 4, chapter 3, where a filter was needed in the description of its semantics. The second aspect is not mathematical, but the aspect of style and of the naturalness of the representation. Can the presentation within classical logic cater for all the natural features and non-mathematical properties of temporal logics arising in application areas?

Part of our task is to make the above vague notion of 'naturalness' more precise. We know, for example, that in classical logic with $\sim$ and $\vee$, implication $A \to B$ is definable as $\sim A \vee B$. Thus the logic of $\sim$ and $\to$ is reducible to (or can be represented within) the logic of $\vee$ and $\sim$ by a simple translation. It can also be described in predicate logic by first Gödel numbering all wffs with $\to$ and $\sim$, then using a proper proof predicate. This is a roundabout way of doing a simple reduction of $\to$ and $\sim$ to $\vee$ and $\sim$ but it is mathematically valid. Yet, we intuitively know that the reduction

$$A \to B \equiv \sim A \vee B$$

is a more direct one and, indeed, a better one.

There are more examples where the representation may be on the borderline of being natural. Our task is therefore to attempt to make our intuitions more precise. In algebra, for example, it is generally accepted that although a finite dimensional vector space V is isomorphic both to its dual V^* and to its dual V^{**}, it is the latter isomorphism $V \leftrightarrow V^{**}$, which is the 'natural' one. The attempts to make this notion more precise led to the development of category theory.

The best way of dealing with our problems is first to translate some temporal logics into the predicate calculus and then, in the light of what we have done, examine the general notions of reduction and translation involved. We need to perform some translations first, to see all our mathematical options. For this purpose we have to build up some extensions of the predicate calculus, i.e. by adding temporal connectives to predicate logic. We also look at temporal logics built (or simulated) within predicate logic. We need a notation system which will help distinguish between these two approaches. We thus write **APL1, APL2**, ... to name temporal logics using temporal connectives as we have done in earlier chapters. Technically these are proper extensions of predicate logic ('A' for addition) and write **IPL1, IPL2**, ..., to name two-sorted systems of predicate logic with a special sort for time ('I' for internal). Schematically we want to check whether, and how, we can say that **IPL=APL**.

Let us first see how the **IPL** languages can be built. We begin by isolating from among the predicates and variables of the general predicate calculus some special symbols which will have temporal meaning. These symbols become distinguished in the same way that the equality symbol $=$ is distinguished in predicate logic. These include:

- a unary predicate T for the flow of time;
- a transitive binary relation $<$ for the earlier–later ordering of time.

We thus have a two-sorted predicate logic with special predicates and variables for each sort.

Example 5.2.1

(a) The property that time is linear can be written as:
$(\forall t, s)(t < s \vee t = s \vee s < t)$.

(b) 'John loves Mary at time t' can be written as:
Love(t,John,Mary).

(c) The truth table for 'A will be true' is represented by the formula
$\psi(\text{ will})(t, T_1) = \exists s > tT_1(s)$,
where T_1 is supposed to be $\{s' \mid A \text{ is true at } s'\}$.

Example 5.2.2 Assume that we want to formalize the properties of the English progressive tense, say in the context of a natural language processing model which can recognize certain temporal features, such as the difference between:

1. 'John walks to school', and
2. 'John is walking to school'.

The **APL** (temporal logic) way of talking about the progressive is to add a new unary connective Nq to the language of predicate logic and represent 'John is walking' as: N Walk(John). The properties of the progressive can be captured by the axioms below:

1. $Nq \to q$.
   If John is walking around now, then he is actually walking now.

2. $Nq \to NNq$.
   The above axiom really says that if John is walking now, then he has been walking a little bit before now, and will be walking a little bit after now. Axiom 2 forces the above property.

3. $N(p \wedge q) \leftrightarrow Np \wedge Nq$.
   John is walking and drinking now, if and only if John is walking now and is drinking now.

4. $\vdash q \Rightarrow \vdash Nq$.
   For example, if we are walking at all times, then we are walking around at any time.

Axiom 2 is needed because we want to force Nq to mean what it means and we want to use axioms involving only N.

If we allow the connectives Since and Until, we can define N in terms of S and U by:

$Nq =$ (definition) $q \wedge S(\mathit{truth}, q) \wedge U(\mathit{truth}, q)$.

But then, of course, we need axioms on S and U.

The **IPL** representation of the progressive is to associate with it the formula $\mathbf{N}(t, Q)$ of predicate logic where:

$\mathbf{N}(t, Q)=$ definition $\exists t_1 \exists t_2 (t_1 < t < t_2 \wedge \forall s (t_1 < s < t_2 \to Q(s))$

The nature of the progressive is determined by this formula. This formula $\mathbf{N}$ is said to be the *table* for N in our terminology of chapter 1, or of the definitions in section 5.3 below.

The axioms for N given in the **APL** style of presentation are all valid and follow in predicate logic from $\mathbf{N}$. However, the axioms do not determine the formula N uniquely. Imagine reading Nq as $N_1 q$ below:

$N_1 q = q$ is true from now onwards.

N_1 satisfies the same axioms as N. The **IPL** presentation of N_1 is $\mathbf{N}_1$, namely:

$\mathbf{N}_1(t, Q) =$ (definition) $\forall s (t < s \to Q(s))$.

It can be proved mathematically that over, e.g., rational numbers (or real numbers) flow of time, N and N_1 satisfy the same logical axioms. Over the integers time, Nq becomes q, i.e. $N(t, Q) = Q(t)$ while $\mathbf{N}_1$ is not. Thus we see that the two methods of representation, i.e. **APL** and **IPL**, are really stylistically different. There are more examples in the next section.

The moral of the N, N_1 example is that the axiomatic presentation of the connective N (or N_1) does not admit a unique meaning and allows for a family of meanings in one axiomatic presentation. To present N or N_1 within classical predicate logic we have to use one of its meanings. We thus lose the other meanings. Of course, one can always use predicate logic as a metalanguage (use Gödel numbers) and simulate the axiom system for N (or N_1). This will indeed yield the axiomatic system for N (or N_1) exactly, but it is perverse to say that we have 'expressed' N or N_1 in predicate logic. We use predicate logic instead of English to describe the system for N_1. We use no properties of predicate logic. We may as well use Latin for the purpose.

5.3 Formal definition of APL and IPL languages

Definition 5.3.1 We define the fragment **IPL** of classical logic capable of handling time as follows. (We use $T, <, t, s$ as agreed above.) The logic is two-sorted with $s, t, t_1, t_2, ...$ being time variable symbols and $x_1, x_2, ...$ being variable symbols for elements of a domain. $Q(t_1, ..., t_k, x_1, ..., x_n)$ means that $Q(x_1, ..., x_n)$ holds at time $(t_1, ..., t_k)$.

(a) The atomic predicates of the fragment have the form either a_1 or a_2 but *not* a_3.

a_1 $Q(t_1, \ldots, t_k, x_1, \ldots, x_n)$ where Q is atomic symbol, $x_1, \ldots, x_n$ are object (non-time) variables, and t_i are time variables. t_1 must appear in Q: so $k \geq 1$. $x_1, \ldots, x_n$ might not appear in Q (i.e. $Q = Q(t_1)$ is allowed). Q is a two-sorted predicate.

a_2 $t < s, t = s$, for t, s time variables.

a_3 $t < x, t = x, x < y, x < t$, are not allowed.

(b) The fragment is closed under $\wedge, \vee, \sim, \rightarrow$ and under quantifications $\forall$ and $\exists$.

Definition 5.3.2 (an m-dimensional table with a parameter R) Any wff built up from atoms of the form $R(s)$, where s is a time variable, and $T_i(t_1, \ldots, t_m), i = 1, \ldots, k$, using quantifiers over $t \in T$ only, and containing m free time variables, is called an m-dimensional table for a k-place connective with a parameter R.

Definition 5.3.3 (the general form of the language APL) Let $\sharp_1, \ldots, \sharp_k$ be symbols for $m_1, \ldots, m_k$ -place connectives. The syntactical temporal extension of predicate logic with the connectives $\sharp_1, \ldots, \sharp_k$ is defined as follows. It is called **APL**($\sharp_1, \ldots, \sharp_k$).

(a) Any predicate wff is a temporal wff.

(b) If $A_1, \ldots, A_{m_i}$ are temporal wffs, so is $\sharp_i(A_1, \ldots, A_{m_i})$.

(c) The set of temporal wffs is closed under the classical connectives and quantifiers.

Definition 5.3.4 The first-order semantical presentation of any **APL** logic of the syntactical form **APL**($\sharp_i$) (as defined above) is obtained by associating with each $\sharp_i$ a table A_i with a parameter R in the sense of Definition 5.3.2. The tables should all be of the same dimension n and for a connective $\sharp_i$ the table is m_i place. The semantics we are giving is in this case n-dimensional. Thus with each $\sharp_i$ we associate a formula $A_i(t_1, \ldots, t_n, R, T_1, \ldots, T_{m_i})$.

Example 5.3.5

(a) Consider the connective F^* of example 1.3.4. The dimension is 2. the number of places is 1. Thus the table for F^* has the form:

$$\Psi_{\mathbf{F}^*}(t, s, T_1) = [t = s \wedge (\exists t' > t)T_1(t', s)] \vee [t > s \wedge T_1(t, t)]$$
$$\vee[t < s \wedge \forall t'(t < t' < s \rightarrow T_1(t', t'))].$$

Here we do not use the parameter R.

(b) Now consider the one-dimensional connective S whose truth conditions on an integer flow of time are given by $\|S(A)\|_n = 1$ if and only if A holds at the next square after n. Here we can use R with extension $\{n^2 \mid n \text{ an integer}\}$ and the table of S is

$$\psi_S(t, T, R) = \forall x((t < x) \wedge R(x) \wedge {\sim}\exists y((t < y < x) \wedge R(y)) \rightarrow T(x)).$$

Definition 5.3.6

(a) The axiomatic presentation of an **APL**($\sharp_i$) logic is obtained by writing axioms and rules for the connectives $\{\sharp_i\}$.

(b) In case we can prove that an axiomatic **APL** presentation and a semantical **APL** presentation define the same logic (i.e the same set of valid wffs), then we say we have proved the completeness of the axioms for the semantics. Note that we haven't yet defined the notion of valid sentences precisely.

Example 5.3.7 Consider the example Nq of the progressive from the preceding section. There are four temporal logics involved with this connective:

(a) An axiomatic presentation:

  (1) $Nq \to q$
  (2) $Nq \to NNq$
  (3) $N(p \wedge q) \leftrightarrow Np \wedge Nq$
  (4) $\vdash q \Rightarrow \vdash Nq$.

(b) A first-order semantical presentation via the table:

$$\mathbf{N}(t, Q) = \exists t_1, t_2(t_1 < t < t_2 \wedge \forall s(t_1 < s < t_2 \to Q(s))).$$

(c) A first-order semantical presentation via the table:

$$\mathbf{N}_1(t, Q) = \forall s(t < s \to Q(s)).$$

(d) A topological semantical presentation as follows:
Let each atom q have a set in a topological space. Let Nq mean Nq=(definition) the topological interior of q.
Thus the table $\mathbf{N}_2$ for N is:
$\mathbf{N}_2(t, Q)$iff $t \in \text{Interior}(Q)$.
The table is not even first-order and involves a topology over sets. This topology may be connected with some natural ordering on the topological space.

We don't think there is any sense in saying that the above four logics for the connective N are the same. It is true that the set of valid sentences in all four logics is the same, but this is all they have in common.

Thus to represent the axiomatic presentation (a) above in first-order logic, we have to use one of its interpretations (e.g. (b) or (c)). We are thus forced to commit ourselves to one interpretation, while the **APL** axiomatic presentation gives no commitment. This is a similar situation to what we have in logical representation of legal rules. The presentation (translation) of the rules into logic already imposes a reading on them and thus deprives the rules from any other interpretation. One's tendency in the legal context is not to translate. The same tendency should apply here.

To examine the difference between the logics further, try and see what is the meaning of some extensions of each logic in terms of the other logics. Consider the following:

Example 5.3.8

(a') Extend the logic (a) of the preceding example with Dummett's axiom as simplified by Geach:

$$N(N(q \to Nq) \to q) \to (\sim N(\sim Nq) \to q).$$

This axiom makes sense in terms of axioms and rules but what does it mean in terms of tables or topologies?

(b', c') Add the condition that the flow of time is dense and linear, to (b) and (c) of the preceding example. Does this correspond to an axiom? How does it affect the table $\mathbf{N}_2$ in case the topology is obtained from some natural ordering.

(d') Add the condition that the topological space of (d) is discrete. Does this have any counterpart in the other systems?

(e') Suppose we restrict our atomic formula to being true only at a finite set of points or its complement (i.e. give them finite sets or complements of finite sets in the topological interpretation). What is the effect of this restriction on the various logics?

Exercise 5.3.9 Work out what is the effect of conditions (a')–(e') on the respective logics.

Given the topological interpretation of N, it is obviously not first order. It has the same valid theorems as the logic in example 5.3.7b and 5.3.7c, which can obviously be represented as first-order. Do we choose one of them, say example 5.3.7b, and claim that we have represented the logic (d') within the frame work of the predicate calculus? Haven't we represented something completely different?

We need standards and protocols for translations from one language to another. The definitions below give a standard translation of sematically presented temporal logics whose table is first-order. Such examples are (b) and (c) of example 5.3.7.

We saw that an $\mathbf{APL}(\sharp_i)$ logic can be presented semantically via a table. The table does not yet define the logic uniquely, it just defines the meaning of the connectives. Thus the table $\mathbf{N}$ for N defines one logic for the case that time is the rational numbers but reduces to $Nq = q$ in the case that time is the integers. Thus to define completely the set of valid sentences of a logic presented to us we must also specify exactly the class of temporal models for this logic. The relevant notions will be defined later in this section. We have enough information at our disposal to be able to reduce semantically defined **APL** logics to **IPL**, provided the **APL** tables are first-order formulae.

Definition 5.3.10 We define a translation $*$ from any m-dimensional semantical (first-order) presentation of an **APL**($\sharp_i$) logic for $\sharp_i$ into **IPL** as follows:

(a) Any atomic $Q(x_1, \ldots, x_n)$ of **APL** is reduced to $Q^*(t_1, \ldots, t_m, x_1, \ldots, x_n)$ of **IPL** (i.e. time co-ordinates are added). In the propositional case, the atom q will translate into $q^* = Q$.

(b) The translation $*$ commutes with the classical connectives and quantifiers.
$(A \wedge B)^* = A^* \wedge B^*$
$(A \vee B)^* = A^* \vee B^*$
$(\sim A)^* = \sim A^*$
$(\forall x A(x))^* = \forall x A^*(x)$
$(\exists x A(x))^* = \exists x A^*(x)$.

(c) The m-dimensional connective $\sharp_i(A_1, \ldots, A_{m_i})$ with table α_i is translated using the table:

$$\sharp_i(A_1, \ldots, A^*_{m_i}) = \alpha_i(t_1, \ldots, t_m, R, A^*_1, \ldots, A^*_{m_i}).$$

Example 5.3.11 The translation of $N(Nq \to q)$ of the preceding example 5.3.7 with table:

$$\mathbf{N}(t, Q) = \exists t_1, t_2, \forall s(t_1 < s < t_2 \to Q(s))$$

is:

$$\exists t_1, t_2(t_1 < t < t_2) \wedge \forall s(t_1 < s < t_2 \to [Nq \to q]^*(s)),$$

where

$$(Nq \to q)^*(s) = (Nq)^*(s) \to q^*(s) = \exists x_1, x_2(x_1 < s < x_2$$
$$\wedge \forall y(x_1 < y < x_2 \to Q(y))) \to Q(s).$$

Thus the translation is:

$$(\exists t_1, t_2)[(t_1 < t < t_2) \wedge \forall s(t_1 < s < t_2 \to (\exists x_1, x_2)((x_1 < s < x_2)$$
$$\wedge \forall y(x_1 < y < x_2 \to Q(y))) \to Q(s))].$$

The semantical presentation of definition 5.3.6 and the translation of definition 5.3.10 are good for any flow of time. To define any specific temporal logic, we need to fix the flow of time. Thus we are led to the following:

Definition 5.3.12

(a) An **APL** semantical presentation of an m-dimensional temporal logic with constant domains and with connectives $\sharp_1, ..., \sharp_k$, is determined by the following components:

(a1) m-dimensional tables $\mathbf{A}_i$ for each connective $\sharp_i, i = 1, ..., i = k$, in the sense of definitions 5.3.4 and 5.3.10.

(a2) A family $\mathbf{K}$ of flows of time with restrictions. The form of each flow of time with restrictions in $\mathbf{K}$ is $(T, <, \mathbf{H}, \mathbf{R})$ where $(T, <)$ is some particular flow of time, $\mathbf{H}$ is a formula (usually first-order) restricting the truth sets which can be assigned to the atoms and $\mathbf{R}$ is a subset of T. $\mathbf{R}$ serves as the parameter $\mathbf{R}$ in the tables $\mathbf{A}_i$. $\mathbf{R}$ may be defined using T.

(b) A temporal model has the form $(T, <, D, h, R) \in \mathbf{K}$ and D, R and h satisfy the restrictions $\mathbf{H}$ and $\mathbf{R}$. D is the non-empty domain of the model and h is an assigment giving for each $t_1, \ldots, t_m \in T$, $R \subseteq T$ and each n-place atomic predicate Q of the temporal language a subset $h(t_1, \ldots, t_m, Q) \subseteq D_n$.

(c) The truth set $\|B\|$ of each wff $B(x_i)$ of the temporal language is defined by induction, for each substitution $x_i/y_i \in D$ (i.e. we are defining $\|B(y_i)\|$ for each $y_i \in D$):

(1) $\|Q(y_i)\| = \{t | (y_1, \ldots, y_n) \in h(t, Q)$, where $t = (t_1, \ldots, t_m)\}$.

(2) $\|(B_1 \wedge B_2)\| = \|B_1\| \cap \|B_2\|$.

(3) $\|(\sim B)\| = T^m - \|B\|$.

(4) $\|(B_1 \vee B_2)\| = \|B_1\| \cup \|B_2\|$.

(5) $\|(B_1 \to B_2)\| = \|(T^m - B_1)\| \cup \|B_2\|$.

(6) $\|(\exists x B(x))\| = \{t^* |$ for some $d \in D, t^* \in \|B(d)\|$, $t^* = (t_1, \ldots, t_m)\}$.

(7) $\|\forall x B(x)\| = \{t^* |$ for all $d \in D, t^* \in \|B(d)\|$, $t^* = (t_1, \ldots, t_m)\}$.

(8) $\|\sharp_i(B_1, \ldots, B_{m_i})\| = \{t | A_i(t, B_1, \ldots, B_{m_i})$ holds in $(T, <)\}$.

(9) We say $B(x_i)$ holds at $(T, <, D, h)$ if for any $d_i \in D, B(d_i)$ contains the set $\{t | R(t)\}$.

(10) We say $B(x_i)$ is valid in the logic if $B(x_i)$ holds at any $(T, <, D, h, R)$, for any $(T, <, \mathbf{H}, \mathbf{R})$ and any h satisfying the restrictions $\mathbf{H}, \mathbf{R}$.

Example 5.3.13

(a) Take the connective N with the table of example 5.3.7(a) and a constant c. Take as the flows of time in $\mathbf{K}$ all linearly ordered $(T, <)$. Take a subset $C \subseteq T$ as a truth table subset for C. Take as restrictions $\mathbf{H}$,$\mathbf{R}$ the condition $\mathbf{H}(Q)$ for any atom Q, and $\mathbf{R}(t)$ below:

For $q \neq c$(i.e. $Q \neq C$) let:

(1) $\mathbf{H}(Q) = \exists t \forall x (\forall s > t)(Q(s,x) \leftrightarrow Q(t,x))$, for $Q \neq C$(i.e. $q \neq c$).

(2) $\mathbf{H}(C) = \exists t{\sim}C(t) \wedge \exists t C(t) \wedge \forall s, s'(C(s) \wedge s < s' \rightarrow C(s'))$.

(3) Let $\mathbf{R}(t) = {\sim}C(t)$.

Thus $\mathbf{H}(Q)$ says that after some time t, Q keeps on giving the same values to any x. $\mathbf{H}(C)$ says that C is initially false and then becomes true. $\mathbf{R}$ says the domain is determined by C being false.

(b) Take as flows of time all finite chains for the connective N of example 5.3.7a, which is the same connective taken in (a) above (without c). Take no restrictions $\mathbf{H}$ or $\mathbf{R}$.

One can show that the logic of the connective N_b (i.e. N of example 5.3.7b above) is related (can be translated) into the logic of the connective N_a of 5.3.7a by the following translation (more or less):

$$N_b q \equiv {\sim}c \wedge N_a(q \vee c).$$

We can now say what it means, as a first approximation, for a temporal logic to be first-order expressible in **IPL**. The temporal logic must have tables A_i for its connectives $\sharp_i$, and first-order conditions defining the class $\mathbf{K}$ of flows of time, and first-order conditions defining the restrictions $\mathbf{R}$,$\mathbf{H}$ on the assignment. If a logic $\mathbf{L}_1$ is defined by other means, not by semantical means, then we can say that if the set of valid theorems of this logic is the same as the set of another logic $\mathbf{L}_2$ which is expressible in **IPL**, then $\mathbf{L}_1$ is also considered expressible in **IPL**. We are not yet ready to give the above as a formal definition, partly because we are going to reject it, as it is too restrictive and misses out on some intuitions. Let us see some more examples before we continue the discussion.

5.4 Examples

Example 5.4.1 Let $\mathbf{L}_1$ be the extension of classical logic with a modality operator $\Box$. Let us read $\Box A$ to mean A *is necessarily true*, and endow $\Box$ with axioms to make it a transitive modality. Thus the axioms for $\mathbf{L}_1$ are (besides all classical logic theorems and rules):

(a) $\dfrac{\vdash A}{\vdash \Box A}$.

(b) $\vdash \Box(A \wedge B) \leftrightarrow \Box A \wedge \Box B$.

(c) $\vdash A \Rightarrow \vdash \Box A$.

(d) $\vdash \forall x \Box P(x) \rightarrow \Box \forall x P(x)$.

(e) $\mathbf{L}_1$ does not prove the reflexivity axiom, $\Box A \rightarrow A$, i.e. $\mathbf{L}_1 \nvdash \Box A \rightarrow A$.

Example 5.4.2 We now define a temporal logic $\mathbf{L}_2$ with a 'progressive' operator NA reading: A *is true now and around now.* This is the logic in example 5.2.2.

Recall the semantics of NA; imagine a flow of time like the rationals or real numbers:

NA is true at t if and only if there is an interval around t in which A is true, i.e.

$$\exists x, y(x < t < y \wedge \forall u(x < u < y \rightarrow A \text{ true at } u)).$$

It can be shown that $\mathbf{L}_2$ entails exactly all the axioms of $\mathbf{L}_1$ together with the reflexivity axiom.

In other words, the modal axioms (a)–(e) axiomatize all the theorems of $\mathbf{L}_2$.

We begin by pointing to three different ways of translating $\mathbf{L}_2$ into $\mathbf{L}_1$. We mean here a mapping $\mathbf{T}$, translating each wff of $\mathbf{L}_2$ into a wff of $\mathbf{L}_1$:

$$\mathbf{T} : A \mapsto \mathbf{T}(A)$$

such that for some set of wffs $\mathbf{S}$ of $\mathbf{L}_1$ we have for all A

$$\mathbf{L}_2 \vdash A \text{ iff } \mathbf{S} \vdash_{\mathbf{L}_1} \mathbf{T}(A),$$

Definition 5.4.3 We define three different translations $\mathbf{T}$, denoted by $\mathbf{T}_1$,$\mathbf{T}_2$,$\mathbf{T}_3$.

The translations $\mathbf{T}_1$ and $\mathbf{T}_3$ are defined by induction on the structure of the formulae of $\mathbf{L}_2$, as follows (compare with definition 5.3.10):

(a) $\mathbf{T}(atom) =$ as described by the particular translation.

(b) $\mathbf{T}(A \wedge B) = \mathbf{T}(A) \wedge \mathbf{T}(B)$.

(c) $\mathbf{T}({\sim}A) = {\sim}\mathbf{T}(A)$.

(d_1) $\mathbf{T}(A \vee B) = \mathbf{T}(A) \vee \mathbf{T}(B)$.

(d_2) $\mathbf{T}(A \to B) = \mathbf{T}(A) \to \mathbf{T}(B)$.

(e) $\mathbf{T}(\exists x A(x)) = (\exists x \in D)\mathbf{T}(A(x))$.

(f) $\mathbf{T}(\forall x A(x)) = (\forall x \in D)\mathbf{T}(A(x))$,
where $A(x)$ is a formula of $\mathbf{L}_1$.

(g) $\mathbf{T}(MA) = M(\mathbf{T}(A))$
where $M(Q)$ is a formula of $\mathbf{L}_1$ with the predicate Q.

The above is not the most general way of translating. Firstly, $\wedge, \vee, {\sim}, \to$ were translated to themselves; this is not always so, see, for example, translation $\mathbf{T}_2$ below. Secondly, each A was translated into a single $\mathbf{T}(A)$. In the general case, $\mathbf{T}(A)$ can be a set of wff of $\mathbf{T}_1$ and we have

$$\mathbf{L}_2 \vdash A \text{ iff } \forall B \in \mathbf{T}(A)(S \vdash_{\mathbf{L}_1} B).$$

The following are the three different ways of translating $\mathbf{T}_2$ into $\mathbf{T}_1$.

1. **Direct translation**
   $\mathbf{T}_1(Q) = Q$, for Q atomic.
   $\mathbf{T}_1(NA) = A \wedge \Box A$.
   It can be shown that
   $\mathbf{L}_2 \vdash A$ iff $\mathbf{L}_1 \vdash \mathbf{T}_1(A)$.

2. **Simulation**
   Since $\mathbf{L}_1$ extends predicate logic, a provability predicate *Demo* (Δ, B) can be defined in $\mathbf{L}_1$ for the provability process in $\mathbf{L}_2$ of B from the data P. Suitable naming of the form $\mathbf{T}_2(A) = name(A) = n(A)$ can be given and we can have, for a suitable set of axioms $\mathbf{S}$ about the translation, that $\mathbf{L}_2 \vdash A$ iff $\mathbf{S} \vdash$ *Demo* $(\Delta, n(A))$.

3. **Semantic translation**
   In $\mathbf{L}_1$ a special variable $t \in T$, and a special set T for time can be allocated, as well as time relation $<$. Any atom $Q(x_i)$ of $\mathbf{L}_2$ can be translated as $Q^*(t, x_i)$ of $\mathbf{L}_1$ reading: $Q(x_i)$ is true at time t, and $\mathbf{T}_3$ can be defined as

   - $\mathbf{T}_3(Q) = Q^*(t), Q$ atomic and

- $\mathbf{T}_3(NA) = \exists x \exists y(x < t < y \wedge \forall u[x < u < y \to \mathbf{T}_3(A)(u)])$.

The translation $\mathbf{T}_3$ actually defines the semantical interpretation of $\mathbf{L}_2$ inside $\mathbf{L}_1$.
We again have that there exists a theory **S** of $\mathbf{L}_1$ such that for any A of $\mathbf{L}_2$,

$$\mathbf{L}_2 \vdash A \text{ iff } S \vdash_{\mathbf{L}_1} \mathbf{T}_3(A).$$

S is needed to say that we are dealing with time in $\mathbf{L}_1$, and that $D \cup T$ is the entire universe, and that some other organizational properties, such as those to do with **Hold**, are satisfied, etc.

The reader may think that any translation $\mathbf{T}_1, \mathbf{T}_2, \mathbf{T}_3$ is no better than the other, but this is not the case. Consider the following points:
Point 1:
In translation $\mathbf{T}_1$ the addition of the axiom $\Box A \leftrightarrow A$ in $\mathbf{L}_1$ 'collapses' the meaning of $\mathbf{T}(NA)$ into A. In translation $\mathbf{T}_2$ and $\mathbf{T}_3$ there is no effect on $\mathbf{T}(NA)$. In other words, in a direct translation the image is not a 'module' and there are reflections from the target logic to the image.
Point 2:
In translation $\mathbf{T}_2$ (the provability simulation) for any extension of $\mathbf{L}_2$ the source can be *automatically* translated. We just modify the provability predicate. This is not true for $\mathbf{T}_1$ and $\mathbf{T}_3$. We may in the case of $\mathbf{T}_1$ or $\mathbf{T}_3$ find new translations defined from scratch, but even this can not always be done.
Point 3:
$\mathbf{T}_1$ and $\mathbf{T}_2$, when applied in the case of $\mathbf{L}_1 = \mathbf{L}_2$, almost yield the identity, not so for $\mathbf{T}_3$.

We see that properties of translations can be studied in terms of:

(a) extendibility;

(b) reflection principles;

(c) compositions;

(d) other properties.

We may be able to give a semi-categorical definition of what a direct translation is and how it is different from a simulation.

Exercise 5.4.4 Write out the conditions **S** for the translation $\mathbf{T}_3$.

Example 5.4.5 Consider example 1.4.6, the logic $Y1$, also discussed in section 3.4. This logic is complete for *finite* flows of time of a certain form. We claim, this logic cannot be directly translated into predicate logic because a set **S** of conditions cannot be found.

In first-order predicate logic the notion of finiteness can not be expressed. We can always fix the flow of time beforehand, and claim as a higher-order condition that the flow is finite. Certainly from the point of view of computer science, finiteness is a very comfortable property.

Furthermore, if one really insists on first-order conditions, one can make use of definition 4.2.13; one can always take all the axioms of Löb's logic and adopt the predicate logic translation of them. Thus in predicate logic we have to add the following flow of time axiom schema, $\mathbf{H}_1$ and $\mathbf{H}_2$.

1. $\mathbf{H}_1$ = Time is linear and discrete.

2. $\mathbf{H}_2$ = Condition corresponding to finiteness.
   $(\exists s < t)\forall u_1, u_2(u_1 < s \wedge u_2 < s \wedge A(u_1) \rightarrow A(u_2))$.

We adopt the schema for all *atoms A* only. It will *automatically* hold for any wff.

This makes the translation into predicate logic more acceptable, because to each wff B containing the atoms $Q_1, \ldots, Q_n$, we can translate B into predicate logic and append the schema for $Q_1, \ldots, Q_n$ to the translation. Thus the translation of any single wff is a single wff.

Definition 5.4.6 A schema has the *transfer property* iff whenever it holds for all the atoms of a wff B it holds for B as well.

Remark 5.4.7 The concepts we have encountered so far are:

1. We can use properties of the time parameters in predicate logic; even so they are not first-order.

2. In translating into predicate logic we prefer that single formulae be translated to single wffs and finitely axiomatizable theories be translated into finitely axiomatizable translations.

5.5 Linked predicate languages

In section 5.1 we discussed the possibility of two interpreters I_1 and I_2 being linked in some way and through that link one can implement some metalevel features about the other. Obviously, a lot depends on the mechanism of 'linking'. We are going to clarify this notion for the case where I_1 and I_2 are two predicate logic theories. This will suffice for the handling of temporal logic.

Sections 5.2 and 5.3 showed how temporal connectives can be 'talked about' from within predicate logic. The temporal metapredicate

$$\text{`}A(x_1, \ldots, x_n) \text{ holds at time } t\text{'}$$

is translated into predicate logic as $A^*(t, x_1, \ldots, x_n)$, where A^* is a two-sorted $n+1$ place predicate, with t a variable of the time sort and $x_1, \ldots, x_n$ of the element sort. The relation $<$ on time also has to be introduced as well as suitable axioms to ensure that the interpretation of the temporal system within predicate logic functions properly.

There is another way of looking at what we are doing, which is quite independent of temporal logic. It has to do with linking two languages via the sharing of variables. The task of this section is to develop this theme.

Consider the first-order langauge **G** with a binary relation $<$ and variables $\{t, s, \ldots\}$. Think of it as a theory of the flow of time. Consider the predicate language **L**, with variables $\{x_1, x_2, \ldots\}$ and predicates $A(x_1, \ldots, x_n)$, $B(y_1, \ldots, y_m)$. The formal operation of what we have done was to 'combine' these two languages by replacing **L** by $\mathbf{L}^*$, where $\mathbf{L}^*$ is a two-sorted language, with variables $\{t, s, \ldots\}$ and $\{x_1, x_2, \ldots\}$, and taking as atomic predicates the two-sorted predicates

$$A^*(t, x_1, \ldots, x_n), B^*(s, y_1, \ldots, y_n), \text{ etc.}$$

We allow for common quantification and mixed wffs. Thus we can write, for example, the mixed formula $\varphi(t)$:

$$\varphi(t) = \forall s(t < s \to A^*(t, x_1, \ldots, x_n)),$$

which you will recognize as the truth table for '$GA(x_1, \ldots, x_n)$ holds at t'.

Formally, φ is just a formula in the mixed language. Let us give a quick definition to clarify our concepts before we continue our discussion.

Definition 5.5.1 (the two-sorted language $\mathbf{L}_k^*(\mathbf{G})$) Let **L** and **G** be the two languages. Both **L** and **G** have atomic predicates. Let $\mathbf{L}_k^*$ be obtained from **L** by replacing each atomic predicate $R(x_1, \ldots, x_n)$ of **L** by $R^*(t_1, \ldots, t_k, x_1, \ldots, x_n)$, where t_i are variables of a new sort. Thus the resulting $\mathbf{L}_k^*$ is a two-sorted language, with the new sort for t-variables. We allow quantification over t-variables. The intention is that the t-variables will be from the language **G**.

1. The atomic wffs of $\mathbf{L}_k^*(\mathbf{G})$ are of the form $R^*(t_1, \ldots, t_k, x_1, \ldots, x_n)$, where $R(x_1, \ldots, x_n)$ is an atomic wff of **L** and $t_1, \ldots, t_k$ are terms of

G, or of the form $Q(t_1, \ldots, t_m)$, where $Q(t_1, \ldots, t_m)$ is an atomic wff of **G**.

2. If φ and ψ are wffs of the language with t free in φ and x free in φ, then $\varphi \wedge \psi, \varphi \vee \psi, \varphi \to \psi, \sim\varphi, \forall t\varphi, \forall x\varphi, \exists t\varphi, \exists x\varphi$ are wffs of the language.

Example 5.5.2

1. Let **G** be the language of order $<$ and let **L** be classical predicate logic. $\mathbf{L}_1^*(\mathbf{G})$ gives a language with atoms $R^*(t, x_1, \ldots, x_n)$, which we used in the preceding example to mean 'R is true at t'. This is the language which we used in earlier parts of this chapter. We denote it by $\mathbf{L}(<)$.

2. The language **G** can be a metalanguage. Consider a propositional language, e.g. the propositional language with $\sim, \wedge, \vee$, and $\to$. For each wff A of the propositional language introduce a constant τ_A. Introduce the operations $\mathbf{f}_\sim(\tau_A) = \tau_{\sim A}$, $\mathbf{f}_\wedge(\tau_A, \tau_B) = \tau_{A \wedge B}$ and similarly $\mathbf{f}_\vee$ and $\mathbf{f}_\to$. Let **G** be the language with τ_A as terms and $\{\mathbf{f}_\sim, \ldots\}$ as functions. The language $\mathbf{L}_1^*(\mathbf{G})$ talks about the formulae of the propositional language as labels $R^*(\tau_A, x_1, \ldots, x_n)$ represent $A : R(x_1, \ldots, x_n)$, where A acts as a label.

The above linkage is still not the most general case. To obtain an idea of what we need, consider the following set-up. Consider a distributed system of stations $s_1, s_2, s_3, \ldots$ related in some manner. Each station has its own internal structure and language **L**. Assume they all use the same language and logic. We also need a global configuration language to describe how these stations are related. This language, denoted by **G** (for global), need not be the same logic as **L**. For example, **G** can be based on intuitionistic logic, while **L** is based on classical logic. **L** may be a Horn clause logic programming language while **G** could be a form of Petri net langauge. The stations s_i need to communicate, so they would have output and input ports which can be accessed by the other stations. For simplicity assume that there are some variables $t_1, \ldots, t_k$ which can be accessed by the stations. This means that the language **G** can talk about t_i.

If we want to link these two languages into one, we can form the syntactic combination $\mathbf{L}_k(\mathbf{G})$ of the preceding definition. The additional feature to worry about is that the logic of **G** and **L** may not be the same. Consider the following cases:

1. $A^*(t, x) \to B^*(s, y)$.

2. $A^*(t, x) \to B^*(t, y)$.

3. $A^*(t, x) \to B^*(s, x)$.

4. $A^*(t, x) \to B^*(t, x)$.

In cases (2) and (4) we can say that this is an internal statement of station t and hence the implication follows the logic of **L**. In case (3) we can use the logic of **G** as this is a relation between stations.

What do we do in case (1)?

The simplest course of action is to understand it as a relationship between stations and hence use the logic of **G**. We thus have a simple rule. Any formula with *different* station variables in it is to be manipulated logically according to **G** but if the variables are equal, it is to be manipulated according to **L**. In fact, once we agree to that, different stations can have different logics, $\mathbf{L}_t$. They must all be based on the same language.

We have one restriction on the system, that is, the configuration logic must be sound with respect to the station logics. In symbols, $\Delta \vdash_{\mathbf{G}} A$ in the configuration logic implies $\Delta \vdash_{\mathbf{L}} A$ in the station logic. The reason for this is unification. We must be able to reason about two possibly different stations in **G** and if we identify them our reasoning (now in **L**) must still be valid.

We now consider the Horn clause fragment of $\mathbf{L}_k^*(\mathbf{G})$.

Definition 5.5.3 (the Horn clause fragment of $\mathbf{L}_k^*(\mathbf{G})$)

1. Any atom of the form $R^*(t_1, \ldots, t_k, x_1, \ldots, x_n)$ of $\mathbf{L}_k^*$ and $Q(t_1, \ldots, t_m)$ of **G** is in the Horn clause fragment and in the atomic fragment.

2. If A and B are in the Horn clause fragment or in the atomic fragment, then so are $A \wedge B, \forall x A, \exists x A, \forall t A$, and $\exists t A$.

3. If A is in the atomic fragment and B is in the Horn clause fragment, then $A \to B$ is in the Horn clause fragment.

Example 5.5.4

1. If $A(t, x), B(s, y)$ are atomic, then $A' = \forall t \exists x A$ is in the atomic fragment and $C = A' \to \exists s \forall y B$ is in the Horn clause fragment. In classical logic, one can write C as:

$$\exists t \forall x \exists s \forall y (A \to B)$$

because we can pull the quantifiers to the front and hence the preceding definition can be simplified. In intuitionistic logic this is not possible and so we need clause (3) of the preceding definition and the notion of atomic fragment, in order not to be committed to classical logic.

2. Notice that for $k = 1$ and **G** the theory of order $<$, the Horn clause fragment of $\mathbf{L}(<)$ is the same as the Horn fragment of the logic $\mathbf{L}(<)$– the linking of classical predicate logic with the theory of order $<$– found in the papers [Gabbay, 1987b] and [Gabbay, 1990].

The above discussion about linking two languages **L** and **G** generalized the special construction of **IPL** in the preceding section. **G** corresponded to the time order relation $<$ and **L** corresponded to the predicate logic. The generalization allows for arbitrary **G** and **L**. In the case of temporal logic, there is the language **APL**, which has temporal connective in it. What would be the counterpart of **APL**, for arbitrary **G** and **L**? Obviously we must generalize the notion of a temporal connective based on time ordering $<$ to some general modal connective based on some language **G**.

We are now ready to define a modal logic $\mathbf{Modal}(\mathbf{L}, \mathbf{G})$ of **L** and **G**. To explain our options, let us reconsider the basic temporal connectives F and P. The truth table for $FA(x)$ is: $FA(x)$ is true at t iff for some $s > t$, $A(x)$ is true at s.

If $A(x)$ is atomic, we translate it into $A^*(t, x)$, understanding $A^*(t, x)$ as '$A(x)$ is true at t' and thus we obtain:

$$FA(x) \text{ is true at } t \text{ iff } \exists s(t < s \wedge A^*(s, x))$$

or

$$(FA(x))^*_t = \exists s(t < s \wedge A^*(s, x)).$$

In $\mathbf{L}^*_1(<)$, $<$ is the atomic relation of the order. $A^*(t, x)$ is an atomic relation of $\mathbf{L}^*_1$. 'F' represents the existential quantifier on the variable s in $t < s \wedge A^*(s, x)$. The existential quantifier on t would be $\exists t(t < s \wedge A^*(t, x))$ and that would correspond to 'P'.

Let us generalize this situation to $\mathbf{L}^*_1(\mathbf{G})$. Here the atoms $A^*(t, x)$ are still the same, but the atoms of **G** could be any $Q(t_1, \ldots, t_m)$. The parallel construction would be:

$$\exists t_i(Q(t_1, \ldots, t_i, \ldots, t_m) \wedge A^*(t_i, x)).$$

We could introduce for each atom $Q(t_1, \ldots, t_m)$ of **G** and each $1 \leq i \leq m$ the existential connective $\Diamond_{Q,i}A(x)$ to mean

$$(\Diamond_{Q,i}A)^*_t = \text{(def) } \exists t_i[Q(\ldots t_i \ldots) \wedge A^*(t_i, x)].$$

Similarly, we can define $\Box_{Q,i}A$:

$$(\Box_{Q,i}A)^*_t = \text{(def) } \forall t_i[Q(\ldots t_i \ldots) \to A^*(t_i, x)].$$

These correspond to 'G' and 'H' of temporal logic. To summarize we have:

$$\begin{aligned} \Box_{<,2} &= G; \\ \Box_{<,1} &= H; \\ \Diamond_{<,2} &= F; \\ \Diamond_{<,1} &= P. \end{aligned}$$

Let us consider what *Since*(A, B) does in this context.

$S(A, B)$ is true at t iff $\exists s < t[A$ is true at S and $\forall y(s < y \wedge y < t$ implies B is true at $y)]$.

Let ψ_S be the following formula in a language with $<$ and the two *new monadic* predicates Q_1 and Q_2:

$$\psi_S(t, Q_1, Q_2) = \text{(def) } \exists s < t[Q_1(s) \wedge \forall y(s < y \wedge y < t \to Q_2(y))].$$

Then we have $S(A, B)$ is true at t iff $\psi_S(t, A^*, B^*)$, where A^*, B^* are considered monadic predicates in the **G** sort.

To generalize the above, let $\mathbf{G}_1^+$ be the extension of the language of **G** with new monadic predicates $\{Q_1, Q_2, \ldots\}$. More generally, let $\mathbf{G}_m^+$ be the extension of the language of **G** with *new* m-place atomic predicates $\{R_1, R_2, \ldots\}$. Let $\psi(t, Q_1, \ldots, Q_n)$ be a formula of $\mathbf{G}_1^+$ with one free variable t and the n new monadic predicates. Let $A_1^*, \ldots, A_n^*$ be n formulae of $\mathbf{L}_1^*$, which can be considered monadic in the sort t. We can introduce the n place connective $\sharp(A_1, \ldots, A_n)$ with the truth table

$\sharp(A_1, \ldots, A_n)$ true at t iff $\psi(t, A_1^*, \ldots, A_n^*)$.

We will now give the general definition and an example.

Definition 5.5.5 Let **G**, **L** be two languages. We construct the modal language $\mathbf{Modal}_m(\mathbf{L}, \mathbf{G})$ as follows:

1. Let $\mathbf{G}_m^+$ be the language **G** supplemented with an infinite sequence of *new* m place atomic predicates $\{R_1, R_2, \ldots\}$.

2. Let $\mathbf{L}_m^*(\mathbf{G})$ be the language obtained from **L** as in definition 5.5.3. We can regard each atomic formula $A^*(t_1, \ldots, t_m, x_1, \ldots, x_n)$ as m-place atomic in the variables $t_1, \ldots, t_m$.

3. Let $\mathcal{G}$ be any model of **G**. We regard $\mathcal{G}$ as a model of possible worlds. With each $t_1, \ldots, t_m \in \mathcal{G}$, associate a model $\mathcal{L}_{(t_1,\ldots,t_m)}$ of **L**. Assume that all the $\mathcal{L}_{(t_1,\ldots,t_m)}$ models have the *same* domain D (constant domain semantics). We refer to $(\mathcal{G}, \mathcal{L}_{(t_1,\ldots,t_m)})$ as a *modal model.*

4. For each wff $\psi(t_1, \ldots, t_m, R_1, \ldots, R_n)$ of $\mathbf{G}_m^+$ with t_i free and R_i new atomic predicates, associate an n place connective $\sharp_\psi$. The full modal language $\mathbf{Modal}_m(\mathbf{L}, \mathbf{G})$ is $\mathbf{L} \cup \{\sharp_\psi \mid \psi \in \mathbf{G}_m^+\}$.

Definition 5.5.6 Given a modal model $(\mathcal{G}, \mathcal{L}_{(t_1,\ldots,t_m)})$ and an assignment h of $\mathbf{L}$ into D, a model $\mathcal{G}^h$ of $\mathbf{G}_m^+$ is induced on $\mathcal{G}$ as follows.

Let $A^*(t_1, \ldots, t_m, x_1, \ldots, x_n)$ be an atom of $\mathbf{L}_k^*$. It can be viewed, for $x_1, \ldots, x_n$ fixed, as an atomic predicate of $\mathbf{G}_m^+$. Let the extension of this predicate in $\mathcal{G}$ be $\{(t_1, \ldots, t_m) \mid \mathcal{L}_{(t_1,\ldots,t_m)} \vDash A(h(x_1), \ldots, h(x_n))\}$.

We define the notion of a formula $A(x_1, \ldots, x_n)$ of $\mathbf{Modal}_m(\mathbf{L}, \mathbf{G})$ holding at the indices $(t_1, \ldots, t_m)$, notation $\|A\|^h_{t_1,\ldots,t_m} = 1$, as follows:

1. $\|A\|^h_{(t_1,\ldots,t_m)} = 1$ iff (def) $\mathcal{L}_{t_1,\ldots,t_m} \vDash A(h(x_1), \ldots, h(x_n))$.
   $\|A\|^h_{t_1,\ldots,t_m} = 0$ otherwise.

2. $\|A \wedge B\| = 1$ iff $\|A\| = \|B\| = 1$.

3. $\|{\sim}A\| = 1$ iff $\|A\| = 0$.

4. $\|\exists x A(x)\| = 1$ iff for some $d \in D, \|A(d)\| = 1$.

5. $\|\sharp_\psi(A_1, \ldots, A_n\|^h_{t_1,\ldots,t_n} = 1$ iff $\mathcal{G}_m^+ \vDash \psi(t_1, \ldots, t_m, Q_1, \ldots, Q_n)$, where $\mathcal{G}_m^+$ is the extension of $\mathcal{G}$ to a model of the language $\mathbf{G}_m^+$ by the assignment
   $Q_i = \{(t_1, \ldots, t_m) \mid \|A\|^h_{t_1,\ldots,t_m} = 1\}$.

6. Let $\mathcal{K}$ be a class of $\mathbf{G}$ models $(\mathcal{G}, (s_1, \ldots, s_m))$ with a distinguished sequence of $s_i \in \mathcal{G}$. Let φ be a formula of $\mathbf{Modal}_m(\mathbf{L}, \mathbf{G})$. We say $\mathcal{K} \vDash \varphi$ iff for all models $(\mathcal{G}, \mathcal{L}_{(t_1,\ldots,t_m)})$ and all h, $\|A\|^h_{s_1,\ldots,s_m} = 1$.

Example 5.5.7 To show the generality of the preceding definition, we give an example from another area of logic. Let $\mathcal{G}$ be a commutative semigroup with multiplication $*$ and a unit 1, i.e. we have the axioms:

1. $\forall xyz((x * y) * z = x * (y * z))$.

2. $\forall xy(x * y = y * x)$.

3. $\forall x(1 * x = x)$.

Let $\mathbf{L}$ be classical propositional logic. Consider the formulae $\psi(t, Q_1, Q_2)$ and $\varphi(t, Q_1, Q_2)$ where:

$$\psi(t, Q_1, Q_2) = \text{ def } \forall y[Q_1(y) \to Q_2(t * y)],$$

$$\varphi(t, Q_1, Q_2) = \text{ def } \exists xy[Q_1(x) \wedge Q_2(y) \wedge t = x * y).$$

Then the logic with the connectives $\sharp_\psi(A, B)$ and $\sharp_\varphi(A, B)$ is linear logic with linear implication and external conjunction:

$$\sharp_\psi(A, B) = A \multimap B,$$

$$\sharp_\varphi(A, B) = A \otimes B.$$

The semantics induced on these connectives by the general definition can be defined directly as follows.

The set of possible worlds is a semigroup $\mathcal{G}$. The propositional assignment h gives a truth value to every atom at a point, i.e. $h(t, q) \in \{0, 1\}$ for $t \in \mathcal{G}$, q atomic. The truth conditions for $\multimap$ and $\otimes$ are as follows:

$$\begin{array}{lll} \|A \multimap B\|_t^h = 1 & \text{iff} & \forall y(\|A\|_y^h = 1 \text{ implies } \|B\|_{t*y}^h = 1); \\ \|A \otimes B\|_t^h = 1 & \text{iff} & \text{for some } x, y \text{ we have } \|A\|_x^h = 1 \text{ and} \\ & & \|B\|_y^h = 1 \text{ and } t = x * y. \end{array}$$

We have $\vDash A$ iff for all $\mathcal{G}$ and all h, $\|A\|_1^h = 1$.

We will not pursue these topics further. See [Gabbay, 1992a].

5.6 The metalanguage HFP

The preceding two sections studied, first, the possiblity of classical logic serving as a metalanguage, and, second, the general mechanism of linking two languages **G** and **L**, so that the linkage can implement metalanguage features of the modal system based on **G**. In both cases our examples used classical logic for **L**. As the language **L** seems to serve as the object language which implements the metafeatures involved, we ask ourselves, is there a convenient general purpose language **L** which is specially suited for expressing metalanguage features? Having such a language, we could, for example, define a new, more flexible **IPL** (of sections 5.2 and 5.3) using this language instead of classical logic.

The answer is yes. The language we have in mind is the language **HFP**, which we introduce in this section.

We begin by describing the intuition behind the language **HFP**. Consider, first, propositional temporal logic, formalized in the usual way, via the addition of extra connectives to classical propositional logic. Assume that the connectives are FA and PA, 'A will be true' and 'A was true' respectively.

In symbols:

$$\|FA\|_t = 1 \text{ iff } \exists s > t \quad \|A\|_s = 1.$$

$$\|PA\|_t = 1 \text{ iff } \exists s < t \quad \|A\|_s = 1.$$

Suppose we want to choose the first moment s_0 such that $s_0 > t$ and $\|A\|_{s_0} = 1$; assume such a moment exists. This moment is a function of t and A. Let us denote it by $s = f_1^1(A, t)$. Here f_1^1 is a two-place function, taking a formula and a point and yielding a point.

Consider now the connective F itself; this connective takes a formula A and yields a formula FA, thus its function is $R_0^1(A)$.

The functionality of f_1^1 and R_0^1 can be described set theoretically. If X is the set of all points in which A is true, we obtain

$$f_1^1(X, t) = \min\ (X \cap \{s \mid s > t\}).$$

$$R_0^1(X) = \{s \mid \exists t \in X(s < t)\}$$

In general we can have general connectives and functions of the form f_n^m, R_n^m, taking m formulae and n points and yielding a point or a formula respectively.

To give another example, consider the connective $S(A, B)$. Its meaning is that since a point in the past where A was true, B has been true all the time. We can now form a new connective $S_y^x(A, B)$ reading at a point t in the past which is different from x, A is true and B has been continuously true at all points since which are greater than y. This connective has the form $S(A, B, x, y)$.

The language **HFP** is a general language in which tables and connectives can be described. It allows for predicates R_n^m and function symbols f_n^m to take both formulae and terms as arguments.

The use of **HFP** is not restricted to the metalevel of modal and temporal languages. It is also the metalanguage of actual Prolog programs. Although, officially, logic programming deals with the Horn clause fragment of classical logic, namely with clauses of the form

$$\wedge A_i \to B,$$

where A_i and B are atoms, in practice the Prolog programmer uses **HFP** as the language.

Expressions such as *Hold*(A, B) *Demo*$(A(x), B(y))$ are written in Prolog. The above are expressions of **HFP**. It may be thought that *Demo* and *Hold* are metapredicates. This is partially correct. They have special properties. The variables x and y in *Demo*$(A(x), B(y))$ can be quantified and unified. Thus *Demo* is not like the general metapredicates of a pure metalanguage **M** but, rather, special and we claim it is as in **HFP**.

We see in definiton 5.6.3 below that **HFP** is given modal semantics. So the question is whether this semantics is adequate for the way in which Prolog uses the language. We will be in a better position to answer in the next section. The answer in general is no, Prolog predicates are not in general modal connectives, but under certain restrictions they can be viewed as such, though this is not the most convenient way of looking at them.

We are now ready to define the general metalanguage **HFP**, of *Hereditarily Finite Predicates.*

Definition 5.6.1 (the language HFP) The language contains the classical connectives $\sim, \wedge, \vee, \rightarrow$ and the variables $\{x, y, z, \ldots\}$ for terms and variables $\{X, Y, Z, \ldots\}$ for formulae, and the quantifiers $\forall$ and $\exists$. There is a list of atomic predicates R_n^m allowing us to form $R_n^m(\psi_1, \ldots, \psi_m, x_1, \ldots, x_n)$, where ψ_i are formula variables and x_j are term variables. There are function symbols f_n^m allowing us to form $f_n^m(\psi_1, \ldots, \psi_m, x_1, \ldots, x_n)$, where ψ_i are formula variables and x_j are term variables. We define now the notions of a *wff* and a *term* of the language **HFP**:

1. **truth** is a formula with no free variables.

2. x is a term with x free. Y is a formula with Y free.

3. If $\psi_i(x_1^i, \ldots, x_{k(i)}^i), i = 1, \ldots, m$, are formulae with free variables x_j^i, $j = 1, ..., k(i)$ and $F_j(\overline{x}_1, \ldots, \overline{x}_{r(j)})$, $j = 1, \ldots, n$ are terms with $\{\overline{x}_1^j, \ldots, \overline{x}_{r(j)}^j\}$ as free variables, R_n^m is a predicate symbol with (m, n) places, and f_n^m is a function symbol with (m, n) places, then:

    (a) $R_n^m(\psi_1, \ldots, \psi_m, F_1, \ldots, F_n)$ is a formula with the free variables $(\overline{x}_j^i, x_s^t, i = 1, \ldots, m, j = 1, \ldots, k(i), t = 1, \ldots, n, s = 1, \ldots, r(t)\}$.

    (b) $f_n^m(\psi_1, \ldots, \psi_m, F_1, \ldots, F_n)$ is a term with the same free variables as in (a).

    The free variables mentioned can be either terms or formulae.

4. If ψ_1, ψ_2 are wffs, then so are $\sim\psi_1, \psi_1 \wedge \psi_2, \psi_1 \rightarrow \psi_2$, and $\psi_1 \vee \psi_2$. $\sim\psi_1$ has the same free variables as ψ_1, and $\psi_1 \wedge \psi_2, \psi_2 \vee \psi_2$, and $\psi_1 \rightarrow \psi_2$ have a set of free variables which is the union of the sets of free variables of ψ_1 and ψ_2.

5. If $\psi(x, y_i)$ is a formula with free variables $\{x, y_i\}$ being either term or formula variables, then $\forall x\psi, \exists x\psi$ are formulae with the free variables $\{y_i\}$.

Example 5.6.2 To give some examples consider the **Hold** predicate. This has the form **Hold** (ψ, t) = 'ψ is true at time t'.

We can now write **Hold**(**Hold**$(\psi, t), s)$

'At time s it was true that ψ was true at t'.

Compare this sentence with **Thought**(**Thought**$(\psi, 0), 5)$:

'At time 5 people thought that at time 0 it was thought that ψ was true.'.

Consider **Thought-F**(ψ, t) which reads:

'At time t, ψ was considered true in the future.'

Thus we can write a rule: if at any time you expect to get a budget for a research assistant, then hire one. (It may be that you never get the budget!):

$$\forall t[\textbf{Thought-F}(\text{budget}, t) \to \textbf{Hire}(t)].$$

We now define several types of models for this language.

Definition 5.6.3 (modal semantics) Let D be a non-empty set. A function h is an assignment for **HFP** iff the following hold:

1. If x is an individual term variable then $h(x) \in D$. If x is a formula variable then $h(x) \subseteq D$.

2. If R_n^m is a predicate symbol then $h(R_n^m)$ is a function with domain $(2^D)^m \times D^n$ and range 2^D.

3. If f_n^m is a function symbol then $h(f_n^m)$ is a function with domain $(2^D)^m \times D^n$ and range D.

4. Given a formula $\psi(x_1, \ldots, x_n)$ with free variables $x_1, \ldots, x_n$ and a term $F(x_1, \ldots, x_n)$ with free variables $x_1, \ldots, x_n$ we define the value $h^*(F) \in D$ and $h^*(\psi) \subseteq D$ as follows:

   (a) $h^*(x) = h(x)$.

   (b) $h^*(\textbf{ truth }) = D$.

   (c) $h^*(R_n^m)(\psi_1, \ldots, \psi_m, F_1, \ldots, F_m) =$
   $h(R_n^m)(h^*(\psi_1), \ldots, h^*(\psi_m), h^*(F_1), \ldots, h^*(F_n))$.

   (d) $h^*(f_n^m(\psi_1, \ldots, \psi_m, F_1, \ldots, F_m) =$
   $h(f_n^m)(h^*(\psi_1), \ldots, h^*(\psi_m), h^*(F_1), \ldots, h^*(F_n))$.

   (e) $h^*(\sim\varphi) = D - h^*(\varphi)$.
   $h^*(\varphi \wedge \psi) = h^*(\varphi) \cap h^*(\psi)$.

(f) $h^*(\forall x\psi(x)) = \bigcap_{d\in D} h^*_{x/d}(\psi(x))$.
where $h_{x/d}$ is the function which differs from h only by having $h_{x/d}(x) = d$,
i.e.
$h_{x/d}(y) = h(y)$ for $y \neq x$.
$h_{x/d}(y) = d$ for $y = x$.
Similarly:
$h^*(\exists x\psi(x)) = \bigcup_{d\in D} h^*_{x/d}(\psi(x))$.

5. (a) A model is a domain and an assignment, (D, h).
   (b) A formula ψ holds in a model if $h^*(\psi) = D$.
   (c) A formula ψ is valid if it holds in all models.

Example 5.6.4

1. Consider classical propositional temporal logic. We can associate with any atomic proposition q a 0-place predicate Q of **HFP**. The domain D is time and assume an order $<$ on D. Q is a $(0,0)$ predicate. Let $\mathbf{F}$ be the $(1,0)$ predicate $\mathbf{F}(Q)$. Let $h_<(\mathbf{F})(Y)$ for $Y \subseteq D$ be the set $Y^+ = \{s \mid \exists y \in Y s < y\}$. Then if $h_<(Q) =$ set where q is true then $h^*_<(FQ)$ is the set of points where Fq is true.

2. Consider the example in (a) above and consider the connective (predicate) $\mathbf{F}(Q, x)$. $\mathbf{F}$ is a $(1,1)$ predicate. It corresponds in the propositional temporal logic to a labelled connective $F^x q$. We may have, for example, $F^x q$ true at t iff $t < x$ and q true at x. In **HFP** the assignment would be:

$$h_<(\mathbf{F})(Y, x) = \{t \mid t < x \text{ and } x \in Y\}.$$

3. Notice that the **Hold** predicate of 5.1 should be of type $\mathbf{Hold}^1_0$, i.e. $\mathbf{Hold}(A)$ if the modal semantics associates a set with it. Thus $\mathbf{Hold}(A) \equiv A$, i.e. $h(\mathbf{Hold}) \equiv$ identity.

The above discussion gave modal semantics to **HFP**. This semantics is adequate for describing the modal and temporal applications of **HFP**, as it talks about truth conditions of connectives. However, it does not seem satsifactory in explaining the way **HFP** predicates are used in practical Prolog programs. Whenever we write in Prolog a clause of the form

$$\mathit{Hold}(\mathit{Demo}(A(x), B(y)) \wedge A(z)) \rightarrow B(z)$$

we are using **HFP** expressions involving predicates of predicates. Sometimes we even write clauses of the form

$$X \to P(X),$$

which play an important role in metaprogramming.

For example, the clause

$$\mathbf{Hold}(\varphi) \text{ if } \varphi$$

may look more familiar to the reader.

The logical status of these phrases and their semantics have to be explained. The main characteristic feature of such use is that the variables in the clauses can unify and change value. Thus it is not immediately obvious whether one can say that in the expression $\varphi(A(x), x), A(x)$ is the name of a formula. If $A(x)$ represents the wff $A(x)$ (i.e. what we really have is $\varphi(`A(x)', x)$, then the '$x$' in $A(x)$ can not change or unify, certainly not in uniformity with the other $x$. In practice of course when '$x$' is unified to be '$a$', we obtain $\varphi(A(a), a)$ and not $\varphi(`A(x)', a)$.

We thus need to explain the above for the case of **HFP**, because, as in Prolog, **HFP** allows for x to be free in $\varphi(A(x), x)$ in both places.

Notice that in the modal semantics the above are interpreted as connectives. So, for example, $\varphi(B, y)$ can be interpreted as the connective saying: *B will always be true but not later than time y*, and $A(x)$ can be interpreted as saying: *Time x is in the future (of now).* Together we form $\varphi(A(x), x)$ to mean: *Time x will always be in the future but not later than time x*, which is really a temporal tautology.

We now proceed to give the *metalevel semantics* for **HFP**, as opposed to the *modal semantics.* Our strategy is to begin with ordinary predicate logic, describe a metalanguage **M** for it and identify **HFP** as part of the metalevel description of predicate logic in **M**.

Definition 5.6.5 Let **L** be predicate logic with predicate symbols $\{P_i, i = 1, 2, 3, \ldots\}$ which are n_i-place, variables $\{x_i, y_i, \ldots\}$ and constants $\{a_i, b_i, \ldots\}$. We refer to **L** as *the object language.* A language **M** is said to be *a metalanguage* for **L** if it is capable of naming the symbols of **L** and contains the following terms:

1. For each symbol of **L** there exists a unique name which is a term of **M**. The naming function is denoted by quotation marks:
    - variables x_i are named 'x_i';
    - constants a_i are named 'a_i';
    - predicate symbols P_i are named 'P_i'.

2. For each P_i, which is an n_i-place predicate, there exists a function $\mathbf{f}_i$ of the metalanguage **M**, satisfying the following:
$\mathbf{f}_i(`P_i\text{'}, `t_i\text{'}, \ldots, `t_{n_i}\text{'}) = `P_i(t_1, \ldots, t_{n_i})\text{'}$
In fact, we can take the above as a definition of the name of $P_i(t_1, \ldots, t_{n_i})$.

3. There exist functions $\mathbf{f}_\wedge, \mathbf{f}_\vee, \mathbf{f}_\rightarrow, \mathbf{f}_\sim, \mathbf{f}_\forall$ and $\mathbf{f}_\exists$ satisfying equations as follows:

   - $\mathbf{f}_\wedge$ ('A', 'B') = '$A \wedge B$';
   - $\mathbf{f}_\vee$ ('A', 'B') = '$A \vee B$';
   - $\mathbf{f}_\rightarrow$ ('A', 'B') = '$A \rightarrow B$';
   - $\mathbf{f}_\sim$ ('A') = '$\sim A$';
   - $\mathbf{f}_\forall$ ('A', 'x') = '$\forall x A(x)$';
   - $\mathbf{f}_\exists$('A', 'x') = '$\exists x A(x)$'.

4. There are functions **Sub**, **s** and **n** satisfying the following, for terms t:
$\mathbf{n}(t)$ = 't',
$\mathbf{s}$('t') = t,
Sub ('$A(x)$', 'x', $\mathbf{n}(t)$) = '$A(t)$',
where t is any term and 't' is its name.

Note that **Sub** is a substitution function. We can assume that $\mathbf{f}_i$, $\mathbf{f}_\wedge$, $\mathbf{f}_\vee$, $\mathbf{f}_\sim$, $\mathbf{f}_\exists$, and $\mathbf{f}_\forall$ are actual function symbols of **M** and thus actually define the terms which name the formulae. **Sub** cannot be assumed as a function symbol because then a formula will have two names. So we have to assume that **Sub** is somehow definable in **M**. An alternative is to take **Sub** as a function symbol and add axioms to **M** to make things right.

Lemma 5.6.6 *Let $\varphi(x_1, \ldots, x_n)$ be a formula, then there exists a term $\mathbf{f}_\varphi(\mathbf{x}_1, \ldots, \mathbf{x}_n)$ of* **M** *with free variables $\mathbf{x}_1, \ldots, \mathbf{x}_n$ (ranging over names) such that*
$\mathbf{f}_\varphi$ *('a_1', ..., 'a_n') = '$\varphi(a_1, \ldots, a_n)$'*

Proof. By induction on the structure of φ. For atomic $\varphi = P_i(x_1, \ldots, x_{n_i})$, we have

$$\mathbf{f}_\varphi = \text{(def) } \mathbf{f}_i(`P_i\text{'}, \mathbf{x}_1, \ldots, \mathbf{x}_{n_i}).$$

If there exist terms $\mathbf{f}_A$ and $\mathbf{f}_B$ for A and B then $\mathbf{f}_{A\sharp B} = \mathbf{f}_\sharp(\mathbf{f}_A, \mathbf{f}_B)$ for $\sharp$ standing for $\wedge, \vee, \rightarrow$, and similarly for $\sim$.

Let $\varphi = \forall x \psi(x)$. Let $\mathbf{f}_\psi$ be given. Then $\mathbf{f}_\varphi = \mathbf{f}_\forall(\mathbf{f}_\psi, `x\text{'})$. Similarly for $\varphi = \exists x \psi(x)$. ∎

Let $\mathcal{M}$ be a model of **M**. Let **Hold** be a unary predicate of **M**. We can derive from $\mathcal{M}$ and **Hold** a model $\mathcal{A}$ of **L** by defining

$$\mathcal{A} \vDash P(a_1, \ldots, a_n) \text{ iff (def) } \mathcal{M} \vDash \mathbf{Hold}(`P(a_1, \ldots, a_n)').$$

We called **Hold** a Hold predicate because we expect that for any $\varphi(a_1, \ldots, a_n)$ we have;

$$\mathcal{A} \vDash \varphi(a_1, \ldots, a_n) \text{ iff } \mathcal{M} \vDash \mathbf{Hold}(`\varphi(a_1, \ldots, a_n)').$$

This will not be the case unless we require axiomatically in **M** that **Hold** satisfies some axioms, for example:

$$\mathbf{Hold}(`A') \wedge \mathbf{Hold}(`B') \leftrightarrow \mathbf{Hold}(\mathbf{f}_\wedge(`A', `B')).$$

and similarly for the other connectives and quantifiers.

Note that the quantifier axioms are:

$$\forall \mathbf{x}\mathbf{Hold}(\mathbf{f}_{\varphi(x)}(\mathbf{x})) \leftrightarrow \mathbf{Hold}(\mathbf{f}_{\forall x \varphi(x)}).$$

$$\exists x \mathbf{Hold}(\mathbf{f}_{\varphi(x)}(\mathbf{x})) \leftrightarrow \mathbf{Hold}(\mathbf{f}_{\exists x \varphi(x)}).$$

Remark 5.6.7 Note that if we have **Sub** then we can define the diagonal function by:

Diag (‘$A(x)$’, ‘x’) = def **Sub** (‘$A(x)$’, ‘x’, $\mathbf{f}_\sim$ (‘$A(x)$’)).

Recall that using the **Diag** function one can prove Tarski's inconsistency theorem, that there exists no truth predicate for **M** in **M**. Thus we can not have a **Hold** predicate satisfying the axiom

Hold (‘A’) $\leftrightarrow A$,

and allow **Diag** to apply to **Hold** as well, because these conditions yield Tarski's inconsistency theorem. We need not worry, however, about this aspect in this book. First, our aim is to provide semantics for **HFP** and Prolog practice in order to undertake our modal and temporal logic programming, rather than construct self-reflective languages which have their own truth predicates. So we do not mind having the **Hold** predicate not apply to itself. See the Perlis papers, [Perlis, 1985] and [Perlis, 1988], for a good coverage of the subject. Second, in the Horn clause fragment of **HFP** negation is not available and it is possible to have a truth predicate in the object level for the object language itself, without leading to contradiction. The worst we can get are loops, if we use negation as failure. We cannot diagonlize with A(‘$\sim A$’), with $\sim$ being negation as failure because its meaning is procedural. If we define a **fail** connective via a program Δ, then we can only have $\mathbf{fail}_\Delta$ as a connective and we obtain A(‘$\mathbf{fail}_\Delta A$’) for the diagonal function. In this case we obtain stratification of **fail** and we do not have a universal object level diagonalization.

We shall not pursue this topic any further, but will continue with the business at hand.

We are now in a position to explain what a formula of **HFP** stands for.

Suppose we name each letter x, P_i, a by itself. This can technically mean that we identify **L** as part of **M**. Thus P_i, x_i, and a_i are already terms of **M**. We let 'a_i'$= a_i$, 'P_i'$= P_i$, and 'x'$= x$ in **M** for these particular symbols. We can define predicates **Pred**(t), **Const**(t), **Var**(t), **Wff**(t), and **Free** of **M** with the axioms:

- $\mathbf{Const}\ (a_i), i = 1, \ldots.$
- $\mathbf{Var}\ (x_i), i = 1, \ldots.$
- $\mathbf{Pred}\ (P_i), \ldots.$
- $\mathbf{Wff}\ (\mathbf{f}_i(P_i, a_1, \ldots, a_{n_i})).$
- $\mathbf{Wff}\ (\mathbf{f}_i(P_i, , x_1, \ldots, x_{n_i})).$
- $\mathbf{Free}\ (\mathbf{f}_i(P_i, , x_1, \ldots, x_{n_i}), x_j).$
- $\mathbf{Wff}\ (A) \wedge \mathbf{Wff}(B) \rightarrow \mathbf{Wff}(\mathbf{f}_\sharp(A, B)).$
- $\mathbf{Wff}\ (A) \wedge \mathbf{Free}(A, x) \rightarrow \mathbf{Free}(\mathbf{f}_\sharp(A, \mathbf{y}), x),$
  where **y** is a metavariable and $\sharp$ is one of $\wedge, \vee, \rightarrow, \sim$.
- $\mathbf{Wff}\ (A) \wedge \mathbf{Free}(A, x) \rightarrow \mathbf{Wff}(\mathbf{f}_\forall(A, x)).$
- $\mathbf{Wff}(A) \wedge \mathbf{Free}(A, x) \rightarrow \mathbf{Wff}(\mathbf{f}_\exists(A, x)).$
- $\mathbf{Wff}(A) \wedge \mathbf{Free}(A, \mathbf{y}) \wedge x \neq \mathbf{y} \rightarrow \mathbf{Free}(\mathbf{f}_\forall(A, x), \mathbf{y}),$
  where **y** is a metavariable.

Having named formulae by themselves we can essentially regard the functions $\mathbf{f}_i(P_i, x_1, \ldots, x_{n_i}), \mathbf{f}_\wedge, \mathbf{f}_\rightarrow$, etc., to be the connectives themselves, taken as functions. So $\mathbf{f}_\varphi$('x_i', ..., 'x_n') will be '$\varphi(x_1, \ldots, x_n)$', which is $\varphi(x_1, \ldots, x_n)$.

If we consider $\mathbf{f}_\varphi(\mathbf{x}_1, \ldots, \mathbf{x}_n)$ with $\mathbf{x}_i$ metavariables ranging over names, then in the convention of expressions of **L** naming themselves, we obtain that it is equal to

$$\varphi(\mathbf{x}_1, \ldots, \mathbf{x}_n).$$

Thus we can unify and change $\mathbf{x}_i$ since they are metavariables.

Consider $\varphi(\mathbf{x}, \mathbf{y})$. This is really a term of **M**. We can certainly substitute in it the term $A(\mathbf{y})$. We thus obtain $\varphi(A(\mathbf{y}), \mathbf{y})$ which is an expression of **HFP**. It really stands for the term $\mathbf{f}_\varphi(\mathbf{f}_A(\mathbf{y}), \mathbf{y})$ of the metalanguage.

Satisfaction of $\varphi(A(\mathbf{y}), \mathbf{y})$ means satisfaction of

$$\mathbf{Hold}(\mathbf{f}_\varphi(\mathbf{f}_A(\mathbf{y}), \mathbf{y})).$$

Let us examine more closely the connection between the satisfaction of $A(y)$ and $\varphi(A(y), y)$. The model $\mathcal{A}$ has as its domain D terms of the metalanguage **M** which are names of wffs of **L**. $A(y)$ holds in $\mathcal{A}$ iff $\mathbf{Hold}(A(y))$ holds in $\mathcal{M}$.

The set $\{y \in D \mid \mathcal{M} \vDash \mathbf{Hold}(A(y))\}$ is the denotation of 'A' and is a subset of the domain D. The denotation of 'φ' is similarly a subset of $D \times D$. '$A(y)$' is a term in D. $\mathcal{A} \vDash A(A(y))$ may or may not hold, i.e. $\mathcal{M} \vDash \mathbf{Hold}(A(A(y)))$ may or may not hold. There is *no* connection in general between the denotation of 'A' and the term '$A(y)$'.

Thus in the model $\mathcal{M}$, and consequently in $\mathcal{A}$, the predicate 'A' has two 'manifestations' or two 'footprints', one as $\{y \in D \mid \mathcal{M} \vDash \mathbf{Hold}(A(y))\}$ and one as a term 'A' in D. **Hold** ('A') also has a value in $\mathcal{M}$, so does $\mathbf{Hold}$('A('A')'), etc. These values are all independent. An **HFP** axiom of the form $X \rightarrow B(X)$ means $\mathbf{Hold}\ (X) \rightarrow \mathbf{Hold}(B(X))$, which is a connection axiom of the **Hold** in **M**. For example, $\mathbf{Hold}(A(y)) \rightarrow \mathbf{Hold}(A(A(y)))$.

We can in **M** read 'A' as representing the set $\{y \mid \mathcal{A} \vdash A(y)\}$. Terms of the form $\mathbf{f}_\varphi$('A', $\mathbf{y}$) can be written in the metalanguage and we can understand them in the object language as $\varphi(\lambda x A(x), y)$.

Example 5.6.8 The modal and temporal semantics provides us with examples of the above situation. A formula may be true at points of the interval $t = [1, 2]$ of time, but whether it is true at the interval itself is another matter. For example, it is not true that one is crossing the Atlantic at any moment at the interval $[1, 2]$, but one may be crossing the Atlantic at $[1, 2]$. Connections between values at intervals and values at points are usually given as persistence criteria. For example, the criterion *if φ holds at an interval it holds at all subintervals*, can be expressed using our language, through such axioms as:

$$\varphi(\text{'}A\text{'}, y) \rightarrow \forall x \varphi(A(x), y)$$

or

$$\forall x(A(x) \rightarrow B(x)) \rightarrow [\varphi(\text{'}B\text{'}, y) \rightarrow \varphi(\text{'}A\text{'}, y)].$$

The above freedom we still have with terms allows us to extend the expressive power of **HFP**. First let us take any binary atomic predicate of **HFP**; let us call it $P(X,Y)$: X,Y formula variables.

Form the formulae

$$P^2(A_1, A_2) = \text{(def) } P(A_1, A_2).$$

$$P^{n+1}(A_1, \ldots, A_{n+1}) = \text{(def) } P(P^n(A_1, \ldots, A_n), A_{n+1}).$$

In the language **M**, there are no axioms connecting $\mathbf{Hold}(P(X,Y))$ with $\mathbf{Hold}(P(P(X,Y),Z))$ and so in $\mathcal{M}$, $\mathbf{Hold}(P^n(X_1,\ldots,X_n))$ has no connection with $\mathbf{Hold}(P^m(X_1,\ldots,X_m))$ for $m \neq n$. However, this trick allows us to regard 'P' as a sequence predicate (i.e. without a fixed number of places) with $P(X_1,\ldots,X_n)$ holding in $\mathcal{A}$, iff $\mathbf{Hold}(P^n(X_1,\ldots,X_n))$ holds in $\mathcal{M}$.

We can thus write our predicate as a set of sequences $(P, X_1, \ldots, X_n)$, where n is arbitrary.

We can similarly treat function symbols f and write $(f, x_1, \ldots, x_n)$. If we do the above for any pure formula predicate of the language **HFP**, we can identify the following Lisp-like fragment of **HFP**.

Definition 5.6.9 (the sublanguage HFP$_{Lisp}$) The sublanguage contains formula variables $X_1, X_2, \ldots$, term variables and Lisp-predicate letters $P_1, P_2, P_3, \ldots$.

The general notion of a wff of $\mathbf{HFP_{Lisp}}$ with free variables is defined in the same way as in the case of definition 5.6.1 of **HFP**, except that we replace the formation rule for atoms 5.1.6(3) by the following formation rule ($3_{\mathbf{Lisp}}$): If n is arbitrary and $\varphi_1, \ldots, \varphi_n$ are formulae, or predicate letters, or formula variables, and P is a predicate letter, then $(P, \varphi_1, \ldots, \varphi_n)$ is a formula. We allow the case $n = 0$ in which case (P) is a formula.

The reader should note the translation $*$ from $\mathbf{HFP_{Lisp}}$ into **HFP**. Each P of $\mathbf{HFP_{Lisp}}$ is read in **HFP** as a binary predicate:

$$(P)^* = P(\mathbf{truth}, \mathbf{truth}))$$

$$(P, \varphi)^* = P(\mathbf{truth}, \varphi^*) \text{ and}$$

$$(P, \varphi_1, \ldots, \varphi_n)^* = P^n(\varphi_1^*, \ldots, \varphi_n^*).$$

HFP can be translated into a fragment of $\mathbf{HFP_{Lisp}}$ via the translation $\sharp$:

$$P(\varphi_1, \ldots, \varphi_n)^\sharp = (P, \varphi_1^\sharp, \ldots, \varphi_n^\sharp).$$

We are now interested in defining the Horn-clause fragment of **HFP**.

Definition 5.6.10 (Horn clause fragment of HFP) We follow closely the defining clauses of definition 5.6.1. Read 'wff' as 'wff in the the Horn fragment':

1. Same as in definition 5.6.1.

2. Same as in definition 5.6.1.

3. Same as in definition 5.6.1 with the restriction that the formulae involved (i.e. $\varphi_1, \ldots, \varphi_n, \psi_1, \ldots, \psi_m, F_1, \ldots, F_n$) are all taken from the Horn fragment.

4. (a) If ψ_1 and ψ_2 are in the Horn fragment, so is $\psi_1 \wedge \psi_2$.

   (b) If ψ_i, φ are atomic, i.e. the result of applying clause (3), then $\wedge\psi_i \rightarrow \varphi$ is in the Horn fragment.

5. Same as in definition 5.6.1.

To define the Horn clause fragment of $\mathbf{HFP}_{\mathbf{Lisp}}$ we follow definition 5.6.9 and replace clause (3) above by the rule $3_{\mathbf{Lisp}}$ restricting the formulae $\phi_1, \ldots, \phi_n$ to the Horn fragment.

The next question to ask is:

Do we have an implementation of the Horn fragment?

The answer is, surprisingly, yes, we do in fact have an old implementation, as the next remark shows.

Remark 5.6.11 The reader can see Clark and McCabe's book [Clark and McCabe, 1984] on Micro-PROLOG. Chapter 9 describes exactly the Micro-PROLOG system language which is the Horn clause fragment of $\mathbf{HFP}_{\mathbf{Lisp}}$. Thus there exist implementations of this language. Micro-PROLOG on the IBM PC gives access to this language. Later versions of LPA-Prolog concentrated on supporting Edinburgh syntax and therefore the original Micro-PROLOG syntax has been suppressed in these later versions.

Another implementation for the full **HFP** is essentially the rewrite language **PLL** of E. Babb [Babb, 1990].

5.7 Term semantics for HFP

The discussion in the preceding section showed that if we name in **M** elements of **HFP** by themselves, we obtain that all formulae of **HFP** are terms in **M**. The term functions $\mathbf{f}_i, \mathbf{f}_\wedge, \mathbf{f}_\vee, \mathbf{f}_\rightarrow, \mathbf{f}_\forall, \mathbf{f}_\sim, \mathbf{f}_\exists$, and **Sub** of definition 5.6.5 all become familiar operations:

1. The naming function λx'x' is the identity.

2. $\mathbf{f}_i(P_i, t_1, \ldots, t_{n_i}) = P_i(t_1, \ldots, t_n)$.
   Thus $\mathbf{f}_i$ is the P_i application function.

3. $\mathbf{f}_\wedge$ is conjunction $\wedge$
   $\mathbf{f}_\vee$ is disjunction $\vee$
   $\mathbf{f}_\rightarrow$ is implication $\rightarrow$
   $\mathbf{f}_\sim$ is negation $\sim$
   $\mathbf{f}_\forall$ is universal quantification
   $\mathbf{f}_\exists$ is existential quantification.

4. **Sub** is the substitution function.
   s, **n** are the identity function.
   $\mathbf{f}_\varphi$ for a formula φ becomes an application function for φ (just like $\mathbf{f}_i$ is for P_i).

The above functions, however, are functions in **M** and can therefore take any term of **M**. Assume P and Q are two unary predicates of **L**. Their names in **M** are 'P' and 'Q'. Given a term a of **L** its name is 'a'. The application functions $\mathbf{f}_P$ and $\mathbf{f}_Q$ of **M** allow us to form:

$$\mathbf{f}_P(\text{`}P\text{'}, \text{`}a\text{'})$$

and we are assured that:

$$\mathbf{f}_P(\text{`}P\text{'}, \text{`}a\text{'}) = \text{`}P(a)\text{'}.$$

We can, however, formally obtain the following terms of **M**:

1. $\mathbf{f}_P(\text{`}Q\text{'}, \text{`}a\text{'})$.

2. $\mathbf{f}_P(\text{`}P\text{'}, \text{`}Q\text{'})$.

We have no axioms in **M** to tell us what the above should mean. The **Hold** predicate of **M** could be true or false of the above terms without any restrictions. We do have axioms on **Hold** to assure us that, for example,

$$\mathbf{Hold}(\text{`}P(a)\text{'} \wedge \text{`}Q(a)\text{'}) \leftrightarrow \mathbf{Hold}(\text{`}P(a)\text{'}) \wedge \mathbf{Hold}(\text{`}Q(a)\text{'})$$

but nothing more.

To extend our semantic meaning to terms of forms (1) and (2) above, we can first say that $\mathbf{f}_P, \mathbf{f}_Q$ should be the same, namely application functions. We can thus replace the n_i place function $\mathbf{f}_i$ by $\mathbf{App}_i$ and obtain

$$\mathbf{App}_{n_i}(\text{`}P_i\text{'}, t_1, \ldots, t_{n_i}) = \text{`}P_i(t_1, \ldots, t_{n_i})\text{'}.$$

Thus all we need are the functions $\mathbf{App}_1, \mathbf{App}_2, \ldots$. So both $\mathbf{f}_Q$ and $\mathbf{f}_P$ are $\mathbf{App}_1$.

Furthermore, one can code $\mathbf{App}_3(x, y, z)$ as $\mathbf{App}_2(\mathbf{App}_2(x, y), z)$ and thus all one needs is a binary **App** function. Further, there is no reason why we can not take application to be concatenation. Thus $\mathbf{f}_i$ can be read as concatenation function. This also allows us to regard all predicates as unary predicates. $R(x, y)$ is understood either as $R(x)(y)$ or as $R((x, y))$. Probably the former is more convenient.

The above syntactical moves do not necessarily involve new semantical commitments. They can be handled as just a matter of notation. We are still left with the problem of the semantical meaning of $P(Q)$ or $P(P)$. If we understand Q to name the set $\{x \mid Q(x)\}$, then we can understand $P(Q)$ as expressing a property of this set. We can now express persistence conditions like

$$P(Q) \to \forall x(Q(x) \to P(x)).$$

In general, $P(Q)$ is independent of $P(t_i)$, for t_i such that $Q(t_i)$ holds. We can see, however, that we are touching on the problem of λ-calculus models.

We can still go on and ask for the meaning of **App**('a', 'Q'). This question has now more than one option, since we have already agreed to read 'Q' as a name for $\{x \mid Q(x)\}$. We can thus read 'a' as a higher predicate on sets Q. It is true of Q iff it is an element of Q:

$$\mathbf{Hold}(a(Q)) \text{ iff } \mathbf{Hold}(Q(a)).$$

'a' can thus be identified with the set of all its properties: cf. Montague semantics [Dowty *et al.*, 1981], [Thomason, 1974d].

It is now up to us whether to allow in **HFP** (say in an extension which we can call $\mathbf{HFP}^+$) formulae of the form $P(Q)$ or not. If we do, we can still use the functions of the metalanguage **M** but we would need a semantical interpretation, possibly some sort of λ-calculus model. We will not pursue this line of research in this chapter; we leave it to subsequent papers. Prolog practice, which uses **HFP** formulae, does not frequently write such expressions as:

Demo (*Demo*, *Demo*).

If we stick to **HFP** itself and ban the above, then it is easy to construct term models for it.

Let Δ be an **HFP** theory. Formally extend Δ to a saturated theory Δ'. Thus whenever $\Delta' \vdash \exists x A(x)$ then for some terms $t, \Delta \vdash A(t)$ and whenever $\Delta' \vdash A \vee B$ then either $\Delta' \vdash A$ or $\Delta' \vdash B$. We can now define $\mathbf{Hold}_\Delta(A)$ iff $A \in \Delta$ and obtain a model for the **Hold** predicate of **M** as applied to the wffs of **HFP** regarded as **M** terms.

The term interpretation of **HFP** can be understood in another way, as notation for generating predicates. To illustrate our view, consider conjunction. It can generate predicates by letting $[P \wedge Q](x)$ mean $P(x) \wedge Q(x)$. This is done in the same way that $[\sim A](x)$ is $\sim A(x)$ and $[\Box A](x) = \Box A(x)$.

If we push this point of view further we can regard any formula $\varphi(x, y)$ of **HFP** as a functional $\lambda xy\varphi$ which takes formulae A, B to form a new formula $\varphi(A, B)$. This view, although *mathematically* compatible with the term semantics, is not *conceptually* compatible with the intuitive way in which we read Prolog. For example, we read *Demo*(A, B) as B can succeed from A and regard *Demo* (*Demo*, x) as meaningless.

Chapter 6

A general theory of axiomatization for propositional temporal logic

6.1 Introduction

Consider a flow of time $(T, <, =)$ and consider one-dimensional interpretations over this flow. This means that we are considering connectives and tables of the form $\|\sharp(q_1, \ldots, q_n)\|_t^h = 1$ iff $\Psi(t, h(q_1), \ldots, h(q_n))$ holds.

There are many possible connectives and therefore many possible logics. Some connectives are specially distinguished either because they appear in applications or because they represent natural technical features of flows of time. We have mentioned before the connectives Fq, Pq ('It will be the case that q' and 'It was the case that q') as natural connectives to consider. In fact, they should be included in any temporal logic. Other natural connectives are $S(p, q)$ and $U(p, q)$, and the technically important Stavi connectives $S'(p, q)$ and $U'(p, q)$. These are also important from the point of view of expressive power of temporal logic.

The tables for F, P, S, U were given before; they are repeated below together with the tables for the Stavi connectives. Their logical properties differ from one type of flow of time to another. We shall therefore consider axiomatizations of extensions of K_t in the language $\{P, F\}$ and in the languages of $\{U, S\}$ and $\{U, S, U', S'\}$ for a variety of flows of time. This is the task of this chapter.

The flows we consider are ordinary ($<$ irreflexive and transitive), linear, linear dense, rational, integer, real, and other varieties. The axioms are very straightforward when the irreflexivity rule IRR is present, and all completeness proofs use the same process of construction of a canonical model using the notion of IRR-theories.

In the later parts of the chapter we shall consider other connectives besides P, F, S, U, U', S', and will prove a general axiomatization theorem for temporal connectives.

Let us summarize here the main connectives and their tables which will be of use later:

1. Fq reading 'q will be true' and Pq reading 'q was true' have tables:
   $\|Fq\|_t^h = 1$ iff $\exists s > t \|q\|_s^h = 1$.
   $\|Pq\|_t^h = 1$ iff $\exists s < t \|q\|_s^h = 1$.

2. $U(p,q)$ reads: 'q will be true until p is true'.
   $S(p,q)$ reads: 'q has been true since p was true'.

   Tables for U, S are:
   $\|U(p,q)\|_t^h = 1$ iff $\exists s > t[\|p\|_s^h = 1$ and
   $\forall x(t < x < s \to \|q\|_x^h = 1)]$.
   $\|S(p,q)\|_t^h = 1$ iff $\exists s < t[\|p\|_s^h = 1$ and
   $\forall x(s < x < t \to \|q\|_x^h = 1)]$.

3. The connectives F, P are definable from U, S:
   $Fq = U(q, q \to q)$.
   $Pq = S(q, q \to q)$.
   We define G and H by
   $Gq = {\sim}F{\sim}q$.
   $Hq = {\sim}P{\sim}q$.

   The following unary connectives are definable from U and S and are very useful in proofs. They are

   $$K^+q = {\sim}U(q \to q, {\sim}q).$$
   $$K^-q = {\sim}S(q \to q, {\sim}q).$$

   $K^+q =$ 'q will be true in the future at points arbitrarily near':

   $$\|K^+q\|_t^h = 1 \text{ iff } \forall s > t \exists x(t < x < s \wedge (\|q\|_x^h = 1)).$$

   $K^-q =$'q was true in the past at points arbitrarily near':

   $$\|K^-q\|_t^h = 1 \text{ iff } \forall s < t \exists x(s < x < t \wedge (\|q\|_x^h = 1)).$$

Not all connectives are always definable using S and U. We shall see in chapter 11 that both the Stavi connectives and also the following connective $\gamma^{-}q$ are not.

4. Let $\gamma^{-}(q)$ be the connective saying that the situation in figure 6.1 holds in the past of t.

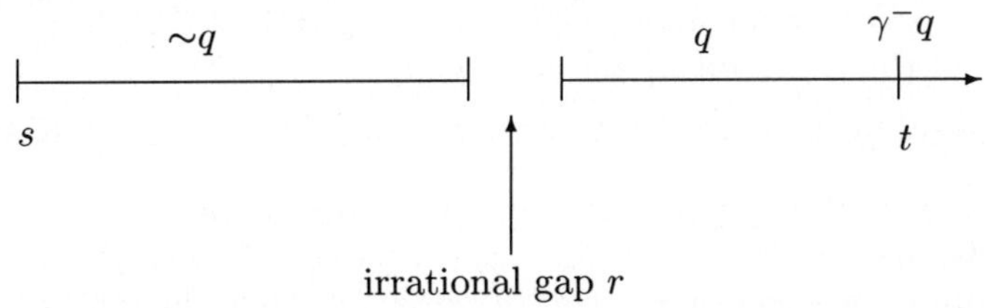

Figure 6.1

We have $\|\gamma^{-}q\|_t^h = 1$ iff

$$\exists s < t\{\exists y(s < y < t \wedge Q(y)) \wedge \exists y(s < y < t \wedge {\sim}Q(y)) \\ \wedge \forall y\{s < y < t \to [Q(y) \leftrightarrow \forall z(y < z < t \to Q(z))] \\ \wedge [{\sim}Q(y) \leftrightarrow \forall z(s < z < y \to {\sim}Q(z))]\}\},$$

where $Q(z) = \{z \mid \|q\|_z^h = 1\}$.

As we shall see in chapter 11, Kamp has shown in [Kamp, 1968b] that over rational time $\gamma^{-}q$ is not definable using $\{U, S\}$. Over real or integer time γ^{-} is definable: it is false.

5. The Stavi connectives. These connectives U' and S' were introduced by Stavi in [Stavi, 1979] and used in a result in [Gabbay *et al.*, 1980], of which we will see more in chapter 12. $U'(p, q)$ is true now iff there is a gap in the future such that

   - q is true until this gap;
   - p is true from the gap into the future for some uninterrupted stretch of time;
   - and in the future of the gap q is false arbitrarily close to the gap.

   If we draw a picture we obtain figure 6.2.

   S' is the mirror image of U', i.e. going into the past instead of into the future. $S'(p, q)$ is true now iff there is a gap in the past such that

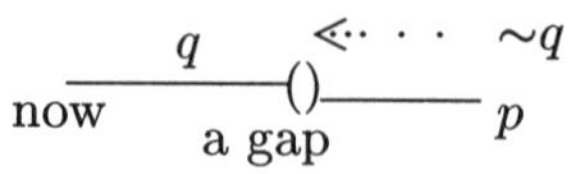

Figure 6.2

- q was true since this gap;
- p was true for some uninterrupted stretch of time from the past of the gap right up until the gap;
- and in the past of the gap q is false arbitrarily close to the gap.

Despite involving gaps, these are first-order connectives. Let $\varphi(t, Q)$ be the first-order table of $U(\top, q) \wedge {\sim}U({\sim}q \vee K^+{\sim}q, q)$. (Evidently, this formula is equivalent to $U'(\top, q) \vee Gq$.) Then the table of $U'(p, q)$ can be written as the conjunction of $\varphi(t, Q)$ and

$$\exists s(s > t \wedge {\sim}Q(s) \wedge \forall v([v < s \wedge \exists s(t < s < v \wedge {\sim}Q(s))] \rightarrow P(v))).$$

Notice that we re-used the variable s; this idea will be important in chapter 13.

6.2 The system K_t formulated with the irreflexivity rule

Consider the following axiom system for a language with F, P only.

Definition 6.2.1 Let K_t be the system of chapter 3, section 3.1 with the additional rule IRR. Recall the following:

1. $K_t \vdash A$ iff for some sequence of wffs $A_1, \ldots, A_n$ we have that each A_i is either an axiom or is obtained from A_j, A_k $(j, k < i)$ via one of the rules of K_t.

2. For Δ a set of wffs, we write $\Delta \vdash_{K_t} A$ iff for some $B_1, \ldots, B_n \in \Delta$ we have $K_t \vdash (\bigwedge_i B_i) \rightarrow A$.

3. Δ is said to be K_t-consistent iff we do not have $\Delta \vdash_{K_t} \bot$. In particular, A is consistent iff $K_t \nvdash {\sim}A$.

4. Δ is said to be complete iff for all A, $A \in \Delta$ or ${\sim}A \in \Delta$.

5. Let Δ, Γ be two (complete) theories. Define
$\Delta < \Gamma$ iff for all A, $GA \in \Delta \rightarrow A \in \Gamma$;
$\Delta > \Gamma$ iff for all A, $HA \in \Delta \rightarrow A \in \Gamma$.

Exercise 6.2.2 Show that the following holds for K_t:

1. If Δ is consistent and q is a new atom not occurring in Δ, then $\Delta \cup \{\sim q \wedge Hq\}$ is consistent. In particular, if A is consistent, then for any atom q not in A, $A \wedge \sim q \wedge Hq$ is also consistent.

2. For any $(T, <)$ and any A, if $K_t \vdash A$, then A is valid in $(T, <)$.

3. Let Δ be a K_t-consistent theory. Show that Δ can be extended to a complete and consistent theory $\Delta' \supseteq \Delta$. The construction for finding Δ' uses classical axioms only.

4. Let Δ, Γ be two complete, consistent theories. Show that $\Delta < \Gamma$ iff $\Gamma > \Delta$ (see definition 6.2.1).

The usual axiomatizations of temporal systems do not take the irreflexivity rule IRR. This rule is not provable, in general, from the other axioms and rules, but in most of the cases we consider it is a conservative addition.

We need it to characterize irreflexivity, i.e. the condition $\forall t \sim(t < t)$ in the time structure. When this rule is available it is inconsistent to add $HA \to A$ as an axiom, since we will obtain $\sim q \wedge Hq \to \bot$, and hence, by IRR, $\vdash \bot$.

What do we mean when we say that IRR is derivable from the other axioms and rules of K_t? It simply happens to be true, being a universal statement of the form 'if $\vdash A$ then $\vdash B$'. But if we add $HA \to A$ to the other axioms (without IRR), it ceases to be true. However, if we adopt IRR officially as a rule (property) of the system, it must be maintained as a rule of the system in any extension (by monotonicity). Thus with IRR as a rule, there is no consistent extension with the axiom schema $HA \to A$.

We are about to see the important rôle that IRR plays in temporal logic. We are justified in adopting it because irreflexivity is an underlying property of all flows of time (the case of transitive circular time is exceptional but then we drop the rule), and also because it makes all proofs much simpler. However, there are some disadvantages to adopting it. In a sense, it introduces variables 'by the back door'. Also, some nice mathematical properties are lost, such as: if K is a finitely axiomatizable class then K is closed under ultraproducts.

We now describe a Henkin-type construction which will be used in all future completeness proofs. For this reason we assume we are dealing with an axiom system Z which contains all the axioms and rules of K_t, and perform the construction for Z. This way we can refer to this construction in the future. The time structure $(T, <, h)$ which will be constructed for Z will be called the *canonical time structure* for Z.

We begin with a few definitions and lemmas. First we define IRR-theories to be those which contain a 'name' of the form $\sim q \wedge Hq$ and also similarly name any other time referred to by some zig-zagging sequence of Fs and Ps.

Definition 6.2.3 A theory Δ is said to be an IRR-theory if (a) and (b) hold:

(a) For some q, $\sim q \wedge Hq \in \Delta$.

(b) Whenever $X = \Diamond_1(A_1 \wedge \Diamond_2(A_2 \wedge \ldots \wedge \Diamond_n A_n)) \cdots) \in \Delta$, then for some new atom q, $X(q) = \Diamond_1(A_1 \wedge \Diamond_2(A_2 \wedge \ldots \wedge \Diamond_n(A_n \wedge \sim q \wedge Hq)) \cdots) \in \Delta$, where $\Diamond_i$ is either P or F.

Lemma 6.2.4 *Let A be any Z-consistent wff. Then there exists a complete, Z-consistent IRR-theory, Δ, such that $A \in \Delta$. In fact, if Δ_0 is any Z-consistent theory such that an infinite number of atomic propositions q_i do not appear in Δ_0, then there exists an IRR Z-consistent and complete theory $\Delta \supseteq \Delta_0$.*

Proof. Let $B_0, B_1, \ldots$ be an enumeration of all wffs of the language, where all wffs of the form X of definition 6.2.3 are especially enumerated as such in the odd places. We construct an increasing sequence of finite Z-consistent Δ_n such that $\bigcup_n \Delta_n = \Delta$ is the desired theory.

First note that by rule IRR and exercise 6.2.2, for any q not appearing in $A, \{A \wedge Hq \wedge \sim q\}$ is consistent. Let Δ_0 be this set.

Assume Δ_n was defined, $n \geq 0$, and consider B_n. Clearly, either $\Delta_n \cup \{B_n\}$ or $\Delta_n \cup \{\sim B_n\}$ is Z-consistent. If $\Delta_n \cup \{\sim B_n\}$ is consistent then take $\Delta_{n+1} = \Delta_n \cup \{\sim B_n\}$. Otherwise, if n is even take $\Delta_{n+1} = \Delta_n \cup \{B_n\}$. Otherwise, B_n has the form X of definition 6.2.3 and $\{B_n\} \cup \Delta_n$ is consistent. Let q be a new atom not appearing in Δ_n or B_n and let $\Delta_{n+1} = \Delta_n \cup \{B_n(q)\} \cup \{B_n\}$. In view of the IRR-rule, exercise 6.2.2 and theorem 3.2.4, Δ_{n+1} is Z-consistent. Note that $\vdash B_n(q) \to B_n$.

$\Delta = \bigcup_n \Delta_n$ is indeed the desired theory. ■

Lemma 6.2.5 *Let Δ be a Z-consistent complete IRR-theory. Let $FA \in \Delta$ ($PA \in \Delta$ respectively). Then there exists a Z-consistent complete IRR-theory Γ such that $A \in \Gamma$ and $\Delta < \Gamma$ ($\Delta > \Gamma$ respectively). [$<, >$ were defined in definition* 6.2.1].

Proof. Since $FA \in \Delta$, for some q we have $F(A \wedge {\sim}q \wedge Hq) \in \Delta$. Let $A' = {\sim}q \wedge Hq \wedge A$ and let $\Gamma_0 = \{A'\} \cup \{X \mid GX \in \Delta\}$; we show that Γ_0 is Z-consistent. For, otherwise, for some X_i,

$$\vdash \bigwedge_i X_i \to {\sim}A',$$

hence

$$\vdash \bigwedge_i GX_i \to G{\sim}A',$$

and hence $G{\sim}A' \in \Delta$, a contradiction, since $FA' \in \Delta$.

We now turn to use the sequence B_n of the preceding lemma to define an increasing sequence of Z-consistent extensions $\Gamma_n \subseteq \Gamma_{n+1}$.

Assume $\Gamma_0 \subseteq \Gamma_n$ has been defined and is Z-consistent. Then either $\Gamma_n \cup \{B_n\}$ or $\Gamma_n \cup \{{\sim}B_n\}$ is Z-consistent. If $\Delta_n \cup \{{\sim}B_n\}$ is consistent then take $\Delta_{n+1} = \Delta_n \cup \{{\sim}B_n\}$. Otherwise, if n is even take $\Delta_{n+1} = \Delta_n \cup \{B_n\}$. Otherwise, again B_n has the form X of the definition 6.2.3, and $\{B_n\} \cup \Delta_n$ is consistent. Clearly,

$$F(A' \wedge \bigwedge(\Gamma_n - \Gamma_0) \wedge B_n) \in \Delta.$$

For otherwise, $G(A' \wedge \bigwedge(\Gamma_n - \Gamma_0) \to {\sim}B_n) \in \Delta$ and hence $\Gamma_n \vdash {\sim}B_n$. Therefore, for some q,

$$F(A' \wedge \bigwedge(\Gamma_n - \Gamma_0) \wedge B_n(q)) \in \Delta$$

(where we use the notation of definition 6.2.3). Let $\Gamma_{n+1} = \Gamma_n \cup \{B_n(q)\} \cup \{B_n\}$.

Thus the sequence Γ_n, $n = 0, 1, \ldots$ is defined and, clearly, $\Gamma = \bigcup_n \Gamma_n$ is the desired Z-consistent and complete IRR-theory Γ.

A similar lemma holds for the case of $PA \in \Delta$. ■

Theorem 6.2.6 (completeness theorem) *For any A, $K_t \vdash A$ iff A is valid in any $(T, <)$.*

Proof. One direction was an exercise. We show that if A is consistent (i.e. $\nvdash {\sim}A$), then for some $(T, < h)$ and $t \in T$, $\|A\|_t^h = 1$. Let A be a consistent sentence. Let T be the set of all IRR-theories. T is ordered by the relation $<$ on theories introduced in definition 6.2.1. Define h by letting for atomic q, $h(q) = \{t \in T \mid q \in t\}$.

We call $(T, <, h)$ the *canonical time structure* for K_t. Clearly, the construction can be made analogously for any system $Z \supseteq K_t$ to give a canonical time structure for Z.

Lemma 6.2.7 *Let $Z \supseteq K_t$ be any consistent system and let $(T, <, h)$ be the canonical time structure for Z. Then for all A we have:*

$$\|A\|_t^h = 1 \text{ iff } A \in t.$$

Proof. By induction on A. The case of atomic q follows from the definition, and the cases of the classical connectives are immediate. If $FA \in t$, then by lemma 6.2.5, for some $s \in T$ we have $t < s$ and $A \in s$; hence $\|A\|_s^h = 1$ by the induction hypothesis and therefore $\|A\|_t^h = 1$.

Let $\sim FA \in t$, i.e. $G{\sim}A \in t$, then for any s, $t < s$ implies by definition $\sim A \in s$, hence by the induction hypothesis $\|A\|_s^h = 0$ and so by definition $\|FA\|_t^h = 0$. We can similarly verify the clause for P, using the fact (exercise 6.2.2(4)) that $\Delta < \Gamma$ iff $\Gamma > \Delta$. ■

Lemma 6.2.8

1. *If $K_t \not\vdash \sim A$ then there is $t \in T$ such that $A \in t$.*

2. *$(T, <)$ for K_t satisfies the conditions of irreflexivity and transitivity.*

Proof.

1. Follows from lemma 6.2.4.

2. Irreflexivity follows from the fact that for all $t \in T$ there is an atom q with $\|{\sim}q \wedge Hq\|_t^h = 1$. Transitivity follows from the axiom $GA \rightarrow GGA$ of K_t and the definition of $<$.

■

The completeness theorem follows from the lemmas. ■

Remark 6.2.9

(a) The above was a Henkin construction for a canonical structure for the logic Z. One can give a semantic tableaux construction. The theory Γ constructed in the proof of lemma 6.2.5 is not unique. We can associate arbitrarily with each Z-consistent and complete IRR-theory and each $FA \in \Delta$ a theory $\Gamma_{FA} = g(\Delta, FA)$, as promised by lemma 6.2.5. Similarly, we can associate with each $PA \in \Delta$ a theory $\Gamma_{PA} = g(\Delta, PA)$. g denotes the arbitrary choice function. We can now build, for any consistent sentence A_0, a canonical Henkin structure of theories (dependent on the arbitrary choice function g). First extend $\{A_0\}$ to a complete and consistent IRR-theory Δ_0, as promised by lemma 6.2.4. Now, as in the 'downward Löwenheim–Skolem theorem', let $T = T_g$ be the smallest set of theories such that $\Delta_0 \in T$ and whenever $\Delta \in T$ and $FB \in \Delta$ (or $PB \in \Delta$), then also $g(\Delta, FB)$ (and $g(\Delta, PB)$) are in T. The resulting $(T, <)$ can be taken as the canonical model for Δ_0. If we do not take a choice function g but let T be the set of all complete and consistent IRR-theories we obtain an uncountable set T.

(b) Note that to any logic Z in which $HA \to A$ is not a theorem (axiom), the irreflexivity rule condition can be satisfied (i.e. IRR-theories can be constructed). In fact, any candidate axiom of the form φ which is not a theorem can be treated in the same manner, and one can assume that all theories contain instances of $\sim\varphi$. For the details, see section 6.6 below.

6.3 Extensions of K_t

We can extend K_t with other axioms, depending on what our conception is of the flow of time. The completeness proof for each of the extensions $A \geq K_t$ runs along the same lines as in the preceding section. We add a condition **C** on the flow of time and find a corresponding axiom A**C** to add to K_t. Thus we obtain a system Z**C**. We verify that the axiom A**C** is valid in any flow of time satisfying **C**. To show the other direction, i.e. that if B is a Z**C**-consistent wff, then for some flow of time $(T, <, h)$ satisfying **C**, and $t \in T$; $\|B\|^h_t = 1$, we proceed in the same way as in the case of K_t. We construct the canonical flow $(T, <, h)$ of IRR-theories for Z**C**. This flow is what we need for B (i.e. $B \in t$ for some $t \in T$). We only have to check that the condition **C** holds for $(T, <)$. To show this we use the fact that T is a set of IRR-theories containing every instance of the axioms A**C**. The axioms are used to show that condition **C** holds.

This procedure works in many cases. In some cases we need to obtain from $(T, <, h)$ another flow, $(T^*, <^*, h^*)$. We use the extra properties of $(T, <)$ to show that $(T^*, <^*)$ fulfils condition **C**.

Definition 6.3.1 Define the following extensions of K_t.

(*a*) Extension for linear time without endpoints:

(a_1) $P\top$;

(a_2) $F\top$;

(a_3) $G(A \wedge GA \to B) \vee G(B \wedge GB \to A)$
$H(A \wedge HA \to B) \vee H(B \wedge HB \to A)$.

(*b*) Extension for linear dense time: add to the axioms of (a) the following:
$FA \to FFA$
$PA \to PPA$
(note that the above axioms yield density only in the presence of the axioms for linearity).

(*c*) Extensions for Dedekind complete real time: add to the system for linear dense time (b) the following axioms:
$FGA \wedge F{\sim}A \wedge G({\sim}A \to F{\sim}A) \to F(GA \wedge {\sim}PGA)$
$PHA \wedge P{\sim}A \wedge H({\sim}A \to P{\sim}A) \to P(HA \wedge {\sim}FHA)$.

(*d*) Extensions for discrete linear time: add to the system for linear time the axioms:
$A \wedge HA \to FHA$
$A \wedge GA \to PGA$.

Theorem 6.3.2 *Each of the axiom systems $Za, Zb, \ldots$ is sound and complete for the flows of time suggested by its name, i.e. all the axioms of $Za, Zb, \ldots$ are valid in any $(T, <)$ satisfying the conditions suggested for them, and whenever A is valid in all such $(T, <)$, then A is provable from the respective axioms.*

Proof. Let Z be one of the systems. It is easy to verify that all axioms are valid in the suggested flows of time. We want to show that if A is Z-consistent then for some $(T, <, h)$ in the suggested class of flows and some $t \in T, \|A\|_t^h = 1$. Let $(T, <, h)$ be the canonical structure constructed for Z in the preceding section.

Clearly, there is a $t \in T$ with $\|A\|_t^h = 1$ (actually $A \in t$). It remains to be shown that $(T, <)$ satisfies the required properties, or, if not, to find a modified structure $(T^*, <^*, h^*)$ that does satisfy the required properties.

The axioms yield the following properties of $(T, <)$:

(a_1) $P\top$ yields $\forall t \exists s (s < t)$.

(a_2) $F\top$ yields $\forall t \exists s (t < s)$.

(a_3) Linearity axiom yields linearity. To show this assume we have the situation in figure 6.3, i.e. $t < s$, $t < s'$, $s \neq s'$, $s \not< s'$, $s' \not< s$.

And so for some x, y, z, $x \in s$, $x \notin s'$, $Gy \in s$, $y \notin s'$, $Gz \in s'$, $z \notin s$.

Then this contradicts the axiom of linearity for we can take

$$A = y \vee x, \quad B = {\sim}x \vee z$$

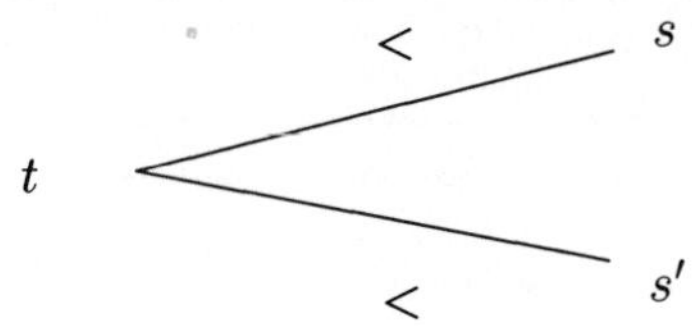

Figure 6.3

and no instance of the axiom for A and B can be in t. A similar argument shows that we cannot have

$$s < t,\ s' < t,\ s \neq s',\ s \not< s',\ s' \not< s.$$

Note that the condition for total linearity, $\forall x, y(x = y \vee x < y \vee y < x)$, is not necessarily valid in a general $(T, <)$. $(T, <)$ may be comprised of several parallel linear flows, but it will be valid in the 'canonical model'. See figure 6.4.

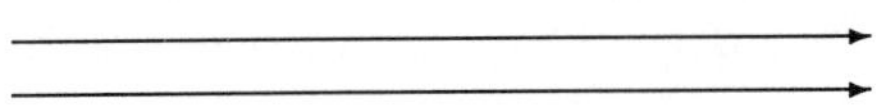

Figure 6.4

(b) The axiom of density $FA \to FFA$ forces $(T, <)$ to be dense. Suppose $t < s$. Let q be such $\sim q \wedge Hq \in s$. Then $F(\sim q \wedge Hq) \in t$ and so $FF(\sim q \wedge Hq) \in t$. Let r be such that $t < r$ and $F(\sim q \wedge Hq) \in r$. Thus we have $t < r \wedge t < s$. We want to show $r < s$. Otherwise $s = r \vee s < r$ by linearity. (Note that we obtain density by using linearity! To obtain density without linearity we need a different axiom.) If $s = r$ then $\sim q \wedge FHq \in s$, a contradiction; and $s < r$ contradicts $\sim q \in s$ and $FHq \in r$.

(c) The axiom system for complete time. Since we have the axioms for linear dense time we know that the canonical time structure $(T, <, h)$ is linear, dense, and without first and last elements, and every $t \in T$ contains every instance of the axiom for complete time. We cannot show that $(T, <)$ itself is complete (in fact for a countable number of atoms T is not complete) but we can add points to T to make it complete (close the gaps) without changing the values, i.e. we construct $T^* \supseteq T$, $<^* \supseteq <$ and h^* extending h such that for any $t \in T$ and any A, the truth value $\|A\|_t^h$ in $(T, <)$ is the same as $\|A\|_t^{h^*}$ in $(T^*, <^*)$. This we now proceed to do.

Let γ denote a gap in $(T, <)$, i.e. we have figure 6.5.

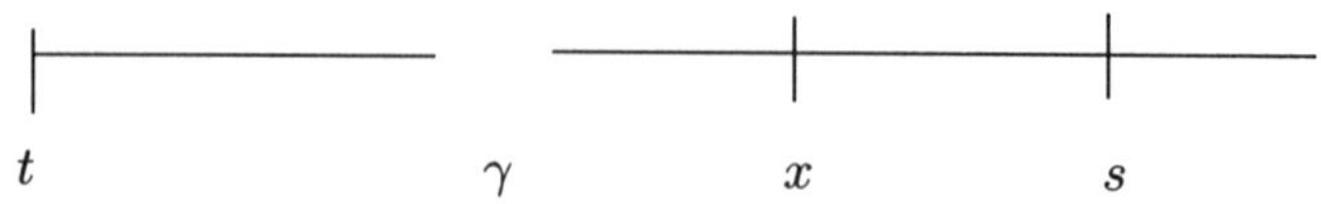

Figure 6.5

First we claim that there cannot be a wff A and a point x as in the diagram such that A is true at all points between γ and x including x and A is false at points arbitrarily close to the gap γ. The reason for the claim is that if it were the case, the axiom schema for complete time would not be valid at any t before γ. For let q be such that $\sim q \wedge Hq \in x$. Let $B = Hq \to A$. Then, clearly, $B \in y$ for every y after γ. On the other hand, for ys arbitrarily near γ from the left we have $\sim B \in y$. Thus the axiom schema for completeness of B does not hold at t, i.e. we have $FGB \wedge F{\sim}B \wedge G({\sim}B \to F{\sim}B) \in t$ but $G(GB \to PGB) \in t$ also. A similar argument by symmetry will not allow the mirror image of this situation to occur.

We thus conclude that for any wff A we have two possibilities only:

(c1) $\exists xy(x < \gamma < y \wedge \forall z(x < z < y \to A \in z))$,

(c2) $\forall xy(x < \gamma < y \to \exists x'y'(x < x' < \gamma < y' < y \wedge {\sim}A \in x' \wedge {\sim}A \in y'))$,

where $x < \gamma$ reads 'x is to the left of γ'. Here $<$ plays the rôle of the order as well as the relation between theories as defined in definition 6.2.1.

Consider now the following theory: $\Gamma_\gamma = \{A \mid$ condition (c 1) holds for $A\}$. Clearly, this theory is consistent. The theory is almost complete. For any sentence of the form FB we have $FB \in \Gamma_\gamma$ or ${\sim}FB \in \Gamma_\gamma$. For if for some $s > \gamma$, $B \in s$, then $FB \in \Gamma_\gamma$. If for no $s > \gamma$, $B \in s$, then for all $s > \gamma$, $G{\sim}B \in s$, and so $G{\sim}B$ must be true at an interval left of γ and hence $G{\sim}B \in \Gamma_\gamma$. We actually proved condition (c3) below:

(c3) $\exists s > \gamma(B \in s) \leftrightarrow FB \in \Gamma_\gamma$ and ${\sim}\exists s > \gamma(B \in s) \leftrightarrow {\sim}FB \in \Gamma_\gamma$.
$\exists t < \gamma(B \in t) \leftrightarrow PB \in \Gamma_\gamma$ and ${\sim}\exists t < \gamma(B \in t) \leftrightarrow {\sim}PB \in \Gamma_\gamma$.

Let Γ^*_γ be a complete consistent extension of Γ_γ. We may not have q with ${\sim}q \wedge Hq \in \Gamma^*_\gamma$. Let $T^* = T \cup \{\Gamma^*_\gamma \mid \gamma$ a gap$\}$. Let $<^*, h^*$ be the extensions of $<, h$ in the obvious way, i.e. also for Γ^*_γ let $\|q\|^{h^*}_\gamma = 1$ iff $q \in \Gamma^*_\gamma$.

Lemma 6.3.3 *For any* $A, t \in T$, γ *a gap:*

(*a*) $\|A\|^h_t$ *in* $(T, <) = \|A\|^{h^*}_t$ *in* $(T^*, <^*)$.

(*b*) $\|A\|^{h^*}_\gamma = 1$ *iff* $A \in \Gamma^*_\gamma$.

Proof. By simultaneous induction on A. The only interesting case is of FA.

(a) Suppose that $\|FA\|^h_t = 1$. Then $\exists s \in T(t < s \wedge \|A\|^h_s = 1)$. Since by the induction hypothesis we have $\|A\|^h_s = \|A\|^{h^*}_s$, we conclude that $\|FA\|^{h^*}_t = 1$.
Conversely, if $\|FA\|^{h^*}_t = 1$, then $\exists x \in T^*(t < x \wedge \|A\|^{h^*}_x = 1)$. If this x is in T then certainly $\|FA\|^h_t = 1$. If it is a gap, say $x = \gamma$, then by the induction hypothesis, condition (2) holds and so $A \in \Gamma^*_\gamma$. Hence condition (c1) can not hold for $\sim A$, so (c2) holds: i.e. $A \in s$ for points s arbitrarily close to γ. Hence $\|FA\|^h_t = 1$.

(b) Follows from (c3) and the induction hypothesis since $FB \in \Gamma^*_\gamma$ iff for some $s > \gamma$, $B \in s$ and so iff $\|FB\|^{h^*}_\gamma = 1$.

Thus the lemma is proved and $(T^*, <^*, h^*)$ is a time structure with no gaps as required.

■

(d) The case of the axioms of discreteness is very simple. It is clear that $A \wedge HA \to FHA$ is valid in any flow such that any point has an immediate successor. Conversely, if we take the canonical time structure $(T, <, h)$ for linear discrete time then any $t \in T$ has a successor. For let q be such that $\sim q \wedge Hq \in t$. Then $Hq \wedge HHq \in t$ and hence $FHHq \in t$ so for some $s > t, HHq \in s$. We cannot have any r with $t < r < s$ since $\sim q \in t \wedge HHq \in s$.

This completes all cases of theorem 6.3.2. ■

Definition 6.3.4 Let US_t be a system based on a general flow of time, just like that of K_t in the language with U, S only. The interpretation for U, S is as before, even though we do not assume that time is linear. So:

$$\|U(p,q)\|^h_t = 1 \quad \text{iff} \quad \exists s > t(\forall y(t < y < s \to \|q\|^h_y = 1) \wedge \|p\|^h_s = 1).$$
$$\|S(p,q)\|^h_t = 1 \quad \text{iff} \quad \exists s < t(\forall y(s < y < t \to \|q\|^h_y = 1) \wedge \|p\|^h_s = 1).$$

We know that F and P are definable by

$$Pq = S(q, \top),$$
$$Fq = U(q, \top).$$

We are going to define axiom systems involving $\{U, S\}$ for a general $(T, <)$, linear $(T, <)$, etc., completely parallel to definition 6.3.1. Remember that F, P are definable from U, S and so axioms involving them really involve U, S.

(a) Let US_t be the axiom system obtained by taking all instances of truth functional tautologies and all axioms and rules of K_t (of definition 6.3.1) for P, F (as defined from U, S) and the following additional axioms:
Axiom UU:

$$\sim r \wedge Hr \rightarrow [U(p, q) \leftrightarrow F(p \wedge H(P{\sim}r \rightarrow q))].$$

Axiom SS:

$$\sim r \wedge Gr \rightarrow [S(p, q) \leftrightarrow P(p \wedge G(F{\sim}r \rightarrow q))].$$

(b) Let the axiom system of U, S for linear time be obtained by adding axioms UU, SS to the axiom system of K_t for linear time of definition 6.3.1a.

(c) The axiom system for U, S for linear dense time is obtained by adding

$$FA \rightarrow FFA,$$
$$PA \rightarrow PPA,$$

as in definition 6.3.1b.

(d) Extension for discrete linear time: add

$$A \wedge HA \rightarrow FHA,$$
$$A \wedge GA \rightarrow PGA,$$

parallel to definition 6.3.1d.

Theorem 6.3.5 *Each of the systems described in the preceding definition is complete for the flow of time suggested by its name.*

Proof.

(a) We show first that US_t is complete for the class of general flows of time $(T, <)$. First note that if $US_t \vdash A$, then A is valid in any flow of time. We have to show that if any A is consistent in US_t then for some $(T, <, h)$ and $t \in T$, $\|A\|_t^h = 1$. The $(T, <, h)$ we are going to take is the canonical model constructed for K_t. We know that P and F, the connectives of K_t, are definable using U and S. So if we formally regard the language of US_t as $\{P, F, U, S\}$ and take the definitions

$$Pq = S(q, q \to q); \quad Fq = U(q, q \to q)$$

as axioms, we will still be working with the logic US_t. In this 'new' system we can pretend that anything of the form $U(A, B)$ or $S(A, B)$ is atomic. So we can pretend we are working with K_t, with P and F. For this 'new' K_t we can construct the canonical structure $(T, <, h)$ of the preceding section. All lemmas go through, especially lemma 6.2.7, namely for any A, $t \in T$:

$$\begin{array}{l} PA \in t \text{ iff } \|PA\|_t^h = 1, \\ FA \in t \text{ iff } \|FA\|_t^h = 1. \end{array}$$

We want to find a similar connection for $U(A, B)$. We shall use the connection we already have, namely the equations above for the wffs of the form FA and PA, to find a connection for $U(A, B)$ and $S(A, B)$. Take $t \in T$ and let q be an atom such that $\sim q \wedge Hq \in t$. Then by axiom UU,

$$U(A, B) \leftrightarrow F(A \wedge H(P{\sim}q \to B)) \in t$$

and hence

$$U(A, B) \in t \text{ iff } F(A \wedge H(P{\sim}q \to B)) \in t$$

iff, by lemma 6.2.7 for the $\{P, F\}$ language,

$$\|F(A \wedge H(P{\sim}q \to B))\|_t^h = 1.$$

Since $\sim q \wedge Hq \in t$ we can continue:

$$\begin{array}{ll} \text{iff} & \exists s > t(\|A\|_s^h = 1 \wedge \forall y(t < y < s \to \|B\|_y^h = 1)) \\ \text{iff} & \|U(A, B)\|_t^h = 1. \end{array}$$

A similar proof will yield the same equation for the case of S. For the case of S we use $r = {\sim}Hq$ (i.e. use $\sim r \wedge Gr \in t$). We can thus summarize:

(*) In $(T, <, h)$ we have for all A, B, t, $U(A, B) \in t$ iff $\|U(A, B)\|_t^h = 1$ and $S(A, B) \in t$ iff $\|S(A, B)\|_t^h = 1$. This gives us a structure validating any consistent A and thus proves (a).

(b, c) The completeness proofs for the axiom system with U, S for linear and/or dense time follow exactly the same lines of proof for (a) above. The canonical structure obtained for the language of $\{P, F\}$ is linear and/or dense. The axioms U and S are available to show that the analogue of lemma 6.2.7 holds for U and S.

(d) The completeness proof for discrete linear time follows similar lines as above and goes through for the same reasons.

■

Definition 6.3.6

(A) The following is an axiom system for the Stavi connectives for linear time:

(1) Axioms of linear time with irreflexivity for $\{P, F\}$.

(2) The following axioms for U', S':
Axiom $U'U'$
$\sim r \wedge Hr \to [U'(p, q) \leftrightarrow (a) \wedge (b) \wedge (c)]$,
where

$$\begin{aligned}
(a) &= F(q \wedge H(P{\sim}r \to q) \wedge G[H(P{\sim}r \to q) \\
&\quad \to F(q \wedge H(P{\sim}r \to q))], \\
(b) &= F{\sim}q \wedge G({\sim}q \to P({\sim}q \wedge P{\sim}r)), \\
(c) &= F({\sim}q \wedge H(P({\sim}q \wedge P{\sim}r) \to p)).
\end{aligned}$$

Axiom $S'S'$
The mirror image of axiom $U'U'$, i.e.
$\sim r \wedge Gr \to [S'(p, q) \leftrightarrow (a*) \wedge (b*) \wedge (c*)]$,
where
$(a*)$ is the mirror image of (a),
$(b*)$ is the mirror image of (b),
$(c*)$ is the mirror image of (c).

(B) The following is an axiom system for the connective $\gamma^{-}q$ of section 6.1:

(1) Axioms and rules for linear time with irreflexivity for $\{P, F\}$.

(2) The following axiom for $\gamma^- q$:

$$\sim r \wedge Hr \to [\gamma^- q \leftrightarrow (a\sim) \wedge (b\sim) \wedge (c\sim)],$$

where
$(a\sim) = (a^*)$ above,
$(b\sim) = (b^*)$ above,
$(c\sim) = (c^*)$ above with $\sim q$ replacing p.

Definition 6.3.7 The following is a suggested axiom system for integer time in the language with P and F.

(a) Axioms and rules for linear time, from definition 6.3.1.

(b) $FGA \wedge G(GA \to A) \to GA$,
$PHA \wedge H(HA \to A) \to HA$.

Theorem 6.3.8 *The suggested axiom system Z for integer time is complete for integer time.*

Proof. One can verify that the axioms and rules suggested are valid in integer time.

We now show the other direction, namely if A_0 is a consistent wff of this system Z, then for some $(T^*, <^*, h^*)$ and $t \in T^*, \|A_0\|_t^{h^*} = 1$, where $(T^*, <^*)$ is the set of integers with $<^*$ the usual order.

Let $(T, <, h)$ be the canonical model for Z. Then for some $t_0 \in T$, $\|A_0\|_{t_0}^h = 1$. $(T, <)$ can be shown to be a linear ordering, since Z contains all axioms of linear order. Furthermore, every instance of axioms (b) of definition 6.3.7 is valid in this canonical model.

We now construct a subset $T^* \subseteq T$ such that $<^* = < \restriction T^*$ (i.e. the restriction of $<$ to T^*) orders T^* like the integers. We shall have $t_0 \in T^*$ and for any subformula B of A_0 and any $x \in T^*$ we have $\|B\|_x^h = \|B\|_x^{h^*}$, where h^* is the restriction of h to T^*, namely, for all q, $h^*(q) = h(q) \cap T^*$.

Let k be the maximal depth of nested occurrences of P, F in A_0. Let Γ_0 be the set of all wffs B built up from the atoms of A_0 such that the maximal depth of nested occurrences of P, F in B is no more than $k + 1$. Γ_0 is finite.

Let B be such that FB, PB are in Γ_0. We consider several possibilities:

(+1) $\|GFB\|_{t_0}^h = 1$.

(−1) $\|HPB\|_{t_0}^h = 1$.

(+2) $\|G{\sim}B\|^h_{t_0} = 1$.

(−2) $\|H{\sim}B\|^h_{t_0} = 1$.

(+3) Possibility (+1) and (+2) do not hold. In this case by the axioms $\|F(B \wedge G{\sim}B)\|^h_{t_0} = 1$. Let $x(FB)$ be a point x in T in which $\|B \wedge G{\sim}B\|^h_x = 1$, and $t_0 < x(FB)$.

(−3) Possibilities (−1) and (−2) are false. Let $x(PB)$ be a point $x(PB) < t_0$ with $\|B \wedge H{\sim}B\|^h_x = 1$.

(−4) Possibility (+2) is true and (−2) is false, then by the temporal theorem $G{\sim}B \wedge {\sim}B \wedge PB \rightarrow P(B \wedge G{\sim}B)$ (see remark 6.3.10 below) we can let $y(PB) \leq t_0$ be a point with $\|B \wedge G{\sim}B\|^h_y = 1$.

(+4) Possibility (−2) is true and (+2) is false. Let $y(FB) \geq t_0$ be a point with $\|B \wedge H{\sim}B\|^h_y = 1$.

Let $T^*_0 = \{t_0\} \cup \{x(PB), x(FB), y(PB), y(FB) \mid FB, PB \in \Gamma_0\}$. T^*_0 is linearly ordered by $<$. Let $a_0 \in T^*_0$ be its first point and $b_0 \in T^*_0$ be its last point. Define by induction $T^*_n \subseteq T^*_{n+1} \cdots$ for all $n \geq 0$.

Assume T^*_n has been defined. T^*_n is finite and linearly ordered by $<$. Let a_n be its first point and b_n be its last point. For any $B \in \Gamma_0$ such that possibility (+1) holds, we have $z = z(n, FB) > b_n, \|B\|^h_z = 1$. For any $B \in \Gamma_0$ such that possibility (−1) holds, there exists $z(n, PB) < a_n$ with $\|B\|^h_z = 1$.

Let $T^*_{n+1} = T^*_n \cup \{z(n, FB), z(n, PB) \mid \text{ all such } B\}$. Let $T^* = \bigcup_n T^*_n$. Let $<^*, h^*$ be the restrictions of $<, h$ to T^*. Clearly, $(T^*, <^*) \cong \mathbb{Z}$.

Lemma 6.3.9 *For any $B \in \Gamma_0, t \in T^*$, we have:*

$$\|B\|^h_t \text{ in } (T, <, h) = \|B\|^{h^*}_t \text{ in } (T^*, <^*, h^*).$$

Proof. By induction on B. The case of atomic B is easy. So is the case of the classical connectives.

Assume $\|FB\|^{h^*}_t = 1$, then for some $s \in T^*, t < s \wedge \|B\|^{h^*}_s = 1$ and hence since $T^* \subseteq T$, by the inductive hypothesis $\|FB\|^h_t = 1$.

Assume $\|FB\|^h_t = 1$. We distinguish several cases:

(a) $\|GFB\|^h_{t_0} = 1$. In this case if $t \in T^*_n$, for any n, then in T^*_{n+1} we put by construction a $z(n, FB) > t$. With $\|B\|^h_z = 1$ and hence by induction hypothesis $\|B\|^{h^*}_z = 1$, and hence $\|FB\|^{h^*}_t = 1$.

(b) $\|GFB\|^h_{t_0} = 0$. If $\|FB\|^h_{t_0} = 1$, then by construction there exists in T^*_0 $x(FB) > t_0$ with $\|B \wedge G{\sim}B\|^h_x = 1$. Since $\|FB\|^h_t = 1$, we must have

$t < x$ and again this yields $\|FB\|_t^{h^*} = 1$. If $\|G{\sim}B\|_{t_0}^h = 1$, then by construction $y(PB) \leq t_0$ exists with $\|B \wedge G{\sim}B\|_y^h = 1$. Again since $\|FB\|_t^h = 1$, we must have $t < y$ and again we obtain $\|FB\|_t^{h^*} = 1$.

The case of P is symmetrical. This proves the lemma. ∎

We then have that $\|A_0\|_{t_0}^{h^*} = 1$ since $\|A_0\|_{t_0}^h = 1$ and this completes the proof of theorem 6.3.8. ∎

Remark 6.3.10

(a) The system of definition 6.3.7 is not equivalent to the system for discrete linear time of definition 6.3.1d. For it is easy to write down a structure with flow of time $\mathbb{Z}+\mathbb{Z}$ (i.e. a copy of $\mathbb{Z}$ followed by another copy of $\mathbb{Z}$) in which definition 6.3.1(d) is valid, but definition 6.3.7 is not.

(b) In the system of definition 6.3.7 (but not in that of definition 6.3.1):
$\vdash G{\sim}B \wedge {\sim}B \wedge PB \to P(B \wedge G{\sim}B)$
$\vdash H{\sim}B \wedge {\sim}B \wedge FB \to F(B \wedge H{\sim}B)$.

6.4 Axiomatization of the logic of an arbitrary connective

The irreflexivity lemma and rule allow us to axiomatize almost any arbitrary logic.

We have already defined the notion of an n-place truth table over a flow of time. Let us give some additional examples of connectives and their tables.

Examples 6.4.1

(a) Fp ('it will be the case that p') has the table $\exists s > tP(s)$.

(b) $U(p,q)$ ('q is true until p is true') and $S(p,q)$ ('q was true since p was true') has the table for 'Until':

$$\exists s > t(P(s) \wedge \forall x(t < x < s \to Q(x)))$$

and similarly for 'Since'.

(c) $K^{-}(p)$ ('p has repeatedly been true') (e.g. as in 'He has been knocking at the door') has the table:

$$\forall s < t \exists x (s < x < t \wedge P(x)).$$

(d) Np ('p is being true', as in 'He is writing') has the table:

$$\exists x, y (x < t < y \wedge \forall z (x < z < y \to P(z))).$$

(e) $P!p$ ('p was true exactly once') has the table:

$$\exists ! x < t P(x).$$

Similarly, $F!p$.

Definition 6.4.2 Let $(T, <)$ be a flow of time and L a language for this flow of time with the connectives P and F (having their usual truth tables) and possibly other temporal connectives (having their respective truth tables). Let $\Psi(t, P_1, \ldots, P_n, <)$ be an n-place truth table. We say that $\Psi(t)$ is parametrically defined in L (using a wff $B(k, p_1, \ldots, p_n)$ of L) iff for any assignment g such that $g(k) = \{t\}$ we have

(*) $\|B\|_t^g = 1$ iff $(T, <) \vDash \Psi(t, g(p_1), \ldots, <)$.

For a connective $\sharp$ with table Ψ, we say that $\sharp$ is parametrically defined iff Ψ is.

Theorem 6.4.3 *Let* **L** *be a temporal language for linear time with* P, F *and a connective* $\sharp$ *with table* Ψ*, which is parameter definable via* B *over linear time. Then the following is an axiomatization of the valid statements of* **L**:

(*a*) *All theorems of linear time in the language with* P *and* F.

(*b*) *The irreflexivity rule.*

(*c*) $k \wedge H{\sim}k \wedge G{\sim}k \to [\sharp \leftrightarrow B(k)]$.

Proof. It is clear that the axiom system is sound. Suppose A is a consistent wff. The assumptions of the irreflexivity lemma 6.2.7 hold. Thus a set S of theories exists satisfying the conditions of the canonical model. $(S, <)$ is linear time by the axioms (a).

Now since every $t \in S$ has a $k = k(t)$ such that $k \wedge G{\sim}k \wedge H{\sim}k \in t$, we obtain $\sharp \leftrightarrow B(k(t)) \in t$ for all $t \in S$. This ensures that $\sharp$ behaves according to the truth table Ψ under any assignment. For let $g(p) = \{t \in S \mid p \in t\}$, for p atomic. The conditions of the irreflexivity lemma ensure that for any wff A with P and F only and any t:

$$(*) \qquad \|A\|(g,t) = 1 \leftrightarrow A \in t,$$

but since $\sharp \leftrightarrow B(k(t)) \in t$ for all t, we obtain $(*)$ for arbitrary A. ■

Example 6.4.4 For each of the connectives of example 6.4.1 let the following be the corresponding B:

1. For $U(p,q)$ let:
   $B = F(p \wedge H(q \vee Fk \vee k))$.

2. For $K^-(p)$ let:
   $B = HF(p \wedge Fk)$.

3. For Np let:
   $B = FH(Pk \rightarrow p) \wedge PG(Fk \rightarrow p) \wedge p$.

4. For $P!p$ let:
   $B = Pp \wedge H(p \rightarrow H{\sim}p \wedge G({\sim}p \vee k \wedge Pk))$.

Remark 6.4.5 Take a language with *Since* and *Until*. Since F and P are definable in such a language (e.g. $Fq = U(q, q \rightarrow q)$), we obtain simple axiomatizations of this logic for the various sorts of linear time.

Example 6.4.6 Consider the connectives $N^+(p)$ and $N^-(p)$ reading $N^+(p)$ iff 'p will be true at some interval completely in the future' and similarly for N^-:

$$\Psi_{N^+}(t) = \exists x, y(t < x < y \wedge \forall u(x < u < y \rightarrow P(u))).$$

Is N^+ parameter definable in terms of P and F? We know that Np is definable; and N^+p is also definable via Np, since

$$k \wedge H{\sim}k \wedge G{\sim}k \rightarrow [N^+p \leftrightarrow FN(p \wedge Pk)].$$

Can N^+p be defined directly? We do not think so. This leads us to the notion of:

Definition 6.4.7 Define the notion of $\sharp$ being weakly parametrically definable from P and F (*wpd*) as follows:

(a) If $\sharp$ is parametrically definable using P, F then it is *wpd*.

(b) If $\sharp_i$ are *wpd* using P and F and $\sharp$ is parametrically definable using P, F, and $\sharp_i$, then $\sharp$ is *wpd* using P and F.

Example 6.4.8 For real numbers flow of time, every connective with table $\Psi(t)$, for any first-order Ψ, is actually definable using S and U. See chapters 9 and 10. Since these two connectives are parametrically definable using P and F, we obtain the result that any connective for any Ψ is weakly pd using P and F in the case of real time.

The proof of theorem 6.4.1 suggests the following theorem.

Theorem 6.4.9 *Let $\sharp_1, \ldots, \sharp_n$, connectives with tables $\Psi_1, \ldots, \Psi_n$, be such that $\sharp_1$ is wpd from P, F and for each $i > 1$, $\sharp_i$ is wpd from $P, F, \sharp_1, \ldots, \sharp_{i-1}$. Then the logic with $P, F, \sharp_i$ $(1 \le i \le n)$ is finitely axiomatizable over linear time.*

Proof. We can obtain the axioms as in the case of theorem 6.4.1, using the parametric definitions as axioms. ■

Theorem 6.4.10 *Let $(T, <)$ be linear time; then any n-table is wpd from P and F.*

Proof. Follows from the general construction in the next section. ■

6.5 Axiomatization of truth tables in irreflexive frames

Definition 6.5.1

1. By an *irreflexive frame* we mean a pair $(T, <)$ satisfying

$$\forall x \sim (x < x) \text{ and } \forall xyz(x < y \wedge y < z \to x < z).$$

2. Given an n-place truth table $B(t, P_1, \ldots, P_n)$ and an n-place connective $\sharp(p_1, \ldots, p_n)$, we can give a semantical interpretation for $\sharp$ by stipulating:

$$(*) \qquad \|\sharp(p_1, \ldots, p_n)\|_t^g = 1 \text{ iff } (T, <) \vDash B(t, g(p_1), \ldots, g(p_n)),$$

where g is an assignment (i.e. $g(p) \subseteq T$ for each atomic p) and $\|A\|_t^g$ is the truth value of A at $t \in T$ under g.

Let L be a language with connectives $\sharp_0, \ldots, \sharp_{m-1}$ having the respective truth tables $B_0, \ldots, B_{m-1}$. Let

$$TH = \{A \mid \|A\|_t^g = 1 \text{ for all irreflexive frames } (T, <) \text{ and all } g, t\}.$$

We can ask the question, is TH axiomatizable? (We know TH is recursively enumerable since all the wff of TH are translatable into the predicate calculus, but we want to see some axioms and rules.)

Definition 6.5.2

1. Let $L \subseteq L'$ be two languages, let L consist of $\sharp_0, \ldots, \sharp_{m-1}$ and L' consist of $\sharp_0, \ldots, \sharp_{m-1}, \sharp_m, \ldots, \sharp_{(m+m'-1)}$. Let $B_0, \ldots, B_{m+m'-1}$ be their respective tables. Let TH and TH' be the sets of valid statements of L, L', respectively.

   It is clear that

   $$(*) \quad \text{for any } A \text{ of } L,\ A \in TH \text{ iff } A \in TH'.$$

   In such a case $(*)$ we say that TH' is a *conservative extension* of TH.

2. We say that TH is *(finitely) pseudo-axiomatizable* iff for some conservative extension TH' of TH as above, TH' is (finitely) axiomatizable.

Our purpose is to prove the following.

Theorem 6.5.3 *Any TH is finitely pseudo-axiomatizable over linear time.*

For simplicity, we will only prove this theorem in the case where L consists of a single connective, $\sharp_0$ (i.e. $m = 1$). The proof for larger L is similar. We need some preliminary lemmas and constructions.

Definition 6.5.4 Let $\Box p$ be the connective with table

$$\|\Box p\|_t^g = 1 \text{ iff } \forall s \neq t (\|p\|_s^g = 1).$$

Let $\Diamond p = {\sim}\Box{\sim}p$, $\Box' p = p \wedge \Box p$, and $\Diamond' p = p \vee \Diamond p$.

Note that for linear time $\Box q = Gq \wedge Hq$, but not for general time (in which the only requirement is that $<$ is irreflexive and transitive).

Definition 6.5.5

1. Let $\Psi(s_0) = \Psi(s_0, P_1, \ldots, P_n)$ be an arbitrary n-table with free variable s_0. Assume $\Psi(s_0)$ is in prenex normal form, so that

$$\Psi(s_0) = (Q_1 s_1)(Q_2 s_2)\ldots(Q_m s_m)\Psi_m(s_0, s_1, \ldots, s_m, P_1, \ldots, P_n),$$

where $s_0, \ldots, s_m$ are distinct variables, $Q_1, \ldots, Q_m \in \{\forall, \exists\}$, and Ψ_m is quantifier free. Note: in order to keep the definition of ψ^* in (4) below as simple as we can, we will assume that the variable s_m does not in fact occur in Ψ_m. (This is in accord with the usual convention that $\varphi(v_1, \ldots, v_n)$ indicates a formula whose free variables lie among $v_1, \ldots, v_n$ but need not include them all.)

2. *Notation.* For brevity, we will write $\bar{P}$ for $P_1, \ldots, P_n$, $\bar{p}$ for $p_1, \ldots, p_n$, and $g(\bar{p})$ for $g(p_1), \ldots, g(p_n)$.

3. For $0 \leq i < m$ let

$$\Psi_j(s_0, s_1, \ldots, s_j, \bar{P}) = (Q_{j+1} s_{j+1})\ldots(Q_m s_m)\Psi_m.$$

So $\Psi_0 = \Psi$, and $\Psi_j(s_0, s_1, \ldots, s_j) = (Q_{j+1} s_{j+1})\Psi_{j+1}(s_0, \ldots, s_{j+1})$ for all $j < m$.

4. Introduce new propositional atoms $s_0^*, \ldots, s_{m-1}^*$. For each subformula ψ of Ψ_m we define a modal formula $\psi^*(s_0^*, \ldots, s_{m-1}^*, \bar{p})$ in the language with connectives $\{G, \Box\}$. ψ^* is defined by induction on ψ. If ψ is atomic, then there are three cases:

$$\begin{array}{rll}
(s_i = s_j)^* = & \Box'(s_i^* \leftrightarrow s_j^*) & (0 \leq i, j < m) \\
(s_i < s_j)^* = & \Box'(s_i^* \rightarrow F s_j^*) & (0 \leq i, j < m) \\
(P_k(s_i))^* = & \Box'(s_i^* \rightarrow p_k) & (0 \leq i < m, 1 \leq k \leq n).
\end{array}$$

Let $(\psi_1 \wedge \psi_2)^* = \psi_1^* \wedge \psi_2^*$, and $(\sim\psi)^* = \sim(\psi^*)$.

Thus we obtain ψ^* for all subformulae ψ of Ψ_m.

Lemma 6.5.6 *For any subformula $\psi(s_0, \ldots, s_m, \bar{p})$ of Ψ_m, ψ has the following property.*

Let $(T, <)$ be an irreflexive linear flow of time. Let g be an assignment to the atoms $\bar{p}$ corresponding to the monadic predicates $\bar{P}$, and to the atoms $s_0^*, \ldots, s_{m-1}^*$, such that for each $i < m$, $g(s_i^*) = \{t_i\}$ for some $t_i \in T$. Then for arbitrary $t_m \in T$,

$$\|\psi^*\|_{t_m}^g = 1 \text{ iff } (T, <) \vDash \psi(t_0, \ldots, t_m, g(\bar{p})).$$

Proof. This we show by induction on the structure of ψ. First, we check the assertion for atomic ψ. We have the following:

$$\begin{array}{rlll} \|(s_i = s_j)^*\|^g_{t_m} = 1 & \text{iff } \|\Box'(s_i^* \leftrightarrow s_j^*)\|^g_{t_m} = 1 & \text{iff } t_i = t_j; \\ \|(s_i < s_j)^*\|^g_{t_m} = 1 & \text{iff } \|\Box'(s_i^* \rightarrow Fs_j^*)\|^g_{t_m} = 1 & \text{iff } t_i < t_j; \\ \|P_k(s_i)^*\|^g_{t_m} = 1 & \text{iff } \|\Box'(s_i^* \rightarrow p_k)\|^g_{t_m} = 1 & \text{iff } t_i \in g(p_k). \end{array}$$

The cases $\psi_1 \wedge \psi_2$ and $\sim\psi$ are easy. This completes the proof. Note that the value of ψ^* does not depend on t_m: that is, ψ^* is either true everywhere in T or false everywhere in T, for a given assignment g. A similar fact holds for ψ, as s_m does not occur in Ψ_m. ■

Definition 6.5.7

1. With each Ψ_j $(0 \leq j \leq m)$ we associate a $(j+n)$-place table A_j, with free variable s_j, as follows:

$$A_j(S_0, S_1, \ldots, S_{j-1}, s_j, \bar{P}) =_{\text{def.}}$$

$$(\exists s_0, s_1, \ldots, s_{j-1})[\Psi_j(s_0, \ldots, s_{j-1}, s_j, \bar{P}) \wedge \bigwedge_{i<j} (S_i(s_i) \wedge \exists! x S_i(x))].$$

   Here, $\exists! x$ means that there exists a unique x. Note that for $j = 0$ the table is $\Psi(s_0, \bar{P})$.

2. Let L' be a language with connectives P, F, and $\sharp_j$ $(j = 0, \ldots, m)$. Let the tables for P, F be as before. Let the number of places of $\sharp_j$ be $j + n$, and let its table be $A_j(S_0, S_1, \ldots, S_{j-1}, s_j, \bar{P})$. (The table of $\sharp_0(\bar{p})$ is $\Psi(s_0, \bar{P})$, the arbitrary table that we started with.) We will regard G, H, $\Box$, $\Diamond$, $\Box'$, and $\Diamond'$ as abbreviations of appropriate F, P-formulae in L' having equivalent tables over linear time: e.g. $\Box q$ abbreviates $Gq \wedge Hq$, etc. Let $TH(L')$ be the set of formulae of this logic that are valid over irreflexive linear time (cf. definition 6.5.2).

3. *Axioms and rules for* $TH(L')$:

   (a) Axioms and rules of linear time for P and F.

   (b) Irreflexivity rule,

$$\frac{q \wedge \Box{\sim}q \rightarrow A}{A}, \quad \text{if } q \text{ does not occur in } A,$$

   plus modus ponens, substitution, and temporal generalization as usual.

(c) $\sharp_m(s_0^*, \ldots, s_{m-1}^*, \bar{p}) \leftrightarrow$

$$[\bigwedge_{i<m} \Diamond'(s_i^* \wedge \Box{\sim}s_i^*) \wedge \Psi_m^*(s_0^*, \ldots, s_{m-1}^*, \bar{p})].$$

(d) For each $0 \leq j < m$,

$$s_j^* \wedge \Box{\sim}s_j^* \to [\sharp_j(s_0^*, \ldots, s_{j-1}^*, \bar{p}) \leftrightarrow \blacksquare_{j+1}\sharp_{j+1}(s_0^*, \ldots, s_j^*, \bar{p})],$$

where $\blacksquare_j = \Box'$ if $Q_j = \forall$, and $\blacksquare_j = \Diamond'$ if $Q_j = \exists$.

We are going to prove that this system axiomatizes $TH(L')$, and hence finitely pseudo-axiomatizes $TH(F, P, \sharp_0)$.

Lemma 6.5.8 *The system of definition 6.5.7(3) is sound for* $TH(L')$.

Proof. We check that these axioms and rules are valid in linear irreflexive frames. Clearly (a) and (b) are valid. Let us check (c) and (d).

To check the validity of (c), let linear $(T, <)$, $t \in T$, and an assignment g into T be given. Recall that the table of $\sharp_m$ is:

$$A_m(S_0, \ldots, S_{m-1}, s_m, \bar{P})$$
$$= \exists s_0, \ldots, s_{m-1}[\Psi_m(s_0, \ldots, s_{m-1}, s_m, \bar{P}) \wedge \bigwedge_{i<m}(S_i(s_i) \wedge \exists! x S_i(x))].$$

So $\|\sharp_m(s_0^*, \ldots, s_{m-1}^*, \bar{p})\|_t^g = 1$ iff for some $s_0, \ldots, s_{m-1} \in T$,

$$g(s_i^*) = \{s_i\} \ (i < m) \text{ and } (T, <) \models \Psi_m(s_0, \ldots, s_{m-1}, t, g(\bar{p})).$$

But by lemma 6.5.6, for any $s_0, \ldots, s_{m-1}$ and g such that $g(s_i^*) = \{s_i\}$ for $i < m$, and for any $t \in T$,

$$(T, <) \models \Psi_m(s_0, \ldots s_{m-1}, t, g(\bar{p})) \iff \|\Psi_m^*(s_0^*, \ldots, s_{m-1}^*, \bar{p})\|_t^g = 1.$$

Thus, $\|\sharp_m(s_0^*, \ldots, s_{m-1}^*, \bar{p})\|_t^g = 1$ iff for some $s_0, \ldots, s_{m-1} \in T$, $g(s_i^*) = \{s_i\}$ for each $i < m$ and $\|\Psi_m^*(s_0^*, \ldots, s_{m-1}^*, \bar{p})\|_t^g = 1$, iff $\| \bigwedge_{i<m} \Diamond'(s_i^* \wedge \Box{\sim}s_i^*) \wedge \Psi_m^*(s_0^*, \ldots, s_{m-1}^*, \bar{p})\|_t^g = 1$. Thus (c) is valid.

To check the validity of (d), fix $0 \leq j < m$, let $s_j \in T$, and assume that g is such that $\|s_j^* \wedge \Box{\sim}s_j^*\|_{s_j}^g = 1$, so that $g(s_j^*) = \{s_j\}$. We need to check that

$$\|\sharp_j(s_0^*, \ldots, s_{j-1}^*, \bar{p})\|_{s_j}^g = \|\blacksquare_{j+1}\sharp_{j+1}(s_0^*, \ldots, s_j^*, \bar{p})\|_{s_j}^g .$$

Recall that the table of $\sharp_j$ is:

$$A_j(S_0, \ldots, S_{j-1}, s_j, \bar{P})$$
$$= \exists s_0, s_1, \ldots, s_{j-1}[\Psi_j(s_0, \ldots, s_{j-1}, s_j, \bar{P}) \wedge \bigwedge_{i<j}(S_i(s_i) \wedge \exists! x(S_i(x)))].$$

Case I: $g(s_i^*)$, $i < j$, are not all one-point sets. Then both $\sharp_j$ and $\blacksquare_{j+1}\sharp_{j+1}$ are false everywhere in T.

Case II: $g(s_i^*) = \{s_i\} \subseteq T$ for all $i < j$. Then $\|\sharp_j(s_0^*, \ldots, s_{j-1}^*, \bar{p})\|_{s_j}^g = 1$
iff $(T, <) \models A_j(g(s_0^*), \ldots, g(s_{j-1}^*), s_j, g(\bar{P}))$
iff $(T, <) \models \Psi_j(s_0, \ldots, s_{j-1}, s_j, g(\bar{p}))$.

Similarly, because we assumed that $g(s_j^*) = \{s_j\}$, we obtain (for arbitrary $t \in T$): $\|\blacksquare_{j+1}\sharp_{j+1}(s_0^*, \ldots, s_j^*, \bar{p})\|_t^g = 1$ iff for all $s_{j+1} \in T$ (or some s_{j+1}, depending on Q_{j+1}),

$$\|\sharp_{j+1}(s_0^*, \ldots, s_j^*, \bar{p})\|_{s_{j+1}}^g = 1,$$

iff for all $s_{j+1} \in T$ (or some s_{j+1}, depending on Q_{j+1}),

$$(T, <) \models \Psi_{j+1}(s_0, \ldots, s_{j+1}, g(\bar{p})),$$

iff $(T, <) \models Q_{j+1}s_{j+1}\Psi_{j+1}(s_0, \ldots, s_{j+1}, g(\bar{p}))$,
iff by definition 6.5.5(3), $(T, <) \models \Psi_j(s_0, \ldots, s_j, g(\bar{p}))$.
So, assuming that $g(s_j^*) = \{s_j\}$, for any $t \in T$ we have

$$\|\sharp_j(s_0^*, \ldots, s_{j-1}^*, \bar{p})\|_{s_j}^g = \|\blacksquare_{j+1}\sharp_{j+1}(s_0^*, \ldots s_j^*, \bar{p})\|_t^g \,.$$

Taking $t = s_j$ shows the validity of axiom (d) and completes the proof of validity of the axioms. ∎

Now let A be a consistent formula of this logic. Let $(T, <)$ be the canonical model. So $(T, <)$ is irreflexive and linear, for all $\Gamma \in T$ there is an atom q such that $q \wedge \Box{\sim}q \in \Gamma$ (we use the irreflexivity rule to get this), and there is $\Gamma_0 \in T$ with $A \in \Gamma_0$. Let g be defined on T by $g(q) = \{\Gamma \in T \mid q \in \Gamma\}$ for q atomic. To show that $(T, <, g)$ is a model of A, we will prove the following lemma.

Lemma 6.5.9 *For all B, Γ, we have $\|B\|_\Gamma^g = 1$ iff $B \in \Gamma$.*

Proof. Let us say that a formula B is *well-behaved* if for all $\Gamma \in T$ we have $\|B\|_\Gamma^g = 1$ iff $B \in \Gamma$. Clearly, any atom is well-behaved. We will prove that for each j with $0 \leq j \leq m + 1$, the following property χ_j holds.

- χ_j If $\mathcal{S}$ is a set of well-behaved formulae, then any formula B made from the formulae in $\mathcal{S}$ using only the boolean connectives, F, P, and $\sharp_i$ for $j \leq i \leq m$, is also well-behaved.

The lemma follows from χ_0, by taking $\mathcal{S}$ to be the set of all atoms.

The proof of χ_j is by downward induction on j. For the base case, χ_{m+1}, we must show that any formula B built from the formulae in $\mathcal{S}$

using only the classical connectives (no $\sharp_j$s) is well-behaved. This can be proved in the accustomed fashion, by induction on the structure of B.

Now assume that $j \le m$, and that $\chi_{j+1}, \ldots, \chi_{m+1}$ all hold. Take any set $\mathcal{S}$ of well-behaved formulae. We will prove that any formula B built from the formulae in $\mathcal{S}$ using only the boolean connectives, F, P, $\sharp_{j+1}, \ldots,$ and $\sharp_m$, is well-behaved.

The proof is again by induction on B. If $B \in \mathcal{S}$ we are given the result; and the cases $B \wedge C$, $\sim B$, FB, PB, and $\sharp_i(B_0, \ldots, B_{i-1}, C_1, \ldots, C_n)$ for $i > j$ follow immediately from χ_{j+1}, letting $\mathcal{S}$ be the set of maximal subformulae of B in each case (i.e. $\{B, C\}$, $\{B\}$, and $\{B_0, \ldots, B_{i-1}, C_1, \ldots, C_n\}$ respectively).

So assume that $B = \sharp_j(B_0, \ldots, B_{j-1}, C_1, \ldots, C_n)$, and (inductively) that $B_0, \ldots, B_{j-1}, C_1, \ldots, C_n$ are well-behaved. We must prove the same for B. For brevity, we will write $\bar{C}$ for $C_1, \ldots, C_n$. We will also write $\gamma(D)$ for $\{\Delta \in T \mid \|D\|^g_\Delta = 1\}$ (for any formula D), and $\gamma(\bar{C})$ for $\gamma(C_1), \ldots, \gamma(C_n)$.

There are two cases.

Case 1: $j = m$. So $B = \sharp_m(B_0, \ldots, B_{m-1}, \bar{C})$. Take $\Gamma \in T$. Let

$$B^* = \bigwedge_{i<m} \Diamond'(B_i \wedge \Box{\sim}B_i) \wedge \Psi^*_m(B_0, \ldots, B_{m-1}, \bar{C}).$$

By axiom (c),

$$B \in \Gamma \quad \text{iff} \quad B^* \in \Gamma.$$

By lemma 6.5.8, axiom (c) is valid in $(T, <)$. Hence

$$\|B\|^g_\Gamma = \|B^*\|^g_\Gamma.$$

So it suffices to prove that B^* is well-behaved. But as $B_0, \ldots, B_{m-1}, \bar{C}$ are well-behaved, this is immediate from χ_{m+1}, by taking $\mathcal{S}$ to be $\{B_0, \ldots, B_{m-1}, \bar{C}\}$.

Case 2: $j < m$. This is similar to the argument of case 1. Take $\Gamma \in T$. Because of the irreflexivity rule, there is an atom q with $q \wedge \Box{\sim}q \in \Gamma$. So by axiom (d),

$$\sharp_j(B_0, \ldots, B_{j-1}, \bar{C}) \in \Gamma \quad \text{iff} \quad \blacksquare_{j+1}\sharp_{j+1}(B_0, \ldots B_{j-1}, q, \bar{C}) \in \Gamma.$$

By lemma 6.5.8, axiom (d) is valid in $(T, <)$; and because, by χ_{m+1}, $q \wedge \Box{\sim}q$ is well-behaved, we have $\|q \wedge \Box{\sim}q\|^g_\Gamma = 1$. Hence

$$\|\sharp_j(B_0, \ldots, B_{j-1}, \bar{C})\|^g_\Gamma = \|\blacksquare_{j+1}\sharp_{j+1}(B_0, \ldots B_{j-1}, q, \bar{C})\|^g_\Gamma,$$

so it suffices to prove that $\blacksquare_{j+1}\sharp_{j+1}(B_0, \ldots B_{j-1}, q, \bar{C})$ is well-behaved. Again, this is immediate from the inductive hypothesis χ_{j+1}, taking $\mathcal{S}$ to be $\{B_0, \ldots B_{j-1}, q, \bar{C}\}$.

We have now shown that $B = \sharp_j(B_0, \ldots, B_{j-1}, \bar{C})$ is well-behaved. This completes the induction on B, and so χ_j is established. Thus, by induction on j, $\chi_0, \ldots, \chi_{m+1}$ all hold, and this completes the proof of lemma 6.5.9. ■

As $A \in \Gamma_0$, it follows from the lemma that $\|A\|^g_{\Gamma_0} = 1$. Thus we have proved completeness, and theorem 6.5.3 follows.

Remark 6.5.10 A general question can be asked here. Given a flow of time $(T, <)$ and a table $B(t, P_1, \ldots, P_m)$ of the kind mentioned above, we can define an m-place temporal connective $\sharp(p_1, \ldots, p_m)$ via the truth table:

$$\|\sharp(p_1, \ldots, p_m)\|^g_t = 1 \text{ iff } (T, <) \vDash B(t, g(p_1), \ldots, g(p_m)).$$

We can now ask: under what conditions can $\sharp$ be defined using unary connectives? For the case of $(T, <)$ = the real numbers, we know from chapter 10 that any $\sharp$ can be defined using Since and Until, and therefore using P and F of our system. But the question remains open for arbitrary $(T, <)$, e.g. the rationals.

6.6 The rule IϕR

The irreflexivity rule played two rôles in temporal logic. The first rôle was to characterize, in a way, the condition of irreflexivity on temporal frames, namely the condition $\forall x{\sim}(x < x)$.

The second rôle of IRR, which is perhaps accidental, is to allow us to give neat and nice axiomatizations of temporal logical systems. In many cases the use of IRR in axiomatization is essential.

This section examines generalizations of the rule IRR as a rule characterizing irreflexivity. If $\rho(x) =_{\text{def.}} x < x$, we read IRR as:

$$\forall x{\sim}\rho(x).$$

We can therefore look for other rules of the form IϕR reading 'not ϕ rule'. For example take linearity; the axiom is $\forall x, y, z\phi$, where:

$$\phi(x, y, z) \equiv (x < y \wedge x < z \rightarrow y = z \vee y < z \vee z < y).$$

Suppose we do not require linearity. Can we write an axiom for non-linearity at each point, i.e. for the condition

$$\forall x \exists y, z {\sim}\phi(x, y, z)?$$

The situation here is the same as in the case of an axiom for irreflexivity. There is no such axiom but one can add a rule. To formulate the rule we need a wff of temporal logic that can do the job. Let us look at the reflexivity axiom:

$$R_\rho(q) = Hq \to q.$$

Our rule $I\phi R$ was

$$\frac{\vdash {\sim}R_\rho(q) \to A}{\vdash A}, \quad q \text{ not in } A.$$

The linearity axiom is $R_\phi(p, q) = G(q \wedge Gq \to p) \vee G(p \wedge Gp \to q)$. The corresponding rule for non-linearity is

$$\frac{\vdash {\sim}R_\phi(p, q) \to A}{\vdash A}, \quad p, q \text{ not in } A.$$

The choice of ϕ, R_ϕ is a matter of a good guess and should satisfy certain conditions specified below.

Definition 6.6.1

1. Let $A(q_1, \ldots, q_n)$ be a wff; A is said to be '*ready-for-a-rule*' iff there exists a wff of $(T, <, =)$ of the form $\phi(x)$ such that for all $t \in T$:

   $$(T, <, =) \vDash \phi(t) \text{ iff } \forall h(\|A\|_t^h = 1).$$

   We say that there is a *local correspondence* between A and ϕ.

2. Let A be ready for a rule. A theory Δ is said to be an $I\phi R$ -theory iff the following holds:

   (a) For some q_i, ${\sim}A(q_i) \in \Delta$.

   (b) Whenever $X = \mathbf{M}_1(A_1 \wedge \mathbf{M}_2(A_2 \wedge \ldots \wedge \mathbf{M}_n A_n)) \cdots) \in \Delta$, then for some q_i, $X(q_i) = \mathbf{M}_1(A_1 \wedge \ldots \mathbf{M}_n(A_n \wedge {\sim}A(q_i)) \cdots) \in \Delta$, where $\mathbf{M}_i \in \{P, F\}$.

3. Let $K_t{\sim}\phi$ be an extension of K_t with the rule $I\phi R$, i.e. with

   $$\frac{\vdash {\sim}A \to B}{\vdash B} \quad \text{where no atoms of } A \text{ appear in } B.$$

Lemma 6.6.2 *Let B be a $K_t{\sim}\phi$ consistent wff. Then for some complete and consistent $I\phi R$-theory Δ we have $B \in \Delta$.*

Proof. Same as lemma 6.2.4. ■

Lemma 6.6.3 *Let Δ be a complete and consistent $I\phi R$-theory and let $FA \in \Delta$ ($PA \in \Delta$). Then for some complete and consistent $I\phi R$ -theory Γ, we have $A \in \Gamma$ and $\Delta < \Gamma$ ($\Gamma < \Delta$ respectively).*

Proof. Same as lemma 6.2.5. ■

Using the above lemmas we obtain a model $(T, <, h)$ of canonical theories for B, such that for each $t \in T$ there are q_i such that $\|A(q_i)\|_t^h = 0$.

Example 6.6.4 Take, for example, the following axioms $A(q)$:

1. $A_1 = HHq \to Hq$.
2. $A_2 = G(q \wedge Gq \to p) \vee G(p \wedge Gp \to q)$.

The corresponding rules $I\phi_1 R$ and $I\phi_2 R$ characterize non-density and non-linearity. In the canonical model for $I\phi_1 R$, each t has a q with $\|HHq \to Hq\|_t^h = 0$; hence $\exists y[{\sim}\exists x(t > x \wedge x > y) \wedge (t > y)]$. In the canonical model for $I\phi_2 R$, for each t there are p, q such that

$$\|G(q \wedge Gq \to p) \vee G(p \wedge Gp \to q)\|_t^h = 0.$$

Hence

$$\|F(q \wedge Gq \wedge {\sim}p) \wedge F(p \wedge Gp \wedge {\sim}q)\|_t^h = 1$$

and hence

$$\exists x, y(t < x \wedge t < y \wedge x \neq y \wedge {\sim}x < y \wedge {\sim}y < x)$$

holds, which is non-linearity.

We can thus state the following theorem:

Theorem 6.6.5 *Let A be an axiom such that for any flow $(T, <)$:*

$$\forall t \in T[\forall h(\|A\|_t^h = 1) \leftrightarrow {\sim}v(t)]$$

holds for some condition $v(t)$.

Then $K_t{\sim}v$, which is K_t with the rule IvR, is complete for the class of all flows $(T, <)$ satisfying $\forall t v(t)$.

Proof. Exercise. ■

Exercise 6.6.6 Take ϕ as the axiom of completeness of definition 6.3.1c. Then the rule $I\phi R$ characterizes the condition that above any point there is a gap.

6.7 Generalized irreflexivity lemma

The irreflexivity rule is especially simple for the case of temporal logic because of theorem 6.3.2, which is a consequence of the axioms

$$q \to GPq$$

$$q \to HFq$$

of K_t. These axioms essentially say that one can go back and forth in the flow of time (time travel). If we are dealing with future temporal logic only, with possibly other connectives, then we can go up to the future but we may not have the suitable connectives to come back. Thus if, for example, we have $F(A \wedge FB)$ and we want to show the consistency of adding a new atom x to where B is true and have $F(A \wedge F(B \wedge \sim x \wedge Gx))$ consistent, we need a special rule for this. We are thus led to the general irreflexivity rule and the general irreflexivity lemma. The rule is no longer simple and nice as in F, P temporal logics.

To achieve generality we move our irreflexivity results to the setting of modal logic. To fix our notation, (S, R, g) denotes a propositional Kripke frame with relation $R \subseteq S^2$ and an assignment g (i.e. $g(p) \subseteq S$, for any atom p).

Consider the modal system K, in a language with one necessity operator and the modal axiom schemas:

1. $\Box(A \to B) \to (\Box A \to \Box B)$,

2. $\dfrac{\vdash A}{\vdash \Box A}$,

besides all substitution instances of all truth functional tautologies.

It is known that K can be characterized by the class of all frames (S, R), i.e.

3. $K \vdash A$ iff for all (S, R) and all g into (S, R) and all $x \in S$, $\|A\|_x^g = 1$.

The condition of reflexivity of R can be characterized by the axiom schema

4. $\Box A \to A$
   in the sense that

5. R is reflexive in (S, R) iff under all g
   $\|\Box A \to A\|_x^g = 1$, for all $x \in S$.

We know that the condition of irreflexivity cannot be characterized in the same way, as shown in chapter 2.

In temporal logics, all flows of time are required to be irreflexive. Take, for example, future time, with the future operator FA reading 'it will be the case that A'. Its truth table is, in a flow of time $(T, <, g)$, where $<$ is irreflexive and transitive:

6. $\|FA\|_t^g = 1$ iff $\exists s > t \|A\|_s^g = 1$.

This shows that the system $G4$ of future time is no other than modal $K4$, i.e. with axioms (1), (2), and the axiom of transitivity

7. $\Box A \to \Box\Box A$

taken for the operator $GA =_{\text{def.}} {\sim}F{\sim}A$, replacing $\Box$. We have no special axiom for the irreflexivity of $<$, and in order to prove completeness, we resort to special constructions.

In this particular case the following lemma was used (cf. section 3.2).

Lemma 6.7.1 *For any frame (S, R, g) in which R is transitive, there exists a frame (S', R', g'), in which R' is transitive and irreflexive, and a function $\sigma : S' \to S$, such that*

1. *σ is an order homomorphism onto,*
2. *$\sigma(x) = y \wedge yRy' \to \exists x'(xR'x' \wedge \sigma(x') = y')$,*
3. *for all x and A, $\|A\|_x^{g'} = \|A\|_{\sigma(x)}^g$.*

Such a σ is called a *p-morphism*.

Similar constructions are used for other temporal systems. All basic constructions proceed roughly as follows:

1. Given a temporal system X which is expected to be complete for a certain kind of flow of time, and a wff A such that $X \nvdash A$, obtain a structure (S, R, g) (usually by taking S to be a set of complete consistent X-theories) in which X is valid and
$$\exists x \in S, \|A\|_x^g = 0.$$
2. Transform (S, R, g) to some $(T, <, g*)$ in which $<$ is irreflexive and all theorems of X are valid and A is not; arrange that $(T, <)$ is a desired flow of time.

A practical point should be mentioned. (2) above is usually achieved by duplicating the points x for which xRx holds. This procedure is sometimes impossible to follow in the case of two-dimensional temporal logics studied in chapter 7. In the two-dimensional case, we evaluate at a pair of points

$\|A\|_s^t$. Duplicating a point x into x_1, x_2 forces us to say not only what $\|A\|_y^{x_i}$ is for $y \neq x$ but also what $\|A\|_{x_2}^{x_1}$ is, and this can prove difficult to do.

We propose a lemma that allows us to build structures with R irreflexive in the first place.

Lemma 6.7.2 (irreflexivity lemma) *Let* $\mathbf{L}$ *be a language containing, besides the classical connectives, two finite sets* $\mathbf{C}_1$ *and* $\mathbf{C}_2 \supseteq \mathbf{C}_1$ *of modal connectives. Let* X *be a modal (l-ary) logic for this language satisfying, among other axioms and rules, the following:*

1. *X proves all substitution instances of truth functional tautologies.*

2. *The axiom* (1) *and rule* (2) *above are provable in* X *for any modal connective in* $\mathbf{C}_2$.

3. *The following rule is a rule of* X:

$$\frac{A_0 \to \Box_1(A_1 \to \cdots \Box_n(A_n \to \sim \bigwedge_{i=1}^m (\sim q_i \wedge \blacksquare_i q_i)) \cdots)}{A_0 \to \Box_1(A_1 \to \cdots \Box_n {\sim} A_n)) \cdots)},$$

   where n and A_i are arbitrary, $\Box_1, \ldots, \Box_n$ are any (not necessarily distinct) operators from $\mathbf{C}_2$, $\blacksquare_i$, $i = 1, \ldots, m$, are all the operators of $\mathbf{C}_1$, and $q_1, \ldots, q_m$ are atoms not appearing in $A_0, \ldots, A_n$.

Then, if A is a consistent wff of X, there exists a set T of X-complete and consistent theories, and a $\Theta \in T$ such that:

4. $A \in \Theta$.

5. *For each $\Gamma \in T$, there exist atoms $q_1, \ldots, q_m$ such that*

$$\bigwedge_{i=1}^{m} {\sim} q_i \wedge \blacksquare_i q_i \in \Gamma, \text{ where } \mathbf{C}_1 = \{\blacksquare_1, \ldots \blacksquare_m\}.$$

6. *For any $\Box \in \mathbf{C}_2$, let $\mathbf{R}_\Box$ be the relation: $\Gamma \mathbf{R}_\Box \Gamma'$ iff for all $A, \Box A \in \Gamma \to A \in \Gamma'$. Then for any Γ, B, if ${\sim}\Box B \in \Gamma$, then for some Γ', ${\sim} B \in \Gamma'$ and $\Gamma \mathbf{R}_\Box \Gamma'$.*

7. *If $\Gamma \mathbf{R} \Gamma'$ is defined as $\Gamma \mathbf{R} \Gamma' = \bigvee_\Box \Gamma \mathbf{R}_\Box \Gamma'$ then for any $\Gamma \in T$ there exists a number $k = k(\Gamma)$ such that $\theta \mathbf{R}^k \Gamma$, where $x \mathbf{R}^0 y$ is $x = y$ and $x \mathbf{R}^{k+1} y$ is $\exists z (x \mathbf{R}^k z \wedge z \mathbf{R} y)$.*

Proof. Section 6.8 is devoted to the proof of this lemma. ■

Remark 6.7.3 To understand what this lemma says, let us look again at future transitive irreflexive time. Let G be $\sim F \sim$. Then certainly (1) and (2) are valid for G. Rule (3) boils down to the rule

$$(a) \qquad \frac{A_0 \to G(A_1 \to G(\cdots G(A_m \to \sim(\sim q \wedge Gq))\cdots)}{A_0 \to G(A_1 \to \cdots G\sim A_m)\cdots)}$$

if we take $\mathbf{C}_1 = \mathbf{C}_2 = \{G\}$, and where q is an atom not in any of $A_1, \ldots, A_m$. This rule is certainly valid, since $<$ is irreflexive. So let $G4*$ be the axiom system with axioms (1), (2), (7), and (a) for G. The lemma yields a T such that for each $\Gamma \in T$, there exists a q with $\sim q \wedge Gq \in \Gamma$. Obviously (6) and (7) assure us that $< = \mathbf{R}_G$ is transitive and irreflexive. A structure of the desired type is the immediate result.

It should be noted that the axiom system $G4$ of (6) is different from the axiom system $G4*$ above. For $G4$ is the same as $K4$ and the system $S4$ is an extension of $G4$ (i.e. $GA \to A$ can be consistently added to $G4$). But this is not the case with $G4*$, as the axiom $GA \to A$, when added to $G4*$, gives an inconsistent logic in view of (a).

Therefore, one may say that axiom (a) characterizes, in a certain sense, the condition of irreflexivity.

We shall explain the use of rule (a) further. This rule simply allows us to carry out a semantic tableaux construction for the logic, allowing us to build models with the property that in each possible world there is an atom true exactly in the world.

To give an example, take the system G4 and let Γ be a complete and consistent G4 theory. If Γ is satisfiable (and this is really the completeness theorem), then there exists a structure (S, R, g) which satisfies Γ. That is, for $0 \in S$ we have: $\Gamma = \{A \mid \|A\|_0^g = 1\}$. So to find (S, R, g) we start with Γ and say that at first we know that a point 0 should be given in which all members of Γ must be true. Describe this situation by writing:

$$0 \bullet \ \Gamma \text{ true}$$

Now if we have point 0 and $\sim GA \in \Gamma$, since $\sim GA$ is true at 0 we must also have a point a such that $0Ra$ in which $\sim A$ is true and $\{B \mid GB \in \Gamma\}$ is also true. So we throw in a point a with $0Ra$, show that $\{\sim A\} \cup \{B \mid GB \in \Gamma\}$ is consistent, and extend the above set to a complete theory Γ_a. Now we have the situation shown in figure 6.6.

We can do this for each $\sim GA_i \in \Gamma$ and then go on treating the newly obtained Γ_{a_i} in the same manner, adding another level of points, etc., etc. In the end this gives a G4-structure for Γ.

In the case where we want the structure to be irreflexive, we want to put in Γ_a an atom q with $Gq \wedge \sim q \in \Gamma_a$. Thus we need to show that the

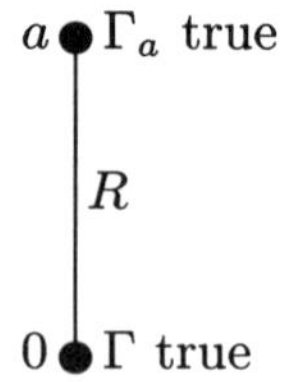

Figure 6.6

following set is consistent:

$$\{\sim A\} \cup \{B \mid GB \in \Gamma\} \cup \{Gq \wedge \sim q\}.$$

In general, we may have $Gp \in \Gamma$ for all atoms p and we may have to assume that q is a completely new atom not appearing in Γ. Now suppose the above set is not consistent; then

$$\vdash \bigwedge B_i \to (\sim A \to \sim(\sim q \wedge Gq)),$$

so

$$\vdash GB \to G(\sim A \to \sim(\sim q \wedge Gq)),$$

where $\bigwedge B_i = B$. We need a special rule to tell us that we cannot have the above situation. This rule is (a). It gives us

$$\vdash GB \to GA$$

and since $GB \in \Gamma$ we obtain $GA \in \Gamma$, a contradiction.

Here we used (a) for depth 1. When we go on building the model we are going to have situations of higher depth, as shown in figure 6.7.

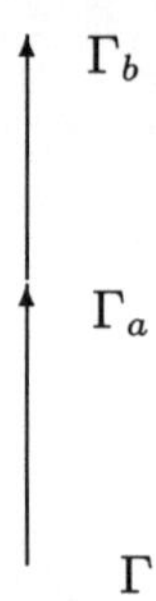

Figure 6.7

The reader who takes a look at the proof of the irreflexivity lemma in the next section will get the impression that the proof is a little bit more complicated than we described. The reason for all the extra work is technical. We saw above that we can add to Γ_a a new atom q with $\sim q \wedge Gq \in \Gamma_a$. The technical problem is that q is new to Γ. Since we are dealing with propositional calculi, we must have Γ and Γ_a in the same language, based on the same set of atomic propositions.

So we use the method of semantic tableaux with the additional assumption that $\Gamma_0, \Gamma_a, \ldots$, etc., are all *finite*, and go up and down the construction tree filling up the tableaux. This requires careful labelling, consistency conditions, etc. This is what is being done in the next section, and this is why the formulation of the irreflexivity lemma is the way it is. It simply says that there exists a successful construction of tableaux.

In P, F temporal logic, e.g. K_t, rule (a) for depth 1 is sufficient because we can go up and down the flow of time and rule (a) for higher depths can be proved. This is theorem 3.1.4.

6.8 A proof of the irreflexivity lemma

Definition 6.8.1 Let $\mathbf{C}_2 = \{\Box_1, \ldots, \Box_m\}$ and $\mathbf{C}_1 = \{\Box_1 \ldots \Box_{m+m'}\}$.

1. Let S^* be the set of all finite sequences of natural numbers including the empty sequence ε. Let, for $t, s \in S^*, t * s$ denote concatenation, i.e. if $t = (x_1, \ldots, x_n), s = (y_1, \ldots, y_k)$, then $t * s = (x_1, \ldots, x_n, y_1, \ldots, y_k)$.
   Let $t^* = x_n$ and let $t' = (x_1, \ldots, x_{n-1})$. If $m \leq n$ define $t^{(m)} = (x_1, \ldots, x_{n-m})$. So $t^{(n)} = \varepsilon$. Write $t \leq s$ iff t is an initial segment of s or $t = s$.
   Write $t\mathbf{R}_i^+ s$ iff for some x, $s = t * (2^{i+1}.3^x)$.
   Write $t\mathbf{R}^+ s$, iff for some $1 \leq i \leq m + m'$, $t\mathbf{R}_i^+ s$.

2. By a *table*, we mean a pair $(S, \mathbf{d})$ where $S \subseteq S^*$, such that

   (2.1) $\varepsilon \in S$.

   (2.2) If $t \in S$ then $t' \in S$.

   (2.3) Every element $t \neq \varepsilon$, has the property that for some x and $1 \leq i \leq m + m'$, $t^* = 2^{i+1}3^x$.

   (2.4) $\mathbf{d}$ is a function such that $\mathbf{d}(t), t \in S$ is a consistent wff with the property that if $t^* = 2^{i+1}3^x$ then $\mathbf{d}(t') \vdash \Diamond_i \mathbf{d}(t)$, where $\Diamond_i = \sim\Box_i\sim$.

3. Let $(S, \mathbf{d})$ be a table, z a wff, and $t \in S$. Assume that $t = (x_1, \ldots, x_n)$. Define a $\mathbf{d}_{z,t}$ on S by
   (a) $\mathbf{d}_{z,t}(s) = \mathbf{d}(s)$ if $s \not\leq t$;
   (b) $\mathbf{d}_{z,t}(t) = \mathbf{d}(t) \wedge z$;
   (c) inductively, for each $m \leq n$, define $\mathbf{d}_{z,t}(t^{(m+1)}) = \mathbf{d}(t^{(m+1)}) \wedge \Diamond_i \mathbf{d}_{z,t}(t^{(m)})$, where $(t^{(m)})^* = 2^{i+1}3^x$ for some x.

Lemma 6.8.2 *Let $(S, \mathbf{d})$ be a table and z a wff and $t \in S$ as above. Then*

1. $\frac{\vdash \sim \mathbf{d}_{z,t}(t^{(m)})}{\vdash \sim \mathbf{d}_{z,t}(t^{(m+1)})}$ *for all $m < n$.*
2. *$\mathbf{d}_{z \vee \sim z,t}(s)$ is always consistent.*
3. $\vdash \sim \mathbf{d}_{z,t}(t^{(m)}) \wedge \sim \mathbf{d}_{\sim z,t}(t^{(m)}) \rightarrow \sim \mathbf{d}_{z \vee \sim z,t}(t^{(m)})$ *for all $m \leq n$.*

Proof. Simply by induction. ■

Lemma 6.8.3 *For $(S, \mathbf{d})$ a table and z a wff and $t \in S$, either $(S, \mathbf{d}_{z,t})$ or $(S, \mathbf{d}_{\sim z,t})$ is a table.*

Proof. From the preceding lemma. ■

Lemma 6.8.4 *Let $(S, \mathbf{d})$ be a table and $t \in S$ be as above, and assume*

$$\mathbf{d}(t) \vdash \Diamond_i z, \text{ for some } i \leq m + m'.$$

Let $q_1, \ldots, q_\ell$ be new atoms not appearing in any $\mathbf{d}(s)$, and let

$$C = z \wedge \bigwedge_{j=1}^{\ell} \sim q_j \wedge \Box_j q_j.$$

Let r be the least number such that $t * (2^{i+1}3^r) \notin S$. Let

$$S' = S \cup \{t * (2^{i+1}3^r)\},$$

and let $\mathbf{d}'(s) = \mathbf{d}(s)$ for s not below t. Let $\mathbf{d}'(t * (2^{i+1}3^r)) = C$. Let $\mathbf{d}'(t) = \mathbf{d}(t) \wedge \Diamond_i C$ and let $\mathbf{d}'(t^{(m+1)}) = \mathbf{d}(t^{(m+1)}) \wedge \Diamond_j \mathbf{d}'(t^{(m)})$ for each $m < n$, where $(t^{(m)})^* = 2^{(j+1)}(3^x)$ for some x.

Then $(S', \mathbf{d}')$ is a table.

Proof. Clearly, since $\mathbf{d}(t) \vdash \Diamond_i z$, then C is consistent. Now if some $\mathbf{d}'(t^{(n)}), n \geq 0$, were inconsistent, then $\vdash \sim \mathbf{d}'(t^{(n)})$ and hence by rule 3 and the fact that $\mathbf{d}(t) \vdash \Diamond_i z$, we obtain that $\mathbf{d}(t^{(n)})$ is not consistent, which is impossible. ■

Definition 6.8.5 A sequence of tables $(S_n, \mathbf{d}_n)$ is called a regular sequence iff:

1. $S_n \subseteq S_{n+1}$.
2. For $t \in S_n, \mathbf{d}_{n+1}(t) \vdash \mathbf{d}_n(t)$.
3. If $\mathbf{d}_n(t) \vdash \Diamond_i Z$, then for some $n' \geq n$ and $s \in S_{n'}$, $t\mathbf{R}_i^+ s$ and $d_{n'}(s) \vdash Z$.
4. For any $t \in S_n$ and any Z, there exists an $m \geq n$, for which $\mathbf{d}_m(t) \vdash Z$ or $\mathbf{d}_m(t) \vdash {\sim}Z$.

Lemma 6.8.6 *If A is a consistent wff, then there exists a regular sequence $(S_n, \mathbf{d}_n)$ with $\mathbf{d}_0(\varepsilon) \vdash A$.*

Proof. Let h be a function giving, for each n, a pair of the form (t, Z), where $t \in S^*$ and Z is a wff. We can assume that each pair (t, Z) is obtained an infinite number of times both as value for even n and as value for odd n.

Define the sequence $(S_n, \mathbf{d}_n)$ by induction.
Case $n = 0$: let $S_0 = \{\varepsilon\}, \mathbf{d}_0(\varepsilon) = A$.
Case $n + 1$: Consider $h(n) = (t, z)$:

1. If $t \in S_n$ and n is even, then if $(S_n, \mathbf{d}_{z,t})$ is a table, let $(S_{n+1}, \mathbf{d}_{n+1})$ be this table, otherwise let $(S_{n+1}, \mathbf{d}_{n+1}) = (S_n, \mathbf{d}_{\sim z,t})$.
2. If $t \in S_n, n$ is odd and $z = \Diamond_i z'$ and $\mathbf{d}_n(t) \vdash z$, then let $(S_{n+1}, \mathbf{d}_{n+1})$ be the table of lemma 6.8.4.
3. In all other cases let $S_{n+1} = S_n, \mathbf{d}_{n+1} = \mathbf{d}_n$.

It is easy to show that we obtain a regular sequence. ■

Proof of the irreflexivity lemma. Since A is consistent, let $(S_n, \mathbf{d}_n)$ be a regular sequence for A. Let $S = \bigcup S_n$. For $t \in S$, let $\Gamma_t = \{B \mid \exists n \mathbf{d}_n(t) \vdash B\}$. Each Γ_t is a complete and consistent theory. Let $\theta = \Gamma_\varepsilon$ and let $T = \{\Gamma_t \mid t \in S\}$. All the required properties can be verified for this T. ■

6.9 Completeness of U and S over the reals

6.9.1 Introduction

A completeness theorem for flow of time the real numbers $\mathbb{R}$, with syntax the Priorean connectives F and P ('in the future', 'in the past', respectively), is given in theorem 6.3.2 above. Notice that the proof actually shows that any *consistent set of formulae* has a model with the real numbers as flow of time. The Prior axiom of definition 6.3.1(c) enforced that any bounded set of time points definable by a temporal formula had an infimum and a supremum (see proposition 6.9.3 below). Since Dedekind completeness is fundamental to the reals, one might expect this property to be enough for completeness in a stronger logic, such as that with connectives U (Until) and S (Since). But in fact, the insistence on $\mathbb{R}$ as the flow of time renders a completeness theorem for infinite sets of *'US'* formulae impossible to attain. The reason is that the compactness theorem fails with U and S over the reals (proposition 6.9.2 below). Let us examine this issue.

The *compactness therorem* — that any theory (set of formulae) has a model if each of its finite subsets does — is a result that holds for classical first-order logic. We can say that a temporal logic $\mathcal{L}$ (such as '*FP*' with connectives F and P, or '*US*' with connectives U and S) has the *compactness property* over a class $\mathcal{K}$ of flows of time, if any $\mathcal{L}$-theory has a model with flow of time in $\mathcal{K}$ whenever every finite subset of it does. Now with our finitary rules of inference (each rule uses only finitely many premisses), a set of formulae is consistent if and only if every finite subset is consistent. The importance of compactness here is that if we have a strong completeness theorem for a class $\mathcal{K}$ of flows of time in a logic $\mathcal{L}$ — any consistent $\mathcal{L}$-theory has a model with flow of time in $\mathcal{K}$ — and the proof rules are finitary, then the compactness property holds for $\mathcal{L}$ over $\mathcal{K}$. Hence, for example, the temporal logics FP and US have the compactness property when the flow of time is taken to be the ordered rationals; and FP has the compactness property over the real numbers. So when compactness fails for $\mathcal{L}, \mathcal{K}$, we cannot hope for a strong completeness theorem for $\mathcal{L}$ over $\mathcal{K}$.

It is not difficult to see that the compactness property over integer or natural number time fails even in FP. For let us define $A_n = F(p \wedge F(p \wedge F(\ldots)) \ldots)$, with n ps (so formally we can let $A_0 = \top$ and $A_{n+1} = F(p \wedge A_n)$ for all $n < \omega$). Now let us consider the theory $\{A_n : n < \omega\} \cup \{FG{\sim}p\}$. Every finite subset of this theory has a model with integer or natural number flow of time, but the theory as a whole does not. As for the reals, although there is a completeness theorem for FP over the class $\{\mathbb{R}\}$, we will now show that compactness fails in $\mathbb{R}$ for US.

Notation 6.9.1 Recall that K^+A abbreviates $\sim U(truth, \sim A)$ and K^-A $\sim S(truth, \sim A)$. So K^+A holds in a structure at a point t if A is *dense to the right of* t: for all $u > t$ there is v with $t < v < u$ and A holding at v. Similarly, K^-A says that A is dense to the left.

Proposition 6.9.2 *The compactness theorem fails in the logic US for $(\mathbb{R}, <)$.*

Proof. Let q and r be atoms, and define formulae $A_n (n < \omega)$ by: $A_0 = F(r \wedge H{\sim}r)$, $A_{n+1} = F(q \wedge A_n)$. So A_n holds at t iff there is $u > t$ where $r \wedge H{\sim}r$ holds, and there are at least n distinct points strictly between t and u where q is true. Let $C = {\sim}K^+q \wedge G({\sim}K^+q \wedge {\sim}K^-q)$.

Now let $\Sigma = \{A_n : n < \omega\} \cup \{C\}$. Clearly, every finite subset of Σ has a model with flow of time isomorphic to $\mathbb{R}$. But Σ does not. For if $N = (\mathbb{R}, <, h)$ is one, so for some $t \in \mathbb{R}$ we have $\|B\|_t^h = 1$ for all $B \in \Sigma$, then there is $u > t$ in $\mathbb{R}$ with $\|r \wedge H{\sim}r\|_u^h = 1$. As $\|A_n\|_t^h = 1$ for all n, there are infinitely many points in the interval (t, u) where q is true. As the flow of time is $\mathbb{R}$, this implies that there is $v \in (t, u)$ where $K^+q \vee K^-q$ holds, or that K^+q holds at t. This contradicts the fact that C holds at t. ■

It follows that there is no set Z of axioms and finitary rules such that any Z-consistent set of US-formulae has a model with the real numbers as flow of time. However, we can still try to find axioms such that any *single* consistent formula has a model with the reals as flow of time. This is what we did in the case of integer time (theorem 6.3.8), and we will do the same for $\mathbb{R}$ in this section. The proof that we present is really just the observation that the elegant argument of Gurevich and Burgess in [Burgess and Gurevich, 1985], which goes back to [Läuchli and Leonard, 1966], can be linked with the usual completeness theorems for U and S proved above, to establish this weak completeness result for the reals. In addition to the Prior 'Dedekind completeness' axiom of definition 6.3.1c, we adopt a new axiom that roughly enforces separability (the reals have a countable dense subset, $\mathbb{Q}$).

The proof works as follows. With an appropriate set of axioms and rules, given a consistent formula A we first use theorem 6.3.2 to construct a model with flow of time isomorphic to the rationals in which all instances of our axioms are valid. Using the techniques of [Burgess and Gurevich, 1985], we then show that with respect to formulae at most as complex as A, this model is equivalent to one with the correct flow of time. This last part is done in two stages.

The argument of [Läuchli and Leonard, 1966] has been further developed by, e.g., Shelah [Shelah, 1975]. It has been applied fruitfully to decidability questions in temporal logic: [Burgess and Gurevich, 1985] used it to prove that the theory of the reals using U and S is decidable. It has also been used with other logics having similarly fixed semantics, such as the monadic second-order theory of one successor function (S1S). We believe that it will find further uses.

6.9.2 Axioms and rules for the real numbers

Recall that our rules of inference in theorem 6.3.2 were:
modus ponens

$$\frac{A \quad A \rightarrow B}{B} \text{ for all formulae } A, B,$$

generalization

$$\frac{A}{GA \wedge HA} \text{ for all formulae } A,$$

substitution

$$\frac{A}{A(q/B)} \text{ for any atom } q \text{ and formula } B.$$

Here, $A(q/B)$ is A with B substituted for q throughout.
IRR (irreflexivity)

$$\frac{q \wedge H{\sim}q \rightarrow A}{A} \text{ for all formulae } A \text{ and atoms } q \text{ not appearing in } A.$$

By choosing our set Z of axioms correctly we obtained several completeness theorems. In particular, let $Z_{\mathbb{Q}}$ consist of:

- $P\top, F\top$,
- $G(A \wedge GA \rightarrow B) \vee G(B \wedge GB \rightarrow A)$,
- $H(A \wedge HA \rightarrow B) \vee H(B \wedge HB \rightarrow A)$,
- $FA \rightarrow FFA$,

in addition to the axioms of K_t. Then $Z_{\mathbb{Q}}$ is complete for the flow of time $(\mathbb{Q}, <)$ (the ordered rationals) in FP, and also in US if we add the axioms of definition 6.4.3(a). And $Z_{\mathbb{Q}}$ plus:

- $FGA \wedge F{\sim}A \rightarrow F(GA \wedge {\sim}PGA)$ (future),
- $PHA \wedge P{\sim}A \rightarrow P(HA \wedge {\sim}FHA)$ (past),

is complete for $(\mathbb{R}, <)$ (the ordered real numbers) in FP only. These last two axioms are due to Prior.

We will define a set $Z_{\mathbb{R}}$ of axioms, and show that any US-formula A is consistent with $Z_{\mathbb{R}}$ iff it has a model with flow of time $\mathbb{R}$. We take $Z_{\mathbb{R}}$ to be $Z_{\mathbb{Q}}$ plus two further axioms: the Prior axiom, and another, which roughly speaking will let us enforce separability of the flow of time.

Let us discuss the Prior axiom first. It is really two axioms:

$$FGq \wedge F{\sim}q \to F(Gq \wedge {\sim}PGq) \quad \text{(future).}$$
$$PHq \wedge P{\sim}q \to P(Hq \wedge {\sim}FHq) \quad \text{(past).}$$

In fact we can usually dispense with either the past or the future axiom, relying instead on the other. The Prior axiom reflects the Dedekind completeness of the reals, but it can only mention *definable sets:* sets of time points of the form $\{t : A \text{ holds at } t\}$, for some formula A. We have the following proposition.

Proposition 6.9.3 *Let $N = (T, <, h)$ be a structure having a dense linear flow of time T without endpoints. Then the following are equivalent:*

1. *All instances of the Prior axiom are valid in N.*

2. *If S is a non-empty definable subset of N, then if S is bounded above it has a supremum, and if S is bounded below it has an infimum.*

Proof. Assume (1). Let A be a formula, and assume that $S = \{t \in T : \|A\|_t^h = 1\}$ is non-empty and bounded above. Consider the instance I of the future Prior axiom for $q = {\sim}A$: I is equivalent to $FG{\sim}A \wedge FA \to F(G{\sim}A \wedge HFA)$. Take $t \in T$ such that there is $s \in S$ with $s > t$. We have $\|I\|_t^h = 1$ as I is valid in N. Hence there is $u > t$ with $\|G{\sim}A \wedge HFA\|_u^h = 1$. Clearly, $u \geq s$ for all $s \in S$, and u is least such. So S has supremum u. The case where S is bounded below is similar and uses the past axiom.

Now assume (2). Let A be a formula and assume that $t \in T$ and

$$\|FGA \wedge F{\sim}A\|_t^h = 1.$$

Then $\{u \in T \mid \|GA\|_u^h = 1\}$ is non-empty and bounded below by t. Let u be its infimum. Then $u > t$, and as N has dense flow of time, $\|GA \wedge {\sim}PGA\|_u^h = 1$. So $\|F(GA \wedge {\sim}PGA)\|_t^h = 1$. As A was arbitrary, it follows that the future Prior axiom is valid in N. The proof for the other axiom is similar. ■

It follows that all instances of the Prior axiom are valid in any structure with Dedekind complete flow of time (e.g. $\mathbb{R}$).

In fact the Prior axiom is not sufficient for a completeness result for single formulae of US in $\mathbb{R}$. The formula (sep) below is valid in any structure with flow of time $\mathbb{R}$, but is not provable from $Z_{\mathbb{Q}}$ plus the Prior axiom.

$$\text{(sep)} \qquad K^+A \wedge \sim K^+(A \wedge U(A, \sim A)) \to K^+(K^+A \wedge K^-A).$$

The name 'sep' stands for 'separability' — we will see the connection with the separability of $\mathbb{R}$ later. Note that (sep) is not expressible in FP, as U, K^+, K^- are not definable.

Proposition 6.9.4

1. *Any substitution instance of (sep) is valid in any temporal structure N with flow of time isomorphic to $\mathbb{R}$.*

2. *(sep) is not provable from $Z_{\mathbb{Q}}$ plus the Prior axiom.*

Proof. (1). If not, there is $(\mathbb{R}, <, h)$, an atom q, and an element $t \in \mathbb{R}$ such that $\|K^+q \wedge \sim K^+(q \wedge U(q, \sim q))\|_t^h = 1$ and $\|K^+(K^+q \wedge K^-q)\|_t^h = 0$. Hence by taking $u \in h(q)$ with $u > t$ and u close enough to t, we can arrange that $S = \{v \in h(q) \mid t < v < u\} \subseteq \mathbb{R}$ has order type $\mathbb{Q}$, and for all $r \in \mathbb{R}$ there is some non-empty open interval I_r of $\mathbb{R}$ abutting r on the left or right (i.e. r is in its closure), and disjoint from S.

Now S has order type $\mathbb{Q}$, so it has gaps corresponding to the gaps (irrationals) of $\mathbb{Q}$. Let γ be a gap of S, defined by the Dedekind cut (X, Y) of S: so X, Y are unbounded open intervals of S, $X \cup Y = S$, and $x < y$ for all $x \in X$, $y \in Y$. Then $X \subseteq S \subseteq \mathbb{R}$; define $r(\gamma) \in \mathbb{R}$ to be the supremum in $\mathbb{R}$ of X. We order the gaps of S in the usual way. Then, clearly, if $\gamma < \delta$ are gaps of S there are points of S between $r(\gamma)$ and $r(\delta)$. Hence $I_{r(\gamma)} \cap I_{r(\delta)} = \varnothing$. It also follows that $r : \{\text{gaps of } S\} \to \mathbb{R}$ is an order-preserving (1-1) map. So $\{I_{r(\gamma)} \mid \gamma \text{ a gap of } S\}$ is an uncountable set of pairwise disjoint intervals of $\mathbb{R}$. This is impossible, as any non-empty interval in $\mathbb{R}$ contains a point of $\mathbb{Q}$, and $\mathbb{Q}$ is countable.

(2) We only sketch the proof, since we will not use it. If $M = (T, <, h)$ is a structure, we say for the duration of this proof that a formula $A(p_1, \ldots, p_n)$ is *valid* in M if for all formulas $B_1, \ldots, B_n$ using the atoms 'known to h' (in the domain of h), and all $t \in T$, we have $\|A(B_1, \ldots, B_n)\|_t^h = 1$. We also say that a rule of inference is *valid* in M if any formula derived by it from M-valid formulae is also M-valid.

It is enough to exhibit a structure M with irreflexive linear flow of time in which every point is 'named' by an atom, the axioms of $Z_{\mathbb{Q}}$ and the Prior axiom are valid, but (sep) is not valid. This is because all the rules of inference (modus ponens, substitution, generalisation and IRR) are valid in M. (In the case of IRR, for example, suppose that $r \wedge H{\sim}r \to A(p_1, \ldots, p_n)$ is M-valid, where r is an atom not occurring in A. As we can derive $A(p_1, \ldots, p_n)$ from this by the IRR rule, we must show that $A(p_1, \ldots, p_n)$ is M-valid. So let $B_1, \ldots, B_n$ be formulas known to M, and t be a point of M. Choose an atom r' true only at t. Then $r' \wedge H{\sim}r' \to A(B_1, \ldots, B_n)$ is a substitution instance of $r \wedge H{\sim}r \to A(p_1, \ldots, p_n)$, and so is true at t. So $A(B_1, \ldots, B_n)$ is true at t too. This proves that $A(p_1, \ldots, p_n)$ is M-valid, as required.) An induction on the length of proofs now shows that if A is provable from $Z_{\mathbb{Q}}$ plus the Prior axiom then A is valid in M. Hence if (sep) is not valid in M, it cannot be a theorem of $Z_{\mathbb{Q}}$ plus the Prior axiom.

We obtain M essentially by replacing each $x \in \mathbb{Q}$ by a copy of $\mathbb{Q} \cap [0, 1]$ in which the special atom q is true at 0 only. (See figure 6.8.)

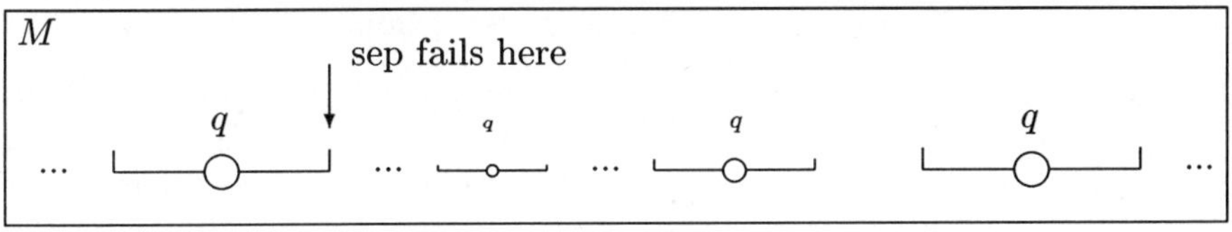

Figure 6.8

The flow of time here is countable, so by introducing a new atom for each of its points we can arrange that for each point t of M there is an atom that is true only at t. M knows only about these atoms and q. This completes the definition of M.

Note that the flow of time of M is dense and without endpoints, so by Cantor's theorem is order-isomorphic to $\mathbb{Q}$. Hence the axioms of $Z_{\mathbb{Q}}$ are valid in M. To see that the Prior axiom is also valid in M is a little harder. Let $A = A(q, r_1, \ldots, r_n)$ be a formula, where q is the special atom and $r_1, \ldots, r_n$ are 'naming' atoms, true only at the points $t_1, \ldots, t_n$ of M. Let t be a point of M, and suppose that $FGA \wedge F{\sim}A$ holds at t. Let S be the set of points in M where GA holds. We wish to find an infimum of S in M.

Let X be the finite set consisting of $t_1, \ldots, t_n$ plus those points 0, 1 and -1 lying in any copy of $\mathbb{Q} \cap [0, 1]$ that contains some t_i. We will prove that $\inf(S) \in X$. It suffices to show:

Claim: If u, v are points of M standing in the same order-relationship to the points of X (i.e. $u < x \iff v < x$ for all $x \in X$, and similarly for '=' and '>'), then $u \in S \iff v \in S$.

Proof of claim: We may suppose that u and v lie in copies U, V of $\mathbb{Q} \cap [0, 1]$, respectively. Let M' be the structure M restricted to the atoms $q, r_1, \ldots, r_n$ occurring in A. It is easily seen by induction on A that any automorphism of M' preserves S.

The proof has two cases. First assume that $U = V$. Then any order-preserving permutation φ of M that fixes (a) the points of $U \cap X$, and (b) the points of M that are not in U, will be an automorphism of M', and hence will satisfy $\varphi(S) = S$. Since we may choose such a φ with $\varphi(u) = v$, we see that $u \in S \iff v \in S$ in this case.

Next, assume that $U \neq V$. Let θ be any order-automorphism of $\mathbb{Q}$ that fixes the indices of all copies of $\mathbb{Q} \cap [0,1]$ containing points of X. Then θ induces an automorphism θ^* of M' in a natural way, by taking the kth copy of $\mathbb{Q} \cap [0,1]$ to the $\theta(k)$th copy. Since u, v lie in the same relation to the points of X, there are no points of X between U and V (inclusive). Hence there is θ with $\theta^*(U) = V$. Also, by the first case, all points of U agree on GA, so $U \subseteq S$ or $U \cap S = \varnothing$. Since $\theta^*(S) = S$, it now follows that $u \in S \iff v \in S$, as required. This proves the claim.

Now we show that (sep) is not M-valid. Each point t of M corresponds to ('is') some element of $\mathbb{Q} \cap [0,1]$. Choose any t corresponding to 1. Then $K^+q \wedge {\sim}K^+(q \wedge U(q, {\sim}q))$ is true at t. But any point u of M lies in a copy of $\mathbb{Q} \cap [0,1]$, so, clearly, there is an interval, abutting u on one side or the other, throughout which q is false. Hence $K^+q \wedge K^-q$ is false at u, for any u in M. This shows that (sep) fails at t, so that (sep) is not M-valid. ■

As (sep) is valid in $\mathbb{R}$, we want to be able to prove it! We will therefore adopt (sep) as a second axiom, in addition to the Prior axiom. We therefore make the following definition:

Definition 6.9.5 Let $Z_{\mathbb{R}}$ be $Z_{\mathbb{Q}}$, plus the Prior axiom, plus (sep). So $Z_{\mathbb{R}}$ is:

- K_t,
- $P\top$, $F\top$,
- $G(A \wedge GA \to B) \vee G(B \wedge GB \to A)$,
- $H(A \wedge HA \to B) \vee H(B \wedge HB \to A)$,
- $FA \to FFA$,
- (Prior) $FGA \wedge F{\sim}A \to F(GA \wedge {\sim}PGA)$ (future)
  $PHA \wedge P{\sim}A \to P(HA \wedge {\sim}FHA)$ (past),
- (sep) $K^+A \wedge {\sim}K^+(A \wedge U(A, {\sim}A)) \to K^+(K^+A \wedge K^-A)$.

Our main result is then:

Theorem 6.9.6 *Let A be a formula of US. Then A is consistent with $Z_{\mathbb{R}}$ iff it has a model with flow of time $\mathbb{R}$.*

The soundness part has already been proved in theorem 6.3.2 and propositions 6.9.3 and 6.9.4. We leave the details as an exercise.

Before proving completeness we need two preliminaries.

6.9.3 Expressive completeness

Let $\varphi(x, P_1, \ldots, P_k)$ be a monadic first-order formula of the signature $\{=, <\}$, where the P_i are unary relation symbols. Let $p_1, \ldots, p_k$ be atoms. As usual we will view these atoms as 'corresponding' to the monadic relation symbols P_i of φ; this association will not usually be stated explicitly.

We can then interpret φ in any temporal structure $N = (T, <, h)$ having interpretations for the p_i. We interpret $<$ as the earlier–later relation and P_i as $h(p_i)$ $(i = 1, \ldots, k)$.

Now if A is a temporal formula written with more or less any of the connectives that we have mentioned, there is an 'equivalent' monadic first-order formula $\alpha(x)$, with sole free variable x. For all $N = (T, <, h)$ and all $t \in T$, we have $\|A\|_t^h = 1$ iff $N \models \alpha(t)$. We can build a canonical $\alpha(x)$ equivalent to a given temporal formula A by induction on the structure of A. This α is called the *table* of A. See lemma 9.1.5 for details of its construction.

Clearly, the table of A is 'equivalent' to A in *any* temporal structure. So there is a monadic first-order formula equivalent to any temporal formula in US (also for FP and other logics). The converse question is of interest: is there a temporal formula equivalent to an arbitrary monadic first-order formula? Chapters 9–13 consider this question.

The answer clearly depends on the available connectives. However, it is known that for any finite set of connectives, there is a monadic formula φ such that no temporal formula is equivalent to φ in all temporal structures. This is because the theory of arbitrary partial orders has infinite 'Henkin dimension' — see chapter 13.

As we will see, more positive results can be obtained if we consider a particular flow of time. For instance, given a monadic formula we can find a US-formula equivalent to it in all temporal structures with (linear) Dedekind complete flow of time ($\mathbb{N}$, $\mathbb{Z}$, or $\mathbb{R}$, for example). We say that Until and Since are *expressively complete* for Dedekind complete time. See chapter 10 below.

However, Until and Since are not expressively complete for the rationals $\mathbb{Q}$. The connective γ^- mentioned in section 6.1 provides an example. This negative result prompted Jonathan Stavi to add further temporal connectives to gain expressive completeness for $\mathbb{Q}$.

Recall the definitions of the binary Stavi connectives U', S' from section 6.1. Informally, $U'(A, B)$ holds (in a structure with flow of time $\mathbb{Q}$, say) at t if there is an irrational ('gap') $\gamma > t$ such that (a) B holds at every rational in the open interval (t, γ), (b) for no interval (γ, u) with $u > \gamma$ does B hold at every rational in (γ, u), and (c) there is $u > \gamma$ such that A holds at every rational in (γ, u). S' is defined symmetrically for the past. In fact U' and S' have first-order tables: see section 6.1.

We borrow the chief result of chapter 11.

Theorem 6.9.7 *Let $\varphi(x)$ be a monadic first-order formula in the signature $\{=, <\}$. Then there is a temporal formula A_φ using connectives U, S, U', S' such that in any temporal structure $N = (T, <, h)$ with linearly ordered flow of time T, for any $t \in T$ we have*

$$N \models \varphi(t) \textit{ iff } \|A_\varphi\|_t^h = 1.$$

So the temporal logic with U, S, U', and S' is expressively complete *for linear time.*

By proposition 6.9.3, in the presence of the Prior axiom we can dispense with the Stavi connectives.

Definition 6.9.8 Call a temporal structure Q a *π-structure* if its flow of time is isomorphic to $(\mathbb{Q}, <)$, for each $t \in \mathbb{Q}$ there is an atom r such that $r \wedge H{\sim}r$ holds at t, and every substitution instance of the Prior axiom by US-formulae is valid in Q.

Proposition 6.9.9 *Let $\varphi(x)$ be a monadic first-order formula. Then there is a temporal formula A of US such that whenever $Q = (\mathbb{Q}, <, h)$ is a π-structure we have*

$$Q \models \varphi(t) \textit{ iff } \|A(t)\|_t^h = 1 \textit{ for all } t \in \mathbb{Q}.$$

Proof. By theorem 6.9.7, U, S, and the Stavi connectives are expressively complete for $\mathbb{Q}$. So there is a temporal formula B using U, S, U', S' such that for any Q as above, $Q \models \varphi(t)$ iff $\|B\|_t^h = 1$ for all $t \in Q$. Clearly, it suffices to show the following:

Claim: For any $USU'S'$ formula B there is a US formula B' such that $B \leftrightarrow B'$ is valid in any π-structure Q.

Proof of Claim: By induction on the structure of B. If B is atomic or $\top$ (truth) we let $B' = B$. We also let $({\sim}B)' = {\sim}B'$, $(C \wedge D)' = C' \wedge D'$,

$U(C,D)' = U(C',D')$, and similarly for $S(C,D)$. The proof that these definitions work is no problem. Assume that B is $U'(C,D)$. Inductively, there are US-formulae C', D' equivalent to C and D in any π-structure Q. Let Q be given. Assume for contradiction that $U'(C,D)$ holds at some $t \in Q$. Choose r so that $r \wedge H{\sim}r$ holds at t. Since D' is a formula of US, we can apply proposition 6.9.3 to the formula $H(D' \vee H{\sim}r)$. This shows that $\{x \in Q : \|H(D' \vee H{\sim}r)\|_x^h = 1\}$ has a rational supremum, which contradicts our assumption that $\|U'(C',D')\|_t^h = 1$. Hence ${\sim}U'(C,D)$ is valid in any π-structure Q, and we can set $(U'(C,D))' = \bot$ (falsity). We can set $(S'(C,D))' = \bot$ in a similar way. This completes the induction and proves the claim, and with it the proposition. ■

6.9.4 Lexicographic sums

We now turn to the second piece of material that we need. This is the notion of a lexicographic sum of structures. For reference, see [Rosenstein, 1982].

Definition 6.9.10 Let $(I,<)$ be a poset, and for each $i \in I$ let $N_i = (J_i, <, h_i)$ be a temporal structure. The *lexicographic sum* $\sum_{i \in I} N_i$ is defined to be the temporal structure $(X, \sqsubset, h)$, where:

- X is the disjoint union of the J_i $(i \in I)$,
- if $x, y \in X$, then $x \sqsubset y$ iff
  - for some $i \in I$, $x, y \in J_i$ and $N_i \models x < y$, or
  - for some $i_1, i_2 \in I$, $x \in J_{i_1}, y \in J_{i_2}$ and $i_1 < i_2$,
- $h(p) = \bigcup\{h_i(p) : i \in I\}$.

I is called the index set of the sum.

Notation 6.9.11 If I is finite (e.g., $I = \{i_1, i_2\}$ with $i_1 < i_2$), then we write $N_{i_1} + N_{i_2}$ for the lexicographic sum. We would write $N_{i_2} + N_{i_1}$ for the sum if we had $i_1 > i_2$. If the N_i are all isomorphic to N, say, we can write $N * I$ for the sum, rather than $\sum_{i \in I} N$.

Let C be a finite set of temporal structures. Let D be a dense linear ordering without endpoints, and let $\sigma : D \to C$ be a map such that

(δ) for each $N \in C$, the set $\sigma^{-1}(N) = \{d \in D : \sigma(d) = N\}$ is dense in D.

Then we define the *shuffle of C over D* to be the sum $\sum_{d \in D} \sigma(d)$. We are replacing each $d \in D$ by a copy of some $N \in C$ in such a way that each N 'occurs' densely in D.

Fact 6.9.12 The isomorphism type of a shuffle $\sum_{q \in \mathbb{Q}} \sigma(q)$ over $\mathbb{Q}$ does not depend on σ, given that σ satisfies (δ). (The proof uses a back and forth argument; see [Rosenstein, 1982] or [Gurevich, 1985].)

Indeed, we can go further.

Definition 6.9.13

1. From now on, fix a US-formula A that is consistent with $Z_{\mathbb{R}}$.

2. Also fix $k < \omega$ with $k \geq 2$ so that the table $\alpha(x)$ of A has quantifier depth less than k.

3. Let $k < \omega$ and let N be any temporal structure. The *k-theory of N* is the set of all monadic first-order sentences σ involving (at most) only those monadic predicates corresponding to the atoms occurring in A, and of quantifier depth at most k, such that $N \models \sigma$.

Now we quote:

Fact 6.9.14

1. Let I be a linear ordering and M_i, N_i $(i \in I)$ be linear temporal structures such that for each i, M_i and N_i have the same k-theory. Then the lexicographic sum $\sum_{i \in I} M_i$ has the same k-theory as $\sum_{i \in I} N_i$.

2. If C is a finite set of temporal structures, and D_1, D_2 are any dense linear orderings without endpoints, then the shuffle of C over D_1 has the same k-theory as the shuffle of C over D_2.

Proof. Since we only have a finite number of monadic relation symbols, these statements can be proved using Ehrenfeucht–Fraïssé games of length k (see [Gurevich, 1977], section 14). Note that as a corollary, if the M_i, N_i have the same full monadic theory (the same k-theory for all $k < \omega$), then so do $\sum_{i \in I} M_i$ and $\sum_{i \in I} N_i$. A similar result holds for the shuffle. ∎

6.9.5 Completeness for the reals

We can now prove completeness, so finishing theorem 6.9.6. The technique is as in [Burgess and Gurevich, 1985], following [Läuchli and Leonard, 1966]. See also [Doets, 1989], whose proof we follow closely.

Definition 6.9.15

1. We define an *R-structure* to be a π-structure satisfying the separability axiom (sep).

2. A linear temporal structure M is said to be *good* if there is a temporal structure R with flow of time (order-isomorphic to) an interval of $\mathbb{R}$ and having the same k-theory as M.

An interval of $\mathbb{R}$ must be one of: $\{a\}$, $(a,b) =_{\text{def}} \{x \in \mathbb{R} \mid a < x < b\}$, $[a,b] =_{\text{def}} \{x \in \mathbb{R} \mid a \leq x \leq b\}$, $[a,b) =_{\text{def}} \{x \in \mathbb{R} \mid a \leq x < b\}$, or $(a,b] =_{\text{def}} \{x \in \mathbb{R} \mid a < x \leq b\}$, where $a < b$ in $\mathbb{R}$.

We will show that every R-structure is good. By theorem 6.3.2 applied to definition 6.3.1c, A has an 'R-model', Q. Hence Q is good, so there is R, k-equivalent to Q, with flow of time an interval of $\mathbb{R}$. By considering the sentence $\forall x \exists y (y < x)$, which holds in Q and therefore (as $k \geq 2$) holds in R, we see that the flow of time of R has no left endpoint. Similarly, it has no right endpoint, and so is isomorphic to $\mathbb{R}$. Also, as $\exists x \alpha(x)$ has quantifier depth at most k and holds in Q, there will be $x \in R$ such that A holds in R at x. Hence A has a model R with flow of time $\mathbb{R}$, and we are done.

So we must prove:

Theorem 6.9.16 *Any R-structure $N = (\mathbb{Q}, <, h)$ is good.*

Proof. The idea of the proof is as follows. We would like to consider the maximal good subintervals of N (i.e. of $\mathbb{Q}$), but it could be that two such intervals overlap. We avoid this complication by considering the 'very good' intervals: roughly, those with the property that each of their subintervals is good. The union of countably many very good intervals of N, if an interval, is very good, so the maximal very good ('best') intervals of N do not overlap. In fact, we will obtain the best intervals as the equivalence classes of a certain equivalence relation. So the best intervals partition N. Using the Prior axiom, which is valid in N, we find that the best intervals are closed intervals of $\mathbb{Q}$, and using proposition 6.9.9 we build a US formula B that holds exactly at the left-hand endpoints of the best intervals.

Now if N is not good, there must be at least two best intervals in N. It follows that the best intervals occur densely in N — between any two there is a third. We then use the B-instance of the separability axiom (sep) to show that between any two best intervals there is a *one-point best interval.* Because of this, we can find in N an interval that is a suitable shuffle of best intervals. Because of the presence of the one-point intervals, this shuffle turns out to be very good. This is a contradiction, since no very good interval can properly contain a best interval. So N is good, as required.

So let $N = (\mathbb{Q}, <, h)$ be any R-structure.

Definition 6.9.17

1. If $S \subseteq \mathbb{Q}$ is non-empty, we write $N\lceil S$ for the temporal structure $(S, <, h')$ where $<$ is the ordering on S induced from $\mathbb{Q}$, and h' is given by $h'(p) = h(p) \cap S$.

2. Define a binary relation $\equiv$ on $\mathbb{Q}$ by: $a \equiv b$ iff (i) $a = b$, or (ii) for every $c < d$ between or equal to a, b, $N\lceil(c, d)$ is good.

Lemma 6.9.18

1. *If $a \equiv b$ and $a < c < b$ in $\mathbb{Q}$ then $a \equiv c$ and $c \equiv b$.*

2. *$\equiv$ is an equivalence relation on $\mathbb{Q}$.*

Proof. (1) This is clear from the definition.

(2) Clearly, $\equiv$ is reflexive and symmetric. For transitivity, suppose that $a \equiv b \equiv c$ in $\mathbb{Q}$; we show that $a \equiv c$. By (1), it suffices to consider the case where $a < b < c$. Let $a \leq a' < c' \leq c$ in $\mathbb{Q}$; we must show that $N\lceil(a', c')$ is good. We can suppose that $a' < b < c'$, since otherwise the result follows immediately from the definition of $\equiv$. By (1), $a' \equiv b \equiv c'$, so there are structures R_1 and R_2, k-equivalent to $N\lceil(a', b)$ and $N\lceil(b, c')$, respectively, and each with flow of time an interval of $\mathbb{R}$. As before, we see that R_1 has no right endpoint and R_2 no left endpoint. So the structure $R_1 + N\lceil\{b\} + R_2$ has flow of time isomorphic to an interval of $\mathbb{R}$. By fact 6.9.13, this structure is k-equivalent to $N\lceil(a', c')$, which is therefore good. Thus $a \equiv c$, as required. ∎

Lemma 6.9.19 $\equiv$ *is definable in the monadic first-order logic over the atoms of A.*

Proof. Since k is finite, there are finitely many k-theories τ in the monadic language over the atoms of A, and each is finite. So each τ can be written as a single sentence, $\bigwedge \tau$. Let ρ_i $(i < g)$ be a list of all these sentences that have a model with flow of time isomorphic to $\mathbb{R}$. Let x, y be new point variables not occurring in the ρ_i. We relativize each ρ to the interval (x, y), by replacing all quantifiers $\exists z \varphi$ by $\exists z(x < z \wedge z < y \wedge \varphi)$. We obtain formulae $\rho_i(x, y)$. So if $a < b$ in $\mathbb{Q}$, we have $N \models \bigvee_{i<g} \rho_i(a, b)$ iff $N\lceil(a, b)$ is good. So if we define

$$\alpha(x, y) = \forall zt[x \leq z < t \leq y \rightarrow \bigvee_i \rho_i(z, t)],$$

then $x \equiv y$ is definable by the formula $x = y \vee (x < y \wedge \alpha(x, y)) \vee (y > x \wedge \alpha(y, x))$. ∎

A *closed* interval of $\mathbb{Q}$ is one that contains its endpoints (if any). Thus, it has the form $\mathbb{Q}$, $(-\infty, a]$, $[a, \infty)$ or $[a, b]$, for $a \leq b$ in $\mathbb{Q}$.

Corollary 6.9.20

1. *Each $\equiv$-class E is a closed interval of $\mathbb{Q}$, and $N\lceil E$ is a good structure.*

2. *If N is not good then there are at least two $\equiv$-classes, and $\mathbb{Q}/\equiv$ is a dense ordering.*

Proof. (1) Let E be a $\equiv$-class with at least two points. Choose $e \in E$. As N is a π-structure there is an atom r with $\|r \wedge H{\sim}r\|_e^h = 1$. Regarding r as a monadic unary predicate symbol, let $\varepsilon(x)$ be the first-order formula $\exists y(x \equiv y \wedge r(y) \wedge \forall z(z < y \rightarrow {\sim}r(z)))$. Then $N \models \varepsilon(s)$ iff $s \in E$. By proposition 6.9.9, there is a temporal formula D of US such that if $a \in \mathbb{Q}$, $\|D\|_a^h = 1$ iff $N \models \varepsilon(a)$. Now by lemma 6.9.18, E can be regarded as an interval with endpoints $a, b \in \mathbb{R} \cup \{\infty, -\infty\}$. So by proposition 6.9.3 we obtain $a, b \in \mathbb{Q} \cup \{\infty, -\infty\}$.

It is now enough to show that for all $c \in (a, b)$, $N\lceil(a, c)$ and $N\lceil(c, b)$ are good. For, granted this, it follows that if $a \in \mathbb{Q}$ then $a \equiv c$, so $a \in E$; and similarly for b. Moreover, assuming that $a, b \in \mathbb{Q}$, we have $N\lceil E = N\lceil\{a\} + N\lceil(a, e) + N\lceil\{e\} + N\lceil(e, b) + N\lceil\{b\}$, and each component of this sum is good. If $a \notin \mathbb{Q}$ or $b \notin \mathbb{Q}$, we merely omit $N\lceil\{a\}$, $N\lceil\{b\}$ from the sum, as appropriate. So we can deduce from fact 6.9.13 as before that $N\lceil E$ is good.

So take $c \in (a, b)$. We will prove that $N\lceil(c, b)$ is good; the case $N\lceil(a, c)$ is similar. Take elements $c = c_0 < c_1 < \ldots < b$ in E, so that for all $d < b$ there is $i < \omega$ such that $c_i > d$. Then $c_i \equiv c_{i+1}$, so that $N\lceil(c_i, c_{i+1})$ is good for all i. Take R_i with flow of time an (open) interval of $\mathbb{R}$ and having the same k-theory as $N\lceil(c_i, c_{i+1})$. Then by fact 6.9.13, the structure $R = \sum_{i<\omega}[(N\lceil\{c_i\}) + R_i]$ has the same k-theory as $N\lceil(c, b)$. Clearly, R also has flow of time isomorphic to an interval of $\mathbb{R}$. So $N\lceil(c, b)$ is good.

(2) By (1), each $\equiv$-class is an interval of $\mathbb{Q}$, so that the set $\mathbb{Q}/\equiv$ of equivalence classes is itself canonically linearly ordered. We write '$<$' for this ordering; it is induced from $<$ on $\mathbb{Q}$. If $\mathbb{Q}$ itself is a $\equiv$-class, then by the above, N is good. So if N is not good then there are at least two classes. Denseness of $\mathbb{Q}/\equiv$ follows trivially from the denseness of $\mathbb{Q}$ and the fact that the $\equiv$-classes are closed intervals. ■

Assume for contradiction that N is not good. By the corollary, $(\mathbb{Q}/\equiv, <)$ is densely ordered. If E is a $\equiv$-class, write $\tau(N\lceil E)$ for the k-theory of $N\lceil E$ in the monadic language over the atoms of A. There are only finitely many k-theories, so for all $E < E'$ in $\mathbb{Q}/\equiv$, the set

$$K(E, E') = \{\tau(N\lceil E^*) \mid E^* \in \mathbb{Q}/\equiv, E < E^* < E'\}$$

is finite. Choose $E < E'$ so that its cardinality n is as small as possible. Let a be the right-hand endpoint of E, and b the left-hand endpoint of E'.

Lemma 6.9.21 *$N\lceil(a,b)$ is good.*

Proof. We begin with:

Claim: There is a one-point $\equiv$-class E_1 with $E < E_1 < E'$.

Proof of Claim: Let B be a US-formula equivalent to $\sim\exists y(y < x \wedge y \equiv x)$. Then B holds exactly at the left-hand endpoint of each $\equiv$-class, and at c if $\{c\}$ is a class. Because $\mathbb{Q}/\equiv$ is densely ordered, $\|K^+B \wedge \sim K^+(B \wedge U(B, \sim B))\|_a^h = 1$. By the separability axiom (sep), $\|K^+(K^+B \wedge K^-B)\|_a^h = 1$, so there is $c \in (a,b)$ with $\|K^+B \wedge K^-B\|_c^h = 1$. Let E_1 be the class of c. Then $E < E_1 < E'$. If there is $d \in E_1$ with $d \neq c$, then B fails at all points between c and d. This is a contradiction to the choice of c. Hence $E_1 = \{c\}$, proving the claim.

By corollary 6.9.20(1), for each $\tau \in K(E,E')$ we may take a structure $R_\tau \models \tau$ having flow of time isomorphic to a (closed) interval of $\mathbb{R}$. Because $k \geq 2$, by considering the sentence $\exists x \forall y(x = y)$ we see that $R_{\tau(N\lceil E_1)}$ also has flow of time consisting of a single point, so we can assume that $R_{\tau(N\lceil E_1)} = N\lceil E_1$ in this case.

By corollary 6.9.20(2), the interval (E,E') of $\mathbb{Q}/\equiv$ has order type $\mathbb{Q}$. Let D be its Dedekind completion: so D is order-isomorphic to $\mathbb{R}$. Define $\sigma : D \to \{R_\tau : \tau \in K(E,E')\}$ by

$$\sigma(d) = R_{\tau(N\lceil d)} \text{ if } d \in \mathbb{Q}/\equiv, \text{ and } N\lceil E_1 \text{ otherwise.}$$

As $|K(E,E')|$ is minimal, it follows that for all $\tau \in K(E,E')$ the set $\{E^* \in \mathbb{Q}/\equiv \mid E < E^* < E', \tau(N\lceil E^*) = \tau\}$ is dense in (E,E'), and hence dense in D. So the sum $R = \sum_{d \in D} \sigma(d)$ is a shuffle. R is easily seen to be dense without endpoints. Because the index set D and each $\sigma(d)$ are Dedekind complete, the latter with endpoints, R is also Dedekind complete. Each R_τ has a countable dense subset $S(R_\tau)$, say; then because $|\sigma(d)| = 1$ for $d \in D \setminus (E,E')$, $\sum_{d \in (E,E')} S(\sigma(d))$ is a countable dense subset of R. Hence R is separable. As is well known (see [Rosenstein, 1982]), this means that R has flow of time isomorphic to $\mathbb{R}$.

By fact 6.9.13(1), $N\lceil(a,b)$ has the same k-theory as $\sum_{d \in (E,E')} \sigma(d)$. By (2) of the fact, this last has the same k-theory as R. Hence $N\lceil(a,b)$ is good. ∎

Corollary 6.9.22 *$a \equiv b$.*

Proof. Let $a \leq c < d \leq b$; we must show that $N\lceil(c,d)$ is good. If $c \equiv d$ then this follows by definition of $\equiv$; so assume $c \not\equiv d$. Let the $\equiv$-classes of

c, d be E_c, E_d respectively. Let c' be the right-hand endpoint of E_c, and d' the left-hand endpoint of E_d. Then as $c \equiv c'$, $N\lceil(c, c')$ is good. Similarly, $N\lceil(d', d)$ is good. But by lemma 6.9.21, which applies since $K(E_c, E_d)$ also has cardinality n, $N\lceil(c', d')$ is also good. Hence by fact 6.9.13 the sum

$$N\lceil(c,d) = N\lceil(c,c') + N\lceil\{c'\} + N\lceil(c',d') + N\lceil\{d'\} + N\lceil(d',d)$$

is good. ■

But we assumed that $a \in E$, $b \in E'$ and $E < E'$. This contradicts the corollary, and completes the proof of theorems 6.9.16 and 6.9.6. ■

Remark 6.9.23

1. Similar arguments will prove completeness theorems for appropriate axioms of US over $\mathbb{N}$ and $\mathbb{Z}$. See [Reynolds, 1993a; Venema, 1991a].
2. Reynolds [Reynolds, 1992c] has eliminated the use of the IRR-rule in the axiomatization of US over $\mathbb{R}$. To do this, he adds axioms such as those of Burgess [Burgess, 1982a] to describe the semantics of U and S, and proves corollary 6.9.20(1) by a longer argument.

6.10 Sahlqvist's theorem

This theorem has two aspects. The first is to do with preservation of certain modal formulae (persistence), and the second is about finding first-order equivalents for them (correspondence). The formulae considered (the so-called *Sahlqvist formulae)* have a restricted syntactic form, which is none the less general enough to be useful and includes several well-known examples, such as $Gp \to p$ (reflexivity), $Gq \to GGq$ (transitivity), $Fp \wedge Fq \to F(p \wedge q) \vee F(p \wedge Fq) \vee F(Fp \wedge q)$ (linearity), $GGq \to Gq$ (density), $FGp \to GFp$ (the so-called Church–Rosser formula), etc.

Roughly speaking, a formula is said to be persistent if whenever there is an assignment making it false, then we can choose the assignment to be 'nice' in a certain sense. The notion of 'nice' is related to the assignments h^* in *canonical structures* (consisting of all maximal consistent theories, as in chapter 3), where $h^*(q) = \{\Gamma \mid q \in \Gamma\}$ for each atom q. Specifically, an assignment h is nice if for all atoms q there is a formula A such that $h(q) = \{t : \|A\|_t^{h^*} = 1\}$. So if all substitution instances of a persistent formula are true at $t \in T$ in a canonical structure $(T, <, h^*)$, then the formula itself is true at t in $(T, <, h)$ for any assignment h into T. We will show that the Sahlqvist formulae are persistent in this sense.

The validity of a formula at a time point t under all assignments sometimes corresponds to a first-order (non-monadic) statement about t. We saw that the formula $A = Gq \to q$ corresponds to reflexivity, in that for any T and $t \in T, t < t$ iff $\|A\|_t^h = 1$ for all assignments h into T. A consequence of the proof of persistence for Sahlqvist formulae is that each one corresponds to a first-order statement in this way.

The theorem was originally proved in [Sahlqvist, 1975]. The proof was complicated. A simpler proof of the correspondence part of the theorem was later given by van Benthem [van Benthem, 1985], and later still, Sambin and Vaccaro gave a simple proof of both halves [Sambin and Vaccaro, 1989]. We present a stylistic variant of the latter proof. In section 6.11 we will give an argument similar to that of Venema [Venema, 1992] to apply the theorem to situations where the assignment is not so nice.

Setting

The modal language will have atoms, $\top$ (true), $\bot$ (falsity), the boolean connectives $\wedge, \vee, \sim$, and a set of box operators and their corresponding diamonds. Note that $\to$ is an abbreviation. Unless otherwise stated we will always assume that formulae are written with atoms from a fixed infinite set.

Semantics will be given by a *Kripke frame* $(T, \bar{r})$ where T is a non-empty set and $\bar{r} = (r_1, r_2, \ldots)$ is a tuple of accessibility relations, one for each box/diamond. We sometimes write $\Box_r, \Diamond_r$ for the box and diamond whose semantic accessibility relations are r. An *assignment* into T is a map h assigning a subset $h(p)$ of T to each atom p. A *structure* is a triple $(T, \bar{r}, h)$, where $(T, \bar{r})$ is a Kripke frame and h an assignment into T. The semantics of formulae are defined as usual; we will write $h(A)$ for $\{t \in T \mid A \text{ is true at } t \text{ under } h\}$. We write $\|A\|_t^h = 1$ iff $t \in h(A)$, and $\|A\|_t^h = 0$ otherwise.

6.10.1 Descriptive assignments

To state and prove Sahlqvist's theorem we need some preliminaries. We begin with *types*, which will allow us to define the *descriptive (nice) assignments*. Then in section 6.10.2 we examine the syntactic restrictions on formulae required for Sahlqvist's theorem.

Definition 6.10.1

1. A *type* is a set of modal formulae.[1] We write Γ, Δ, etc., for types. If

[1]The terms 'type' and 'theory' are equivalent, but we use 'theory' more usually for an element of a canonical structure.

$(T, \overline{r}, h)$ is a structure, an element $t \in T$ is said to *realize* Γ (*over* h, or *in* (T, h)) if $\|A\|_t^h = 1$ for all $A \in \Gamma$. We write $h(\Gamma) \subseteq T$ for the set of realizations of Γ over h. A type Γ is said to be (h)-consistent if every finite subset of Γ is realized in (T, h). A subset $S \subseteq T$ is said to be *type-definable* (*over* h) if there is a type Γ with $S = h(\Gamma)$.

2. Let $(T, \overline{r}, h^*)$ be a structure. h^* is said to be *descriptive* (for $(T, \overline{r})$) if the following three conditions hold.

   (separation) If $t \neq u$ in T then there is A such that $\|A\|_t^{h^*} \neq \|A\|_u^{h^*}$.

   (saturation) Any h^*-consistent type is realized over h^* in T.

   (type-definable accessibility, tda) For every $t \in T$, and every accessibility relation r, the set $\{u \in T \mid r(t, u)\}$ is type-definable over h^*.

For any h we can obtain a topology on T by taking the set of basic closed and open sets to be $\{h(A) \mid A \text{ a formula}\}$. The separation property above corresponds to the Hausdorff property of the topology; saturation corresponds to its compactness, and (tda) corresponds to its being 'point-closed'. In fact, h^* is descriptive iff $(T, \overline{r}, \{h^*(A) \mid A \text{ a formula}\})$ is a *descriptive general frame,* and all such frames arise in this way. See [Sambin and Vaccaro, 1989] for more details.

Example 6.10.2 Let $(T, \overline{r}, h)$ be a *canonical structure*, so that:

- T is the set of all maximal consistent sets Γ in some consistent modal logic Z.
- For each r in $\overline{r}, r(\Gamma, \Delta)$ holds iff for all A, $A \in \Delta$ whenever $\Box_r A \in \Gamma$.
- $h(q) = \{\Gamma \mid q \in \Gamma\}$ for all atoms q.

Recall that:

($*$) For all formulae $A, h(A) = \{\Gamma \mid A \in \Gamma\}$.

This was proved by induction on A.

We check that h is descriptive. Separation is clear, as T is made of sets of formulae; and (tda) holds by definition of the accessibility relations. Saturation holds because if Γ is a consistent type in the sense defined above, then it is Z-consistent, so is contained in some $\Delta \in T$. By ($*$), Δ realizes Γ.

Properties of types

For types Γ and descriptive assignments h^* into a frame $(T, \overline{r})$, the sets $h^*(\Gamma)$ have nice properties. We will examine these first, as they are of general use. Let $(T, \overline{r})$ be a frame. We will write h, h^*, etc., for assignments into T, and h^* will usually be descriptive.

Definition 6.10.3

1. A type Γ is said to be $(\wedge)$-*closed* if whenever $B, C \in \Gamma$ then $B \wedge C \in \Gamma$. We write $cl(\Gamma)$ for the set of all finite conjunctions of formulae in Γ. Clearly, $cl(\Gamma)$ is a closed type, and $h(cl(\Gamma)) = h(\Gamma)$ for all h.

2. Let Γ, Δ be types. We define the following types:
   $\Gamma \wedge \Delta = \{B \wedge C \mid B \in \Gamma, C \in \Delta\}$
   $\Gamma \vee \Delta = \{B \vee C \mid B \in \Gamma, C \in \Delta\}$
   $\Box\Gamma = \{\Box B \mid B \in \Gamma\}$
   $\Diamond\Gamma = \{\Diamond B \mid B \in \Gamma\}$.

We show first that under *descriptive* assignments h^*, the 'semantics' of these types behave as we expect.

Lemma 6.10.4 *Let Γ, Δ be types, and let p, q be atoms. Suppose that h, h^* are assignments into T such that $h(p) = h^*(\Gamma)$ and $h(q) = h^*(\Delta)$. Then:*

1. *$h(p \wedge q) = h^*(\Gamma \wedge \Delta)$.*

2. *$h(p \vee q) = h^*(\Gamma \vee \Delta)$.*

3. *$h(\Box p) = h^*(\Box\Gamma)$ for each $\Box$ in the language.*

4. *If Γ is a $\wedge$-closed type and h^* is descriptive, then also $h(\Diamond p) = h^*(\Diamond\Gamma)$ for each $\Diamond$ in the language.*

Proof.

1. This says that $h^*(\Gamma \wedge \Delta) = h^*(\Gamma) \cap h^*(\Delta)$. But obviously, an element $t \in T$ realizes $\Gamma \wedge \Delta$ over h^* iff $\|B \wedge C\|_t^{h^*} = 1$ for all $B \in \Gamma$ and $C \in \Delta$, iff t realizes Γ and Δ over h^*.

2. Here we have to prove that $h^*(\Gamma \vee \Delta) = h^*(\Gamma) \cup h^*(\Delta)$. If $t \in T$ realizes Γ over h^* then it must realize $\Gamma \vee \Delta$, and similarly for Δ, so '$\supseteq$' is clear. For '$\subseteq$', note that if $t \in T$ realizes neither Γ nor Δ (over h^*), then there is $B \in \Gamma$ and $C \in \Delta$ such that $\|B\|_t^{h^*} = \|C\|_t^{h^*} = 0$. Thus $\|B \vee C\|_t^{h^*} = 0$, and as $B \vee C \in \Gamma \vee \Delta, t$ does not realize $\Gamma \vee \Delta$ over h^*.

3. Let $\Box$ have accessibility relation r, and let $t \in T$. Then $t \in h(\Box p)$ iff $u \in h(p)$ for all u such that $r(t,u)$, iff u realizes Γ over h^* whenever $r(t,u)$ holds. This is equivalent to saying that t realizes $\Box\Gamma$ over h^*.

4. Let $\Diamond$ have accessibility relation r. First assume that $t \in h(\Diamond p)$. So there is $u \in h(p)$ with $r(t,u)$, so that u realizes Γ (over h^*). Hence, for each $B \in \Gamma$, $\|\Diamond B\|_t^{h^*} = 1$, so t realizes $\Diamond\Gamma$ over h^*.

   Conversely, suppose that $t \in h^*(\Diamond\Gamma)$. Let $\Box$ be '$\sim\Diamond\sim$', and using (tda) choose a type Δ such that $h^*(\Delta) = \{u \in T \mid r(t,u)\}$. We claim that the type $\Gamma \cup \Delta$ is h^*-consistent. For let $X_1, \ldots, X_n \in \Gamma$. Then as Γ is $\wedge$-closed, $X = X_1 \wedge \ldots \wedge X_n \in \Gamma$. As t realizes $\Diamond\Gamma$, $\|\Diamond X\|_t^{h^*} = 1$, so there is $u \in h^*(\Delta)$ with $\|X\|_u^{h^*} = 1$. It follows that any finite subset of $\Gamma \cup \Delta$ is realized over h^*, so $\Gamma \cup \Delta$ is h^*-consistent, as claimed.

   As h^* is descriptive, by saturation there is u realizing $\Gamma \cup \Delta$ over h^*. Then $u \in h^*(\Gamma) = h(p)$, and as u realizes Δ we have $r(t,u)$. Thus $t \in h(\Diamond p)$ as required.

■

We now consider iterations of the accessibility relations.

Definition 6.10.5

1. Let β be any sequence of boxes and diamonds. We define the binary relation r_β on T by induction. Let $t, u \in T$. If β is empty then $r_\beta(t,u)$ holds iff $t = u$. If $\beta = \gamma \cdot \Box_r$ or $\gamma \cdot \Diamond_r$, then $r_\beta(t,u)$ holds iff there exists $v \in T$ with $r_\gamma(t,v)$ and $r(v,u)$.

2. If h is an assignment into $T, t \in T$, and β is a sequence of boxes, write $tp_h(t,\beta)$ for $\{A \mid \|\beta A\|_t^h = 1\}$. We will write simply $tp_h(t)$ if β is the empty sequence, and we will omit the h if it is clear from the context. If $M = (T, \overline{r}, h)$ is a structure, and $t \in T$, we may write $\mathrm{tp}_M(t)$ for $\mathrm{tp}_h(t)$.

Remark 6.10.6 For any sequence δ of diamonds and type Γ, define the type $\delta\Gamma$ by induction on the length of δ. If δ is empty then let $\delta\Gamma = \Gamma$; and for any sequence δ and diamond $\Diamond$, define $(\Diamond\delta)\Gamma$ to be $\Diamond(cl(\delta\Gamma))$. Then by lemma 6.10.4 and induction on the length of δ, if h^* is descriptive then for all $u \in T$,

$$\{t \in T \mid r_\delta(t,u)\} = h^*(\delta \cdot \mathrm{tp}_{h^*}(u))$$

is type-definable. See figure 6.9.

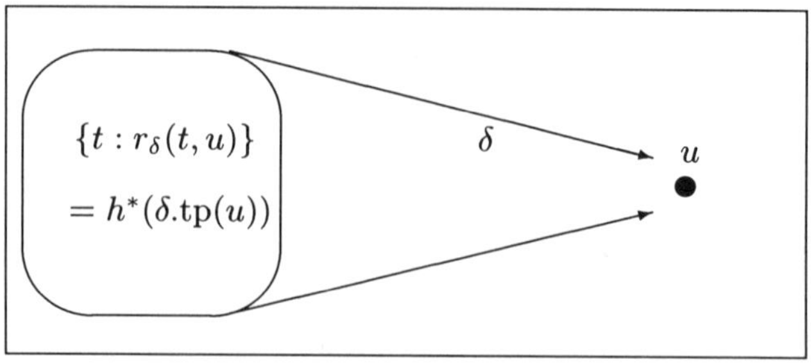

Figure 6.9

What about the 'inverse situation'? In fact, if $t \in T$ and h^* is descriptive then $\{u \in T \mid r_\beta(t,u)\}$ is also type-definable over h^*. We have this for β of length 0 or 1 by separation and (tda); the general case is proved below. Here, rather than add diamonds to $\mathrm{tp}(u)$, we will remove boxes from $\mathrm{tp}(t)$.

Lemma 6.10.7 *Let $M = (T, \overline{r}, h)$ be arbitrary, let β be a sequence of boxes, and let $t \in T$. Then the following are equivalent:*

1. *$\{u \in T \mid r_\beta(t,u)\}$ is type-definable over h.*
2. *$\{u \in T \mid r_\beta(t,u)\} = h(tp(t,\beta))$.*

In particular, if h^* is descriptive then $\{u \in T \mid r(t,u)\}$ is type-definable over h^* by $tp_{h^*}(t, \Box_r)$.

Proof. Only (i) $\Rightarrow$ (ii) requires proof. If $u \in T$ and $r_\beta(t,u)$, then clearly u realizes $\mathrm{tp}(t,\beta)$, so that $\{u \in T \mid r_\beta(t,u)\} \subseteq h(\mathrm{tp}(t,\beta))$. Conversely, suppose that Γ is a type such that $r_\beta(t,u)$ holds iff u realizes Γ over h. If $A \in \Gamma$ then, clearly, $\|\beta A\|_t^h = 1$, so that $\Gamma \subseteq \mathrm{tp}(t,\beta)$. Hence $\{u \in T \mid r_\beta(t,u)\} = h(\Gamma) \supseteq h(\mathrm{tp}(t,\beta))$. ■

Lemma 6.10.8 (generalized tda, 'gtda') *Assume that h^* is descriptive, and let $t \in T$. Then for any sequence β of boxes, $\{u \in T \mid r_\beta(t,u)\}$ is type-definable over h^*, by $tp(t,\beta)$.*

Proof. By induction on the length of β. If this is 0 then the result follows by the separation property of h^*. Assume the result for γ, and suppose that $\Box$ is a box with accessibility relation r. If $r_{\gamma\cdot\Box}(t,u)$ then there is v such that $r_\gamma(t,v)$ and $r(v,u)$. So for all A such that $\|\gamma\cdot\Box A\|_t^{h^*} = 1$, $\|\Box A\|_v^{h^*} = 1$, so $\|A\|_u^{h^*} = 1$. Hence u realizes $\mathrm{tp}(t, \gamma\cdot\Box)$.

Conversely, suppose u realizes $\mathrm{tp}(t,\gamma\cdot\Box)$ (over h^*). We must prove that there is $v \in T$ with $r_\gamma(t,v)$ and $r(v,u)$. As h^* is saturated, by the remark above it suffices to prove that $\mathrm{tp}(t,\gamma) \cup \Diamond\mathrm{tp}(u)$ is consistent. So let $A_1,\ldots,A_n \in \mathrm{tp}(t,\gamma)$ and $\Diamond B_1,\ldots,\Diamond B_m$ be such that $\|B_1\|_u^{h^*} = \ldots = \|B_m\|_u^{h^*} = 1$. Clearly, $B = B_1 \wedge \ldots \wedge B_m$ is also true at u. By assumption on u, $\|\gamma\Box{\sim}B\|_t^{h^*} = 0$, so there is $v \in T$ with $r_\gamma(t,v)$ at which $\Box{\sim}B$ is false, i.e. $\|\Diamond B\|_v^{h^*} = 1$. As $\|\gamma A_i\|_t^{h^*} = 1$, we see that $\|A_i\|_v^{h^*} = 1$ (all i). So $\|A_1 \wedge \ldots \wedge A_n \wedge \Diamond B\|_v^{h^*} = 1$, and v witnesses the consistency of the above type. This completes the proof. ∎

Figure 6.10 illustrates this.

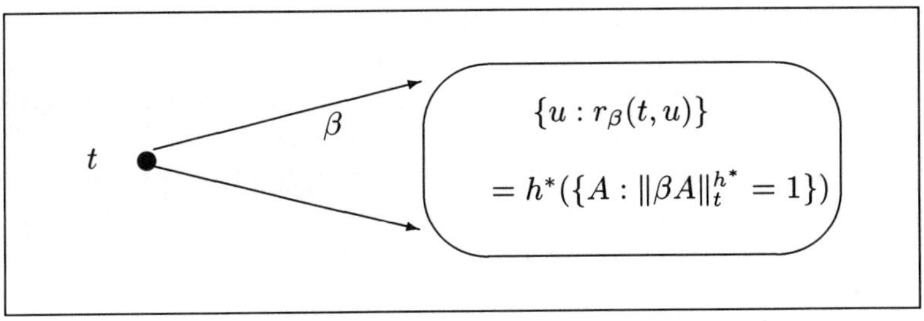

Figure 6.10

We round off our look at types with another corollary of lemma 6.10.4, showing that for a fixed language and set V of valid formulae there is essentially at most one structure with a descriptive assignment. If V is axiomatizable, the canonical structure is descriptive (example 6.10.3), so there is exactly one.

Definition 6.10.9 Let $M = (T,\overline{r},h), M' = (T',\overline{r},h')$ be structures, where h, h' are defined on the same set of atoms.

1. A formula A is said to be *valid in M* if $h(A) = T$.

2. We write $M \equiv M'$ if for all formulae A, A is valid in M iff A is valid in M'.

3. A map $\alpha : T \to T'$ is said to be an *isomorphism* from M to M' if α is a bijection from T on to T' such that:

   (a) for all accessibility relations r and all $t,u \in T, r(t,u)$ holds iff $r(\alpha(t),\alpha(u))$ holds;

   (b) for all atoms p and all $t \in T, t \in h(p)$ iff $\alpha(t) \in h'(p)$.
   We write $M \cong M'$ if there is an isomorphism from M to M'.

Corollary 6.10.10 *Let L be any infinite set of atoms, and suppose that $M = (T, \bar{r}, h)$ and $M' = (T', \bar{r}, h')$ are L-structures such that h and h' are descriptive. Assume that $M \equiv M'$. Then there is a unique isomorphism $\alpha : T \to T'$.*

Proof. Let Γ be any type. Clearly, Γ is M-consistent iff for all $A_1, \ldots, A_n \in \Gamma$, $\sim(A_1 \wedge \ldots \wedge A_n)$ is not valid in M. Since $M \equiv M'$, Γ is M-consistent iff Γ is M'-consistent. By saturation, a type is M-consistent iff it is realized in M; and similarly for M'. Thus M and M' realize the same types.

By separation, $t = u$ in T iff $\mathrm{tp}_M(t) = \mathrm{tp}_M(u)$, and similarly for M'. Define a bijection $\alpha : T \to T'$ by letting $\alpha(t)$ be the unique element $u \in T'$ with $\mathrm{tp}_{M'}(u) = \mathrm{tp}_M(t)$. Then for all $t \in T$, $\mathrm{tp}_M(t) = \mathrm{tp}_{M'}(\alpha(t))$. So by lemma 6.10.7, for all t, u in T and all $r, r(t, u)$ holds iff $\mathrm{tp}_M(t, \Box_r) \subseteq \mathrm{tp}_M(u)$, iff $\mathrm{tp}_{M'}(\alpha(t), \Box_r) \subseteq \mathrm{tp}_{M'}(\alpha(u))$, iff $r(\alpha(t), \alpha(u))$ holds. Thus α preserves the accessibility relations; and it clearly preserves the atoms. So α is an isomorphism from M to M'.

It is easily seen by induction on A that if α' is any isomorphism from M to M', then for all $t \in T$, $\|A\|_t^h = \|A\|_{\alpha'(t)}^{h'}$. Thus $\mathrm{tp}_M(t) = \mathrm{tp}_{M'}(\alpha'(t))$ for all t. It follows that $\alpha' = \alpha$. Hence α is unique. ■

6.10.2 Sahlqvist formulae and untied formulae

Now we define the Sahlqvist formulae — the ones whose syntactic form makes it possible to prove persistence and correspondence. The so-called untied formulae are the main building blocks of Sahlqvist formulae, and we will study the relationship between the two.

Definition 6.10.11

1. Two formulae A, B are said to be (*semantically*) *equivalent* if for all structures $(T, \bar{r}, h)$ and $t \in T$, $\|A\|_t^h = \|B\|_t^h$.

2. A formula A is said to be in *negation-normal form (nnf)* if it is built from atoms, $\top, \bot$, and their negations without using $\sim$. Note that for any A, $\sim\Diamond A$ is equivalent to $\Box\sim A$, and $\sim(A \wedge B)$ to $\sim A \vee \sim B$, and similarly for the duals. Hence any formula is equivalent to one in nnf.

3. A formula is *positive* if it is (semantically) equivalent to one that does not involve $\sim$: so it is built from atoms, $\top$ and $\bot$ using only $\wedge, \vee, \Box$, and $\Diamond$. It is *negative* if it is equivalent to $\sim A$ for a positive formula A.

4. A *Sahlqvist formula* is one of the form $\beta(A \to B)$, where β is a finite sequence of boxes (possibly empty), B is positive, and A is a formula in nnf such that for any box $\Box$ and diamond $\Diamond'$, if $\Box C$ is a subformula of A then no positive occurrence of an atom in C is in a subformula of C of the form $\Diamond' D$ or $D \vee E$. That is, the formation tree of A has no branch containing the pattern

$$\text{——} \Box \text{——} (\Diamond' \text{ or } \vee) \text{——} +p.$$

5. A *strongly positive formula* is a conjunction of formulae of the form βq, where q is an atom, $\top$ or $\bot$, and β is a (possibly empty) sequence of boxes.

6. An *untied formula* is one built from strongly positive and negative formulae using only $\wedge$ and $\Diamond$.

We will prove:

Proposition 6.10.12

1. *Any Sahlqvist formula is (semantically) equivalent to a conjunction of negations of untied formulae.*

2. *Conversely, any conjunction of negations of untied formulae is equivalent to a Sahlqvist formula.*

Part (2) of the proposition is easily seen. If $A_1, \ldots, A_n$ are untied then $\bigwedge_i \sim A_i$ is equivalent to $(\bigvee_i A_i) \to \bot$, which is evidently a Sahlqvist formula. For part (1) we need some definitions and lemmas.

Definition 6.10.13 A formula A in nnf is said to be *good* if all subformulae of A of the form $\Diamond C$ or $C \vee D$ are negative.

Lemma 6.10.14 *Let A be a formula in nnf. If A is good then it is (semantically) equivalent to $C \wedge N$ where C is strongly positive and N is negative.*

Proof. By induction on A. If A is atomic or negated atomic (including $\top$ and $\bot$) the result is clear. Assume that it holds for A_1, A_2. If $A_1 \wedge A_2$ is good, then so are A_1, A_2. Hence $A_1 \wedge A_2$ is equivalent to $C_1 \wedge N_1 \wedge C_2 \wedge N_2$ where the C_i are strongly positive and the N_i negative. As $C_1 \wedge C_2$ is strongly positive and $N_1 \wedge N_2$ is negative, the result holds for $A_1 \wedge A_2$. If $A_1 \vee A_2$ is good then it is already negative. Similarly, if $\Diamond A$ is good then it is negative.

Finally, if $\Box A$ is good then A is good, and so is equivalent to $C \wedge N$ as above; thus $\Box A$ is equivalent to $\Box C \wedge \Box N$. But $\Box N$ is negative, and $\Box C$ is equivalent to a strongly positive formula. This completes the proof. ■

Corollary 6.10.15 *Let A be any nnf formula whose formation tree has no branch containing the pattern*

$$\text{---}\ \Box\ \text{---}\ (\Diamond'\ or\ \vee)\ \text{---}\ +p$$

(as in Sahlqvist formulae). Then A is semantically equivalent to a disjunction of untied formulae.

Proof. By induction on A. If A is atomic or negated atomic, it is already untied. Assume that the result holds for A and B. If $A \vee B$ satisfies the condition of the corollary, then so do A and B, so that each is equivalent to a disjunction of untied formulae. Hence so is $A \vee B$. If $A \wedge B$ satisfies the condition, then in the same way, $A \wedge B$ is equivalent to $\bigvee_{i \leq n} D_i \wedge \bigvee_{j \leq m} E_j$ where D_i, E_j are untied. This is equivalent to $\bigvee_{i,j} D_i \wedge E_j$, a disjunction of untied formulae. If $\Diamond A$ satisfies the condition for some $\Diamond$, then so does A. Hence A is equivalent to $\bigvee_{i \leq n} D_i$ for untied D_i. Then $\Diamond A$ is equivalent to $\bigvee_{i \leq n} \Diamond D_i$ as required.

Finally, if $\Box A$ satisfies the condition on branches, then A, and so $\Box A$, are good. By the lemma, $\Box A$ is equivalent to $C \wedge N$ for strongly positive C and negative N; and this is untied. ■

Now we can prove proposition 6.10.12(1).

Proof. Let $S = \beta(A \to B)$ as in the definition. By corollary 6.10.14, A is equivalent to a disjunction $\bigvee_i U_i$ of untied formulae. Thus S is equivalent to $\beta{\sim}(\bigvee_i U_i \wedge {\sim}B)$, i.e. to $\beta{\sim}\bigvee_i[U_i \wedge {\sim}B]$, and so to $\bigwedge_i {\sim}\delta(U_i \wedge {\sim}B)$, where δ is obtained from β by replacing each $\Box$ of β by its corresponding $\Diamond$. Since for each $i, \delta(U_i \wedge {\sim}B)$ is untied, the proof is complete. ■

6.10.3 Sahlqvist's persistence theorem

We will first prove the persistence part of Sahlqvist's theorem. The correspondence part will follow by inspecting the proof. We will also prove a similar theorem that will be needed in the next section.

Definition 6.10.16 Let $(T, \overline{r}, h^*)$ be any structure.

1. A formula $A(q_1, \ldots, q_n)$ is said to be *locally persistent (over h^*)* if for each $t \in T$, if all substitution instances $A(B_1, \ldots, B_n)$ of A (for arbitrary formulae $B_1, \ldots, B_n$) are true at t in $(T, \overline{r}, h^*)$, then $\|A\|_t^h = 1$ for all assignments h into T.

2. h^* is said to be *discrete* if for each $t \in T$ and sequence β of boxes, there is a formula $W_{t,\beta}$ such that $\{u \in T : r_\beta(t,u)\} = h^*(W_{t,\beta})$.

Note that if h^* is discrete, separation and (tda) hold. Saturation fails if T is infinite, for $\{\sim W_{t,\varnothing} \mid t \in T\}$ is consistent, but is not realized over h^*.

We will prove:

Theorem 6.10.17 (persistence theorem) *Let $(T, \overline{r}, h^*)$ be any structure.*

1. *(Sahlqvist) If h^* is descriptive, then any Sahlqvist formula is locally persistent over h^*.*

2. *(Venema,* [Venema, 1992]*) If h^* is discrete, then any Sahlqvist formula is locally persistent over h^*.*

Remark 6.10.18 By example 6.10.2, part (1) of this theorem applies to canonical structures $(T, \overline{r}, h^*)$ consisting of all consistent theories in some logic Z. Thus if S is a Sahlqvist formula that is an axiom of Z, then S is valid in $(T, \overline{r}, h)$ for all assignments h. But the failure of saturation means that the theorem does not apply directly to 'IRR-structures'. In the next section, inspired by [Venema, 1992], we show how to apply (2) of the theorem to this situation.

For the rest of this section we fix a structure $(T, \overline{r}, h^*)$, where h^* is either descriptive or discrete. *Local persistence* will mean local persistence over h^*.

Since local persistence is preserved under semantic equivalence, and (from the definition) if A_i are locally persistent formulae then so is $\bigwedge_i A_i$, it is enough if we show that

(∗) if A is untied then $\sim A$ is locally persistent over h^*.

In fact, the following theorem proves (∗).

Theorem 6.10.19 (distributed persistence) *Let $k, n, m < \omega$, and let $A_1, \ldots, A_n$ be strongly positive formulae and $N_1, \ldots, N_m$ negative formulae, all using at most the atoms $q_1, \ldots, q_k$. Let $a_1, \ldots, a_n$, $b_1, \ldots, b_m \in T$. Suppose that there exists an assignment h such that:*

$$\|A_i\|^h_{a_i} = \|N_j\|^h_{b_j} = 1 \text{ for all } i \le n \text{ and } j \le m.$$

Then there are formulae $C_1, \ldots, C_k$ such that

$$\|A_i(C_1, \ldots, C_k)\|^{h^*}_{a_i} = \|N_j(C_1, \ldots, C_k)\|^{h^*}_{b_j} = 1 \text{ for all } i \le n, j \le m.$$

For let $A(q_1, \ldots, q_k)$ be an untied formula. It is easily seen (by induction on the number of $\wedge$s and $\Diamond$s needed to construct A from negative and strongly positive formulae) that there are A_i, N_j as in theorem 6.10.19 such that for all assignments h and all $t \in T$, the following are equivalent:

1. $\|A\|_t^h = 1$.
2. There exist $a_i, b_j \in T$ $(i \leq n, j \leq m)$ satisfying certain relations of the form $r_\gamma(x, y)$ for $x, y \in \{t, a_i, b_j \mid i \leq n, j \leq m\}$ and sequences δ of diamonds, such that $\|A_i\|_{a_i}^h = \|N_j\|_{b_j}^h = 1$ for all $i \leq n, j \leq m$.

Assuming theorem 6.10.19, by using this equivalence twice we see that if $\|A\|_t^h = 1$ for some h, t, then there are formulae $C_1, \ldots, C_k$ such that $\|A(C_1, \ldots, C_k)\|_t^{h^*} = 1$. This proves that $\sim A$ is locally persistent, so $(*)$ is established.

So it only remains to prove theorem 6.10.19. As the first step, we will show that if h exists as in the theorem, then there is a *type-definable* such h: one such that for each $i, h(q_i)$ is type-definable over h^*. To do this we will consider the strongly positive formulae A_i. Since repetitions in the a_i are permitted, we can assume that each A_i has the form $\beta_i p_i$ for some (possibly empty) sequence β_i of boxes and $p_i \in \{q_1, \ldots, q_k\}$.

Definition 6.10.20

1. If $i \leq n$, let Γ_i be the following type:
   - $\text{tp}_{h^*}(a_i, \beta_i) = \{B \mid \|\beta_i B\|_{a_i}^{h^*} = 1\}$ if h^* is descriptive;
   - $\{W_{a_i, \beta_i}\}$ if h^* is discrete.
2. For each $j \leq k$ let $\Gamma(q_j)$ be the $\wedge$-closure of $\bigvee\{\Gamma_i \mid i \leq n, p_i = q_j\}$ (see definition 6.10.3). Thus if q does not occur in any A_i we have $\Gamma(q) = \{\bot\}$.

Note that the Γs depend on h^*.

Proposition 6.10.21 *Let h be any assignment. Then if $i \leq n$, the following three conditions are equivalent.*

1. $\|A_i\|_{a_i}^h = 1$.
2. *For all $t \in T, t \in h(p_i)$ whenever $r_{\beta i}(a_i, t)$.*
3. $h^*(\Gamma_i) \subseteq h(p_i)$.

Moreover, $\|A_i\|_{a_i}^h = 1$ for all $i \leq n$ iff $h^*(\Gamma(q_j)) \subseteq h(q_j)$ for all $j \leq k$.

Proof. As $A_i = \beta_i p_i$, the equivalence of (1) and (2) is trivial. If h^* is descriptive, then by lemma 6.10.8, $r_{\beta i}(a_i, t)$ is equivalent to $t \in h^*(\Gamma_i)$. If it is discrete, then $r_{\beta i}(a_i, t)$ is equivalent to $t \in h^*(W_{a_i,\beta_i}) = h^*(\Gamma_i)$. Thus (2) is equivalent to (3).

By the equivalence of (1) and (3), $\|A_i\|^h_{a_i} = 1$ for all $i \leq n$ iff $h^*(\Gamma_i) \subseteq h(q_j)$ whenever $p_i = q_j$, i.e., $\bigcup\{h^*(\Gamma_i) \mid i \leq n, p_i = q_j\} \subseteq h(q_j)$ for all $j \leq k$. But by lemma 6.10.4(2), $\bigcup\{h^*(\Gamma_i) \mid p_i = q_j\} = h^*(\Gamma(q_j))$. This completes the proof. ■

A formula $C(q_1, \ldots, q_k)$ is said to be *monotonic* if whenever h, h' are assignments such that $h(q_j) \subseteq h'(q_j)$ for all $j \leq k$, then $h(C) \subseteq h'(C)$. Clearly, all positive formulae are monotonic.

Assume that h exists as in theorem 6.10.19. Define an assignment h_0 by $h_0(q_j) = h^*(\Gamma(q_j))$ for all $j \leq k$. Evidently, by proposition 6.10.21 we have $\|A_i\|^{h_0}_{a_i} = 1$ for all $i \leq n$. Moreover, by the same proposition, h_0 is the 'smallest' such assignment, so that $h_0(q_j) \subseteq h(q_j)$ for all $j \leq k$. Hence by monotonicity of $\sim N_j$, $\|N_j\|^{h_0}_{b_j} = \|N_j\|^h_{b_j} = 1$ for all $j \leq m$. Thus by replacing h by h_0 we can assume that h is type-definable, by: $h(q) = h^*(\Gamma(q))$ for all $q \in \{q_1, \ldots, q_k\}$.

In the second step we turn our attention to the negative formulae N_j.

Definition 6.10.22

1. If $B(q_1, \ldots, q_k)$ is any formula, define the type $\Gamma(B)$ to be the type $\{B(D_1, \ldots, D_k) \mid D_1 \in \Gamma(q_1), \ldots, D_k \in \Gamma(q_k)\}$. $\Gamma(B)$ is not $\wedge$-closed in general.

2. Define for any positive formula B, the type $\Gamma^*(B)$ by induction:

   - $\Gamma^*(q) = \Gamma(q)$, $\Gamma^*(\top) = \{\top\}$ and $\Gamma^*(\bot) = \{\bot\}$;
   - $\Gamma^*(B \wedge C) = cl(\Gamma^*(B) \wedge \Gamma^*(C))$;
   - $\Gamma^*(B \vee C) = cl(\Gamma^*(B) \vee \Gamma^*(C))$;
   - $\Gamma^*(\Box B) = cl(\Box\Gamma^*(B))$;
   - $\Gamma^*(\Diamond B) = cl(\Diamond\Gamma^*(B))$.

   So for each B, the type $\Gamma^*(B)$ is $\wedge$-closed.

Remark 6.10.23 It is easy to see that $\Gamma(B) \subseteq \Gamma^*(B)$ for all B. If h^* is discrete then $\Gamma(B) = \Gamma^*(B)$. But if not, then regrettably we do not have equality in general. Suppose, for instance, that B is $q \wedge \Box q$. Then if $C, D \in \Gamma(q)$ are distinct, $C \wedge \Box D \in \Gamma^*(B) \setminus \Gamma(B)$.

However, we will prove that $h(\Gamma(B)) = h(\Gamma^*(B))$ for all h, which is good enough for us.

Definition 6.10.24

1. Let B and C be formulae. We write $T \models B \to C$ if $h(B) \subseteq h(C)$ for all assignments h into T.
2. Let $\Gamma \subseteq \Delta$ be types. We say that Γ is a *base* for Δ if for all $C \in \mathrm{cl}(\Delta)$ there is $B \in \Gamma$ with $T \models B \to C$.

Clearly, if Γ is a base for Δ then $h(\Gamma) = h(\Delta)$ for all h.

Lemma 6.10.25 *Let $B(q_1, \ldots, q_k)$ be a positive formula involving at most the atoms $q_1, \ldots, q_k$. Then $\Gamma(B)$ is a base for $\Gamma^*(B)$.*

Proof. Clearly, $\Gamma(B) \subseteq \Gamma^*(B)$. We show that $\Gamma(B)$ is a base for $\Gamma^*(B)$ by induction on B. Of course, this is easy if h^* is discrete, since in this case $\Gamma(B) = \Gamma^*(B)$ for all B. So we can assume that h^* is descriptive instead.

If B is atomic we have $\Gamma(B) = \Gamma^*(B)$. Assume that $B \vee C$ is positive and that the result holds for B and C. Let $X \in \Gamma^*(B \vee C)$, so that for some $m < \omega$ and formulae $Y_i \in \Gamma^*(B)$, $Z_i \in \Gamma^*(C)$ (for each $i \leq m$), $X = \bigwedge_{i \leq m} Y_i \vee Z_i$. By the inductive hypothesis there are $D_{ij}, D'_{ij} \in \Gamma(q_j)$ for each $i \leq m$ and $j \leq k$, with $T \models B(D_{i1}, \ldots, D_{ik}) \to Y_i$ and $T \models C(D'_{i1}, \ldots, D'_{ik}) \to Z_i$ for each $i \leq m$. Let $E_j = \bigwedge_{i \leq m} D_{ij} \wedge D'_{ij}$ for each $j \leq k$. Then $E_j \in \Gamma(q_j)$ as this type is $\wedge$-closed. As B is positive, it is monotonic, so that $T \models B(E_1, \ldots, E_k) \to Y_i$ for each $i \leq m$. Similarly, $T \models C(E_1, \ldots, E_k) \to Z_i$ for each i. Thus $T \models [B \vee C](E_1, \ldots, E_k) \to Y_i \vee Z_i$ for each i, so that $T \models [B \vee C](E_1, \ldots, E_k) \to X$. This completes the case for $B \vee C$. The case for $B \wedge C$ is similar and indeed simpler, as $\Gamma^*(B) \wedge \Gamma^*(C)$ is already $\wedge$-closed. We leave it to the reader.

Now assume that $\Diamond B$ is positive and that the result holds for B. Let $X = \bigwedge_{i \leq m} \Diamond Y_i \in \Gamma^*(\Box B)$, where each $Y_i \in \Gamma^*(B)$. Inductively, for each $i \leq m, j \leq k$ there is $D_{ij} \in \Gamma(q_j)$ such that $T \models B(D_{i1}, \ldots, D_{ik}) \to Y_i$. Let $E_j = \bigwedge_{i \leq m} D_{ij} \in \Gamma(q_j)$. As before, $T \models B(E_1, \ldots, E_k) \to Y_i$, so that $T \models \Diamond B(E_1, \ldots, E_k) \to \Diamond Y_i$ for each i. Hence $T \models \Diamond B(E_1, \ldots, E_k) \to X$, as required. The case for $\Box$ is similar. ■

Corollary 6.10.26 *Let h be an assignment into T with $h(q) = h^*(\Gamma(q))$ for each q. Then $h(B) = h^*(\Gamma(B))$ for all positive formulae B.*

Proof. Again, this is obvious if h^* is discrete. Assume then that h^* is descriptive. By the lemma, $h^*(\Gamma(B)) = h^*(\Gamma^*(B))$. So it suffices to prove that $h(B) = h^*(\Gamma^*(B))$ for all positive formulae B. But as the types $\Gamma^*(B)$ are $\wedge$-closed and h^* is descriptive, this follows easily from lemma 6.10.4 by induction on B. ∎

We can now prove theorem 6.10.19. Recall that we assumed h type-definable: $h(q_j) = h^*(\Gamma(q_j))$ for each $j \leq k$. So as each $\sim N_i$ is positive, by the corollary we obtain $h(\sim N_i) = h^*(\Gamma(\sim N_i))$ for all $i \leq m$. But $\|N_i\|_{b_i}^h = 1$ for all $i \leq m$, so there are formulae $C_{ij} \in \Gamma(q_j)$ for each $i \leq m, j \leq k$, such that $\|N_i(C_{i1}, \ldots, C_{ik})\|_{b_i}^{h^*} = 1$. Let $C_j = \bigwedge_{i \leq m} C_{ij}$. As $\Gamma(q_j)$ is $\wedge$-closed, $C_j \in \Gamma(q_j)$ for all $j \leq k$; and by monotonicity of $\sim N_i$, $\|N_i(C_1, \ldots, C_k)\|_{b_i}^{h^*} = 1$ for all i. As $h^*(C_j) \supseteq h^*(\Gamma(q_j))$ for all j, by proposition 6.10.21 we also have $\|A_i(C_1, \ldots, C_k)\|_{a_i}^{h^*} = 1$ for all $i \leq n$. This completes the proof of theorem 6.10.19, and with it the persistence theorem (6.10.17). ∎

6.10.4 Sahlqvist's correspondence theorem

We now move on to the correspondence part of Sahlqvist's theorem.

Definition 6.10.27
Let $A(q_1, \ldots, q_k)$ be a modal formula with table $\alpha(t, Q_1, \ldots, Q_k)$, and let $\varphi(t)$ be any formula in the monadic second-order language with monadic predicates $Q_1, \ldots, Q_k$ and binary relation symbols $=$ and r for each r in $\overline{r}$. Quantification over the Q_i is allowed. We say that A *locally corresponds* to φ if the formula

$$\forall t[(\forall Q_1 \ldots \forall Q_k \alpha(t, Q_1, \ldots, Q_k)) \leftrightarrow \varphi(t)]$$

is valid in every frame $(T, \overline{r})$. That is, for all frames $(T, \overline{r})$ and $t \in T$, $\|A\|_t^h = 1$ for all h iff $(T, \overline{r}) \models \varphi(t)$.

Certainly, A locally corresponds to $\forall Q_1 \ldots \forall Q_k \alpha(t, Q_1, \ldots, Q_k)$. But if A is a Sahlqvist formula, we can say more.

Theorem 6.10.28 (correspondence theorem) *Let $S(q_1, \ldots, q_k)$ be a Sahlqvist formula. Then there is a (non-monadic) first-order formula $\varphi(t)$ that locally corresponds to S. Moreover, φ is obtainable effectively from S.*

Note that this theorem concerns only frames. Descriptive assignments are not involved.

Proof. This theorem can be read off quite simply from the proofs above. Again, it is enough to show that if A is untied then $\sim A$ locally corresponds to some effectively obtainable first-order φ. For by proposition 6.10.12 we can assume that S has the form $\bigwedge_i \sim A_i$, where each A_i is an untied formula with table $\alpha_i(t, Q_1, \ldots, Q_k)$. Assume that $\sim A_i$ locally corresponds to $\varphi_i(t)$, so that $\forall Q_1 \ldots \forall Q_k {\sim} \alpha_i(t, Q_1, \ldots, Q_k)$ is frame-equivalent to $\varphi_i(t)$. Then S locally corresponds to $\forall Q_1 \ldots \forall Q_k \bigwedge_i \sim \alpha_i(t, Q_1, \ldots, Q_k)$, i.e. to

$$\bigwedge_i \forall Q_1 \ldots \forall Q_k {\sim} \alpha_i(t, Q_1, \ldots, Q_k),$$

and so to the first-order formula $\bigwedge_i \varphi_i(t)$, which is effectively obtainable from the φ_i.

So let $A(q_1, \ldots, q_k)$ be an untied formula with table $\alpha(t, Q_1, \ldots, Q_k)$. We want to find (by an algorithm) a first-order equivalent of

$$\forall Q_1 \ldots \forall Q_k {\sim} \alpha(t, Q_1, \ldots, Q_k),$$

or equally, ${\sim}\exists Q_1 \ldots \exists Q_k \alpha(t, Q_1, \ldots, Q_k)$. By the argument following theorem 6.10.19 (a simple induction on A), it can be seen that $\alpha(t, Q_1, \ldots, Q_k)$ has the form

$$\exists a_1 \ldots \exists a_n \exists b_1 \ldots \exists b_m (\rho(t, \overline{a}, \overline{b}) \wedge \bigwedge_{i \leq n} \alpha_i(a_i, P_i) \wedge \bigwedge_{j \leq m} \nu_j(b_j, Q_1, \ldots, Q_k)),$$

where ρ is a first-order, non-monadic formula expressing the $\overline{r}$-relationships between t, the a_i, and the b_j; α_i is the table of an appropriate formula of the form $\beta_i p_i$ for some sequence β_i of boxes and $p_i \in \{q_1, \ldots, q_k\}$; and ν_j is the table of an appropriate negative formula written with atoms from $q_1, \ldots, q_k$. So $\exists Q_1 \ldots \exists Q_k \alpha$ is equivalent to:

$$\begin{aligned}&\exists a_1 \ldots \exists a_n \exists b_1 \ldots \exists b_m \\ &(\rho(t, \overline{a}, \overline{b}) \wedge \exists Q_1 \ldots \exists Q_k [\textstyle\bigwedge_{i \leq n} \alpha_i(a_i, P_i) \wedge \bigwedge_{j \leq m} \nu_j(b_j, Q_1, \ldots, Q_k)])\end{aligned}$$

(where each P_i is some Q_j), since the second-order $\exists$s commute with the first-order $\exists$s and ρ is non-monadic.

Now by (1)–(2) of proposition 6.10.21[2] and the monotonicity of negations of negative formulae, there exist first-order non-monadic formulae $\delta_j(x, \overline{a})$ for each $j \leq k$ such that

$$\exists Q_1 \ldots Q_k [\bigwedge_{i \leq n} \alpha_i(a_i, P_i) \wedge \bigwedge_{j \leq m} \nu_j(b_j, Q_1, \ldots, Q_k)]$$

is equivalent to

$$\psi(\bar{a}, \bar{b}) = \bigwedge_{i \leq n} \alpha_i(a_i, \varepsilon_i(x, \overline{a})) \wedge \bigwedge_{j \leq m} \nu_j(b_j, \delta_1(x, \overline{a}), \ldots, \delta_k(x, \overline{a})),$$

[2]This does not mention h^*, and so does not require it to be descriptive or discrete.

where if P_i is Q_j then ε_i is δ_j. In fact, by the proof of the final clause of proposition 6.10.21, δ_j has the form:

$$(*) \qquad \delta_j(x, \bar{a}) = \bigvee_{i \leq n, P_i = Q_j} r_{\beta_i}(a_i, x).$$

So $\forall Q_1 \ldots \forall Q_k {\sim}\alpha(t, Q_1, \ldots, Q_k)$ is equivalent to the first-order formula

$${\sim}\exists a_1 \ldots a_n \exists b_1 \ldots \exists b_m [\rho(t, \bar{a}, \bar{b}) \wedge \psi(\bar{a}, \bar{b})].$$

Clearly the construction of this formula is effective (cf. the examples below). This completes the proof of theorem 6.10.28. ■

Thus all of Sahlqvist's theorem is proved.

6.10.5 Examples of correspondence

The algorithm of theorem 6.10.28 is simple enough to implement on a personal computer (as pointed out in [Sambin and Vaccaro, 1989]). Let us see some examples of its use. We let $\Box$ be a box with accessibility relation r.

1. The 'reflexivity formula' $\Box q \to q$ is a Sahlqvist formula; it is equivalent to ${\sim}(\Box q \wedge {\sim}q)$, and $\Box q \wedge {\sim}q$ is untied with table $\alpha(t, Q) = \forall u(r(t, u) \to Q(u)) \wedge {\sim}Q(t)$. Here, $n = m = 0$. Then $\exists Q\alpha$ is equivalent to $\forall u(r(t, u) \to \delta(u)) \wedge {\sim}\delta(t)$, where $\delta(x)$ is $r(t, x)$ in this case. This is obviously equivalent to ${\sim}r(t, t)$. Thus ${\sim}\exists Q\alpha$ is equivalent to $r(t, t)$, i.e. reflexivity of r at t.

2. Density: $\Box\Box q \to \Box q$. It is equivalent to ${\sim}(\Box\Box q \wedge {\sim}\Box q)$. $\Box\Box q \wedge {\sim}\Box q$ is untied and has table

$$\alpha(t, Q) = \forall u(\exists y(r(t, y) \wedge r(y, u)) \to Q(u)) \wedge {\sim}\forall v(r(t, v) \to Q(v)).$$

In this case $\delta(x)$ is clearly $\exists y(r(t, y) \wedge r(y, x))$, so that $\exists Q\alpha$ is equivalent to

$$\forall u(\exists y(r(t, y) \wedge r(y, u)) \to \delta(u)) \wedge {\sim}\forall v(r(t, v) \to \delta(v)),$$

i.e. to

$${\sim}\forall v(r(t, v) \to \delta(v)).$$

Thus, ${\sim}\exists Q\alpha(t, Q)$ is equivalent to

$$\forall v(r(t, v) \to \exists y(r(t, y) \wedge r(y, v))),$$

justifying the soubriquet 'density'.

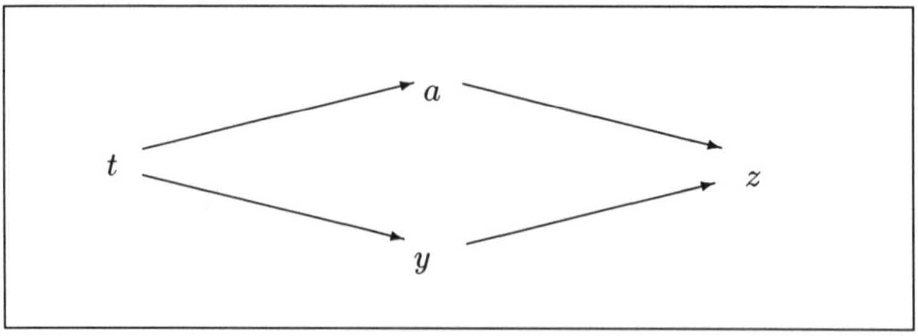

Figure 6.11

3. The 'Church–Rosser formula' $\Diamond\Box q \to \Box\Diamond q$ is a Sahlqvist formula, equivalent to $\sim(\Diamond\Box q \wedge \sim\Box\Diamond q)$, the negation of an untied formula ($\Box q$ is strongly positive and $\sim\Box\Diamond q$ is negative). The table of the untied part is $\alpha(t,Q) = \exists a[r(t,a) \wedge \forall u(r(a,u) \to Q(u)) \wedge \beta(t,Q)]$, where $\beta(t,Q)$ is the table of $\sim\Box\Diamond q$. Here, $m = 0$ but now $n = 1$. So in this case $\delta(x,a)$ depends on a, and using $(*)$ from the proof of theorem 6.10.28 we see that it is $r(a,x)$. Substituting it for Q in α gives $\exists a[r(t,a) \wedge[\forall u\ (r(a,u) \to r(a,u))$ $\wedge\beta(t,r(a,u))]]$, or equivalently,

$$\exists a[r(t,a) \wedge \beta(t,r(a,u))].$$

Now we need to evaluate $\beta(t,Q)$. It is $\exists y(r(t,y) \wedge \forall z(r(y,z) \to \sim Q(z)))$. Substituting $\delta(z,a)$ for $Q(z)$ yields

$$\exists y(r(t,y) \wedge \forall z(r(y,z) \to \sim r(a,z))).$$

Thus $\sim\exists Q\alpha(t,Q)$ is equivalent to

$$\sim\exists a[r(t,a) \wedge \exists y(r(t,y) \wedge \forall z(r(y,z) \to \sim r(a,z)))].$$

Simplifying, we obtain

$$\forall a\forall y[r(t,a) \wedge r(t,y) \to \exists z(r(y,z) \wedge r(a,z))],$$

the well-known 'amalgamation diamond' equivalent of the Church–Rosser formula. See the figure 6.11.

4. Linearity, $\Diamond p \wedge \Diamond q \to \Diamond(p \wedge q) \vee \Diamond(p \wedge \Diamond q) \vee \Diamond(\Diamond p \wedge q)$.

Let $N = \sim[\Diamond(p\wedge q)\vee\Diamond(p\wedge\Diamond q)\vee\Diamond(\Diamond p\wedge q)]$, a negative formula. Thus the orginal linearity formula is equivalent to the negation of the untied formula $\Diamond p\wedge\Diamond q\wedge N$. The table $\alpha(t,P,Q)$ of this is $\exists a_1\exists a_2(r(t,a_1)\wedge P(a_1)\wedge r(t,a_2)\wedge Q(a_2) \wedge \nu(t,P,Q))$, where ν is the usual table of N. Thus by $(*)$ above, $\delta_p(x,a_1,a_2)$ is $x = a_1$, and $\delta_q(x,a_1,a_2)$ is $x = a_2$.

Substituting δ_p for P and δ_q for Q in ν yields the formula $\sim(\exists y[r(t,y)\wedge y=a_1\wedge y=a_2]\vee\exists y[r(t,y)\wedge y=a_1\wedge\exists z(r(y,z)\wedge z=a_2)]\vee\exists y[r(t,y)\wedge y=a_2\wedge\exists z(r(y,z)\wedge z=a_1)])$, which simplifies to

$$\sim[(r(t,a_1)\wedge a_1=a_2)\vee(r(t,a_1)\wedge r(a_1,a_2))\vee(r(t,a_2)\wedge r(a_2,a_1))].$$

So $\sim\exists P\exists Q\alpha(t,P,Q)$ is equivalent to

$$\begin{aligned}\sim\exists a_1\exists a_2\quad & [r(t,a_1)\wedge r(t,a_2)\\ & \wedge\sim[r(t,a_1)\wedge a_1=a_2\\ & \qquad\vee r(t,a_1)\wedge r(a_1,a_2)\vee r(t,a_2)\wedge r(a_2,a_1)]].\end{aligned}$$

Deleting redundant $r(t,a_1)$s and $r(t,a_2)$s, we obtain

$$\forall a_1a_2[r(t,a_1)\wedge r(t,a_2)\to a_1=a_2\vee r(a_1,a_2)\vee r(a_2,a_1)],$$

transparently a linearity axiom.

6.11 Sahlqvist's theorem with irreflexivity rules

In this section we indicate briefly how to use Venema's version of Sahlqvist's theorem (where h^* is discrete) in a case where saturation fails. The main situation that we have in mind is that of a canonical structure consisting of 'IRR-theories'. As in section 6.10, the language will in general contain many boxes $\Box$ and their corresponding diamonds $\Diamond$.

Definition 6.11.1

1. Let B be any formula and let $n<\omega$. The inference rule $\rho_n(B)$ is:

$$\frac{\vdash\sim[A_n\wedge\Diamond_n(A_{n-1}\wedge\Diamond_{n-1}(\ldots\wedge\Diamond_1(A_0\wedge B(\overline{q})))\cdots)]}{\vdash\sim[A_n\wedge\Diamond_n(A_{n-1}\wedge\Diamond_{n-1}(\ldots\wedge\Diamond_1A_0))\cdots)]},$$

   where $A_0,\ldots,A_n$ are any formulae, $B(\overline{q})$ is a substitution instance of B by any atoms $\overline{q}$ not occurring in the A_i, and $\Diamond_1,\ldots,\Diamond_n$ are any diamonds.

2. The 'rule' $\rho(B)$ is the set of all the rules $\rho_n(B)$, for all $n<\omega$.

3. Assume we are given a set of rules $\rho(B)$ as above. A *ρ-theory* is a theory Γ such that whenever $\rho(B)$ is a given rule, $n<\omega$ and $A_n\wedge\Diamond_n(A_{n-1}\wedge\Diamond_{n-1}(A_{n-2}\wedge\ldots\wedge\Diamond_1A_0))\cdots)\in\Gamma$ then $A_n\wedge\Diamond_n(A_{n-1}\ldots\wedge\Diamond_1(A_0\wedge B(\overline{q})))\cdots)\in\Gamma$ for some atoms $\overline{q}$.

The first rule of this form that we saw was the IRR-rule of chapter 3:

$$\frac{\vdash q \wedge H{\sim}q \to A}{\vdash A}, \text{ if } q \text{ does not occur in } A.$$

This rule is just $\rho_0(q \wedge H{\sim}q)$. We later saw two others: the $I\varphi R$-rule of section 6.6, and the generalised IRR-rules of sections 6.7 and 6.8. In the case of tense logic, where for every $\Diamond$ (with accessibility relation $r(x, y)$) the language has a dual connective $\Diamond^*$ with accessibility relation $r(y, x)$, the full rule ρ follows from ρ_0. See theorem 3.2.4.

Let Z be a consistent modal logic (i.e. a set of axioms and rules) with (at least) the axioms and rules of the basic modal logic K for each $\Box, \Diamond$ in the language. Assume that the given rules $\rho(B)$ are countable in number and are valid in Z, so that each $\rho(B)$ is either already a rule of Z, or is merely a valid statement about $\vdash_Z$. The same style of proof as in section 6.2 shows that any Z-consistent set Γ can be extended to a maximal consistent ρ-theory, possibly in a language containing more atoms. If T is the set of all ρ-theories in this larger language, we can define the accessibility relations on T as usual, by: if $\Diamond, \Box$ have accessibility relation r, then for all $\Gamma, \Delta \in T, r(\Gamma, \Delta)$ holds iff for all formulae A, $\Box A \in \Gamma \Rightarrow A \in \Delta$, or, equivalently, $A \in \Delta \Rightarrow \Diamond A \in \Gamma$. We then define an assignment h^* into T by $h^*(p) = \{\Gamma \in T \mid p \in \Gamma\}$. The argument of section 6.2 then shows that for all formulae A and all $\Gamma \in T$,

$$A \in \Gamma \text{ iff } \|A\|_{\Gamma}^{h^*} = 1.$$

We may even restrict T to a single 'component' by removing all theories not accessible from a chosen 'home theory' by the $\overline{r}$ in some way.

The structure $M = (T, \overline{r}, h^*)$ so obtained is in many ways better behaved than the set of all maximal consistent Z-theories. For example, it may be irreflexive (see section 6.2). However, from the point of view of Sahlqvist's theorem it is not so good, because although h^* satisfies (sep) and (tda) of definition 6.10.1, it is not in general saturated. For example, assuming that the rule ρ is $\rho(B)$ where $\nvdash_Z B$, the type $\{{\sim}B(\overline{q}) \mid \text{all atoms } \overline{q}\}$ is consistent with M in the sense of definition 6.10.1, but is not realized in M. Thus we cannot apply theorem 6.10.17(1).

But we would like to use the Sahlqvist conclusions for M. For example, suppose that S is a Sahlqvist formula, and the algorithm of theorem 6.10.28 yields the first-order formula $\varphi(t)$. Then φ locally corresponds to S. For example, if S is $\Box_r q \to q$, then $\varphi(t)$ will be logically equivalent to $r(t, t)$. If Z should contain S as an axiom,[3] then all substitution instances of S will be valid in M. If, further, S *were locally persistent in* M, then for all

[3] Of course, Z can not then contain the rule $\rho(q \wedge \Box_r{\sim}q)$.

$t \in T$ and all h, we would have $\|S\|_t^h = 1$. So as φ locally corresponds to S, it would follow that $(T, \bar{r}) \models \forall t \varphi(t)$. Thus we would have a general way of establishing certain properties of $(T, \bar{r})$, namely by adding corresponding Sahlqvist formulae as axioms to Z.

Thus we would like to apply theorem 6.10.17(2) instead. So we would like to show that in the case of ρ-theories, h^* can be made discrete. In this section we will describe briefly two situations in which this can be done. Roughly, the first case is when the language is a *tense language* — as in [Venema, 1992] — so that each $\Diamond$ with accessibility relation $r(x, y)$ has a *dual* diamond $\Diamond^*$ with accessibility relation $r(y, x)$. The second case is when each diamond $\Diamond$ with accessibility relation $r(x, y)$, has a *complementary* diamond $\Diamond'$ with accessibility relation $\sim r(x, y)$. The proof in the second case was also inspired by [Venema, 1992], and the arguments in the two cases are very similar.

6.11.1 A tense language with the O connective

Firstly, we follow [Venema, 1992] and thus assume the following:

1. For each $\Diamond$ (with accessibility relation $r(x, y)$) there is a dual connective $\Diamond^*$ whose accessibility relation is $r(y, x)$. In temporal logic, for example, F is dual to P.

2. There is in the language a special unary connective O, the table $\alpha(t, Q)$ of Oq being $\forall x(Q(x) \leftrightarrow x = t)$. We read Oq as 'q is only true here'.

3. $(T, \bar{r}, h^*)$ is a structure, and for each $t \in T$ there is an atom q such that $\|Oq\|_t^{h^*} = 1$.

Examples of such $(T, \bar{r}, h^*)$ include the set T of all ρ-theories related by $<, =$ or $>$ to a given Γ_0, in the logic $K_t + IRR$ plus a linearity axiom. In this case, the connective O is *definable* by $Oq = q \wedge H{\sim}q \wedge G{\sim}q$, and there is no need to add it to the language. [Venema, 1992] shows how to axiomatize O using the rule $\rho_0(Oq)$, and to build a set T of ρ-theories in which O has the right semantics.

It is easy to see that h^* is discrete. If β is a sequence of boxes, define β^* by induction: if β is empty then so is β^*; and if $\beta = \gamma \cdot \Box$ then $\beta^* = \Diamond^* \cdot \gamma^*$, where $\Diamond$ is $\sim\Box\sim$. Then if $t \in T$ and β is a sequence as above, let q be an atom such that $\|Oq\|_t^{h^*} = 1$. We can now take $W_{t,\beta}$ to be $\beta^* q$. Clearly, $h^*(W_{t,\beta}) = \{u \in T \mid r_\beta(t, u)\}$, so that h^* is discrete, as required.

6.11.2 A language with complements

Secondly, although we do not require that diamonds have duals (as in tense logics), we put other, rather heavy, restrictions on the language.

Definition 6.11.2 Let $\mathcal{B}$ be the set of boxes of the language.

1. $\mathcal{B}$ is said to be *standardized* if the following hold:

   (*a*) For each $\Box \in \mathcal{B}$ there is a *complementary* box $\Box' \in \mathcal{B}$. We require that in each case, $\Box' \neq \Box$ and $\Box'' = \Box$.

   (*b*) There is a distinguished box $I \in \mathcal{B}$.

   (*c*) There is an associative binary operation $*$ on $\mathcal{B}$ such that $I * \Box = \Box * I = \Box$ for all $\Box \in \mathcal{B}$.

   (*d*) $\mathcal{B}$ is countable (though our proofs work in more general cases).

   If $\Box \in \mathcal{B}$ and $\Diamond = {\sim}\Box{\sim}$ is the diamond corresponding to $\Box$, we write $\Diamond'$ for the diamond ${\sim}\Box'{\sim}$ corresponding to $\Box'$.

2. Let $\mathcal{B}$ be a standardized set of boxes, and let $\bar{r}$ be a tuple of accessibility relations corresponding to the boxes in $\mathcal{B}$. A Kripke frame $(T, \bar{r})$ is said to be *standard* if the following three conditions hold:

   (a) Iq has the same semantics as q (i.e. its accessibility relation r_I is $=$).

   (b) If $\Diamond, \Box$ have accessibility relation $r(x, y)$, then $\Diamond', \Box'$ have accessibility relation ${\sim}r(x, y)$.

   (c) If the boxes $\Box_1, \Box_2 \in \mathcal{B}$ have accessibility relations $r_1(x, y)$, $r_2(x, y)$, respectively, then the box $\Box_1 * \Box_2$ has accessibility relation $\exists z(r_1(x, z) \wedge r_2(z, y))$ (i.e. it is $r_{\Box_1 \Box_2}$ of definition 6.10.6).

Note that the connective O of section 6.11.1 is definable in a standard frame, by $Oq = Iq \wedge I'{\sim}q$.

Example 6.11.3

1. For any set $\mathcal{B}$ of boxes let $\mathcal{B}'$ be a set $\{\Box' \mid \Box \in \mathcal{B}\}$ of distinct new boxes disjoint from $\mathcal{B}$. Let $\mathcal{B}^*$ be the set of all finite sequences of elements of $\mathcal{B}$. We identify each $\Box$ with the sequence $\langle \Box \rangle$ of length 1.

   Now let $\mathcal{B}_0$ be any countable set of boxes. Define $\mathcal{B}_{2n+1} = \mathcal{B}_{2n} \cup \mathcal{B}'_{2n}$ and $\mathcal{B}_{2n+2} = \mathcal{B}^*_{2n+1}$ for each $n < \omega$. Then if $\mathcal{B} = \bigcup_{n<\omega} \mathcal{B}_n$, we can define I to be the empty sequence, and $*$ to be concatenation of sequences. If $b \in \mathcal{B}$ let $n < \omega$ be least such that $b \in \mathcal{B}_n$. If n is even then, already, $b' \in \mathcal{B}_{n+1} \subseteq \mathcal{B}$ is defined. If n is odd then $b = c'$ for some $c \in \mathcal{B}_{n-1}$; we define $b' = c$. Then $\mathcal{B}$ is standardized. Thus any set $\mathcal{B}_0$ of boxes can be standardized in a canonical manner.

2. Let $\mathbb{Z}$ be the set of non-zero integers (we could use rationals, etc.). Define $I = 1, z' = -z$ ($z \in \mathbb{Z}$), and let $*$ be integer multiplication. Then $(\mathbb{Z}, I, ', *)$ is standardized. In this case $*$ is commutative.

3. If T is any set, let $\mathcal{B} \subseteq 2^{T \times T}$ be any (countable) non-empty set satisfying:

   - $I = \{(t,t) \mid t \in T\} \in \mathcal{B}$;
   - if $S \in \mathcal{B}$ then $S' =_{\text{def.}} T^2 \setminus S \in \mathcal{B}$;
   - if $S_1, S_2 \in \mathcal{B}$ then:
     $S_1 * S_2 =_{\text{def.}} \{(t,u) \mid \exists v \in T((t,v) \in S_1 \wedge (v,u) \in S_2)\} \in \mathcal{B}$.

   Then each $S \in \mathcal{B}$ can be viewed as a box with accessibility relation S. The frame $(T, \mathcal{B})$ is standard.

The intended semantics of a standardized set $\mathcal{B}$ of boxes are a structure $(T, \overline{r}, h)$ where $(T, \overline{r})$ is a standard Kripke frame. We should check that standard frames exist. This is easy in some cases.

Remark 6.11.4 Let $\mathcal{B}$ be canonically constructed from a given set $\mathcal{B}_0$ of boxes, as in (1) of the example. Then any Kripke frame $K_0 = (T, \overline{r}_0)$, where $\overline{r}_0$ are accessibility relations corresponding to the boxes in $\mathcal{B}_0$, can be extended canonically to a standard frame $K = (T, \overline{r})$ for $\mathcal{B}$. We define frames $K_n = (T, \overline{r}_n)$ $(n < \omega)$ by induction. If n is even, we add to $\overline{r}_n$ a relation r' for each $\Box' \in \mathcal{B}'_n$ and interpret it in T as $\sim r$, where r is inductively the accessibility relation of $\Box$. If n is odd, then for each sequence $\beta = \langle \Box_1, \ldots, \Box_k \rangle \in \mathcal{B}_{n+1}$, where $\Box_1, \ldots, \Box_k \in \mathcal{B}_n$ have accessibility relations $r_1, \ldots, r_k$, respectively, we add to $\overline{r}_n$ the accessibility relation $r_\beta(x,y)$ on T given by: $\exists x_0 \ldots \exists x_k (x_0 = x \wedge x_k = y \wedge \bigwedge_{i<k} r_{i+1}(x_i, x_{i+1}))$. Letting $\overline{r}$ be the collection of all the $\overline{r}_n$, the frame $(T, \overline{r})$ is clearly standard.

If the accessibility relations of the boxes in $\mathcal{B}_0$ are all different, we can use example 6.11.3(3) instead: identifying each box in $\mathcal{B}_0$ with the interpretation of its accessibility relation in T, we let $\mathcal{B}$ be the smallest subset of 2^{T^2} that contains $\mathcal{B}_0$ and is closed under the operations of example 6.11.3(3). Each $\Box \in \mathcal{B}$ is then its own accessibility relation.

The frame $T = \{1, -1\}$, where $r_n(x,y)$ holds for $n \in \mathbb{Z}$ and $x, y \in T$ iff xy and n have the same sign, is standard for the set of boxes in example 6.11.3(2) above. But since $I' * I' = I$, there is no standard frame of size greater than 2.

An interesting problem, leading into the field of *relation algebras,* is to characterize those standardized sets of boxes for which standard frames exist. By theorem 6.11.6(2) below, standard frames exist for $\mathcal{B}$ iff the minimal logic K_ν is consistent.

In what follows, $(\mathcal{B}, I, ', *)$ will be a standardized set of boxes.

Definition 6.11.5 K_ν is the logic based on $\mathcal{B}$, with the following axioms[4] and rules:

- the axioms and rules of K *for each* $\Box \in \mathcal{B}$*:* so, e.g., we have $\Box$-generalization and $\Box(p \to q) \to (\Box p \to \Box q)$ for each $\Box \in \mathcal{B}$;
- (identity) $q \leftrightarrow Iq$;
- (composition) $\Box\blacksquare q \leftrightarrow (\Box * \blacksquare)q$ for all $\Box, \blacksquare \in \mathcal{B}$;
- (normality) $\Box q \wedge \Box' q \to \blacksquare q$ for all $\Box, \blacksquare \in \mathcal{B}$;
- (symmetry) $\Diamond(\Box q \wedge \Box' q) \to q$ for all $\Box \in \mathcal{B}$, where $\Diamond = \sim\Box\sim$;
- the rule $\rho(\Box q \wedge \Box' \sim q)$, for every $\Box \in \mathcal{B}$.

K_ν is recursive if $\mathcal{B}$ can be presented recursively. We will prove:

Theorem 6.11.6

1. *K_ν is sound in the class of standard frames.*

   Let Z be any consistent system of axioms and rules containing those of K_ν. Then:

2. *Z is complete for the class of standard frames.*

3. *If $S \in Z$ is a Sahlqvist axiom that locally corresponds to $\varphi(x)$, then Z is complete for the class of those standard frames $(T, \overline{r})$ with $(T, \overline{r}) \models \forall x\varphi(x)$.*

Proof. Part (1) is easy to prove. If there are no standard frames then there is nothing to prove. Assume otherwise. Soundness of the axioms is easy to check, and the proof that $\rho(\Box q \wedge \Box' \sim q)$ is sound is similar to that of chapter 3 for the simple IRR-rule, with the condition of 'irreflexivity' of the accessibility relation being replaced by the assumption of standardness. (If irreflexivity of a relation is also required, an appropriate axiom or rule can be added to Z.)

[4] Using the identity axiom we could combine normality and symmetry into the single axiom $\Diamond(\Box q \wedge \Box' q) \to \blacksquare q$ for all $\Box, \blacksquare \in \mathcal{B}$, where $\Diamond = \sim\Box\sim$.

Assume from now on that $Z \supseteq K_\nu$ is a logical system, and Γ_0 a Z-consistent theory written with a countable set of atoms. We will prove that there is a canonical structure $(T, \overline{r}, h)$ consisting of ρ-theories that is a model of every formula in Γ_0. This will be sufficient to deduce (2) and (3) of the theorem.

Many of the arguments are as in the early part of this chapter. For example:

Proposition 6.11.7 *There is a maximal Z-consistent ρ-theory $\Gamma_1 \supseteq \Gamma_0$, using a set L of at most countably many more atoms than Γ_0.*

Proof. As in section 6.2. ■

Let T^* be the set of all maximal Z-consistent ρ-theories written in L. From now on, all formulae will be L-formulae. Let $\Box$ be any box of the language. We define its accessibility relation r in T^* by: if $\Gamma, \Delta \in T^*$, then $r(\Gamma, \Delta)$ holds iff for all A, $\Box A \in \Gamma$ implies $A \in \Delta$.

Proposition 6.11.8 *If $\Gamma, \Delta \in T^*$, $r(\Gamma, \Delta)$ holds iff for all A, $A \in \Delta$ implies $\Diamond A \in \Gamma$, where $\Diamond$ is the diamond $\sim\Box\sim$.*

Proof. Easy. ■

Write r' for the accessibility relation of $\Box'$, where $\Box$ has accessibility relation r. So $r'' = r$ always. We want r' to have the same interpretation as $\sim r$. This will not be true without deleting some theories from T^*, since there may exist $\Gamma, \Delta \in T^*$ with $\sim r(\Gamma, \Delta)$ and $\sim r'(\Gamma, \Delta)$. But we can already show that not both r, r' hold.

Proposition 6.11.9 *Let $\Gamma, \Delta \in T^*$ and let r be any accessibility relation. Then we do not have both $r(\Gamma, \Delta)$ and $r'(\Gamma, \Delta)$.*

Proof. Let $\Box$ have accessibility relation r. As Γ is a ρ-theory, there is an atom q such that $\Box q \wedge \Box' \sim q \in \Gamma$. If $r(\Gamma, \Delta)$ and also $r'(\Gamma, \Delta)$ held, we would have $q \wedge \sim q \in \Delta$, rendering Δ inconsistent. ■

Definition 6.11.10 Define a binary relation $\approx$ on T^* by:
$\Gamma \approx \Delta$ iff there is an accessibility relation r such that $r(\Gamma, \Delta)$ holds.

Lemma 6.11.11

1. *For any r and $\Gamma, \Delta \in T^*$, if $r(\Gamma, \Delta)$ and $r'(\Gamma, \Delta)$ both fail then there is A such that $\Box A \in \Gamma$ for all $\Box$, and $\sim A \in \Delta$.*

2. *Suppose that $\Gamma, \Delta \in T^*$ and that $\Gamma \approx \Delta$. Then for all accessibility relations r, either $r(\Gamma, \Delta)$ or $r'(\Gamma, \Delta)$ holds.*

Proof. 1. Let r have accessibility relation $\blacksquare$. There are $\blacksquare B, \blacksquare' C \in \Gamma$ with $B, C \notin \Delta$. Let $A = B \vee C$. Then by the usual modal axioms and rules, $\blacksquare A \wedge \blacksquare' A \in \Gamma$ and $\sim A \in \Delta$. By the normality axiom, $\Box A \in \Gamma$ for all $\Box$.

2. Assume that $\Gamma \approx \Delta$, and that $r(\Gamma, \Delta)$ and $r'(\Gamma, \Delta)$ both fail. There is r_1 such that $r_1(\Gamma, \Delta)$ holds. Let $\Box$ be its box. By (1) there is A with $\Box A \in \Gamma$ and $A \notin \Delta$. This is a contradiction. ■

Proposition 6.11.12 *$\approx$ is an equivalence relation on T^*.*

Proof. Let $\Gamma, \Delta, \Phi \in T^*$.

1. Reflexivity: It is enough to show that for all A, $IA \in \Gamma \Rightarrow A \in \Gamma$. But this is clear by the identity axiom.

2. Symmetry: Assume that $\Gamma \approx \Delta$, so that for some $\Box$ with accessibility relation r, $r(\Gamma, \Delta)$ holds. We show that either $r(\Delta, \Gamma)$ or $r'(\Delta, \Gamma)$ holds. If neither holds, then, as in the lemma, there is a formula A such that $\Box A \wedge \Box' A \in \Delta$ and $\sim A \in \Gamma$. But $r(\Gamma, \Delta)$, so by proposition 6.11.8, $\Diamond(\Box A \wedge \Box' A) \in \Gamma$. By the symmetry axiom $\Diamond(\Box q \wedge \Box' q) \to q$ we obtain $A \in \Gamma$, a contradiction.

3. Transitivity: Assume that $\Gamma \approx \Delta \approx \Phi$. Choose boxes $\Box, \blacksquare$ such that $\Box A \in \Gamma \Rightarrow A \in \Delta$ and $\blacksquare A \in \Delta \Rightarrow A \in \Phi$. Then if $(\Box * \blacksquare) A \in \Gamma$, $\Box \blacksquare A \in \Gamma$ by the composition axiom, so that $\blacksquare A \in \Delta$ and $A \in \Phi$. Hence $r_{\Box * \blacksquare}(\Gamma, \Phi)$ holds, so that $\Gamma \approx \Phi$.

■

Now let $T \subseteq T^*$ be the $\approx$-equivalence class of the theory Γ_1 of proposition 6.11.7. Define a structure $(T, \overline{r}, h)$ by: the interpretation of each accessibility relation r in T is the restriction to T of its interpretation in T^*; and $h(q) = \{\Gamma \in T \mid q \in \Gamma\}$.

Lemma 6.11.13 *For all $\Gamma \in T$ and all formulae A, $A \in \Gamma$ iff $\|A\|_t^h = 1$.*

Proof. This is proved by induction on A in the same way as in section 6.2. We will only treat the case $\Diamond A$. Assume $\Diamond$ has accessibility relation r. If $\|\Diamond A\|_\Gamma^h = 1$ then for some $\Delta \in T$ with $r(\Gamma, \Delta)$ we have $\|A\|_\Delta^h = 1$. As, clearly, $\Delta \in T$, by the inductive hypothesis we have $A \in \Delta$. By proposition 6.11.8, $\Diamond A \in \Gamma$.

Conversely, assume that $\Diamond A \in \Gamma$. We must prove that $\|\Diamond A\|^h_\Gamma = 1$. Write $\Box$ for $\sim\Diamond\sim$, where $\Diamond$ is the diamond in $\Diamond A$.

Claim: For all formulae D, $\Diamond D \in \Gamma$ iff the set $\Sigma = \{D\} \cup \{B \mid \Box B \in \Gamma\}$ is (Z-)consistent.

Proof of Claim:

$\Rightarrow$: If Σ is inconsistent, there are $\Box B_1, \ldots, \Box B_n \in \Gamma$ such that $\vdash_Z \bigwedge_i B_i \to \sim D$. By generalization and axioms of K, $\vdash_Z \bigwedge_i \Box B_i \to \sim\Diamond D$. But $\bigwedge_i \Box B_i \in \Gamma$, so that $\sim\Diamond D \in \Gamma$.

$\Leftarrow$: Conversely, if $\Diamond D \notin \Gamma$ then $\Box\sim D \in \Gamma$ so that $\sim D$ is in the second component of Σ. This proves the claim.

By the claim, $\Delta_0 = \{A\} \cup \{B \mid \Box B \in \Gamma\}$ is consistent. As in section 6.2, we extend it to a maximal consistent ρ-theory Δ. List all the formula–box pairs as $(B_0, \Box_0), (B_1, \Box_1), \ldots$ in such a way that each pair comes up infinitely often in the list. Define Δ_i ($i < \omega$) by induction. Δ_0 is already defined. If Δ_i is defined, we set $\Delta_{i+1} = \Delta_i \cup \{\sim B_i\}$ if this is consistent. Otherwise, $\Delta_i \cup \{B_i\}$ is consistent. If B_i is not of the form

$$(*) \qquad C_m \wedge \Diamond_m(C_{m-1} \wedge \ldots \wedge \Diamond_1 C_0)) \cdots)$$

for some $m < \omega$, formulae $C_0, \ldots, C_m$ and arbitrary diamonds $\Diamond_1, \ldots, \Diamond_m$, we let $\Delta_{i+1} = \Delta_i \cup \{B_i\}$.

Suppose then that B_i has this form. The decomposition of B_i in this way is not unique — for example, if $m > 0$ we also have $B_i = C_m \wedge \Diamond_m C'$ where $C' = C_{m-1} \wedge \Diamond_{m-1}(C_{m-2} \wedge \ldots \wedge \Diamond_1 C_0)) \cdots)$ — but as $B_i = B_j$ for infinitely many $j < \omega$, we can deal with each possible decomposition as the construction proceeds. Write D for $A \wedge \bigwedge(\Delta_i \setminus \Delta_0)$. D is a well-defined formula as $\Delta_i \setminus \Delta_0$ is finite.

By assumption, $D \wedge B_i$ is consistent with $\{B \mid \Box B \in \Gamma\}$. So by the claim, we must have $\Diamond(D \wedge B_i) \in \Gamma$. Now $\Diamond(D \wedge B_i)$ is itself of the form $(*)$ above. So as Γ is a ρ-theory, there is already an atom q such that, writing

$$B_i' = C_m \wedge \Diamond_m(C_{m-1} \wedge \ldots \wedge \Diamond_1(C_0 \wedge \Box_i q \wedge \Box_i' \sim q)) \cdots),$$

we have $\Diamond(D \wedge B_i') \in \Gamma$. So by the claim again, $\Delta_i \cup \{B_i'\}$ is Z-consistent. We define Δ_{i+1} to be $\Delta_i \cup \{B_i, B_i'\}$. Since, clearly, $\vdash_Z B_i' \to B_i$, Δ_{i+1} is Z-consistent. This completes the definition of the Δ_i.

Setting $\Delta = \bigcup_{i<\omega} \Delta_i$, we see that Δ is a complete Z-consistent ρ-theory, i.e. $\Delta \in T^*$. Clearly, we have $r(\Gamma, \Delta)$, so $\Gamma \approx \Delta$ and $\Delta \in T$. Moreover, $A \in \Delta$, so by the inductive hypothesis $\|A\|^h_\Delta = 1$. Thus $\|\Diamond A\|^h_\Gamma = 1$, completing the proof. ■

Corollary 6.11.14 *$(T, \overline{r})$ is a standard frame.*

Proof.

1. We check that the accessibility relation r_I of I is '='. If $\Gamma, \Delta \in T$, then by definition, $r_I(\Gamma, \Delta)$ holds iff for all A, $IA \in \Gamma$ implies $A' \in \Delta$. But as $Iq \leftrightarrow q$ is an axiom, this says that $A \in \Gamma$ implies $A \in \Delta$, i.e. that $\Gamma = \Delta$.

2. By proposition 6.11.9 and lemma 6.11.11, for each $\Gamma, \Delta \in T$ and each accessibility relation r, we have $r'(\Gamma, \Delta) \iff {\sim}r(\Gamma, \Delta)$. Thus the accessibility relation of each $\Box'$ is ${\sim}r_\Box$.

3. Finally, we prove that for all $\Box, \blacksquare \in B$, the accessibility relation r of the box $\Box * \blacksquare$ is $r_{\Box . \blacksquare}$ in the sense of definition 6.10.5. Let $\Gamma, \Delta \in T$ and assume first that $r_{\Box . \blacksquare}(\Gamma, \Delta)$ holds. Thus there is $\Phi \in T$ with $r_\Box(\Gamma, \Phi)$ and $r_\blacksquare(\Phi, \Delta)$. Then if $(\Box * \blacksquare)A \in \Gamma$, we have $\Box\blacksquare A \in \Gamma$ by the composition axiom, so that $\blacksquare A \in \Phi$ and $A \in \Delta$. Hence $r_{(\Box * \blacksquare)}(\Gamma, \Delta)$ holds.

   Conversely, assume that $r_{(\Box * \blacksquare)}(\Gamma, \Delta)$ holds. As Δ is a ρ-theory we can choose an atom q such that $q \wedge I'{\sim}q \in \Delta$. By proposition 6.11.8, $\Diamond * \blacklozenge q \in \Gamma$, so by the composition axiom, $\Diamond\blacklozenge q \in \Gamma$. By lemma 6.11.13, $\|\Diamond\blacklozenge q\|_\Gamma^h = 1$, so there is $\Delta' \in T$ with $r_{\Box . \blacksquare}(\Gamma, \Delta')$ and $q \in \Delta'$. If $\Delta \neq \Delta'$, then as $I'{\sim}q \in \Delta$, by (1) and (2) above we would have ${\sim}q \in \Delta'$, a contradiction. Thus $\Delta = \Delta'$, and so $r_{\Box . \blacksquare}(\Gamma, \Delta)$, as required.

■

The results of proposition 6.11.7 and lemma 6.11.13 show that for all $A \in \Gamma_0$, $\|A\|_{\Gamma_1}^h = 1$, so $(T, \overline{r}, h)$ is a model of each formula in Γ_0. By corollary 6.11.14, $(T, \overline{r})$ is standard, so theorem 6.11.6(2) is proved. Theorem 6.11.6(3) will follow from:

Proposition 6.11.15 *Let $(T, \overline{r}, h)$ be as above. Let β be a sequence of boxes, and let $\Gamma \in T$. Then there is an atom q such that $\{\Delta \in T \mid r_\beta(\Gamma, \Delta)\} = h(q)$. Thus h is discrete.*

Proof. By corollary 6.11.14, $(T, \overline{r})$ is standard. So as I is a box, and for all $\Box, \blacksquare$, $\Box * \blacksquare$ is a box, we can assume that β is a box. As Γ is a ρ-theory, there is an atom q such that $\beta q \wedge \beta'{\sim}q \in \Gamma$. By lemma 6.11.13 and the standardness of $(T, \overline{r})$, $\{\Delta \in T \mid r_\beta(\Gamma, \Delta)\} = h(q)$, as required. ■

Now if $S \in Z$ is a Sahlqvist axiom that locally corresponds to $\varphi(x)$, then any substitution instance of S is a theorem of Z. Hence each $\Gamma \in T$ contains all substitution instances of S, so by lemma 6.11.13 they are all valid in $(T, \overline{r}, h)$. By proposition 6.11.15 and theorem 6.10.17(2), S is locally persistent over h, so S is valid in $(T, \overline{r}, h)$ for all h. Thus $(T, \overline{r}) \models \forall x \varphi(x)$, as required. The proof of theorem 6.11.6 is complete. ■

Chapter 7

Basic many-dimensional systems

7.1 Introduction and discussion

Traditional temporal logic is one-dimensional in the sense that the semantic evaluation of formulae involves only one point of reference in an indexical way. The standard truth definition, for instance, characterizes the truth of a temporal formula with respect to a time structure (or flow of time) and a moment t in that structure. The moment t is thus taken to be the (temporal) context where the formula as a whole holds. The semantic account of temporal operators such as P and F involves changing the initial point and reduces the truth of a formula at t to the truth of its subformulae at other moments. Except for certain tricks associated with the IRR-rule, there is really no way to 'get back' to t — to 'keep track' of the initial point of evaluation.

In many-dimensional temporal logic we have both the idea that the initial point of evaluation must be available during the course of evaluation of a formula in order, for example, to deal in an adequate way with the connective *now*, and also that other points (introduced by explicit or implicit reference devices) must remain accessible as complex formulae (and texts) are processed semantically.

In this chapter we survey the technical side of many-dimensional temporal logic, in particular, as far as the basic options in the interpretation of atomic formulae as well as the expressive power of the various languages are concerned. In this introduction we look at some more general issues.

Definition 7.1.1 Consider the first-order signature $\{<, R_1, \ldots, R_n\}$, where $<$ is a binary relation symbol and the R_i are m-place relation symbols. (Formulae of the signature may involve '$=$', but we treat '$=$' as a logical symbol and so do not include it in the signature.) Any formula $\varphi(x_1, \ldots, x_m, R_1, \ldots, R_n)$ of this signature, with m free variables $x_1, \ldots, x_m$, is called an *m-truth table for an n-place connective,* or an *$(m;n)$-truth table.*

Definition 7.1.2 Consider a propositional temporal language with the atomic propositions $q_1, \ldots, q_k, \ldots$, the classical (boolean) connectives, and k additional connectives $\#_1, \ldots, \#_k$, where $\#_i$ $(1 \leq i \leq k)$ is an n_i-place connective. Let $\varphi_1, \ldots, \varphi_k$ be truth tables such that φ_i is an $(m; n_i)$-truth table.

We can define, using the φ_i, a 'strong' m-dimensional interpretation for our language as follows. Consider a flow of time $(T, <)$, where $<$ is a transitive irreflexive binary relation on T. Let h be a function from atomic propositions into subsets of T^m (i.e. $h(q) \subseteq T^m$ for each propositional atom q). We define the value of $\|A\|^h_{t_1,\ldots,t_m}$ for $t_1, \ldots, t_m \in T$ by induction on A:

- $\|q\|^h_{t_1,\ldots,t_m} = 1$ iff $(t_1, \ldots, t_m) \in h(q)$ for q atomic;
- as usual for the boolean connectives;
- $\|\#_i(A_1, \ldots, A_{n_i})\|^h_{t_1,\ldots,t_m} = 1$ iff
  $$(T, <) \vDash \varphi_i(t_1, \ldots, t_m, h(A_1), \ldots, h(A_{n_i})),$$
  where inductively $h(A_j) = \{(u_1, \ldots, u_m) \in T^m \mid \|A_j\|^h_{u_1,\ldots,u_m} = 1\}$.

Definition 7.1.3 We can distinguish two notions of *strong validity* in $(T, <)$:

1. A is strongly valid iff for all h, $t_1, \ldots, t_m$, $\|A\|^h_{t_1,\ldots,t_m} = 1$;
2. A is strongly diagonally valid iff for all t, h, $\|A\|^h_{t,\ldots,t} = 1$.

Of course we can also define validities over a *class* of flows of time, but we will not do that here.

Definition 7.1.4 There is another notion, that of *weak interpretation,* where $h(q) \subseteq T$ (instead of $h(q) \subseteq T^m$). The notion of $\|A\|^h_{t_1,\ldots,t_m}$ has to be redefined for this case. We assign a special role to the coordinate t_1 and stipulate that

- $\|q\|^h_{t_1,\ldots,t_m} = 1$ iff $t_1 \in h(q)$ for q atomic.

In this case t_1 is called the *evaluation* point and $t_2, \ldots, t_m$ are called *reference* points. The inductive clauses for the truth value proceed as before. We can define notions of weak validity and of weak diagonal validity in $(T, <)$, analogously to the strong case.

It is obvious that for a fixed flow of time $(T, <)$ and a given m-dimensional interpretation, validity implies diagonal validity and strong validity implies weak validity. Two nice consequences of restricting attention to the weak two-dimensional logic are that validity may become decidable and that we may be able to find expressively complete connectives (see chapters 9 and 15). This need not hold in the strong case — see, for example, [Halpern and Shoham, 1986] and [Venema, 1990b]. The notion of weak validity, where we have one evaluation point and the rest are reference points, is more applicable to the study of the tenses in English.

Examples 7.1.5 Among the best-known many-dimensional systems are the following:

- Kamp invented one with a two-dimensional *Now* operator. The interpretation is weak (in the sense defined above) and the principal notion of validity is diagonal validity.
- Åqvist and Guenthner extended Kamp's system and incorporated a form of operator corresponding to the *then* of Vlach. We study this sytem in section 7.2.
- In section 7.2 we study a system which is many-dimensional with respect to all operators, in particular also for P and F, the interpretation of which depends on the particular two-dimensional configuration at which the operator is evaluated.
- Two-dimensional systems where atomic formulae have a strong interpretation can be found in [Segerberg, 1973]. In the setting of temporal logic such interpretations are perhaps best regarded as *interval logics,* at least as far as the atomic sentences are concerned. Interval logics have been extensively studied — see, for example, [Allen, 1984], [van Bentham, 1983], [Burgess, 1982b], [Halpern *et al.*, 1983b], [Halpern and Shoham, 1986], [Hamblin, 1972], [Humberstone, 1979], [Röper, 1980], [Schwartz *et al.*, 1983b], [Venema, 1990b], [Venema, 1991b]. We will defer their consideration until volume 2.

The notion of dimension

Here we want to examine the notion of dimension. To simplify matters, we will take the notion of validity to be *weak diagonal validity,* and we fix an arbitrary linear flow of time $(T, <)$.

Definition 7.1.6 Consider a language with connectives $\sim$, $\wedge$, $\vee$, $\rightarrow$, $\#_1$, $\ldots, \#_k$, where $\#_i$ is n_i-place. This language can be given many interpretations — in particular, interpretations of differing dimension — by choosing various different truth tables for the connectives. Assume that a set τ of formulae of this language is given. We say that τ is a *logical system* if it is closed under modus ponens and contains all substitution instances of truth functional tautologies. We say that τ is *characterized* by $(T, <)$, or *complete* for $(T, <)$, iff τ is the set of all formulae that are valid in $(T, <)$. We will often write $\tau \vdash A$, rather than $A \in \tau$, to recall its logical nature.

We want to show now that the notion of a dimension is non-trivial. We will exhibit a logical system τ which is complete for a notion of two-dimensional weak diagonal validity (for $(T, <)$) but is not complete for any one-dimensional validity over $(T, <)$. This will show that a two-dimensional semantics is essentially richer than a one-dimensional one. The example can be modified to show that an $m+1$-dimensional semantics is 'stronger' than an m-dimensional one.

Definition 7.1.7 To define τ consider the language with unary operators P, F, J^-, J^+. Truth value is two-dimensional, $\|A\|^h_{t,u}$, with t the evaluation point and u the reference point. The value is computed as follows:

$$
\begin{array}{ll}
\|q\|^h_{t,u} = 1 & \text{iff } t \in h(q), \text{ for } q \text{ atomic;} \\
\|FA\|^h_{t,u} = 1 & \text{iff there is } s > t \text{ with } \|A\|^h_{s,u} = 1; \\
\|PA\|^h_{t,u} = 1 & \text{iff there is } s < t \text{ with } \|A\|^h_{s,u} = 1; \\
\|J^+A\|^h_{t,u} & = \|A\|^h_{u,u}; \\
\|J^-A\|^h_{t,u} & = \|A\|^h_{t,t}.
\end{array}
$$

We let GA, HA abbreviate $\sim F \sim A$ and $\sim P \sim A$ respectively. We also define G^- and H^- by letting $G^-A = GJ^-A$ and $H^-A = HJ^-A$. We let

$$\tau = \{A \mid A \text{ is diagonally valid over } (T, <)\}.$$

Theorem 7.1.8

1. *Let A be a formula without any occurrences of J^+ or J^-. Let A^- be obtained from A by replacing any G by G^- and H by H^-. Then $\tau \vdash A \leftrightarrow A^-$.*

2. *$\tau \vdash G^-(J^+q \leftrightarrow q)$ and $\tau \nvdash G(J^+q \leftrightarrow q)$ for atomic q.*

3. *There is no one-dimensional interpretation over* $(T, <)$ *for which* τ *is complete.*

Proof. (1) and (2) are clear from the semantics given above, assuming that $(T, <)$ is non-trivial. For (3), suppose for contradiction that there is a one-dimensional interpretation for which τ is complete. Let φ_G, φ_H, φ_{J^+}, φ_{J^-} be the respective one-dimensional tables. The table φ_{G^-} for G^- can be obtained immediately from φ_G and φ_{J^-} by substitution, since we defined G^- to be GJ^-. In view of (1), $\tau \vdash Gq \leftrightarrow G^-q$ for any atom q, and it follows that φ_{G^-} and φ_G are equivalent formulae over $(T, <)$. Therefore, $G^- = G$ semantically; but this is impossible in view of (2). ■

To gain a better understanding of the τ of definition 7.1.7, let us prove a theorem of Kamp and Vlach:

Theorem 7.1.9 *Let* B *be any formula of the language with* P, F, J^-, J^+. *Then there exists a formula* $B^\#$ *without* J^+, J^- *such that for all* t, h *over linear time, we have* $\|B\|^h_{t,t} = \|B^\#\|^h_{t,t}$. *Hence* $\tau \vdash B \leftrightarrow B^\#$.

Proof. First prove the theorem for formulae B without J^-. Let m be the maximal number of nested J^+s in B, and let $C_1, \ldots, C_n$ be all subformulae of B of the form $C_i = J^+C'_i$ such that C'_i is a formula with $m-1$ nested occurrences of J^+. We prove the theorem by induction on m and n. It follows from the semantics of J^+ that

$$\tau \vdash B \leftrightarrow (C'_n \wedge B) \vee (\sim C'_n \wedge B) \leftrightarrow (C'_n \wedge B^\top) \vee (\sim C'_n \wedge B^\perp),$$

where $B^\top$ (respectively, $B^\perp$) is obtained from B by replacing every occurrence of $C_n = J^+C'_n$ by truth (respectively, falsity). But $C'_n \wedge B^\top$ and $\sim C'_n \wedge B^\perp$ have $n-1$ maximal J^+-nested subformulae, and hence by the inductive hypothesis B is equivalent to some $B^\#$ without $J^\pm$.

In case the formula B contains J^- as well, let k be the maximal number of nested J^-s in B, and $D_1, \ldots, D_n$ be all subformulae of B of the form $D_i = J^-D'_i$ such that D'_i is a formula with $k-1$ nested occurrences of J^-. Assume inductively that any formula D with smaller k, or the same k and smaller n, has an equivalent $D^\#$ without any occurrences of $J^\pm$. Let B^* be obtained from B by replacing the subformula D_n by $(D'_n)^\#$, which exists by the inductive hypothesis. Since $(D'_n)^\#$ has no Js, for any h and any t, u we have

$$\|D_n\|^h_{t,u} = \|D'_n\|^h_{t,t} = \|(D'_n)^\#\|^h_{t,t} = \|(D'_n)^\#\|^h_{t,u}.$$

Hence $\tau \vdash B \leftrightarrow B^*$. But the inductive hypothesis applies to B^*, so there is a formula $(B^*)^\#$ without occurrences of $J^\pm$ and with $\tau \vdash B^* \leftrightarrow (B^*)^\#$. So if we let $B^\# = (B^*)^\#$, we have $\tau \vdash B \leftrightarrow B^\#$, as required. ■

So although τ does not have a one-dimensional interpretation, any formula A of the language of τ is equivalent to an $A^{\#}$, which does have a one-dimensional interpretation. What fails is the principle of substitutivity. Although $\tau \vdash A \leftrightarrow A^{\#}$, we cannot obtain $\tau \vdash C(A) \leftrightarrow C(A^{\#})$ for arbitrary C. For example, if q is an atom then $\tau \vdash q \leftrightarrow J^{+}q$, but $\tau \nvdash Gq \leftrightarrow GJ^{+}q$.

One may be led to think that perhaps one of the notions in definition 7.1.10(6–8) below is the correct one. Recall that we have a fixed linear flow of time $(T, <)$ in mind, and that we are considering weak diagonal validity.

Definition 7.1.10 Let $\mathcal{L}, \mathcal{L}'$ be languages with finitely many connectives, the tables of the connectives of $\mathcal{L}$ being k-dimensional, and those for $\mathcal{L}'$, m-dimensional. Let the logical systems τ, τ' be the set of valid formulae in $\mathcal{L}, \mathcal{L}'$ respectively.

1. Let $\mathcal{F}$ be a class of maps. We write $\mathcal{L} \leq_{\mathcal{F}} \mathcal{L}'$ if there is a map $f \in \mathcal{F}$ such that for all formulae A of $\mathcal{L}$, $f(A)$ is defined and is a formula of $\mathcal{L}'$, and for all assignments h into T and all $t \in T$, $\|A\|^h_{t^k} = \|f(A)\|^h_{t^m}$, where $t^k = (t, \ldots, t) \in T^k$ and $t^m = (t, \ldots, t) \in T^m$.

   We write $\tau \leq_{\mathcal{F}} \tau'$ if there is a map $f \in \mathcal{F}$ such that for all formulae A of $\mathcal{L}$, $f(A)$ is defined and is a formula of $\mathcal{L}'$, and $\tau \vdash A$ iff $\tau' \vdash f(A)$.

2. We let $\mathcal{V}$ be the class of all maps.

3. We let e be the class of effective (recursive) maps: $f \in e$ if and only if there is an algorithm (i.e. a Turing machine) that computes $f(A)$ from A (for all A of $\mathcal{L}$).

4. We let p be the class of maps f such that there is an algorithm that computes $f(A)$ in time polynomial in the length of A, for all A of $\mathcal{L}$.

5. We let s be the class of 'substitutive' maps $A \mapsto A^*$ such that for all formulae $A, B_1, \ldots, B_n$ of $\mathcal{L}$:

   - if A involves atoms $q_1, \ldots, q_n$ then so does A^*;
   - in that case, $A(B_1, \ldots, B_n)^* = A^*(B_1^*, \ldots, B_n^*)$.

6. We say that $\mathcal{L}$ is *pseudo-m-dimensional* if there exists a language $\mathcal{L}'$ as above (i.e. m-dimensional) such that $\mathcal{L} \leq_{\mathcal{V}} \mathcal{L}'$.

7. We say that τ is *pseudo-m-dimensional* if there exists a logical system τ', in an m-dimensional language, which is complete for $(T, <)$ and such that $\tau \leq_{\mathcal{V}} \tau'$.

8. We define *effectively m-dimensional, polynomially m-dimensional,* and *structurally m-dimensional* similarly, replacing $\leq_{\mathcal{V}}$ by $\leq_e$, $\leq_p$, and $\leq_s$, respectively.

Remarks 7.1.11

1. In the notation of the definition, we have $\mathcal{L} \leq_{\mathcal{F}} \mathcal{L}' \Rightarrow \tau \leq_{\mathcal{F}} \tau'$ for any $\mathcal{F}$. So, for example, if $\mathcal{L}$ is polynomially m-dimensional then so is τ.

2. It is not difficult to see that $p \subseteq e \subseteq \mathcal{V}$. Hence

$$\begin{array}{rl} & \mathcal{L} \leq_p \mathcal{L}' \Rightarrow \mathcal{L} \leq_e \mathcal{L}' \Rightarrow \mathcal{L} \leq_{\mathcal{V}} \mathcal{L}' \\ \text{and} & \tau \leq_p \tau' \Rightarrow \tau \leq_e \tau' \Rightarrow \tau \leq_{\mathcal{V}} \tau'. \end{array}$$

3. For any τ, τ', as $\top \in \tau'$ and $\bot \notin \tau'$ we may define f by $f(A) = \top$ if $A \in \tau$, and $f(A) = \bot$ otherwise. Clearly, this shows that $\tau \leq_{\mathcal{V}} \tau'$, so the notion $\leq_{\mathcal{V}}$ for logical systems is not interesting.

4. The argument of theorem 7.1.8 above can be adapted to show that for all m there is an $m+1$-dimensional logical system τ_{m+1} that is not structurally m-dimensional.

Theorem 7.1.12 *Over linear time, any $\mathcal{L}$ is effectively one-dimensional.*

Proof. Let A be any formula of $\mathcal{L}$; then A has a table, $\varphi(t)$, such that in any linear flow of time, for arbitrary h, t we have $\|A\|^h_{t,\ldots,t} = 1$ iff $\varphi(t)$ is h-valid. φ can be obtained effectively from A in the usual way. Now Kamp defined a one-dimensional logic US which is *functionally complete* for the real numbers or integers — see chapter 10. Stavi added two more connectives U', S' to obtain a functionally complete logic over arbitrary linear time — see chapters 11 and 12. This means that there is an $A^{\#}$ of $USU'S'$, in fact, obtainable effectively from φ, such that for all h,t, $\|A^{\#}\|^h_t = 1$ iff $\varphi(t)$ is h-valid (see theorem 11.5.4). So $\|A\|^h_{t,\ldots,t} = \|A^{\#}\|^h_t$ for all h, t. Hence $\mathcal{L} \leq_e USU'S'$, as required. ■

Local validity

Let $(T, <, 0)$ be a flow of time with $0 \in T$ a distinguished point. 0 can be taken as the actual moment or 'now', in analogy to the notion of actual world in modal logic. Using this notion we can define validity differently.

Consider the notion of weak m-dimensional diagonal validity in $(T,<)$. According to our definition, A is weakly m-dimensionally diagonally valid in $(T,<)$ iff for all h and all $t \in T$,

$$\|A\|^h_{t,\ldots,t} = 1.$$

Alternatively, we could say that A is weakly m-dimensionally diagonally valid in $(T,<,0)$ iff for all h, $\|A\|^h_{0,\ldots,0} = 1$.

These two notions are not the same. For example, Kamp's logic with P, F, J^- can be shown to be one-dimensional in linear $(T,<,0)$ (where we let $\|J^-A\|_t = \|A\|_0$), but we know (from an argument similar to the one above) that this logic is not one-dimensional in $(T,<)$. The example with P, F, J^-, J^+ is not one-dimensional in $(T,<)$ either.

7.2 The two-dimensional system of Vlach and Åqvist

This is the basic system containing all the 'existential' quantifiers.

Definition 7.2.1 Consider a linear irreflexive flow of time $(T,<)$. Consider a language with the operators H, G, J^+, J^-, J, and the constants $C_=, C_>, C_<$. For an assignment g, giving each atomic p a set $g(p) \subseteq T$, and elements $t, s \in T$, consider the following two-dimensional tables:

(a)	$\|p\|^s_t = 1$	iff	$t \in g(p)$, for atomic p.
(b)	$\|C_=\|^s_t = 1$	iff	$t = s$.
	$\|C_>\|^s_t = 1$	iff	$t > s$.
	$\|C_<\|^s_t = 1$	iff	$t < s$.
(c)	$\|GA\|^s_t = 1$	iff	$\forall u > t, \|A\|^s_u = 1$.
	$\|HA\|^s_t = 1$	iff	$\forall u < t, \|A\|^s_u = 1$.
(d)	$\|JA\|^s_t = \|A\|^t_s$.		
	$\|J^+A\|^s_t = \|A\|^s_s$.		
	$\|J^-A\|^s_t = \|A\|^t_t$.		

Let s be $\{A \mid \|A\|^s_t = 1$, for all linear $(T,<,g)$ and all $t,s \in T\}$. The problem is to axiomatize s.

Notation 7.2.2 FA, PA will abbreviate $\sim G\sim A$, $\sim H\sim A$ respectively. Or will abbreviate $r \wedge G\sim r \wedge H\sim r$, meaning that r holds Only here. We also let $\Box A$ abbreviate $A \wedge GA \wedge HA$, and $\Diamond A$ abbreviate $A \vee FA \vee PA$.

Definition 7.2.3 (Axiomatization of s) We take the following axioms and rules.

1. The axioms and rules cited in lemma 6.7.2 of section 6.7, with **C2**$=\{G, H, J^+, J^-, J\}$ and **C1** $= \{G, H\}$.

2. Linear time axioms for G, H.

3. $J{\sim}A \leftrightarrow {\sim}JA$
   $J(A \wedge B) \leftrightarrow JA \wedge JB$
   Similarly for J^+, J^-.

4. $JJA \leftrightarrow A$
   $JJ^-A \leftrightarrow J^+A$
   $J^-A \leftrightarrow J^-J^-A$.

5. $A \rightarrow J^+\Diamond A$
   $A \rightarrow J^-\Diamond JA$
   $\#\Diamond J^-A \rightarrow \Diamond J^-A$ for $\# \in \{J, J^+, J^-\}$.

6. $FJ^-A \leftrightarrow J^-FJ^-A$
   $J^+A \rightarrow GJ^+A$.
   Similarly for P, H.

7. Let x be built up using G, H and the classical connectives; then take the axiom $J^-x \leftrightarrow x$.

8. $C_< \vee C_= \vee C_>$
   $C_< \leftrightarrow {\sim}(C_= \vee C_>)$
   $C_= \leftrightarrow {\sim}(C_< \vee C_>)$
   $C_> \leftrightarrow {\sim}(C_= \vee C_<)$
   $C_= \rightarrow GC_> \wedge HC_<$
   $C_= \rightarrow (A \rightarrow (\Box J^+A \wedge \Diamond J^-A))$
   $C_= \rightarrow (A \leftrightarrow J^-A)$
   $J^+C_= \wedge J^-C_=$.

9. $\Diamond(A \wedge Or) \rightarrow \Box(Or \rightarrow A)$.

Theorem 7.2.4 *The axiom system is complete for the semantical interpretation.*

Proof. For soundness, verify that all axioms and rules are valid in the interpretation.

For the completeness part, let A_0 be a consistent formula. We will build a linear irreflexive flow of time $(T, <)$ such that $\|A_0\|_t^s = 1$ for some $t, s \in T$ (with respect to some assignment of atoms to subsets of T).

Using the irreflexivity lemma 6.7.2 of chapter 6 we obtain a set S of theories Γ that are complete and consistent with respect to the axioms and rules listed, and a $\Theta \in S$ with $A_0 \in \Theta$, the whole with all the properties promised by the lemma. In particular, we can assume:

- For each $\Gamma \in S$ there is an atom r such that $Or \in \Gamma$.
- For each $\# \in \{F, P, J, J^+, J^-, \Diamond\}$ and $\Gamma, \Delta \in S$, we write

  $$\Gamma \; r_\# \; \Delta$$

  if for all $A \in \Delta$ we have $\#A \in \Gamma$. Then if $\#B \in \Gamma \in S$, there is $\Delta \in S$ with $\Gamma \; r_\# \; \Delta$ and $B \in \Delta$.
- For each $\Gamma, \Delta \in S$ there are $\Gamma = \Gamma_0, \Gamma_1, \ldots, \Gamma_n = \Delta$ such that for all $i < n$ there is $\# \in \{F, P, J, J^+, J^-\}$ with $\Gamma_i \; r_\# \; \Gamma_{i+1}$.

We want to make S into a model of A_0. Our problem is that S is simply a set of theories, whereas the logic **s** is two-dimensional. To obtain a two-dimensional interpretation of **s** from S, we let

$$T = \{\Gamma \in S \mid C_= \in \Gamma\}$$

(the 'diagonal' of S). We will use t, u to denote elements of T. For $t, u \in T$, let

$$t < u \quad \iff \quad \text{for all } A,\ GJ^-A \in t \Rightarrow A \in u.$$

Lemma 7.2.5 $(T, <)$ *is linear irreflexive time.*

Proof. Let $t \in T$. Choose an atom r such that $Or \in t$. Hence $G{\sim}r \in t$. By axiom (7) and G-generalization, $\vdash G({\sim}r \leftrightarrow J^-{\sim}r)$. But axiom (1) includes $G(A \to B) \to (GA \to GB)$ (G-normality), so $\vdash G{\sim}r \leftrightarrow GJ^-{\sim}r$. Hence $GJ^-{\sim}r \in t$. If $t < t$, then we would have ${\sim}r \in t$, a contradiction to the consistency of t. Hence $<$ is irreflexive.

By axiom (2) we have $\vdash GJ^-A \to GGJ^-A$. Applying G-generalization and G-normality to axiom (6) yields $\vdash GGJ^-A \to GJ^-GJ^-A$. Hence $\vdash GJ^-A \to GJ^-GJ^-A$, and it follows that $<$ is transitive.

For linearity, we first show that for all $t \in T$, $\Gamma \in S$, and all A,

$$(*) \qquad A \in t \Rightarrow \Diamond J^-A \in \Gamma.$$

By lemma 6.7.2, there are $\Gamma_0 = t, \Gamma_1, \ldots, \Gamma_n = \Gamma$ in S such that for all $i < n$ there is $\# \in \{F, P, J, J^+, J^-\}$ with $\Gamma_i \; r_\# \; \Gamma_{i+1}$. We prove $(*)$ by induction on n. If $n = 0$ then $(*)$ follows from axiom (8), since $C_= \in t$. Assume the result for smaller n, and let $A \in t$. Then inductively, $\Diamond J^-A \in \Gamma_{n-1}$, and hence $\#\Diamond J^- A \in \Gamma$ for some $\# \in \{F, P, J, J^+, J^-\}$. If $\#$ is F or P, we have $\vdash \#\Diamond J^- A \to \Diamond J^- A$ by axiom (2); the same holds by axiom (5) if $\#$ is J, J^+, or J^-. So $\Diamond J^- A \in \Gamma$, completing the induction. Hence (*) is proved.

Next, observe that $\vdash C_= \to (A \to GJ^-PJ^-A)$. For by axiom (2), $\vdash J^-A \to GPJ^-A$. By axiom (6) and G-generalization and normality, $\vdash GPJ^-A \to GJ^-PJ^-A$. By axiom (8), $\vdash C_= \to (A \leftrightarrow J^-A)$, and the result follows. It can now be shown in the usual way that if $t < u$ in T and $HJ^-A \in u$ then $A \in t$.

Now let $t, u \in T$ be distinct, and suppose for contradiction that $t \not< u$ and $u \not< t$. Then there are A, B, C such that

- $J^-A \in t$ and $\sim A \in u$ (use axiom (8));
- $GJ^-B \in t$ and $\sim B \in u$;
- $HJ^-C \in t$ and $\sim C \in u$.

Let $D = A \vee B \vee C$. Then $\Box J^-D \in t$, and yet $\sim D \in u$. This contradicts (*). Hence $(T, <)$ is linear, as required. ■

Remark 7.2.6 The proof shows that for all $t, u \in T$, $A \in t \Rightarrow \Diamond J^-A \in u$. If A only involves at most G, H and the boolean connectives, then by axiom (7) we have $A \in t \Rightarrow \Diamond A \in u$.

Definition 7.2.7 If t, u in T, we define $\alpha(t, u)$ to be the unique $\Gamma \in S$ such that $u\ r_\Diamond\ \Gamma$ and $Or \in \Gamma \cap t$ for some atom r.

Lemma 7.2.8 *For all $t, u \in T$, $\alpha(t, u)$ exists and is unique.*

Proof. First we check existence. Using lemma 6.7.2 choose an atom r such that $Or \in t$. So by remark 7.2.6, $\Diamond Or \in u$. By lemma 6.7.2 again, there is $\Gamma \in S$ with $u\ r_\Diamond\ \Gamma$ and $Or \in \Gamma$, as required.

For uniqueness, suppose Γ is as above, and $\Delta \in S$ is such that $u\ r_\Diamond\ \Delta$ and $Os \in \Delta \cap t$ for some atom s. Let $A \in \Gamma$. Then $A \wedge Or \in \Gamma$, so $\Diamond(A \wedge Or) \in u$. Now $Or \wedge Os \in t$, so by the remark, $\Diamond(Or \wedge Os) \in u$. By axiom (9), $\Box(Or \to Os) \in u$, and so $\Diamond(A \wedge Os) \in u$. Hence $\Box(Os \to A) \in u$ by axiom (9) again, so that $Os \to A \in \Delta$. Since $Os \in \Delta$ we have $A \in \Delta$. Thus $\Gamma \subseteq \Delta$; and as Γ and Δ are maximal consistent we have $\Gamma = \Delta$, as required. ■

The following lemma gives handy criteria for a formula's being in $\alpha(t, u)$.

Lemma 7.2.9 *Let $t, u \in T$. Then:*

1. *If x is built using only G, H and the classical connectives, then $x \in t \iff x \in \alpha(t,u)$.*

2. *For all formulae A, $A \in \alpha(t,u)$ iff $\Diamond(A \wedge Or) \in u$, where r is any atom such that $Or \in t$.*

Proof. 1. Choose an atom r such that $Or \in t$. Suppose that $x \in t$. Then by the remark above, $\Diamond(x \wedge Or) \in u$. So there is $\Gamma \in S$ with $u\ r_\Diamond\ \Gamma$ and $x \wedge Or \in \Gamma$. By the uniqueness part of lemma 7.2.8, $\Gamma = \alpha(t,u)$. Hence $x \in \alpha(t,u)$. Similarly, $\sim x \in t \Rightarrow \sim x \in \alpha(t,u)$. The result now follows by maximal consistency of t and $\alpha(t,u)$.

2. Choose an atom r such that $Or \in t$. Note that by (1), $Or \in \alpha(t,u)$ also. So if $A \in \alpha(t,u)$ then $A \wedge Or \in \alpha(t,u)$, so $\Diamond(A \wedge Or) \in u$ by definition of α. Conversely, suppose that $\Diamond(A \wedge Or) \in u$. By axiom (9), this implies that $\Box(Or \to A) \in u$. But $u\ r_\Diamond\ \alpha(t,u)$ by definition, so we have $Or \to A \in \alpha(t,u)$. As $Or \in \alpha(t,u)$, we obtain $A \in \alpha(t,u)$. ■

Definition 7.2.10 Let $\Gamma \in S$. Define

$$\begin{aligned} \Gamma^+ &= \{A \mid J^+A \in \Gamma\}, \\ \text{and} \quad \Gamma^- &= \{A \mid J^-A \in \Gamma\}. \end{aligned}$$

Lemma 7.2.11 *For all $\Gamma \in S$ we have:*

1. *$\Gamma^-, \Gamma^+ \in T$;*

2. *$\alpha(\Gamma^-, \Gamma^+) = \Gamma$;*

3. *If $\Gamma = \alpha(t,u)$ for $t, u \in T$ then $t = \Gamma^-$ and $u = \Gamma^+$;*

4. *$\alpha : T \times T \to S$ is bijective.*

Proof. 1. By axiom (8), $J^-C_= \in \Gamma$, so by lemma 6.7.2 there is $t \in T$ with $\Gamma\ r_{J^-}\ t$. Clearly, $t \subseteq \Gamma^-$. But by axiom (3), Γ^- is maximal consistent. Hence $t = \Gamma^-$, as required. A similar argument works for Γ^+.

2. We show that $\Gamma \subseteq \alpha(\Gamma^-, \Gamma^+)$; this is sufficient since Γ and $\alpha(\Gamma^-, \Gamma^+)$ are maximal consistent. So let $A \in \Gamma$. Choose an atom r such that $Or \in \Gamma^-$. Then $J^-Or \in \Gamma$, and by axiom (7), $Or \in \Gamma$. Hence $A \wedge Or \in \Gamma$, and by axiom (5) we obtain $J^+\Diamond(A \wedge Or) \in \Gamma$, and so $\Diamond(A \wedge Or) \in \Gamma^+$. So by lemma 7.2.9, $A \in \alpha(\Gamma^-, \Gamma^+)$, as required.

3. Choose an atom r such that $Or \in t$. If $A \in t$ then by remark 7.2.6, $\Diamond J^-(A \wedge Or) \in u$. By axioms (3) and (7), $\Diamond(J^-A \wedge Or) \in u$, so by lemma

7.2.9, $J^- A \in \Gamma$, and so $A \in \Gamma^-$ by definition of Γ^-. Hence $t \subseteq \Gamma^-$, so $t = \Gamma^-$ as both sets are maximal consistent. The argument for u is similar: $A \in u$ implies $\Box J^+ A \in u$ by axiom (8), so $J^+ A \in \alpha(t,u) = \Gamma$, by definition of α, and $A \in \Gamma^+$. Hence $u \subseteq \Gamma^+$, and so we have equality as before.

4. Immediate from (2) and (3). ■

Now define an assignment g into T by $g(q) = \{t \in T \mid q \in t\}$ for each atom q, as usual.

Theorem 7.2.12 *With respect to this assignment g, for all $t, u \in T$ and all formulae A,*

$$\|A\|_t^u = 1 \iff A \in \alpha(t,u).$$

Proof. We go by induction on the structure of A.

Case A atomic: For an atom q, we have $\|q\|_t^u = 1$ iff $t \in g(q)$, iff $q \in t$, iff $q \in \alpha(t,u)$ by lemma 7.2.9(1).

The boolean cases are clear, which leaves $C_=$, etc., the J-cases, and F and P.

Case $C_=$, etc.: Let r be an atom such that $Or \in t$. If $t = u$ then for all $A \in t$, $A \wedge Or \in u$, so certainly $\Diamond(A \wedge Or) \in u$, yielding $A \in \alpha(t,u)$ by lemma 7.2.9. Hence $\alpha(t,u) = t = u$, so $C_= \in \alpha(t,u)$. If $t > u$ then $C_= \wedge FOr \in u$. By axiom (8), $GC_> \in u$, and it follows that $\Diamond(C_> \wedge Or) \in u$. Hence $C_> \in \alpha(t,u)$ by lemma 7.2.9. The case $t < u$ is similar. By axiom (8) these possibilities are mutually exclusive, so the cases C are complete.

Case J^+, J^-: By lemma 7.2.11(3), $J^+ A \in \alpha(t,u)$ iff $A \in [\alpha(t,u)]^+ = u = \alpha(u,u)$. The result for J^+ now follows using the inductive hypothesis, and the proof for J^- is similar.

Case J: By the inductive hypothesis, it is enough to show that for all $t, u \in T$ we have

$$JA \in \alpha(t,u) \Rightarrow A \in \alpha(u,t).$$

Choose an atom s such that $Os \in u$. We claim that $JOs \in \alpha(t,u)$. Well, $Os \in u$, so by lemma 7.2.11, $J^+ Os \in \alpha(t,u)$. By axiom (4), $JJ^- Os \in \alpha(t,u)$. By axiom (7) and J-generalization, $\vdash JOs \leftrightarrow JJ^- Os$, so $JOs \in \alpha(t,u)$, as claimed.

Now assume that $JA \in \alpha(t,u)$. By axiom (5), $J^- \Diamond J(JA \wedge JOs) \in \alpha(t,u)$, so by lemma 7.2.11 again, $\Diamond J(JA \wedge JOs) \in t$, i.e. $\Diamond(A \wedge Os) \in t$. Hence $A \in \alpha(u,t)$, by lemma 7.2.9.

Case F: We show that $FA \in \alpha(t,u)$ iff $A \in \alpha(t',u)$ for some $t' \in T$ with $t' > t$. By lemma 7.2.11, it suffices to show that for all $\Gamma, \Delta \in S$,

$$(*) \qquad \Gamma\, r_F\, \Delta \quad \iff \quad \Gamma^+ = \Delta^+ \text{ and } \Gamma^- < \Delta^-.$$

For left-to-right of (*), suppose that $\Gamma\ r_F\ \Delta$. Assume that $A \in \Gamma^+$, so that $J^+A \in \Gamma$. By axiom (6), $GJ^+A \in \Gamma$, so that $J^+A \in \Delta$. Hence $A \in \Delta^+$, so that $\Gamma^+ \subseteq \Delta^+$. As these sets are maximal consistent, we obtain $\Gamma^+ = \Delta^+$. Similarly, if $A \in \Delta^-$ then $J^-A \in \Delta$, so $FJ^-A \in \Gamma$. By axiom (6), $J^-FJ^-A \in \Gamma$, so $FJ^-A \in \Gamma^-$. This proves that $\Gamma^- < \Delta^-$, as required.

Conversely, assume the right-hand side of (*) and let $GA \in \Gamma$. We show that $A \in \Delta$. Choose atoms r, s such that $Or \in \Gamma^-$ and $Os \in \Delta^-$. Because $\Gamma^- < \Delta^-$, by axiom (7) we have $FOs \in \Gamma^-$, and so by remark 7.2.6 we obtain $\Diamond(FOs \wedge Or) \in \Gamma^+$. By axiom (9) we obtain $\Box(Or \to FOs) \in \Gamma^+$. Now $GA \in \Gamma$, so $\Diamond(GA \wedge Or) \in \Gamma^+$ by lemma 7.2.9. So we obtain $\Diamond F(A \wedge Os) \in \Gamma^+ = \Delta^+$. Hence by axiom (2), $\Diamond(A \wedge Os) \in \Delta^+$, so by lemma 7.2.9, $A \in \Delta$, as required.

Case P: This is similar to F. ∎

Recall that our consistent formula A_0 lies in some $\Theta \in S$. Choose $t, u \in T$ such that $\alpha(t, u) = \Theta$. By the theorem, ${}^g\|A_0\|_t^u = 1$, and so $(T, <, g)$ is a model for A_0. This completes the proof of theorem 7.2.4. ∎

7.3 A system suggested by the future perfect

Consider the sentence

John will come and ask for the money
but by that time it will have been stolen.

We have three uses of the future here. We start with our reference point and evaluation point being the same. We denote this by (n, n) (n for now).

1. One application of F (will come) sends the evaluation point into the future. So we have (n, x) with $n < x$.

2. The 'will have' is really P, but it gives a new evaluation point $y, n < y < x$. We take the reference point also to be y. (Later, in section 7.3, we investigate what happens if we don't do that.) Thus we obtain (y, y) with $n < y < x$ from the application of P to (n, x), if $n < x$.

3. What about an application of F ('and ask') to (n, x) for the case $n < x$? We have the option of taking $x' \geq x$ as evaluation point. What do we take as reference? Looking at the example we see that $x' = x$ and possibly the reference becomes x also. Thus we take F applied to (n, x) with $n < x$ to be (x, x).

There are other possibilities. This is not the only way of representing the perfect tense.

The truth table given for P and F above is not arbitrary. P and F do occur in English in other connections.

1. Clause (1) captures the behaviour of 'will' in 'He will come'.
2. Clause (2) captures the meaning of 'would' in: 'He said he would come'. Or better, its mirror image captures the future perfect.
3. Clause (3) captures the behaviour of the second 'will' in 'I will come and bring the parcel', 'He will say he is coming'.

To make our connectives P and F more plausible to the reader, here is an example of how to express 'until' using 'will' and 'will have':

When the postman delivers the parcel he will ask what is in it and add that he has been wondering what is in it all the time.

This has the form

$$F(Fp \wedge Hq),$$

or U(deliver, ask) (i.e. asks until delivers).

Note that here we have two-dimensional unary connectives, and we want to stress that they are not the same as the unary connectives P and F of preceding chapters. To avoid further confusion, the representation of unary P and F in terms of the binary P and F will be denoted by F^+, P^+ (actually $F^+ = FF$, $P^+ = PP$), and similarly for U, S. To summarize we have the following:

- One-dimensional future F, past P, since, and until, are denoted by F, P, S, U in preceding chapters.
- Their interpretation in terms of two-dimensional connectives will be indicated by F^+, P^+, U^+, S^+ in this chapter.
- We will continue to let U, S denote the one-dimensional until and since in this chapter.
- The two-dimensional *will* and *was* will be denoted by F, P in this section.
- The one-dimensional evaluation at points will be denoted by $\|A\|_t$ and the two-dimensional evaluation by $[A]^t_s$.

We can now give the table for the two-dimensional P, F. It will turn out that they are equivalent in strength to the one-dimensional system $\{U, S\}$. Recall that:

- $\|S(A,B)\|_t = 1$ iff for some $s < t, \|A\|_s = 1$ and for all t', if $s < t' < t$ then $\|B\|_{t'} = 1$.
- $\|U(A,B)\|_t = 1$ iff for some $s > t, \|A\|_s = 1$ and for all t', if $t < t' < s$ then $\|B\|_{t'} = 1$.

Axiomatizations for **US** (the system with U, S) were given in chapter 6 for various flows of time.

Definition 7.3.1 Consider now the following truth table, for a system **FP** in a language with unary two-dimensional P and F only. The notation

$$^g[A]^t_s$$

will represent the truth value of A at the point s, with respect to the reference point t, and assignment g. We omit g when possible.

1. $[p]^t_s = 1$ iff $s \in g(p)$, for p atomic.
2. The conditions for $\wedge, \vee, \sim, \rightarrow$ are the usual.
3. $[FA]^t_s = 1$ iff
   (a) $t = s$ and for some $s' > s$, $[A]^t_{s'} = 1$, or
   (b) $t < s$ and $[A]^s_s = 1$, or
   (c) $t > s$ and for some $s < u < t$, $[A]^u_u = 1$.
4. $[PA]^t_s = 1$ iff
   (a) $t = s$ and for some $s' < s$, $[A]^t_{s'} = 1$, or
   (b) $t < s$ and for some $t < u < s$, $[A]^u_u = 1$, or
   (c) $t > s$ and $[A]^s_s = 1$.

We let $Hx = \sim P \sim x$ and $Gx = \sim F \sim x$.

Let **FP** be the set of all formulae A such that $[A]^t_t = 1$ for all t in linear time and all assignments g. Thus we have, for any formula A:

$$A \in \mathbf{FP} \text{ iff for all linear } (T,<) \text{ and all } g, t, {}^g[A]^t_t = 1.$$

In terms of definition 7.1.4, we are dealing with *weak diagonal* validity. Similarly, we define **US** to be the set of all US-formulae that are valid over linear time. The main purpose of this section is to show that **US** and **FP** are inter-translatable into each other. We will prove:

Theorem 7.3.2

1. *For each formula A of the language of* **US** *there exists a formula A^* of the language of* **FP** *such that for all t in linear time and all assignments g, we have $\|A\|_t^g = {}^g[A^*]_t^t$. Hence $A \in$* **US** *iff $A^* \in$* **FP**; *or equivalently, if* **a(US)** *is any axiomatization of* **US** *then* **a(US)** *$\vdash A$ iff $A^* \in$* **FP**.

2. *For each B of the language of* **FP** *there exists a $B\sharp$ of the language of* **US** *such that for all t in linear time and all assignments g, we have $\|A\sharp\|_t^g = {}^g[A]_t^t$. Hence $B \in$* **FP** *iff $B\sharp \in$* **US** *(or* **a(US)** *$\vdash B$).*

Remark 7.3.3 The preceding theorem yields an axiomatization of **FP**, obtained from the axiomatization of **US**. We shall see that later. The two functions $* : A \to A^*$ and $\sharp : B \to B\sharp$ are defined effectively and inductively on the complexity of the formulae. In the terminology of definition 7.1.10, we have **US** $\leq_s$ **FP**, and **FP** $\leq_e$ **US**, and the same for the respective languages. **FP** is of particular interest, as its connectives are unary (P and F) yet they have the same expressive power as the binary connectives U and S. Furthermore, **US** is a very strong system in the sense that it is expressively complete for integer and real time (see chapter 10).

We begin with the proof of theorem 7.3.2(1).

Definition 7.3.4 Let $*$ be the following translation from **US** into **FP**.

(a)	p^*	$= p$, for atomic p.
(b)	$(A \wedge B)^*$	$= A^* \wedge B^*$
	$(\sim A)^*$	$= \sim A^*$
	$(A \vee B)^*$	$= A^* \vee B^*$
	$(A \to B)^*$	$= A^* \to B^*$.
(c)	$S(A,B)^*$	$= P(PA^* \wedge GB^*)$.
(d)	$U(A,B)^*$	$= F(FA^* \wedge HB^*)$.

Lemma 7.3.5 *For all formulae A of the language of* **US**, *for any assignment g and any t, $\|A\|_t^g = {}^g[A^*]_t^t$.*

Proof. Obvious from the semantical conditions. ■

Hence also $A \in$ **US** iff $A^* \in$ **FP**. Note in particular that F^+ of **US** is translated into

$$(F^+A)^* = U(A, \top)^* = F(FA^* \wedge H\top),$$

which is equivalent to FFA^*, and similarly for P. This shows what is semantically evident: that F^+ can be taken as FF and P^+ as PP in **FP**, and similarly for G^+ and H^+.

The proof of theorem 7.3.2(1) is complete. We now proceed to define the translation $\sharp$ from **FP** into **US**, and prove theorem 7.3.2(2). For this purpose we need some auxiliary results.

Remark 7.3.6 Below are some axioms and rules valid in **FP**.

A. All formulae of the form below and their mirror images (obtained by interchanging P and F):

1. $A \wedge HA \wedge GA$, for any A that is a substitution instance of a truth functional tautology.

2. Let $U^+(x,y) = F(Fx \wedge Hy)$ and $S^+(x,y) = P(Px \wedge Gy)$. For any A of **US** such that $A \in$ **US**, let A^* be the formula obtained from A by replacing each U by U^+ and S by S^+. Then $A^* \in$ **FP**.

3. $G(A \wedge B) \leftrightarrow GA \wedge GB$

   $GF\top$, $GH\top$

   $F(p \wedge A) \leftrightarrow F(Fp \wedge A)$, for p atomic or negated atomic

   $G(FA \wedge FB \leftrightarrow F(A \wedge B))$

   $G(F{\sim}A \leftrightarrow {\sim}FA)$

   $G(HA \wedge HB \leftrightarrow H(A \wedge B))$.

4. $A \wedge H^+A \rightarrow G(HA \rightarrow FH^+A)$.

5. $G(FA \rightarrow \bigvee_{i=1}^m HB_i) \rightarrow G^+({\sim}B_1 \rightarrow G(FA \rightarrow \bigvee_{i=2}^m HB_i))$.

B. The following rules are valid in **FP**:

1. modus ponens,

2. $$\frac{(P^+B \wedge F^+A) \rightarrow C}{A \rightarrow H(PB \rightarrow GC)},$$

3. $$\frac{A \leftrightarrow B}{E(A) \leftrightarrow E(B)},$$

   where $E(x)$ is any formula built up using U^+ and S^+ only (i.e. $E = B(x)^*$ for some $B(x)$ of the language of **US**), and x is atomic.

Lemma 7.3.7

1. *In* **FP**, *any formula of the form FD is equivalent to a disjunction*

$$\bigvee_{i=1}^{n} Q_i,$$

each Q_i having the form $\mathbf{y}_F(A, \{B_j\}, C)$, where

$$\mathbf{y}_F\{A, \{B_1, \ldots, B_m\}, C) = F(FA \wedge HC \wedge \bigwedge_{j=1}^{m} PB_j).$$

2. *Similarly, in the mirror image case of PD we can obtain an equivalent* $\mathbf{y}_P$.

Proof. We only prove (1), as (2) is similar. Write D in disjunctive normal form, regarding anything of the form FX or PX as atoms. Thus D is equivalent to a disjunction $\bigvee D_i$, where each D_i has the form $\bigwedge_j X_j$, X_j being either atomic, or of the form FA_j, or of the form PB_j, or a negation of these forms. FD is equivalent to $\bigvee FD_i$.

We know (see remark 7.3.6(3)) that $F(p \wedge A)$ is equivalent to $F(Fp \wedge A)$, whenever p is an atomic or negated atomic formula. Therefore, each FD_i is equivalent to $F \bigwedge_j X_j$, where each X_j is of the form FA, PB, or their negations. Furthermore, since $G({\sim}FA \leftrightarrow F{\sim}A) \in$ **FP**, we can assume that the form ${\sim}FA$ does not occur; and since $G(FA \wedge FB \leftrightarrow F(A \wedge B)) \in$ **FP**, we can assume that there exists at most one X_j of the form FA. Also, since ${\sim}PA$ is the same as $H{\sim}A$ and $G(HA \wedge HB \leftrightarrow H(A \wedge B)) \in$ **FP** we can assume that there exists at most one formula X_j of the form ${\sim}PB$, which is, in effect, of the form HC.

Thus we are left with at most one FA, at most one HC, and possibly several PB_i. We can assume that at least one FA and HC appear because by the remark, $GF\top$, $GH\top$, and $G(A \wedge B) \leftrightarrow (GA \wedge GB)$ are all in **FP**. We can conclude that each FD_i is equivalent to a formula of the form $\mathbf{y}_F(A, \{B_j\}, C)$. ■

Definition 7.3.8 For any formulae A, C, and any finite set Γ of formulae, we define by induction on the size of Γ a formula $\mathbf{y}_U(A, \Gamma, C)$ of the language of **US**. We let $\mathbf{y}_U(A, \varnothing, C) = U(A, C)$, and

$$\mathbf{y}_U(A, \Gamma, C) = \bigvee_{\varnothing \neq \Delta \subseteq \Gamma} U(C \wedge (\bigwedge \Delta) \wedge \mathbf{y}_U(A, \Delta^-, C), C),$$

where $\Delta^- = \Gamma - \Delta$. We define $\mathbf{y}_S$ similarly, transposing U and S above.

Lemma 7.3.9 *For any assignment g and any moment of time t, any atomic A, C and any finite set Γ of atomic formulae,*

$$^g[\mathbf{y}_F(A,\Gamma,C)]^t_t = \|\mathbf{y}_U(A,\Gamma,C)\|^g_t,$$

where $\mathbf{y}_F$ is as in lemma 7.3.7.

Proof. By induction on the size n of Γ. If this is 0, then by the semantic truth condition and the fact that A and C are atomic,

$$[\mathbf{y}_F(A,\varnothing,C)]^t_t = [F(FA \wedge HC)]^t_t = \|U(A,C)\|_t = \|\mathbf{y}_U(A,\varnothing,C)\|_t$$

(with respect to any assignment).

Assume the result for n, and suppose that $|\Gamma| = n+1$. Let $\mathcal{B}$ be the set of all non-empty $\Delta \subseteq \Gamma$, and for any $\Delta \in \mathcal{B}$ let Δ^- be $\Gamma - \Delta$. Then $\|\mathbf{y}_U(A,\Gamma,C)\|_t = 1$ iff $\| \bigvee_{\Delta\in\mathcal{B}} U(C \wedge \bigwedge\Delta \wedge \mathbf{y}_U(A,\Delta^-,C),C)\|_t = 1$, which holds iff for some $t' > t$ and some non-empty $\Delta \subseteq \Gamma$,

(a) $\|C\|_x = 1$ for all $t < x \leq t'$;

(b) $\| \bigwedge\Delta\|_{t'} = 1$;

(c) $\|\mathbf{y}_U(A,\Delta^-,C)\|_{t'} = 1$.

Since A, C and the formulae in Γ are all atomic and Δ^- has fewer than $n+1$ elements, by the induction hypothesis we obtain equivalent conditions:

(a′) $[C]^x_x = 1$ for all $t < x \leq t'$;

(b′) $[\bigwedge\Delta]^{t'}_{t'} = 1$;

(c′) $[\mathbf{y}_F(A,\Delta^-,C)]^{t'}_{t'} = 1$.

Thus we have established that $\|\mathbf{y}_U(A,\Gamma,C)\|_t = 1$ iff for some $t' > t$ and $\Delta \in \mathcal{B}$, conditions (a′), (b′), and (c′) hold.

Now consider the matter from the other direction. By definition, we see that $[\mathbf{y}_F(A,\Gamma,C)]^t_t = 1$ iff $[F(FA \wedge HC \wedge \bigwedge_{B\in\Gamma} PB)]^t_t = 1$, iff there exist $s > t$, and $u(B)$ with $t < u(B) < s$ for each $B \in \Gamma$, such that $[A]^s_s = 1$, $[B]^{u(B)}_{u(B)} = 1$ for each B, and $[C]^y_y = 1$ for all y with $t < y < s$. The $u(B)$ need not necessarily be different. Let $t' \in \{u(B) \mid B \in \Gamma\}$ be the smallest of them (i.e. the nearest to t), and let $\Delta = \{B \in \Gamma \mid u(B) = t'\}$, $\Delta^- = \Gamma - \Delta$. Then (a′), (b′), and (c′) hold at t'. Conversely, if (a′), (b′), and (c′) hold at some $t' > t$ for some non-empty $\Delta \subseteq \Gamma$, then certainly $[\mathbf{y}_F(A,\Gamma,C)]^t_t = 1$. We can thus conclude that

$$[\mathbf{y}_F(A,\Gamma,C)]^t_t = \|\mathbf{y}_U(A,\Gamma,C)\|_t,$$

since both sides of the equation are equal to 1 iff (a′), (b′), and (c′) all hold. Thus the induction step is completed and the lemma follows. ■

Remark 7.3.10 In the lemma we assumed that A, C, and the formulae in Γ are atomic. This assumption was used only to show that $g[X]^t_t = {}^g\|X\|_t$ for all such X. Therefore, the following version of lemma 7.3.9 is true.

Let $A, A\sharp, B_i, B_i\sharp, C, C\sharp$ be such that for any $X \in \{A, B_i, C\}$ any g and any t, ${}^g[X]^t_t = \|X\sharp\|^g_t$. Then for any g and t,

$${}^g[\mathbf{y}_F(A, \{B_i\}, C)]^t_t = \|\mathbf{y}_U(A\sharp, \{B_i\sharp\}, C\sharp)\|^g_t.$$

Definition 7.3.11 We can now define the translation $\sharp$ of **FP** into **US**.

1. $A\sharp = A$, for A atomic.
2. $(A \wedge B)\sharp = A\sharp \wedge B\sharp$
   $(\sim A)\sharp = \sim A\sharp$
   $(A \vee B)\sharp = A\sharp \vee B\sharp$
   $(A \to B)\sharp = A\sharp \to B\sharp$.
3. For the case of FD, we have only to define the translation for formulae of the form $\mathbf{y}_F(A, \{B_j\}, C)$, in view of lemma 7.3.7. Hence let
$$\mathbf{y}_F(A, \{B_j\}, C)\sharp = \mathbf{y}_U(A\sharp, \{B_j\sharp\}, C\sharp).$$
4. Similar to (3) for the case of PD.

Lemma 7.3.12 *For any A of* **FP** *and any assignment g and moment t, we have* ${}^g[A]^t_t = \|A\sharp\|^g_t$.

Proof. By induction, in view of remark 7.3.10. ■

Lemma 7.3.12 yields theorem 7.3.1(2). ■

Axiomatizing FP

We propose an axiomatization η(**FP**) of **FP** over rational time and prove a direct completeness theorem for η(**FP**).

Definition 7.3.13 (The system η(FP)) We write η(**FP**) for the axiom system with the following axioms and rules and their mirror images (recall that $F^+ = FF$, $P^+ = PP$, $G = \sim F \sim$, and $H = \sim P \sim$).

1. All substitution instances of truth functional tautologies.

2. The usual axioms for rational time with P^+ and F^+, i.e.
   (a) $G^+(A \to B) \to (G^+A \to G^+B)$;
   (b) $G^+A \to G^+G^+A$;
   (c) $A \to G^+P^+A$;
   (d) $G^+(A \wedge G^+A \to B) \vee G^+(B \wedge G^+B \to A)$;
   (e) $G^+G^+A \to G^+A$;
   (f) $F^+\top, P^+\top$.
3. (a) $G(p \leftrightarrow Fp)$, p atomic;
   (b) $G({\sim}FA \leftrightarrow F{\sim}A)$;
   (c) $G(F(A \wedge B) \leftrightarrow FA \wedge FB)$;
   (d) $G(HA \wedge HB \leftrightarrow H(A \wedge B))$;
   (e) GA, for all A that are substitutions of truth functional tautologies;
   (f) $G(A \to B) \to (GA \to GB)$.
4. $x \wedge H^+x \wedge G^+{\sim}x$

$$\to [F(FC \wedge \bigwedge_i PB_i \wedge HD) \leftrightarrow F^+(C \wedge H^+(x \vee D) \wedge \bigwedge_i P^+(B_i \wedge {\sim}x))].$$

5. The rule of modus ponens.
6.

$$\frac{\vdash A}{\vdash G^+A}.$$

7.

$$\frac{\vdash A \leftrightarrow A', \quad \vdash B_i \leftrightarrow B_i', \quad \vdash C \leftrightarrow C'}{\vdash F(FA \wedge HC \wedge \bigwedge_i PB_i) \leftrightarrow F(FA' \wedge HC' \wedge \bigwedge_i PB_i')}.$$

8. The irreflexivity rule for H^+, G^+.

We will prove the following theorem.

Theorem 7.3.14 $\eta(\mathbf{FP}) \vdash A$ *iff* $A \in \mathbf{FP}$.

Proof. To show one direction, check every axiom and rule of $\eta(\mathbf{FP})$ and verify they are all valid and are in **FP**.

For the other direction, namely, $A \in \mathbf{FP} \Rightarrow \eta(\mathbf{FP}) \vdash A$, we need a series of constructions, culminating in theorem 7.3.23 below. We must follow a different strategy to that of theorem 7.2.4, because the validity here is *diagonal* in the sense of section 7.1. Thus, for example, in a structure $(T, <, g)$, if $t \neq u$ in T then $\{A \mid {}^g[A]_t^u = 1\}$ may contain $\sim G(q \leftrightarrow Fq)$, and so may not be **FP**-consistent. Therefore, we cannot simply build $T \times T$ and pick out the diagonal T, as in theorem 7.2.4; we must build the diagonal T first and then construct the rest.

Lemma 7.3.15 *Let A_0 be a consistent formula of the logic $\eta(\mathbf{FP})$. Then there exists a set T of $\eta(\mathbf{FP})$-complete and consistent theories, and a relation $<$ on T given by $\Delta < \Gamma$ iff for all formulae B, $G^+B \in \Delta$ implies $B \in \Gamma$, such that the following hold:*

1. *There is $\Gamma_0 \in T$ with $A_0 \in \Gamma_0$.*
2. *Each Δ in T is a complete and consistent $\eta(\mathbf{FP})$-theory.*
3. *For each Δ in T there exists a q such that $\sim q \wedge G^+q \in \Delta$ and r such that $\sim r \wedge H^+r \in \Delta$.*
4. *$\Delta_1 < \Delta_2$ iff for all A, $H^+A \in \Delta_2$ implies $A \in \Delta_1$.*
5. *If $F^+A \in \Delta \in T$ then for some $\Delta' \in T$ with $\Delta < \Delta'$ we have $A \in \Delta'$. Similarly, the mirror condition for P^+.*
6. *$(T, <)$ is isomorphic to the rationals in order, with $<$ irreflexive.*

Proof. Using the argument of theorem 6.3.2, we can obtain a set T of theories in the language of **FP** with countably many atoms, satisfying (1)–(4). Transitivity of '$<$' follows from axiom (2b) of definition 7.3.13. Irreflexivity follows from (3) above. Linearity follows from axiom (2d) and density from axiom (2e) of definition 7.3.13. The order is non-ending since axiom (2f) is available. It is countable since by (3), every point is 'named' by an atom; so by Cantor's theorem it must be isomorphic to $(\mathbb{Q}, <)$. So (6) also holds. For (5) we copy the proof of lemma 6.2.5. ■

Note that the version of lemma 7.3.15(5) obtained by replacing F^+ by F need not hold. For the first step of the construction of Δ' would be to show that $\{A\} \cup \{B \mid GB \in \Delta\}$ is consistent. If it were not, we would have $\vdash \bigwedge_i B_i \to \sim A$ for some $GB_1, \ldots, GB_n \in \Delta$. We would now like to obtain $\vdash \bigwedge_i GB_i \to \sim FA$ by G-generalization and normality; but G-generalization is not valid and not available.

From now on we deal with the $(T, <)$ obtained above.

Definition 7.3.16

1. Let Θ be a set of formulae. We say Θ is $\eta(\mathbf{FP})G$-consistent iff for no $A_1, .., A_n \in \Theta$ do we have $\eta(\mathbf{FP}) \vdash G{\sim}\bigwedge A_i$. (Notice we use G here and not G^+.) Similarly, define the notion of $\eta(\mathbf{FP})H$-consistency.

2. We say that an $\eta(\mathbf{FP})G$-consistent set Θ of formulae is *complete* if $A \in \Theta$ or ${\sim}A \in \Theta$ for all formulae A.

Lemma 7.3.17 *Any $\eta(\mathbf{FP})G$-consistent set can be extended to a complete $\eta(\mathbf{FP})G$-consistent set.*

Proof. Assume that Θ is $\eta(\mathbf{FP})G$-consistent and let A be any formula. It is sufficient to show that $\Theta \cup \{A\}$ or $\Theta \cup \{{\sim}A\}$ is $\eta(\mathbf{FP})G$-consistent. If not, then for some $B_i, C_j \in \Theta$,

$$\vdash G(\bigwedge B_i \to {\sim}A) \quad \text{and} \quad \vdash G(\bigwedge C_j \to A).$$

From the axioms of definition 7.3.13 it follows that $\vdash G(\bigwedge_j C_j \wedge \bigwedge_i B_i \to A \wedge {\sim}A)$, and hence $\vdash G{\sim}(\bigwedge_j C_j \wedge \bigwedge_i B_i)$, a contradiction. ■

Lemma 7.3.18 *Let Θ be a complete $\eta(\mathbf{FP})G$-consistent set. Then:*

1. *If $\vdash GA$ then $A \in \Theta$.*

2. *Θ is propositionally consistent.*

3. *Θ contains all substitution instances of propositional tautologies.*

4. *Θ is closed under the rule of modus ponens: i.e. if A, $A \to B \in \Theta$ then $B \in \Theta$.*

5. *$A \in \Theta$ iff ${\sim}A \notin \Theta$.*

6. *For all A, B, $A \wedge B \in \Theta$ iff $A, B \in \Theta$.*

Proof.

1. If $\vdash GA$ then by $\eta(\mathbf{FP})G$-consistency, $\sim A \notin \Theta$, so by completeness, $A \in \Theta$.

2. If not, there are $A_1, \ldots, A_n \in \Theta$ such that $\sim \bigwedge_{i=1}^{n} A_i$ is a substitution instance of a propositional tautology. Hence by axiom (3e), $\vdash G(\sim \bigwedge_{i=1}^{n} A_i)$, contradicting the $\eta(\mathbf{FP})G$-consistency of Θ.

3. This follows from (2) and completeness. Or use axiom (3e) and (1).

4. If not, then by completeness, $A, A \to B, \sim B \in \Theta$, contradicting (2).

5. Let A be given. Θ is complete so contains at least one of $A, \sim A$, and by (2) it cannot contain both.

6. This is immediate from (2) and (5).

■

Lemma 7.3.19 *Let $\Delta, \Gamma \in T$ and $\Delta < \Gamma$. Then there exists a unique $\eta(\mathbf{FP})G$-consistent and complete $\Theta = \alpha(\Delta, \Gamma)$ such that:*

1. $GA \in \Delta \Rightarrow A \in \Theta$.

2. $C \in \Gamma \Rightarrow FC \in \Theta$.

3. If B is such that for all $\Phi \in T$, $\Delta < \Phi < \Gamma \Rightarrow B \in \Phi$, then $HB \in \Theta$.

4. If B is such that for some $\Phi \in T$, $\Delta < \Phi < \Gamma$ and $B \in \Phi$, then $PB \in \Theta$.

Proof. To show uniqueness, we first prove:

Claim: For any Θ with the above properties, and any formula A, we have:

1. If A is an atom p, then $p \in \Theta \iff p \in \Gamma$.

2. $PA \in \Theta$ iff for some $\Phi \in T$, $\Delta < \Phi < \Gamma$ and $A \in \Phi$.

3. $FA \in \Theta \iff A \in \Gamma$.

Proof of Claim:

1. For atomic p, since $\vdash G(p \leftrightarrow Fp)$, by lemma 7.3.18(1) we obtain $p \leftrightarrow Fp \in \Theta$. The same holds for negated atoms. Hence by lemma 7.3.18(4),

$$p \in \Gamma \Rightarrow Fp \in \Theta \Rightarrow p \in \Theta, \text{ and}$$
$$\sim p \in \Gamma \Rightarrow F\sim p \in \Theta \Rightarrow \sim p \in \Theta.$$

So by lemma 7.3.18(5), $p \in \Theta \iff p \in \Gamma$ for all atoms p.

2. If there is $\Phi \in T$ with $\Delta < \Phi < \Gamma$ and $A \in \Phi$, then by (4), $PA \in \Theta$. Otherwise, by (3) we have $H{\sim}A \in \Theta$, i.e. ${\sim}PA \in \Theta$. So by lemma 7.3.18(5), $PA \notin \Theta$, as required.

3. If $A \in \Gamma$ then by (2), $FA \in \Theta$. If instead ${\sim}A \in \Gamma$ then $F{\sim}A \in \Theta$, so by axiom (3b) and lemma 7.3.18(1,4,5), $FA \notin \Theta$.

Now assume that Θ_1, Θ_2 have the above properties. We show that for all formulae A, $A \in \Theta_1 \iff A \in \Theta_2$. The proof is by induction on A. The atomic case follows from the claim, as do the cases of PA and FA. The boolean cases follow easily from lemma 7.3.18(5,6) using the inductive hypothesis. Hence $\Theta_1 = \Theta_2$, as required.

To show existence, let Θ_0 be the set

$$\begin{gathered}\{A \mid GA \in \Delta\} \cup \{FC \mid C \in \Gamma\} \\ \cup \{PB \mid B \in \Phi \text{ for some } \Phi \in T \text{ such that } \Delta < \Phi < \Gamma\} \\ \cup \{HD \mid \text{ for all } \Psi \in T,\ \Delta < \Psi < \Gamma \Rightarrow D \in \Psi\}.\end{gathered}$$

We claim Θ_0 is $\eta(\mathbf{FP})G$-consistent. Otherwise for some A_j, B_i, C_k, D_m,

$$\vdash G{\sim}(\bigwedge A_j \wedge \bigwedge PB_i \wedge \bigwedge FC_k \wedge \bigwedge HD_m).$$

From the axioms of definition 7.3.13, we obtain

$$\vdash G{\sim}(HD \wedge A \wedge FC \wedge \bigwedge PB_i),$$

where $A = \bigwedge A_j$, $C = \bigwedge C_k$ and $D = \bigwedge D_m$. So by axiom (3f),

$$\vdash GA \to G{\sim}(HD \wedge FC \wedge \bigwedge_i PB_i).$$

As $GA_j \in \Delta$ for all j, by modus ponens and axioms (3e) and (f), we obtain $GA \in \Delta$. So

$${\sim}F(HD \wedge FC \wedge \bigwedge_i PB_i) \in \Delta.$$

Since we can find an atom x such that $x \wedge H^+x \wedge G^+{\sim}x \in \Delta$, by axiom (4) of definition 7.3.13 we obtain ${\sim}F^+(H^+(D \vee x) \wedge C \wedge \bigwedge_i P^+(B_i \wedge {\sim}x)) \in \Delta$. But this is impossible since by lemma 7.3.15(4,5),

$$C \wedge H^+(D \vee x) \wedge P^+(B_i \wedge {\sim}x) \in \Gamma \text{ and } \Delta < \Gamma.$$

Thus Θ_0 is $\eta(\mathbf{FP})G$-consistent and can be extended to an $\alpha(\Delta, \Gamma)$ as required. A similar mirror lemma holds for the existence of $\alpha(\Delta, \Gamma)$ for $\Delta > \Gamma$. ∎

Lemma 7.3.20 *For every formula A, there exists a formula A' of the form*

$$A' = \bigvee_j (FC_j \wedge HD_j \wedge \bigwedge_i PB_{ij}),$$

such that $\eta(\mathbf{FP}) \vdash G(A \leftrightarrow A')$.

Proof. Follows the steps of the proof of lemma 7.3.7 and note that the axioms of definition 7.3.13 allow you to prove the equivalence in $\eta(\mathbf{FP})$. A similar lemma holds for the case of H. ■

Lemma 7.3.21 *For each $\Delta \in T$, $A = F(FC \wedge HD \wedge \bigwedge_i PB_i) \in \Delta$ if and only if for some $\Gamma \in T$, $\Delta < \Gamma$ and $(FC \wedge HD \wedge \bigwedge_i PB_i) \in \alpha(\Delta, \Gamma)$.*

Proof. For some atom x we have $x \wedge H^+x \wedge G^+{\sim}x \in \Delta$. From axiom (4) of definition 7.3.13, it follows that

$$A \leftrightarrow F^+(C \wedge H^+(x \vee D) \wedge \bigwedge_i P^+(B_i \wedge {\sim}x)) \in \Delta.$$

Hence $A \in \Delta$ iff for some $\Gamma > \Delta$ with $C \in \Gamma$, there are sets $\Phi_i \in T$ with $\Delta < \Phi_i < \Gamma$ and $B_i \in \Phi_i$, and also $D \in \Psi$ for all $\Delta < \Psi < \Gamma$. By construction of $\alpha(\Delta, \Gamma)$, this last holds iff

$$(FC \wedge HD \wedge \bigwedge_i PB_i) \in \alpha(\Delta, \Gamma).$$

A similar lemma holds for the mirror case. ■

Lemma 7.3.22 *Define $\alpha(\Delta, \Delta) = \Delta$, and let the assignment g be given by $g(q) = \{\Delta \in T \mid q \in \Delta\}$, for q atomic. Then for $(T, <, g)$ and any formula A,*

$$^g[A]^\Delta_\Gamma = 1 \iff A \in \alpha(\Delta, \Gamma).$$

Proof. By induction on A. For atomic A, this holds by definition of g and the fact that axiom (3a) of definition 7.3.14 is available. The cases of $\wedge$, $\vee$, ${\sim}$, and $\rightarrow$ present no difficulty. We now tackle the case of FA.

Case 1: $\Delta = \Gamma$.

Let A' be as in lemma 7.3.20. Then by that lemma, $FA \in \alpha(\Delta, \Gamma) = \Delta$ iff $FA' \in \Delta$, iff $\bigvee_j F(FC_j \wedge HD_j \wedge \bigwedge_i PB_{ij}) \in \Delta$, iff some disjunct j is in Δ. By lemma 7.3.21, this is to say that for some j and some Γ with $\Delta < \Gamma$ we have $FC_j \wedge HD_j \wedge \bigwedge_i PB_{ij} \in \alpha(\Delta, \Gamma)$, iff $A' \in \alpha(\Delta, \Gamma)$ for some $\Gamma > \Delta$. By lemmas 7.3.18(1) and 7.3.20, $A \leftrightarrow A' \in \alpha(\Delta, \Gamma)$, so the above holds iff there is Γ with $\Delta < \Gamma$ and $A \in \alpha(\Delta, \Gamma)$. By the induction hypothesis, this holds iff $^g[A]^\Delta_\Gamma = 1$ for some Γ such that $\Delta < \Gamma$, i.e. iff $^g[FA]^\Delta_\Delta = 1$.

Case 2: $\Delta < \Gamma$.
$FA \in \alpha(\Delta, \Gamma)$ iff (by the claim of lemma 7.3.19) $A \in \Gamma$, iff (by the induction hypothesis) ${}^g[A]^\Gamma_\Gamma = 1$, iff ${}^g[FA]^\Delta_\Gamma = 1$.

Case 3: $\Gamma < \Delta$:
$FA \in \alpha(\Gamma, \Delta)$ iff (by the claim of lemma 7.3.19 again) $A \in \Phi$ for some Φ with $\Gamma < \Phi < \Delta$, iff (by the induction hypothesis) ${}^g[A]^\Phi_\Phi = 1$, iff ${}^g[FA]^\Delta_\Gamma = 1$.

The case of P is the mirror of the case of F. Thus lemma 7.3.22 is proved. ■

What we have proved is the following:

Theorem 7.3.23 *Let A_0 be an $\eta(\mathbf{FP})$-consistent formula. Then there exists a structure of the form $(T, <, g)$ with $(T, <)$ isomorphic to the rationals, such that:*

1. *$\eta(\mathbf{FP}) \vdash B$ implies ${}^g[B]^t_t = 1$, for all $t \in T$ and all B.*
2. *For some $t_0 \in T$, ${}^g[A_0]^{t_0}_{t_0} = 1$.*

This completes the proof of theorem 7.3.14. ■

7.4 More two-dimensional systems

In the system of the preceding section we had the connectives P, F, and the table for ${}^g[Pq]^t_s = 1$ (for $t < s$) was that $\exists u(t < u < s \wedge {}^g[q]^u_u = 1)$. Here the reference point is identical to the evaluation point.

There are two other possibilities for the reference point, namely t or s. That is, if $t < s$ then

$$\|Pq\|^t_s = 1 \quad \Longleftrightarrow \quad \exists u(t < u < s \text{ and } \|q\|^x_u = 1),$$

where $x = t$ or $x = s$ give the two other possibilities.

Let us summarize the three available systems:

Definition 7.4.1 Consider the following three systems for $\{P_\mathbf{a}, F_\mathbf{a}\}$ with two dimensions.

1. Evaluation is at two indices $\| - \|^t_s$, t being the reference point and s the evaluation point.
2. The table for $F_\mathbf{a}$ is the following:

$$\begin{array}{rl} \|F_{\mathbf{a}}q\|_s^t = 1 \text{ iff} & t = s \wedge \exists u > t(\|q\|_u^t = 1), \text{ or} \\ & t < s \wedge (\|q\|_s^s = 1), \text{ or} \\ & t > s \wedge \exists u, x[\mathbf{a}(x) \wedge (s < u < t) \wedge (\|q\|_u^x = 1)]; \\ \|P_{\mathbf{a}}q\|_s^t = 1 \text{ iff} & t = s \wedge \exists u < t(\|q\|_u^t = 1), \text{ or} \\ & t > s \wedge (\|q\|_s^s = 1), \text{ or} \\ & t < s \wedge \exists u, x[\mathbf{a}(x) \wedge (t < u < s) \wedge (\|q\|_u^x = 1)]. \end{array}$$

The first system is obtained as $\{F_{\mathbf{a}}, P_{\mathbf{a}}\} = \{F, P\}$ by taking $\mathbf{a}(x)$ to be $x = u$. This gives the system of the preceding section.

The second system with $\{F_{\mathbf{a}}, P_{\mathbf{a}}\} = \{F', P'\}$ is obtained by taking $\mathbf{a}(x)$ to be $x = t$.

The third system with $\{F_{\mathbf{a}}, P_{\mathbf{a}}\} = \{F'', P''\}$ is obtained by taking $\mathbf{a}(x)$ to be $x = s$.

Remark 7.4.2 Any one-dimensional connective can be viewed as a two-dimensional one in two ways.

1. One index — the reference point — can simply be ignored. This way we have the same reference all the time.

2. The reference point can be taken as equal to the evaluation point all the time. Thus take, for example, the one-dimensional F and P. Their representation as two-dimensional connectives is denoted by F^+, P^+ and are always evaluated at pairs of the form (t, t). Thus,

$$\|F^+q\|_t^t = 1 \text{ iff } \exists s > t(\|q\|_s^s = 1).$$

Similarly, $U^+(p, q)$ is:

$$\|U^+(p,q)\|_t^t = 1 \text{ iff } \exists s > t[\|p\|_s^s = 1, \text{ and } \forall u(t < u < s \Rightarrow \|q\|_u^u = 1)].$$

Remark 7.4.3 Let $H_{\mathbf{a}} = {\sim}P_{\mathbf{a}}{\sim}$ and $G_{\mathbf{a}} = {\sim}F_{\mathbf{a}}{\sim}$. Then $U^+(p, q)$ can be defined in each of the systems of definition 7.4.1 as follows:

$$\begin{array}{c} U^+(p,q) = F(Fp \wedge Hq) \\ U^+(p,q) = F'(F'p \wedge H'F'q) \\ U^+(p,q) = F''(F''p \wedge H''P''q). \end{array}$$

We saw in the preceding section that over linear time, the system $\{P, F\}$ of definition 7.4.1 is essentially equivalent to a $\{U^+, S^+\}$-system. It turns out that the other two systems (namely $\{F', P'\}$, $\{F'', P''\}$) are essentially

$\{U^+, S^+, U'^+, S'^+\}$-systems, where U' and S' are the Stavi connectives. These are more powerful in expressive power than Kamp's $\{U, S\}$. U'^+ is not definable using $\{P, F\}$ of definition 7.4.1.

We can express U' in these systems as follows:

$$\begin{aligned} U'^+(p,q) = \; & U^+(q,q) \wedge {\sim}U^+({\sim}q \vee {\sim}U^+(\top, q), q) \\ & \wedge F'[F'({\sim}q \wedge p) \wedge H'(P'F'{\sim}q \rightarrow F'p)]; \\ U'^+(p,q) = \; & U^+(q,q) \wedge {\sim}U^+({\sim}q \vee {\sim}U^+(\top, q), q) \\ & \wedge F''[F''({\sim}q \wedge p) \wedge H''(P''{\sim}q \rightarrow G''F''p)]. \end{aligned}$$

7.5 The three-dimensional system F^oP^o

Consider the following five examples from English:

1. I knew him after he had been to prison.
2. I knew him before he had been to prison.

Example (2) is problematic. It does not conform to the rules of the past perfect.

3. Mary waited until Bill came.
4. When Bill arrived Mary was reading.

There are several problems with the formal representation of (3) and (4). We know that Bill's coming is in the past of now. If we write the obvious, i.e. PU(*Bill comes,Mary waits*), we do not guarantee that *Bill comes* in the past of now. It seems that one has the feeling that the 'entire story' of (3) took place in the past.

A similar analysis applies to sentence (4). Again, if we write the obvious, $P[p \wedge q \wedge S(\top, q) \wedge U(\top, q)]$, where p = *Bill arrives* and q = *Mary reads,* we cannot be sure that Mary stops reading before now.

So our problem is the following. Given a story or, formally, a formula $A(q_i, P, F)$ built up from atoms q_i and the connectives P and F, we want to put A into the past, i.e. write PA but in such a way that the 'entire story' A is restricted into the past. So in any Fq appearing inside this PA, the 'F' refers to a point in the future but still in the past of 'now'.

Let us denote this special past connective by P^o. For example, ordinary $P[Fq]$ has the table (cf. figure 7.1)

$$\|PFq\|_t = 1 \text{ iff } (\exists x < t)(\exists y > x)(\|q\|_y = 1).$$

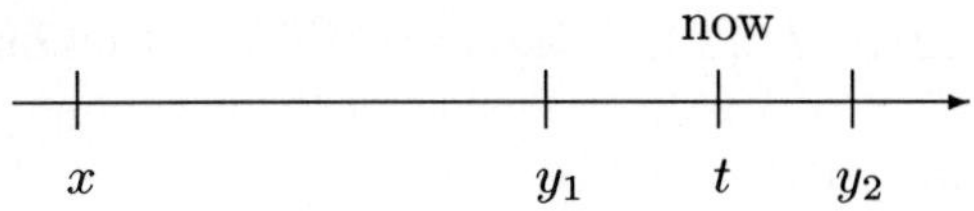

Figure 7.1

There is no restriction on y to be in the past of t. So we may have $y = y_1 < t$ or $y = y_2 > t$. Both are possible.

The connective P^o tells us that all Fs within its range are restricted by 'now's'. Thus we have

$$\|P^o[Fq]\|_t = 1 \text{ iff } (\exists x < t)(\exists y > x)(y < t \wedge (\|q\|_y = 1)).$$

The y is restricted to $y < t$.

The usual method of evaluation is inductive. Thus, when we evaluate

$$\|PA\|_t = 1 \text{ iff } \exists x < t(\|A\|_x = 1),$$

we write $\|A\|_x = 1$, and in this expression t is no longer mentioned. So in the case of P^oA we must put a sign in $\|A\|_x = 1$ to remind us that any F inside A is restricted by t. This means that we must carry t as a reference, i.e.

$$\|P^oA\|_t^t = 1 \text{ iff } \exists x < t(\|A\|_x^t = 1).$$

A similar discussion will lead to F^o, which puts the entire story into the future:

$$\|F^oA\|_t^t = 1 \text{ iff } \exists x > t(\|A\|_x^t = 1).$$

Here any P inside A is restricted by now (i.e. t), because the entire story should be in the future. If we have both F^o, P^o we need two reference points, one for the P^o and one for the F^o.

The surprising fact about this system (with F^o, P^o) is that it is expressively complete over linear time. For any $\psi(t, Q_i)$ there exists an $A(q_i, F^o, P^o)$ such that for any h, t $\|A\|_t^h = 1$ iff $\psi(t, h(q_i))$. Perhaps it is not so surprising after all: whenever we write P^oA we are putting A entirely into the past, and that is the *separation* of chapter 9, is it not?

Before we give the formal definitions, we have to explain the oddity in (2). If we start with

2′. I know him before he goes to prison,

we can represent (2) as $P^o(2')$. The odd tense construction ('had been to') is a result of the P^o. So (2) is no ordinary construction but a P^o application.

Another example of this sort is

5. The slippers had been found.

This is P^o(slippers are found).

The system b

Consider the system **b** with two unary connectives F^o and P^o and three-dimensional tables for them. We evaluate at a point t with two reference points x and y with $x < t < y$.

F^oA is read as *It will be the case that* A, but its value at t is restricted by y. F^oA is true at t iff A is true at the future of t but before y. We also allow $x = -\infty$ and $y = +\infty$ to indicate that there are no restrictions. Another feature of F^o is that once you go forward into the future from t you cannot go back past t.

Definition 7.5.1

1. Let **b** be the three-dimensional system with F^o, P^o defined as follows. Let h be an assignment, t the evaluation point, and x, y the reference points. We allow $x = -\infty$ and $y = +\infty$ and require $x < t < y$. The table for $\|A\|_t^{h,x,y}$ is given below:

$$\begin{aligned} \|q\|_t^{h,x,y} &= 1 \text{ iff } t \in h(q) \text{ for } q \text{ atomic;} \\ \|F^oA\|_t^{h,x,y} &= 1 \text{ iff } \exists x'(t < x' < y \wedge \|A\|_{x'}^{h,t,y} = 1); \\ \|P^oA\|_t^{h,x,y} &= 1 \text{ iff } \exists y'(x < y' < t \wedge \|A\|_{y'}^{h,x,t} = 1). \end{aligned}$$

   Notice that the table for F^o, P^o has three free variables, x, y and t.

2. Let $H^o = \sim P^o \sim$ and $G^o = \sim F^o \sim$.

3. Functional completeness in this case (over linear time) means the following. Let $\psi(t, P_1, \ldots, P_n)$ be any monadic formula with one free variable, t. So ψ is of the signature with $<$ but may also include the n unary predicate variables $P_1, \ldots, P_n$. Then there exists a formula A of **b** such that for any linear flow of time $(T, <)$, and for all h, t,

$$\|A\|_t^{h,-\infty,+\infty} = 1 \text{ iff } (T, <) \vDash \psi(t, h(P_1), \ldots, h(P_n)).$$

We proceed to prove expressive completeness for **b** over linear time, using some of Kamp's lemmas.

Definition 7.5.2 We define four classes of formulae in the monadic language with free variables x, y or both: namely,

$$\texttt{At}(x), \texttt{Before}(x), \texttt{Between}(x,y) \text{ and } \texttt{After}(x).$$

The definition is by induction on n.

1. $\texttt{At}(x)$ contains all boolean combinations of formulae of the form $Q(x)$, for all unary predicates Q.

2. $\texttt{Before}_0(x) = \texttt{Between}_0(x,y) = \texttt{After}_0(x) = \varnothing$.

3. $\texttt{Between}_{n+1}(x,y) = \texttt{Between}_n(x,y) \cup$ $\{\exists z(x < z < y \wedge \psi_1 \wedge \ldots \wedge \psi_m), \forall z(x < z < y \to \psi_1 \vee \ldots \vee \psi_m)$ $\mid \psi_i \in \texttt{At}(z) \cup \texttt{Between}_n(x,z) \cup \texttt{Between}_n(z,y)$, m arbitrary, z a new variable, $z \neq x, y\}$.

4. $\texttt{Before}_{n+1}(x) = \texttt{Before}_n(x) \cup$ $\{\exists z(z < x \wedge \psi_1 \wedge \ldots \wedge \psi_m), \forall z(z < x \to \psi_1 \vee \ldots \vee \psi_m)$ $\mid \psi_i \in \texttt{At}(z) \cup \texttt{Before}_n(z) \cup \texttt{Between}_n(z,x)$, z a new variable, $z \neq x\}$.

5. $\texttt{After}_{n+1}(x) = \texttt{After}_n(x) \cup$ $\{\exists z(x < z \wedge \psi_1 \wedge \ldots \wedge \psi_m), \forall z(x < z \to \psi_1 \vee \ldots \vee \psi_m)$ $\mid \psi_i \in \texttt{At}(z) \cup \texttt{After}_n(z) \cup \texttt{Between}_n(x,z)$, z a new variable, $z \neq x\}$.

6. $\texttt{Before}(x) = \bigcup_n \texttt{Before}_n(x)$,
   $\texttt{After}(x) = \bigcup_n \texttt{After}_n(x)$ and
   $\texttt{Between}(x,y) = \bigcup_n \texttt{Between}_n(x,y)$.

Theorem 7.5.3 (*See [Kamp, 1968b]*). *Every monadic formula $\psi(y)$ with y as the only free variable is equivalent over linear time to a boolean combination of members of* $\texttt{Before}(y) \cup \texttt{At}(y) \cup \texttt{After}(y)$.

Proof. This follows from the separation results of chapter 9. See especially lemma 9.3.2. ■

Remark 7.5.4 In view of theorem 7.5.3, if we show that any table $\psi(y) \in \texttt{Before}(y) \cup \texttt{At}(y) \cup \texttt{After}(y)$ can be represented in **b**; this yields the functional completeness of **b**.

Lemma 7.5.5 *Let* $\alpha(x) \in \texttt{At}(x)$, $\varphi(x,y) \in \texttt{Between}(x,y)$, $\psi(x) \in \texttt{Before}(x)$ *and* $\psi'(x) \in \texttt{After}(x)$.

1. *There exists a formula* $W(\alpha)$ *of* $\mathbf{b}$ *such that for all* $y < x < y'$ *and all* h, $\|W(\alpha)\|_x^{h,y,y'} = 1$ *iff* $\alpha(x)$ *is* h*-valid.*

2. *There exist formulae* $W^+(\varphi)$ *and* $W^-(\varphi)$ *of* $\mathbf{b}$ *such that for all* h, *whenever* $-\infty \leq x' < x < y < y' \leq +\infty$ *we have* $\|W^+(\varphi)\|_x^{h,x',y} = \|W^-(\varphi)\|_y^{h,x,y'} = 1$ *iff* $\varphi(x,y)$ *is* h*-valid.*

3. *There exists a formula* $W(\psi)$ *of* $\mathbf{b}$ *such that for all* $x' > x$ *and all* h, $\|W(\psi)\|_x^{h,-\infty,x'} = 1$ *iff* $\psi(x)$ *is* h*-valid.*

4. *There exists a* $W(\psi')$ *of* $\mathbf{b}$ *such that for all* $x' < x$ *and all* h, $\|W(\psi')\|_x^{h,x',+\infty} = 1$ *iff* $\psi'(x)$ *is* h*-valid.*

Proof. For a formula α from $\mathtt{At}(x)$ the lemma is clearly valid. The other parts are proved by simultaneous induction on n in the definition of $\mathtt{Between}_n(x,y)$, $\mathtt{After}_n(x)$ and $\mathtt{Before}_n(x)$. Assume the result for stage n of the definition of the classes.

1. Let $\varphi(x,y) \in \mathtt{Between}_{n+1}(x,y)$. Then it has the form (a_1) or (a_2):

$$\begin{array}{ll}(a_1) & \exists z(x < z < y \wedge \alpha \wedge \bigwedge_i \psi_{i,1} \wedge \bigwedge_i \psi_{i,2}),\\ (a_2) & \forall z(x < z < y \to \alpha \vee \bigvee_i \psi_{i,1} \vee \bigvee_i \psi_{i,2}),\end{array}$$

where $\alpha \in \mathtt{At}(z)$, $\psi_{i,1} \in \mathtt{Between}_n(x,z)$ and $\psi_{i,2} \in \mathtt{Between}_n(z,y)$. Let

$$\begin{array}{l}A = W(\alpha) \wedge \bigwedge_i W^-(\psi_{i,1}) \wedge \bigwedge_i W^+(\psi_{i,2}),\\ B = W(\alpha) \vee \bigvee_i W^-(\psi_{i,1}) \vee \bigvee_i W^+(\psi_{i,2}).\end{array}$$

These formulae exist by the inductive hypothesis. Let $W^+(\varphi) = F^oA$, $W^-(\varphi) = P^oA$ in case φ is of form (a_1), and let $W^+(\varphi) = G^oB$, $W^-(\varphi) = H^oB$ in case φ is of form (a_2). If we abbreviate $\bigwedge_i \psi_{i,j} = \psi_j$ for $j = 1, 2$, we obtain, for example, for the case (a_1):
$\|W^+(\varphi)\|_x^{h,x',y} = 1$
iff $\exists z(x < z < y$ and $\|W(\alpha) \wedge W^-(\psi_1) \wedge W^+(\psi_2)\|_z^{h,x,y} = 1)$,
iff $\exists z(x < z < y \wedge \alpha(z) \wedge \psi_1(x,z) \wedge \psi_2(z,y))$ is h-valid,
iff $\varphi(x,y)$ is h-valid; and
$\|W^-(\varphi)\|_y^{h,x,y'} = 1$
iff $\exists z(x < z < y$ and $\|W(\alpha) \wedge W^-(\psi_1) \wedge W^+(\psi_2)\|_z^{h,x,y} = 1)$,
iff $\varphi(x,y)$ is h-valid, as before.

2. Let $\psi(x)$ be in $\mathtt{Before}_{n+1}(x)$. Suppose that $\psi(x)$ has the form

$$(a_1) \quad (\exists z < x)(\alpha \wedge \psi_1 \wedge \psi_2),$$

where $\alpha \in \texttt{At}(z)$, $\psi_1(z)$ is a conjunction of formulae in $\texttt{Before}_n(z)$ and ψ_2 is a conjunction of formulae in $\texttt{Between}_n(z,x)$. Let

$$W(\psi) = P^o(W(\alpha) \wedge W^-(\psi_1) \wedge W^+(\psi_2)) \quad \text{in case } (a_1)$$
$$W(\psi) = H^o(W(\alpha) \wedge W^-(\psi_1) \wedge W^+(\psi_2)) \quad \text{in case } (a_2).$$

These formulae exist by the inductive hypothesis. Let $x' > x$. Then:
$\|W(\psi)\|_x^{h,-\infty,x'} = 1$
iff $(\exists z < x)(\|W(\alpha) \wedge W^-(\psi_1) \wedge W^+(\psi_2)\|_z^{h,-\infty,x} = 1)$,
iff $(\exists z < x)(\alpha(z) \wedge \psi_1(z) \wedge \psi_2(z,x))$ is h-valid,
iff $\psi(x)$ is h-valid.

The other case,

$$(a_2) \quad (\forall z < x)(\alpha \vee \psi_1 \vee \psi_2),$$

is similar.

3. This is the mirror case of (2).

Thus the induction step is complete and lemma 7.5.5 is proved. ■

Theorem 7.5.6 *For any table $\psi(t)$ there exists a formula A of* **b** *such that for all h, t,*

$$\|A\|_t^{h,-\infty,+\infty} = 1 \textit{ iff } \psi(t) \textit{ is } h\textit{-valid.}$$

Proof. By remark 7.5.4 it is sufficient to prove the theorem for formulae ψ in $\texttt{Before}(t) \cup \texttt{At}(t) \cup \texttt{After}(t)$. By lemma 7.5.5, such an A exists. ■

Remark 7.5.7

1. Even for the case of the real numbers, the systems **US** and **b** are not the same, despite the fact that they are both functionally complete. F^o, P^o are not definable as connectives in **US**.

2. The system **b** is not functionally complete for an arbitrary partially ordered flow of time. The table $\psi(t) = \exists x(x \neq t)$ (which is not expressible as $\exists x(x < t \vee t < x)$ unless time is linear) cannot be represented in **b**, as H. Kamp has observed.

   To see this, consider a flow of time $(T, <) = (\{0,1\}, \varnothing)$. This is certainly an irreflexive and transitive flow of time. Let $(T', <') = (\{0\}, \varnothing)$. Then for any h and A of **b**, and $h' = h\lceil T'$,

$$\|A\|_0^{h,-\infty,\infty} \text{ in } T = \|A\|_0^{h',-\infty,\infty} \text{ in } T'.$$

   But $(T, <) \vDash \psi(0)$ and $(T', <') \vDash \sim\psi(0)$.

7.6 Another expressively complete logic

Consider the two-dimensional, two-place connective $E(p,q)$ and the two, one-place connectives J^+ and J^-. Evaluation is of the form $\|A\|_x^{h,y}$ for $x \leq y$. We allow either $x = -\infty$ or $y = +\infty$ but not both.

Definition 7.6.1 The evaluation is as follows:

1. $\|p\|_x^{h,y} = 1$ iff $(x \in h(p) \wedge x \neq -\infty) \vee (y \in h(p) \wedge x = -\infty)$.
2. $\|J^+p\|_x^{h,y} = 1$ iff $(x = -\infty \wedge \|p\|_x^{h,y} = 1) \vee (x \neq -\infty \wedge \|p\|_x^{h,+\infty} = 1)$.
3. $\|J^-p\|_x^{h,y} = 1$ iff $(y = +\infty \wedge \|p\|_x^{h,y} = 1) \vee (y \neq +\infty \wedge \|p\|_{-\infty}^{h,y} = 1)$.
4. $\|E(p,q)\|_x^{h,y} = 1$ iff $\exists z(x < z < y \wedge \|p\|_x^{h,z} = 1 \wedge \|q\|_z^{h,y} = 1)$.

Let $A(p,q)$ be defined as $\sim E(\sim p, \sim q)$. Then

$$\|A(p,q)\|_x^{h,y} = 1 \text{ iff } \forall z(x < x < y \Rightarrow \|p\|_x^{h,z} = 1 \vee \|q\|_z^{h,y} = 1).$$

To show the functional completeness of this system (over linear time) we have to prove the analogue of lemma 7.5.5 of the preceding section.

Lemma 7.6.2 *Suppose that* $\alpha(x) \in \mathtt{At}(x)$, $\varphi(x,y) \in \mathtt{Between}(x,y)$, $\psi(x) \in \mathtt{Before}(x)$, *and* $\psi'(x) \in \mathtt{After}(x)$. *Then there exist formulae* $W(\alpha)$, $W(\varphi)$, $W(\psi)$, $W(\psi')$ *(in the language of* E*) such that for all* h, x, y *with* $-\infty < x < y < +\infty$*:*

1. $\|W(\alpha)\|_x^{h,y} = \|W(\alpha)\|_x^{h,+\infty} = 1$ *iff* $(T, <, h) \vDash \alpha$*;*
2. $\|W(\varphi)\|_x^{h,y} = 1$ *iff* $(T, <, h) \vDash \varphi$*;*
3. $\|W(\psi)\|_{-\infty}^{h,x} = 1$ *iff* $(T, <, h) \vDash \psi$*;*
4. $\|W(\psi')\|_x^{h,+\infty} = 1$ *iff* $(T, <, h) \vDash \psi'$.

Proof. (1) is clearly valid. We prove (2–4) by induction on n of the definition of $\mathtt{Between}_n(x,y)$, $\mathtt{Before}_n(x)$, and $\mathtt{After}_n(x)$. Assume the lemma holds for stage n of the definition of the classes.

Let $\varphi(x,y) \in \mathtt{Between}_{n+1}(x,y)$. Hence it has the form (a_1) or (a_2):

$$(a_1) \quad \exists z(x < z < y \wedge \alpha \wedge \textstyle\bigwedge_i \varphi_{i,1} \wedge \bigwedge_i \varphi_{i,2}),$$
$$(a_2) \quad \forall z(x < z < y \rightarrow \alpha \vee \textstyle\bigvee_i \varphi_{i,1} \vee \bigvee_i \varphi_{i,2}),$$

where $\alpha \in \mathtt{At}(z)$, $\varphi_{i,1} \in \mathtt{Between}_n(x,z)$, and $\varphi_{i,2} \in \mathtt{Between}_n(z,y)$. Let $W(\varphi) = E(\bigwedge_i W(\varphi_{i,1}), W(\alpha) \wedge \bigwedge_i W(\varphi_{i,2}))$ for case (a_1), and $W(\varphi) =$

$A(\bigvee_i W(\varphi_{i,1}), W(\alpha) \vee \bigvee_i W(\varphi_{i,2}))$ in case (a_2). Then in case (a_1) we have
$\|W(\varphi)\|_x^{h,y} = 1$
iff $\|E(\bigwedge_i W(\varphi_{i,1}), W(\alpha) \wedge \bigwedge_i W(\varphi_{i,2}))\|_x^{h,y} = 1$,
iff there is z with $x < z < y$ such that
$\|\bigwedge_i W(\varphi_{i,1})\|_x^{h,z} = \|W(\alpha) \wedge \bigwedge_i W(\varphi_{i,2})\|_z^{h,y} = 1$,
iff (by the inductive hypothesis) there is z with $x < z < y$ such that
$(T, <, h) \models \bigwedge_i \varphi_{i,1}(x, z)$, $(T, <, h) \models \alpha(z)$ and $(T, <, h) \models \bigwedge_i \varphi_{i,2}(z, y)$,
iff $(T, <, h) \models \varphi(x, y)$.

We can treat the set $\texttt{Before}(x)$ as $\texttt{Between}(-\infty, x)$, and similarly $\texttt{After}(x)$ as $\texttt{Between}(x, +\infty)$. ■

The lemma yields functional completeness.

Theorem 7.6.3 *For any $\psi(t)$ there exists an A such that for all h, t,*

$$\|A\|_t^{h,t} = 1 \textit{ iff } (T, <, h) \models \psi(t).$$

Proof. It is sufficient to prove the theorem for $\psi \in \texttt{Before}(t)$, $\psi \in \texttt{After}(t)$ and $\psi \in \texttt{At}(t)$. For $\psi \in \texttt{At}(t)$, such an A exists by the preceding lemma. For $\psi \in \texttt{Before}(t)$, there exists by the lemma an A' with the property that for all h, t,

$$\|A'\|_{-\infty}^{h,t} = 1 \text{ iff } (T, <, h) \models \psi(t).$$

Let $A = J^- A'$. Similarly, for $\psi \in \texttt{After}(t)$. ■

Another natural two-dimensional expressively complete logic, CDT, is defined by Venema [Venema, 1991b]. A complete axiomatization is given. The expressive completeness of a logic similar to Venema's $CC\lambda$ over linear time is proved in theorem 13.6.3.

7.7 Logics of historical necessity

We have already discussed in chapter 2, example 2.1.3, a logic for historical necessity. We considered a flow of time of the form $(T, <, T_i, g), i \in J$, where $(T, <)$ is a flow of time with linear past and branching future, $T_i \subseteq T$ are (usually) maximal linear subsets of T such that $T = \bigcup T_i$. Each T_i is considered a history. For a point $t \in T$, any T_i such that $t \in T_i$ is considered a history through t. g is an assignment giving each atom q the set of points $g(q) \subseteq T$ in which it is true. Evaluation is at a point and history, i.e. provided $t \in T_i$, we write

$$\|A\|_{t,i}^g = ?$$

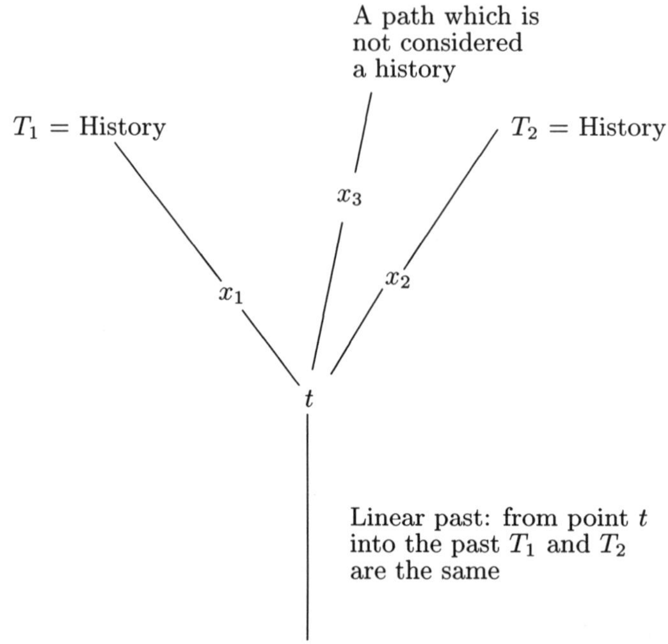

Figure 7.2

The connectives we have are F and P ranging over the same history and M ranging over different histories. Note that often M is seen as $\Box$ in the literature, and its derivative $L = {\sim}M{\sim}$ is seen as $\Diamond$.

The tables are:

$$\begin{array}{ll} \|Fq\|_{t,i} = 1 & \text{iff } \exists x > t(x \in T_i \wedge (\|q\|_{x,i} = 1)) \\ \|Pq\|_{t,i} = 1 & \text{iff } \exists x < t(x \in T_i \wedge (\|q\|_{x,i} = 1)) \\ \|Mq\|_{t,i} = 1 & \text{iff } \exists j \in J(x \in T_i \wedge (\|q\|_{t,j} = 1)). \end{array}$$

In figure 7.2, for example, if q holds at x_1 (in T_1) and nowhere else then Fq holds at t in T_1 but not at t in T_2. Instead, MFq holds at t in T_2.

Given a point $t \in T$ there are several histories passing through t. For example, see figure 7.3.

The history T_i goes from t through x_i and up to y_i in figure 7.3 ($i = 1, 2, \ldots$). We have no requirement that forces the existence of a history T_0 going through the axis $(t, x_1, x_2, \ldots, x_n, \ldots)$. In other words, there is no requirement that forces us to make one history out of an increasing sequence of segments of history. (We consider having this extra requirement later in the section.) If the logic we are considering is a logic which follows the executions of programs, then T_i, passing through (x_i, y_i), is a possible

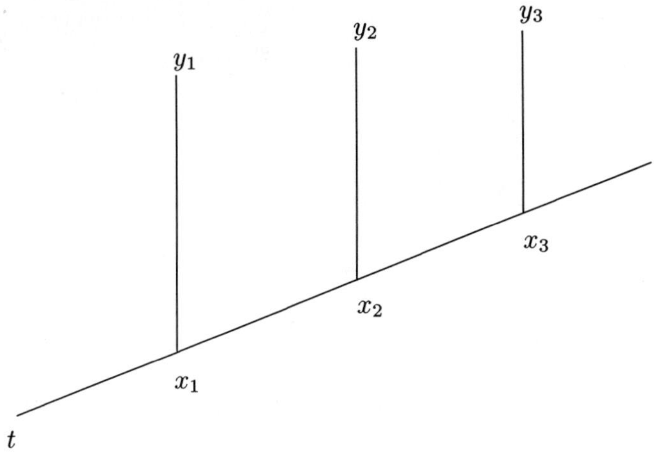

Figure 7.3

execution course for the program which may result in termination. The sequence $T = (t, x_1, x_2, \ldots)$ may indicate a loop and is therefore not counted as a history.

Another philosophical question concerns whether histories should be maximal linear subsets of T. This is usually assumed, and we will proceed under this assumption in considering a logic **HN2** later in the section. For now, though, we will just require that histories are downward closed linear subsets in the sense made precise below. This question affects whether such formulae as $G\bot \to LG\bot$ are valid or not.

For our convenience we will present a logic **HN1** of historical necessity in a slightly different form to that described above. Consider a set of histories $\mathcal{H} = \{T_i \mid i \in J\}$. Above, we said that the point t may appear in several histories, but it is more convenient to make the histories disjoint and take a relation $t \approx s$ to show that t and s are really the same point.

For example, in figure 7.4 we have two histories $T_1 \cap T_2 \doteq [-\infty, t]$. We split them into two separate disjoint histories, as in figure 7.5.

We have $h_1 \cap h_2 = \varnothing$ and indicate the fact that really h_1 and h_2 have a common past by writing $t \approx t'$, $x \approx x'$, $y \approx y'$. We add the requirement (3) below to capture the idea of the necessity of the past including its 'downward closure'.

Definition 7.7.1 Here is the formal description:

The structures consist of

- a set $\mathcal{H}$ of *histories,*

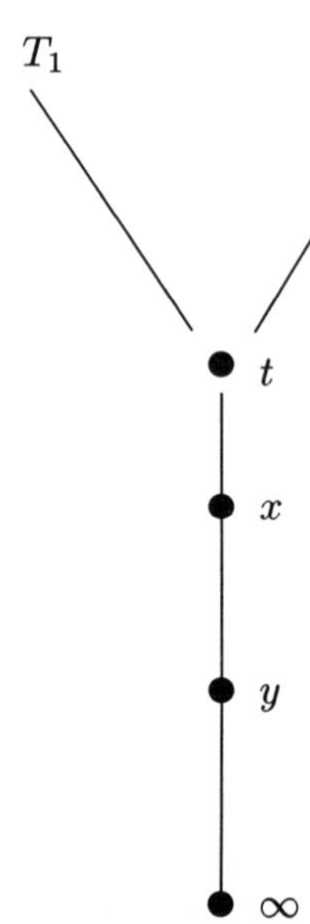

Figure 7.4

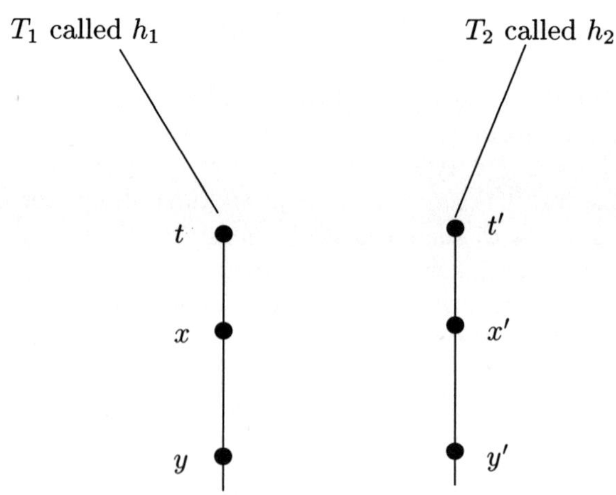

Figure 7.5

- a linear irreflexive order $(T^h, <^h)$ for each history h,
- an equivalence relation $\approx$ on $T = \bigcup_{h \in \mathcal{H}} T^h$
- and an assignment g giving subsets $g(q) \subseteq T$,

such that:

1. For histories $h_1 \neq h_2$, T^{h_1} and T^{h_2} are disjoint.
2. If $t \neq s$ are from T^h then $t \not\approx s$.
3. If $t \in T^h$, $t' \in T^{h'}$, and $t \approx t'$, then there is an isomorphism f between $\{u \in h \mid u < t\}$ and $\{u' \in h' \mid u' < t'\}$ such that for all u, $u \approx f(u)$. (Note: in the above $<$ is understood to be $<_h$ for the appropriate h. From now on we are going to write $<$ only.)
4. If $t \in g(q)$ and $t \approx s$ then $s \in g(q)$.

Evaluation is at two indices t and h for $t \in T^h$, and the table is as follows:

- $\|q\|^g_{t,h} = 1$ iff $t \in g(q), q$ atomic;
- $\|FA\|^g_{t,h} = 1$ iff $\exists s \in h(t < s \wedge \|A\|^g_{s,h} = 1)$;
  $\|PA\|^g_{t,h} = 1$ iff $\exists s \in h(s < t \wedge \|A\|^g_{s,h} = 1)$;
- $\|MA\|^g_{t,h} = 1$ iff $\exists h' \exists s \in h'(s \approx t \wedge \|A\|^g_{s,h'} = 1)$.

A formula A is a *theorem* of the logic of historical necessity iff $\|A\|^g_{t,h} = 1$ for all structures and all t, h.

An axiomatization for a similar semantics is given in [Zanardo, 1985] but here we present one using the IRR-rule.

Definition 7.7.2 Consider the following axiom system **HNl** for historical necessity; where $L = {\sim}M{\sim}$

1. Axioms for linear temporal logic involving $\{F, P\}$ (see chapter 6).
2. Axioms for S5 modality involving $\{M\}$:
   $L(A \wedge B) \leftrightarrow LA \wedge LB$
   $LA \rightarrow LLA$
   $LA \rightarrow A$
   $A \rightarrow LMA$.
3. $A \rightarrow LA$, for any A not containing F.
4. ${\sim}r \wedge Hr \wedge LA \rightarrow GLH({\sim}r \wedge Hr \rightarrow A)$.

5. $A \to GLPMA$.

6. The rules of modus ponens and $\vdash A \Rightarrow \vdash LA, \vdash GA, \vdash HA$.

7. The rule IRR, for q, an atom not in A:

$$\frac{\vdash {\sim} q \wedge Hq \to A}{\vdash A}.$$

One can verify that all axioms and rules are valid.

Definition 7.7.3 For any $\Diamond \in \{P, F, M\}$ let $\Diamond'$ be defined by $F' = P$, $P' = F$, and $M' = M$.
Let $\Box = {\sim}\Diamond{\sim}$, and $\Box' = {\sim}\Diamond'{\sim}$.

Lemma 7.7.4 *Theorem 3.2.4 holds for any $\Box$ in $\{H, G, L\}$.*

Proof. We use induction and axiom (5) and the same proof as given in chapter 3 since the conditions hold for $\Box, \Box'$. ∎

Definition 7.7.5

1. A theory is said to be IRR iff it satisfies the conditions of definition 6.2.3 of section 2 of chapter 6 for $\Diamond_i \in \{P, F, M\}$.

2. For Δ, Γ theories, let $\Delta < \Gamma$ be
$\forall B(GB \in \Delta \to B \in \Gamma)$.
Let $\Delta \approx \Gamma$ iff $\forall B(LB \in \Delta \to B \in \Gamma)$.

Lemma 7.7.6

1. *Let A be a consistent formula. Then for some IRR complete and consistent theory Δ, $A \in \Delta$.*

2. *Let FA (respectively PA) be in Δ, Δ a complete consistent IRR-theory. Then for some consistent and complete IRR-theory $\Gamma, \Delta < \Gamma (\Gamma < \Delta$ resp.) $A \in \Gamma$.*

3. *Let $MA \in \Delta$, a complete and consistent IRR-theory. Then for some complete and consistent IRR-theory $\Gamma, A \in \Gamma$ and $\Delta \approx \Gamma$.*

Proof. As in section 2 of chapter 6. ∎

Lemma 7.7.7 *For IRR-theories:*
$\Delta \approx \Gamma$ *and* $\Gamma' < \Gamma$ *imply not* $\Delta \approx \Gamma'$.

Proof. There is an atom q with $\sim q \wedge Hq \in \Delta$. Hence by axiom (3), $\sim q \wedge Hq \in \Gamma$ and hence $q \in \Gamma'$. Therefore, $\Delta \not\approx \Gamma'$. ■

Lemma 7.7.8 *$<$ is a linear ordering and $\approx$ is an equivalence relation, where by linear ordering we mean the conditions:*
$\Delta < \Gamma$ and $\Delta < \Gamma'$ imply $\Gamma = \Gamma' \vee \Gamma < \Gamma' \vee \Gamma' < \Gamma$.
$\Delta > \Gamma \wedge \Delta > \Gamma'$ imply $\Gamma = \Gamma' \vee \Gamma < \Gamma' \vee \Gamma' < \Gamma$.

Proof. Follows from the axioms. ■

Lemma 7.7.9 *Let Δ', Γ', Γ be IRR-theories such that $\Delta < \Gamma$ and $\Gamma \approx \Gamma'$. Then for some $\Delta', \Delta' \approx \Delta$ and $\Delta' < \Gamma'$.*

Proof. For some atom $q, \sim q \wedge Hq \in \Delta$. Hence $P(\sim q \wedge Hq) \in \Gamma$ and hence $P(\sim q \wedge Hq) \in \Gamma'$ and hence for some $\Delta, \Delta' < \Gamma'$ and $\sim q \wedge Hq \in \Delta'$.

We show now that $\Delta \approx \Delta'$. Assume that $LA \in \Delta$. Then by axiom (4), $GLH(\sim q \wedge Hq \to A) \in \Delta$. Hence $LH(\sim q \wedge Hq \to A) \in \Gamma$ and so $H(\sim q \wedge Hq \to A) \in \Gamma'$ and $\sim q \wedge Hq \to A \in \Delta'$, and hence $A \in \Delta'$. ■

Definition 7.7.10 Let $\mathcal{M}$ be the set of all maximal consistent IRR-theories in some fixed language. Let $<$ and $\approx$ be as defined above. For each $\Gamma \in \mathcal{M}$ let the history containing Γ be $h_\Gamma = \{\Delta \mid \Delta = \Gamma, \Delta < \Gamma \text{ or } \Gamma < \Delta\}$. Let $\mathcal{H}$ be the set of all such histories, let $T^h = h$, and let g be defined by $g(q) = \{\Gamma \in T \mid q \in \Gamma\}$.

Theorem 7.7.11 *1. $(\mathcal{H}, T, <, \approx, g)$ is a historical necessity structure.*

2. For any A, for any $\Gamma \in T$,
$\|A\|^g_{\Gamma, h_\Gamma} = 1$ *iff* $A \in \Gamma$.

Proof. (1) follows from preceding lemmas. (2) follows by induction using the lemmas. ■

Remark 7.7.12 One can combine the completeness results of preceding sections and obtain axioms for various conditions on the order of the histories $\mathcal{H}$. For example, the axioms

$$A \wedge HA \to FHA \qquad A \wedge GA \to PGA$$

for discrete linear time will give us the semantics where each history is discretely ordered.

Remark 7.7.13 As we mentioned it is usual for the semantics of historical necessity to include the requirement that histories are maximal linear subsets of T. This can be enforced by adding to definition 7.7.1 the requirement

5. For *no* histories $h \neq h'$ is there an isomorphism f from T^h into $T^{h'}$ such that for all u, $u \approx f(u)$.

Zanardo [Zanardo, 1992] has axiomatized this logic which we may call **HN2** by simply adding the following axiom to an axiomatization of **HN1**:

$$G\bot \to LG\bot.$$

To prove completeness one may continue from our proof above but separate out many paths through time built from piecing together bits of histories. The proof is quite intricate so we refer readers to [Zanardo, 1992] for details.

Remark 7.7.14 One should investigate the system **HN3** with the added condition that piecing together parts of histories is also a history. We want the following condition on $(T^h, <^h), h \in \mathcal{H}$:

Let $x_n \in \cup_{h\in\mathcal{H}} T^h$,a sequence such that for each n there exists a history T^h and elements $y_1, \ldots, y_n$ in T^h such that $y_1 <^h \ldots <^h y_{n-1} <^h y_n$ and $x_1 \approx y_1, \ldots, x_n \approx y_n$. Then there exists a history $h' \in \mathcal{H}$ with $x_i \in T^{h'}$ for all i.

We are looking for an axiomatization of this system. We believe that here we need an additional rule IAR for a suitable **A** and some additional axioms, and all this besides the irreflexivity rule IRR, which we will keep.

The problems are unsolved.

Remark 7.7.15 We would like to thank Alberto Zanardo for pointing out several important problems with an earlier version of this section.

Chapter 8

Propositional quantifiers and fixed point operators

8.1 Introduction

In this chapter we investigate the two related topics of propositional quantifiers and fixed point operators. In sections 8.1–8.3 we look at languages in which there are formulae $\forall xA$ and $\exists xA$ where x behaves in A as a proposition. There are different semantics depending on what we mean by a 'proposition', i.e. what we let q range over. In any case we obtain such expressive languages that they are not directly useful for computation. In the rest of the chapter we look at some very useful sublanguages in which quantifiers of the form $LFP_q A$ are used to build formulae and in which their semantics are defined as the least fixed point of certain operators. As we will see, these fixed point languages allow a sort of inductive definition of new propositions.

Consider the propositional logic K_t, studied in chapter 3. The wffs of the system are built up from atomic propositions $q_1, q_2, \ldots$, using the classical connectives $\sim, \wedge$ and the temporal connectives F and P. Let $A(x)$ be such a wff with the atom x appearing in A. We may want to say that there exists a proposition q which makes $A(q)$ true or we may want to to say that any q would make $A(q)$ true. For example, $A_1(x)$ may be $p \wedge x \to (\sim x \to C)$ and $A_2(x) = p \wedge x \to C$. In the first case we may want to point out that any q would make $A_1(q)$ true, and in the second case we may want to say that some q could make $A_2(q)$ true. To be able to express the above notions we must enrich the system by adding two quantifier symbols $\forall$ and $\exists$ and allowing the formation of the wffs $\forall xA$

or $\exists x A$ where x is atomic. This is propositional quantification but at the moment it is ambiguous. What do we mean by a 'proposition'?

There seem to be at least two natural definitions which we present below.

Perhaps proposition means already existing atomic proposition. We define PQ(atoms) to be the propositionally quantified temporal logic with F and P over the class of all flows of time with the meaning of $\exists x A(x)$ being that there exists an atom q which makes $A(q)$ true and that of $\forall x A(x)$ being that $A(q)$ is true for all atoms q.

Example 8.1.1 Let $(T, <, h)$ be a model where $(T, <)$ is the integers time and h is an assignment making each atom p_i true exactly at point i. Thus, $h(p_i) = \{i\}$. So for each $t \in T$, $\|\exists x(Fx)\|_t^h = 1$, and $\|\forall x(Fx)\|_t^h = 1$ iff $t \leq 0$.

Alternatively, perhaps we should quantify over all possible propositions: i.e. over all subsets of time. We define *the full second-order propositional temporal logic,* PQ(full), to be the propositionally quantified temporal logic with F and P over the class of all flows of time with the meaning of $\exists x A(x)$ being that there exists a subset $h(q) \subseteq T$ which makes $A(q)$ true, and that of $\forall x A(x)$ being that $A(q)$ is true for all subsets $h(q)$.

Example 8.1.2 Let $(T, <, h)$ be as above in example 8.1.1.
Then for each $t \in T$, $\|\exists x(Fx)\|_t^h = 1$ and $\|\forall x(Fx)\|_t^h = 0$.

Both of these logics can be fitted into a more general formalism as follows.

First let us define the syntax. Because there are proposition variables we will have the notion of free variables and have to distinguish between formulae and sentences.

Definition 8.1.3 Consider a language with a set $\mathbf{X} = \{x_1, x_2, \ldots\}$ of propositional atomic variables as well as a set $\mathbf{P} = \{p_1, p_2, \ldots\}$ of atomic propositions, the classical connectives, the temporal connectives P, F, and the quantifier symbols $\forall, \exists$. We define by induction the notion of a formula $A(x_1, \ldots, x_n)$ with free variables from the set $\{x_1, \ldots, x_n\}$.

1. An atom $x \in \mathbf{X}$ is a formula with x free.

2. $p \in \mathbf{P}$ is a formula with no free variables.

3. If A is a formula with the variables $x_1, \ldots, x_n$ free and B is a formula with $y_1, \ldots, y_m$ free then $A \wedge B$ are formulae with $\{x_1, \ldots, x_n\} \cup \{y_1, \ldots, y_n\}$ free; and $\sim A$, FA, PA are formulae with $x_1, \ldots, x_n$ free.

4. If $A(y, x_1, \ldots, x_n)$ is a formula with $y, x_1, \ldots, x_n$ free and $y \in \mathbf{X}$ then $\exists y A$ are formulae with $x_1, \ldots, x_n$ free.

A sentence is just a formula with no free variables. We have the usual abbreviations $\vee$, GA, etc., as well as the new $\forall x A$ for $\sim\exists x(\sim A)$.

Now the semantics.

Definition 8.1.4 Let $(T, <)$ be a flow of time, h a valuation of the atoms (i.e. a map from $\mathbf{P}$ into the power set of T), and $\mathbf{S}$ be a family of subsets of T.

1. h is an $\mathbf{S}$-valuation iff for all $p \in \mathbf{P}$, $h(p) \in \mathbf{S}$.
2. Any map g which takes each atom in $\mathbf{P} \cup \mathbf{X}$ into $\mathbf{S}$ is called an $\mathbf{S}$-assignment.
3. An $\mathbf{S}$-assignment is called an $(h, \mathbf{S})$-*assignment* iff it agrees with h on $\mathbf{P}$.
4. Let g, g' be two assignments and x an atom from $\mathbf{X}$. We write $g =_x g'$ to mean that for all variables $y \neq x$ we have $g(y) = g'(y)$.

Definition 8.1.5 For each formula A of the language with propositional quantifiers we can define the value $\|A\|_t^{g,\mathbf{S}}$ as follows, where g is an $(h, \mathbf{S})$-assignment:

1. $\|p\|_t^{g,\mathbf{S}} = 1$ iff $t \in g(p)$ iff $t \in h(p)$.
2. $\|x\|_t^{g,\mathbf{S}} = 1$ iff $t \in g(x)$.
3. For the recursive clauses of the connectives $\sim, \wedge$, F, and P take the usual defining clauses.
4. $\|\exists x A(x)\|_t^{g,\mathbf{S}} = 1$ iff for some $(h, \mathbf{S})$-assignment g' such that $g' =_x g$ we have $\|A(x)\|_t^{g',\mathbf{S}} = 1$.

For some $\mathbf{S}$-valuation h, and some sentence A, define $\|A\|_t^{h,\mathbf{S}} = 1$ iff for some (equivalently all) $(h, \mathbf{S})$-assignments g, $\|A\|_t^{g,\mathbf{S}} = 1$.

The logic PQ has validity defined by $\models A$ iff for any flow of time $(T, <)$, for any set $\mathbf{S}$ of subsets of T, for any $\mathbf{S}$-valuation h, for any $t \in T$, we have $\|A\|_t^{h,\mathbf{S}} = 1$.

We can obtain many stronger logics $PQ(R)$ by restricting the structures $(T, <, h, \mathbf{S})$ which we consider:

1. $R =$ atoms: we have met this logic in which we put $\mathbf{S} = \{h(p) \mid p \in \mathbf{P}\}$.

2. $R =$ full: again we have met this where $\mathbf{S} = 2^T$.

3. $R =$ named: here we only consider structures in which for every $t \in T$, there is an atom p such that $t \in h(p)$ and for all $s < t$, $s \notin h(p)$.

4. $R =$ irr: when $(T, <)$ is irreflexive.

5. $R =$ linear: when $(T, <)$ is a linear irreflexive order.

6. $R =$ comprehensive: when we have for each sentence A an atom p such that $A \leftrightarrow p$ holds everywhere.

7. Or any sensible conjunction of the above restrictions.

We will see that there is a big difference in the behaviour of these logics: for example, $PQ($ atoms, linear$)$ is axiomatizable while $PQ($ full, linear$)$ is not even arithmetical. In any case, these languages are very expressive.

This is a very good reason for studying them. In our studies of both pure temporal systems and various applied system we have encountered a variety of temporal connectives, temporal axioms, and temporal rules. Quantified propositional temporal logic with its rich expressive power can provide a general background against which all of this variety of systems can be compared, if not expressed, using propositional quantifiers.

There are many other reasons for studying propositional quantifiers. The need for a stronger propositional language appears in many areas of application. The need to quantify over propositions or moments of time appears everywhere. It is natural of course to study quantified predicate temporal logic, and quantified propositional temporal logic is an important and natural intermediary step.

The computer language Tempura uses propositional quantifiers, as do many computer language with time and fixed point operators. Systems with propositional quantifiers have strong expressive power. We shall see later that they are as strong as classical predicate logic or the monadic predicate logic of $(T, <, =)$.

The expressive power of quantification over propositions allows one to define many new connectives in a natural way. Suppose we want to say that every other day (every second day) q is true. For example, John takes a pill every other day. How can we express that?

Suppose we have integer time with Tq for q true tomorrow as well as F. We can write something like the following:

$$TTq \wedge G(TTq \to TTTTq) \text{ i.e. } T^2q \wedge G(T^2q \to T^4q).$$

We can read the above as 'John will take a pill the day after tomorrow and always if he will take a pill in two days time then he will also take the pill in four days time.' This, however, is a bit stronger than we had in mind. All we want to say is that John will take the pill every other day, no more and no less. So if he also happens to take the pill three days from now we do not intend this to mean that because of this fact, he will also take the pill five, seven, nine, etc., days from now. The above statement $G(T^2q \to T^4q)$ forces this meaning. What we really want to say is that if a proposition x is true every two days exactly, beginning the day after tomorrow, then $G(x \to \text{John takes the pill})$ is true. Thus what we need is the connective $(TT^*)(q)$ defined as:

$$(TT^*)(q) = \forall x[T({\sim}x) \wedge T^2x \wedge G(T({\sim}x) \to T^3({\sim}x)) \wedge G(T^2x \to T^4x) \\ \to G(x \to q)]$$

Of course we must use the full second-order propositional temporal logic here so as to guarantee that there is at least one proposition x: quantifying over existing atomic propositions might leave us with no such xs.

Note that $(TT^*)(q)$ is then the transitive closure of TT, i.e. $(TT^*)(q) = \bigwedge_n (TT)^n(q)$. The transitive closure of (TT) is not first-order definable over $(\mathbb{N}, <)$. However, it is an example of a least fixed point operator.

After we see the formal definitions of the semantics of least fixed point languages in later sections of this chapter it will be clear that we can similarly define all fixed point expressions in our full second-order propositional temporal logic. There it is more convenient, though, to regard the least fixed point as a specialized new propositional quantifier in its own right. However it is defined, it is a very powerful, interesting, and useful way of producing new languages.

Before we look at fixed points in more detail let us prove some results about the propositional quantifiers $\forall x$ and $\exists x$.

8.2 Axiomatizing propositional logics

In this section we will axiomatize the temporal logic $PQ($ atoms, linear, named$)$ in which we restrict our attention to structures $(T, <, h, S)$ where $(T, <)$ is a linear irreflexive flow, for each $t \in T$ there is an atom p such that t is minimal in $h(p)$, and $S = \{h(p) \mid p \text{ is an atom}\}$.

The reason we consider only what we have called 'named' structures is that we want to use the very helpful IRR-rule in our axiomatization. Recall that the IRR-rule,

IRR-rule $\dfrac{\vdash {\sim}q \wedge Hq \to A}{\vdash A}$, where q does not appear in A, was shown to be valid in most of the logics we have been dealing with and that it was very helpful in constructing natural axiomatizations and then proving their completeness.

Surprisingly, in the case of propositional quantification this rule is not actually generally valid. Consider the following example in the unrestricted logic PQ:

$$\frac{\vdash q \wedge H{\sim}q \rightarrow \exists x(x \wedge H{\sim}x)}{\vdash \exists x(x \wedge H{\sim}x)}\,.$$

It is clear that the antecedent of this application of the IRR-rule is valid. However, the consequent is not valid as can be seen by considering any structure with an uncountable number of time points.

The example also shows that with propositional quantifier logics, the IRR-rule is only valid if the following sentence, which we will call the IRR-axiom, is also valid:

IRR-axiom $\exists x(x \wedge H{\sim}x)$.

This explains our restriction to 'named' structures which, incidentally, also means that we are only dealing with countable flows of time.

But now if we make this restriction then we might as well use the IRR-axiom in our axiomatization. In fact, it is clear that the IRR-axiom will do just as well instead of the IRR-rule in establishing that a canonical model is irreflexive, and this is just what we want the rule to do. The axiom will be seen to be much easier to use in this way. It is a pity that it is not available to use more generally for these purposes.

So let us present an axiomatization. We define a Hilbert system QL for the propositionally quantified temporal logic $PQ($ atoms, linear, named $)$ as defined in definition 8.1.5.

Definition 8.2.1 QL consists of exactly the following rules and axiom schemas:

- Modus ponens and the rules of temporal generalization.
- Any substitution instance of an axiom schema of K_t These are:

  any classical tautology;

  $G(p \rightarrow q) \rightarrow (Gp \rightarrow Gq)$ and dual;

  $q \rightarrow GPq$ and dual;

  $Gp \rightarrow GGp$ and dual.
- Any substitution instance of the linearity axiom $G(p \wedge Gp \rightarrow q) \vee G(q \wedge Gq \rightarrow p)$ or its dual.

- The IRR-axiom $\exists x(x \wedge H{\sim}x)$.
- The usual quantifier rules as follows. Suppose that A is a formula with the propositional variable x its only free variable, that B is a sentence (so B has no free variables), and that q is an atom from P. Then:

  1. $\forall x A \to A(q)$;
  2. $A(q) \to \exists x A$;
  3. $$\frac{\vdash A(q) \to B}{\vdash \exists x A \to B} \quad q \text{ does not appear in } B;$$
  4. $$\frac{\vdash B \to A(q)}{\vdash B \to \forall x A} \quad q \text{ does not appear in } B.$$

- The Barcan sentence: $\forall x G A(x) \leftrightarrow G \forall x A(x)$ and its dual.

Checking the soundness of this system is not as easy as usual due to the strange nature of quantification over already existing propositions. It is as though we were quantifying over constant symbols in a predicate language (instead of over elements of the domain), and this means that it is more difficult than usual to check the validity of the quantifier rules. Nevertheless...

Lemma 8.2.2 *The system QL is sound for the logic PQ(atoms, linear , named).*

Proof. We proceed by induction on the length n of a proof of A as usual but what we show is the following:

- A is valid in the logic and for any atoms p and q from $\mathbf{P}$, if we replace all appearances of p in A by q to get B then B is also valid.

That the base case holds can easily be seen from a quick check of the axioms. The only difficult part of the inductive step concerns the quantifier rules.

Consider number (3). Suppose that we have proved $A(q) \to B$ in n steps, that q does not appear in B, and that we use rule (3) to deduce $\exists x A(x) \to B$. Let us check that this is valid first.

So $A(q) \to B$ is valid as is any result of substituting q by another atom in it. Given a structure $(T, <, h, S)$ in which $\|\exists x A(x)\|_t^{h,S} = 1$ we know by the definition of the semantics that there is an atom p such that $\|A(p)\|_t^{h,S} = 1$.

But by the inductive hypothesis we have $\|A(p) \to B\|_t^{h,S} = 1$ and so we have $\|B\|_t^{h,S} = 1$. This proves the validity of $\exists x A(x) \to B$.

To prove the other part of the inductive claim we simply replace q by p in every step of the proof of $C = \exists x A(x) \to B$ to obtain a proof of the C with q replaced by p.

The other quantifier rules are similar or easier to deal with. ■

It can be shown that when, as in QL, we have the quantifier rules in a system, then the IRR-axiom and the IRR-rule are equivalent, i.e. either can be used.

Theorem 8.2.3 *Suppose that we have an axiom system for a propositionally quantified logic which includes the usual four quantifier rules above.*

Then if either the irreflexivity rule IRR or the IRR-axiom are valid in the system then they are both valid.

Proof.

1. Assume the irreflexivity axiom. We show that IRR is valid. Suppose that we have $\vdash q \wedge H{\sim}q \to A$, with q not in A. Then using the usual quantifier rules, which are valid for $\exists$, we can obtain, since q does not appear in A, that $\vdash \exists x(x \wedge H{\sim}x) \to A$, and by modus ponens since $\vdash \exists x(x \wedge H{\sim}x)$ we obtain $\vdash A$.

2. Assume IRR and prove the irreflexivity axiom. By the available quantifier rules we have $\vdash q \wedge H{\sim}q \to \exists x(x \wedge H{\sim}x)$. Since q is not in the consequent we can conclude by IRR, $\vdash \exists x(x \wedge H{\sim}x)$.

■

Now let us check the completeness of the system:

Theorem 8.2.4 *The system QL above is sound and complete for the logic PQ(atoms, linear, named).*

Proof. We have already done one direction.

To prove the other direction we need some lemmas and definitions.

Definition 8.2.5

1. A propositional quantifier theory Δ is said to be saturated iff it is consistent and whenever $\exists x A(x) \in \Delta$, then for some atom y, $A(y) \in \Delta$.

2. Δ is said to be Henkin iff whenever $\Delta \vdash A(y)$ for all atoms y, then $\Delta \vdash \forall x A(x)$.

Lemma 8.2.6 *Every consistent theory can be extended to a complete and saturated theory in a language with countably infinitely more atoms.*

Proof. As in classical logic. A theory Δ is saturated if whenever $\exists x A(x) \in \Delta$ then for some atom q, $A(q) \in \Delta$. ■

Lemma 8.2.7 *Every Henkin theory can be extended to a complete and saturated theory in the same language (i.e. without adding new atoms).*

Proof. Let Δ be saturated. Enumerate all sentences of the language as $B_1, B_2, \ldots$. Define the finite extensions, $\Delta_m \supseteq \Delta_{m-1} \supseteq \ldots \supseteq \Delta$ of Δ for each n as follows:

1. Let $\Delta_0 = \Delta$.

2. Assume Δ_m is defined and consider $\Delta_m \cup \{B_m\}$ and $\Delta_m \cup \{\sim B_m\}$.

One of the above theories is consistent. Assume $\Delta_m \cup \{B'_m\}$ is consistent, where B'_m is either B_m or is $\sim B_m$.

If B'_m is not of the form $\exists x D_m(x)$ then let $\Delta_{m+1} = \Delta_m \cup \{B'_m\}$.

Otherwise, i.e. if $B'_m = \exists x D_m(x)$ we want to add more to Δ_{m+1}. We claim that for some atom $y, \Delta_m \cup \{\exists x D_m(x), D_m(y)\}$ is consistent. For if not then $\Delta_m \cup \{\exists x D_m(x)\} \vdash \sim D_m(y)$ for all y. Hence $\Delta \vdash [\bigwedge(\Delta_m - \Delta) \wedge \exists x D_m(x) \to \sim D_m(y)]$ for all y. Since Δ is Henkin we obtain that $\Delta \vdash \bigwedge(\Delta_m - \Delta) \wedge \exists x D_m(x) \to \forall y {\sim} D_m(y)$, and hence $\Delta_m \vdash \forall y {\sim} D_m(y)$, which contradicts the assumption that $\Delta_m \cup \{\exists x D_m(x)\}$ is consistent. We can therefore conclude from the above that for some y_0, $\Delta_m \cup \{\exists x D(x), D(y_o)\}$ is consistent. Let Δ_{m+1} be this theory. Thus Δ_m for all m is defined.

Let $\Delta^\infty = \bigcup_m \Delta_m$. Δ^∞ is a complete consistent saturated theory containing Δ in the same language as Δ, as required by lemma 8.2.7. ■

Lemma 8.2.8 *Let Δ be a consistent, complete, and saturated theory and let $FA \in \Delta$ (let $PA \in \Delta$ respectively). Then $\Delta_0(FA) = \{B \mid GB \in \Delta\} \cup \{A\}$ ($\Delta_0(PA) = \{B \mid HB \in \Delta\} \cup \{A\}$ respectively), is a Henkin consistent theory, and can be completed to a complete and consistent theory $\Delta(FA)$ ($\Delta(PA)$ respectively) in the same language as Δ.*

Proof. We show that the above set $\Delta_0(FA)$ is Henkin. Assume $\{B \mid GB \in \Delta\} \cup \{A\} \vdash C(y)$ for all y. We can assume y is a variable not appearing in A. Hence for each y there exists a finite set of $B_{y,i}$, $i = 1, \ldots, n(y)$, such that $\vdash \bigwedge B_{y,i} \to (A \to C(y))$.

Hence

$$\vdash G[\bigwedge_i B_{y,i} \to (A \to C(y))]$$

Hence

$$\Delta \vdash G(A \to C(y))$$

for all y.

Since Δ is saturated and complete, it is Henkin and so

$$\Delta \vdash \forall y(G(A \to C(y)),$$

$$\Delta \vdash G\forall y(A \to C(y)),$$

$$\Delta \vdash G(A \to \forall y C(y)),$$

and hence $\forall y C(y)$ is provable from $\{B \mid GB \in \Delta\} \cup \{A\}$.

The above shows that $\Delta_o(FA)$ is Henkin and can therefore be extended to a saturated complete theory in the same language. A similar theorem will yield a theory $\Delta(PA)$ for $PA \in \Delta$. ■

We can prove the other direction of the completeness theorem now. Let B be a consistent sentence. It can be extended to a saturated complete theory Δ_B with $B \in \Delta_B$. Let T be the set of all complete and consistent and saturated theories. Define $<$ on T by:

$$\Delta < \Gamma \text{ iff for all } A,\ GA \in \Delta \text{ implies } A \in \Gamma.$$

Just as in the case of linear time we obtain that $<$ is irreflexive (since $\exists x({\sim}x \wedge Hx) \in \Gamma$ for all $\Gamma \in T$) and linear. We further have for all $\Delta < \Gamma$ and $HA \in \Gamma$ that $A \in \Delta$. Moreover, for each Γ and each $FA, PA' \in \Gamma$, the theories $\Gamma(FA) \in T$ and $\Gamma(PA') \in T$ exist, $\Gamma(PA') < \Gamma < \Gamma(FA)$ by construction, and $A' \in \Gamma(PA')$ and $A \in \Gamma(FA)$.

Finally, for each atom x let $h(x) = \{\Gamma \mid x \in \Gamma\}$.

Lemma 8.2.9 *In $(T, <, h)$ defined above we have for any sentence A and any Γ, $\|A\|^h_\Gamma = 1$ iff $A \in \Gamma$.*

Proof. By induction on A, using the properties of $(T, <, h)$. ■

This completes theorem 8.2.3.

■

Other axiomatizations

Another interesting propositionally quantified logic is *PQ*(linear , named). The semantics are defined here without restricting S at all. In fact, it turns out that we have met this logic already:

Theorem 8.2.10 *The theorems of PQ(atoms, linear, named) are exactly those of PQ(linear, named).*

Thus QL axiomatizes both logics.

Proof. The equivalence of the logics can be shown in several ways. Perhaps the easiest is to show first the soundness of QL for *PQ*(linear, named). This is easier than for *PQ*(atoms, linear, named).

Then one simply observes that a model $(T, <, h, S), t$ for a consistent sentence as constructed in the preceding theorem will also do as a model for the logic *PQ*(linear, named). ■

So restricting the semantics of *PQ* (linear, named) to $S = h(\mathcal{L})$ does not change the logic. In the next section, on the other hand, we will see that restricting our semantics to $S = 2^T$ is very different and results in a logic which is not recursively axiomatizable.

Another way of changing the logic is to consider other connectives. However, this is not very interesting in the context of propositional quantification and the IRR-axiom. Let us see why.

If $<$ is irreflexive and linear then the notion atom $(x) = \Diamond(x \wedge H{\sim}x \wedge G{\sim}x)$, where $\Diamond q = q \vee Pq \vee Fq$ defines the fact that x characterizes a point. The notion $<$ between atoms can be defined as $x < y = \Diamond(x \wedge Fy)$.

For a system with irreflexivity, this $<$ is identical to the $<$ of the linear flow of time itself, and the system is stronger than the predicate monadic language of $(T, <)$. For example $U(p, q)$ is definable as $\exists x[\text{ atom } x \wedge F(p \wedge x) \wedge G(Fx \rightarrow q)]$. In fact all connectives with first-order tables are definable since $<$ and atom (x) are available. The table can be translated directly into a connective.

8.3 The strength of propositional quantification

In this section we will consider the decidability of some of the logics with propositional qunatification.

Let us start by showing that *PQ*(atoms) is undecidable. We do this by interpreting the first-order logic of a reflexive symmetric relation in this logic. The former is well known to be undecidable.

An RS-structure is any non-empty set on which is defined a reflexive symmetric binary relation R. For any sentence A in the first-order language with (equality and) a binary relation symbol R, write $RS \models A$ iff A is valid in RS-structures. There is no decision procedure which when given any such A will answer whether or not $RS \models A$.

Suppose that $\mathcal{L}$ is our set of propositions for temporal structures. To interpret the one logic in the other we are going to need to be able to code structures of one in structures of the other. This we do by *associating* a pair of structures. First, some definitions.

Suppose that t is a point in the temporal structure $\mathcal{T}$. We define the set $\mathcal{A}(\mathcal{T}, t)$ of *atoms not too far above* t to contain exactly those $p \in \mathcal{L}$ for which there exists a point $u > t$ in T such that p is true at u and u has a future.

Say that atoms p, q from $\mathcal{A}(\mathcal{T}, t)$ are *indistinguishable* and write $p \approx q$ iff $GH(p \leftrightarrow q)$ holds at t. This is clearly an equivalence relation.

With $\mathcal{T}$ and t determined by the context let $\mathcal{A}/\approx$ be the set of $\approx$-classes of $\mathcal{A}(\mathcal{T}, t)$. Write $[q]$ for the $\approx$-class containing q.

Then we say that $(\mathcal{T}, t)$ and the RS-structure (S, R) are *associated* iff there is a map $a \mapsto q_a$ between S and the set $\mathcal{A}(\mathcal{T}, t)$ such that

- $a \mapsto [q_a]$ is a bijection between S and the set $\mathcal{A}/\approx$,
- for all a, b from S aRb iff there is a point $u > t$ where $Pq_a \wedge Pq_b$ holds.

To see how these mechanisms work let us look at how to code an RS-structure in a temporal structure. Because of the lack of atoms we can only do this for countable structures.

Lemma 8.3.1 *For any countable RS-structure there is an associated temporal structure* $((T, <, h), t)$.

Proof. Given (S, R) which is countable, define T to consist of all the points of S along with a new point t and new points $r(a, b)$ for every pair $a, b \in S$ such that aRb. Define $<$ by:

- for all $a \in S$, $t < a$, and
- for all $a, b \in S$ such that aRb, $t < r(a, b)$, $a < r(a, b)$ and $b < r(a, b)$.

Finally choose a distinct atom q_a for each of the countable number of $a \in S$ and define $h(q_a) = \{a\}$.

It is clear that we have our result. ■

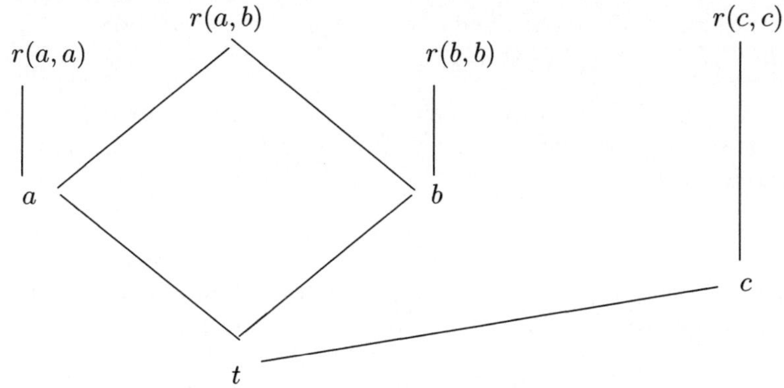

Figure 8.1

For example, if $S = \{a, b, c\}$ and $R = \{(a, a), (a, b), (b, a), (b, b), (c, c)\}$ then we have figure 8.1.

Of course it is not much use having such a coding if we cannot translate formulae between the two languages and preserve truth.

For any formula $A(\overline{x})$ in the first-order language with (equality and) a binary relation symbol R, define a sentence A^* in the temporal language with quantification over propositions by recursion as follows:

- $(x_i = x_j)^*$ is $GH(x_i \leftrightarrow x_j)$,
- $(x_i R x_j)^*$ is $F(Px_i \wedge Px_j)$,
- $(A \wedge B)^*$ is $A^* \wedge B^*$,
- $(\sim A)^*$ is $\sim(A^*)$,
- $(\exists x_i A)^*$ is $\exists x_i(F(x_i \wedge F\top) \wedge A^*)$.

Lemma 8.3.2 *If* $((T, <, h), t)$ *and* (S, R) *are associated then* $(S, R) \models A$ *iff* A^* *is true at time* t.

Proof. We show that for all formulae A in the language with binary R, for all n so that $\overline{x} = (x_1, ..., x_n)$ includes all the free variables of A, and for all n-tuples $\overline{a}$ from S, $(S, R) \models A(\overline{a})$ iff β is true at time t in $(T, <, h)$ where β is obtained by replacing each free occurrence of x_i in A^* by the atom q_{a_i}.

We proceed by induction on the construction of A.

$x_i = x_j$: By definition of $\approx$ and the fact that the range of the map $a \mapsto q_a$ contains only one representative of each $\approx$-class, we know that $a_i = a_j$ iff $GH(q_{a_i} \leftrightarrow q_{a_j})$ is true at t.

$x_i R x_j$: By the definition of association, $a_i R a_j$ iff there is a point $u > t$ where Pq_{a_i} and Pq_{a_j} are true. This is iff $F(Pq_{a_i} \wedge Pq_{a_j})$ is true at t as required.

$A \wedge B$ and $\sim A$: Straightforward.

$\exists x_i C$, $\Rightarrow$: Without loss of generality, $i = 1$ so we know that there is $c \in S$ such that $(S, R) \models C(c, a_2, ..., a_n)$. By the inductive hypothesis,

$$(Ii) \quad \delta \text{ is true at } t \text{ in } (T, <, h),$$

where δ is obtained from C^* by replacing each free occurrence of x_j by q_{a_j} for $j > 1$ and replacing each free occurrence of x_1 by q_c.

Now since $c \in S$, definition of association tells us there is a point $u > t$ with q_c and $F\top$ true at u. So

$$(Iii) \quad F(q_c \wedge F\top) \text{ is true at } t \text{ in } (T, <, h).$$

From (Ii) and (Iii), we deduce that β is true at time t in $(T, <, h)$ where β is obtained by replacing each free occurrence of x_j in $\exists x_1(F(x_1 \wedge F\top) \wedge C^*)$ by the atom q_{a_j}.

$\exists x_i C$, $\Leftarrow$: Without loss of generality, $i = 1$. We know that β is true at time t in $(T, <, h)$ where β is obtained by replacing each free occurrence of x_j in $\exists x_1(F(x_1 \wedge F\top) \wedge C^*)$ by the atom q_{a_j}. So there is some atom q such that $F(q \wedge F\top)$ is true at t and δ is true at t where δ is obtained from C^* by replacing each occurrence of x_j by q_{a_j} when $j > 1$ and x_1 by q.

By definition of $\mathcal{A}(\mathcal{T}, t)$, q is in this set so that by the surjectivity of the map $a \mapsto [q_a]$ there is some $a \in S$ such that $q_a \approx q$. The lemma below shows that $\delta[q_a/q]$ holds at t so, by the inductive hypothesis, $(S, R) \models C(a, a_2, ..., a_n)$.

Thus $(S, R) \models \exists x_1 C$ as required.

So we have a lemma to finish:

Lemma 8.3.3 *Let A be any formula of the first-order language with binary R with free variables all among $x_1, ..., x_n$, and let $q, q_1, ..., q_n$ be atoms.*

Let $\mathcal{T}$ be any temporal structure.

Suppose that δ is A^* with each free occurrence of x_i replaced by q_i, and that B is obtained from A^* by replacing all occurrences of q_1 by q.

Finally, suppose that q_1 and q are both from $\mathcal{A}(\mathcal{T}, t)$ and satisfy $q \approx q_1$.

Then A^* is true at t iff B is true at t.

Proof. This is a very straightforward induction on the construction of A. It is supported by the two base cases which themselves follow from the observations that $GH(q_1 \leftrightarrow q)$ entails both $GH(q_i \leftrightarrow q_1) \rightarrow GH(q_i \leftrightarrow q)$ and $F(Pq_i \wedge Pq_1) \rightarrow F(Pq_i \wedge Pq)$. ■

This completes the proof of the lemma. ■

We now show

Theorem 8.3.4 *For any sentence A in the first-order language with a binary relation symbol R (and equality), $RS \models A$ iff $(\exists x F(x \wedge F\top)) \rightarrow A^*$ is a theorem in the logic $PQ(atoms)$.*

Proof.

⇒: Assume $RS \models A$. Now given any temporal structure $(T, <, h, h(\mathcal{L}))$ which satisfies $\exists x F(x \wedge F\top)$ at some time t, find, by lemma 8.3.5, an associated RS-structure (S, R). By assumption, $(S, R) \models A$ so by lemma 8.3.2, A^* is true at time t in $(T, <, h, h(\mathcal{L}))$, as required.

⇐: Assume $(\exists x F(x \wedge F\top)) \rightarrow A^*$ is a theorem and that (S, R) is an RS-structure. By the Löwenheim–Skolem theorem there is a countable structure (S', R') which is elementarily equivalent to (S, R). By lemma 8.3.1, there is a temporal structure $((T, <, h), t)$ which is associated with (S', R'). Since S is non-empty, $\exists x F(x \wedge F\top)$ is true at time $t \in T$. So A^* is also true at time t. By lemma 8.3.2, $(S', R') \models A$ and elementary equivalence gives us the result.

■

Lemma 8.3.5 *For any temporal structure which satisfies $\exists x F(x \wedge F\top)$ at some time, there is an associated RS-structure.*

Proof. Given $(T, <, h)$ in which $\exists x F(x \wedge F\top)$ is true at time t, we select one representative from each $\approx$-class in $\mathcal{A}(\mathcal{T}, t)$. Now simply define S to be the set of these representatives. Define R to relate those elements of S which share a future in the future of t, i.e. pRq iff there is $u > t$ in T where $Pp \wedge Pq$ holds.

Since $\exists pF(p \wedge F\top)$ is true at t, S is non-empty.

R is symmetric by definition and R is reflexive as the atoms in S have a future in the future of t.

It is immediate that $((T, <, h), t)$ and (S, R) are associated. ■

Corollary 8.3.6 *The logic PQ(atoms, named) is undecidable.*

Note that with a little more work on coding we could have interpreted RS-structures in linear temporal structures and have concluded that PQ(atoms, linear, named) is undecidable.

We now continue to examine the strength of the propositional quantifiers logic. We show now that it is as strong as the monadic theory with $<$. Let us show this, by way of example, for the case of linear time and with names for points. We need to define the subset T of points, the relation $<$ on points, and indicate which are the monadic predicates and what points satisfy them.

(a) Let $\mathbf{J}(x) = \Diamond(x \wedge H{\sim}x)$.
$\mathbf{J}(x)$ says x is a point.

(b) Let $x < y$ be $\Box(Hx \rightarrow Hy)$.
$x < y$ reads x is earlier than y.

(c) Any atom q is considered a monadic predicate Q. The relation can be expressed by $x \in Q = \Box(x \rightarrow q)$.

(d) The quantifiers $\forall x, \exists x$ over points are translated in the logic as $\forall x \in \mathbf{J}, \exists x \in \mathbf{J}$.

Theorem 8.3.7 *Let ψ be a wff of the monadic theory of $(T, <, =)$ and let $\mathbf{B}$ be the translation of ψ obtained from ψ by translating as above and restricting the quantifiers to the symbol $\mathbf{J}$. Then*

$$\mathbf{QL} \vdash \mathbf{B} \textit{ iff}$$

ψ is valid in linear time.

Proof. Follows from the above discussion. ■

As promised, we must also consider the strength of the logic $PQ($ full, linear). However it is clear how to interpret the full monadic second-order logic in this logic — just code time points as atoms which are true at exactly that time — so it follows from the results reported in section 15.3.2 that $PQ($ full, linear) is also not decidable.

8.4 Fixed point operators

It is known that conventional temporal logic is insufficiently expressive to handle issues arising in concurrency. Several extended temporal logic systems, with second-order capability, now exist in the literature, including the quantified propositional temporal logics of sections 8.1–8.3, Wolper's ETL [Wolper, 1983], and Bannieqbal and Barringer's calculus [Bannieqbal and Barringer, 1986] using minimal and maximal fixed points. The system *USF*, introduced in [Gabbay, 1989a], and of interest here, involves a fixed point operator with recursively defined semantics. Over the natural numbers, all these systems are as expressive as the monadic second-order logic $S1S$, in which quantification over subsets as well as elements is allowed.

$S1S$ has been studied extensively; its decidability was proved by Büchi in 1962, using automata, and as *USF* is closely related to automata we can easily elicit the relationship between the two logics, and show that *USF* is also decidable and has the same expressive power as $S1S$. None the less, *USF* is itself surprisingly well-behaved, with elegant properties that can be studied without recourse to automata theory. In this vein we will prove that unbounded depth of nesting of the fixed point operator is not required for full expressive power. For the 'past' fragment, this can indeed be done within *USF*, and in fact *no* nested fixed point operators are needed. Our proof for full *USF* goes via the automata connection, and converts any formula effectively into one in which no fixed point operator is nested inside more than one other. We would like to find a more direct proof that avoids the use of automata; such a proof might yield a more efficient conversion algorithm. We give an example of our construction. We can also show that if we allow restrictions on the assignments to atoms, the fixed point operators can be eliminated completely.

We should mention that we will be using the logic *UYF* instead of *USF*. This has technical advantages and leads to no loss in expressive power (see below).

Notation

$\mathbb{N}$ will be the set $\{0, 1, 2, \ldots\}$ of natural numbers. 2^S will denote the set of all subsets of the set S. We often write $\bar{x}, \bar{a}, \ldots$, for *tuples* — finite sequences of variables, atoms, elements of a structure, etc. Other notations will be defined when required.

8.4.1 Syntax of *UYF*

We start by developing the syntax and semantics of the fixed point operator. This is not entirely a trivial task. We will fix an infinite set of propositional atoms, with which our formulae will be written; we write $p, q, r, s, \ldots$ for atoms.

Definition 8.4.1

1. The set of formulae of *UYF* is the smallest class closed under the following:

   (a) Any atom q is a formula of *UYF*, as is $\top$.

   (b) If A is a formula so is $\sim A$.

   (c) If A is a formula so is YA. We read Y as 'yesterday'.

   (d) If A and B are formulae, so are $A \wedge B$ and $U(A, B)$. (The latter is read as 'Until'. $A \vee B$ and $A \to B$ are regarded as abbreviations.)

   (e) Suppose that A is a formula such that every occurrence of the atom q in A not within the scope of a φq is within the scope of a Y but not within the scope of a U. Then $\varphi q A$ is a formula. (The conditions ensure that $\varphi q A$ has fixed point semantics.)

2. The *depth of nesting* of φs in a formula A is defined by induction on its formation: formulae formed by clause (a) have depth 0, clause (e) adds 1 to the depth of nesting, clauses (b) and (c) leave it unchanged, and in clause (d), the depth of nesting of $U(A, B)$ and $A \wedge B$ is the maximum of the depths of nesting of A and B. So, for example, $\sim\varphi r(\sim Yr \wedge \varphi q Y(q \to r))$ has depth of nesting of 2.

3. A *UYF*-formula is said to be a *YF-formula* if it does not involve U.

4. Let A be a formula and q an atom. A *bound occurrence* of q in A is one in a subformula of A of the form $\varphi q B$. All other occurrences of q in A are said to be *free*. An occurrence of q in A is said to be *pure past* in A if it is in a subformula of A of the form YB but not in a

subformula of the form $U(B, C)$. So $\varphi q A$ is well-formed if and only if all free occurrences of q in A are pure past.

8.4.2 Semantics of *UYF*

An *assignment* is a map h providing a subset $h(q)$ of $\mathbb{N}$ for each atom q. If h, h' are assignments, and $\bar{q}$ a tuple of atoms, we write $h =_{\bar{q}} h'$ if $h(r) = h'(r)$ for all atoms r not occurring in $\bar{q}$. If $S \subseteq \mathbb{N}$ and q is an atom, we write $h_{q/S}$ for the unique assignment h' satisfying: $h' =_q h$, $h'(q) = S$.

For each assignment h and formula A of *UYF* we will define a subset $h(A)$ of $\mathbb{N}$, the interpretation of A in $\mathbb{N}$. Intuitively, $h(A) = \{n \in \mathbb{N} \mid A$ is true at n under $h\} = \{n \in \mathbb{N} \mid \|A\|^h_n = 1\}$. We will ensure that, whenever $\varphi q A$ is well-formed,

(∗) $\qquad h(\varphi q A)$ is the unique $S \subseteq \mathbb{N}$ such that $S = h_{q/S}(A)$.

Notation 8.4.2 If $S \subseteq \mathbb{N}$, we write $S + 1$ (or $1 + S$) for $\{s + 1 \mid s \in S\}$.

Definition 8.4.3 We define the semantics of *UYF* by induction. Let h be an assignment. If A is atomic then $h(A)$ is already defined. We set:

- $h(\top) = \mathbb{N}$.
- $h(\sim A) = \mathbb{N} \setminus h(A)$.
- $h(YA) = h(A) + 1$.
- $h(A \wedge B) = h(A) \cap h(B)$.
- $h(U(A, B)) = \{n \in \mathbb{N} \mid \exists m > n(m \in h(A) \wedge \forall m'(n < m' < m \rightarrow m' \in h(B)))\}$.
- Finally, if $\varphi q A$ is well-formed we define $h(\varphi q A)$ as follows. First define assignments h_n ($n \in \mathbb{N}$) by induction: $h_0 = h, h_{n+1} = (h_n)_{q/h_n(A)}$. Then $h(\varphi q A) \stackrel{\text{def}}{=} \{n \in \mathbb{N} \mid n \in h_n(A)\} = \{n \in \mathbb{N} \mid n \in h_{n+1}(q)\}$.

To establish (∗) we need some definitions and lemmas.

Definition 8.4.4

1. If $n \in \mathbb{N}$, we say that subsets $S_1, S_2 \subseteq \mathbb{N}$ *agree before* n if for all $m < n$, $m \in S_1$ if and only if $m \in S_2$. We say that S_1 and S_2 *agree up to* n if they agree before $n + 1$.

2. Assume that A is a formula of *UYF*. If $n \in \mathbb{N}$, we say that assignments g, h are *A-similar up to n* if for all atoms q:
   - if all free occurrences of q in A are pure past, then $g(q)$ and $h(q)$ agree before n;
   - if not all free occurrences of q in A are pure past, but still no free occurrence of q in A is within the scope of a U, then $g(q)$ and $h(q)$ agree up to n;
   - otherwise, $g(q) = h(q)$.
3. A *UYF*-formula A is said to be *local* if $g(A)$ and $h(A)$ agree up to n whenever g, h are assignments that are A-similar up to n.

Remark 8.4.5 From the definitions, if g, h are φqA-similar up to n and $g(q)$, $h(q)$ agree before n, then g, h are A-similar up to n.

Lemma 8.4.6 *Assume that A is a local UYF formula, and that φqA is well-formed. Then:*

1. *If g, h are assignments with $g =_q h$, then $g(\varphi qA) = h(\varphi qA)$.*
2. *If $S \subseteq \mathbb{N}$ and h is an assignment, then $h_{q/S}(A) = S$ if and only if $S = h(\varphi qA)$.*
3. *φqA is local.*

Proof.
1. By definition 8.4.3 it suffices to show that for all $n \in \mathbb{N}$, $g_n(q)$ and $h_n(q)$ (as in the definition) agree before n. We do so by induction on n. If $n = 0$ there is nothing to prove. Assume the statement for n. Clearly, $g_n =_q h_n$. By remark 8.4.5 and the inductive hypothesis, g_n and h_n are A-similar up to n. As A is local, $g_n(A)$ and $h_n(A)$ agree up to n, i.e. $g_{n+1}(q)$ and $h_{n+1}(q)$ agree before $n + 1$. This completes the induction.

2. By (1) we can replace h by $h_{q/S}$. So it is enough to show that for any h, $h(A) = h(q) \iff h(\varphi qA) = h(q)$.

$\Rightarrow$: Suppose that $h(A) = h(q)$. First observe that $h_1(q) = h(A) = h(q)$, so that $h = h_1$. It follows by the definition of the h_n and induction that $h = h_n$ for all n. But now for all n, $n \in h(\varphi qA)$ if and only if $n \in h_{n+1}(q) = h(q)$, so that $h(\varphi qA) = h(q)$, as required.

$\Leftarrow$: Assuming that $h(\varphi qA) = h(q)$, we show that for all n:

$(a)_n$ $h(A)$ and $h_n(A)$ agree up to n;

$(b)_n$ $h(q)$ and $h_n(A)$ agree up to n.

It will clearly follow that $h(A) = h(q)$. We proceed by induction on n. $(a)_0$ holds because $h_0 = h$. We now show that $((a)_m \mid m \leq n) \Rightarrow (b)_n$. Let $m \leq n$. Then $m \in h(q) = h(\varphi q A)$ if and only if $m \in h_m(A)$ by definition 8.4.3, if and only if $m \in h(A)$ by $(a)_m$, if and only if $m \in h_n(A)$ by $(a)_n$. This proves $(b)_n$.

We now show that $(b)_n \Rightarrow (a)_{n+1}$. $(b)_n$ says that $h(q)$ and $h_{n+1}(q)$ agree before $n+1$. Thus, h and h_{n+1} are A-similar up to $n+1$. As A is local, $h(A)$ and $h_{n+1}(A)$ agree up to $n+1$, proving $(a)_{n+1}$.

3. To prove $\varphi q A$ local, we must show that whenever g, h are assignments that are $\varphi q A$-similar up to n, then $g(\varphi q A)$ and $h(\varphi q A)$ agree up to n. By (1) and (2) we can assume that $g(q) = g(A) = g(\varphi q A)$ and $h(q) = h(A) = h(\varphi q A)$; the hypothesis and what we have to prove are unchanged.

The proof is by induction on n. Assume the result for all $m < n$ and suppose that g, h are $\varphi q A$-similar up to n. If $m < n$ then g and h are $\varphi q A$-similar up to m, so by the inductive hypothesis, $g(\varphi q A)$ and $h(\varphi q A)$ agree up to m. Hence $g(A)$ and $h(A)$ agree before n. By remark 8.4.5, g and h are A-similar up to n, so as A is local, $g(A)$ and $h(A)$ agree up to n, as required. This completes the proof. ∎

Proposition 8.4.7 *Every formula A of UYF is local, and for any h, $h(A)$ depends only on $h(q)$ for those atoms q that have free occurrences in A.*

Proof. We show by induction on A that A is local, and that whenever g, h are assignments agreeing on the atoms occurring free in A, then $g(A) = h(A)$. In the atomic case and the cases of the boolean connectives the proof is simple, and the case of $\varphi q A$ is covered by the lemma. For $A = U(B, C)$, if assignments g, h are A-similar up to n then $g(q) = h(q)$ for all atoms q with free occurrences in A. Clearly, g and h agree on the free atoms of B and of C, so by the inductive hypothesis, $g(B) = h(B)$ and $g(C) = h(C)$, yielding $g(A) = h(A)$ — so these two sets certainly agree up to n. This proves both claims for A. Now consider the case of $A = YB$. Pick $n \in \mathbb{N}$ and a pair g, h of assignments that are YB-similar up to n. We claim that g and h are B-similar up to $n-1$. Let q be an atom. If all free occurrences of q in B are pure past, then the same holds for YB, so that $g(q), h(q)$ already agree up to $n-1$. Otherwise, if no free occurrence of q in A is under a U, then all free occurrences of q in YB are pure past. Hence again, $g(q), h(q)$ agree up to $n-1$. If none of these apply to q then $g(q) = h(q)$. It follows that g and h are B-similar up to $n-1$, as claimed. By the inductive hypothesis, $g(B)$ and $h(B)$ agree up to $n-1$. So by definition of the semantics of Y, $g(YB)$ and $h(YB)$ agree up to n. The proof that $h(A)$ depends only on $h(q)$ for q occurring free in A is straightforward. ∎

Combining lemma 8.4.6 and proposition 8.4.7 yields:

Theorem 8.4.8 (fixed point theorem)

1. *Suppose that A is any UYF-formula and φqA is well formed. Then if h is any assignment, there is a unique subset $S = h(\varphi qA)$ of $\mathbb{N}$ such that $S = h_{q/S}(A)$. Thus, regarding $S \mapsto h_{q/S}(A)$ as a map $\alpha : 2^{\mathbb{N}} \to 2^{\mathbb{N}}$ (depending on h, A), α has a unique fixed point $S \subseteq \mathbb{N}$, and we have $S = h(\varphi qA)$. For any h, $h(A) = h(q) \iff h(\varphi qA) = h(q)$.*

2. *If q has no free occurrence in a formula A and $g =_q h$, then $g(A) = h(A)$.*

3. *If φqA is well-formed and r is an atom not occurring in A, then for all assignments h, $h(\varphi qA) = h(\varphi rA(q/r))$, where $A(q/r)$ denotes substitution by r for all free occurrences of q in A.*

Proof. By the proposition, A is local, so (1) follows from the lemma. Part (2) of the theorem is proved in the proposition, and (3) is clear from (1). ■

We can also define a logic *USF* using the first-order connectives Until and Since as well as the fixed point operator. The logic *UYF* is just as expressive as *USF*: Yq is definable in *USF* by the formula $S(q, \bot)$, while $S(p, q)$ is definable in *UYF* by $\varphi rY(p \vee (q \wedge r))$. Using *UYF* allows easier proofs and stronger results.

8.5 Elementary results and recursive systems

Here we establish some simple results on the way the fixed point operator interacts with the other connectives of the logic. They are proved using the fixed point theorem, and some will be needed later.

Definition 8.5.1 Two *UYF*-formulae A, B are said to be *equivalent* if for all assignments h we have $h(A) = h(B)$. We write $A \equiv B$ if A and B are equivalent.

Proposition 8.5.2 *Let $\varphi qA(q)$ be a UYF-formula. Then $\sim\varphi qA(q)$ is equivalent to $\varphi q{\sim}A(\sim q)$. Here, $A(\sim q)$ denotes the result of replacing each free occurrence of q by $\sim q$ throughout A.*

Proof. Let h be any assignment and assume that $h(\varphi q A(q)) = S \subseteq \mathbb{N}$. We wish to show that $h(\varphi q{\sim}A({\sim}q)) = \mathbb{N} \setminus S$. By the fixed point theorem, it suffices to show that $h_{q/\mathbb{N}\setminus S}({\sim}A({\sim}q)) = \mathbb{N} \setminus S$. But $h_{q/\mathbb{N}\setminus S}({\sim}A({\sim}q)) = h_{q/S}({\sim}A(q)) = \mathbb{N} \setminus h_{q/S}(A(q))$, and this last is equal to $\mathbb{N} \setminus S$ by the fixed point theorem again. ∎

In a similar way we can show:

Proposition 8.5.3 *Let q, r be distinct atoms and suppose that $\varphi q \varphi r A$ is well-formed. Then $\varphi q \varphi r A \equiv \varphi r \varphi q A \equiv \varphi q A(r/q)$. Here, $A(r/q)$ denotes substitution of q for all free occurrences of r in A.*

Proof. Choose any assignment h. By the fixed point theorem, for all $S \subseteq \mathbb{N}$ we have $h(\varphi q \varphi r A) = S$ if and only if $h_{q/S}(\varphi r A) = S$. Using the theorem again, this is if and only if $h_{q/S,r/S}(A) = S$. So by symmetry, $\varphi q \varphi r A \equiv \varphi r \varphi q A$. Moreover, as we clearly have $h_{q/S,r/S}(A) = S$ if and only if $h_{q/S}(A(r/q)) = S$, the last part follows. ∎

Now we examine how φ interacts with the *yesterday* connective.

Proposition 8.5.4 *Let $B(q)$ be any formula and write $B(Yq)$ for the result of replacing every free occurrence of q in B by the formula Yq. Then $Y\varphi q B(Yq) \equiv \varphi q Y B(q)$.*

Proof. Let h be given, and let $S = h(\varphi q B(Yq))$, so that the interpretation of the left-hand-side under h is just $S + 1$. By the fixed point theorem, it suffices to show that $h_{q/S+1}(YB(q)) = S + 1$. But $h_{q/S+1}(YB(q)) = h_{q/S}(YB(Yq)) = 1 + h_{q/S}(B(Yq))$, and the latter is equal to $1 + S$, by choice of S and using the fixed point theorem once more. ∎

This result allows us to normalise *UYF*-formulae, by pushing all Ys inwards until they are next to atoms.

Definition 8.5.5

1. If A is any *UYF*-formula, and $n < \omega$, we define $Y^n A$ by induction: $Y^0 A = A$ and $Y^{n+1} A = Y(Y^n A)$. Formulae of the form $Y^n q$ for atomic q, or $Y^n \top$, are called *basic.*

2. A *UYF*-formula A is said to be *normal* if it is built up from basic formulae (the *basic subformulae* of A) using only $\sim$, $\wedge$, U, and φ (no Y). A subformula B of A is said to be a *normal subformula* of A if every basic subformula of B is a basic subformula of A.

So a 'basic subformula' of A is a subformula that is maximal among those subformulae of A that are basic. For example, the basic subformulae of $A = {\sim}Yq \to YYq$ are the first Yq and the YYq; neither the two occurrences of q nor the subformula Yq of the YYq are basic subformulae of A. The normal subformulae of A are just the first Yq, ${\sim}Yq$, YYq, and A.

Lemma 8.5.6 *Let A be any normal formula. Then YA is equivalent to a normal formula B.*

Proof. We will show in addition that for any atom q, if all free occurrences of q in YA are pure past then the same holds for B.

We go by induction on the number of ${\sim}$s, $\wedge$s, Us, and φs in A. If this is 0 then A is basic, so YA is already normal. The condition on atoms is trivially valid. It is easily seen that $Y{\sim}A \equiv Y\top \wedge {\sim}YA$, $Y(A \wedge B) \equiv YA \wedge YB$, and $YU(A, B) \equiv Y\top \wedge (A \vee (B \wedge U(A, B)))$, so the result follows immediately from the inductive hypothesis in these cases. Now assume the result for all formulae with no more $\wedge$s, ${\sim}$s, Us and φs than A, and consider φqA (assumed well-formed and normal). By theorem 8.4.8(3) we can rename bound atoms of A if necessary, so we can suppose that all occurrences of q in A are free and (necessarily) pure past. Now A is normal, so it has the form $B(Yq)$ where $B(q)$ is normal and with the same number of ${\sim}$s, $\wedge$s, Us, and φs as A. By the inductive hypothesis, $YB(q)$ is equivalent to a normal formula C. As all free occurrences of q in $YB(q)$ are pure past, the same holds for C, so that φqC is well-formed. It is clearly normal, and by proposition 8.5.4 is equivalent to $Y\varphi qA$. Finally, note that if r is any atom, all of whose free occurrences in $Y\varphi qA$ are pure past, then the same holds for YB and so (inductively) for C. Hence all free occurrences of r in φqC are pure past, as required. ∎

Theorem 8.5.7 *Any UYF-formula A is equivalent to a normal formula.*

Proof. By induction on A. The only difficult case is YA, which is dealt with by the preceding lemma. ∎

8.5.1 Recursive systems

The semantics of the fixed point operator were defined by nested recursion. We will now see how to unravel the nesting. This will be a main step in our proof of elimination of nested φs for *YF*.

Definition 8.5.8 Let $n > 0$. A *recursive system (of width n)* is a pair, $\rho = (\bar{r}, \bar{B})$, where $\bar{r} = (r_1, \ldots, r_n)$ and $\bar{B} = (B_1, \ldots, B_n)$, the r_i are distinct atoms, and the B_i are *UYF*-formulae. We require that for all $i, j \leq n$, every free occurrence of r_i in B_j is pure past.

A formula A is said to be *φ-free* if it contains no φ-operator. The recursive system $(\bar{r}, \bar{B})$ is said to be *φ-free* if each B_j is φ-free.

Semantics of recursive systems

We give them a fixed point semantics, as for φ. Let $\rho = (\bar{r}, \bar{B})$ be a recursive system. It can be shown using the technique of lemma 8.4.6 that for any assignment h there is a unique assignment h_ρ with $h_\rho =_{\bar{r}} h$ and $h_\rho(r_i) = h_\rho(B_i)$ for all $i \leq n$. h_ρ can be defined by recursion as before.

Definition 8.5.9

1. Let $\rho = ((r_1, \ldots, r_n),(B_1, \ldots, B_n))$ be a recursive system, and let A be a *YF*-formula. We say that A and ρ are *equivalent* if for all h we have $h(A) = h_\rho(B_1)$ $(= h_\rho(r_1))$.
2. If $\rho' = ((r'_1, \ldots, r'_m),(B'_1, \ldots, B'_m))$ is another recursive system, we say that ρ and ρ' are *equivalent* if for all assignments h, $h_\rho(r_i) = h_{\rho'}(r'_i)$ for all $i \leq min(n, m)$. Note that, in general, equivalence is not transitive; but the definition is no less useful for that.

We begin by showing that *UYF* is at least as expressive as recursive systems.

Proposition 8.5.10 *Let $\rho = ((r_1, \ldots, r_n), (B_1, \ldots, B_n))$ be any recursive system. Then there is a UYF-formula A that is equivalent to ρ. If the B_i are YF-formulae then such an A can be found in YFalso.*

Proof. By induction on the width n. If $n = 1$ we let $A = \varphi r_1 B_1$. Assume the result for n and let $\rho = ((r_1, \ldots, r_{n+1}), (B_1, \ldots, B_{n+1}))$ be given. Let $\rho^* = ((r_1, \ldots, r_n), (B_1^*, \ldots, B_n^*))$, where $B_i^* = B_i(r_{n+1}/\varphi r_{n+1} B_{n+1})$ for each $i \leq n$. We claim that ρ and ρ^* are equivalent; the result will then follow by induction.

Let h be any assignment, and let $S_i = h_\rho(r_i)$ for each $i \leq n+1$. Since $h_\rho(r_{n+1}) = h_\rho(B_{n+1})$, it follows from the fixed point theorem (8.4.8) that $h_\rho(\varphi r_{n+1} B_{n+1}) = h_\rho(r_{n+1})$ $(= S_{n+1})$. Hence from h_ρ's point of view, replacing r_{n+1} by $\varphi r_{n+1} B_{n+1}$ in B_i makes no difference, and we have $h_\rho(B_i^*) = h_\rho(B_i)$ for each i. But r_{n+1} does not occur free in the B_i^*, so letting $h^* = h_{(r_i/S_i \mid i \leq n)}$, we see that $h^*(B_i^*) = h_\rho(B_i^*)$ for each i. Thus $h^*(B_i^*) = h_\rho(B_i^*) = h_\rho(B_i) = S_i = h^*(r_i)$ for each i. It follows by uniqueness of the fixed point that $h^* = h_{\rho^*}$, so that ρ and ρ^* are equivalent.

■

The fact that formulae B_i in a recursive system may contain U will be needed later, but our current aim is to show that nesting of fixed point operators in *YF* can be eliminated. So from now until the end of section 5 we restrict attention to *YF*: all formulae will be *YF*-formulae. We first prove a converse to the preceding proposition.

Theorem 8.5.11 *Let A be any YF-formula. Then there is a φ-free recursive system $\rho = ((r_1, \ldots, r_n), (B_1, \ldots, B_n))$ (for some n) that is equivalent to A.*

Proof. By theorem 8.5.7 we can assume that A is normal. By renaming bound atoms if need be (cf. theorem 8.4.8(3)), we can also assume that for any subformula φqB of A, the only occurrences of q in A are in B. If A is a formula, we write $\hat{A}$ for the formula obtained from A by omitting all φs. Formally, $\hat{q} = q$ for any atom q, $(\top)^\wedge = \top$, $\widehat{\sim A} = \sim\hat{A}$, $(A \wedge B)^\wedge = \hat{A} \wedge \hat{B}$, $\widehat{YA} = Y\hat{A}$, and $\widehat{\varphi qA} = \hat{A}$. (Later we will also use $\hat{A}$ for *UYF*-formulae; we then include the clause $U(A, B)^\wedge = U(\hat{A}, \hat{B})$.)

We will find ρ as above, with the additional property:

(∗) No r_i occurs free in A, and each B_i is a normal subformula of $\hat{A}$.

We go by induction on A. If A is basic we let $\rho = (r, A)$ where r is any atom not occurring in A. Clearly, ρ is equivalent to A, and (∗) holds. If the recursive system $(\bar{r}, \bar{B}) = ((r_1, \ldots, r_n),(B_1, \ldots, B_n))$ is equivalent to A, and (∗) holds, then $((r_0, \bar{r}), (\sim B_1, \bar{B}))$ is equivalent to $\sim A$, where r_0 is a new atom; and (∗) still holds.

Assume that $A \wedge A'$ satisfies the condition on bound atoms, and that the recursive systems $((r_1, \ldots, r_n), (B_1, \ldots, B_n))$ and $((r'_1, \ldots, r'_m), (B'_1, \ldots, B'_m))$ are equivalent to A, A', respectively, (∗) holding for each. Then no r_i occurs free in A. If some r_i occurs free in A', then by the condition on bound atoms it cannot occur bound in A. Hence it does not occur at all in A, nor in any B_j (since they are subformulae of $\hat{A}$). In consequence, the functionality of $(\bar{r}, \bar{B})$ is unaffected if we replace r_i by a new atom not occurring at all in $A \wedge A'$. If we do this for all r_i where necessary, and undertake similar modifications for the r'_i, then the recursive system $((r_0, \bar{r}, \bar{r'}), (B_1 \wedge B'_1, \bar{B}, \bar{B'}))$, where r_0 is a new atom, satisfies (∗) for $A \wedge A'$; and it is certainly equivalent to $A \wedge A'$. This completes the case of $\wedge$.

Finally, we consider the case φqA (as A is normal, the case YA does not arise). Assume $\rho = ((r_1, \ldots, r_n), (B_1, \ldots, B_n))$ is equivalent to A, and that (∗) holds. We let $\rho^* = ((q, r_2, \ldots, r_n), (B_1, \ldots, B_n))$. By (∗), all occurrences of q in all the B_i are pure past, so ρ^* is well-formed. (This is where the assumption that A is normal is used.)

Evidently, (∗) holds for ρ^*, φqA. We claim that ρ^* is equivalent to φqA. Let h be an arbitrary assignment, suppose that $h(\varphi qA) = S$, say, and let $h' = h_{q/S}$. Obviously, $h'_\rho(B_i) = h'_\rho(r_i)$ for each $i \geq 2$. But, further, $h'_\rho(B_1) = h'(A)$ since A and ρ are equivalent; and by the fixed point theorem, $h'(A) = S = h'(q)$. If q does occur in A then by $(*)$, $q \notin \{r_1, \ldots, r_n\}$, and if not, then we can certainly assume this; so $h'(q) = h'_\rho(q)$. We have shown that $h'_\rho(B_1) = h'_\rho(q)$. Hence h'_ρ yields a fixed point of ρ^*, and by uniqueness of fixed points, $h_{\rho^*}(q) = h'_\rho(q) = S = h(\varphi q A)$. So A and ρ^* are equivalent, as claimed. This completes the induction, and with it the proof of the theorem. ■

We can restrict recursive systems further without losing any of their expressive power.

Definition 8.5.12 A recursive system $\rho = ((r_1, \ldots, r_n), (B_1, \ldots, B_n))$ is said to be *simple* if (a) it is φ-free, (b) the B_i are normal, and (c) all occurrences of the r_j in the B_i lie under the same number of Ys, i.e. there is $d > 0$ such that every basic subformula of any B_i of the form $Y^k r_j$ (for any j) is such that $k = d$. This unique d is called the *depth* of ρ.

Proposition 8.5.13 *Suppose that $\rho = ((r_1, \ldots, r_n), (B_1, \ldots, B_n))$ is a φ-free recursive system. Then ρ is equivalent to a simple recursive system of depth* 1.

Proof. By theorem 8.5.7 we can assume that the B_i are normal. Let k be maximal such that for some j, $Y^k r_j$ occurs as a basic subformula of some B_i. Introduce new atoms s_j^ℓ for $j \leq n$ and $1 \leq \ell \leq k$, and define S_j^1 to be the (normal) formula obtained from B_j by replacing each basic subformula $Y^\ell r_j$ by $Y s_j^\ell$ (all ℓ), and $S_j^\ell = Y s_j^{\ell-1}$ for each $\ell \geq 2$. Then let $\rho' = (\bar{s}, \bar{S})$, where $\bar{s} = ((s_j^\ell \mid j \leq n) \mid \ell \leq k)$, and similarly for $\bar{S}$.

Certainly ρ' is well-formed and simple of depth 1. We claim that it is equivalent to ρ. Let h be given, and define h' by:

$$h' =_{\bar{s}} h, \qquad h'(s_j^\ell) = h_\rho(Y^{\ell-1} r_j) \text{ for each } j, \ell.$$

Then for each j, $h'(s_j^\ell) = h'(Y s_j^{\ell-1}) = h'(S_j^\ell)$ if $\ell \geq 2$, and

$$h'(S_j^1) = h_\rho(S_j^1(s_i^\ell / Y^{\ell-1} r_i \mid i \leq n, \ell \leq k)) = h_\rho(B_j) = h_\rho(r_j) = h'(s_j^1).$$

Hence h' is a fixed point assignment for ρ', and by uniqueness of fixed points, $h' = h_{\rho'}$. As $h_{\rho'}(s_j^1) = h_\rho(r_j)$ for each j, ρ' and ρ are equivalent, as claimed. ■

Unfolding

We need a final result on syntactic manipulation of recursive systems.

Definition 8.5.14 Let $\rho = ((r_1, \ldots, r_n), (B_1, \ldots, B_n))$ be a simple recursive system of depth 1. Define formulae B_i^k for $k < \omega$ by induction:

- $B_i^0 = r_i$ for all i,
- B_i^{k+1} is obtained by *normalizing* $B_i^k(r_j/B_j \mid j \leq n)$ (i.e. replacing it by a normal equivalent).

Also define recursive systems $\rho^k = ((r_1, \ldots, r_n), (B_1^k, \ldots, B_n^k))$ for each $k \geq 1$.

Lemma 8.5.15 *Let $\rho = ((r_1, \ldots, r_n), (B_1, \ldots, B_n))$ be a simple recursive system of depth 1, as above. Then for each $k \geq 1$, the system ρ^k is simple, of depth k, and equivalent to ρ — we have $h_\rho = h_{\rho^k}$ for all h, k.*

Proof. By induction on k. Assume the result for $k \geq 1$ and let h be any assignment. Clearly, ρ^{k+1} is φ-free. To normalize a φ-free formula we move all Ys in through the $\wedge$s and $\sim$s using the rules $Y(A \wedge B) \equiv YA \wedge YB$ and $Y{\sim}A \equiv Y\top \wedge {\sim}YA$. These rewrite rules clearly preserve the total number of Ys above each atom, and it follows that ρ^{k+1} is simple and of depth $k+1$.

We claim that h_ρ gives a fixed point of ρ^{k+1}; the lemma will then follow by uniqueness of fixed points. But as $h_\rho(r_j) = h_\rho(B_j)$ for each j, the substitution of B_j for r_j in B_i^k makes no difference from h_ρ's point of view, so that $h_\rho(B_i^{k+1}) = h_\rho(B_i^k)$. By inductive hypothesis this last is equal to $h_\rho(r_i)$, which completes the proof. ■

8.6 Elimination of fixed point operators

Theorem 8.6.1 (elimination of fixed point nesting) *Let A be any YF-formula. Then A is equivalent to a formula A' without nested φs — a boolean combination of YF-formulae of the form φqB, where B is φ-free. Moreover, A' is obtainable effectively from A (in time polynomial in the length of A).*

The proof is rather technical. Nesting of the fixed point operator corresponds to recursive systems of width greater than 1. The idea of the proof is (roughly) to express A as a recursive system of width n, reduce n to 1 by coding the truth values of its formulae at regular intervals by a single atom, and obtain the value of A at intermediate times by interpolation.

Notation 8.6.2 If $k, p \in \mathbb{N}, p > 0$, we write $k \bmod p$ for the unique i with $0 \leq i < p$ and $i \equiv k \pmod p$. We write k_p for $k - (k \bmod p)$; k_p is the largest multiple of p not greater than k.

Let A be any *YF*-formula. By theorem 8.5.11 there is a φ-free recursive system $\rho = ((r_1, \ldots, r_n), (B_1, \ldots, B_n))$ that is equivalent to A. By proposition 8.5.13, ρ can be taken to be simple of depth 1. Study of the proofs in sections 3–4 shows that ρ is obtainable effectively from A in polynomial time.

Let $p = 2n + 3$. To prove theorem 8.6.1 it suffices to show:

Lemma 8.6.3 *For each $j \leq n$ there is a YF-formula D_j of the form $\varphi q C$, where C is φ-free, such that for any assignment h and any multiple m of p, $m \in h(D_j) \iff m \in h_\rho(B_j)$. D_j is obtainable effectively from ρ in polynomial time.*

For assuming the lemma, we can interpolate to obtain the values of A at any time. Observe first that the 'clock' formula $\tau = \varphi q {\sim} Y {\sim} Y^{p-1} q$ satisfies $h(\tau) = \{mp \mid m \in \mathbb{N}\}$ for any assignment h. We will show:

Claim: A is equivalent to $A' = \bigwedge_{0 \leq i < p} Y^i \tau \to B_1^i(r_j/D_j \mid j \leq n)$, where B_1^i is as in definition 8.5.14.

Proof of Claim: Let h be any assignment, let $m \in \mathbb{N}$, and let $i = m \bmod p$. Then $m \in h(A')$ if and only if $m \in h(B_1^i(r_j/D_j \mid j \leq n))$. Now by definition 8.5.14 and lemma 8.5.15, B_1^i is normal and every occurrence of each r_j in it is in a basic subformula of the form $Y^i r_j$. Moreover, as m_p is divisible by p, $m_p \in h(D_j)$ if and only if $m_p \in h_\rho(B_j) = h_\rho(r_j)$. So as $h =_{\bar{r}} h_\rho$, it follows that $m \in h(B_1^i(r_j/D_j \mid j \leq n))$ if and only if $m \in h_\rho(B_1^i)$. But by lemma 8.5.15, ρ^i is equivalent to ρ and hence to A, so this is if and only if $m \in h(A)$, proving the claim.

Evidently A' has no nested fixed point operators. Moreover, it is obtainable effectively from ρ in polynomial time, assuming that the D_j are. So theorem 8.6.1 follows from the lemma. ■

Proof of lemma 8.6.3 Assume for simplicity of notation that $j = 1$; the proof for other j is the same. By lemma 8.5.15 we can replace ρ by ρ^p (effectively and in polynomial time), and thus assume that ρ is a simple recursive system of depth p. Choose a surjective function $\chi : \mathbb{N} \to \{\top, \bot, B_i \mid i \leq n\}$ with the following properties:

- $\chi(m) = \chi(m+p)$ for all $m \in \mathbb{N}$;
- $\chi(0) = B_1$;
- if $k < p-3$ is odd then $\chi(k) = \bot$;
- $\chi(p-3) = \chi(p-2) = \top$;
- $\chi(p-1) = \bot$.

We can find such a χ since p is large enough. χ has the important property that for any assignment h, $h(\chi(m))$, $h(\chi(m+1))$ are both non-empty if and only if $m \equiv p-3 \pmod{p}$. We will design a φ-free recursive system $\sigma = (s, C)$ satisfying, for every assignment h,

$$(*) \qquad \forall m \in \mathbb{N}\ (m \in h_\sigma(s) \iff m \in h_\rho(Y^{m \bmod p}\chi(m))).$$

In particular, if $p \mid m$ then $m \in h_\sigma(s) \iff m \in h_\rho(B_1)$. The formula $D_1 = \varphi sC$ then satisfies the conclusion of the proposition, as, by the fixed point theorem, $h(D_1) = h_\sigma(s)$ for any h.

We let the formula C be:

$$\bigwedge_{0 \le i < p} Y^i[{\sim}Y\top \vee (Y^2 s \wedge Y^3 s)] \to Y^i[\chi(i)(Y^p r_j / Y^{p-\ell_j} s \mid j \le n)].$$

Here, $0 \le \ell_j < p$ is such that $\chi(\ell_j) = B_j$; ℓ_j exists because χ is onto. It is clear that C is obtainable effectively in polynomial time from the B_j. Hence the lemma will be established if we can prove that $(*)$ holds for all m.

We do this by induction on m. Fix an assignment h. Let $m \in \mathbb{N}$ and assume that $(*)$ holds for all $m' < m$. We will show that it holds for m.

Claim: For all $i < p$, $Y^i[{\sim}Y\top \vee (Y^2 s \wedge Y^3 s)]$ is true at m under the assignment h_σ if and only if $i = m \bmod p$.

Proof of claim: If $i > m$ then $Y^i A$ is false at m for any A under any assignment, and $i \neq m \bmod p$. So we can assume $i \le m$, in which case the left-hand side holds if and only if

$$(\dagger) \qquad m - i \in h_\sigma({\sim}Y\top \vee (Y^2 s \wedge Y^3 s)).$$

But clearly $h'({\sim}Y\top) = \{0\}$ for any assignment h', so $(\dagger)$ holds if and only if $m = i$ or $m-i-2$, $m-i-3 \in h_\sigma(s)$. By the inductive hypothesis, this holds if and only if $m = i$ or $m-i-2 \in h_\rho(Y^{(m-i-2) \bmod p}\chi(m-i-2))$, and similarly for $m-i-3$: i.e. if and only if

- $m = i$, or
- $(m-i-2)_p \in h_\rho(\chi(m-i-2))$ and $(m-i-3)_p \in h_\rho(\chi(m-i-3))$.

But by choice of χ, this holds if and only if $m = i$ or $m - i - 3 \equiv p - 3 \pmod{p}$, i.e. if and only if $m \equiv i \pmod{p}$. This proves the claim.

We now prove (*) for m. Let $i = m \bmod p$. Now $m \in h_\sigma(s)$ if and only if $m \in h_\sigma(C)$, so by the claim, we see that we must prove:

$$m \in h_\sigma(Y^i[\chi(i)(Y^p r_j / Y^{p-\ell_j} s \mid j \leq n)]) \iff m \in h_\rho(Y^i \chi(m)),$$

or, equivalently,

$$m_p \in h_\sigma(\chi(m)(Y^p r_j / Y^{p-\ell_j} s \mid j \leq n)) \iff m_p \in h_\rho(\chi(m)).$$

Now $\chi(m)$ is $\top$, $\perp$ or a normal formula B_k in which all occurrences of the r_j are in normal subformulae of the form $Y^p r_j$, which in C are replaced by $Y^{p-\ell_j} s$. So as $h_\sigma =_{\bar{r},s} h_\rho$, we need only check that:

$$(\ddagger) \qquad m_p \in h_\sigma(Y^{p-\ell_j} s) \iff m_p \in h_\rho(Y^p r_j) \text{ for each } j \leq n.$$

So let $1 \leq j \leq n$. First assume that $m < p$, so that $m_p = 0$. Now $p - \ell_j \geq 1$, so that $0 \notin h_\sigma(Y^{p-\ell_j} s)$; and clearly $0 \notin h_\rho(Y^p r_j)$. Hence (‡) holds in this case. Now assume $m \geq p$, so that $m_p \geq p$. Then $m_p \in h_\sigma(Y^{p-\ell_j} s)$ if and only if $m_p - p + \ell_j \in h_\sigma(s)$. As $\ell_j < p$, $m_p - p + \ell_j < m$, so (using (*) inductively) this holds if and only if $m_p - p + \ell_j \in h_\rho(Y^{\ell_j} \chi(m_p - p + \ell_j))$, if and only if $m_p - p \in h_\rho(\chi(\ell_j)) = h_\rho(B_j) = h_\rho(r_j)$, if and only if $m_p \in h_\rho(Y^p r_j)$. Thus (‡) is proved.

So (*) holds for m, and hence by induction, it holds for all $m \in \mathbb{N}$, completing the proof of the lemma. ■

8.7 Decidability and expressive power

Now we return to the full logic *UYF*. We will prove that it has the same expressive power as the monadic second-order logic $S1S$, *the second-order theory of one successor function.* Decidability of *UYF* will then follow from the known decidability of $S1S$. [Gurevich, 1985] surveys $S1S$ and automata.

We first explain the need to invoke second-order logic. Recall the temporal logic *'US'*, which involves in addition to the boolean connectives the binary temporal connective Until and its dual Since. There is no fixed point operator in *US*. Now consider the monadic first-order logic over $\mathbb{N}$. Its signature consists of the first-order signature $\{=, <\}$, augmented with monadic predicate variables $Q(x), R(x), \ldots$ associated with the atoms $q, r, \ldots$. The semantics are those of first-order logic in the structure $(\mathbb{N}, <, h)$. The assignment h provides the semantics of the unary predicate variables, so that if $Q(x)$ is associated with the atom q, and $n \in \mathbb{N}$, then $(\mathbb{N}, <, h) \models Q(n)$ if and only if $n \in h(q)$.

It is easily seen by induction on A that for any formula $A(q_1, \ldots, q_k)$ of US there is an *equivalent* monadic first-order formula $\psi_A(x, Q_1, \ldots, Q_k)$: for all h and $n \in \mathbb{N}$, $n \in h(A)$ if and only if $(\mathbb{N}, <, h) \models \psi_A(n)$. We will prove in chapters 9 and 10 that the converse also holds: for each monadic first-order formula there is an ($\mathbb{N}$-)equivalent US-formula. Thus we say that US is *fully expressive* with respect to monadic first-order logic over $\mathbb{N}$.

The need for second-order logic when treating the fixed point operator is prompted by the following observation by Wolper:

Proposition 8.7.1

1. *Assume that for every atom q, either $h(q)$ is finite or $\mathbb{N} \setminus h(q)$ is finite (in the latter case we say: $h(q)$ is cofinite). Then for all formulae A of US, $h(A)$ is either finite or cofinite.*

2. *If $\psi(x)$ is a monadic first-order formula, and $h(q)$ is finite or cofinite for each Q occurring in ψ, then so is $\{n \in \mathbb{N} \mid (\mathbb{N}, <, h) \models \psi(n)\}$.*

3. *There is no first-order formula $\epsilon(x)$ in the signature $\{=, <\}$ such that for all $n \in \mathbb{N}$, $(\mathbb{N}, <) \models \epsilon(n)$ if and only if n is even.*

Proof. 1. By induction on the complexity of A. For atomic A we are given the result. The set of finite and cofinite subsets of $\mathbb{N}$ is closed under the boolean operations, so the only remaining cases are $U(A, B)$ and $S(A, B)$. Assume inductively that $h(A)$ and $h(B)$ are finite or cofinite. Inspection of the semantics of U shows that if $n \in h(U(A, B))$ then there is $m > n$ in $h(A)$, and $n + 1 \in h(A)$ implies $n \in h(U(A, B))$. Hence $h(U(A, B))$ is finite or cofinite according as $h(A)$ is. For S, choose $m \in \mathbb{N}$ so large that the truth values of A and B at all $n \geq m$ are constant. Now in $\mathbb{N}$ we have $n + 1 \in h(S(A, B))$ if and only if $n \in h(A \vee (B \wedge S(A, B)))$ for all n. So if $m < k \in h(S(A, B))$ then either $k - 1 \in h(A)$, so that $k \in h(A)$ and $k + 1 \in S(A, B)$, or $k - 1 \in h(B \wedge S(A, B))$, which implies $k \in h(B)$; and as already $k \in h(S(A, B))$, we obtain $k + 1 \in S(A, B)$ again. It follows by induction that if there is $k > m$ in $h(S(A, B))$ then $k' \in h(S(A, B))$ for all $k' > k$. Hence if $h(S(A, B))$ is infinite, it is cofinite.

2. This is immediate from (1) and the full expressiveness of U and S over $\mathbb{N}$.

3. This is a special case of (2). ∎

But the *UYF*-formula $\tau = \varphi q {\sim} Y q$ satisfies $h(\tau) = \{0, 2, 4, \ldots\}$ for all h, so the expressive power of *UYF* goes beyond that of first-order logic over $\mathbb{N}$.

Recall the definition of $S1S$. The signature of this logic is as for monadic first-order logic (above). The formation rules for $S1S$-formulae are also as for first-order logic, but with the additional clause: if ψ is a formula and Q a monadic predicate variable then $\exists Q\psi$ is a formula. The semantics of $S1S$ are as for monadic first-order logic in $(\mathbb{N}, <, h)$, with the additional second-order clause: if $\psi(x_1, \ldots, x_n)$ is a formula, and $\bar{a} = (a_1, \ldots, a_n) \in \mathbb{N}^n$, then $(\mathbb{N}, <, h) \models \exists Q\psi(\bar{a})$ if and only if $(\mathbb{N}, <, h') \models \psi(\bar{a})$ for some $h' =_q h$.

We want to compare the expressive power of *UYF* and $S1S$. The natural definition to make is the following.

Definition 8.7.2 Let A be a *UYF*-formula and $\psi(x)$ a formula of $S1S$ with a single free variable, x. We say that A and ψ are *equivalent* if for all assignments h, $h(A) = \{n \in \mathbb{N} \mid (\mathbb{N}, <, h) \models \psi(n)\}$.

It is easy to show that $S1S$ is at least as expressive as *UYF*.

Proposition 8.7.3 *Let A be any UYF-formula. Then there is a formula $\psi_A(x)$ of $S1S$ that is equivalent to A. There is an algorithm that constructs ψ_A from A.*

Proof. By induction on A. If A is the atom q we let $\psi_A(x)$ be $Q(x)$, and we let $\psi_\top(x)$ be $x = x$. The boolean clauses are as expected, $\psi_{YA}(x)$ is $\exists y < x(\psi_A(y) \wedge {\sim}\exists z(y < z < x))$, and $\psi_{U(A,B)}(x)$ is $\exists y > x(\psi_A(y) \wedge \forall z(x < z < y \to \psi_B(z)))$. Finally, assume that $\varphi q A$ is well-formed and suppose that we have defined $\psi_A(x)$. The fixed point theorem shows that for any h, $h(\varphi q A)$ is the unique $S \subseteq \mathbb{N}$ such that $h_{q/S}(A) = S$. So we can define $\psi_{\varphi q A}(x)$ as $\exists Q[Q(x) \wedge \forall y(Q(y) \leftrightarrow \psi_A(y))]$. A standard induction on A now shows that ψ_A is always equivalent to A, and clearly the construction of ψ_A from A is effective. ■

We will prove the converse of proposition 8.7.3:

Theorem 8.7.4 *For any formula of $S1S$ with a single free variable, there is an equivalent UYF-formula, which is effectively constructable.*

This will establish that the system *UYF* is 'fully expressive' with respect to monadic second-order logic. The proof will use automata.

Definition 8.7.5

1. Let L be a finite set of atoms. We define a (*Muller*) *L-automaton* to be a 4-tuple $M = (S, s_0, T, F)$ where S is a finite non-empty set (of *states*), $s_0 \in S$ is the *initial state*, $T \subseteq S \times 2^L \times S$ is the *transition table*, and $F \subseteq 2^S$ is the set of *accepting conditions*.

2. M (as above) is said to be *deterministic* if for all $s \in S$ and $X \subseteq L$ there is a unique $s' \in S$ with $(s, X, s') \in T$. In this case we will often regard T as a function : $S \times 2^L \to S$.

3. If M is an automaton as above, and h is an assignment, a *run* of M over h is a sequence $(s_n \mid n < \omega)$ of states, such that for all n, $(s_n, \{q \in L \mid n \in h(q)\}, s_{n+1}) \in T$. The run is said to be *accepting* if $\{s \in S \mid s = s_n$ for infinitely many $n < \omega\} \in F$. Otherwise it is said to be *rejecting*.

4. M is said to *accept* an assignment h if there exists an accepting run of M over h.

5. Two L-automata are said to be *equivalent* if they accept exactly the same assignments.

An L-automaton is thought of as running along $\mathbb{N}$: at each $n \in \mathbb{N}$ it 'reads' which atoms of L are true at n under h, and chooses its next state in the light of this and its current state. It then advances to $n + 1$ and the process repeats.

The main lemma that we need follows after a definition.

Definition 8.7.6 If h is an assignment and $m \in \mathbb{N}$, we write $h_{\geq m}$ for the assignment given by $h_{\geq m}(q) = \{n \in \mathbb{N} \mid m + n \in h(q)\}$.

Intuitively, an automaton M accepts $h_{\geq m}$ if and only if M would accept h if it started its run at m instead of at 0.

Lemma 8.7.7 *Let $M = (S, s_0, T, F)$ be a deterministic automaton. Then there is a formula A_M of UYF such that for any assignment h and $n \in \mathbb{N}$, $n \in h(A_M)$ if and only if M accepts $h_{\geq n}$. A_M is obtainable effectively from M.*

Proof. The idea is taken from [McNaughton, 1966], and we will only sketch it. Suppose that M has k states. We first describe a deterministic automaton M^* involving $k+1$ 'copies' $M_0, \ldots, M_k$ of M. We describe the run of M^*. At each time $t \in \mathbb{N}$, M^* releases a dormant copy of M. Then for each $s \in S$ it checks to see if more than one currently active copy of M is in state s; if so, it renders dormant all but the longest-running copy (the one that was released first). (Thus M^* must keep track of the order of launch of the currently active copies of M — it can do this with finitely many extra states.) After this check, at most k copies of M can be active, so there will always be at least one dormant copy. M^* then advances to $t + 1$ by allowing all surviving copies of M so to advance, and the process repeats.

Consider a copy M_{i_0} of M that is released at time t_0. If it is later deactivated, this is because at some time $t_1 \geq t_0$ it arrives in the same state as another copy M_{i_1} of M released before t_0. At all times $t \geq t_1$, M_{i_1} (if not itself deactivated) will be in the same state as M_{i_0} would have been in, had it survived. If M_{i_1} is later deactivated then it itself will be replaced by another copy M_{i_2}, launched earlier than M_{i_1}. The resulting sequence $M_{i_0}, M_{i_1}, \ldots$ of 'descendants' of M_{i_0} is of length at most $k+1$, since clearly all the M_{i_j} were already active at time t_0. Let M_{i_ℓ} be the final descendant of M_{i_0}. Then M_{i_ℓ} is never deactivated, and as it is eventually always in the same state as M_{i_0} would have been in, we see that M accepts $h_{\geq t_0}$ if and only if the 'run' of M_{i_0}, had it lasted, would have been accepting, i.e. if and only if the 'run' of M_{i_ℓ} is accepting. This is the condition that we have to check with *UYF*-formulae.

We can simulate M^* by a recursive system $\rho = (\bar{r}, \bar{B})$. The atoms $\bar{r}$ involve:

- atoms r_{is} for each $i \leq k, s \in S$: r_{is} will be true at n if and only if the copy M_i of M is active and in state s at n;

- atoms o_{ij} $(i, j \leq k)$: o_{ij} being true at some point will mean that copies M_i and M_j of M are both currently active, and M_i was released first;

- atoms p_{ij} $(i, j \leq k)$: p_{ij} being true at a point will mean that at that point, copy M_i was made dormant because it was in the same state as copy M_j, which was not made dormant (i.e. M_j was the oldest copy in that state).

It is clear that, knowing the transition table of M, the values of these atoms at time $t+1$ are a fixed boolean combination of their values and of the values of the atoms of L at time t. Thus ρ can in fact be taken to be simple of depth 1; the formulae $\bar{B}$ will not involve Until. By proposition 8.5.10, for each atom r of $\bar{r}$ there is a *YF*-formula A_r such that for all h, $h(A_r) = h_\rho(r)$.

Let $D^0_{ii} = \top, D^0_{ij} = \bot$ if $i \neq j$, and for $d \geq 0$,

$$D^{d+1}_{ij} = \bigvee_{i' \neq i} U((p_{i,i'} \wedge D^d_{i',j}), \bigwedge_{i'' \neq i} {\sim} p_{i,i''}).$$

D^d_{ij} says that the dth descendant of M_i is M_j. Then let

$$C_{ij} = \bigvee_{s \in S} r_{is} \wedge \bigvee_{d \leq k} [D^d_{ij} \wedge \bigwedge_{j' \neq i} {\sim} D^{d+1}_{i,j'}].$$

Then $n \in h_\rho(C_{ij})$ if and only if M_i is active at time n and its final descendant is M_j. We can also express that the run of M_j is accepting, by

$$B_j = \bigvee_{X \in F} \big(\bigwedge_{s \in X} I(r_{js}) \wedge \bigwedge_{s' \in S \setminus X} {\sim} I(r_{j,s'}) \big).$$

Here, $I(q)$ abbreviates the formula $GFq = {\sim}U({\sim}U(q, \top), \top)$, saying that q holds infinitely often in the future. The formula A_M is now obtained from $\bigwedge_{i \leq k}(r_{i,s_0} \to \bigwedge_{j \leq k}(C_{ij} \to B_j))$ by substituting A_r for r (for each atom r of $\bar{\bar{r}}$). ■

To complete the argument we will need the following standard results about automata.

Fact 8.7.8

1. Every automaton is equivalent to a deterministic automaton, which can be constructed effectively. This was proved in [McNaughton, 1966].

2. For every sentence σ of $S1S$ whose free monadic predicates correspond to atoms in the finite set L, there is an L-automaton M such that for all assignments h, M accepts h if and only if $(\mathbb{N}, <, h) \models \sigma$. M can be constructed effectively from σ. See, e.g., [Gurevich, 1985].

Corollary 8.7.9 *UYF is decidable: there is an algorithm that, given a formula $A(\bar{q})$ of UYF, decides whether or not there is an assignment h such that $h(A) \neq \varnothing$.*

Proof. Consider the sentence $\sigma = \exists x \exists \bar{Q} \psi_A(x)$ of $S1S$, which by the proposition is effectively constructable from A. Clearly, $\mathbb{N} \models \sigma$ if and only if A has a model. By fact 8.7.8, from σ we can construct a deterministic $\varnothing$-automaton $M = (S, s_0, T, F)$ that accepts an assignment h if and only if $(\mathbb{N}, <, h) \models \sigma$. As σ has no atoms, the behaviour of M is independent of h, and T is a function from S to S. Thus M has a unique run $s_0, s_1, \ldots$, where $s_{n+1} = T(s_n)$ for all n; and A has a model just when this run is accepting.

Since M has finitely many states, there are n, k with $k > 0$ and $n+k \leq r$, such that $s_n = s_{n+k}$. As M's next state is determined by its current state, we have $s_m = s_{m+k}$ for all $m \geq n$. We can construct the part $s_0, \ldots, s_{n+k}$ of the run by an algorithm, since $s_m = T^m(s_0)$ for all m. Then the run is accepting if and only if $\{s_n, \ldots, s_{n+k}\} \in F$, since clearly these are the states that occur infinitely often in the run. We can decide this question algorithmically. Note that we have actually proved that $S1S$ is decidable. ∎

Now let $\psi(x)$ be an $S1S$-formula with a single free variable, x. We want to find a *UYF*-formula A equivalent to ψ. Let L be the set of atoms occurring in ψ and let Q be a monadic predicate variable not occurring in ψ. By fact 8.7.8 we can find a deterministic $L \cup \{q\}$-automaton $M = (S, s_0, T, F)$ that accepts an assignment h if and only if $(\mathbb{N}, <, h) \models \forall x(Q(x) \leftrightarrow \psi(x))$.

Definition 8.7.10

1. If $s \in S$ we write M_s for any deterministic equivalent of the version of (S, s, T, F) that 'guesses' values of $h(q)$: formally, M_s is a deterministic equivalent of the L-automaton (S, s, T', F), where $T' = \{(s_1, X \cap L, s_2) \mid (s_1, X, s_2) \in T\}$.
2. We write A_s for the formula A_{M_s} of lemma 8.7.7.

Clearly, for any h there is a unique $h^* =_q h$ that M accepts, namely the h^* satisfying $h^*(q) = \{n \in \mathbb{N} \mid (\mathbb{N}, <, h) \models \psi(n)\}$. We can simulate the run of M over h^* from within h, obtaining the following result.

Lemma 8.7.11 *For each $s \in S$ there is a UYF-formula C_s such that for all assignments h and all $n \in \mathbb{N}$, $n \in h(C_s)$ if and only if the state of M at time n during its run on h^* is s.*

Proof. We define a recursive system $\rho = ((r_s \mid s \in S), (B_s \mid s \in S))$ as follows. If $s \in S$ let $\pi(s) = \{(s', X) \in S \times 2^L \mid T(s', X) = s$ or $T(s', X \cup \{q\}) = s\}$. We let

$$B_s = A_s \wedge \bigvee_{(s',X)\in\pi(s)} Y(r_{s'} \wedge \bigwedge_{x\in X} x \wedge \bigwedge_{y\in L\setminus X} {\sim}y)$$

for each $s \in S \setminus \{s_0\}$, and

$$B_{s_0} = {\sim}Y\top \vee [A_{s_0} \wedge \bigvee_{(s,X)\in\pi(s_0)} Y(r_s \wedge \bigwedge_{x\in X} x \wedge \bigwedge_{y\in L\setminus X} {\sim}y)].$$

Note that the B_s are formulae of *UYF* (and not *YF*). None the less, by proposition 8.5.10 we can find for each $s \in S$ a formula C_s such that for all h, $h(C_s) = h_\rho(r_s)$. Thus it suffices to show that for any assignment h, if $s_0, s_1, \ldots$ is the (accepting) run of M on h^* then for all n and s, $n \in h_\rho(r_s)$ if and only if $s = s_n$.

This is clear if $n = 0$. Assume the result for n. Let $X = \{p \in L \mid n \in h(p)\}$, and define $s^+ = T(s_n, X \cup \{q\})$, $s^- = T(s_n, X)$. By definition of the B_s and the inductive hypothesis, for each $s \in S$ we have: $n+1 \in h_\rho(r_s)$ if and only if

1. $s = s^+$ or $s = s^-$, and

2. $n + 1 \in h(A_s)$.

And (2) holds if and only if M_s accepts $h_{\geq n+1}$, if and only if there is $h' =_q h$ that M accepts 'when started at $n+1$ in state s'.

Claim: (1) and (2) hold if and only if $s = s_{n+1}$.

Proof of Claim: Certainly, $s_{n+1} \in \{s^+, s^-\}$; and M obviously accepts h^* when started in state s_{n+1} at $n+1$ (its run is just $s_{n+1}, s_{n+2}, \ldots$). Conversely, suppose we are given $s \in \{s^+, s^-\}$ and h' as above. We can assume that $h'(q)$ and $h^*(q)$ agree before n, and that $T(s_n, h'^{-1}(n) \cap (L \cup \{q\})) = s$. An accepting run of M on $h'_{\geq n+1}$ when started in state s, when preceded by $s_0, \ldots, s_n$, gives an accepting run of M on h'. By uniqueness, $h' = h^*$, and so $s = T(s_n, h^{*-1}(n) \cap (L \cup \{q\})) = s_{n+1}$, as required. This proves the claim, and with it the lemma. ■

Corollary 8.7.12 *There is a UYF-formula D equivalent to $\psi(x)$.*

Proof. We let D be

$$\bigwedge_{s \in S, X \subseteq L} (C_s \wedge \bigwedge_{x \in X} x \wedge \bigwedge_{y \in L \setminus X} {\sim} y) \to U(A_{T(s, X \cup \{q\})}, \bot).$$

Let h be given, and let $s_0, s_1, \ldots$ be the run of M on h^*, as before. Let $n \in \mathbb{N}$, and write X for $\{p \in L \mid n \in h(p)\}$. Then $n \in h(D)$ if and only if $n \in h(U(A_{T(s_n, X \cup \{q\})}, \bot))$, if and only if $n+1 \in h(A_{T(s_n, X\cup\{q\})})$. By the claim of lemma 8.7.11, this holds if and only if $s_{n+1} = T(s_n, X \cup \{q\})$. Now $s_{n+1} = T(s_n, X \cup \{q\})$ or $T(s_n, X)$, according as $n \in h^*(q)$ or not. If we had $T(s_n, X \cup \{q\}) = T(s_n, X)$, we could replace h^* by h', identical with h^* except that $n \in h'(q) \iff n \notin h^*(q)$; then $s_0, s_1, \ldots$ would be an accepting run of M over h', contradicting the uniqueness of h^*. Hence $T(s_n, X \cup \{q\}) \neq T(s_n, X)$. So $n \in h(D)$ if and only if $n \in h^*(q)$, i.e. if and only if $(\mathbb{N}, <, h) \models \psi(n)$, as required. ■

Thus the proof of theorem 8.7.4 is complete. We remark that the formula D of the theorem is effectively obtainable from ψ (since M and the M_s are). By theorem 8.6.1, D can in fact be taken to have depth of nesting of fixed point operators of 2 (i.e. no branch of the formation tree of D

contains more than two φs). Thus, combining proposition 8.7.3 and theorem 8.7.4, we see that any *UYF*-formula is effectively equivalent to such a formula.

Example 8.7.13 We give a very simple example of our construction. We will construct the formula D equivalent to the $S1S$-formula $\epsilon(x) = \exists P[P(x) \wedge \forall y(P(y) \leftrightarrow \sim\exists z(z < y \wedge \sim\exists t(z < t < y) \wedge P(z)))]$, defining the even numbers in $\mathbb{N}$. The automaton M for $\forall x(Q(x) \leftrightarrow \epsilon(x))$ can be taken to have three states, 0 (initial state), 1 and 2, as in figure 8.2.

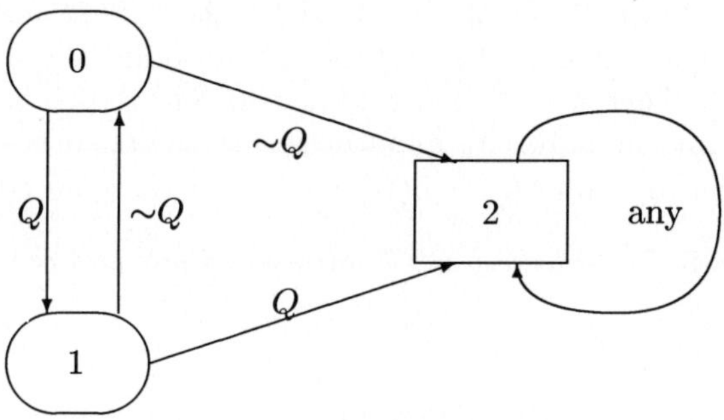

Figure 8.2

The set F of accepting conditions is $\{\{0,1\}\}$. The version of M that 'guesses' Q can accept if and only if it is initiated in state 0 or 1. Hence we can take $A_0 = A_1 = \top$ and $A_2 = \bot$. As $L = \varnothing$, we obtain $B_0 = \sim Y\top \vee Yr_1$, $B_1 = Yr_0$, and $B_2 = \bot$. Applying proposition 8.5.10 to the recursive system $(\bar{r}, \bar{B})$, we obtain (up to equivalence) $C_0 = E$, $C_1 = YE$, and $C_2 = \bot$, where $E = \varphi r_0(\sim Y\top \vee Y^2 r_0)$. Observe that E is true at even numbers only. The formula D is thus $\bigwedge_{i \leq 2}(C_i \rightarrow TA_{i'})$, where Tq ('tomorrow, q') abbreviates $U(q, \bot)$, and i' is the next state of M when currently it is in state i and Q is true. Since $0' = 1$ and $1' = 2' = 2$, we obtain $D = (E \rightarrow T\top) \wedge (YE \rightarrow T\bot) \wedge (\bot \rightarrow T\bot)$, which is equivalent to $\top \wedge \sim YE \wedge \top$, i.e. to E. Hence our construction yields the formula E as equivalent to $\epsilon(x)$.

8.8 Eliminating φ altogether

Finally, we prove proposition 8.8.2, a variant of which was also shown by Strulo [Strulo, 1990]. It shows how to eliminate all fixed point operators from a *UYF*-formula A, at the cost of restricting the values assigned to the bound atoms of A.

We can assume that the bound atoms of A are distinct and do not occur free in A. We let $A\dagger = \bigwedge(q \leftrightarrow \hat{B})$, where the conjunction is taken over all subformulae of A of the form φqB, and $\hat{A}$ is as defined in section 8.5. Equivalently we can define $A\dagger$ by induction: $q\dagger = \top\dagger = \top$ for any atom q, $(\sim A)\dagger = (YA)\dagger = A\dagger$, $(A \wedge B)\dagger = (U(A,B))\dagger = A\dagger \wedge B\dagger$, and $(\varphi qA)\dagger = A\dagger \wedge (q \leftrightarrow \hat{A})$.

Lemma 8.8.1 *Let A be a formula of UYF, and let h be an assignment such that $h(A\dagger) = \mathbb{N}$. Then for all subformulae B of A, $h(B) = h(\hat{B})$.*

Proof. By induction on B. We need only check the case $B = \varphi qC$, as the other cases are simple. Assume that φqC is a subformula of A, and that the result holds for C. Then $q \leftrightarrow \hat{C}$ is a conjunct of $A\dagger$. Hence $h(q) = h(\hat{C})$, and $h(\hat{C}) = h(C)$ by the inductive hypothesis. By the fixed point theorem (8.4.8), $h(\varphi qC) = h(q)$, and $h(q) = h(\hat{C}) = h(\widehat{\varphi qC})$, as required. ∎

Proposition 8.8.2 *Let A be any UYF-formula with bound atoms $\bar{r}$. Then for any assignment h there is a unique assignment $h^A =_{\bar{r}} h$ satisfying $h^A(A\dagger) = \mathbb{N}$. We have $h^A(\hat{A}) = h(A)$ for all h.*

Proof. By induction on A. If A is atomic or $\top$, then $\bar{r}$ is empty, $A\dagger = \top$ and $\hat{A} = A$, so $h^A = h$ and the result is trivial. Assume the result for A. As YA has the same bound atoms as A, and $(YA)\dagger = A\dagger$, we must have $h^{YA} = h^A$ by the uniqueness part of the inductive hypothesis. But then, $h^{YA}(\widehat{YA}) = h^A(Y\hat{A}) = 1 + h^A(\hat{A}) = 1 + h(A)$ by the inductive hypothesis, and this last is $h(YA)$, as required. The proof for $\sim A$ is similar. Now consider $U(A,B)$. Suppose the result holds for A, B, having bound atoms $\bar{a}, \bar{b}$ respectively. There are unique $h^A =_{\bar{a}} h$, $h^B =_{\bar{b}} h$ with $h^A(A\dagger) = h^B(B\dagger) = \mathbb{N}$. As $\bar{a}$ and $\bar{b}$ are disjoint, we can define $h^* =_{\bar{a}\bar{b}} h$ by $h^* =_{\bar{b}} h^A$, $h^* =_{\bar{a}} h^B$. Then $h^*(U(A,B)\dagger) = \mathbb{N}$, and h^* is unique given this condition. Moreover, we have $h^*(U(A,B)^\wedge) = h^*(U(\hat{A},\hat{B}))$, which by the inductive hypothesis is $h(U(A,B))$, as required. The argument for $A \wedge B$ is similar.

Finally, assume the result for A and consider φqA. Let $h(\varphi qA) = S$, and $h^* = (h_{q/S})^A$. Then (i) $h^*(\widehat{\varphi qA}) = (h_{q/S})^A(\hat{A})$; by the inductive hypothesis this is $h_{q/S}(A)$, which is $h(\varphi qA)$, by the fixed point theorem. Also, (ii) $h^*(A\dagger) = \mathbb{N}$ since h^* is of the form h'^A. It follows from (i) that (iii) $h^*(q) = S = h^*(\hat{A})$. ¿From (ii) and (iii) we obtain (iv) $h^*((\varphi qA)\dagger) = \mathbb{N}$. Hence by (i) and (iv) we can let $h^{\varphi qA} = h^*$.

It remains to check uniqueness. Suppose h' is any assignment such that $h' =_{\bar{r},q} h$ (where $\bar{r}$ are the bound atoms of A) and $h'((\varphi qA)\dagger) = \mathbb{N}$. Then in particular, $h'(q) = h'(\hat{A})$. As $h'(A\dagger) = \mathbb{N}$, lemma 8.8.1 applies, and we obtain $h'(\hat{A}) = h'(A)$. But $h'(q) = h'(A)$ yields $h'(q) = S$ by the uniqueness part of the fixed point theorem. Hence $h' =_{\bar{r}} h_{q/S}$, and since

$h'(A\dagger) = \mathbb{N}$, we obtain $h' = (h_{q/S})^A = h^{\varphi q A}$ by the uniqueness part of the inductive hypothesis. ■

We say that a formula A is *valid* if $h(A) = \mathbb{N}$ for all assignments h.

Theorem 8.8.3 *Let A be any UYF-formula. Then A is valid if and only if the UY-formula $\Box(A\dagger) \to \hat{A}$ is valid, where $\Box(A\dagger)$ abbreviates $A\dagger \wedge \sim U(\sim A\dagger, \top) \wedge \sim S(\sim A\dagger, \top)$.*

Proof. Let h be any assignment, and let $t \in \mathbb{N}$. Assume first that A is valid. If $t \in h(\Box(A\dagger))$, we want to show that $t \in h(\hat{A})$. Clearly, $h(A\dagger) = \mathbb{N}$, so by lemma 8.8.1, $h(\hat{A}) = h(A)$, which is $\mathbb{N}$ since A is valid. Hence $t \in h(\hat{A})$, as required.

Conversely, if $\Box(A\dagger) \to \hat{A}$ is valid, we need to show that $t \in h(A)$. By validity, $t \in h^A(\Box(A\dagger) \to \hat{A})$. By proposition 8.8.2, $h^A(A\dagger) = \mathbb{N}$, so $t \in h^A(\Box(A\dagger))$. Hence $t \in h^A(\hat{A})$, and as the proposition proved that $h^A(\hat{A}) = h(A)$, we obtain the result. ■

Thus, in a sense, any *UYF*-formula A is 'equivalent' to the UY-formula $\Box(A\dagger) \to \hat{A}$, which does not involve φ at all.

Chapter 9

Expressive power of one-dimensional temporal connectives: basic concepts

9.1 Introduction

This chapter deals with the basic concepts and theorems involved in measuring the expressive power of a temporal system. We deal here with one-dimensional temporal logics, those in which the unit for truth value is a point of the flow of time. Below we define the notions and connectives involved.

These definitions are a special case of the definition in chapter 4. We repeat them here for the sake of clarity and relative independence of chapters.

Some of the connectives involved in this and the next four chapters are *since*(S), *until*(U), *will*(F), and *was*(P). Given a flow of time $(T, <, h)$, the truth value of each of the above connectives at a point $t \in T$ is determined as follows:

$$\begin{array}{lll}
\|Fp\|_t^h = 1 & \text{iff} & (\exists s > t)\|p\|_s^h = 1, \\
\|Pp\|_t^h = 1 & \text{iff} & (\exists s < t)\|p\|_s^h = 1, \\
\|U(p,q)\|_t^h = 1 & \text{iff} & (\exists s > t)(\|p\|_s^h = 1 \wedge \forall y(t < y < s \rightarrow \|q\|_y^h = 1), \\
\|S(p,q)\|_t^h = 1 & \text{iff} & (\exists s < t)(\|p\|_s^h = 1 \wedge \forall y(s < y < t) \rightarrow \|q\|_y^h = 1).
\end{array}$$

Thus:

- $\|Fp\|_t^h = 1$ iff $\chi_F(t, h(p))$ holds in $(T, <)$,
- $\|Pp\|_t^h = 1$ iff $\chi_P(t, h(p))$ holds in $(T, <)$,
- $\|U(p, q)\|_t^h = 1$ iff $\chi_U(t, h(p), h(q))$ holds in $(T, <)$, and
- $\|S(p, q)\|_t^h = 1$ iff $\chi_S(t, h(p), h(q))$ holds in $(T, <)$,

where

$\chi_F(t, P) = (\exists s > t)P(s)$,

$\chi_P(t, P) = (\exists s < t)P(s)$,

$\chi_U(t, P, Q) = (\exists s > t)(P(s) \wedge \forall y(t < y < s \rightarrow Q(y))$,

$\chi_S(t, P, Q) = (\exists s < t)(P(s) \wedge \forall y(s < y < t \rightarrow Q(y)))$.

$\chi_\#$ is called the *truth table* for the connective #. It is a wff in the monadic first-order language—the language allowing for quantification over elements of the domain, with relation symbols $<$ and $=$ and unary predicates (i.e. subsets of T) to be used as parameters. (This is the language **IPL** of chapter 4.)

Example 9.1.1 Consider $\chi(t, P) = \exists xy(t < x < y \wedge \forall s(x < s < y \rightarrow P(s)))$. χ says 'There is an interval in the future of t inside which P is true.' This is a table for a connective Fint: $\|\text{Fint}(p)\|_t^h = 1$ iff $\chi(t, h(p))$ holds in $(T, <)$.

Definition 9.1.2

1. Any wff $\chi(t, Q_1, ..., Q_m)$ with one free variable t, in the monadic first-order language with predicate variable symbols Q_i, is called an m-place truth table (in one dimension). (See chapter 4.)

2. Given a syntactic symbol # for an m-place connective, we say it has a truth table $\chi(t, Q_1, ..., Q_m)$ iff for any T, h and t, $(*)$ holds:

$$(*): \quad \|\#(q_1, ..., q_m)\|_t^h = 1 \quad \text{iff} \quad (T, <) \models \psi(t, h(q_1), ..., h(q_m)).$$

This way we can define as many connectives as we want. The complexity of the connective is determined by its truth table. Some connectives are definable using other connectives, for example:

Example 9.1.3 $Fp \equiv U(p, \top)$ and $\delta \equiv U(\top, \bot) \equiv$ there exists a next moment.

However, U, for example, is not definable using F and P. In fact, δ is not definable using F and P over linear time. Whether a connective is definable or not from other connectives depends on the flow of time, as the following example shows.

Example 9.1.4

1. Over dense linear time δ is equivalent to $\bot$ and hence is definable from P and F.

2. Let $T = \{... -2, -1, 0, 1, 2, ...\} \cup \{(1/n) \mid n = 1, 2, 3...\}$, with $<$ being the order by magnitude, then δ is not definable using P and/or F.

Proof.

1. It is simple.

2. Suppose δ is equivalent to A where A is written with P and F and, maybe, atoms. Replace all appearances of atoms by $\bot$ to obtain A'. Since $\delta \leftrightarrow A$ holds in the structure $(T, <, h')$ with all atoms always false, in this structure $\delta \leftrightarrow A \leftrightarrow A'$ holds. As neither δ nor A' contain atoms $\delta \leftrightarrow A'$ holds in all other $(T, <, h)$ as well. Now A' contains only P and F, $\top$, and $\bot$ and the classical connectives. Since $F\top \equiv P\top \equiv \top$ and $F\bot \equiv P\bot \equiv \bot$, at every point, A' must be equivalent (in $(T, <)$) to either $\top$ or $\bot$ and so cannot equal δ which is true at 1 and false at 0.

■

Given a family of connectives, e.g.$\{F, P\}$ or $\{U, S\}$, we can define (using them) as many new connectives as we want. Previously we have seen how to define H and G using F and P, for example. In fact, any wff $A = A(q_1, ..., q_m)$ of the language of the given initial connectives can be considered as a new connective. For example, FHq has the table $\psi_{FHq} = (\exists s > t)(\forall y < s)Q(y)$ and is definable from F and P. So any wff A has a *table* ψ_A:

Lemma 9.1.5 *Let $\#_1(q_1, ..., q_{m_1}), ..., \#_n(q_1, ..., q_{m_n})$ be n connectives with tables $\chi_1, ..., \chi_n$. Let A be any wff built up from atoms $q_1, ..., q_m$, the classical connectives, and these connectives. Then there exists a monadic $\psi_A(t, Q_1, ..., Q_m)$ such that for all T and h,*

$$\|A\|_t^h = 1 \text{ iff } (T, <) \models \psi_A(t, h(q_1), ..., h(q_m)).$$

Proof. By induction on A.

- $\psi_{q_j} = Q_j(t)$,
- $\psi_{\sim A} = \sim\psi_A$,
- $\psi_{A\wedge B} = \psi_A \wedge \psi_B$,
- $\psi_{\#_i(A_1,\ldots,A_{m_i})} = \chi_i(t, \psi_{A_1}, \ldots, \psi_{A_{m_i}})$, the right-hand side being notation for the wff obtained, after renaming to avoid variable symbol clashes, by substituting $\psi_{A_j}(x)$ in wherever $Q_j(x)$ appears in χ_i.

■

Example 9.1.6 We have $\psi_{FU(p,q)} = \psi_{Fp}(\psi_{U(p,q)}) = \chi_F(\psi_{U(p,q)}) = \exists s > t\psi_{U(p,q)}(s, P, Q)$.

The expressive power of a family of connectives over a flow of time is measured by how many ψ_As it can express over the flow of time. A family of connectives is said to be fully expressive (or functionally complete) over a flow of time iff all $\psi(t, Q_1, \ldots, Q_m)$ can be obtained as ψ_A (up to equivalence) for suitable $A(q_1, \ldots, q_m)$.

Definition 9.1.7 A temporal language with one-dimensional connectives is said to be *expressively complete* or, equivalently, *functionally complete*, in one dimension over a class $\mathcal{T}$ of partial orders iff for any monadic $\psi(t, Q_1, \ldots, Q_m)$, there exists an A of the language such that for any $(T, <)$ from $\mathcal{T}$, for any interpretation h for $q_1, \ldots, q_m$,

$$(T, <) \models \forall t(\psi \leftrightarrow \psi_A)(t, h(q_1), \ldots, h(q_m)).$$

In the cases where $\mathcal{T} = \{(T, <)\}$ we talk of expressive completeness over $(T, <)$. For example, the language of Since and Until is expressively complete over integer time and real number flow of time but not over rational numbers time.

Definition 9.1.8 A flow of time $(T, <)$ is said to be expressively complete (or functionally complete) (in one dimension) iff there exists a *finite* set of (one-dimensional) connectives which is expressively complete over $(T, <)$, in one dimension.

The qualification of one-dimensionality in the definitions above will be explained in the more general terminology of chapter 13.

These notions parallel the definability and expressive completeness of classical logic. We know that in classical logic $\{\sim, \rightarrow\}$ is sufficient to define all other connectives. Furthermore, for any n-place truth table $\psi : 2^n \to 2$ there exists an $A(q_1, ..., q_n)$ of classical logic such that for any h,

$$h(A) = \psi(h(q_1), ..., h(q_n)).$$

This is the expressive completeness of $\{\sim, \to\}$ in classical logic.

The question of the existence of a *finite* expressively complete set of connectives is not trivial. Some flows of time have one and some do not. For example, the class of all branching flows of time (i.e. $<$ irreflexive and transitive) does not allow for any finite expressively complete set of connectives (see chapter 13). The question of expressive completeness is important for applications to computing. In an application area such as the study of execution sequences of parallel programs using temporal logic one may choose to use certain 'natural' temporal connectives. If these can be supplemented to an expressively complete set of connectives then one is assured that one can say whatever one wants on this application area.

9.2 The separation property for linear time

This subsection deals with linear flows of time. So 'equivalent' will mean 'equivalent over linear time', etc.

Consider the wff Fq. This formula 'talks' about the future only. Its truth value at any t, in any $(T, <, h)$ does not depend on the past or the present. We call such a wff *pure future.* We can similarly talk about *pure past wffs* such as, for example, Pq. Some wffs are mixed.

Example 9.2.1 Consider $F(p \wedge Hq)$ being true now as pictured in figure 9.1.

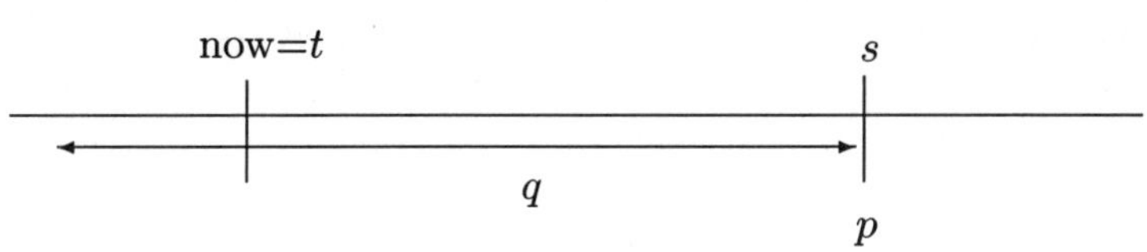

Figure 9.1

$F(p \wedge Hq)$ is true at t iff $\exists s > t$ such that p is true at s and q is true in all points in the past of s. This formula is not pure future because its truth value depends on the past (for it to be true, q must be true at all points in the past of t) and so if we change the past of t we may change the value of $F(p \wedge Hq)$ at t. It is also dependent on the future because a point s must be found in the future of t. So $F(p \wedge Hq)$ is neither pure future nor pure past.

It is equivalent (in linear flows), however, to a boolean combination of wffs, each of which is pure future, past, or atomic. In fact, $F(p \wedge Hq) \equiv Hq \wedge q \wedge U(p,q)$.

It seems reasonable of us to conjecture that any wff talking about t should be equivalent to some boolean combination of wffs talking about the future of t only (pure future), past of t only (pure past), and t itself only (pure present). After all, we don't expect that the descriptions of the future and past are dependent. However, we can only do this decomposition if the language is sufficiently expressive. For take a language with P, F only. As we have seen, this language cannot express U (for otherwise it could express δ). Can we rewrite $F(p \wedge Hq)$ in a different, separated way using only F and P and *not* using U?

No. The following lemma shows that if we could do that then F and P alone could express U. Since we know that they cannot we can deduce that $F(p \wedge Hq)$ cannot be decomposed into a purely future bit, purely past bit, and some atoms.

Lemma 9.2.2 *If in a temporal logic with P, F and possibly other connectives, $F(p \wedge Hq)$ is equivalent over linear time to a boolean combination of pure future wffs, pure past wffs, and atoms, then U is definable in the logic.*

Proof. Assume then that $F(p \wedge Hq) \equiv \beta(A_j^+(p,q), p, q, A_j^-(p,q))$ where β stands for some boolean combination, the $A_j^+(p,q)$ are pure future, and $A_i^-(p,q)$ are pure past wffs. All wffs are built from F,P, atoms, and logical connectives only.

Let us replace p and q by truth in each A_i^- to obtain D_i^- and let $B = \beta(D_i^-, \top, \top, A_j^+)$. We claim that B is equivalent to $U(p,q)$.

We must show that the equivalence holds at all times t in all linear $(T, <, h)$. To see this we define two new interpretations: h', which agrees with h in the future of t but makes p and q true on $\{s \mid s \leq t\}$; and $h*$, which makes p and q true everywhere.

Evaluating D_j^- under h is the same as evaluating A_j^- under $h*$ but, since A_j^- is pure past and $h*$ and h' agree on the past of t, then that in turn is the same as evaluating A_j^- under h'. Similarly, p and q under h' are just $\top$. Finally, A_i^+ must agree under h' and h as they reflect the same future. Thus, at t, B under h must agree with $\beta(A_j^-, p, q, A_i^+)$ under h' and we have $\|B\|_t^h = \|F(p \wedge Hq)\|_t^{h'}$.

Now it is not difficult to see that under an interpretation like h' the right-hand side is just $U(p,q)$ under h'. But $U(p,q)$, being pure future, agrees under h' and h.

We have $\|B\|_t^h = \|U(p,q)\|_t^h$, as required. ■

Let us now give formal definitions to the notions we have used.

Definition 9.2.3

- Let h, h' be two assignments and $t \in T$. We say that h, h' agree on the past of t, $h =_{<t} h'$, iff for any atom q and any $s < t$,

$$s \in h(q) \text{ iff } s \in h'(q).$$

  We similarly define $h' =_{=t} h$ and $h' =_{>t} h$.

- Let $\mathcal{T}$ be a class of linear flows of time and A be a wff in a temporal language over $(T, <)$. We say that A is a *pure past wff over* $\mathcal{T}$, iff for all $(T, <)$ in $\mathcal{T}$, for all $t \in T$,

$$\forall h, h', (h =_{<t} h') \text{ implies that } \|A\|_t^h = \|A\|_t^{h'}.$$

  Similarly, we define *pure future* and *pure present* wffs.

- Let $\mathcal{T}$ be a class of flows of time and A be a wff in a temporal language $\mathbf{L}$. We say A is *separable* in $\mathbf{L}$ over $\mathcal{T}$ iff there exists a wff in $\mathbf{L}$ which is a boolean combination of pure past, pure future, and atomic wffs and is equivalent to A everywhere in any $(T, <)$ from $\mathcal{T}$.

- A set of temporal connectives is said to have the separation property over $\mathcal{T}$ iff every wff in the temporal language of these connectives is separable in the language (over $\mathcal{T}$).

Note that all these definitions depend on (the connectives of) the language and on the flow of time.

We show in the next section the main theorem that for a given set of connectives over a linear flow of time, the property of separation and the property of expressive completeness are essentially equivalent. In other words, you can have full separation for every wff over linear time iff you have full expressive power iff the temporal language is essentially the full monadic predicate logic of the linear flow of time.

In chapter 10 we prove separation (syntactically by actually rewriting each A) for $\{U, S\}$ over integer time and real number time. Then, in chapter 11, we show separation of a more expressive language over the class of all linear flows of time. Thus we obtain the expressive completeness of these languages for the respective flows.

9.3 Separation equals expressive completeness over linear flows

Theorem 9.3.1 *Let* **L** *be a temporal language built from any number (finite or infinite) of connectives in which P and F are definable over a class $\mathcal{T}$ of linear flows of time. Then* **L** *has the separation property over $\mathcal{T}$ implies that* **L** *is expressively complete over $\mathcal{T}$.*

Proof. Assume **L** has separation. We will show that **L** is expressively complete. This is trivial if $\mathcal{T}$ is empty so suppose not. We have to show that for any $\varphi(t, \overline{Q})$ in the monadic theory of linear order with predicate variable symbols $\overline{Q} = (Q_1, ..., Q_n)$, there exists a wff $A = A(q_1, ..., q_n)$ in the temporal language such that for all flows of time $(T, <)$ from $\mathcal{T}$, for all h, t, $\|A\|_t^h = 1$ iff $(T, <) \models \varphi(t, h(q_1), ..., h(q_n))$.

We denote this wff by $A[\varphi]$ and proceed by induction on the depth m of nested quantifiers in φ.

$m = 0$: In this case, $\varphi(t)$ is quantifier free. Just replace each appearance of $t = t$ by $\top$, $t < t$ by $\bot$, and each $Q_j(t)$ by q_j to obtain $A[\varphi]$.

$m > 0$: Assume now that for any wff $\phi(t, \overline{Q})$, with any number of Q_i and depth of nested quantifiers at most m, there exists wff $A[\phi]$. This is the induction hypothesis. We now want to show that the hypothesis holds for $m + 1$. It is enough to treat the case of $\exists x \psi(t, x, \overline{Q})$ where ψ has quantifier depth $\leq m$.

Assuming that we do not use t as a bound variable symbol in ψ and that we have replaced all appearances of $t = t$ by $\top$ and $t < t$ by $\bot$ then the atomic wffs in ψ which involve t have one of the following forms: $Q_i(t)$, $t < y$, $t = y$, or $y < t$, where y could be x or any other variable letter occuring in ψ.

If we regard t as fixed, the relations $t < y, t = y, t > y$ become unary. Introduce new n-ary predicates

- $R_=(y) = (t = y)$,
- $R_>(y) = (t > y)$, and
- $R_<(y) = (t < y)$.

Then ψ can be rewritten equivalently as

$$\psi_0^t(x, \overline{Q}, R_=, R_>, R_<),$$

in which t appears only in the form $Q_i(t)$. Since t is free in ψ, we can go further and prove (by induction on the quantifier depth of ψ) that ψ_0^t can be equivalently rewritten as

$$\psi^t = \bigvee_j [\alpha_j(t) \wedge \psi_j^t(x, \overline{Q}, R_=, R_>, R_<)],$$

where

- $Q_i(t)$ appear only in $\alpha_j(t)$ and not at all in ψ_j^t,
- $\alpha_j(t)$ is quantifier free,
- and each ψ_j^t has quantifier depth $\leq m$.

Each ψ_j^t is a wff with at most m nested quantifers and hence by the induction hypothesis there exists a temporal wff $A_j = A_j(\overline{q}, r_=, r_>, r_<)$ such that for any h, x,

$$\begin{aligned} & \|A_j\|_x^h = 1 \\ \text{iff} \quad & (T, <) \models \psi_j^t(x, h(q_1), ..., h(q_n), h(r_=), h(r_>), h(r_<)). \end{aligned}$$

Now let $\Diamond q$ be an abbreviation for temporal wffs equivalent (over $\mathcal{T}$) to $Pq \vee q \vee Fq$ whose existence in **L** is guaranteed in the statement of the theorem. Then let $B(\overline{q}, r_=, r_>, r_<) = \bigvee_j (A[\alpha_j] \wedge \Diamond A_j)$. $A[\alpha_j]$ can be obtained from the quantifier free case.

In any structure $(T, <)$ from $\mathcal{T}$ for any h interpreting the atoms $\overline{q}$,$r_=$, $r_>$ and $r_<$, we have

$$\begin{aligned} & \|B\|_t^h = 1 \\ \text{iff} \quad & \| \bigvee_j (A[\alpha_j] \wedge \Diamond A_j)\|_t^h = 1 \\ \text{iff} \quad & \bigvee_j (\|A[\alpha_j]\|_t^h = 1 \wedge \| \Diamond A_j\|_t^h = 1) \\ \text{iff} \quad & \bigvee_j (\alpha_j(t) \wedge \exists x (\|A_j\|_x^h = 1)) \\ \text{iff} \quad & \bigvee_j (\alpha_j(t) \wedge \exists x \psi_j^t(x, h(q_1), ..., h(q_n), h(r_=), h(r_>), h(r_<))) \\ \text{iff} \quad & \exists x \bigvee_j (\alpha_j(t) \wedge \psi_j^t(x, h(q_1), ..., h(q_n), h(r_=), h(r_>), h(r_<))) \\ \text{iff} \quad & \exists x \psi_0^t(x, h(q_1), ..., h(q_n), h(r_=), h(r_>), h(r_<)). \end{aligned}$$

Now provided we interpret the r atoms as the appropriate R predicates, i.e. provided that:

- $h^*(r_=) = \{t\}$,
- $h^*(r_<) = \{s \mid t < s\}$, and
- $h^*(r_>) = \{s \mid s < t\}$,

we can continue and obtain

$$\begin{array}{rl} & \|B\|_t^{h^*} = 1 \\ \text{iff} & \exists x \psi(t, x, h^*(q_1), ..., h^*(q_n)) \\ \text{iff} & \varphi(t, h^*(q_1), ..., h^*(q_n)). \end{array}$$

B is almost the $A[\varphi]$ we need except for one problem. B contains, besides the q_i, also three other atoms, $r_=, r_>$, and $r_<$, and equation $(*)$ from definition 9.1.1 above is valid for any h^* which is arbitrary on the q_i but very special on $r_=, r_>, r_<$. We are now ready to use the separation property (which we haven't used so far in the proof). We use separation to eliminate $r_=, r_>, r_<$ from B. Since we have separation B is equivalent to a boolean combination of atoms, pure past wffs, and pure future wffs.

So there is a boolean combination $\beta = \beta(\overline{p}_+, \overline{p}_-, p_0)$ such that

$$B \leftrightarrow \quad \beta(\overline{B}_+, \overline{B}_-, B_0),$$

where B_0 is a combination of atoms, B_{+i} are pure future, and B_{-j} are pure past wffs.

Therefore,

$$\|B\|_t^{h^*} = \|\beta(\overline{B}_+, \overline{B}_-, B_0)\|_t^{h^*}.$$

Consider now $B_{-j}(\overline{q}, r_>, r_=, r_<)$.

Let the interpretation g^* be defined as h^* on the q_i but also by

- $g^*(r_>) = \top$
- $g^*(r_<) = g^*(r_=) = \bot$.

So g^* agrees with h^* on the past of t. Therefore,

$$\|B_{-j}(\overline{q}, r_>, r_=, r_<)\|_t^{g^*} = \|B_{-j}(\overline{q}, r_>, r_=, r_<)\|_t^{h^*},$$

but under g^*,$r_>$ is $\top$, and $r_=, r_<$ are $\bot$. Substitute these values in B_{-j} and obtain B^*_{-j}, i.e. $B^*_{-j} = B_{-j}(\overline{q}, \top, \bot, \bot)$, and obtain that

$$\|B_{-j}\|_t^{h^*} = \|B_{-j}\|_t^{g^*} = \|B^*_{-j}\|_t^{h^*}.$$

We can similarly argue this for $B^*_{+i} = B_{+i}(\overline{q}, \bot, \top, \bot)$ where we substitute $r_> = \bot, r_< = \top, r_= = \bot$ in B_{+i} and for B^*_0, where we obtain a similar equation by letting $r_= = \top$ and $r_< = r_> = \bot$.

Let $B^* = \beta(B^*_{+i}, B^*_{-j}, B^*_0)$. Then we obtain for any h^*,

$$\begin{array}{rc} & \|B\|_t^{h^*} \\ = & \beta(\|B_{+i}\|_t^{h^*}, \|B_{-j}\|_t^{h^*}, \|B_0\|_t^{h^*}) \\ = & \beta(\|B^*_{+i}\|_t^{h^*}, \|B^*_{-j}\|_t^{h^*}, \|B_0\|_t^{h^*}) \\ = & \|B^*\|_t^{h^*}. \end{array}$$

Hence

$$\|B^*\|_t^{h^*} = 1 \quad \text{iff} \quad (T, <) \models \varphi(t, h^*(\overline{q})).$$

This equation holds for any h^* arbitrary on $\overline{q}$, but restricted on $r_<, r_>, r_=$. But $r_<, r_>, r_=$ do not appear in it at all and hence we obtain that for any h, $\|B^*\|_t^h = 1$ iff $(T, <) \models \varphi(t, h^*(\overline{q}))$.

Let $A[\varphi] = B^*$ and we have completed the induction step and the proof of the theorem.

■

Now we want to prove the other direction, namely that if **L** is expressively complete it is separated. The idea is to use lemma 9.1.5.

Given A, we take ψ_A(its table), which is a wff of the monadic theory of $(T, <)$. We separate ψ_A, using properties of the first-order theory of $(T, <)$ and then we use the expressive completeness to find suitable $A_i^{\pm}$ for the separated parts. Thus we achieve separation for A. We therefore need some lemmas on the monadic first-order language of $(T, <)$ that will show how to separate any ψ.

Lemma 9.3.2 *Let ψ be a wff of the monadic language. Then for any set $(x_1, ..., x_n)$ of variable symbols containing all of ψ's free ones, under the assumption $x_1 < x_2 < ... < x_n$ we have ψ equivalent to a disjunction of conjunctions of wffs of the form*

$$\bigvee_i (\varphi[i, < x_1] \wedge \varphi[i, x_1] \wedge \varphi[i, x_1, x_2] \wedge \varphi[i, x_2] \wedge ... \varphi[i, x_n] \wedge \varphi[i, > x_n]) .$$

where

- *in $\varphi[i, y]$ only y is free and φ is quantifier free.*
- *$\varphi[i, < y]$ is a wff in which only y is free and all quantifiers are relativized to $< y$.*
- *Similarly, $\varphi[i, > y]$.*

- *$\varphi[i, x < y]$ is a wff with only x, y free and all quantifiers are relativized to between x and y.*

Proof. Proceed by induction on the quantifier depth of ψ. We assume $x_1 < x_2 < ... < x_n$ is some ordering of distinct variables including all of ψ's free ones.

The quantifier free case: Here, by putting ψ in disjunctive normal form, we see that ψ is equivalent to a disjunction (over i) of conjunctions of the form $\varphi[i, x_1] \wedge \varphi[i, x_2] \wedge ... \wedge \varphi[i, x_n]$.

Inductive case: Here, we may assume that ψ is an existentially quantified wff. Otherwise write ψ as a boolean combination of such wffs (just rearrange the top level boolean connectives so as not to change any quantifier depths), use the proof below to rewrite each of the boolean constituents, and put the whole resulting wff into disjunctive normal form. This will be in the right form as $\sim\varphi$ is an acceptable conjunct exactly when φ is.

So suppose $\psi = \exists y \varphi(y, x_1, ..., x_n)$. Since we can assume $x_1 < x_2 < ... < x_n$ we have

$$\begin{array}{rl} \psi \leftrightarrow & \\ & (\exists y < x_1)\varphi \\ \vee & \bigvee_{i=1}^{n} \varphi(x_i, x_1, ..., x_i, ..., x_n) \\ \vee & \bigvee_{i=1}^{n-1} \exists y((x_i < y < x_{i+1}) \wedge \varphi) \\ \vee & (\exists y > x_n)\varphi. \end{array}$$

Assume $x_1 < ... < x_i < y < x_{i+1} < ... < x_n$. Then $\varphi(y, x_1, ..., x_n)$ is equivalent by the induction hypothesis to $\bigvee_k \varphi_k$ where

$$\begin{array}{rl} \varphi_k = & \varphi[k, < x_1] \wedge \varphi[k, x_1] \wedge \varphi[k, x_1, x_2] \\ & \wedge ... \wedge \varphi[k, x_i, y] \wedge \varphi[k, y] \wedge \varphi[k, y, x_{(i+1)}] \wedge .... \end{array}$$

Hence $\exists y(x_i < y < x_{i+1} \wedge \bigvee_k \varphi_k)$ is equivalent to

$$\begin{array}{rl} \bigvee_k & [\varphi[k, < x_1] \wedge \varphi[k, x_1] \wedge .. \wedge \varphi[k, x_i] \\ & \wedge \exists y(x_i < y < x_{i+1} \wedge \varphi[k, x_i, y] \wedge \varphi[k, y] \wedge \varphi[k, y, x_{i+1}]) \\ & \wedge \varphi[k, x_{i+1}, x_{i+2}] \wedge ...]. \end{array}$$

We can obtain a similar expression for $(\exists y < x_1)\varphi$, $(\exists y > x_n)\varphi$ and for the cases when y is equal to some x_i.

We thus obtain ψ equivalent to a disjunction of the required form. This completes the induction step.

■

Corollary 9.3.3 *Let $\psi(t)$ be a wff with t free only. Then ψ is equivalent to a disjunction of the conjunctions of the form*

$$\bigvee_k (\psi[k, < t] \wedge \psi[k, t] \wedge \psi[k, > t]).$$

Theorem 9.3.4 *Let* **L** *be expressively complete over linear time. Then* **L** *is separated.*

Proof. Assume **L** is expressively complete and show **L** has separation. Let A be any wff of **L**. Then there exists, by lemma 9.1.5 of section 1, a monadic $\psi(t, \overline{Q})$ such that for any $(T, <, h)$,

$$\|A\|_t^h = 1 \text{ iff } (T, <) \models \psi(t, h(q_1), ..., h(q_n)).$$

We have shown above that the ψ can be separated. Thus ψ is equivalent to

$$\bigvee_k \psi[k, < t] \wedge \psi[k, t] \wedge \psi[k, > t].$$

By the expressive completeness of **L** let

$$A^* = \bigvee_k (A[\psi[k, < t]] \wedge A[\psi[k, t]] \wedge A[\psi[k, > t]]),$$

which is equivalent to A, since for all h and t,

$$\|A\|_t^h = 1 \text{ iff } (T, <) \models \psi(t) \text{ iff } \|A^*\|_t^h = 1.$$

By examining the forms of their tables (e.g. $\psi[k, < t]$) we deduce that each $A[\psi[k, < t]]$ is pure past, each $A[\psi[k, t]]$ pure present, and each $A[\psi[k, > t]]$ pure future. We thus also have that A^* is separated. ■

9.4 The generalized separation property

By careful examination of the separation theorem, we can extract the relevant ideas and prove a general separation theorem which will allow us to deduce expressive completeness of a language over any class of—even non-linear— flows of time provided that we can demonstrate a generalized separation property. A proof of the result below appears in [Amir, 1985].

To do this, we will have to have some relations which, playing the role of $<$, $>$ and $=$ in the linear case, separate time into disjoint regions like past, present and future.

Definition 9.4.1 Let $\varphi_i(x, y), i = 1, ..., n$ be n given formulae in the monadic language with $<$ and let $\mathcal{T}$ be a class of flows of time.

Assume the following:

- for every t in each $(T,<)$ in $\mathcal{T}$ the sets $T(i,t) = \{s \in T \mid \varphi_i(s,t)\}$ for $i = 1,...,n$ are mutually exclusive and $\bigcup_i T(i,t) = T$;
- for each i there is a formula $B(t,x)$ such that $\varphi_i(t,x)$ and $B(t,x)$ are equivalent over $\mathcal{T}$ and B is a boolean combination of some $\varphi_j(x,t)$; and
- $<$ and $=$ can be equivalently expressed (over $\mathcal{T}$) as boolean combinations of the φ_i.

Then we have the following series of definitions:

- For any t from any $(T,<)$ in $\mathcal{T}$, for any $i = 1,...,n$, we say that truth functions h and h' agree on $T(i,t)$ if and only if $h(q)(s) = h'(q)(s)$ for all s in $T(i,t)$ and all atoms q.
- We say that a formula A is *pure* φ_i over $\mathcal{T}$ if for any $(T,<)$ in $\mathcal{T}$, any $t \in T$ and any two truth functions h and h' which agree on $T(i,t)$, we have
$$\|A\|_t^h = \|A\|_t^{h'}.$$
- The logic $\mathbf{L}$ has the *generalized separation property* over $\mathcal{T}$ iff every formula A of $\mathbf{L}$ is equivalent over $\mathcal{T}$ to a boolean combination of pure formula.

Theorem 9.4.2 (generalized separation theorem) *If* $\mathbf{L}$ *has the generalized separation property over a class* $\mathcal{T}$ *of flows of time and if each of the* 1*-place connectives* $\#_i$ *defined below can be expressed equivalently in* $\mathbf{L}$ *then* $\mathbf{L}$ *is expressively complete over* $\mathcal{T}$.

For $i = 1,...,n$ *define* $\#_i$ *by*

$$\|\#_i(p)\|_t^h = 1 \quad \text{iff} \quad \exists s \quad \varphi_i(s,t) \text{ holds in } (T,<) \text{ and } \|p\|_s^h = 1.$$

Proof. In the proof let 'equivalent' be read as 'equivalent over $\mathcal{T}$'.

It is clear that because of the first condition we have assumed holds on the ϕ_i, any boolean combination of these formulae is equivalent to a disjunction of them. It follows that there are subsets $K_=$ and $K_<$ and J_i for each i of $\{1,...,n\}$ such that $x = y$ is equivalent to $\bigvee_{k \in K_=} \phi_k$, $x < y$ is equivalent to $\bigvee_{k \in K_<} \phi_k$ and for each i, $\phi_i(t,s)$ is equivalent to $\bigvee_{j \in J_i} \phi_j(s,t)$.

In fact, without loss of generality we may assume that $\varphi_1(x,y)$ is $x = y$. To see this just throw away all the ϕ_k for $k \in K_=$ and use $x = y$ for $\phi_1(x,y)$. The generalized separation property is preserved as any formula which was pure ϕ_k for some old $k \in K_=$ is now pure ϕ_1.

Next we note that we are best using a slightly different language than the usual monadic one. We suppose that instead of $=$, $<$ and $Q_i(x)$ our atomic formula are now ϕ_i or $Q_i(x)$. To prevent confusion we use a binary symbol c_i instead of each ϕ_i. $(T, <)$ becomes a $\{c_1, ..., c_n\}$-structure just by interpreting c_i as ϕ_i. It is clear that appearences of $=$ and $<$ can be equivalently rewritten in the new language and vice versa.

We are going to show that any monadic formula in $\{=, <\}$ with one free variable t has a temporal equivalent in L by induction on m using the following hypothesis:

- for every formula $\phi(t, Q_1, ..., Q_k)$ using the language with $c_1, ..., c_n$ but having quantifier depth at most m there is a temporal formula A in L such that $\|A\|_y^h = 1$ if and only if $(T, <) \models \psi(t, h(q_1), ..., h(q_n))$.

Base Case: First assume $\varphi(t)$ is quantifier free. Just replace each appearance of $c_1(t,t)$ by $\top$, $c_i(t,t)$ for $i > 1$ by $\bot$, and $Q_j(t)$ by q_j and we have A as required. There are no other atomic formulae.

Induction Case: Assume that the induction hypothesis holds for m. We need only consider the case when ϕ is $\exists x \psi(t, x, Q_1, ..., Q_k)$ and when the depth of nesting of quantifiers in ψ is not greater than m.

Since t is free in ψ, we can prove (by induction on the quantifier depth of ψ) that $\exists x \psi(t, x)$ is equivalent to

$$\bigvee_l [\alpha_l \wedge \exists x \psi_l(t, x)],$$

where

- $Q_i(t)$ appear only in $\alpha_l(t)$ and not at all in ψ_l,
- $\alpha_l(t)$ is quantifier free,
- and each ψ_l has quantifier depth $\leq m$.

We can also suppose that t is not used as a bound variable symbol. Further, by replacing $c_1(t,t)$ by $\top$ and $c_i(t,t)$ for $i > 1$ by $\bot$ and by rewriting each appearance of $c_i(t, y)$ as a boolean combination (hence not effecting the quantifier depth) of some $c_j(y, t)$ we can guarantee that the only atomic formulae in ψ_l are

- $Q_i(y)$ for $i = 1, ..., k$,
- $c_i(y, t)$ for $i = 1, ..., n$, and
- $c_i(y, y')$ for $i = 1, ..., n$,

where y and y' are variables different from t. So the only atoms in ψ_l which include the variable t are $c_i(y,t)$ for $i = 1, ..., n$.

We are going to proceed for a while now as if t is fixed. This will allow us to pretend that we only have a free variable x and to use the induction hypothesis. To do this, introduce monadic symbols $R_1, ..., R_n$ and substitute $R_i(y)$ for each $c_i(y,t)$ in ψ_l to obtain $\psi_l^*(x, \overline{Q}, R_1, ..., R_n)$. As the quantifier depth in ψ_l^* is not greater than m the induction hypothesis tells us that there is a formula

$$B_l(\overline{q}, r_1, ..., r_n)$$

in **L** such that

$$\|B_l\|_x^g = 1 \text{ iff } (T,<) \models \psi_j^*(x, g(q_1), ..., g(q_k), g(r_1), ..., g(r_n)).$$

If each $g(r_i)$ is just the extension of $c_i(\_, t)$ which is $(T(i,t)$ then we have $\|B_l\|_x^g = 1$ iff $(T,<) \models \psi(t, x, g(q_1), ..., g(q_k))$.

Now, as in the linear case, we are going to use separation to show that we do not really need the atoms $r_1, ..., r_n$ anyway.

First we reduce our considerations to one new atom r. Replace r_1 in B_l by r and each r_i for $i > 1$ by $\bigvee_{j \in J_i} \#_j(r)$ to obtain A_l where we recall that $\phi_i(t,s) \equiv \bigvee_{j \in J_i} \phi_j(s,t)$.

Now, given a valuation h of the original atoms and r, and a $t \in T \in \mathcal{T}$ such that

$$(*) \qquad h(\{r\}) = \{t\}$$

then we can extend h to g on $r_1, ..., r_n$ by $g(r_i) = T(i,t)$. Thus

$$\begin{array}{ll} & \|r_i\|_s^g = 1 \\ \text{iff} & s \in T(i,t) \\ \text{iff} & \phi_i(s,t) \text{ holds in } (T,<) \\ \text{iff} & \bigvee_{j \in J_i} \phi_j(t,s) \text{ holds in } (T,<) \\ \text{iff} & \text{for some } j \in J_i, \phi_j(t,s) \text{ holds in } (T,<) \\ \text{iff} & \exists j \in J_i, \exists u \in T, \phi_j(t,s) \text{ holds in } (T,<) \text{ and } \|r\|_u^h = 1 \\ \text{iff} & \exists j \in J_i, \|\#_j(r)\|_s^h = 1 \\ \text{iff} & \|\bigvee_{j \in J_i} \#_j(r)\|_s^h = 1. \end{array}$$

A simple induction on the construction of B_l then shows that $\|B_l\|_s^g = 1$ if and only if $\|A_l\|_s^h = 1$.

If we now take $A = \bigvee_l D_l \wedge (\bigvee_{i=1}^n \#_i(A_l))$ where D_l is obtained from α_l by replacing $Q_i(t)$ with q_i.

Then, for all h, t such that $h(r) = \{t\}$,

$$\begin{array}{ll} & \|A\|^h_t = 1 \\ \text{iff} & \exists l, \|D_l\|^h_t = 1 \text{ and } \exists i = 1, ..., n \text{ s.t. } \|\#_i(A_l)\|^h_t = 1 \\ \text{iff} & \exists l, (T, <) \models \alpha_l(t, h(q_1), ..., h(q_k)) \\ & \quad \text{and } \exists i = 1, ..., n, \exists s \in T, \phi_i(s, t) \text{holds in} (T, <) \text{ and } \|A_l\|^h_s = 1 \\ \text{iff} & \exists l, (T, <) \models \alpha_l(t, h(\overline{q})) \\ & \quad \text{and } \exists i, \exists s \in T, \phi_i(s, t) \text{ and } \|B_l\|^g_s = 1 \\ \text{iff} & \exists l, (T, <) \models \alpha_l(t, h(\overline{q})) \\ & \quad \text{and } \exists i, \exists s \phi_i(s, t) \text{and} (T, <) \models \psi_l(t, s, h(\overline{q})) \\ \text{iff} & (T, <) \models \bigvee_l (\alpha_l \wedge \exists x \psi_l(t, x)) \\ \text{iff} & (T, <) \models \phi \end{array}$$

as required.

Finally we get rid of r.

Because of the separation property, A is equivalent to a boolean combination $\beta(E_1, ..., E_L)$ of pure formulae E_i.

We define a new formula E'_i for each E_i. There are two cases. If E_i is pure ϕ_1 then make E_i by replacing each occurrence of r by $\top$. If E_i is pure ϕ_j for some $j > 1$ then replace r by $\bot$ to obtain E'_i. It is easy to see that because of their purity we have in either case $\|E_i\|^h_t = 1$ iff $\|E'_i\|^h_t = 1$ for any valuation h which satisfies $(*)$. For such a valuation we have $\|\beta(E'_1, ..., E'_L)\|^h_t = 1$ iff $(T, <) \models \phi(t, h(\overline{q}))$. But now the temporal formula $\beta(E'_1, ..., E'_L)$ does not contain r and so we can drop the restriction on valuations and we have finished our induction step.

This completes the proof of the theorem. ■

9.5 Separation for general time

As an example of the use of the generalized separation theorem of the last section let us prove a more specific result after choosing certain φ_i. For each $i = 1, ..., 4$, let φ_i be given as follows:

- $\varphi_1(x, y) = (x = y)$,
- $\varphi_2(x, y) = (x < y)$,
- $\varphi_3(x, y) = (y < x)$, and
- $\varphi_4(x, y) = (x \# y) = \sim [(x = y) \vee (x < y) \vee (y < x)]$.

Definition 9.5.1 Let D be the connective given by

$$\|Dq\|_t^h = 1 \text{ iff } \exists x \# t \|q\|_x^h = 1.$$

and $\mathcal{T}$ be any class of flows of time. Then A is said to be *pure parallel* over a class $\mathcal{T}$ of flows of time iff for all t from any $(T, <)$ from $\mathcal{T}$, for all $h =_{\#t} h'$,

$$\|A\|_t^h = \|A\|_t^{h'},$$

where $h =_{\#t} h'$ iff $\forall x \# t \forall q (x \in h(q) \leftrightarrow x \in h'(q))$.

It is clear what separation means in the context of pure present, past, future, and parallel.

Corollary 9.5.2 *Let* **L** *be a language with* F, P, D *over any class of flows of time. Then* **L** *has a separation implies* **L** *is expressively complete.*

Proof. It is simple to check that the φ_i satisfy the general separation property and other preconditions for using the generalized separation theorem. Thus the theorem gives us our result immediately.

■

Chapter 10

Expressive completeness of Since and Until over integer and real time

10.1 Introduction

In this chapter we will show how the separation theorem — theorem 9.3.1 — can be used to prove the expressive completeness of the language with Until and Since over some very important flows of time. To do this we need to show the separation of the language over the flows. First, in section 10.2, we examine the discrete Dedekind complete flows of time, concentrating on the integers. This serves as a simple introduction to the techniques which we then use in section 10.3 to show separation of this language over any class of Dedekind complete flows of time. The separation procedure is performed in two stages: first some basic equivalences, called *eliminations*, are shown; then we show how to put eliminations together in the right order to achieve separation.

10.2 Separation for S, U over integer time

In this section we prove that the language with the connectives S and U has the separation property over the flow of time of the integers. This proof appeared in [Gabbay, 1989a]. It can be seen that the proof will also work for the natural numbers.

Wffs of the language may contain nested occurrences of S and U, one inside the other, as does, for example, $S(a \wedge U(b, c \vee S(x,y)), d)$. We shall describe a syntactic procedure for pulling out the Us systematically from inside the Ss (and vice versa depending on the case at hand) until we arrive at an equivalent wff with the property that no U is in the scope of an S and no S is in the scope of a U. A wff with this property is already separated, because any wff beginning with U and containing only atoms and Us is pure future. Anything with S only is pure past and the atoms are of course pure present. The process is very simple. As we shall see there are eight cases of nested occurrences of U within an S to worry about. Each case is not difficult to deal with and this should practically finish the proof since for any given A, we can keep on eliminating Us and Ss as required.

First we need some equivalence of relations over flow of time.

Lemma 10.2.1 *The following are valid over linear time:*

- $U(A \vee B, C) \leftrightarrow U(A, C) \vee U(B, C)$;
- $S(A \vee B, C) \leftrightarrow S(A, C) \vee S(B, C)$;
- $U(A, B \wedge C) \leftrightarrow U(A, B) \wedge U(A, C)$;
- $S(A, B \wedge C) \leftrightarrow S(A, B) \wedge S(A, C)$.

Proof. Simple. ■

Lemma 10.2.2 *The following hold over integer time:*

- $\sim U(A, B) \leftrightarrow G(\sim A) \vee U(\sim A \wedge \sim B, \sim A)$;
- $\sim S(A, B) \leftrightarrow H(\sim A) \vee S(\sim A \wedge \sim B, \sim A)$;
- $\sim U(A, B) \leftrightarrow G(\sim A) \vee U(\sim A \wedge \sim B, B \wedge \sim A)$;
- $\sim S(A, B) \leftrightarrow H(\sim A) \vee S(\sim A \wedge \sim B, B \wedge \sim A)$.

Proof. Simple. ■

Now we describe some more equivalences which we will call *eliminations* because they allow us to eliminate a nested occurrence of a U from beneath an S.

It should be noted that all the results in the rest of this section have dual results in which U and S exchange roles.

Lemma 10.2.3 *Let a, q, A, and B be atoms. Consider the following wffs:*

1. $S(a \wedge U(A, B), q)$,

2. $S(a \wedge \sim U(A, B), q)$,

3. $S(a, q \vee U(A, B))$,

4. $S(a, q \vee \sim U(A, B))$,

5. $S(a \wedge U(A, B), q \vee U(A, B))$,

6. $S(a \wedge \sim U(A, B), q \vee U(A, B))$,

7. $S(a \wedge U(A, B), q \vee \sim U(A, B))$, *and*

8. $S(a \wedge \sim U(A, B), q \vee \sim U(A, B))$.

Each of the wffs above is equivalent, over integer time, to another wff in which the only appearances of the until connective are as the wff $U(A, B)$ and no appearance of that wff is in the scope of an S.

Proof. Some proofs here involve semantic argument, some involve just the use of other eliminations and equivalences from preceding lemmas and some involve both types of argument. We will only present some of the details.

1. It is very easy to show that $S(a \wedge U(A, B), q)$ is equivalent to

$$\begin{array}{cc} & S(a, q) \wedge S(a, B) \wedge B \wedge U(A, B) \\ \vee & [A \wedge S(a, B) \wedge S(a, q)] \\ \vee & S(A \wedge q \wedge S(a, B) \wedge S(a, q), q). \end{array}$$

   To see this notice that the original wff holds at t iff there is $s < t$ and $u > s$ such that a holds at s, A at u, B everywhere between s and u, and q everywhere between s and t. The three disjuncts correspond to the cases $u > t$,$u = t$, and $u < t$ respectively. Note that we make essential use of the linearity of time.

2. $S(a \wedge \sim U(A, B), q)$ can be rewritten by using lemmas 10.2.1 and 10.2.2, noting that $S(a \wedge G(\sim A), q)$ is equivalent to $S(a, \sim A \wedge q) \wedge (\sim A) \wedge G(\sim A)$, using the first elimination and then using lemma 10.2.2 again (in reverse).

   We end up with $S(a \wedge \sim U(A, B), q)$ equivalent to

$$\begin{array}{cc} & [S(a, q \wedge \sim A) \wedge \sim A \wedge \sim U(A, B)] \\ \vee & [\sim A \wedge \sim B \wedge S(a, \sim A \wedge q)] \\ \vee & S(\sim A \wedge \sim B \wedge q \wedge S(a, \sim A \wedge q), q). \end{array}$$

3. To separate $S(a, q \vee U(A, B))$ we look at its negation and use lemma 10.2.2 and then elimination (2) above to obtain:

$$S(a, q \vee U(A, B))$$

$$\begin{array}{lll} \leftrightarrow & \sim(& H(\sim a) \\ & \vee & [S(\sim a \wedge \sim q, \sim a \wedge \sim A) \wedge \sim A \wedge (\sim U(A, B) \vee \sim B)] \\ & \vee & S(\sim A \wedge \sim B \wedge \sim a \wedge S(\sim a \wedge \sim q, \sim A \wedge \sim a), \sim a) \qquad). \end{array}$$

4. The case of $S(a, q \vee \sim U(A, B))$ is very straightforward to do semantically. It is equivalent to

$$\begin{array}{ll} & S(a, \sim a \wedge [S(\sim q \wedge \sim a, \sim a \wedge B) \rightarrow \sim A]) \\ \wedge & (S(\sim q \wedge \sim a, \sim a \wedge B) \rightarrow \sim[A \vee (B \wedge U(A, B))]). \end{array}$$

5. $S(a \wedge U(A, B), q \vee U(A, B))$ is equivalent to

$$\begin{array}{ll} & S(a, B) \wedge [A \vee (B \wedge U(A, B))] \\ \vee & S(A \wedge S(a, B), A \vee B \vee \sim S(\sim q, \sim A)) \\ & \wedge [A \vee (B \wedge U(A, B))] \wedge \sim S(\sim q, \sim A). \end{array}$$

The first disjunct holds when the A from $U(A, B)$ is true in the future or present and the second when it is true in the past.

6. The case of $S(a \wedge \sim U(A, B), q \vee U(A, B))$. Let s be the time in the past indicated by the S. The wff is separated by considering when the first occurrence (if any) of $\sim B$ after s is. We obtain an equivalent wff with two disjuncts:

$$\begin{array}{ll} & [S(a, q \wedge \sim A) \wedge \sim A \wedge \sim(B \wedge U(A, B))] \\ \vee & S(\sim B \wedge \sim A \wedge (q \vee U(A, B)) \wedge S(a, q \wedge \sim A), q \vee U(A, B)). \end{array}$$

Eliminations (3) and (5) can be used to finish the separating.

7. $S(a \wedge U(A, B), q \vee \sim U(A, B))$ By considering when A is true we deduce that our formula is equivalent to the disjunction:

$$\begin{array}{ll} & [S(A \wedge (q \vee \sim U(A, B)) \wedge S(a, B \wedge q), q \vee \sim U(A, B))] \\ \vee & [S(a, B \wedge q) \wedge A] \\ \vee & [S(a, B \wedge q) \wedge B \wedge U(A, B)]. \end{array}$$

The first disjunct can be further eliminated by eliminations (8) and (4).

8. The case $D = S(a \wedge \sim U(A,B), q \vee \sim U(A,B))$ can be reduced to cases already discussed since

$$\begin{aligned}\sim S(a \wedge x, q \vee y) \leftrightarrow\ & H(\sim a \vee \sim x) \\ \vee\ & S(\sim q \wedge \sim y \wedge \sim a, \sim a \vee \sim x) \\ \vee\ & S(\sim q \wedge \sim y \wedge \sim x, \sim a \vee \sim x).\end{aligned}$$

Substituting $y = x = \sim U(A,B)$ we obtain

$$\begin{aligned}\sim D \leftrightarrow & \\ & H(\sim a \vee U(A,B)) \\ \vee\ & S(\sim q \wedge \sim a \wedge U(A,B), \sim a \vee U(A,B)) \\ \vee\ & S(\sim q \wedge U(A,B), \sim a \vee U(A,B)).\end{aligned}$$

Notice the last disjunct in $\sim D$ is redundant.

These are cases we can handle by other eliminations, especially elimination (5).

■

We now know the basic steps in our proof of separation. We simply keep pulling out Us from under the scopes of Ss and vice versa until there are no more. Given a wff A, this process will eventually leave us with a *syntactically separated* wff, i.e. a wff B which is a boolean combination of atoms, wffs $U(E,F)$ with E and F built without using S and wffs $S(E,F)$ with E and F built without using U. Clearly, such a B is separated.

Lemma 10.2.4 *Suppose that A and B are wffs in which U and S do not appear and that both wffs C and F are such that each (if any) appearance of U in either of them is as $U(A,B)$ and is not nested under any Ss.*

Then $S(C,F)$ is equivalent to a syntactically separated wff in which U only appears as the formula $U(A,B)$.

Proof. If $U(A,B)$ does not appear then we are done. Otherwise, by rearrangement of C and F into disjunctive and conjunctive normal form, respectively, and repeated use of lemma 10.2.1 we can rewrite $S(C,F)$ equivalently as a boolean combination of wffs $S(C_1, C_2)$ with no U appearing and wffs of the form either:

- $S(C_1 \wedge \pm U(A,B), C_2)$,
- $S(C_1, C_2 \vee \pm U(A,B))$, or
- $S(C_1 \wedge \pm U(A,B), C_2 \vee \pm U(A,B))$,

for some boolean combinations C_1 and C_2 of atoms and pure past formulae. Now the preceding lemma shows that each such boolean constituent is equivalent to a boolean combination of formulae of the following three forms:

- $S(X, Y)$ where X and Y are built from C_1, C_2, A, and B just using S and not U,
- C_1, C_2, A, and B, and
- $U(A, B)$.

Thus we have a separated equivalent. ■

Next let us begin the inductive process of removing Us from under Ss. We break the induction into a series of step-by-step lemmas.

First consider the case of one level of U beneath Ss and where U only appears in copies of one subformula $U(A, B)$.

Lemma 10.2.5 *Suppose that A and B are wffs built without S or U and that the only appearance of U in the wff D is as $U(A, B)$.*

Then D is equivalent to a syntactically separated wff in which U only appears as the formula $U(A, B)$.

Proof. By induction on the maximum number k of nested Ss above any $U(A, B)$.

Case $k = 0$: In this case D is already separated.

Case $k > 0$: Apply the preceding lemma to each of the most deeply nested $S(C, F)$ in which $U(A, B)$ appear and then we have an equivalent wff in which the maximum depth of nesting of $U(A, B)$ is reduced. Note also that U still only appears in the form $U(A, B)$. The induction hypothesis then gives us the result.

■

Now consider one level of U beneath Ss.

Lemma 10.2.6 *For each $i = 1, ..., n$ let A_i and B_i be wffs built without S or U. Suppose that the only appearances of U in the wff D are in the form $U(A_i, B_i)$.*

Then D is syntactically separable.

Proof. Proceed by induction on n:

Case $n = 1$: This is the preceding lemma.

Case $n > 1$: First we separate only for $U(A_n, B_n)$. For each $i = 1, ..., n-1$, we start this by replacing each appearance of $U(A_i, B_i)$ by a new atom q_i to obtain a wff D'. The preceding lemma gives us syntactically separated E' equivalent to D' with U appearing only as $U(A_n, B_n)$.

E' is separated and so is a boolean combination of atoms, of pure future wffs (i.e. $U(A_n, B_n)$) and pure past wffs D_j which are built from atoms including $q_1, ..., q_{n-1}$, those in A_n and B_n, and others of D. Note that $U(A_n, B_n)$ does not appear in any D_j. Now substitute $U(A_i, B_i)$ for each q_i ($i = 1, ..., n-1$) in each D_j and, using the induction hypothesis, separate them. Also substituting $U(A_i, B_i)$ for any other q_is gives us our result.

■

Next we allow nestings of Us beneath nestings of Ss but no S within a U.

Lemma 10.2.7 *Suppose that wff D contains no S nested within a U. Then D is syntactically separable.*

Proof. By induction on the maximum depth n of nesting of Us beneath an S.

Case $n = 1$: This is the case of the preceding lemma.

Case $n > 1$: Let $U(A_i, B_i)$ ($i = 1, ..., N$) be some subformulae of D such that every appearance of U in D is as a subformula of an appearance of one of the $U(A_i, B_i)$. Each A_i and B_i are built up as a boolean combination from wffs of the form $U(X_{ij}, Y_{ij})$ and atoms. Replace each $U(X_{ij}, Y_{ij})$ in A_i and B_i by the new atom x_{ij} to form wffs A_i' and B_i' which are just boolean combinations of atoms. Thus when we substitute x_{ij} by $U(X_{ij}, Y_{ij})$ in A_i' and B_i' we obtain A_i and B_i respectively.

Replace each occurrence of $U(A_i, B_i)$ (which is not contained within another $U(A_j, B_j)$) in D by $U(A_i', B_i')$ to obtain D', which can be separated by the preceding lemma. Let E' be its separated form. E' will be a boolean combination of atoms (including the x_{ij}), pure future formulae (like $U(A_i', B_i')$) and pure past formulae (for example A_i's and B_i's nested under Ss). Furthermore, when we substitute in x_{ij} by $U(X_{ij}, Y_{ij})$ we obtain a wff E_1 equivalent to D.

Unfortunately, E_1 is not separated: what were pure past formulae in E' have become, on replacement of x_{ij} by $U(X_{ij}, Y_{ij})$, impure. To correct this we use the induction hypothesis on each of these pure past subformula of E_1. It is clear that we can do so as the level of nesting of U in $U(A_i, B_i)$ must be strictly greater than that in its subformula $U(X_{ij}, Y_{ij})$.

■

Lemma 10.2.8 *Any wff of the language with U and S is syntactically separable over the integer flow of time.*

Proof. We first introduce some notation. Suppose that B is a subformula of wff A in the language with U and S. We can define the *junction depth* (≥ 0) of an appearance B in A as follows. If $C_1, ..., C_n$ are subformulae of A such that

- B is a subformula of C_1,
- each C_i is a subformula of C_{i+1},
- each C_i is either an Until (of the form $U(D, E)$) or a Since (of the form $S(D, E)$), and
- the C_is alternate between Until's and Since's

then the junction depth of the appropriate appearance of B in A is at least n. If the junction depth of a certain appearance of B in A is at least n but not at least $n+1$, then it is n. The *junction depth* of a wff is the maximum junction depth of any of its appearances of subformulae.

For example, in

$$S(a \wedge U(A, S(C, D)), S(S(C, D), E))$$

the junction depth of the first appearance of C is 3 while its second appearance has junction depth 1. The junction depth of the whole formula is 3.

Now we begin the proof, being given a wff D. We proceed by induction on the junction depth of D.

If it is zero or one then D is already syntactically separated. So assume the induction hypothesis and that the junction depth of D is at least two.

D is a boolean combination of atoms, wffs of the form $S(D_1, D_2)$ and wffs of the form $U(D_1, D_2)$. We are done when we syntactically separate the latter two forms. Because of the dual nature of the results so far we need only demonstrate the syntactic separation of a wff of the form $D = S(D_1, D_2)$.

Let $U(A_i, B_i)$ $(i = 1, ..., N)$ be the subformula covering the maximal appearances of U, i.e. every appearance of U in D is as a subformula of an appearance of one of the $U(A_i, B_i)$.

Since the junction depth of D is at least two, there are some subformula of some $U(A_i, B_i)$ which are of the form $S(E, F)$. Replace each maximal such subformula in $U(A_i, B_i)$ by its own new atom x_{ij} to obtain $U(A_i', B_i')$. Change $S(D_1, D_2)$ into E' by replacing each $U(A_i, B_i)$ by $U(A_i', B_i')$. The preceding lemma now tells us how to separate E' into a wff E_1'.

If we resubstitute the original wffs for each x_{ij} then we will have a formula equivalent to $S(D_1, D_2)$ but of one less junction depth and we may use the induction hypothesis.

■

Theorem 10.2.9 (separation theorem) *Each wff in the language with $\{U, S\}$ is equivalent, over the integer flow of time, to a separated wff.*

Proof. This follows directly from the preceding lemma because, as we have already noted, syntactic separation implies separation. ■

Theorem 10.2.10 *The language $\{U, S\}$ is expressively complete over integer time.*

Proof. This follows from the separation theorem and the results of chapter 9. ■

10.3 Separation for U and S over Dedekind complete time

10.3.1 Introduction

To show separation over Dedekind complete time, we follow a similar procedure for eliminating occurrences of S from under the scope of a U and U from under S to that we have just finished above. Here the elimination and induction is slightly more complicated.

We will need to consider some new connectives:

- $K^+q = \sim U(\top, \sim q)$, and
- $K^-q = \sim S(\top, \sim q)$.

We have K^+q true at t iff q is true arbitrarily close to t from the future, i.e.:

$$\|K^+q\|_t^h = 1 \text{ iff } \forall x > t, \exists y(t < y < x \wedge \|q\|_y^h = 1).$$

Similarly,

$$\|K^-q\|_t^h = 1 \text{ iff } \forall x > t, \exists y(x < y < t \wedge \|q\|_y^h = 1).$$

In integer time, these connectives are not very interesting for $K^+q \equiv K^-q \equiv \bot$.

Now consider the wff PK^+q. This wff is actually pure past—its truth at time t depends only on the behaviour of q in the past of t. More generally, if we build any wff out of S, K^-, K^+ only, we will obtain a pure past wff as long as any K^+ is under the scope of an S(or a K^-). The reason for this last restriction is that K^+q itself is pure future, but under the scope of any S it does no harm since it expresses 'local' properties.

Thus, in doing eliminations as we did in the last section, we may leave K^+ within the scope of an S, and still have a separated wff. This is just as well for in the language with just U and S it is impossible to rewrite such wffs as PK^+q in such a way that there is no U in the scope of an S and vice versa.

In order to make our lives easier we will extend our temporal language to include K^+ and K^- as connectives. It is clear that, since these new connectives can be defined in terms of U and S, separation and expressive completeness either hold for both or fail for both the temporal language with just U and S and its extension, which includes the $K^\pm$.

Now it is possible to define syntactic separation in this extended language and then go through the eliminations and inductions to prove that any wff can be syntactically separated. However, for the benefit of the next chapter, we are going to make things slightly more complicated.

We will introduce still more connectives ($\Gamma^\pm$), use a special atom (c) whose interpretation is restricted, and keep a close watch on how we go about rewriting wffs when we separate them. All this extra machinery means more work in this chapter (for example the $\Gamma^\pm$ eliminations are really redundant here) but saves us work in chapter 11.

First we define

Definition 10.3.1

$$\Gamma^+(B) = {\sim}K^+({\sim}B) \wedge K^-({\sim}B)$$

and

$$\Gamma^-(B) = {\sim}K^-({\sim}B) \wedge K^+({\sim}B).$$

Again note that the expressive power has clearly not been increased from that of U and S.

Next we introduce a new atom c and require that it is always interpreted *relatively densely* in the linear order, i.e. c is true somewhere in each interval of the form

- $(a, b] = \{x \in T \mid a < x \leq b\}$, or
- $[a, b) = \{x \in T \mid a \leq x < b\}$,

for $a < b$ from T. This seeming restriction on the structures which we are looking at is not really one in this chapter, as to any temporal structure we can add c to the language and interpret it as true everywhere. We do have

Lemma 10.3.2 *Over Dedekind complete time with c interpreted relatively densely $K^+(A) \leftrightarrow {\sim}U(c, {\sim}A)$ and the dual result.*

In the rest of this chapter we will work in the language with connectives $\{U, S, K^\pm, \Gamma^\pm\}$ and a special atom c. All equivalences will be valid in Dedekind complete structures with c interpreted relatively densely.

Definition 10.3.3 A wff of $\{U, S, K^\pm, \Gamma^\pm\}$is *syntactically separated* iff it is a boolean combination of wffs of the form:

- atoms,
- $U(A, B)$ without S,
- $S(A, B)$ without U,
- $K^+(A)$ without S, and
- $K^-(A)$ without U.

For example, $K^+(S(p, q))$ is not syntactically separated but $S(K^+(S(p, q)), r)$ is.

Note that $\Gamma^\pm$ must only appear nested below other temporal connectives.

Lemma 10.3.4 *Syntactic separation implies separation.*

Proof. It is not difficult to prove by induction on the construction of any wff A of $\{U, S, K^\pm, \Gamma^\pm\}$that

- if A does not contain S then its truth at $t \in T$ is determined by any $s < t$ along with the interpretation of the atoms on $\{u \in T \mid s < u\}$,

- and dually for U.

Then, given any wff $U(A,B)$ with A and B not containing S, we know that its truth at $t \in T$ depends just on the valuations of A and B at points $> t$ but by the above these are determined by the interpretation of the atoms on $\{u \in T \mid t < u\}$, i.e. on the future of t.

And similarly for the other cases. ■

10.3.2 Pre-eliminations

As before we have a useful negation lemma:

Lemma 10.3.5 *The following hold over Dedekind complete flows of time:*

$$\begin{array}{rl} {\sim}U(A,B) \leftrightarrow & K^+({\sim}B) \vee {\sim}U(A,\top) \\ \vee & U({\sim}A \wedge {\sim}B, {\sim}A) \vee U({\sim}A \wedge \Gamma^-(B), {\sim}A); \\ {\sim}S(A,B) \leftrightarrow & K^-({\sim}B) \vee {\sim}S(A,\top) \\ \vee & S({\sim}A \wedge {\sim}B, {\sim}A) \vee S({\sim}A \wedge \Gamma^+(B), {\sim}A). \end{array}$$

Now let us look more closely at the basic eliminations which we will need to carry out. Our elimination procedure may disregard any K^+ in the scope of S or K^-. However, since $K^\pm$ are allowed in the scope of S, a U may be under the scope of a K^+ or a K^- which is itself in the scope of an S. Thus we must pull the U out of the scope of the $K^\pm$, in order to pull it out further from under the scope of S. For example, in $S(a \wedge K^-(p \wedge U(A,B)), q)$ we must obtain the U out of the scope of K^- and then out of the scope of S.

Because of facts like $K^+(p \vee q)$ being equivalent to $K^+(p) \vee K^+(q)$ it turns out that, in addition to the eight cases of eliminations needed for the integers, for this section we have to consider the four additional elimination cases seen in lemma 10.3.8.

We need similar results for $\Gamma^\pm$ but first a technical lemma:

Lemma 10.3.6 *Let $Q(A,B,C)$ be the abbreviation*

$$\begin{array}{rl} & [C \rightarrow {\sim}K^+({\sim}B)] \\ \wedge & [({\sim}B \vee \Gamma^-(B)) \rightarrow (S(C,{\sim}A) \rightarrow A)]. \end{array}$$

1. If

- $\forall x \in (t_0,t_1), \|C \rightarrow U(A,B)\|_x = 1,$
- *and* $\|(A \wedge {\sim}C) \vee K^+(A) \vee U(A,B)\|_{t_0} = 1,$

then $\forall x \in (t_0,t_1), \|Q(A,B,C)\|_x = 1.$

2. If

- $\forall x \in (t_0, t_1), \|Q(A, B, C)\|_x = 1$,
- *and* $\|A \vee K^-(A) \vee (B \wedge U(A, B))\|_{t_1} = 1$,

then $\forall x \in (t_0, t_1), \|C \to U(A, B)\|_x = 1$.

Proof. Let $Q = Q(A, B, C)$.

1. Suppose the conditions are true and $x \in (t_0, t_1)$. If $\|C\|_x = 1$ then $\|U(A, B)\|_x = 1$ so we certainly have $\|{\sim}K^+({\sim}B)\|_x = 1$.

   Assume $\|({\sim}B \vee \Gamma^-(B))\|_x = 1$ and $\|S(C, {\sim}A)\|_x = 1$. We show that $\|A\|_x = 1$.

   There is $z < x$ with $\|C\|_z = 1$ and for all $w \in (z, x)$, $\|{\sim}A\|_w = 1$. There are two cases:

   $z \leq t_0$ when the initial conditions on t_0 imply that $\|U(A, B)\|_{t_0} = 1$. Put $u = t_0$.

   $t_0 < z < x$ when C being true at $u = z$ implies that $\|U(A, B)\|_u = 1$.

   In each case, since A is not true between u (where $U(A, B)$ holds) and x, and B does not remain true past x, we must have A true at x, as required.

2. Suppose the conditions are true and $x \in (t_0, t_1)$. Assume that $\|C\|_x = 1$ and we show that $\|U(A, B)\|_x = 1$.

   Let $y = \sup\{z \in (x, t_1) \mid \text{for all } u \in (x, z), \|B\|_u = 1\}$. Since $\|{\sim}K^+({\sim}B)\|_x = 1$ we know that y exists and is $> x$. There are two cases:

   $x < y < t_1$ Now we have either ${\sim}B$ or $\Gamma^-(B)$ holding at y and thus, since Q also holds there we have $\|S(C, {\sim}A) \to A\|_y = 1$. If A is true anywhere between x and y then we are done (as B holds everywhere on that interval) but otherwise we have A holding at y (since $S(C, {\sim}A)$ does) and we are again finished.

   $y = t_1$ when it follows directly from the initial conditions on t_1 that $U(A, B)$ is true at x as required.

■

Lemma 10.3.7 *The following hold over Dedekind complete flows of time:*

- $K^+(A \vee B) \leftrightarrow K^+A \vee K^+B$,

- $K^-(A \vee B) \leftrightarrow K^-A \vee K^-B$.

Let us perform the four eliminations for U beneath $K^\pm$ first:

Lemma 10.3.8 *Over Dedekind complete time we have the following equivalences in which the abbreviation $Q(A,B,p)$ is as defined in the Q lemma. Note that in the results U only appears as $U(A,B)$ and not under S,$K^\pm$ or $\Gamma^\pm$.*

1. $K^+(p \wedge U(A,B))$

$$\leftrightarrow [K^+p \wedge U(A,B)] \vee [K^+p \wedge K^+(A \wedge S(p,B))].$$

2. $K^-(p \wedge U(A,B)$

$$\begin{array}{rr} \leftrightarrow & [K^-p \wedge A \wedge {\sim}K^-({\sim}B)] \\ \vee & [K^-p \wedge {\sim}K^-({\sim}B) \wedge B \wedge U(A,B)] \\ \vee & [K^-p \wedge K^-(S(p,B) \wedge A)]. \end{array}$$

3. $K^+(p \wedge {\sim}U(A,B))$

$$\leftrightarrow K^+p \wedge {\sim}U(A,B) \wedge [{\sim}K^+A \vee K^+({\sim}Q(A,B,p))].$$

4. $K^-(p \wedge {\sim}U(A,B))$

$$\begin{array}{rr} \leftrightarrow & [K^-p \wedge K^-({\sim}Q(A,B,p))] \\ \vee & [K^-p \wedge {\sim}(A \vee K^-A \vee (B \wedge U(A,B)))]. \end{array}$$

Proof. These are all very straightforward semantic arguments, the last two of which are simplified somewhat by use of the Q lemma. ■

Notice that we sometimes don't end up with a syntactically separated wff after using the above equivalences: for example S appears (in Q) beneath K^+ in the third. This doesn't matter in the proof where we just require the $U(A,B)$ to be removed from beneath the $K^\pm$.

Next the Γs.

Lemma 10.3.9

$$\Gamma^+(A \wedge B) \leftrightarrow [\Gamma^+(A) \wedge {\sim}K^+({\sim}B)] \vee [\Gamma^+(B) \wedge {\sim}K^+({\sim}A)],$$

and similarly $\Gamma^-(A \wedge B)$.

Lemma 10.3.10 *Each wff below is equivalent (over Dedekind complete time) to a wff in which U only appears as $U(A,B)$ and is not nested beneath S, $K^\pm$, or $\Gamma^\pm$:*

1. $\Gamma^-(q \vee U(A,B))$;

2. $\Gamma^+(q \vee U(A,B))$;

3. $\Gamma^-(q \vee \sim U(A,B))$;

4. $\Gamma^+(q \vee \sim U(A,B))$.

Dual results with Γ^+ *exchanged with* Γ^- *and* U *exchanged with* S *hold. In fact, we have:*

$$\begin{array}{ll}
1. & \Gamma^-(q \vee U(A,B)) \\
\leftrightarrow & [\Gamma^-(q) \wedge \sim U(A,B) \wedge (\sim K^+(A) \vee K^+(\sim Q))] \\
\vee & [K^+(\sim q) \wedge \sim U(A,B) \wedge (\sim K^+(A) \vee K^+(\sim Q)) \wedge \sim K^-(\sim Q) \wedge A] \\
\vee & [K^+(\sim q) \wedge \sim U(A,B) \wedge \Gamma^+(\sim A) \wedge \sim K^-(\sim Q)] \\
\vee & [K^+(\sim q) \wedge \sim U(A,B) \wedge \Gamma^-(Q) \wedge K^-(A)].
\end{array}$$

$$\begin{array}{ll}
2. & \Gamma^+(q \vee U(A,B)) \\
\leftrightarrow & [\Gamma^+(q) \wedge K^-(\sim Q)] \\
\vee & [\Gamma^+(q) \wedge \sim A \wedge \sim K^-(A) \wedge (\sim B \vee \sim U(A,B))] \\
\vee & [U(A,B) \wedge K^-(\sim q) \wedge K^-(\sim Q) \wedge (\sim B \vee \Gamma^+(B))] \\
\vee & [U(A,B) \wedge K^-(\sim q) \wedge \sim A \wedge \sim K^-(A) \wedge \sim B] \\
\vee & [K^+(A) \wedge \Gamma^+(Q) \wedge K^-(\sim q)] \\
\vee & [\Gamma^-(\sim A) \wedge \sim K^+(\sim Q) \wedge K^-(\sim q) \wedge \sim A \wedge (\sim B \vee \sim U(A,B))].
\end{array}$$

$$\begin{array}{lll}
3. & & \Gamma^-(q \vee \sim U(A,B)) \\
\leftrightarrow & & [\Gamma^-(q) \wedge U(A,B)] \\
\vee & & [\Gamma^-(q) \wedge K^+(A \wedge S(\sim q, B))] \\
\vee[& & (\sim A \vee K^-(\sim B)) \\
 & \wedge & (((\Gamma^+(B) \vee \sim B) \wedge \sim K^-(S(\sim q, B) \wedge A) \wedge K^+(\sim q) \wedge U(A,B)) \\
 & & \vee(\sim(\sim K^-(\sim B) \wedge B \wedge U(A,B)) \\
 & & \wedge \Gamma^-(\sim A \vee \sim S(\sim q, B)) \wedge K^+(\sim q)))].
\end{array}$$

$$\begin{array}{ll}
4. & \Gamma^+(q \vee \sim U(A,B)) \\
\leftrightarrow & [\Gamma^+(q) \wedge A \wedge \sim K^-(\sim B)] \\
\vee & [\Gamma^+(q) \wedge \sim K^-(\sim B) \wedge B \wedge U(A,B)] \\
\vee & [\Gamma^+(q) \wedge K^-(S(\sim q, B) \wedge A)] \\
\vee & [K^-(\sim q) \wedge \sim U(A,B) \wedge \sim K^+(A \wedge S(\sim q, B) \wedge A \wedge \sim K^-(\sim B))] \\
\vee & [K^-(\sim q) \wedge \sim U(A,B) \wedge \Gamma^+(\sim A \vee \sim S(\sim q, B))].
\end{array}$$

Proof. Rewrite $\Gamma^\pm$ in terms of $K^\pm$ and use the $K^\pm$ eliminations. Then rearrange and reform some $\Gamma^\pm$s. ∎

10.3.3 Eliminations

Remember that the simple equivalences of lemma 10.2.1, such as that involving $S(A \vee B, C)$, hold in any linear order.

Now we consider the cases of elimination which were also seen in the integer separations.

Lemma 10.3.11 *Consider the following wffs with a, q, A, and B being atoms:*

1. $S(a \wedge U(A,B), q)$,
2. $S(a \wedge {\sim}U(A,B), q)$,
3. $S(a, q \vee U(A,B))$,
4. $S(a, q \vee {\sim}U(A,B))$,
5. $S(a \wedge U(A,B), q \vee U(A,B))$,
6. $S(a \wedge {\sim}U(A,B), q \vee U(A,B))$,
7. $S(a \wedge U(A,B), q \vee {\sim}U(A,B))$, *and*
8. $S(a \wedge {\sim}U(A,B), q \vee {\sim}U(A,B))$.

Over Dedekind complete time each of the above wffs can be syntactically separated in such a way that the only appearance of U in the separated wff is as $U(A,B)$.

Proof.

1. In this case, the wff here is equivalent to the same wff obtained in the case of the integers in lemma 10.2.3, i.e. $S(a \wedge U(A,B), q)$ is equivalent to

$$\begin{array}{cc} & [S(a,q) \wedge S(a,B) \wedge B \wedge U(A,B)] \\ \vee & [A \wedge S(a,B) \wedge S(a,q)] \\ \vee & S(A \wedge q \wedge S(a,B) \wedge S(a,q), q). \end{array}$$

2. It is straightforward to show that $S(a \wedge {\sim}U(A,B), q)$

$$\begin{array}{cc} \leftrightarrow & S(K^{+}({\sim}B) \wedge a, q) \\ \vee & S({\sim}A \wedge {\sim}B \wedge q \wedge S(a, {\sim}A \wedge q), q) \\ \vee & S({\sim}A \wedge \Gamma^{-}(B) \wedge q \wedge S(a, {\sim}A \wedge q), q) \\ \vee & [S(a, {\sim}A \wedge q) \wedge {\sim}A \wedge ({\sim}B \vee {\sim}U(A,B))]. \end{array}$$

3. Below we show that $S(a, U(A,B) \vee q)$

$$\begin{array}{cc} \leftrightarrow & S(a,q) \\ \vee & [S(\alpha,Q) \wedge \beta] \\ \vee & S(A \wedge (q \vee U(A,B)) \wedge S(\alpha,Q), q) \\ \vee & S(\Gamma^+(q) \wedge q \wedge (A \vee K^-(A)) \wedge S(\alpha,Q), q), \end{array}$$

where $Q = Q(A, B, \sim q)$,

$$\alpha = a \vee ((\sim q \vee \Gamma^-(q)) \wedge S(a,q) \wedge (q \vee U(A,B))),$$

and

$$\beta = A \vee K^-(A) \vee [B \wedge U(A,B)].$$

Elimination (1) used several times will give us separation.

Let us outline the proof of the equivalence above:

($\Rightarrow$) Assume that $S(a, U(A,B) \vee q)$ holds at t. So there is $s < t$ such that $\|a\|_s = 1$ and $U(A,B) \vee q$ is true everywhere between s and t.

Let

$$L = \{x \in (s,t) \mid \forall y \in (s,x), \|q\|_y = 1\},$$

$l = \sup L$ (or $l = s$ if L is empty),

$$R = \{x \in (s,t) \mid \forall y \in (x,t), \|q\|_y = 1\}.$$

and $r = \inf R$ (or $r = t$ if R is empty).

If $L = (s,t)$ then $\|S(a,q)\|_t = 1$ and we are done. So suppose that $l < t$.

Clearly, $l \leq r \leq t$. Now either $s = l$ so the first disjunct of α holds at r or $s < l$ so the second holds. In each case $K^+(A) \vee U(A,B)$ is true at l. We thus have $S(\alpha, Q)$ true at r.

There are three cases:

$r = t$ Now we must have $K^-(\sim q)$ true at t so $K^-(U(A,B))$ also. This implies β and hence $S(\alpha,Q) \wedge \beta$ holds at t.

$r < t$ and $U(A,B)$ is false at r: We must have q true at r but since $r = \inf R$ we need also have $\Gamma^+(q)$ true here. Because $K^-(\sim q)$ so $K^-(U(A,B))$ holds and we have $A \vee K^-(A)$. Thus

$$S(\Gamma^+(q) \wedge q \wedge (A \vee K^-(A)) \wedge S(\alpha,Q), q)$$

holds at t and we finish.

$r < t$ and $U(A, B)$ holds at r: Suppose that $w > r$ witnesses the until. It is clear then that Q is true at r and up until w so we have further sub-cases:

$r < w < t$: $S(A \wedge S(\alpha, Q), q)$ holds at t.

$t \leq w$: Now $A \vee (B \wedge U(A, B))$ holds at t so we have $S(\alpha, Q) \wedge \beta$ there.

($\Leftarrow$) Suppose that one of the four disjuncts holds at t. It is clear that if the first does then $S(a, U(A, B) \vee q)$ holds at t, as required. Now consider the cases of the other disjuncts holding.

If the second disjunct holds put $u = t$ and β holds here.

If the third or fourth disjuncts hold let $u < t$ witness the since so we have either A or $K^{-}(A)$ holding at u and q true between u and t.

In each case we have a point $u \leq t$ with $\|S(\alpha, Q)\|_u = 1$ and various other truth conditions depending on the case. Thus there is $v < u$ where α holds and Q is true everywhere on (v, u). Either a holds at v when we put $s = v$ or, if the other disjunct of α holds at v, there is $s < v$ witnessing $S(a, q)$.

We are done when we show that $q \vee U(A, B)$ holds everywhere in (s, t). Consider x from this interval.

If $s < x < v$ then q holds at x.

If $s < x = v$ then $q \vee U(A, B)$ holds.

If $v < x < u$ then since Q holds on (v, u) and β holds at u we have the Q lemma implying $q \vee U(A, B)$ at x.

If $x = u < t$ then we have disjuncts three or four holding and the desired result.

If $u < x < t$ then we have q true at x and we are done.

4. $S(a, {\sim}U(A, B) \vee q)$

$$
\begin{array}{rrr}
\leftrightarrow {\sim}[& & K^{-}(U(A, B) \wedge {\sim}q) \\
& \vee & {\sim}S(a, \top) \\
& \vee & S({\sim}a \wedge U(A, B) \wedge {\sim}q, {\sim}a) \\
& \vee & S({\sim}a \wedge \Gamma^{+}({\sim}U(A, B) \vee q), {\sim}a) \quad].
\end{array}
$$

By lemma 10.3.3 the first disjunct can be separated. By the first elimination the third disjunct can be separated and by lemmas Γ and elimination (1) or (2) the fourth disjunct can be separated.

5. To separate $S(a \wedge U(A, B), q \vee U(A, B))$ use elimination (3) to rewrite

it equivalently as

$$\begin{array}{rc} & S(a \wedge U(A,B),q) \\ \vee & [S(\alpha,Q) \wedge \beta] \\ \vee & S(A \wedge (q \vee U(A,B)) \wedge S(\alpha,Q),q) \\ \vee & S(\Gamma^{+}(q) \wedge q \wedge (A \vee K^{-}(A)) \wedge S(\alpha,Q),q). \end{array}$$

where $Q = Q(A, B, \sim q)$,

$$\begin{array}{rl} \alpha = & (a \wedge U(A,B)) \\ & \vee((\sim q \vee \Gamma^{-}(q)) \wedge S(a \wedge U(A,B),q) \wedge (q \vee U(A,B))), \end{array}$$

and

$$\beta = A \vee K^{-}(A) \vee [B \wedge U(A,B)].$$

The first elimination will separate the first disjunct and the expression α. Further use of that elimination will separate $S(\alpha, Q)$ and finally also the expression which the latter is nested within.

6. To separate $S(a \wedge \sim U(A,B), q \vee U(A,B))$ use elimination (3) and then elimination (2) in a similar manner to the preceding elimination.

7. $S(U(A,B) \wedge a, \sim U(A,B) \vee q)$

$$\begin{array}{rc} \leftrightarrow & S(a, B \wedge q) \wedge (A \vee (B \wedge U(A,B))) \\ \vee & S(S(a, B \wedge q) \wedge A \wedge (q \vee \sim U(A,B)), \sim U(A,B) \vee q). \end{array}$$

We then use the eighth and fourth eliminations.

8. $S(\sim U(A,B) \wedge a, \sim U(A,B) \vee q)$

$$\begin{array}{rrcl} \leftrightarrow & \sim[& K^{-}(U(A,B) \wedge \sim q) & \\ & \vee & \sim S(\sim U(A,B) \wedge a, \top) & \\ & \vee & S((U(A,B) \vee \sim a) \wedge U(A,B) \wedge \sim q, U(A,B) \vee \sim a) & \\ & \vee & S((U(A,B) \vee \sim a) \wedge \Gamma^{+}(\sim U(A,B) \wedge q), & \\ & & U(A,B) \vee \sim a) &]. \end{array}$$

This can be separated by lemmas 10.3.3, and eliminations (2) and (5).

■

10.3.4 Induction

For the benefit of the next chapter we need to account for the rewrites we use in separating wffs.

Definition 10.3.12 An *acceptable rewrite* of wff α is any change accomplished by replacing subwff γ of α by δ where γ is got from δ by:

- boolean rearrangement,
- replacing $K^+(x)$ by $\sim U(c, \sim x)$ and similarly K^-,
- one of the simple $U/S(A \vee B, C)$, $U/S(A, B \wedge C)$, $K^{\pm}(A \vee B)$ and $\Gamma^{\pm}(A \wedge B)$ equivalences,
- one of the four $K^{\pm}$ eliminations,
- replacing a boolean constituent $\Gamma^{\pm}(A)$ of α by
$$K^{\mp}(\sim A) \wedge \sim K^{\pm}(\sim A),$$
- one of the four $\Gamma^{\pm}$ eliminations,
- one of the eight U/S eliminations,
- or several of such steps.

Definition 10.3.13 Write *acceptably separated* for acceptably rewritten as a syntactically separated wff.

Note that acceptable rewriting preserves equivalence over Dedekind complete linear orders in which c is relatively dense.

We now have the necessary tools to prove the separation theorem of Since and Until over Dedekind complete time.

Lemma 10.3.14 *Suppose that A and B are wffs of $\{U, S, K^{\pm}, \Gamma^{\pm}\}$ in which U and S do not appear and that E and F are wffs in which U only appears as $U(A, B)$ and never nested within S, $K^{\pm}$ or $\Gamma^{\pm}$.*

Then each of $S(E, F)$, $K^{\pm}(E)$, and $\Gamma^{\pm}(F)$ can be acceptably rewritten as a boolean combination of wffs of the following forms:

- *atoms,*
- *$U(A, B)$, and*
- *$S(E_1, F_1)$, $K^{\pm}(E_1)$ and $\Gamma^{\pm}(E_1)$ in which U does not appear.*

Furthermore, we do not have to use the acceptable rewrite of replacing $\Gamma^{\pm}$ by $K^{\pm}$s in doing so.

Proof. By using boolean rearrangement, which is an acceptable rewrite, and rules about taking disjuncts (in the case of E) or conjuncts (in the case of F) out of connectives, we can without loss suppose that $U(A, B)$(possibly negated) only appears as a conjunct in E and/or a disjunct of F.

Now use the appropriate one of the 16 eliminations to remove the $U(A, B)$ from within the $K^{\pm}$,$\Gamma^{\pm}$, or S. An inspection of the results of these eliminations will reveal that we have our proof.

Remember that A and B are not necessarily atoms when doing the inspection. ■

Lemma 10.3.15 *Suppose that A and B are wffs of $\{U, S, K^{\pm}, \Gamma^{\pm}\}$in which U and S do not appear and that in both wffs C and F the only appearances of U, if any, are as the subformula $U(A, B)$ and are never nested below S,$K^{\pm}$, or $\Gamma^{\pm}$.*

Then $S(C, F)$ can be acceptably separated as a wff in which U only appears as the formula $U(A, B)$.

Proof. So C and F are boolean combinations of atoms, of wffs (built from $K^{\pm}$,$\Gamma^{\pm}$, or S) not containing U and of appearances of $U(A, B)$ itself. As in the integer case we rewrite in normal forms and use lemma 10.2.1 to acceptably rewrite $S(C, F)$ as a boolean combination of wffs $S(C_1, C_2)$ with no U appearing and of wffs of the form either:

- $S(C_1 \wedge \pm U(A, B), C_2)$,
- $S(C_1, C_2 \vee \pm U(A, B))$, or
- $S(C_1 \wedge \pm U(A, B), C_2 \vee \pm U(A, B))$.

for some wffs C_1 and C_2 not containing U.

Using the S eliminations it can be seen that we can acceptably rewrite wffs of these latter three forms as boolean combinations of:

- A and B,
- $S(C', F')$ containing no U,
- $U(A, B)$, and
- $K^{-}(C')$ containing no U.

Checking each elimination will reveal that there are no other possibilities.

Now we are almost finished as all these subformulae are pure except possibly boolean constituents of A and B. The only possible offenders are constituents of the form $\Gamma^{\pm}(C')$ where C', like A and B, contains neither U nor S. We simply rewrite such Γs as Ks and we are done. ■

Lemma 10.3.16 *Suppose that A and B are wffs of $\{U, S, K^{\pm}, \Gamma^{\pm}\}$ in which U and S do not appear and that in both wffs C and F the only appearances of U, if any, are as the subformula $U(A, B)$.*

Then $S(C, F)$ can be acceptably separated as a wff in which U only appears as the formula $U(A, B)$.

Proof. If $U(A, B)$ does not appear then we are done. Assume otherwise. The proof is by induction on the maximum depth k of nested occurrences of $S, K^{\pm}$ or $\Gamma^{\pm}$ within $S(C, F)$ which contain U (i.e. $U(A, B)$) in their scope.

$k = 1$ This is just the last lemma.

$k > 1$**:** Let $D_j, j = 1, \ldots, N$ be all the wffs of the forms $S(C_1, C_2)$,$K^{\pm}(C_1)$, $\Gamma^{\pm}(C_1)$ such that C_1 and C_2 may contain $U(A, B)$ but contain no nested occurrence of $U(A, B)$ within $K^{\pm}$,$\Gamma^{\pm}$ or S.

By the moving-up-of-Us lemma these can be acceptably rewritten not using the Γ to K rewrite so that U still only appears as $U(A, B)$ but no longer nested within S,$K^{\pm}$, or $\Gamma^{\pm}$. Replacing these D_js by their rewritten form is an acceptable rewrite of $S(C, F)$ but reduces the depth of nesting and allows us to use the induction hypothesis to finish.

■

Lemma 10.3.17 *For each $i = 1, ..., n$ let A_i and B_i be wffs built without S or U. Suppose that the only appearances of U in the wff D are in the form $U(A_i, B_i)$.*

Then D is acceptably separable.

Proof. Proceed by induction on n as in the integers case, lemma 10.2.6.

When $n = 1$ we have the preceding lemma so suppose that $n > 1$. First replace each appearance of $U(A_i, B_i)$ by a new atom q_i for each $i = 1, ..., n-1$ to obtain a new wff D'. Now acceptably separate D' to E' using the last lemma. Substitute the $U(A_i, B_i)$ back in for the q_i to obtain E and it is clear that this is an acceptable rewrite of D.

Remembering that E' is syntactically separated let us show how to acceptably separate each boolean constituent of E. These will be of the following forms:

- (old) atoms which are already pure,
- $U(A_i, B_i)$ which are pure future,
- $K^+(E_1)$ containing no S (so already separated), and
- $K^-(E_1)$ and $S(E_1, F_1)$ containing U only in the forms $U(A_i, B_i)$ for $i \leq n-1$.

Rewriting the $K^-(E_1)$s as $\sim S(c, \sim E_1)$s we note that in these latter K^- and S cases we can use the induction hypothesis to finish. ■

Lemma 10.3.18 *Suppose that wff $S(C, F)$ contains no S nested within a U.*

Then $S(C, F)$ is acceptably separable.

Proof. By induction on the maximal depth n of nesting of Us beneath any S.

Case $n = 1$: This is the case of the preceding lemma.

Case $n > 1$: Let $U(A_i, B_i)$ $(i = 1, ..., N)$ be the subformula covering the least deeply nested appearances of U, i.e. every appearance of U in $S(C, F)$ is as a subformula of an appearance of one of the $U(A_i, B_i)$. Each A_i and B_i are built up as a combination from wffs of the form $U(X_{ij}, Y_{ij})$ and atoms using boolean connectives $\Gamma^\pm$ and $K^\pm$. Replace each $U(X_{ij}, Y_{ij})$ in A_i and B_i by the new atom x_{ij} to form wffs A_i' and B_i' which are now just combinations of atoms. Thus when we substitute in A_i' and B_i' the value $x_{ij} = U(X_{ij}, Y_{ij})$ we obtain A_i and B_i respectively.

Replace each occurrence of $U(A_i, B_i)$ in D by $U(A_i', B_i')$ to obtain D'. D' can be acceptably separated by the preceding lemma. Let E' be its separated form. E' will be a boolean combination of

- atoms (including the x_{ij}),
- pure future formulae (like $U(A_i', B_i')$), and
- pure past formulae (for example A_i's and B_i's nested under Ss).

Furthermore, when we substitute in $x_{ij} = U(X_{ij}, Y_{ij})$ we obtain a wff E which is an acceptable rewrite of D.

Unfortunately, E is not necessarily separated: what were pure past formulae in E' have become, on replacement of x_{ij} by $U(X_{ij}, Y_{ij})$, impure. To correct this we use the induction hypothesis on each of these pure past subformula of E. Wff $K^-(C)$ will have to be rewritten as $\sim S(c, \sim C)$ in order to do so.

This can be done since the level of nesting of U in $U(A_i, B_i)$ must be strictly greater than that in its subformula $U(X_{ij}, Y_{ij})$.

■

Let us bring in all the assumptions when we write out our final lemma:

Lemma 10.3.19 *Suppose that D_1 and D_2 are wffs from the language with $\{U, S, K^\pm, \Gamma^\pm\}$ and special atom c.*

Then any wff of the form $D = U(D_1, D_2)$ or $D = S(D_1, D_2)$ can be acceptably rewritten as a wff which is

- *equivalent to D over any Dedekind complete flow of time in which c is interpreted relatively densely, and*
- *syntactically separated.*

Proof. Again this is similar to lemma 10.2.8. We define *junction depth* just as we did there: i.e. we ignore the $K^\pm$ and $\Gamma^\pm$ as we do the classical connectives.

Now we begin the proof being given a wff D of one of the above forms. We proceed by induction on the junction depth d of D.

If it is zero or one then D is already syntactically separated. So assume the induction hypothesis and that the junction depth of D is at least two.

Again because of the dual nature of the results so far we need only demonstrate the syntactic separation of a wff of the form $D = S(D_1, D_2)$.

Let $U(A_i, B_i)$ $(i = 1, ..., N)$ be the subformula covering the least deeply nested appearances of U, i.e. every appearance of U in $S(D_1, D_2)$ is as a subformula of an appearance of one of the $U(A_i, B_i)$.

There are some subformula of $U(A_i, B_i)$ which are of the form $S(E, F)$. We note that the junction depth of any of these subformula is at most $d-2$. Replace each least deeply nested such subformula $S(E_{ij}, F_{ij})$ in $U(A_i, B_i)$ by its own new atom x_{ij} to obtain $U(A_i', B_i')$. Change $S(D_1, D_2)$ into D' by replacing each $U(A_i, B_i)$ by $U(A_i', B_i')$. The preceding lemma now tells us how to acceptably separate D' into a wff E'.

Now E' will be a boolean combination of subformula of the form:

- x_{ij},
- other atoms,
- $S(C_1, C_2)$ not containing U,

- $U(C_1, C_2)$ not containing S,
- $K^+(C_1)$ not containing S, and
- $K^-(C_1)$ not containing U.

To finish we resubstitute in the original $S(E_{ij}, F_{ij})$ (of junction depth $\leq d-2$) for each x_{ij} in each of the forms of subformulae above to obtain E which is an acceptable rewrite of $S(D_1, D_2)$. We still need to acceptably separate each subformula.

What was x_{ij} in E' has become $S(E_{ij}, F_{ij})$ in E but this has junction depth at most $d-2$ so we can use the induction hypothesis.

Atoms are pure.

Those of the form $S(C_1, C_2)$ now have junction depth at most $d-2$ while those of the form $U(C_1, C_2)$ have depth less than or equal to $d-1$ so our induction hypothesis completes the proof.

Consider $K^+(C)$ not containing S changed into $K^+(C')$ by resubstitutions for each x_{ij}. We can rewrite this as $\sim U(c, \sim C')$ and we see that the junction depth is one greater than that of the greatest of the $S(E_{ij}, F_{ij})$ used. This is still at most $d-1$ and we can use the induction hypothesis.

Finally, consider $K^-(C')$ which we rewrite as $\sim S(c, \sim C')$. Here, resubstitution doesn't increase the depth above the maximum of the $S(E_{ij}, F_{ij})$ and we can apply the induction hypothesis.

■

Theorem 10.3.20 (separation theorem) *Over Dedekind complete time, each wff in the language with $\{U, S\}$ is equivalent to a separated wff.*

Proof. All the separation equivalences hold with c interpreted as true everywhere so we can ignore this special atom.

The wff is a boolean combination of atoms and wffs of the form $U(A, B)$ and $S(A, B)$. The latter two types of subformula can be syntactically separated in the extended language with $K^{\pm}$ by the preceding lemma. This leaves us with a syntactically separated, and hence separated, formula. By rewriting any appearances of $K^{\pm}$ or $\Gamma^{\pm}$ using only U and S we can preserve the purity of the appropriate subformulae (which is a semantic property), and hence separation. ■

Theorem 10.3.21 *The language $\{U, S\}$ is expressively complete over Dedekind complete time.*

Proof. This follows from the separation theorem and the results of chapter 9. ■

Chapter 11

Expressive completeness of Stavi connectives over general linear time

11.1 Introduction

In the preceding chapter we saw that the connectives S and U have the separation property and hence expressive completeness over any Dedekind complete flow of time.

In the next section we will prove that $\{U, S\}$ is not expressively complete over a general linear flow of time so that Since and Until alone cannot have the separation property over the class of all such flows.

In section 11.3 we reintroduce the Stavi connectives denoted by U' and S'. We then show that the set $\{U, S, U', S'\}$ does indeed have the separation property over all linear flows of time and so is expressively complete for linear time.

11.2 Failure of Separation for S, U over linear time

The failure of the separation property for $\{U, S\}$ over general linear time is due to the failure of these connectives to describe the behaviour of truth values in the neighbourhood of gaps. This can be intuitively seen as follows.

Example 11.2.1 Consider the wff $D = P{\sim}U(A, B)$. This holds at t iff one of the diagrams in figure 11.1 applies where y may be in the past of t (as shown), equal to t, or in the future of t.

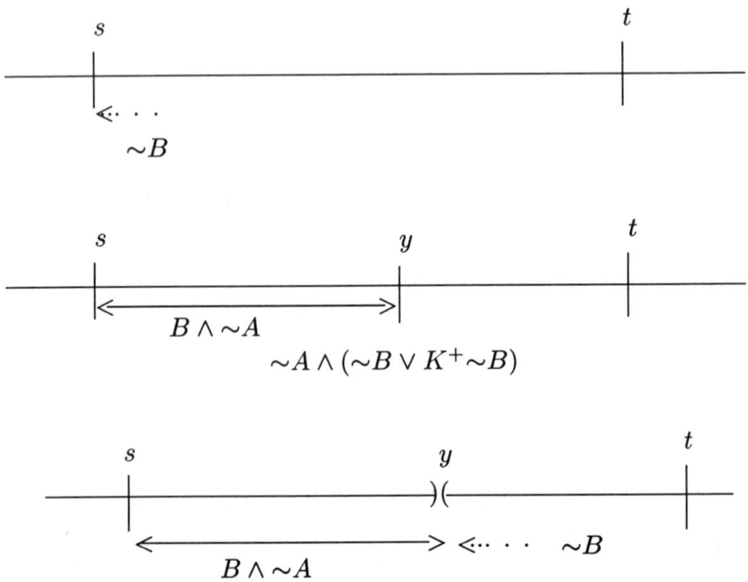

Figure 11.1

In the third diagram, y is a gap and hence cannot be equal to t. $B \wedge {\sim}A$ is true at all points between s and y, and ${\sim}B$ is true at a sequence of points approaching the gap from the right-hand side. This we denote in the diagram by $< \cdots ({\sim}B)$. We could have written $K^+{\sim}B$ true at y. The gap could be either in the past or in the future of t.

Separation can be obtained for the first two cases as in the case of real number time. Namely, for the first diagram, D is equivalent to $D_I = PK^+({\sim}B)$. For the second, D is equivalent to

$$\begin{aligned} D_{II} &= [P({\sim}A \wedge ({\sim}B \vee K^+{\sim}B) \wedge S(\top, B \wedge {\sim}A))] \\ &\quad \vee[{\sim}A \wedge ({\sim}B \vee K^+{\sim}B) \wedge S(\top, B \wedge {\sim}A)] \\ &\quad \vee[S(\top, B \wedge {\sim}A) \wedge {\sim}A \wedge {\sim}U(A, B)]. \end{aligned}$$

The three disjuncts of D_{II} reflect the situation of $y < t$, $y = t$, and $y > t$ respectively.

It should be clear that we cannot separate the gap case so easily. In fact we cannot do so as we have no connectives which talk about gaps.

The third diagram does not arise in a complete flow of time but for general linear flows we need additional connectives.

Let us prove formally that, indeed, $\{U, S\}$ fails to have separation over linear time. We shall accomplish this by showing that these connectives are not expressively complete over linear time and relying on theorem 9.3.1. Below we show that a first-order connective, η^+, is not definable from $\{U, S\}$.

Definition 11.2.2 Let $\eta^+(q)$ mean that q is true from now until a gap in time after which it stays false for a while.

It is easy to see that η^+ is in fact first order.

Theorem 11.2.3 *The connective η^+ is not definable from $\{U, S\}$ over general linear time.*

Proof. Suppose, for contradiction, that $\eta^+(q)$ is equivalent, over all linear flows of time, to a wff $G(q)$ built only from U and S.

Let $\mathcal{Q}$ be the rationals. Let I be a union of intervals of the form $\{r \in \mathcal{Q} \mid \gamma_1 < r < \gamma_2\}$ where the γ_i are gaps, such that I and $\mathcal{Q}\backslash I$ both contain arbitrarily large positive and negative numbers.

It is easy to show, by induction on its construction, that any wff built up from q in the language of $\{U, S\}$ is equivalent over all structures $(\mathcal{Q}, <, h)$ with $h(q) = I$ as above to either $\top$, $\bot$, q, or $\sim q$. Thus, on such structures, $\eta^+(q)$ must be equivalent to one of these four simple wffs. We now show that this leads to a contradiction.

Let $(x_n)_{n\geq 0}$ be a decreasing sequence of irrationals all less than π and approaching another irrational $x > 0$. Let I contain only the following rationals:

- those between an odd multiple of π and the next, even, multiple,
- those between $-\pi$ and x, and
- those r such that $x_{2n+1} < r < x_{2n}$ for some n.

It is clear that I has the form required above but since $\|\eta^+(q)\|_0^h = 0$ we can deduce that $\eta^+(q)$ can only be equivalent to either $\bot$ or $\sim q$ of the four simple wffs.

Now consider another structure defined by the subset I' of $\mathcal{Q}$. Let I' contain only the following rationals:

- those between an odd multiple of π and the next even multiple, and
- those between $-\pi$ and x.

Again we have I' of the appropriate form but now $\|\eta^+(q)\|_0^h = 1$. Since neither $\bot$ nor $\sim q$ have changed their truth values at 0, $\eta^+(q)$, which has, cannot be equivalent to either and we have the desired contradiction.

■

11.3 A language $L*$ for evaluation at gaps

In the preceding section we saw that, because of the problem with talking about gaps, we need additional connectives besides U and S if we want to obtain separation and expressive completeness over general linear time. Eventually we are going to show that $\{U, S, U', S'\}$ will do where U' and S' are the Stavi connectives that we saw in section 6.1. Recall that $U'(A, B)$ is as pictured in figure 11.2.

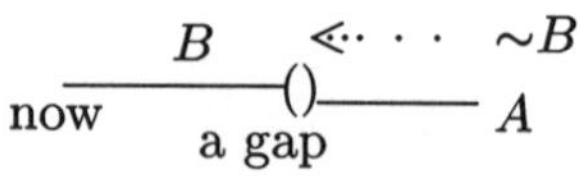

Figure 11.2

S' is defined dually, i.e. with past and future swapped. Despite involving gaps we saw that U' and S' are first-order connectives.

Throughout the proof we will make use of the concept of a definable gap. These are gaps which are defined or indicated by a temporal formula. There are two types of gaps defined by the wff B:

- gaps defined by B on the right where B is true for a while after the gap but false arbitrarily soon before it, and
- gaps defined by B on the left where B is true for a while up until the gap but false arbitrarily soon afterwards.

Our strategy is to obtain another set of connectives for which separation holds. To find such a set, one may try to carry out separation with U and S as in the preceding chapter and in the process discover the need for new connectives, something like: 'B is true until a gap definable by A and at this gap C is true' and 'until A, C is true at every gap definable by B', and combinations of them.

We will shortly define a new language $L*$ from such connectives and then show that the new system is intertranslatable with $\{U, U', S, S'\}$. By then showing, in the next section, that $L*$ has the separation property we thus obtain separation for $\{U, U', S, S'\}$.

In order to define the connectives in $L*$ we need to consider the completion of the time flow. Let $(T, <, h)$ be a linear structure. Let $(T*, <)$ be the completion of $(T, <)$. That is, we add to T all of its gaps. We do not extend h to $T*$. To keep track of all the points of T let c be a new propositional constant which is true at all points of T and false at all gaps (i.e. in $T*$, $\|c\|_x = 1$ iff $x \in T$).

Given an assignment h into T, for any atomic q, we can evaluate $\|q\|_x^h$ at any $x \in T$. For $x \in T* - T$, we have no value for $\|q\|_x^h$. However, this does not mean that we cannot evaluate wffs at $x \in T* - T$. For example, $U(p \wedge c, c \rightarrow q)$ can be evaluated at a gap $x \in T* - T$, because it describes the situation in figure 11.3.

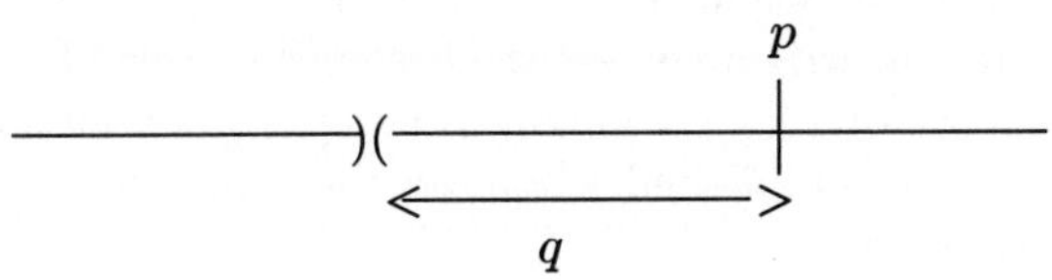

Figure 11.3

If we build a set $L*$ of wffs such that we evaluate at points and gaps in such a way that we are *never* required to evaluate an atom q at a gap x, then for any $A \in L*$ and any $x \in T*$, $\|A\|_x^h$ can be computed even though h is an assignment to T and not to $T*$.

For example, we know how to evaluate $X = U(p \wedge c, c \rightarrow q)$ at a gap y. Suppose we want to say there is a sequence of gaps $y_n \rightarrow x$ in which X is true.

This would be

$$\alpha = K^{-}(\sim c \wedge X).$$

The truth value of α can be computed at any $x \in T*$ including x in T. We write, as usual, $\|\alpha\|_t^h = 1$ when α is true at t even though this is a generalization of our previous use of the notation. Truth then was defined inductively in a straightforward manner. Unfortunately, here things are not quite so simple—defining truth involves a little detour via the completion of T. The problem with α is that it is not first order in $(T, <, h)$: we cannot find a wff $\psi_\alpha(t)$ such that for all $t \in T$

$$(T, <) \models \psi_\alpha(t) \text{ iff } \|\alpha\|_t^h = 1.$$

The reason is that we have to look for gaps around t, and we cannot say 'y is a gap' in a first-order way.

Since we are looking for a language that is intertranslatable with the $\{U, S, U, S'\}$ whose wffs all have first-order tables we have gone too far.

Suppose that, instead, we were asking to look at gaps definable by B from the left only. In other words, we look at:

$$\beta = K^-(\sim c \wedge S(\top, B) \wedge K^+(\sim B) \wedge X).$$

As the reader can check, the addition of B makes β first order because we can equivalently evaluate

$$K^-(U'(q \wedge X, B))$$

in $(T, <, h)$ instead.

It turns out that if whenever we speak of a gap we say which wff B defines it, then all the wffs involved are first order over $(T, <)$. A rigorous proof that we have not strayed beyond first-order connectives in defining our language $L*$ is accomplished when we show that it can be translated into the Stavi connectives.

The above suggests that we should look at wffs involving $\sim c$ (gaps) such that with any mention of $\sim c$, a wff B defining the gap is also mentioned.

So we will not have one of the usual connective based inductively constructed temporal languages. Instead we use only some of the wffs from such a language and for the semantics on $(T, <, h)$ we actually evaluate these wffs in $T*$.

Let us proceed to formally define such a language $L*$.

Suppose that we have some set of atoms. Formulae of $L*$ will be given truth values on temporal structures $(T, <, h)$ with valuations h of these atoms. Since $L*$ looks as though it uses different atoms we will call these original atoms the *normal atoms.* For each normal atom p we will use two new and distinct atoms p^+ and p^-. We will also use a completely different atom c.

Now consider the temporal language *Big*, obtained by taking our new set of atoms, and then building wffs in the usual way from the connectives $\{U, S, K^\pm, \Gamma^\pm\}$ above. We define, by simultaneous induction, *legitimate*, *active*, and *passive* wff. For convenience we consider $\sim$ and $\wedge$ as the only two boolean connectives and regard the rest as abbreviations.

Legitimate wffs in *Big*:

- c is legitimate as are $p^\pm$ for any normal atom p;
- $U(\alpha, \beta)$,$S(\alpha, \beta)$, $K^\pm(\alpha)$ and $\Gamma^\pm(\beta)$ are legitimate iff α is active and β is passive;
- $\top$ and $\bot$ are legitimate;

- $\alpha \wedge \beta$ is legitimate iff both α and β are; and
- $\sim\alpha$ is legitimate iff α is.

Active wffs in *Big*:

- c is active as are $p^{\pm}$ for any normal atom p;
- $\Gamma^{\pm}(\beta)$ is active iff β is passive;
- $\bot$ is active;
- $\alpha \wedge \beta$ is active iff both α and β are legitimate and at least one is active; and
- $\sim\alpha$ is active iff α is passive.

Passive wffs in *Big*:

- $\top$ is passive;
- $\alpha \wedge \beta$ is passive iff both α and β are; and
- $\sim\beta$ is passive iff β is active.

Clearly all active and passive wffs are legitimate.

Definition 11.3.1 We define the language $L*$:

Syntax: The wffs of $L*$ are just those active wffs of *Big*.

Semantics: Given any linear order $(T, <)$ and an interpretation h for the normal atoms we proceed as follows. First construct the completion $(T*, <)$ of $(T, <)$. Now define an interpretation g for the new atoms on the frame $(T*, <)$ as follows:

- the interpretation of c is $T \subseteq T*$;
- for each normal atom p, the interpretation of p^+ includes exactly the points of T where h makes p true; and
- for each normal atom p, the interpretation of p^- includes exactly the points of T where h makes p false.

For any A of $L*$, for any $t \in T$, put $\|A\|_t^h$ in $(T, <)$ equal to $\|A\|_t^g$ in $(T*, <)$.

We call a temporal structure in the signature with atoms $\{c, p^{\pm}\}$ *nice* iff

- the flow is Dedekind complete,
- c is interpreted relatively densely, and
- for each normal atom p,

$$(\sim c \wedge \sim p^{+} \wedge \sim p^{-}) \vee (c \wedge p^{+} \wedge \sim p^{-}) \vee (c \wedge \sim p^{+} \wedge p^{-})$$

is valid.

Given any linear temporal structure $(T, <, h)$, we can construct a nice temporal structure $(T*, <, g)$ such that the truth value of each wff of $L*$ agrees at $t \in T \subseteq T*$ whether we evaluate it as a wff of $L*$ in $(T, <, h)$ or as a wff of *Big* in $(T*, <, g)$. To see this we need only show that the points of a linear order T are relatively dense in the completion $T*$. This follows from basic properties of infimums and supremums.

Definition 11.3.2 A wff of *Big* is in *active normal form* iff it is a disjunction of conjunctions of wffs of the following form:

- $c \wedge p^{\pm}$ for some normal atom p,
- $c \wedge \pm U/S(\alpha_1, \sim\alpha_2)$ for some wffs α_1 and α_2 in active normal form, or
- $\sim c \wedge \Gamma^{\pm}(\sim\alpha_0) \wedge \pm U/S(\alpha_1, \sim\alpha_2)$ for wffs α_i in active normal form.

Let *passive normal form* be active normal form under a single negation symbol.

Lemma 11.3.3 *Each active or passive wff of Big is equivalent over nice flows of time to one in the appropriate normal form.*

Proof.

Proceed by simultaneous induction on the construction of active and passive wffs. We write $\langle\alpha\rangle$ for the rewritten form of active or passive α. Choose any normal atom p. The base cases follow when we put

- $\langle c\rangle = (c \wedge p^{+}) \vee (c \wedge p^{-})$,
- $\langle\bot\rangle = (c \wedge p^{+}) \wedge (c \wedge p^{-})$, and
- $\langle\top\rangle = \sim\langle\bot\rangle$.

Supposing that we have already shown how to rewrite passive β then

$$\begin{aligned}\langle \Gamma^{+}(\beta)\rangle = \quad & [\sim c \wedge \Gamma + (\langle\beta\rangle) \wedge \sim U(\langle\bot\rangle, \langle\top\rangle)] \\ & \vee[(c \wedge U(\langle c\rangle, \langle\beta\rangle)) \wedge (c \wedge \sim S(\langle c\rangle, \langle\beta\rangle))]\end{aligned}$$

and $\Gamma^{-}(\beta)$ is dual.

Clearly, negations present no problem so we are finished when we show how to rewrite $\alpha \wedge \gamma$ for active α and legitimate γ. By rewriting α first and then doing a little induction on conjuncts and disjuncts we can without loss of generality suppose that α is one of the three forms

- $c \wedge p^{\pm}$ for some normal atom p,
- $c \wedge \pm U/S(\alpha_1, \sim\alpha_2)$ for some wffs α_1 and α_2 in active normal form, or
- $\sim c \wedge \Gamma^{\pm}(\sim\alpha_0) \wedge \pm U/S(\alpha_1, \sim\alpha_2)$ for wff α_i in active normal form.

Also note that we can do similar things to γ. We can rewrite $\Gamma^{\pm}(\beta)$ as $K^{-}(\sim\beta) \wedge U(c,\beta)$, $K^{+}(\alpha)$ as $\sim U(c, \sim\alpha)$ and similarly for their duals. Since it is clear that any legitimate wff can be equivalently rearranged into disjunctive normal form, we know that we can equivalently rewrite any legitimate wff as a disjunct of conjuncts of the following forms:

- $c \wedge p^{\pm}$ for a normal atom p, and
- $\pm U/S(\alpha, \beta)$.

Now by doing a little induction on the number of disjuncts and conjuncts in γ we can suppose that it is just $c \wedge p^{\pm}$ for a normal atom p or $\pm U/S(\alpha', \beta)$ for active α' and passive β.

In each of the six cases of α and γ it is clear how to rewrite $\alpha \wedge \gamma$ in active normal form.

We should also consider the conjunction of passive wffs but this is straightforward. ■

11.4 Equivalence of $L*$ and the Stavi connectives

To make use of $L*$ we need to prove two results:

- that it is intertranslatable with the language of $\{U, S, U', S'\}$, and
- that $L*$ has a separation property.

We leave the second result until the next section. To show the first we need to present two translations: one, $*$, from from the language of $\{U, U', S, S'\}$ into the language $L*$ and one, $\Box$, from the language $L*$ into the language of $\{U, U', S, S'\}$. To be translations these two maps must of course preserve meaning.

The existence of the $*$ map will follow from later results about the expressive power of $L*$ but we can explicitly present the translation with minimal effort:

Definition 11.4.1 We define a map $*$ which takes each wff of the language of $\{U, U', S, S'\}$ to a wff of the language $L*$ by first defining another map $\star$ by induction.

$$\begin{array}{lcl}
\star[U'(A,B)] & = & U(\sim c \wedge \Gamma^-(c \to \star[B]) \wedge U(c, c \to \star[A]), c \to \star[B]) \\
\star[S'(A,B)] & = & S(\sim c \wedge \Gamma^+(c \to \star[B]) \wedge S(c, c \to \star[A]), c \to \star[B]) \\
\star[S(A,B)] & = & S(c \wedge \star[A], c \to \star[B]) \\
\star[U(A,B)] & = & U(c \wedge \star[A], c \to \star[B]) \\
\star[\sim A] & = & \sim(\star[A]), \text{ if } A \text{ not atomic} \\
\star[A \wedge B] & = & \star[A] \wedge \star[B] \\
\star[\pm p] & = & p^{\pm}, \text{ if } p \text{ is atomic.}
\end{array}$$

Now we simply put $A* = c \wedge \star A$ for any wff A of the language with $\{U, U', S, S'\}$.

It is now simple to show, by induction, that $\star A$ is always a legitimate wff of *Big* and hence that $A*$ is indeed in $L*$ and further that:

Lemma 11.4.2 *The translation $*$ satisfies for any $T, h, t \in T$,*

$$\|A*\|_t^h = \|A\|_t^h.$$

Before considering the $\Box$ translation we introduce some new (abbreviation) connectives, in the language of $\{U, U', S, S'\}$.

Two simple unary ones to start with:

- $I^+q = U(\top, q)$, and similarly
- $I^-q = S(\top, q)$.

Definition 11.4.3 We define the following four binary connectives as abbreviations in the language with $\{U, S, U', S'\}$:

- $BU\|I^+A = U'(A, B)$ reading B until a future gap in which I^+A holds is pictured in figure 11.4.

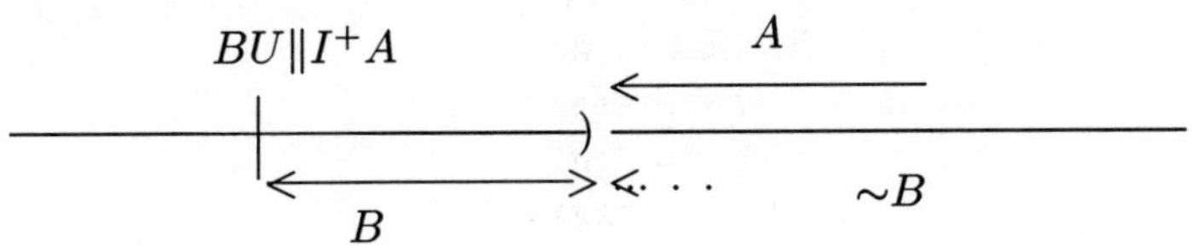

Figure 11.4

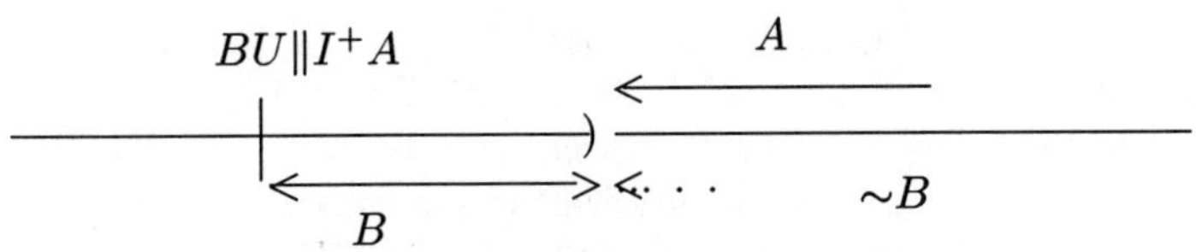

Figure 11.5

- $BU \| K^+ A = U'(\top, B) \wedge {\sim}U'({\sim}A, B)$ reading B until a future gap in which K^+A holds is pictured in figure 11.5.
- $BU \| I^- A = U(A \wedge B \wedge U'(\top, A \wedge B) \wedge {\sim}U'(B, A \wedge B), B)$ reading B until a future gap in which I^-A holds is pictured in figure 11.6.

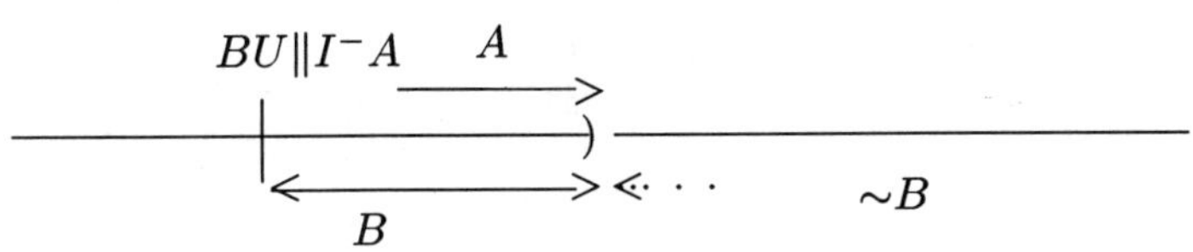

Figure 11.6

- $BU \| K^- A = U'(\top, B) \wedge {\sim}(BU \| I^- A)$ reading B until a future gap in which K^-A holds is pictured in figure 11.7.

We can define duals:

$BS \| I^+ A$ reads B since a gap in which I^+A,

$BS \| K^+ A$ reads B since a gap in which K^+A,

$BS \| I^- A$ reads B since a gap in which I^-A, and

$BS \| K^- A$ reads B since a gap in which K^-A.

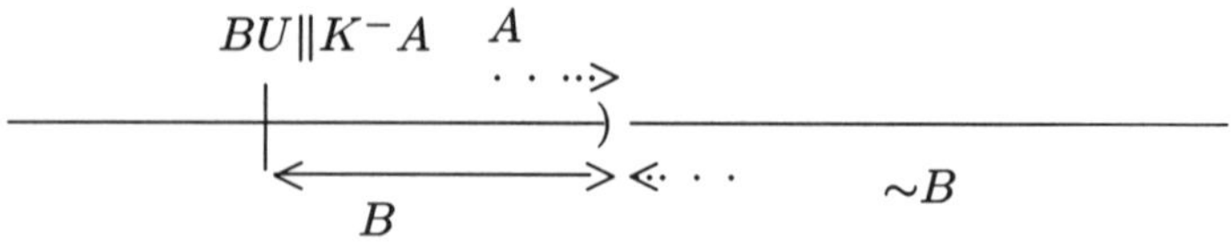

Figure 11.7

We want to define the connective $BU\|M$ to read B until a gap in which M holds. We will only need to do so when M is a boolean combination of wffs of the form $S(A,B)$ or $U(A,B)$. In this case we can write M in the form $\vee \wedge M_{ij}$, where M_{ij} is either of the forms $\pm U/S(A,B)$. Clearly, $BU\|M \leftrightarrow \bigvee_i \bigwedge_j BU\|M_{ij}$.

Thus it is sufficient to define (from $\{U, U', S, S'\}$) the four connectives following:

Definition 11.4.4

$BU\|S(p,q)$ holds iff the situation in figure 11.8 holds.

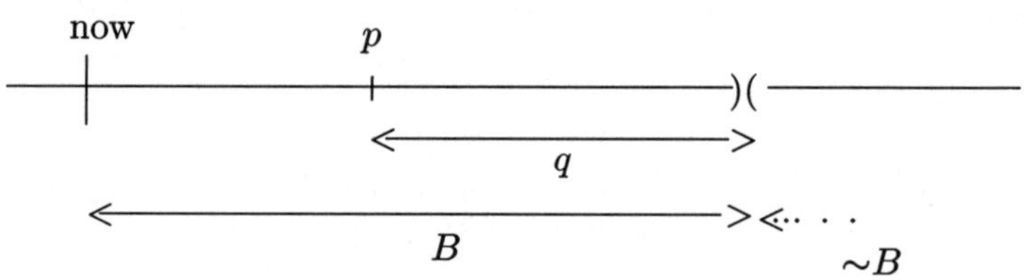

Figure 11.8

In figure 11.8, the point y can be before, equal to, or after t.

We therefore define:

$$BU\|S(p,q) = BU\|I^-(q \wedge S(p,q)).$$

Just substitute $A = q \wedge S(p,q)$ in the definition of $BU\|I^-A$.

$$BU\|{\sim}S(p,q) = (BU\|K^-{\sim}q) \vee BU\|I^-(q \wedge {\sim}S(p,q)),$$

$$BU\|U(p,q) = BU\|I^+(q \wedge U(p,q)), \text{ and}$$

$$BU\|{\sim}U(p,q) = BU\|K^+({\sim}q) \vee BU\|I^+(q \wedge {\sim}U(p,q)).$$

We can similarly define $BS\|M$ for M a boolean combination of $U/S(A,B)$s.

We now want to define the connective $U(A,B\|M)$ reading until A, M is true at any gap defined by B (right gap or left gap).

Definition 11.4.5 Let $\beta = \beta(M,B)$

$$= (BU\|\top \to BU\|M) \wedge (BS\|\top \to BS\|M)$$

and put

$$\begin{array}{rl} R = & U(A,\beta) \\ \vee & U((\sim B \vee K^-\sim B) \wedge U(A,B),\beta) \\ \vee & (\beta U\|(M \wedge U(A,B)). \end{array}$$

Then

$$\begin{array}{rl} U(A,B\|M) = & U(A,B) \\ \vee & [(\sim B \vee K^-(\sim B) \vee K^+(\sim B)) \wedge R] \\ \vee & U((\sim B \vee K^+\sim B) \wedge R, B) \\ \vee & BU\|(M \wedge R). \end{array}$$

We similarly define $S(A,B\|M)$.

Definition 11.4.6 We now define the connective $U(B\|M,q)$ reading q is true until a gap definable by B (either from the right or from the left) in which gap M is true.

$$\begin{array}{rl} U(B\|M,q) = & U(q \wedge B \wedge [(q \wedge B)U\|M \wedge K^+(\sim B)],q) \\ \vee & U(\sim B \wedge q \wedge U(BS\|M,q),q) \\ \vee & qU\|(M \wedge I^+B \wedge K^-(\sim B)). \end{array}$$

The first disjunct in the definition describes the case of a left B-gap. The second describes the case of a right B-gap in which q continues to be true in an interval beyond the gap and the third disjunct describes the case of a right B-gap with q continuously true exactly until the gap and no further.

$S(B\|M,q)$ is similarly defined.

Definition 11.4.7 Suppose that M and N are boolean combinations of Until and Since. We define the connective

$$U(A\|N,B\|M)$$

to mean that until a gap of A in which N is true, M is true at any gap of B as follows:

Let

$$\beta = \beta(M, B) = (BU\|\top \to BU\|M) \wedge (BS\|\top \to BS\|M)$$

and put

$$\begin{array}{rcl} R = & & U(A\|N, \beta) \\ & \vee & U((\sim B \vee K^-(\sim B)) \wedge U(A\|N, B), \beta) \\ & \vee & U(B\|(M \wedge U(A\|N, B)), B). \end{array}$$

Then

$$\begin{array}{rcl} U(A\|N, B\|M) = & & U(A\|N, B) \\ & \vee & [(\sim B \vee K^-(\sim B) \vee K^+(\sim B)) \wedge R] \\ & \vee & U((\sim B \vee K^+ \sim B) \wedge R, B) \\ & \vee & BU\|(M \wedge R). \end{array}$$

$S(A\|N, B\|M)$ is defined similarly.

Important Note:

The notation $B\|M$ or $A\|N$ does not distinguish between right and left gaps. If we want to describe a right gap specifically, for example, we write $B\|(I^+B \to M)$ or $B\|(I^+B \wedge M)$ depending on which side of the gap we are on. Thus, for example,

$$U(A\|(N \wedge I^-A), B\|(I^+B \to M))$$

says that until a left A-gap in which N holds, M holds in any right B-gap.

Lemma 11.4.8 *Let D be any wff of $L*$. Then there exists a wff $D^\Box$ in the language of $\{U, U', S, S'\}$ such that for any linear order $(T, <)$, any valuation h, and any $t \in T$,*

$$\|D^\Box\|_t^h = \|D\|_t^h.$$

The map $\Box$ is defined in such a way that it commutes with the boolean connectives.

Proof. We are going to define some complicated maps $\sqcap$, GDEF, gap, and point inductively on the construction of some wffs in *Big* and then use them to define $\Box$.

We say that a legitimate wff A of *Big* is in active conjunctive form iff it is a conjunction of wff of the following three forms:

- $c \wedge p^\pm$ for some normal atom p,

- $c \wedge \pm U/S(\alpha, {\sim}\delta)$, or

- ${\sim}c \wedge \Gamma^{\pm}({\sim}\zeta) \wedge \pm U/S(\alpha, {\sim}\delta)$,

where α,δ and ζ are themselves in active conjunctive form.

It is not difficult to show that any active wff of *Big* can be equivalently rewritten as a wff in active conjunctive form. We first write it in active normal form and then inductively remove disjunctions from the first place of Untils and Sinces, conjunctions from the second place, and conjunctions from within $\Gamma^{\pm}$s. To do this we need the simple equivalences from chapter 10.

Next we classify wff $\theta_1 \wedge \ldots \wedge \theta_k$ in active conjunctive form as *point* form if θ_1 is of either of the first two forms above and of $\pm$ right (${\sim}\zeta$) defined *gap* form iff θ_1 is of the third form (where $-$ right $=$ left).

Now by simultaneous recursion on the construction of wffs in active conjunctive form we define our four maps.

$$
\begin{array}{lcl}
\text{point } [c \wedge p^{\pm}] & = & \pm p \\
\text{point } [c \wedge \pm U/S(\alpha, {\sim}\delta)] & = & \pm \sqcap U/S(\alpha, {\sim}\delta) \\
\text{point } [{\sim}c \wedge \Gamma^{\pm}({\sim}\zeta) \wedge \pm U/S(\alpha, {\sim}\delta)] & = & \bot \\
\text{point } [\alpha \wedge \delta] & = & \text{point } [\alpha] \wedge \text{ point } [\delta]
\end{array}
$$

$$
\begin{array}{lcl}
\text{gap } [c \wedge p^{\pm}] & = & \bot \\
\text{gap } [c \wedge \pm U/S(\alpha, {\sim}\delta)] & = & \bot \\
\text{gap } [{\sim}c \wedge \Gamma^{+}({\sim}\zeta) \wedge \pm U/S(\alpha, {\sim}\delta)] & = & \\
\quad {\sim}\sqcap S(c, {\sim}\zeta) \wedge \sqcap U(c, {\sim}\zeta) \wedge \pm \sqcap U/S(\alpha, {\sim}\delta) & & \\
\text{gap } [{\sim}c \wedge \Gamma^{-}({\sim}\zeta) \wedge \pm U/S(\alpha, {\sim}\delta)] & = & \\
\quad {\sim}\sqcap U(c, {\sim}\zeta) \wedge \sqcap S(c, {\sim}\zeta) \wedge \pm \sqcap U/S(\alpha, {\sim}\delta) & & \\
\text{gap } [\alpha \wedge \delta] & = & \text{gap } [\alpha] \wedge \text{ gap } [\delta].
\end{array}
$$

Notice that gap α is always a boolean combination of Untils and Sinces.

For each gap formula we need to isolate the wff which defines the gap. For this purpose we introduce the map GDEF which again is defined on the wff in active conjunctive form. For such a wff ζ, GDEF (ζ) is a set of formulae from *Big*. Because gaps can be located by the behaviour of other gaps this map is recursive and quite complicated.

- If ζ is in point form then GDEF (ζ) $=$ { point (ζ)},

- if ζ is in gap form and the first conjunct of ζ is

$$ {\sim}c \wedge \Gamma^{+}({\sim}\theta) \wedge C, $$

then

$$\text{GDEF }(\zeta) = \quad \text{GDEF }(\theta) \\ \cup\{S({\sim}B\|\text{ gap }(\zeta), {\sim}B) \mid B \in \text{ GDEF }(\theta)\}.$$

and

- if ζ is in gap form and the first conjunct of ζ is

$${\sim}c \wedge \Gamma^{-}({\sim}\theta) \wedge C,$$

then

$$\text{GDEF }(\zeta) = \quad \text{GDEF }(\theta) \\ \cup\{U({\sim}B\|\text{ gap }(\zeta), {\sim}B) \mid B \in \text{ GDEF }(\theta)\}.$$

Suppose that α, δ, θ and ζ are in active conjunctive form. We define $\sqcap U/S(\alpha, {\sim}\delta)$ differently depending on the form of the subwff.

- α and δ are point form. Put

$$\sqcap U/S(\alpha, {\sim}\delta) = U/S(\text{ point }(\alpha), {\sim}\text{ point }(\delta)).$$

- α point and δ is $\pm$ right ${\sim}\theta$ gap. Put

$$\sqcap U/S(\alpha, {\sim}\delta) = \bigwedge_{C \in \text{ GDEF }(\theta)} U/S(\text{ point }(\alpha), ({\sim}C)\|{\sim}\text{ gap }(\delta)).$$

- α $\pm$ right ${\sim}\zeta$ gap and δ is point. Put

$$\sqcap U/S(\alpha, {\sim}\delta) \\ = \bigvee\nolimits_{B \in \text{ GDEF }(\zeta)} U/S(({\sim}B)\|\text{ gap }(\alpha), {\sim}\text{ point }(\delta)).$$

- α $\pm_1$ right ${\sim}\zeta$ gap and δ is $\pm_2$ right ${\sim}\theta$ gap. Put

$$\sqcap U/S(\alpha, {\sim}\delta) = \bigvee\nolimits_{B \in \text{ GDEF }(\zeta)} \bigwedge\nolimits_{C \in \text{ GDEF }(\theta)} \\ U/S(({\sim}B\|\text{ gap }(\alpha)), ({\sim}C)\|{\sim}\text{ gap }(\delta)).$$

Now we show by simultaneous induction on construction of wffs α,δ of *Big* in active constructive form the following for all linear orders T and for all valuations h on T.

Suppose that we know how to evaluate a wff $U/S(A, B)$ of the language with $\{U, S, U', S'\}$ at a gap, i.e. at an element of $T * -T$.

Then

point : For all points $t \in T$,

$$\|\alpha\|_t^h = \| \text{ point } \alpha\|_t^h.$$

gap : For all gaps $g \in T* - T$,

$$\|\alpha\|_g^h = \| \text{ gap } \alpha\|_g^h.$$

GDEF : Suppose that α is in $\pm$ right $(\sim\zeta)$ gap form and that $g \in T* - T$. Then $\|\alpha\|_g^h = 1$ iff there is $B \in$ GDEF (ζ) such that g is a gap defined by $\sim B$ (on the $\pm$ right) at which gap (α) holds.

$\sqcap$: For all elements $e \in T*$,

$$\|U/S(\alpha, \sim\delta)\|_e^h = \| \sqcap U/S(\alpha, \sim\delta)\|_e^h.$$

Let us sketch the proof of this result:

point and gap: are straightforward.

GDEF : The only if direction follows directly from the result for gap as does the second half of the if direction. So consider α and ζ being in active conjunctive form, that α is in $\pm$ right $(\sim\zeta)$ gap form, that $g \in T* - T$, and that $\|\alpha\|_g^h = 1$. We must show that there is B from GDEF (ζ) such that g is a gap of T defined by $\sim B$ on the $\pm$ right.

So we know that $\alpha = \sim c \wedge \Gamma^{\pm}(\sim\zeta) \wedge \beta$ holds at g. Without loss of generality deal with the Γ^+ case.

There are two cases depending on the form of ζ.

ζ **in point form:** Arbitrarily soon before g there are elements of $T*$ where ζ holds and for a while after g there are no such elements. Since ζ is in point form these elements can only be points of T and by the result above for point they must be exactly the points of T where point (ζ) holds. Thus g is a right gap of T defined by $\sim$ point (ζ) and since GDEF $(\zeta) = \{$ point $(\zeta)\}$ we are done.

ζ **in gap form:** Assume that $\zeta = \sim c \wedge \Gamma^{\pm}(\sim\theta) \wedge \eta$. Wlog work with the Γ^+ case.

Again, arbitrarily soon before g there are elements of $T*$ where ζ holds and for a while after g there are no such elements. This time these elements are exactly those gaps γ in T for which there exists C in GDEF (θ) such that

- γ is defined on the right by $\sim C$, and
- gap (ζ) is true at γ.

Thus there must be some C in GDEF (θ) whose corresponding gaps appear arbitrarily soon before g but not for a while after.

We will show that either g itself is a gap defined by $\sim C$ on the right or that it is a gap defined by $\sim S(\sim C \|$ gap $(\zeta), \sim C)$ on the right. This will establish the result.

Assume that g is not defined by $\sim C$ on the right.

Clearly, there are points of T where $S(\sim C \|$ gap $(\zeta), \sim C)$ holds arbitrarily close before g. Just look soon enough after any of those gaps preceding g.

Now there is some $a \in T$ after g such that within the interval (g, a) there are no gaps defined by $\sim C$ on the right where gap(ζ) holds. So if $S(\sim C \|$ gap $(\zeta), \sim C)$ holds at any point t within this interval—and we are going to show that this cannot be—then the indicated $\sim C$ gap must be before or at g. Being before is impossible as there are other $\sim C$ gaps closer to g and so $\sim C$ could not remain true from then until t.

We have ruled out the possibility of g being itself a $\sim C$ defined gap so we are done.

$\sqcap$: is straightforward.

Now to define the map $\Box$. First extend $\sqcap$ to $\sqcap_1$ taking any legitimate wff $D = U/S(A, B)$, rewriting D as a boolean combination of $U/S(\alpha, \sim\delta)$ where α and δ are in active conjunctive form, commuting with the boolean connectives, and being the same as $\sqcap$ on the restricted domain.

Define $D^{\Box}$ to commute with the boolean connectives and evaluate

$$\begin{aligned} c^{\Box} &= \top \\ p^{\pm}\,\Box &= \pm p \\ U/S(A, B)^{\Box} &= \sqcap_1 U/S(A, B). \end{aligned}$$

We can translate the $\sim c$ to falsity here because we only evaluate $\Box$ at points of the order. The result follows easily. ■

11.5 Separation results

In this section we use the separation results of chapter 10 to show that $L*$ has a separation property and hence that it and the expressively equivalent language with Until, Since, and the Stavi connectives have expressive completeness over general linear time. First we show that the separating procedures—the acceptable rewriting—of chapter 10 preserve the legitimacy of formulae from the language with *Big*.

Lemma 11.5.1 *Let A,B,C, and D be formulae of Big. Suppose that A is a subwff of the legitimate wff B and that C is a wff which is equivalent to A over Dedekind complete time. Let D be the result of replacing any (i.e. none,some, or all) appearances of A in B by C.*

Clearly, D is equivalent to B over Dedekind complete time but also, under any of the following conditions, D is legitimate:

- *A is of the form $U(A_1, A_2)$, $S(A_1, A_2)$ or $K^{\pm}(A_1)$ and C is legitimate;*
- *A is of the form $\Gamma^{\pm}(A_1)$ and C is active;*
- *A is a boolean constituent of B of the form $\Gamma^{\pm}(A_1)$ and C is legitimate; or*
- *C is a boolean rearrangement of A.*

Proof. Since the legitimacy, activity, and passivity of a wff depends constructively on those of its subwff, most of the above replacements clearly make D legitimate. The last two require more investigation.

In the case of $A = \Gamma^{\pm}(A_1)$, a boolean constituent of legitimate B, we must have all the rest of the consituents legitimate. Replacing A by the legitimate C thus makes the new D legitimate.

Because of the remarks above we need only show that boolean rearrangement preserves legitimacy, activity, and passivity to establish the last part of the lemma. Clearly, legitimacy is preserved.

To show that boolean rearrangement preserves activity we need to check that the following three rewrites do:

- $\sim(\alpha \wedge \beta) \equiv \sim\alpha \vee \sim\beta$;
- $\sim(\alpha \vee \beta) \equiv \sim\alpha \wedge \sim\beta$; and
- $\alpha \wedge (\beta \vee \gamma) \equiv (\alpha \wedge \beta) \vee (\alpha \wedge \gamma)$.

These follow easily from the definition of the three adjectives, remembering that $\vee$ is thought of as an abbreviation there.

It follows that passivity is also preserved and we are done. ∎

Lemma 11.5.2 *Suppose that the legitimate formula A from Big can be acceptably rewritten as B. Then B is also legitimate.*

Proof. By the lemma above we need only observe that each of the acceptable rewrite rules listed in definition 10.3.4 preserve legitimacy and that the Γ eliminations preserve activity.

This can be done by inspection of each rule.

One useful result is to show that if A and C are active and B passive then $Q(A, B, C)$ of lemma 10.3.4 is passive. Each of A,B, and C is legitimate. C is active and $\sim A$ passive so $S(C, \sim A) \to A$ is legitimate. A formula of the form $D \to E$ for D active and E legitimate is passive so both disjuncts of Q are passive. Hence Q itself is.

In inspecting the Until/Since eliminations it will be noticed that in many cases we do not present the final separated result but only say which other eliminations to use to finish. If these other eliminations preserve legitimacy then in each case it can be seen that the original elimination will also. ■

Theorem 11.5.3 *Every wff of $L*$ is equivalent, over linear time, to a boolean combination of pure wffs of $L*$.*

Proof. By the active normal form lemma, any wff from $L*$ will be equivalent to a boolean combination of the wffs of the following forms:

- $c \wedge p^{\pm}$ for some atom p,
- $c \wedge \pm U/S(\alpha_1, \sim\alpha_2)$ for some wffs α_1 and α_2 in active normal form, or
- $\sim c \wedge \Gamma^{\pm}(\sim\alpha_0) \wedge \pm U/S(\alpha_1, \sim\alpha_2)$ for wff α_i in active normal form.

We can further rewrite constituents of the third form as $\perp$ since we will evaluate the *Big* wffs at a point of $T \subseteq T*$ where c is true.

Wffs in the first category are already pure so we need only show how to rewrite those wffs in the second category in an equivalent but separated form.

So suppose we look at $A = c \wedge \pm U/S(\alpha, \sim\delta)$ for active α and δ. Now $U/S(\alpha, \sim\delta)$ is a legitimate wff of *Big* and so, by lemma 10.3.13, can be acceptably rewritten as a boolean combination B of pure wffs which is equivalent to it over nice temporal structures. Without loss of generality we can suppose that B is a disjunction of conjunctions

$$B = \bigvee_i \bigwedge_j B_{ij}$$

of wffs from *Big*.

By the preservation of legitimacy lemma B is also legitimate and it follows from the definition of legitimacy that the B_{ij} are legitimate.

But then put $C = \bigvee_i \bigwedge_j (c \wedge B_{ij})$. Then

- C and A are equivalent wffs of *Big* over all nice temporal structures and
- C and each $(c \wedge B_{ij})$ is active and so is a wff of $L*$.

We still need to show that A and C are equivalent formulae of $L*$ over any linear temporal structure. This follows because, as remarked earlier, we can find a nice temporal structure in which truth of *Big* formulae corresponds exactly to truth of $L*$ formulae in any such structure. ■

Theorem 11.5.4 *The connectives* $\{U, U', S, S'\}$ *have the separation property over linear time and are therefore expressively complete.*

Proof. Given any A of the language L of $\{U, U', S, S'\}$, translate A into equivalent $A*$ of the language $L*$. $A*$ is equivalent to a wff B of $L*$ which is separated, i.e. a boolean combination of pure subformulae. Using the translation $\Box$, B is equivalent to $B^{\Box}$ in the language L. Because the translation $\Box$ preserves boolean combinations and because it preserves truth tables, so it preserves purity, $B^{\Box}$ is separated. Since $B^{\Box}$ is equivalent to A, and obviously F and P are definable in L, we can use the separation theorem for linear time. ■

Chapter 12

Further expressive completeness results

12.1 Introduction

In this chapter we prove the same result as we did in the last. First we will examine the idea of gaps in the flow of time in some detail. Then we will introduce some new connectives to talk about such gaps and use the expressive completeness of the Stavi connectives to deduce expressive completeness results for these new connectives. We also present some axiomatizations of these connectives.

Finally, we prove the expressive completeness of the Stavi connectives again but this time we do not use separation. Instead we present a more direct proof using games. Generalizations of such a proof may be useful in cases where separation is difficult to show.

12.2 Gaps in the flow of time

We identify gaps in a flow of time with supremum-less, non-empty, proper initial segments of the order and insert the gap in the appropriate place in the order. Dedekind complete orders, then, are those without gaps. The completion $T*$ of an order T is another order consisting of T and all the gaps in the right places, and is Dedekind complete.

The simplest kind of gap imaginable is an *isolated gap* which exists in an open interval of time which is otherwise gap-free. Taking one point out of the reals or sticking two copies of the integers together are two straightforward ways of producing an isolated gap.

We are going to define a hierarchy of kinds of gaps. For any (zero, successor or limit) ordinal α, an αth-order gap is a gap which is not of lesser order but lies in an open interval which contains, apart from itself, only gaps of order less than α. So a zero-order gap is just an isolated gap.

Of course this hierarchy does not include all the possible gaps. For example, nowhere in the rationals is there a gap of any order at all.

We will use the game characterization of unranked gaps. Let γ_0 be a gap of T. Players $\forall$ and $\exists$ move alternately, defining a sequence $\gamma_i (0 < i < \omega)$ of gaps. In each round, $\forall$ chooses an open interval I_i containing γ_i, and $\exists$ chooses $\gamma_{i+1} \in I_i$ with $\gamma_{i+1} \neq \gamma_i$. $\exists$ wins iff the game goes on for ω moves. γ_0 is unranked iff $\exists$ has a winning strategy for the game.

To see this, one can employ a straightforward transfinite induction to show that if γ_0 is ranked then $\forall$ has a winning strategy. This simply involves continually choosing open intervals around gap γ_i which contain, apart from γ_i itself, only gaps of lesser ranks. Conversely it can be seen that if γ_0 is unranked then every open interval containing it also contains other unranked gaps. $\exists$ can win by always choosing unranked gaps.

It is interesting to note that if all gaps in a flow of time have ordinal order then the cardinality of the flow is at least as great as the cardinality of any of those orders, and for every infinite ordinal α, there exists a flow of time of cardinality the same as α with a gap of order α.

Definition 12.2.1 Let T be a linear order of cardinality κ, and suppose that Γ is a set of gaps of T. A gap γ of T is said to be Γ-rich if every open interval I of T containing γ contains $\geq \kappa^+$ gaps from Γ. Here, κ^+ is the next largest cardinal after κ.

Proposition 12.2.2 *Let T be a linear order of cardinality κ. Suppose that Γ is a set of gaps of cardinality $\geq \kappa^+$. Then there is a Γ-rich gap $\gamma \in \Gamma$.*

Proof. If not, for each $\gamma \in \Gamma$ choose an open interval I_γ with endpoints $a_\gamma < b_\gamma$ in T, such that $\gamma \in I_\gamma$ and $|I_\gamma \cap \Gamma| \leq \kappa$. As $|\Gamma| > \kappa$, there is $\Gamma' \subseteq \Gamma$ with $|\Gamma'| > \kappa$ and $I_\gamma = I$, say, for all $\gamma \in \Gamma'$. Then $\Gamma' \subseteq I$, a contradiction. ■

Corollary 12.2.3 *Let T be a linear order of cardinality κ. Then T has at most κ ranked gaps.*

Proof. Assume not. Let Γ be a set of κ^+ ranked gaps of T. We will show that any Γ-rich gap is unranked; this will contradict the proposition.

Let γ_0 be an Γ-rich gap. $\forall$ and $\exists$ will play the game above, starting with γ_0. $\exists$ will privately construct sets $\Gamma_i (i < \omega)$ of κ^+ ranked gaps, so that each γ_i is Γ_i-rich. She begins by defining $\Gamma_0 = \Gamma$.

Inductively assume that $i < \omega$, and γ_i is a Γ_i-rich gap. $\forall$ chooses an interval $I_i = (a, b)$, say, around γ_i. I_i contains κ^+ gaps from Γ_i. Let Γ_{i+1} be the gaps from Γ_i contained in (a, γ_i) if this set has cardinality κ^+; otherwise let Γ_{i+1} be the gaps from Γ_i contained in (γ_i, b). So in any case, $|\Gamma_{i+1}| = \kappa^+$. By the proposition, $\exists$ can choose a Γ_{i+1}-rich gap $\gamma_{i+1} \in \Gamma_{i+1}$. If she does this, the game goes on forever and she wins. Hence γ_0 was unranked, as required. ■

Corollary 12.2.4 *Let T be a linear order of cardinality κ and let γ be a ranked gap of T. Then $|rank\ (\gamma)| \leq \kappa$.*

Proof. Any gap γ of rank α has gaps of rank β arbitrarily close, for all $\beta < \alpha$. So if T has a gap of rank α with $|\alpha| > \kappa$, then T has more than κ ranked gaps. The result now follows from the preceding corollary. ■

12.3 Connectives to talk about gaps

Recall from section 6.1 that the Stavi connectives U' and S', which, as we have seen in chapter 11, do talk about gaps, have first-order tables. By presenting our new connectives below in terms of U, S, U', and S' we thus guarantee that they are also first-order.

We start off with some new unary connectives which talk about a single gap located by the vicissitudes of a single temporal formula. First we need to know that there is a gap coming up:

$$\gamma^+(A) \quad = \quad U(\sim A, \top) \wedge U(A, A) \wedge \sim U(\sim A, A) \wedge \sim U(\sim U(\top, A), A).$$

This is true whenever A holds up until a gap but fails to hold arbitrarily soon afterwards. We call such a gap an A left gap: A is true on the left of the gap. See figure 12.1. Dually we can define γ^- and A right gaps. Notice

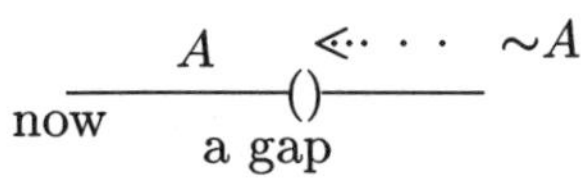

Figure 12.1

that $\gamma^\pm$ are expressible in $\{U, S\}$.

Next we specify that the gap coming up is isolated, as far as gaps definable by the same formula and in the same direction go:

$$\gamma_0^+(A) = \gamma^+(A) \wedge U'(\sim\gamma^+(A), A).$$

Dually we can define γ_0^-. Notice that we use the Stavi connectives here.

Now we can recursively define a hierarchy of connectives. For every $n \geq 0$, define

$$\gamma_{\leq n}^+(A) = \gamma_0^+(A) \vee ... \vee \gamma_n^+(A)$$

and

$$\begin{aligned} \gamma_{n+1}^+(A) &= \gamma^+(A) \\ &\wedge\sim\gamma_{\leq n}^+(A) \\ &\wedge U'(\gamma^+(A) \to \gamma_{\leq n}^+(A), A). \end{aligned}$$

$\gamma_{\leq n}^-$ and γ_n^- are defined dually.

Notice that there is a distinction between gaps in the flow of time and gaps definable by a particular temporal formula or even by any temporal formula. Thus we need to define another hierarchy of gaps—this time within a temporal structure rather than just in a flow of time. Let A be a temporal formula. For any ordinal α, an αth-order A left gap is an A left gap which is not of lesser order but begins an interval containing only A left gaps of lesser order. Dually we can define A right gaps of each order.

For $\alpha < \omega$, gap γ is an αth-order A left gap if and only if $\gamma_\alpha^+(A)$ holds in an interval on the left of γ. We consider the possibility of $\gamma_\alpha^+(A)$ for $\alpha \geq \omega$ later.

We have mentioned the distinction between αth-order A gaps and αth-order gaps in the flow of time. Nevertheless, it is clear that there is only an A gap when there is a gap in time at the right place and that it is an A gap of order α when that gap in time is at least of order α or possibly of non-ordinal order.

12.3.1 Construction of a non-ordinal order definable gap

In this subsection we will demonstrate the existence of definable gaps which do not fit into our scheme of classification. The idea is Robin Hirsch's.

We create a flow of time from a certain subset of the set $\mathcal{Q}^*$ of finite sequences of rational numbers. Let T consist of those non-empty sequences in which every rational number but the last is a power of 1/2 and the last number in the sequence is neither a power of 1/2 nor zero. We order the sequences as follows: $(a_0, a_1, ..., a_m) < (b_0, b_1, ..., b_n)$ iff there is some $k \geq 0$ such that $k \leq n$, $k \leq m$, for all $i < k$, $a_i = b_i$, and $a_k < b_k$.

We turn $(T, \leq)$ into a $\{p\}$-structure by making $T \models p(t)$ if and only if the last number in the sequence t is negative.

Each p left gap in T occurs just after the segment $(a^\wedge(-1), a^\wedge 0)$ where a is a sequence of powers of $1/2$ (possibly the empty sequence).

It is easy to prove that none of these gaps is isolated. Let $(a^\wedge 0, t)$ be any open interval after a gap. If t is not of the form $a^\wedge q^\wedge b$ for some possibly empty sequence b and some rational q, then we show that there is a left gap in $(a^\wedge 0, a^\wedge 1)$ and note that $a^\wedge 1$, which is of that form, must be less than t. So wlog t is $a^\wedge q^\wedge b$. Let r be any power of $1/2$ less than q. Clearly, all elements of T of the form $a^\wedge r^\wedge s$ are in the interval $(a^\wedge 0, t)$ and so is the gap at $a^\wedge r^\wedge 0$.

Now a very straightforward transfinite induction proves that

Lemma 12.3.1 *For any formula A and any non-zero ordinal α, each αth order A left gap is followed arbitrarily soon by zero-order A left gaps.*

So if a flow has no isolated p left gaps then it has no p left gaps of any order at all and we have our result.

12.4 Expressive power

From theorem 11.5.4 stating the expressive completeness of U, S, and the Stavi connectives over linear time, our first result falls out easily:

Lemma 12.4.1 *Over flows of time with only isolated gaps, $\{U, S\}$ is expressively complete.*

This is because, over such flows,

$$U'(A, B) \equiv \gamma^+(B) \wedge U({\sim}B, \gamma^+(B) \vee A).$$

Kamp's pioneering theorem is then a special case of this lemma.

Our next lemma shows that gaps don't have to get much more complicated before Until and Since are not sufficient.

Lemma 12.4.2 *In general linear time, $\{U, S\}$ is not expressively complete. There are even flows of time with a single non-isolated gap on which γ_0^+ is not expressible in terms of $\{U, S\}$.*

Proof. We take a flow of time $(T, \leq)$ with a single non-isolated gap and show that there is no temporal formula built from $\{U, S\}$which is equivalent to $\gamma_0^+(p)$ on all temporal structures over T.

T is constructed in two successive parts: the first one is obtained by taking a copy of $\mathcal{Z}$ for each negative integer and joining them into one long line, and the second has a copy of $\mathcal{Z}$ for each integer arranged in order. There is a gap at the beginning of the whole order and a gap at the end of each copy of $\mathcal{Z}$. The only non-isolated gap is that between the two parts.

A p-structure over T will be called nice iff

- on each little copy of the integers, either p is always true or always false, and
- every point has both p and $\sim p$ true in both its past and future.

An easy induction with several cases shows that for any formula ϕ constructed from p in the language $\{U, S\}$, there is a formula p, $\sim p$, $\top$ or $\bot$ which is uniformly equivalent to ϕ everywhere in all nice structures over T. It is easy to show, though, that these four formulae are all distinct in their truth conditions. For example, $U(p,p)$ is always equivalent to p.

So suppose, for contradiction, that we can express $\gamma_0^+(p)$ in $\{U, S\}$. By the above argument, we have a formula ψ always equivalent to it over nice structures.

Look now at a particular, nice p-structure in which p alternates in truth on copies of $\mathcal{Z}$ but is true in the last copy of the first part. Here $\gamma_0^+(p)$ is false. Thus ψ must be either $\sim p$ or $\bot$.

Look next at a structure in which p alternates in the first part, is true in the last copy of $\mathcal{Z}$ there, is false for an initial segment of the second part, and then alternates again. Here $\gamma_0^+(p)$ is true in the end of the first part. Thus ψ must be either p or $\top$ and we have our contradiction. ■

A similar proof to the above readily shows that

Lemma 12.4.3 *If for all i, $m < n_i$ then $\gamma_m^\pm$ is not expressible over all linear flows by any formula built from U,S and any (finite) number of $\gamma_{n_i}^\pm$.*

It is a little more difficult to prove that any $\gamma_m^\pm$ can be expressed in terms of $\gamma_n^\pm$ (in combination with U and S) for any $n < m$.

Lemma 12.4.4 *For any temporal formula P and any $n \geq 0$,*

$$\gamma_{n+1}^+(P) = \gamma_n^+(P \wedge \gamma^+(P) \wedge \sim\gamma_n^+(P)).$$

The dual result also holds.

Proof. This is immediate from the more informative lemma which follows the next: ■

Lemma 12.4.5 *Let $n \geq 0$ and P be any temporal formula. We write Q for*

$$P \wedge \gamma^+(P) \wedge \sim\gamma_n^+(P)$$

and consider the left P gaps in a structure.

- *Every left Q gap is a left P gap.*

- *No order n left P gap is a left Q gap.*
- *All the other left P gaps are left Q gaps.*

The dual result also holds.

Proof.

- To prove the first claim let us examine a left Q gap α say. Q is true in an interval, containing a point t say, on the left of α and false arbitrarily soon after. P, as a conjunct of Q, is thus true from t until α. If P is false arbitrarily soon after α then we have a left P gap at α as required. Suppose for contradiction that P is instead true for a while after α. Thus, like P, $\gamma^+(P)$ must stay true for a while after α. Finally, look at the third conjunct, $\sim\gamma_n^+(P)$, of Q. Since it is also true at t, β can not be an order n left P gap and again the conjunct remains true after α, at least as far as β. We have shown that Q remains true before and after α and we have our desired contradiction.
- The second observation is clear as $\gamma_n^+(P)$ is true arbitrarily recently before an order n left P gap.
- For the third let us look at a non-nth-order left P gap. For a while, on the left P,$\gamma^+(P)$ and $\sim\gamma_n^+(P)$ are all true. Since P, and hence Q, is false arbitrarily soon after the gap, we have a left Q gap.

■

Now we can actually be more specific about the orders of the gaps involved:

Lemma 12.4.6 *Let k and n be whole numbers and P be any temporal formula. We write*

$$Q \;=\; P \wedge \gamma^+(P) \wedge \sim\gamma_n^+(P)$$

and consider the left P gaps in a structure.

Any order k left P gap is

- *an order k left Q gap if $k < n$,*
- *not a left Q gap at all if $k = n$, and*
- *an order $k - 1$ left Q gap if $k > n$.*

Any order k left Q gap is

- *an order k left P gap if $k < n$, and*
- *an order $k+1$ left P gap if $k \geq n$.*

The dual result with right substituted for left also holds.

Proof. Fix n. Now we proceed by induction on k.

First part Suppose that we have an order k left P gap at α. If $k = n$ then the preceding lemma gives us our result. So suppose not. Thus α is a left Q gap. We will show that α is an order K left Q gap where

$$K = \begin{cases} k & \text{if } k < n \\ k-1 & \text{if } k > n \end{cases}.$$

Now for a while after α any left P gaps are of order less than k. Any left Q gaps which are in this interval are then, by the preceding lemma, left P gaps and so of order less than k as left P gaps. If $k < n$ then these gaps are, by the inductive hypothesis, left Q gaps of the same order less than $k = K$. If $k > n$ then these gaps are, also by the inductive hypothesis, left Q gaps of order one less than their order as P gaps, which is less than $K = k - 1$. In either case, for a while after α, all left Q gaps are of order less than K.

If $K = 0$ then we have shown that α is an isolated left Q gap.

Otherwise, since α is an order k left P gap it must have order $k - 1$ left P gaps arbitrarily soon afterwards. There are three cases:

- if $k < n$ then by the inductive hypothesis these are order $K-1 = k-1$ left Q gaps;
- if $k > n+1$ then these are order $k-2 = K-1$ left Q gaps; and
- if $k = n+1 > 1$ then the order $k-1 = n > 0$ left P gaps are also followed arbitrarily closely by order $k-2$ left P gaps which are, by the inductive hypothesis, also order $k-2 = K-1$ left Q gaps;
- if $k = n+1 = 1$ then $K = k-1 = 0$, which we have supposed not to be the case.

This proves that α is a left Q gap of order K.

Second part Suppose that α is an order k left Q gap. It is also a left P gap. We will show that it is also an order K left P gap where

$$K = \begin{cases} k & \text{if } k < n \\ k+1 & \text{if } k \geq n \end{cases}.$$

For a while after α all left Q gaps are of orders less than k. By our inductive hypothesis they will also be left P gaps of various finite orders. Thus α must be a finite-order left P gap, say, of order l. We know that l is not n for then α wouldn't be a left Q gap at all.

By the first part, if $l < n$ then $k = l$ so $K = k = l$ as required.

If $l > n$ then $k = l - 1 \geq n$ so $K = k + 1 = l$ as required.

■

Now what if we can use $\gamma_0^{\pm}$? Let us consider the new set of connectives $\{U, S, \gamma_0^{\pm}\}$ and ask about its expressive power. In fact, the connectives which talk of higher-order gaps are redundant. In expressive power, the γ_i^+ hierarchy collapses: for each $n \geq 0$,

$$\begin{array}{lc} \gamma_{n+1}^+(p) \equiv & \gamma^+(p) \wedge \sim\gamma_{\leq n}^+(p) \\ \wedge & \gamma_0^+(\gamma^+(p) \wedge \sim\gamma_{\leq n}^+(p)) \\ \wedge & U(\gamma_{\leq n}^{\pm}(p), p \vee U(\gamma_{\leq n}^{\pm}(p), \sim\gamma^+(p) \vee \gamma_{\leq n}^{\pm}(p))). \end{array}$$

Thus one might think that higher-order gaps hold no surprises for $\{U, S, \gamma_0^{\pm}\}$. In fact we do not even need to stop at finite orders.

Lemma 12.4.7 *$\{U, S, \gamma_0^{\pm}\}$ is expressively complete over general linear time.*

Proof. We will exhibit a $\{U, S, \gamma_0^{\pm}\}$ formula which is equivalent to $U'(p, q)$ in any $\{p, q\}$-structure. Because the dual formula will be equivalent to $S'(p, q)$, we will have shown that $\{U, S, \gamma_0^{\pm}\}$ is expressively complete over such structures.

Let ϕ be:

$$\begin{array}{ll} & \gamma^+(q) \wedge U(U(\sim q, p), q) \\ \vee & \gamma^+(q) \\ \wedge & U(\sim q, \gamma^+(\sim U(\sim q, p)) \vee p) \\ \wedge & \sim U(\sim q, \sim U(\sim q, p)) \\ \wedge & \gamma_0^+(\sim U(\sim q, p)). \end{array}$$

Suppose that B is a $\{p, q\}$-structure. We will show that for any $b \in B$,

$$\begin{array}{ll} & B \models U'(p, q)(b) \\ \Leftrightarrow & B \models \phi(b). \end{array}$$

($\Longleftarrow$) Let us assume that ϕ holds at b. We must show that $U'(p, q)$ is true at b. There are two cases.

If the first disjunct holds then it is clear that $U'(p, q)$ does too.

Now suppose that the second disjunct of ϕ holds at b but that the first does not. The first conjunct guarantees that q is true from b up until a gap which we can call β. $\sim q$ is true arbitrarily soon after β.

For contradiction we also suppose that $U'(p,q)$ does not hold at b. Thus p is false arbitrarily soon after β. Since $U(\sim q, \gamma^+(\sim U(\sim q,p)) \vee p)$ holds at b, we have $\gamma^+(\sim U(\sim q,p))$ true arbitrarily soon after β.

Since q is true up until β but p is false arbitrarily soon afterwards, we must have $\sim U(\sim q,p)$ holding from b at least up until β. But

$$\sim U(\sim q, \sim U(\sim q,p))$$

holds at b so $U(\sim q,p)$ must be true arbitrarily soon after β.

So $\sim U(\sim q,p)$ is true up until β but false arbitrarily soon afterwards. Thus $\gamma^+(\sim U(\sim q,p))$ holds at b and β is the $\sim U(\sim q,p)$ left gap involved.

Knowing that both $U(\sim q,p)$ and $\gamma^+(\sim U(\sim q,p))$ are true arbitrarily soon after β tells us that there are $\sim U(\sim q,p)$ left gaps arbitrarily soon after β.

Thus β is not an isolated $\sim U(\sim q,p)$ left gap and this contradicts

$$\gamma_0^+(\sim U(\sim q,p)).$$

holding at b. We are done.

($\Longrightarrow$) Suppose that $B \models U'(p,q)(b)$. So q is true for a while after b up until a gap, called β, say. We must show that ϕ is true at b. There are two cases.

If p is true for a while before β as well as after then it is clear that the first disjunct of ϕ holds at b.

So let us assume that that p is false arbitrarily soon before β.

In this case it is not difficult to see that the second disjunct of ϕ holds at b. To see that those conjuncts involving $\sim U(\sim q,p)$ hold, one need only notice that $\sim U(\sim q,p)$ holds from b up until β and is false for a while afterwards. It is false after the gap, i.e. $U(\sim q,p)$ holds because p is true for a while and $\sim q$ is true arbitrarily soon after β. ■

12.5 Other connectives

The connective

$$\rho^+(q) = U'(\sim q, q)$$

and its dual ρ^- could equally well have been used in this chapter instead of $\gamma_0^\pm$. We can define

$$\gamma_0^+(p) = \rho^+(\gamma^+(p))$$

so that the set $\{U, S, \rho^\pm\}$ is expressively complete.

$Ung(p,q)$ iff q holds until a gap after which it is arbitrarily soon false and after which there are no left p gaps for a while. Dually we define Sng. Note that $\gamma_0^+(p)$ equals $Ung(p,p)$. Thus $\{U, S, Ung, Sng\}$ is expressively complete.

We say that a p left gap is pure if it is not itself an isolated p left gap but there are no isolated p gaps for a while after the gap. Lemma 12.3.1 above shows that pure gaps have non-ordinal order. The non-ordinal order gaps constructed in the example above are pure. We can define a new connective using purity: $\pi^+(q)$ holds iff $\gamma^+(q) \wedge U({\sim}q, {\sim}\gamma_0^+(q))$ does.

An argument similar to the proof of lemma 12.4.2 shows that π^+ is not expressible in terms of U and S. We compare the truth of $\pi^+(p)$ before gaps in two different structures. In one p is true up until a gap, after which p is false for a while. In the other p is true up until a gap, after which open intervals of p being true, and open intervals of p being false replace rational numbers in an interval from that ordering.

The proof of lemma 12.4.2 can also be employed to show that γ_0^+ can not be expressed in terms of U, S, or $\pi^\pm$. Clearly, $\pi^\pm(p)$ is always false in the structures defined there.

12.6 An axiomatization of U, S, $\gamma_0^\pm$ using the irreflexivity rule

We first axiomatize U, S, and $\gamma_0^\pm$ over arbitrary linear flows of time using the irreflexivity rule of [Gabbay, 1981b]. This rule allows simple axiomatizations of many temporal connectives over irreflexive flows of time. We derive some simple consequences and list some open questions. In the next section we will relate some of these questions to the class of scattered flows of time.

In this section, unless otherwise stated, a *temporal formula* will mean one written with the connectives U, S, γ_0^+, and γ_0^-. We will use the standard abbreviations F, P, H, and G: Fp abbreviates $U(p, \top)$, etc. Recall also that $K^+(q)$ abbreviates ${\sim}U(\top, {\sim}q)$ and $\gamma^+(q)$ abbreviates $F{\sim}q \wedge U(q,q) \wedge {\sim}U({\sim}q \vee K^+({\sim}q), q)$; and similarly for K^- and γ^-.

We adopt as axioms the following:

1. All truth functional tautologies.

2. $G(p \to q) \to (Gp \to Gq)$
   $H(p \to q) \to (Hp \to Hq)$.

3. $q \to GPq, q \to HFq$.

4. $FFq \to Fq$ [transitivity].

5. $G(p \land Gp \to q) \lor G(q \land Gq \to p)$
   $H(p \land Hp \to q) \lor H(q \land Hq \to p)$ [linearity].

6. $r \land H{\sim}r \to [U(p,q) \leftrightarrow F(p \land H(Pr \to q))]$
   $r \land H{\sim}r \to [S(p,q) \leftrightarrow P(p \land G(F(r \land H{\sim}r) \to q))]$.

7. $r \land H{\sim}r \to [\gamma_0^+(q) \leftrightarrow (\gamma^+(q) \land F({\sim}q \land H(P({\sim}q \land Pr) \to {\sim}\gamma^+(q))))]$
   $r \land H{\sim}r$
   $\to [\gamma_0^-(q) \leftrightarrow (\gamma^-(q) \land P({\sim}q \land G(F({\sim}q \land F(r \land H{\sim}r)) \to {\sim}\gamma^-(q))))]$.

The rules of inference are:

modus ponens
substitution
generalization: $\vdash A \Rightarrow \vdash GA \land HA$
irreflexivity: $\vdash r \land H{\sim}r \to A \Rightarrow \vdash A$
(for all A and atoms r not occurring in A).

These axioms and rules are valid over irreflexive linear time.

Definition 12.6.1 If A is a temporal formula, N a temporal structure, and t a point of the flow of time of N (for short, '$t \in N$'), we write $N \vDash A(t)$ if A holds at t in N.

Take any set Σ of temporal formulae. A **model** of Σ will be an irreflexive linear temporal structure N such that for some $t \in N, N \vDash A(t)$ for all $A \in \Sigma$.

Theorem 12.6.2 (completeness) *Given any countable consistent set Σ of formulae, there is a countable model N of Σ in which all instances of the axioms are valid at every point.*

Proof (sketch, see chapter 6 for details). Using standard techniques we can obtain a countable irreflexive linear temporal structure N whose points are maximal consistent sets of temporal formulae. The irreflexivity rule allows us to assume that for each $t \in N$ there is an atom r with $r \land H{\sim}r \in t$. Further:

- There is $t_0 \in N$ with $\Sigma \subseteq t_0$.
- For all atoms q and all $t \in N, N \vDash q(t)$ iff $q \in t$.
- For each formula A there is an atom q such that $A \leftrightarrow q \in t$ for all $t \in N$.

- For all formulae A built using only F and P, and all $t \in N$, $A \in t$ iff $N \models A(t)$.

It now easily follows that for all $t \in N$ and all temporal formulae A, $N \models A(t)$ iff $A \in t$. The proof is by induction on the structure of A using axioms (6) and (7) (cf. theorem 6.3.5). Hence as $\Sigma \subseteq t_0$, we have constructed a model of Σ. ■

Question

Is there an axiomatization of U, S, and $\gamma_0^{\pm}$ without using the irreflexivity rule? Burgess axiomatizes U and S over arbitrary linear time in [Burgess, 1982a], without using this rule.

Even if the answer is negative, we still obtain the following corollaries, whose statements do not mention the irreflexivity rule.

Corollary 12.6.3 (compactness) *Let Σ be a set of temporal formulae (of U, S, γ_0^+ and γ_0^-). Suppose that every finite subset of Σ has a model. Then Σ has a model.*

Proof. With the given axioms and finitary rules, no contradiction is derivable from Σ. Hence by theorem 12.6.2 Σ has a model as stated. ■

Corollary 12.6.4

1. *The connective $\gamma^+_{\geq\omega}(-)$, saying that there is a gap of rank at least ω coming up on the right, is not definable by any first-order formula.*

2. *Not both of the connectives $\gamma^+_{\omega}(-)$ and $\gamma^+_{\text{ordinal}}(-)$, saying that coming up on the right is a gap of rank ω, or (respectively) ordinal rank, are first-order definable.*

Proof.

1. Assume for contradiction that $\gamma^+_{\geq\omega}(q)$ has a first-order table. Hence by expressive completeness of $\{U, S, \gamma^+, \gamma^-\}$ (lemma 12.4.8 above) there is already a temporal formula equivalent to $\gamma^+_{\geq\omega}(q)$. So consider $\Sigma = \{\sim\gamma^+_{\geq\omega}(q) \wedge \gamma^+_{\geq n}(q) : n < \omega\}$. Every finite subset of Σ has a model, but Σ does not. This contradicts the preceding corollary.

2. We have $\gamma^+_{\geq\omega}(q) \equiv \gamma^+(q) \wedge [\sim\gamma^+_{\text{ordinal}}(q) \vee \gamma^+_{\omega}(q) \vee \sim U'(\sim\gamma^+_{\omega}(q), q)]$, so the definability of both of γ^+_{ω} and $\gamma^+_{\text{ordinal}}$ would contradict (1).

■

Questions

1. Is γ^+_ω definable? Note that γ^+_ω is definable from $\gamma^+_{\geq\omega}$ by $\gamma^+_\omega(q) \equiv \gamma^+_{\geq\omega}(q) \wedge U'(\sim\gamma^+_{\geq\omega}(q), q)$.

2. Is $\gamma^+_{\text{ordinal}}$ first-order definable?

By corollary 12.6.4(part 2), relevant to the definability of γ_ω is the fact that the flows of time in which there are essentially no unranked gaps are essentially exactly the scattered flows: those that do not embed the rationals. They are our next topic.

12.7 Unranked gaps and scattered flows of time

We will observe that any temporal logic with first-order connectives over the class of all scattered flows of time is decidable. This gives a *weak* recursive axiomatization of the temporal structures with scattered flows of time, though a *strong* axiomatization is not possible (cf. the discussion after proposition 12.7.4).

Recall that a q-definable gap (one where $\gamma^+(q)$ holds on some interval to the left) is of rank ∞ (‘**unranked**’) if it is not of rank α for any ordinal α. An example of such gaps was given in section 12.3.1. They can also be exhibited by first defining $N_i (i = 0, 1)$ to be a structure with flow of time $\mathbb{Q}$, on which q is always true ($i = 1$) or always false ($i = 0$), and then replacing each $i \in \mathbb{Q}$ by a copy of N_0 or N_1 in such a way that any interval of $\mathbb{Q}$ contains copies of both structures. Let Q be the resulting temporal structure. Each $i \in \mathbb{Q}$ that is given a copy of N_1 yields a pure unranked q-gap in Q corresponding to the ‘right-hand end’ of that copy. Note that the flow of time of Q is isomorphic to $\mathbb{Q}$.

We defined unranked gaps of a flow of time in section 12.2. As an example, all gaps in $\mathbb{Q}$ are unranked. Flow-of-time gaps may not be ‘definable’ by a temporal formula (i.e. detectable by γ). However, note that an unranked definable gap is also an unranked flow-of-time gap.

Definition 12.7.1

1. If I is a linear order and $x, y \in I$ we will write $[x, y]$ for the closed interval of I with endpoints x, y. This extends the usual notation to the case where $x \geq y$.

2. An equivalence relation $\equiv$ on a linear ordering I is called a **condensation** if the $\equiv$-classes are convex (i.e. are intervals, but possibly

one-point intervals or with gaps for endpoints). Note that if $\equiv$ is a condensation, the ordering of I induces a canonical linear ordering of $I/\equiv$. Strictly speaking, the condensation is this linear ordering, and not the corresponding relation $\equiv$.

3. Recall that I is said to be **scattered** if $\mathbb{Q}$ does not embed into I. See [Rosenstein, 1982] for general information on scattered orderings.

Proposition 12.7.2 (cf. [Doets, 1989], lemma 2.3) *A linear ordering I is scattered iff whenever $\equiv$ is a condensation of $I, I/\equiv$ is not dense.*

Proof.

$\Rightarrow$ If $\equiv$ is a dense condensation of I, we can use the axiom of choice to choose a set of representatives of the $\equiv$-classes. Some subset of this will have order type $\mathbb{Q}$.

$\Leftarrow$ If $\mathbb{Q} \subseteq I$ define $\equiv$ on I by $x \equiv y$ iff $[x, y] \cap \mathbb{Q}$ is finite. Clearly, $I/\equiv$ is dense.

■

Theorem 12.7.3 *Let I be a linear ordering.*

1. *Suppose that I is scattered. Then there are no unranked flow-of-time gaps in I.*

2. *Assume that I is countable and that no temporal structure M with flow of time I has unranked definable gaps. Then I is scattered.*

Proof.

1. Clearly, $(*)$ any open interval of I containing an unranked gap contains infinitely many unranked gaps. Suppose that γ is an unranked gap of I. We define a chain of finite sets $S_n \subseteq I$ by induction on n so that for all adjacent points $i < j$ in S_n, the open interval (i, j) contains (a) an unranked gap, and (b) a point of S_{n+1}.

   Choose $i_0 < \gamma < i_1$ arbitrarily and let $S_0 = \{i_0, i_1\}$. Let $S_n = \{s_0, \ldots, s_k\}$ be given, satisfying (a) and with $s_0 < s_1 \ldots < s_k$. By $(*)$, for each $i < k$ we can take $s_i < t_i < s_{i+1}$ such that both (s_i, t_i) and (t_i, s_{i+1}) contain unranked gaps. Define $S_{n+1} = S_n \cup \{t_i : i < k\}$. Clearly, (b) holds now for S_n, and (a) holds for S_{n+1}.

   Having defined the S_n, we observe that $\bigcup_{n<\omega} S_n$ has order type $\mathbb{Q} \cap [0, 1]$, so that $\mathbb{Q}$ embeds into I. Hence I is not scattered.

Note that in the case where I is already a temporal structure and γ is a q-definable gap, the same argument shows that the extensions (truth sets) in I of q and of $\sim q$ both embed $\mathbb{Q}$.

2. The example $I = \mathbb{R}$ shows that the theorem can fail if the assumption of countability is discarded. Assume that I is not scattered. Let $\equiv$ be a condensation of I such that $(I/\equiv) \cong \mathbb{Q} \cap [0,1]$ (use proposition 12.7.2, the countability of I, and Cantor's theorem). Let Q^* be obtained from the structure Q made from N_0 and N_1 as above, by adding left and right endpoints at which q is false (say). Hence there is an order isomorphism $\theta : I/\equiv \to Q^*$. Define I as a q-structure M by: if $m \in I, M \models q(m)$ iff $Q^* \models q(\theta(m/\equiv))$. Then each unranked q-definable gap of Q^* gives rise to a similar gap in M.

■

If the compactness theorem held for the scattered orderings, then non-definability of γ_ω (even in the class of scattered orderings) would again follow. For the preceding argument using compactness would show that $\gamma_{\geq\omega}$ is not definable even over the scattered orderings. But $\gamma^+_{\geq\omega}(q) \equiv \gamma^+(q) \wedge [\sim\gamma^+_{\text{ordinal}}(q) \vee \gamma^+_\omega(q) \vee \sim U'(\sim\gamma^+_\omega(q), q)]$, as above. In scattered orderings, because of theorem 12.7.3 we have $\gamma^+_{\geq\omega}(q) \equiv \gamma^+_\omega(q) \vee \gamma^+(\sim\gamma^+_\omega(q))$, so that γ_ωs being definable would force $\gamma_{\geq\omega}$ to be definable, a contradiction.

However, we now show that this is not the case.

Proposition 12.7.4 *The compactness theorem fails for the class of scattered orderings.*

Proof. Introduce propositional atoms $q_i (i \in \mathbb{Q})$. Let $\Sigma = \{P(q_i \wedge H{\sim}q_i \wedge Pq_j) : j < i \text{ in } \mathbb{Q}\}$. Then any finite subset of Σ has a scattered model. But if M were a scattered model of Σ, then $\mathbb{Q}$ would embed into M via $i \mapsto m_i$ where $m_i \in M$ satisfies $M \models (q_i \wedge H{\sim}q_i)(m_i)$. ■

Now the rules of inference are finitary, so completeness implies compactness. Hence, for the class of scattered orderings, there is no completeness theorem of the form: Σ is consistent iff Σ has a scattered model. However, there is a weak completeness theorem that deals with the case where Σ is finite. That is, there is a recursive set of axioms such that $\vdash A$ iff $\models A$ for all temporal formulae A. This follows trivially from the following decidability result.

Proposition 12.7.5

1. *The monadic second-order theory of the class of countable scattered linear orders is decidable.*

2. *Over scattered flows of time, any temporal logic using connectives with first-order tables is decidable.*

Proof.

1. Let σ be a monadic second-order sentence in the signature $\{=, <\}$, where quantification over elements and subsets is allowed. Let Q be a new unary relation symbol and let σ^Q denote the relativization of σ to Q (i.e. the first-order quantifiers $\exists x, \forall x$ are replaced by $\exists x \in Q, \forall x \in Q$ respectively, and the second-order quantifiers $\exists X, \forall X$ by $\exists X \subseteq Q$ and $\forall X \subseteq Q$ respectively. Later we give a formal definition of relativization in the first-order case.) Let $\xi(Q)$ be the formula

$$\forall R \subseteq Q([\exists xy(R(x) \wedge R(y) \wedge x < y)]$$

$$\rightarrow \exists xy(R(x) \wedge R(y) \wedge x < y \wedge \sim\exists z(x < z < y \wedge R(z)))).$$

So $\xi(Q)$ says that the set of points where Q holds is a scattered ordering. Now any countable linear ordering embeds into $(\mathbb{Q}, <)$. So $\mathbb{Q} \models \exists Q(\xi(Q) \wedge \sigma^Q)$ iff σ has a countable scattered model.

It follows from Rabin's celebrated result [Rabin, 1969] that the monadic second-order theory of $\mathbb{Q}$ is decidable: cf. [Burgess and Gurevich, 1985], theorem 2.6] and section 15.4.5. Hence there is an algorithm to decide whether $\mathbb{Q} \models \exists Q(\xi(Q) \wedge \sigma^Q)$. This completes the proof.

2. It follows from the downward Löwenheim–Skolem theorem (see [Chang and Keisler, 1990]) that if A is a temporal formula with a first-order table, then A has a scattered model iff A has a countable scattered model. Let A use atoms $p_1, \ldots, p_n$ and have table $\alpha(x, P_1, \ldots, P_n)$, where the P_i are unary relation symbols corresponding to the atoms. Then A has a scattered model iff the monadic second-order sentence

$$\exists P_1, \ldots, P_n \exists x \alpha(x, P_1, \ldots, P_n)$$

holds in some countable scattered linear order. By (1) there is an algorithm to decide this question.

■

Remarks 12.7.6

1. It follows trivially that given any set of connectives with first order tables, there is a recursive axiomatization of the class K of temporal structures with scattered flow of time. We simply take as axioms $\{A : A \text{ is valid in every structure in } K\}$; this set is recursive by proposition 12.7.5. The only proof rule required is substitution.

2. In section 6.9 a finite (not merely recursive) axiomatization of the temporal logic with Until and Since over the real numbers $\mathbb{R}$ was given. In that proof a certain condensation $\sim_r$ (for $r < \omega$) was defined, and the irreflexivity rule used to show that every $\sim_r$ −class was a closed interval of the flow of time. The temporal translation B of $\sim\exists y < x(y \sim_r x)$ was then true exactly at the left-hand endpoint of each $\sim$-class, so a single axiom could be used to specify properties of the condensation $M/\sim_r$, uniformly in r. In our case the relevant axiom would be $\Diamond(B \wedge FB) \to \Diamond(B \wedge U(B, \sim B))$ (cf. proposition 12.7.4), but we have not found a formula true exactly once in each $\sim_r$ −class (our proof of proposition 12.7.4 uses the axiom of choice). So this method does not appear to be applicable in the scattered case.

12.8 Expressive completeness of U, S and Stavi connectives over linear time

In this section we will prove theorem 11.5.4 again. That is, we establish expressive completeness of U, S, and the Stavi connectives for arbitrary linear flows of time. The formal statement follows after some initial definitions. This time we will not use separation. Our argument was sketched in [Gabbay *et al.*, 1980] for the case of U and S over natural numbers time; the generalization to arbitrary linear time was indicated but not proved.

Definition 12.8.1

1. We fix an arbitrary *finite* set L of propositional atoms. We will consider first-order formulae $\varphi(\overline{x})$ in the 'monadic' language with $=, <$, and a unary relation symbol Q for every atom $q \in L$. We also consider temporal formulae. Unless otherwise stated, a temporal formula will be one built from the atoms of L using the boolean connectives and the binary temporal connectives U, S, U', and S' (standing for *Until* and *Since* and the Stavi connectives).

2. A **temporal (L-)structure** is formally a triple $N = (T, <, h)$, where $(T, <)$ is an irreflexive poset (the **flow of time** of N) and $h : L \to 2^T$ is the assignment map. We will often abuse notation by identifying N with its flow of time T. Moreover, as every temporal formula A defines a subset of a structure—the set of time points $h(A)$ where A is true—we will regard A as an extra atom and use it in monadic first-order formulae as a monadic relation symbol. This simplifies the notation a little. So, for example, $N \models \forall x U(A, B)(x)$ iff $U(A, B)$ is true at every point of N.

3. We will usually use Roman letters for temporal formulae and Greek for classical first-order ones.

In this setting, theorem 11.5.4 becomes:

For all L-formulae $\varphi(x)$ there is a temporal formula A such that if N is a linear temporal structure (i.e. one with linear flow of time) in which the atoms of φ have interpretations then for all $t \in N, N \models \varphi(t)$ iff $N \models A(t)$. Moreover, A is effectively obtainable from φ (i.e . by an algorithm).

This says that the temporal logic with Until, Since, and the Stavi connectives is **expressively (functionally) complete** over linear time. Our proof here is based on the sketch in [Gabbay *et al.*, 1980]. The algorithm resulting from separation is probably more efficient than this one.

We begin with some definitions.

Definition 12.8.2 (rank) The **rank** of a temporal formula A is defined to be the maximum depth of nesting of temporal connectives in A.

Example: if p, q are atoms then

$$rank(p \wedge q) = 0,$$

and

$$rank(\sim U(p, \sim S'(\sim q, q))) = 2.$$

Since L is finite, it is easy to show by induction on r that for each $r < \omega$ there is a finite set of temporal formulae of rank r such that every rank r formula is logically equivalent to one of them.

Definition 12.8.3 (gaps) We will use the definition of a **gap** in a linear order discussed in section 12.2 save that here we also consider there to be a gap at the end (respectively beginning) of the flow of time if there is no last (respectively first) point. We need a few extra notions. Let $M = (M, <, h)$ be any linear temporal structure. If γ is a gap and $S \subseteq M$, we say that $\gamma =$**sup(S)** if for all $t \in M, t \geq s$ for all $s \in S$ iff $t \geq \gamma$. We also say that $\gamma =$**inf(S)** if for all $t \in M, t \leq s$ for all $s \in S$ iff $t \leq \gamma$.

Let γ be a gap and let D be a temporal formula. We say that γ is **definable on the left by** D if D is true at all points of M in some non-empty interval (t, γ) on the left of γ, and not true throughout any non-empty interval (γ, u) on the right. The definition of a gap being **definable (by** D**) on the right** is made in a similar way. If $r < \omega$, an r**-definable gap** is one that is definable (on the left or right) by a formula D of rank at most r. For $r < \omega$ we let $M_r = M \cup \{r\text{-definable gaps of } M\}$, with the induced ordering $<$. So in general $M \subset M_0 \subset M_1 \subset \ldots$. For example, if M has no last element then $+\infty$ is a gap of M definable on the left by $\top$, so that $\infty \in M_0 \backslash M$. The situation for $-\infty$ is similar.

Definition 12.8.4 (relativized connectives) There is a natural way of evaluating temporal formulae of the form $\sharp(A, B)$ for $\sharp \in \{U, S, U', S'\}$ at gaps. For example, $U(A, B)$ holds at a gap γ (i.e. $\gamma \in M_n$ for some n) iff there is a *point* $t > \gamma$ where A holds, with B holding at all *points* $u \in (\gamma, t)$. To formalize this we relativize our connectives to points.

Fix $r < \omega$ and let $\mu \notin L$ be a new propositional atom. We define M_r as a temporal $L \cup \{\mu\}$–structure $(M_r, <, h')$ by:

$$h'(q) = h(q) \subseteq M \text{ for all } q \in L$$
$$h'(\mu) = M.$$

We will relativize U, S, U', and S' to μ.

Let $\varphi(\overline{x})$ be any first-order formula in the signature consisting of $=, <$, and a unary relation symbol for each atom of $L \cup \{\mu\}$. We define the **relativization** φ^μ of φ to μ by induction on φ :

$$\text{if } \varphi \text{ is quantifier free then } \varphi^\mu = \varphi;$$

$$(\sim \varphi)^\mu = \sim (\varphi^\mu);$$
$$(\varphi \wedge \psi)^\mu = \varphi^\mu \wedge \psi^\mu;$$
$$[\exists y \varphi(\overline{x}, y)]^\mu = \exists y(\mu(y) \wedge \varphi^\mu(\overline{x}, y)).$$

We introduce connectives U^μ, S^μ, U'^μ, and S'^μ whose tables are the relativizations to μ of the tables of U, S, U', and S' respectively. We can write formulae using those connectives that are meaningful in any $L \cup \{\mu\}$–structure. In particular we can interpret them in M_r. If A is any formula of $USU'S'$, we let A^μ be the formula obtained by replacing each U in A by U^μ, and similarly for S, U', and S'.

Remark 12.8.5

1. Let $\alpha(x)$ be the canonical **first-order table** of the temporal formula A, as defined in lemma 9.1.5: in any temporal structure $T, \{t \in T : T \models A(t)\} = \{t \in T : T \models \alpha(t)\}$. Then it is easily seen that the table of A^μ is just α^μ :i.e. for all $t \in M_r, M_r \models A^\mu(t)$ iff $M_r \models \alpha^\mu(t)$ (this holds for any $L \cup \{\mu\}$-structure).

2. If $t \in M$ then $M \models A(t)$ iff $M_r \models A^\mu(t)$.

3. Let $A = S'(B, C)$ where C has rank $\leq r$. If $t \in M_r$ and $M_r \models A^\mu(t)$, then the gap that A asserts the existence of actually lies in M_r (as C defines it on the right).

The Stavi connectives can express the existence of gaps, but cannot talk directly about what formulae are 'true' at them. So we need to transform properties of a gap into properties of 'real' points. This is done in the following definition and lemma.

Definition 12.8.6 Let D be any temporal L-formula. We define a temporal L-formula **left**(A, D) by induction on A:

- left$(p, D) = \bot$ for atomic p;
- left$({\sim}A, D) = U'(\top, D) \wedge {\sim}left(A, D)$;
- left$(A \wedge B, D) =$ left$(A, D) \wedge$ left(B, D);
- left$(U(A, B), D) = U'(B \wedge U(A, B), D)$;
- left$(U'(A, B), D) = U'(B \wedge U'(A, B), D)$;
- left$(S(A, B), D) = U(D \wedge B \wedge S(A, B) \wedge U'(\top, B \wedge D) \wedge {\sim}U'(D, B \wedge D), D)$;
- left$(S'(A, B), D) = U(D \wedge B \wedge S'(A, B) \wedge U'(\top, B \wedge D) \wedge {\sim}U'(D, B \wedge D), D)$.

So rank(left$(A, D)) \leq$ max$(rk(A), rk(D)) + 2$. We define **right**(A, D) similarly by swapping each U with S and U' with S' in the definition above.

The point of this definition is given by the following lemma.

Lemma 12.8.7 *Let A, D be temporal formulae with D of rank at most r. Let $m \in M_r$. Then the following are equivalent:*

1. *$M_r \models$ left$(A, D)^\mu(m)$.*

2. *There is $\gamma \in M_r - (M \cup \{\pm\infty\}), \gamma$ a gap of M defined by D to the left, with (a) $\gamma > m$, (b) D holds in M on (m, γ), and (c) $M_r \models A^\mu(\gamma)$.*

Proof. Clear. A corresponding result holds for right(A, D). ■

Definition 12.8.8 (games) We will need some results on Ehrenfeucht-Fraïssé games. Let Σ be any finite first-order signature without function symbols. Let M, N be Σ-structures. If $n < \omega$ we define a game $G^n(M, N)$ between two players, $\forall$ (male) and $\exists$ (female). The game has n rounds. In each round, $\forall$ chooses an element from whichever of M, N he wishes. Then $\exists$ responds by choosing an element of the other structure. After n rounds, two n-tuples $\overline{a}, \overline{b}$ of elements have been chosen from M, N respectively; the order of the elements in each tuple is the order that they were chosen as the game was played. $\exists$ wins this 'play' $(\overline{a}, \overline{b})$ of the game iff for all quantifier-free formulae $\varphi(\overline{x})$ of $\Sigma, M \models \varphi(\overline{a})$ iff $N \models \varphi(\overline{b})$. This is slightly stronger than saying that the map $\overline{a} \mapsto \overline{b}$ is a partial isomorphism, since Σ may have constant symbols.

A **strategy** for $\exists$ in a game is a set of rules (not necessarily deterministic) telling her what to do—this can be formalized as a family of functions. The strategy is said to be **winning** if whenever she uses it she wins.

The following is a well-known result of Ehrenfeucht–Fraïssé game theory.

Proposition 12.8.9 *Let Σ be any signature as above. Let M, N be Σ-structures and let $n < \omega$. The following are equivalent:*

1. *$\exists$ has a winning strategy for $G^n(M, N)$.*

2. *$M \models \sigma$ iff $N \models \sigma$ for all Σ-sentences σ of quantifier depth of nesting at most n.*

Proof (See [Ehrenfeucht, 1961]). As is well known, $\Leftarrow$ can fail if Σ is assumed infinite or to have function symbols. ■

Notation 12.8.10 If $x < y$ in M_r we write (x, y) for $\{t \in M : x < t < y\}$, and if $n \leq r, (x, y)_n$ for $\{t \in M_n : x < t < y\}$. We write $[\mathbf{x}, \mathbf{y}]_\mathbf{n}$ for $\{t \in M_n : x \leq t \leq y\}$, etc. We do not require that $x, y \in M_n$.

Definition 12.8.11 (special games on temporal structures) We now introduce a modified version of the game defined above. Let M and N be linear temporal structures. The game $\mathbf{G_{n;r}(M, xy; N, x'y')}$ for $n, r <$

$\omega, x < y$ in M_r, and $x' < y'$ in N_r, is played as follows. There are only two rounds. $\forall$ begins by choosing n elements $a_1, \ldots, a_n \in [x,y]_r$; $\exists$ responds with elements $a'_1, \ldots, a'_n \in [x', y']_r$. Then $\forall$ chooses one more element $b' \in [x', y']$—so b' must not be a gap—and $\exists$ replies with $b \in [x, y]$.

$\exists$ wins iff:

1. the tuples $xy\overline{a}b$ and $x'y'\overline{a'}b'$ have the same order type, and if $t \in xy\overline{a}b$ and t' is the corresponding element of $x'y'\overline{a'}b'$, then:
2. t is a gap of M iff t' is a gap of N,
3. for each temporal L-formula A of rank at most $r, M_r \models A^\mu(t)$ iff $N_r \models A^\mu(t')$.

Lemma 12.8.12 *Let M, N, etc., be as above. Suppose that $\exists$ has a winning strategy σ for $G_{n;r}(M, xy; N, x'y')$ for some $n, r < \omega$. Let $n' \leq n, r' \leq r$. Then σ gives in the natural way a winning strategy for $G_{n';r'}(M, xy; N, x'y')$ provided that $x, y \in M_{r'}$ and $x', y' \in N_{r'}$.*

Proof. Recall that K^+X abbreviates the formula $\sim U(\top, \sim X)$, and $K^-X = \sim S(\top, \sim X)$.

Suppose in a play of $G_{n';r'}(M, xy; N, x'y')$, $\forall$ chooses $a_1, \ldots, a_{n'} \in [x,y]_{r'}$. Then $\exists$ defines $a_{n'+1}, \ldots, a_n$ to be x, say. So $a_1, \ldots, a_n \in [x, y]_r$. She applies σ to $\overline{a}$ to obtain $\overline{e} \in [x', y']_r$.

We claim that each $e_i \in [x', y']_{r'}$. This is clear if $r' = r$, so assume that $r' < r$. Take i; certainly if $a_i \in M$ then $e_i \in N$. Otherwise a_i is defined by some formula $\sim D$ of rank $\leq r'$. So letting $D' = (K^+D \wedge \sim K^-D) \vee (K^-D \wedge \sim K^+D)$, a formula of rank $\leq r' + 1 \leq r$, we have $M_r \models D'^\mu(a_i)$. As σ is winning, $N_r \models D'^\mu(e_i)$. Hence e_i is also a gap defined by $\sim D$; so $e_i \in N_{r'}$.

If $\forall$ now chooses $a' \in [x', y']$ then $\exists$ simply uses σ to respond with $e' \in [x, y]$. Then $\overline{a}e'$ and $\overline{e}a'$ satisfy the same order relations and rank r temporal formulae, hence also the same temporal formulae of rank r'. Hence $\exists$ has won the play. ■

We want to characterize the formulae associated with these games.

Definition 12.8.13

1. Let $r < \omega$ and $t \in M_r$ be given. Define X_t to be the conjunction of all temporal L-formulae X of rank $\leq r$ with $M_r \models X^\mu(t)$. This conjunction is effectively finite, as because L is finite there are up to logical equivalence only finitely many distinct formulae of any rank. Hence X_t can be taken to be a temporal formula of rank r.

If $t < u$ in M_r, define $X_{(t,u)}$ to be $\bigvee_{v \in (t,u)} X_v$. Again the disjunction is effectively finite, so that $X_{(t,u)}$ can be taken to be a formula of rank r. Note that only non-gaps contribute to the disjunction.

2. (This definition is from [Gabbay *et al.*, 1980].)
An $n;r$-**decomposition** formula is a first-order formula of the form:

$$\psi(x_1, x_2) = \exists y_1, \ldots, y_n[x_1 < y_1 < \ldots < y_n < x_2 \wedge \forall z \chi(x_1, x_2, \overline{y}, z)],$$

where χ is a conjunction of formulae of the following kinds:

(a) $\theta(t)$, where t is an element of $x_1 x_2 \overline{y}$ and θ is either μ, $\sim\mu$, or A^μ for some temporal L-formula A of rank $\leq r$;

(b) $\mu(z) \wedge a < z < b \to B^\mu(z)$, where $a < b$ are adjacent elements of the sequence $x_1 y_1 \ldots y_n x_2$, and B is a temporal formula of rank $\leq r$.

Cf. Definition 12.8.1(2).

Lemma 12.8.14 *Let $M, N, x, y, x,', y'$ be as above. Let $n, r < \omega$. Then the following are equivalent:*

1. *$\exists$ has a winning strategy for $G_{n;r}(M, xy; N, x'y')$.*

2. *For all $n;r$-decomposition formulae $\varphi(x_1, x_2)$, $M_r \models \varphi(x, y) \Rightarrow N_r \models \varphi(x', y')$.*

Proof. (1) $\Rightarrow$ (2)—clear.
(2) $\Rightarrow$ (1) Let $\forall$ choose $a_1, \ldots, a_n \in [x, y]$ in his first move. Assume without loss that $x < a_1 < \ldots < a_n < y$. Write a_0 for x and a_{n+1} for y. Let $\psi(y_0, y_{n+1}) = \exists y_1 \ldots y_n[y_0 < y_1 < \ldots < y_{n+1} \wedge \forall z(\bigwedge_{a_i \in M} \mu(y_i) \wedge \bigwedge_{a_i \notin M} \sim\mu(y_i) \wedge \bigwedge_{i \leq n+1} X_{a_i}(y_i) \wedge \bigwedge_{i \leq n}(\mu(z) \wedge y_i < z < y_{i+1} \to X_{(a_i, a_{i+1})}(z)))]$. Then ψ is an $n;r$-decomposition formula and $M_r \models \psi(x, y)$. Hence by assumption $N_r \models \psi(x', y')$, and so there are $e_i \in (x', y')$ witnessing the $\exists$s in ψ. If $\exists$ chooses the e_i she can easily win the game. ■

The main step in our proof is

Theorem 12.8.15 *Suppose that M, N are linear temporal structures. Then $(*)_n$ holds for all $n < \omega$:*

$(*)_n$ *For all $r < \omega$, if $x < y$ in M_r, $x' < y'$ in N_r, and $\exists$ has a winning strategy for $G_{1+3n;r+4n}(M, xy; N, x'y')$, then $\exists$ has a winning strategy for $G_{n;r}(N, x'y'; M, xy)$.*

This says that if $\exists$ has winning strategies for enough 'forward' games $G(M, xy; N, x'y')$ then she has a winning strategy for a given 'backward' game $G(N, x'y'; M, xy)$. The proof does not use compactness and is really a syntactic result—we could equally prove that a certain class of formulae is closed under negation up to equivalence, which is what is done (for the case of U and S over $\mathbb{N}$ and without full proof) in [Gabbay *et al.*, 1980]. However, the game approach, though still complicated, seems rather simpler to present.

Before we prove theorem 12.8.15 we finish our result on expressive completeness.

Proposition 12.8.16 *Let M, N be linear temporal structures and let $x \in M, y \in N$. Suppose $r < \omega$ and that x and y satisfy the same temporal formulae of rank $r + 4n + 1$ in their respective structures. Then $\exists$ has a winning strategy for $G_{n;r}(M, -\infty x; N, -\infty y)$ and $G_{n;r}(M, x\infty; N, y\infty)$ for all $n < \omega$.*

Proof (sketch). Suppose for simplicity that $\forall$ chooses n points $x < a_1 < \ldots < a_n$ in M in the future of x. Let $a_0 = x$. Define C_n to be $X_{a_n} \wedge \sim U(\sim X_{(a_n,\infty)}, \top)$, and for $i < n, C_i$ to be $X_{a_i} \wedge U(C_{i+1}, X_{(a_i,a_{i+1})})$. So rank $(C_i) = r + n + 1 - i$. Then $M \vDash C_0(x)$, so that $N \vDash C_0(y)$. $\exists$ can use the form of C_0 to choose points $y = e_0 < e_1 < \ldots < e_n$ in N such that $N \vDash X_{a_i}(e_i)$ and $N \vDash X_{(a_i,a_{i+1})}(t)$ for all (non-gaps) $t \in (e_i, e_{i+1})$. If $\forall$ now chooses $t \in (e_i, e_{i+1})$ then $N \vDash X_u(t)$ for some $u \in (a_i, a_{i+1})$. If $\exists$ responds with such a u, she wins the game. The argument for the 'past' game is similar. If some of the a_i are gaps, the idea is the same but the formulae C are more complicated and involve formulae D defining the gaps, together with the formulae left(X_{a_i}, D) or right(X_{a_i}, D)—cf. the proof of Cases III and IV of theorem 12.8.15. In all cases we have rank$(C_0) \leq r + 4n + 1$. ■

Definition 12.8.17 Let f, g be any functions on ω satisfying $f(0) = g(0) = 0, f(n+1) \geq (1+3f(n))(2k_n)+1$, and $g(n+1) \geq g(n)+4f(n)$, where k_n is the number of inequivalent $(1+3f(n)); (g(n)+4f(n))$-decomposition formulae.

Proposition 12.8.18 *For all $n < \omega$ the following holds. Let M, N be linear temporal structures and let $x_1 < \ldots < x_m, y_1 < \ldots < y_m$ be increasing m-tuples of elements of M, N respectively, for arbitrary $m < \omega$. Define $x_0 = -\infty$ and $x_{m+1} = \infty$ in M_0. Define y_0, y_{m+1} similarly.*

Suppose that $\exists$ has winning strategies for

$$G_{f(n);g(n)}(M, x_i, x_{i+1}; N, y_i, y_{i+1})$$

and

$$G_{f(n);g(n)}(N, y_i, y_{i+1}; M, x_i, x_{i+1})$$

for all $0 \leq i \leq m$. *Then* $\exists$ *has a winning strategy for the Ehrenfeucht–Fraïssé game* $G^n((M,\overline{x}),(N,\overline{y}))$.

Proof. By induction on n. If $n = 0$ the result is trivial. Assume it true for n, let $r = g(n) + 4f(n) \leq g(n+1)$, and suppose that $\exists$ has a winning strategy for the games $G_{f(n+1);r}(M, x_i, x_{i+1}; N, y_i, y_{i+1})$ and $G_{f(n+1);r}(N, y_i, y_{i+1}; M, x_i, x_{i+1})$.

Let $\forall$ begin $G^{n+1}((M,\overline{x}),(N,\overline{y}))$ by choosing without loss $a \in M$. (If $\forall$ chooses in N the proof is the same as we have complete symmetry.) If $a \in \{x_1, \ldots, x_m\}$ then $\exists$ chooses the corresponding element of $\overline{y}$, and the result then follows using the induction hypothesis and lemma 12.8.12. So let $i \leq m$ be such that $x_i < a < x_{i+1}$. List as $\varphi_1, \ldots, \varphi_j$ the $[1+3f(n)]$; r- decomposition formulae $\varphi(u,v)$ such that $M_r \vDash \varphi(x_i, a)$, and as $\psi_1, \ldots, \psi_k$, the $[1+3f(n)]$;r-decomposition formulae $\psi(u,v)$ with $M_r \vDash \psi(a, x_{i+1})$.

Let $\exists$ choose witnesses for the existential quantifiers of each φ, ψ, together with a, making at most $n' = (1+3f(n))\ (j+k)+1 \leq f(n+1)$ points in $(x_i, x_{i+1})_r$ in all. She now applies her winning strategy for $G_{f(n+1);r}(M, x_i x_{i+1}; N, y_i y_{i+1})$. Let e be the point she chooses corresponding to a. Clearly (cf. lemma 12.8.12) we have $N_r \vDash \varphi_s(y_i, e)$ for all $s \leq j$ and $N_r \vDash \psi_s(e, y_{i+1})$ for $s \leq k$. By lemma 12.8.14, $\exists$ has a winning strategy for $G_{1+3f(n);r}(M, x_i a; N, y_i e)$ and for $G_{1+3f(n);r}(M, a x_{i+1}; N, e y_{i+1})$. Crucially, by theorem 12.8.15, she also has winning strategies for

$$G_{f(n);g(n)}(N, y_i e; M, x_i a)$$

and

$$G_{f(n);g(n)}(N, e y_{i+1}; M, a x_{i+1}).$$

By the induction hypothesis, $\exists$ has a winning strategy σ for

$$G^n((M, \overline{x}a),(N, \overline{y}e)).$$

So in $G^{n+1}(M,\overline{x}),(N,\overline{y}))$, $\exists$ can choose e in response to $\forall$'s choice of a and then follow σ. This strategy wins the game for her. ■

Corollary 12.8.19 *Let* M, N *be linear temporal structures and let* $x \in M, y \in N$. *Suppose that* x *and* y *satisfy the same temporal formulae of rank* $g(n+1)+1$ *in their respective structures. Then for all monadic first-order formulae* φ *(of* L*) of quantifier depth* $\leq n, M \vDash \varphi(x)$ *iff* $N \vDash \varphi(y)$.

Proof. By propositions 12.8.9, 12.8.16, and 12.8.18. ■

Expressive completeness now follows easily. For given $\varphi(x)$ of quantifier depth n, we may choose a finite L with atoms corresponding to the monadic predicates of φ. Now take a finite set Ψ of temporal formulae of rank $1 + g(n + 1)$ such that (1) if $A, B \in \Psi$ and $A \wedge B$ is consistent then $A = B$; (2) each temporal formula C of rank $1 + g(n + 1)$ is equivalent to a disjunction of formulae in Ψ. Let $\Psi' = \{B \in \Psi :$ for some linear M and $t \in M, M \models B(t)$ and $M \models \varphi(t)\}$. Then by corollary 12.8.19, φ is equivalent over linear time to the rank $1 + g(n + 1)$–formula $\bigvee \Psi'$.

Note that by a result of Gurevich ([Burgess and Gurevich, 1985], 2.7(a)), the universal monadic second-order theory of linear order is decidable (cf. theorem 15.4.6). Hence Ψ' is computable by an algorithm, so that the translation of first-order formulae into temporal ones is effective.

Proof (of theorem 12.8.15). We must prove

$(*)_n$ For all $r < \omega$, if $x < y$ in $M_r, x' < y'$ in N_r, and $\exists$ has a winning strategy for $G_{1+3n;r+4n}(M, xy; N, x'y')$, then $\exists$ has a winning strategy for $G_{n;r}(N, x'y'; M, xy)$.

We prove $(*)_n$ by induction on n. For the case $n = 0$ (r is arbitrary), assume that $\exists$ has a winning strategy σ for $G_{1;r}(M, xy; N, x'y')$ and that $\forall$ chooses $a \in (x, y)$ in the second round of $G_{0;r}(N, x'y'; M, xy)$ (as $n = 0$ the first round is 'empty'). a is not a gap. $\exists$ simply applies σ to choose a response $e \in (x', y')$. Clearly, $\exists$ has won.

Assume $(*)_n$ for $n < \omega$; we prove $(*)_{n+1}$. Fix $r < \omega, x < y$ in M_r and $x' < y'$ in N_r. Assume that $\exists$ has a winning strategy for

$$G_{4+3n;r+4(n+1)}(M, xy; N, x'y').$$

We will construct a winning strategy for $\exists$ in $G_{n+1;r}(N, x'y'; M, xy)$.

Suppose $\forall$ chooses $n + 1$ points $x' < a_0 < \ldots a_n < y'$ in N_r (we may assume that they are all distinct, for otherwise the result follows by the inductive hypothesis and lemma 12.8.12). Define the following rank r temporal formulae:

$$A = X_{(a_{n-1}, a_n)}$$
$$C = X_{(a_n, y')},$$

where if $n = 0$ we take a_{n-1} in A to be x'. Clearly, in N, A holds on (a_{n-1}, a_n) and C on (a_n, y'). Let

- $c = \inf\ \{t \in [x, y] : M \models C(u)$ for all $u \in (t, y)\}$.

If $c \notin M$ then either $c = x \in M_r$ already, or c is a gap definable on the right by C. Hence $c \in M_r$. Define $c' \in N_r$ similarly. See figure 12.2.

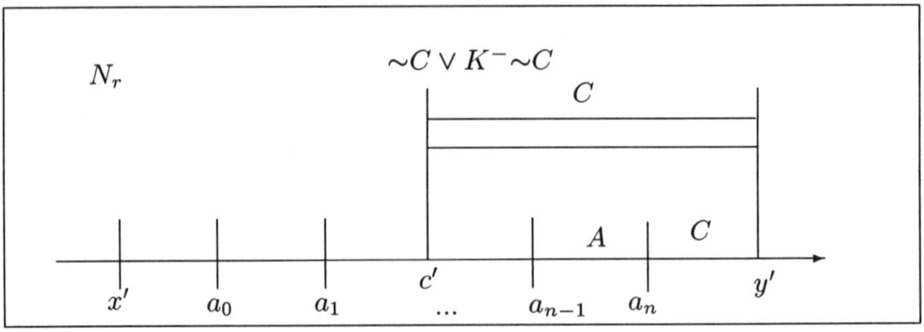

Figure 12.2

Claim 1
Consider a play of the game $G_{m;r'}(M, xy; N, x'y')$ for arbitrary $r' > r, m \geq 1$ in which ∃ uses a winning strategy. Let ∀ begin by choosing c plus $m - 1$ other points, and let ∃'s response to c be d (plus $m - 1$ other points). Then $d = c'$.

Proof of claim
As the strategy is winning, any rank r' temporal formula satisfied by one of ∀'s choices must also be satisfied by the corresponding choice of ∃. Now the rank $r + 1$ formula $C' = \sim C \vee K^- \sim C$ satisfies $M_r \models C'^{\mu}(c)$. Hence also $N_r \models C'^{\mu}(d)$, so $d \leq c'$.

If $d < c'$ then ∀ can choose $d' \in (d, y')$ with $N \models \sim C(d')$. ∃ now has no winning response, a contradiction. Hence $d = c'$. This proves the claim.

Claim 2
∃ has a winning strategy for

$$G_{1+3n;r+4(n+1)}(M, xc; N, x'c')$$

and for

$$G_{1+3n;r+4(n+1)}(M, cy; N, c'y').$$

Proof of claim
Let $r' = r + 4(n + 1)$. Suppose that ∀ chooses $1 + 3n$ elements in the interval $[x, c]_{r'}$. By assumption, ∃ has a winning strategy σ for the game $G_{4+3n;r'}(M, xy; N, x'y')$. ∃ adds c to ∀'s choices and applies σ (cf. lemma 12.8.12). As the order of ∃'s element choices from σ matches the order of ∀'s, Claim 1 ensures that her responses to ∀'s choices all lie in $[x', c']_{r'}$. If ∀ then chooses in $[x', c']$ then again ∃'s strategy will yield an answer in $[x, c]$. The strategy is clearly winning. To sum up, the restriction of σ to games in which ∀ always chooses in $[x, c]_{r'}$ and then in $[x', c']$ can

yield a winning strategy for $G_{1+3n;r+4(n+1)}(M, xc; N, x'c')$. Similarly for the intervals $[c, y], [c', y']$. This establishes the claim. We will use this argument repeatedly.

Hence by inductive hypothesis $(*)_n$, $\exists$ has winning strategies σ, τ for the backward games $G_{n;r+4}(N, x'c'; M, xc)$ and $G_{n;r+4}(N, c'y'; M, cy)$.

Now, clearly, $c \leq a_n$, so that $(x', c')_r$ contains at most n points from $\{a_0, \ldots, a_n\}$. The proof will divide into cases according to whether a_n is a point of N, a left- or a right-definable gap.

Case I: $a_0 \leq c'$: Then $(c', y')_r$ also contains at most n points from $\{a_0, \ldots, a_n\}$. So as $\exists$ is trying to win $G_{n+1;r}(N, x'y'; Mxy)$, she can use σ and τ to choose points $e_0, \ldots, e_n \in M_r$. She applies σ to those a_i in $(x', c')_r$ and τ to the rest using the method of lemma 12.8.12; if an a_i happens to be c' it can be dealt with by either strategy. If $\forall$ then responds in $[x, c)$ she uses σ, and if in $[c, y], \tau$. If she does this then by lemma 12.8.12 she will win the game.

Case II: All the points $a_0, \ldots, a_n$ lie in (c', y'), and $a_n \in N$ is not a gap.

Recall that $\exists$ is trying to win $G_{n+1;r}(N, x'y'; M, xy)$, i.e. to preserve all rank r formulae. Define $B = X_{a_n}$, and $b = \sup\{t \in (x, y) : M \models B(t)\} \in M_r$ (as before, either $b \in M, b = y$ or b is an r-definable gap, defined on the right by $\sim B$). Define $b' \in N_r$ similarly. Then, clearly, $b' \geq a_n$. See figure 12.3.

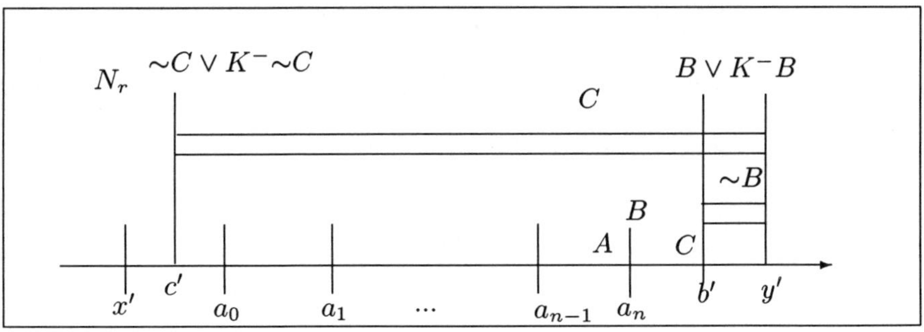

Figure 12.3

As in Claim 1, in any play of $G_{4+3n;r+4(n+1)}(M, xy; N, x'y')$ in which $\exists$ is using her winning strategy and $\forall$ chooses b, c among other points, $\exists$ will respond with b', c' among others. Hence again $\exists$ has a winning strategy for $G_{1+3n;r+4(n+1)}(M, cb; N, c'b')$. So by the induction hypothesis $(*)_n$ she has a winning strategy τ for $G_{n;r+4}(N, c'b'; M, cb)$. She already has a winning strategy σ for $G_{n;r+4}(N, x'c'; M, xc)$.

Let her first use τ in response to $a_0, \ldots, a_{n-1}$. It delivers n points

$e_0, \ldots, e_{n-1} \in (c, b)_r$ (cf Lemma 12.8.12). Now clearly

$$N_r \models U(B, A)^\mu(a_{n-1}):$$

a_n is a witness to this. (This holds even if a_{n-1} is a gap; if $n = 0$ we take $a_{-1}$to be c' and (see below) e_{-1} to be c.) $U(B, A)$ has rank $r + 1$, so as τ preserves formulae up to rank $r + 4, M_r \models U(B, A)^\mu(e_{n-1})$. Hence there is $z > e_{n-1}$ in M with $M \models B(z)$ and $M \models A(t)$ for all $t \in (e_{n-1}, z)$. But $e_{n-1} < b$. Hence we can assume that $z \leq b$. $\exists$ defines e_n to be such a z, completing her move. Clearly, e_n and a_n satisfy the same temporal formulae of rank r, as they both satisfy B.

Suppose that $\forall$ continues by choosing $t \in [x, y]$. Recall that by the game rules, t is not a gap. If $t < c$ then $\exists$ uses σ to respond, and if $c \leq t \leq e_{n-1}$ she uses τ. If $t \in (e_{n-1}, e_n)$ then $M \models A(t)$. By definition of A there is $t' \in (a_{n-1}, a_n)$ with $M \models X_{t'}(t)$. $\exists$ can then choose any such t' as her response. It follows that t and t' agree on all rank r temporal formulae, as required. If $t = e_n$ then $\exists$ responds with a_n. Finally, if $y > t > e_n$ then certainly $t > c$, so $M \models C(t)$. By definition of C there is $t' > a_n$ with $M \models X_{t'}(t)$, and $\exists$ can choose such a t' in response to t. If $\exists$ follows these directions she will win.

The remaining cases are similar to Case II, which gave a response e_n to a_n by letting B describe a_n and making $U(B, A)$ true at e_{n-1}. But a_n will now be a gap, so we must use the Stavi U'—and $U'(B, A)$ does not say that B^μ is true at the gap. So we use the formulae left(-,-) and right(-,-) instead.

Case III: All the points $a_0, \ldots, a_n$ lie in (c', y'), and a_n is a gap defined on the left by some formula D of rank $\leq r$. Clearly, a_n is also defined by $A \wedge D$, so we can assume that $D \vdash A$.

Write B for X_{a_n}, and δ for $A \wedge$ left(B, D). δ is a formula of rank $\leq r+2$, and $N_r \models U(\delta, A)^\mu(a_{n-1})$ (again we set a_{n-1} to be c' if $n = 0$). Define d', g' by:

- $d' = \sup\{t \in (x', y') : N \models {\sim}D(t)\}$;
- $g' = \sup\{t \in (x', d') : N \models \delta(t)\}$.

See figure 12.4.

Define d, g similarly. Note that as before, all these points lie in M_{r+2}, N_{r+2}. Clearly, $a_n < d'$ and the fact that $N_r \models U(\delta, A)^\mu(a_{n-1})$ is witnessed at a point $t' \in N$ where δ holds, with $t' < g'$.

Now if $\exists$ uses a winning strategy for $G_{4+3n;r+4(n+1)}(M, xy; N, x'y')$ and adds c, g, and d to $\forall$'s choices, then, as before, her strategy delivers *inter alia* c', g', and d' in response. So again, $\exists$ has a winning strategy for $G_{1+3n;r+4(n+1)}(M, cg; N, c'g')$ for all m, r'. By $(*)_n$ $\exists$ has a winning

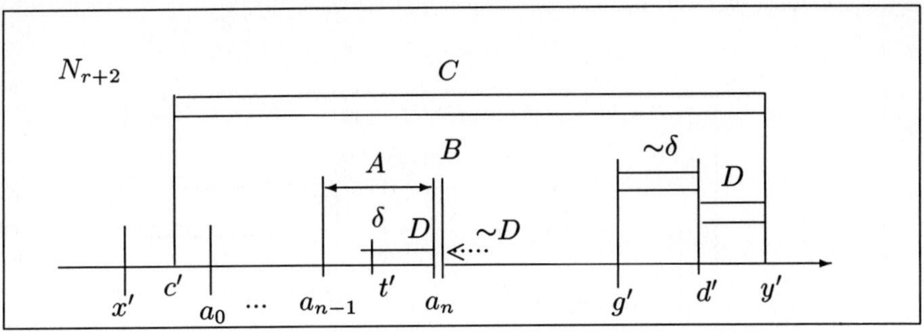

Figure 12.4

strategy for $G_{n;r+4}(N, c'g'; M, cg)$. Let her use it to choose $e_0, \ldots, e_{n-1}$ in response to $a_0, \ldots, a_{n-1}$. Then, as in lemma 12.8.12, $e_0, \ldots, e_{n-1} \in (c, g)_r$, and as rank $r+4$ formulae are preserved, $M_r \models U(\delta, A)^\mu(e_{n-1})$. As $e_{n-1} < g$ we can choose $t \leq g$ in M with $M \models \delta(t)$ and such that A holds at all $u \in (e_{n-1}, t]$.

By definition of δ and lemma 12.8.7, there is a gap $e_n \in (t, d)_r$ defined by D on the left, and such that A holds between t and e_n. Moreover, any rank r formula holds at e_n iff it holds at a_n, as they both satisfy B. ∃ chooses e_n in response to a_n, so completing her move. The same argument as in Case I allows ∃ to complete the remainder of the game, winning it.

Case IV: $a_0, \ldots, a_n \in (c', y'), a_n \in N_r$, and a_n is not definable on the left by any formula of rank $\leq r$.

It follows from the case assumption that A holds throughout some interval containing a_n. Choose D of rank $\leq r$ defining a_n on the right. Define $B = X_{a_n}$ and $\delta = A \wedge {\sim}D \wedge U(\text{right}(B, D), A)$ (rank $r + 3$). Let $d' = \sup\{t \in (x', y') : N \models \text{right}(B, D)(t)\}$, and then $g' = \sup\{t \in (x', d') : N \models \delta(t)\}$. Define $d, g \in M_{r+3}$ similarly.

See figure 12.5.

Clearly, there are $a_{n-1} < t' < a_n < u' < y'$, with $t', u' \in N, N \models \delta(t'), N \models \text{right}(B, D)(u')$, and A holding on (t', u') (if $n = 0$ we take a_{n-1} to be c' as usual). Hence $t' \leq g'$ and $u' \leq d'$. As usual, if ∃ uses a winning strategy for $G_{4+3n;r+4(n+1)}(M, xy; N, x'y')$ and adds c, g, and d to ∀'s choices she can derive a winning strategy for $G_{1+3n;r+4(n+1)}(M, cg; N, c'g')$. So by $(*)_n$ she has a winning strategy for $G_{n,r+4}(N, c'g'; M, cg)$. Let her use it to respond to $a_0, \ldots, a_{n-1}$ with $e_0, \ldots, e_{n-1}$. So as $U(\delta, A)$ has rank $\leq r + 4, M_r \models U(\delta, A)^\mu(e_{n-1})$. We can choose $e_{n-1} < t \leq g$ with $t \in M, M \models \delta(t)$, and A holding on (e_{n-1}, t). Then we can choose $u \in M$

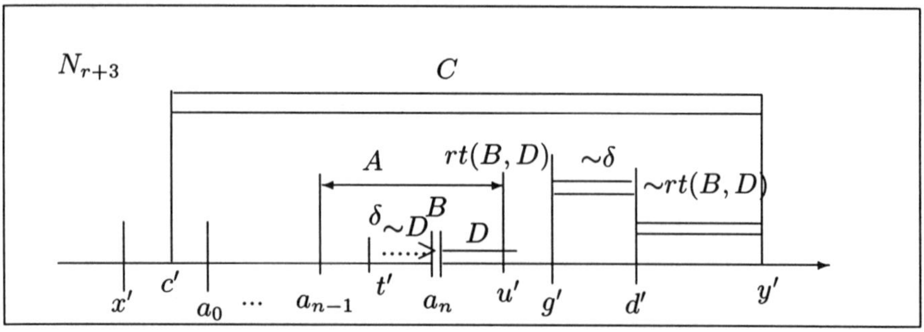

Figure 12.5

with $t < u \leq d$, $M \models \text{right}(B, D)(u)$ and such that A holds in (e_{n-1}, u).

By lemma 12.8.7 there is a gap $e_n \in (t, u)$ defined by D and at which the same relativized rank r formulae hold as at a_n in N_r. (We have $e_n > t$ because $M \models \sim D(t)$.) Then $\exists$ adds e_n to her choices to complete the move. The remainder of the game is as before.

This ends the proof of the theorem. ∎

Chapter 13

H-dimension

13.1 Introduction

We saw that expressive completeness of a set of connectives depends on the flow of time. Kamp's Since and Until are expressively complete over the integers but not over the rationals. The natural question to ask is, given a flow of time, $(T, <, =)$, can one find a finite set of connectives which is expressively complete over $(T, <, =)$?

We shall see in section 13.2 below that a necessary condition for a positive answer to the question is that $(T, <, =)$ has a property known as *finite Henkin* or *H-dimension.* We can use this to show that because some flows of time have infinite H-dimension, they admit no finite expressively complete set of connectives. The property of finite H-dimension is important. In sections 13.4 and 13.5 we give a method of generating more and more flows of time with finite H-dimension. Some of these flows are non-linear. In section 13.6 we examine a notion related to finite H-dimension, the so-called *k-variable property.*

What about a *sufficient* condition for a finite expressively complete set of connectives for a flow of time to exist? If we allow *many-dimensional connectives,* the condition of finite H-dimension is sufficient — we will prove this in section 13.2. However, when we consider *one-dimensional connectives* such as Since and Until, finite H-dimension is no longer a sufficient condition. In section 13.7 we give an example of a class of 'circular' flows of time with H-dimension 3 but without any finite expressively complete set of connectives. We do not know of a necessary and sufficient condition in the one-dimensional case.

We now gather the basic definitions and concepts that have to do with many-dimensional systems. There is no need to restrict ourselves to working only with 'flows of time', and in fact in later sections we will need to consider more general structures. So we will conduct our study of H-dimension in the more general setting of a class of first-order structures. The notion of finite H-dimension involves limiting the number of bound variables employed in formulae, and so many of the proofs in this chapter will involve keeping careful track of how many variables are used. Thus some of them may seem unnatural or have more details than usual.

Notation 13.1.1 L will be a first-order signature, containing constants, relation, and function symbols. When we are concerned with flows of time, L will be simply $\{<\}$, where $<$ is the ordinary binary earlier–later relation. As usual, the equality relation, $=$, is always assumed to be in the logic: we do not need to include it in L. We will often write T for an L-structure, recalling our interest in flows of time. We will write $t \in T$, $\bar{a} \in T$, etc., to denote that t is an element, or $\bar{a}$ a tuple of elements, of the domain of T. Similarly, $S \subseteq T$ (respectively, T^m) will mean that S is a subset of the domain of T (respectively, T^m). We use $\models$ to denote the usual Tarskian validity in T.

Definition 13.1.2

1. An *m-dimensional table for an n-place connective* is a formula $\beta(x_1, \ldots, x_m; R_1, \ldots, R_n)$, where:
   - β is a formula of the first-order predicate language, written with symbols from $L \cup \{R_1, \ldots, R_n\}$, where $R_1, \ldots, R_n$ are special m-place relation symbols not occurring in $L \cup \{=\}$;
   - the free variables of β are exactly $x_1, \ldots, x_m$;
   - any atomic subformula $R(t_1, \ldots, t_m)$ of β (where the t_i are terms of L and $R \notin L \cup \{=\}$) is in fact of the form $R(y_1, \ldots, y_m)$, where $y_1, \ldots, y_m$ are distinct variables.
2. If $\alpha(x_1, \ldots, x_m, R_1, \ldots, R_n)$ is a formula of $L \cup \{R_1, \ldots, R_n\}$, T is an L-structure, $a_1, \ldots, a_m \in T$ and $S_1, \ldots, S_n \subseteq T^m$, we write $T \models \alpha(a_1, \ldots, a_m, S_1, \ldots, S_n)$ if $\alpha(a_1, \ldots, a_m, R_1, \ldots, R_n)$ is valid in T when the relation symbols $R_1, \ldots, R_n$ are interpreted as $S_1, \ldots, S_n$ respectively.

Remark 13.1.3 The reason for this restriction on atomic subformulae in (1) will be found in definition 13.1.5(c). It is not a semantic restriction. For example, a formula that is not a (three-dimensional) table for a two-place connective is

$$\nu(x_1, x_2, x_3; R_1, R_2) = \forall y(\exists z R_1(y, y, z) \to R_2(x_1, y, x_3) \vee R_2(x_1, x_2, y)).$$

The reason is that the atomic subformula $R_1(y,y,z)$ occurs with only two distinct arguments — y and z — rather than three. But we can convert ν into the equivalent table

$$\forall y\big(\exists zv(v = y \wedge R_1(y,v,z)) \rightarrow R_2(x_1,y,x_3) \vee R_2(x_1,x_2,y)\big).$$

To convert an arbitrary formula $\alpha(x_1,\ldots,x_m,R_1,\ldots,R_n)$ into an equivalent table, we replace every atomic subformula $R_k(t_1,\ldots,t_m)$ of α, where the t_j are terms, by the logically equivalent formula

$$\exists y_1,\ldots,y_m \bigwedge_{j\leq m} (y_j = t_j) \wedge R_k(y_1,\ldots,y_m),$$

where $y_1,\ldots,y_m$ are new distinct variables. Since each replacement is logically equivalent to what it replaces and has the same free variables, the resulting table is logically equivalent to α.

Now fix a temporal language $\mathcal{T}$ with atoms $q_1,q_2,\ldots,$ the classical connectives, and the special symbols $\sharp_1,\ldots,\sharp_j$, standing for n_1-,$\ldots$,n_j-place connectives respectively. Let $\beta_1,\ldots,\beta_j$ be m-dimensional n_1-,$\ldots$,n_j-place tables respectively. Since there are finitely many β_i to consider, we can further assume that there is $b \geq m$ such that each β_i is written with variables $x_1,\ldots,x_b$ only.

Definition 13.1.4 Let T be any L-structure. Let h be an assignment into T, i.e. for any atom q, $h(q) \subseteq T$. We define the truth value of each formula A of the language $\mathcal{T}$ at m indices $a_1,\ldots,a_{m-1},t \in T$ under h, as follows:

1. $\|q\|^h_{a_1,\ldots,a_{m-1},t} = 1$ iff $t \in h(q)$, q atomic.

2. $\|A \wedge B\|^h_{a_1,\ldots,a_{m-1},t} = 1$ iff $\|A\|^h_{a_1,\ldots,a_{m-1},t} = \|B\|^h_{a_1,\ldots,a_{m-1},t} = 1$.

3. $\|{\sim}A\|^h_{a_1,\ldots,a_{m-1},t} = 1$ iff $\|A\|^h_{a_1,\ldots,a_{m-1},t} = 0$.

4. Similarly for $\vee, \rightarrow$.

5. For each i $(1 \leq i \leq j)$, $\|\sharp_i(A_1,\ldots,A_{n_i})\|^h_{a_1,\ldots,a_{m-1},t} = 1$ iff
   $T \models \beta_i(a_1,\ldots,a_{m-1},t,h(A_1),\ldots,h(A_{n_i}))$, where

   $$h(A_k) =_{\text{def.}} \{(t_1,\ldots,t_m) \in T^m \mid \|A_k\|^h_{t_1,\ldots,t_m} = 1\}.$$

   Note that the sets $h(A_k)$ are known to us by the induction hypothesis.

While the logic is m-dimensional, the atoms are only evaluated in one dimension. That is, the value of $\|q\|^h_{a_1,\ldots,a_{m-1},t}$ does not depend on $a_1, \ldots, a_{m-1} \in T$, but only on t. (We call t the *evaluation point,* and $a_1, \ldots, a_{m-1}$ *reference points.*) Because of this, we can easily find for each temporal formula a corresponding table in the *monadic* language over T. This is shown in the following definition and theorem.

Definition 13.1.5

1. Let L^M denote the monadic language with the symbols of L, first-order quantifiers over elements, and an arbitrary number of monadic predicate symbols Q_i for subsets of structures. We will regard the Q_i as predicate (subset) variables, implicitly associated with the atoms q_i.

2. For each temporal formula A of the language $\mathcal{T}$ we define a formula δA of L^M, written only with variables from $x_1, \ldots, x_b$ and with m free variables $x_1, \ldots, x_m$. The definition is by induction on the structure of A.

   (a) If A is an atom q_i, we set $\delta A = (x_1 = x_1) \wedge \ldots \wedge (x_{m-1} = x_{m-1}) \wedge Q_i(x_m)$.

   (b) If we have defined δA and δB, we set $\delta(A \wedge B) = \delta A \wedge \delta B$, and $\delta({\sim}A) = {\sim}\delta A$. Similarly for $\vee$ and $\rightarrow$.

   (c) Let $A = \sharp_i(A_1, \ldots, A_{n_i})$, where the A_k are formulae of $\mathcal{T}$. We define δA as follows. First consider the table $\beta_i(x_1, \ldots, x_m; R_1, \ldots, R_{n_i})$ of $\sharp_i$. Recall that we assumed β_i is written using variables $x_1, \ldots, x_b$ only.

   For every atomic subformula of β_i of the form $R_k(y_1, \ldots, y_m)$, where $R_k \notin L \cup \{=\}$, we undertake the following substitution process. We choose a permutation σ of $\{x_1, \ldots, x_b\}$ such that $\sigma(x_1) = y_1, \ldots, \sigma(x_m) = y_m$. This is possible because β is a table, so that $y_1, ..., y_m$ are distinct variables. Now by the inductive hypothesis, δA_k is written only with variables $x_1, \ldots, x_b$, and has $x_1, \ldots, x_m$ free. Simultaneously, for each $j \leq b$ we change every occurrence of x_j in δA_k into $\sigma(x_j)$. We then substitute the resulting formula for this occurrence of $R_k(y_1, \ldots, y_m)$ in β_i. We will write the resulting formula simply as

$$\beta_i(x_1, \ldots, x_m, \delta A_1, \ldots, \delta A_{n_i}).$$

This completes the definition.

Clearly, if A is a temporal formula built up from atoms $q_1, \ldots, q_k$, then δA will contain precisely the monadic predicates $Q_1, \ldots, Q_k$ as free subset variables.

Theorem 13.1.6 *Let B be any formula of $\mathcal{T}$ built up from the atoms $q_1, \ldots, q_k$ using the connectives $\sharp_1, \ldots, \sharp_j$, and the classical connectives. Then for all L-structures T, all assignments h into T, and all $a_1, \ldots, a_{m-1}, t \in T$:*

$$(*) \qquad \|B\|^h_{a_1,\ldots,a_{m-1},t} = 1 \textit{ iff } T \models \delta B(a_1, \ldots, a_{m-1}, t, h(q_1), \ldots, h(q_k)).$$

Proof. By induction on B.

1. For B atomic, $(*)$ is true since atoms are evaluated at points t only (the evaluation point).
2. The case of the classical boolean connectives is trivial.
3. Let $B = \sharp_i(A_1, \ldots, A_{n_i})$ for some $i, 1 \leq i \leq j$. By the induction hypothesis, for all $l \leq n_i$,

   $$h(A_l) = \{(t_1, \ldots, t_m) \mid T \models \delta A_l(t_1, \ldots, t_m, h(\bar{q}))\}.$$

   By definition we have:

   $$\|B\|^h_{t_1,\ldots,t_m} = 1 \text{ iff } T \models \beta_i(t_1, \ldots, t_m, h(A_1), \ldots, h(A_{n_i})),$$

   and hence,

   $$\|B\|^h_{t_1,\ldots,t_m} = 1 \text{ iff } T \models \beta_i(t_1, \ldots, t_m, \delta A_1, \ldots, \delta A_{n_i}).$$

This completes the induction step and the proof of theorem 13.1.6. ■

Definition 13.1.7 Suppose that $\mathcal{K}$ is any class of L-structures, $\bar{x} = x_1, \ldots, x_m$ are variables, and $\bar{Q} = Q_1, \ldots, Q_r$ are monadic predicates.

1. If $\alpha(\bar{x}, \bar{Q})$, $\beta(\bar{x}, \bar{Q})$ are formulae of the monadic language L^M with free variables $\bar{x}$ and free monadic predicates $\bar{Q}$, we say that α and β are *$\mathcal{K}$-equivalent* if for all $T \in \mathcal{K}$ and all subsets $S_1, \ldots, S_r \subseteq T$,

   $$T \models \forall \bar{x}\Big(\alpha(\bar{x}, S_1, \ldots S_r) \leftrightarrow \beta(\bar{x}, S_1, \ldots, S_r)\Big).$$

2. If Ξ is a theory in L, α and β are said to be *Ξ-equivalent* if they are $\mathcal{K}$-equivalent, where $\mathcal{K}$ is the class of all L-structures that are models of Ξ.

3. Let $1 \leq n \leq m$. We say the temporal system $\mathcal{T}$ is *expressively complete* or (equivalent terminology) *functionally complete over $\mathcal{K}$ in n dimensions* if for any $\alpha(x_1, \ldots, x_n, \bar{Q})$ of L^M with free variables $x_1, \ldots, x_n$, there exists a temporal formula $B(\bar{q})$ of $\mathcal{T}$ built up from the atoms $\bar{q} = q_1, \ldots, q_r$, such that $\alpha \wedge \bigwedge_{n<i\leq m} x_i = x_i$ and δB are equivalent in $\mathcal{K}$.

4. We say that an m-dimensional temporal system is expressively complete over a structure T in n dimensions if this holds for the class $\{T\}$.

Example 13.1.8 The m-dimensional temporal system is functionally complete over $\mathcal{K}$ in one dimension iff for any $\alpha(t, \bar{Q})$ there exists a temporal $B(\bar{q})$ such that for any structure $T \in \mathcal{K}$, any $a_1, \ldots, a_m \in T$, and any assignment h into T, we have $\|B\|^h_{a_1,\ldots,a_m} = 1$ iff $T \models \alpha(a_1, h(\bar{q}))$.

Definition 13.1.9 Let $\mathcal{K}$ be a class of structures.

1. Let $1 \leq n \leq m$ again. $\mathcal{K}$ is said to be *m-functionally complete in n dimensions* (symbolically, FC^m_n) if there exists a finite set of m-dimensional connectives that is functionally complete over $\mathcal{K}$ in n dimensions.

2. $\mathcal{K}$ is said to be *functionally complete* if there is m such that $\mathcal{K}$ is m-functionally complete in one dimension (i.e. $\mathcal{K}$ is FC^m_1 for some m).

We say that an L-structure T is functionally complete (etc.) if the property holds for the class $\{T\}$.

The other important notion we need is that of *finite Henkin dimension.*

Definition 13.1.10 Let d be a non-negative integer.

1. We say that a class $\mathcal{K}$ of structures has *Henkin or H-dimension d* if (a) and (b) hold:

   (a) Let $\alpha(x_1, \ldots, x_n, Q_1, \ldots, Q_r)$ be any monadic formula of L^M, with free variables among $x_1, \ldots, x_n$ and monadic predicates Q_1, $\ldots, Q_r$ (with n, r arbitrary). Then there exists an L^M-formula $\alpha'(x_1, \ldots, x_n, Q_1, \ldots, Q_r)$ that is $\mathcal{K}$-equivalent to α and uses no more than d different bound variable letters.

   (b) d is the smallest number satisfying (a) above.

2. We say that an L-theory U has *H-dimension* d if the class of models of U has H-dimension d.

3. We say that a structure T has H-dimension d if the class $\{T\}$ has H-dimension d.

Example 13.1.11 To say that there are at least three different elements one usuallly needs three letters, namely

$$\exists x_0 \exists x_1 \exists x_2 \bigwedge_{i<j<3} (x_i \neq x_j).$$

Over linear time one can manage with two letters:

$$\exists x(\exists y(y > x) \wedge \exists y(y < x)).$$

Remark 13.1.12 If $\mathcal{K} \subseteq \mathcal{K}'$ are classes of L-structures, then:

1. For all $1 \leq n' \leq n \leq m$, if $\mathcal{K}$ is FC_n^m then it is $FC_{n'}^m$.

2. If $\mathcal{K}'$ has H-dimension d then $\mathcal{K}$ has H-dimension at most d.

3. If $\mathcal{K}'$ is FC_n^m then so is $\mathcal{K}$.

We shall see in section 13.2 that the notion of functional completeness (in the sense of FC_1^m for some m) is equivalent to the notion of finite H-dimension. Moreover, some classes of structures have infinite H-dimension and are therefore not functionally complete. Note that it follows from this equivalence that if $\mathcal{K}$ is expressively (functionally) complete in one dimension (FC_1^1) then $\mathcal{K}$ has finite H-dimension. The converse fails: see section 13.7 below.

13.2 Expressive completeness and finite Henkin dimension

We now show that for any class of structures, finite Henkin dimension is equivalent to functional completeness (FC_1^m for some m). So for any class of structures with infinite Henkin dimension, no finite set of connectives can ever be functionally complete. We will give examples of such classes: e.g. the class of all partially ordered flows of time. We can also obtain a bound on the Henkin dimension of some classes known to be functionally complete, simply by counting the variables used in the tables of the expressively complete connectives.

We begin with some examples of classes with infinite H-dimension, and then go on to relate finite H-dimension and functional completeness.

Theorem 13.2.1

1. *Consider the following sentence of the theory of equality:*

$$\sigma_n = \exists x_1, \ldots, x_n \bigwedge_{i \neq j} (x_i \neq x_j).$$

σ_n says that there are at least n different elements. This cannot be expressed using less than n variable symbols, in the pure theory of equality.

2. *Consider the following sentence in the theory of partial order:*

$$\sigma'_n = \exists x_1, \ldots, x_n \bigwedge_{i \neq j} [(x_i \neq x_j) \wedge {\sim}(x_i < x_j)].$$

Then σ'_n can not be equivalently expressed using less than n variable symbols in the theory of partial order.

Proof. First notice that (2) follows from (1), since the pure theory of equality can be obtained from the theory of partial order by the addition of the axiom ${\sim}\exists xy(x < y)$ — in which case, effectively, $\sigma'_n = \sigma_n$.

We now prove (1). To show this we prove that any sentence α of the pure theory of equality is equivalent (modulo the theory of equality) to a boolean combination of units of the form $x = y$ or σ_n, where x, y, and the variable letters of the σ_n all appear in α.

We prove this by induction on the quantifier depth of α. If α has no quantifiers the claim is obvious. The cases ${\sim}\alpha$ and $\alpha \wedge \beta$ are also clear. Assume then that α has the form $\exists y \alpha'$. By the induction hypothesis we can assume that α' is a disjunction of conjunctions of the form

$$\bigwedge_i y = x_i \wedge \bigwedge_j y \neq x'_j \wedge \bigwedge_i v_i = u_i \wedge \bigwedge_j v'_j \neq u'_j \wedge \bigwedge_k \pm\sigma_{m(k)},$$

where all the letters appearing are from α. As any conjuncts $y = y$ or $y \neq y$ can be replaced by $\top$ or $\bot$, we can assume $y \notin \{x_i, x'_j\}$. We can bring the

$\exists y$ inside the disjunction, and ignore it on conjuncts not involving y. So it is sufficient to deal with

$$\exists y(\bigwedge_i y = x_i \wedge \bigwedge_j y \neq x'_j).$$

If the left-hand conjunction is non-empty — say $y = x_0$ appears — we can finish by replacing the above with $\bigwedge_i x_0 = x_i \wedge \bigwedge_j x_0 \neq x'_j$. Otherwise we assume we have to deal with $\beta = \exists y(\bigwedge_j y \neq x'_j)$. Let $X = \{x'_j\}$. Consider atoms of the form $x'_{j_1} = x'_{j_2}$ where $x'_{j_1}, x'_{j_2} \in X$. Let $\mathbf{a}_1, \mathbf{a}_2, \ldots$ be a list of all the possible maximal consistent conjunctions built up from these atoms or their negations. Then $\bigvee_k \mathbf{a}_k = \top$ is the full disjunctive normal form of $\top$ over the above atoms. Hence β is equivalent to $\bigvee_k(\beta \wedge \mathbf{a}_k)$.

But each $\mathbf{a}_k$ says that there are exactly $n(k)$ elements of X that are pairwise different (for some $n(k)$), and the others are equal to some of them. Thus $\mathbf{a}_k \wedge \beta$ is equivalent to $\mathbf{a}_k \wedge \sigma_{n(k)+1}$. Note that $n(k)$ can be at most $|X|$ and so $n(k) + 1$ can be at most $|X \cup \{y\}|$. So we do have enough variable letters from α to express $\sigma_{n(k)+1}$. Thus we have seen that α is equivalent to a sentence of the required form. Note that in the new form of α the same free variables occur.

Now to prove (1) note that σ_n can be equivalent to no α with less than n variable letters since this α would then be equivalent to a disjunction of conjunctions of $\pm\sigma_k$ for $k < n$, which is impossible. ■

Thus the theory of equality and the theory of partial order do not have finite H-dimension.

The following lemma is also useful.

Lemma 13.2.2 *For all n, the formula $\pi_n = \exists x_1, \ldots, x_n \bigwedge_{i \neq j} \sim(x_i < x_j) \wedge (x_i \neq x_j)$ cannot be expressed using less than n variables in a structure where the following axioms hold:*

A. $\exists x \forall y((x \neq y) \to (x < y))$

B. $\sim[\exists x, y, z((x < y) \wedge (y < z))]$.

Explanation: The meaning of the formula π_n in the structure is that there are at least n 'fanning' elements, as pictured in figure 13.1.

Proof. The model where the formula π_n holds true is a structure of at least n elements that are mutually incomparable, and one element that is less than all those n. Thus there is one unique element x such that $\forall y((y \neq x) \to (x < y))$ and for all other x, y, $\sim(x < y)$ and $\sim(y < x)$.

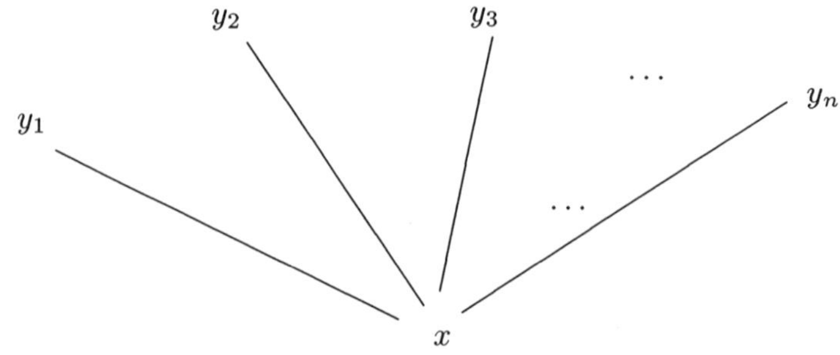

Figure 13.1. Fanning elements.

Introduce a name c for the minimal element and do away with the relation '$<$'. This can be done since '$<$' can be defined in this structure using the constant c, as follows:

$$x < y \text{ iff } x = c \text{ and } y \neq c.$$

Therefore the structure $(R, <, =)$ is definable in the structure $(R, =, c)$.

Assume for contradiction that π_n is expressible in $(R, <, =)$ by a sentence χ using fewer than n variables. Replacing all atomic subformulae $x < y$ of χ by $x = c \wedge y \neq c$, we obtain a sentence $\chi'(c)$ of $(R, =, c)$, expressing with fewer than n variables that there exist at least n elements different from c. Regarding c as a new variable, the sentence $\exists c\chi'$ now expresses in the theory of equality that there are at least $n + 1$ distinct elements. As it only uses at most n variables, this contradicts theorem 13.2.1, so the proof is complete. ■

We now relate functional completeness and finite H-dimension. The following lemma will be very useful.

Lemma 13.2.3 *Let $\mathcal{K}$ be a class of L-structures, and d a whole number. Then the following are equivalent.*

1. *$\mathcal{K}$ has H-dimension at most d;*

2. *For each monadic sentence $\sigma(P_1, \ldots, P_n)$ of L^M, there is a monadic sentence $\sigma^*(P_1, \ldots, P_n)$ $\mathcal{K}$-equivalent to σ and using at most d variables.*

Proof. Only (2) implies (1) requires proof. So assume (2) and let $\alpha(x_1, \ldots, x_n, \bar{Q})$ be any formula of L^M with free variables among $x_1, \ldots, x_n$ and monadic predicates from $\bar{Q} = Q_1, \ldots, Q_r$. We introduce new monadic predicates $\bar{R} = R_1, ..., R_n$, and let $\sigma(\bar{Q}, \bar{R})$ be the sentence

$$\exists x_1, \ldots, x_n (\bigwedge_{1 \leq i \leq n} R_i(x_i) \wedge \alpha).$$

By assumption, there is a sentence $\sigma^*(\bar{Q}, \bar{R})$ of L^M using at most d variables and $\mathcal{K}$-equivalent to σ. We can assume that σ^* is written with variables other than $x_1, \ldots, x_n$. We now obtain the formula $\alpha^*(x_1, \ldots, x_n)$ of L^M as the result of replacing each $R_i(z)$ in σ^* by $x_i = z$ (all $i \leq n$ and all z).

Clearly, α^* has at most d bound variables. We claim that it is $\mathcal{K}$-equivalent to α. Let $T \in \mathcal{K}$, $t_1, \ldots, t_n \in T$ and h be any assignment into T. As is easily seen, $T \models \alpha(t_1, \ldots, t_n, h(\bar{Q}))$ iff $T \models \sigma(h(\bar{Q}), \{t_1\}, ..., \{t_n\})$, iff $T \models \sigma^*(h(\bar{Q}), \{t_1\}, ..., \{t_n\})$, iff $T \models \alpha^*(t_1, \ldots, t_n, h(\bar{Q}))$, proving the claim. It now follows that $\mathcal{K}$ has H-dimension at most d, as required. ■

Theorem 13.2.4 *For any class $\mathcal{K}$ of L-structures, the following are equivalent:*

1. *$\mathcal{K}$ is functionally complete (i.e. there exists a finite dimensional temporal logic with a finite number of connectives which is functionally complete over $\mathcal{K}$ in one dimension: $\mathcal{K}$ is FC_1^m for some m);*

2. *$\mathcal{K}$ has finite H-dimension.*

Proof. First we assume (1) and show (2). Let us assume that it is our m-dimensional logic $\mathcal{T}$ of definition 13.1.4 that is expressively complete over $\mathcal{K}$ in one dimension (definition 13.1.7). Let $\sigma(\bar{Q})$ be a sentence of L^M. So there exists a $B(\bar{q})$ of $\mathcal{T}$ such that the formulae $x_1 = x_1 \wedge ... \wedge x_m = x_m \wedge \sigma(\bar{Q})$ and

$$\delta B(x_1, \ldots, x_m, \bar{Q})$$

are equivalent in $\mathcal{K}$. By definition 13.1.5, δB is written using variables x_1, $\ldots, x_b$ only. Hence the sentence $\sigma^* = \exists x_1 ... \exists x_m \delta B(x_1, \ldots, x_m, \bar{Q})$ has at most b variables, and is clearly $\mathcal{K}$-equivalent to σ. So every sentence of L^M is $\mathcal{K}$-equivalent to one with at most b variables. The result now follows by lemma 13.2.3.

Now assume (2) and prove (1). We will define an $(m+1)$-dimensional logic **d** satisfying (1), where m is the H-dimension of $\mathcal{K}$. Evaluation is of the form $\|A\|^h_{x_0, \ldots, x_m}$, where h is an assignment ($h(q) \subseteq T$ for atomic q), and where each formula A can be built up from the following connectives, the connectives of **d**. Note that the $C^<_{i,j}$ below are constants (0-place connectives).

1. $\|q\|^h_{x_0,\ldots,x_m} = 1$ iff $x_0 \in h(q)$ for q atomic.

2. The tables for $\sim, \wedge, \vee, \to$ are the usual ones.

3. If $i \neq j$ and $0 \leq i, j \leq m$, $\|C^<_{i,j}\|^h_{x_0,\ldots,x_m} = 1$ iff $x_i < x_j$. Similarly we define the semantics of $C^=_{i,j}$.

4. If $0 \leq i \leq m$, $\|J_i A\|^h_{x_0,\ldots,x_m} = 1$ iff $\|A\|^h_{x_i,\ldots,x_i} = 1$.

5. If $0 \leq i \leq m$, $\|\Box_i A\|^h_{x_0,\ldots,x_m} = 1$ iff $\|A\|^h_{x_0,\ldots,x_{i-1},y,x_{i+1},\ldots,x_m} = 1$ for all $y \in T$.

Lemma 13.2.5 *Let β be a formula of L^M written only using the variable letters $u_0, \ldots, u_m$, and having $u_{i_1}, .., u_{i_k}$ free for arbitrary $k \leq m$ and distinct indices $i_1, \ldots, i_k \leq m$. Then there exists a temporal formula A of* **d** *such that for all $h, t_0, \ldots, t_m \in T$,*

$$\|A\|^h_{t_0,\ldots,t_m} = 1 \textit{ iff } \beta(t_{i_1}, \ldots, t_{i_k}) \textit{ is } h\textit{-valid.}$$

Proof. By induction on β. Assume first that β is atomic. If β is $u_i < u_j$ let $A = C^<_{i,j}$ if $i \neq j$, and $\perp$ otherwise. Similarly for $u_i = u_j$. If β is $Q(u_i)$, let A be $J_i(q)$.

The classical connectives present no difficulties. We turn to the case where β is $\forall u_i \alpha(u_{i_1}, .., u_{i_k})$. Let A be the formula suitable for α; then $\Box_i A$ is the formula suitable for β. This proves the lemma. ■

Now to prove functional completeness for **d** we proceed as follows. Let $\beta(u_0)$ be any formula of L^M with one free variable u_0. As $\mathcal{K}$ has H-dimension m, we can suppose that β is written with variables $u_0, ..., u_m$. By lemma 13.2.5 there exists an A of $\mathcal{T}$ such that for any $T \in \mathcal{K}$, $t \in T$, and assignment h into T, $\|A\|^h_{t,\ldots,t} = 1$ iff $\beta(t)$ is h-valid. This completes the proof of theorem 13.2.4. ■

Corollary 13.2.6 *The theory of linear order with monadic predicates has H-dimension at most 3. For the special case of the real numbers and integers the H-dimension is also at most 3.*

Proof. We simply use the fact that we have functional completeness for the one-dimensional system with U', U, S', S, whose tables use only three variables. Certainly the tables of U and S only involve three variables. The table for $U'(p,q)$ is $\alpha \wedge \beta$, where $\alpha(t)$ is the table of:

$$U(q,q) \wedge \sim U(\sim q \vee \sim U(\top, q), q),$$

i.e. $\gamma^+(q)$ in the notation of chapter 12, and $\beta(t)$ is

$$\exists s > t(\sim q(s) \wedge \forall u[t < u < s \wedge \exists s(t < s < u \wedge \sim q(s)) \to p(u)]).$$

Note that s was reused in β. This table $\alpha \wedge \beta$ can be written with only three variables. Of course the table for S' is the mirror image of this. Thus the H-dimension of the class of all linear flows of time is at most 3, by the proof of theorem 13.2.4.

For the reals or integers we need only observe that the tables of Until and Since are written with three variables. See the definitions of the tables of the respective connectives. Of course, the reals and integers are linear flows so have H-dimension at most 3 by the above; but we can observe this directly knowing that U and S alone are expressively complete for these flows. ■

Using the techniques of section 13.6 below we will be able to see that the class of linear flows of time has H-dimension exactly 3.

13.3 Expressive completeness in higher dimensions

We now show that from the existence of a one-dimensional functionally complete system over a class of structures, it follows that there exists a functionally complete system in any dimension. That is, FC_1^1 implies FC_m^m for all m. For simplicity we will only consider an example, *real and integer time,* but the argument for general L-structures is similar.

So in this section we take L to be the signature $\{<\}$ appropriate for flows of time. Let $\sharp(q_1, \ldots, q_n)$ be an n-place connective with table

$$\beta_\sharp(t, <, =, Q_1, \ldots, Q_n) \qquad (Q_i \subseteq T).$$

Each $Q_i(s)$ can be regarded as an m-place relation via $Q_i(x_1, \ldots, x_{m-1}, s) \iff Q_i(s)$. With this in mind we can regard $B_\sharp(t, <, Q_i)$ as a table for an m-dimensional connective $\sharp^m$. For example, to turn $U(p, q)$ into the connective $U^2(p, q)$ we use the same table B_U and obtain

$$\|U^2(p,q)\|^h_{y,t} = 1 \text{ iff } \models B_U(t, h(p), h(q))$$

$$\text{iff } \exists s > t\big(\|p\|^h_{y,s} = 1 \wedge \forall u(t < u < s \Rightarrow \|q\|^h_{y,u} = 1)\big).$$

The y is not touched. In simple words we ignore the reference point y and go on ignoring it.

Thus we can regard the system $\{U^m, S^m\}$ as a system of m-dimensional connectives. We shall use the fact that we have a functionally complete system in dimension 1 to modify the above into a functionally complete system in dimension m.

Definition 13.3.1 Define the following m-dimensional connectives with the following fixed tables:

1. Let E_i $(1 \leq i \leq m-1)$ be the 0-place connective with table $\|E_i\|^h_{x_1,\dots,x_{m-1},t} = 1$ iff $x_i = t$.

2. Let U^m, S^m be the connectives with the following tables (as mentioned above):
   $\|U^m(A,B)\|^h_{x_1,\dots,x_{m-1},t} = 1$ iff for some $s > t$,
   $\|A\|^h_{x_1,\dots,x_{m-1},s} = 1$ and for all $u, t < u < s \Rightarrow \|B\|^h_{x_1,\dots,x_{m-1},u} = 1$;
   $\|S^m(A,B)\|^h_{x_1,\dots,x_{m-1},t} = 1$ iff for some $s < t$,
   $\|A\|^h_{x_1,\dots,x_{m-1},s} = 1$ and for all u, $s < u < t \Rightarrow \|B\|^h_{x_1,\dots,x_{m-1},u} = 1$.

Let US^m be the m-dimensional logic with the above mentioned connectives. Let US be the fragment of US^m with the connectives U^m, S^m only.

Lemma 13.3.2 *US is the logic with Since and Until of chapter 10.*

Proof. The reference points $x_1, \dots, x_{m-1}$ are untouched by the tables of U, S and the reference points are not involved in the evaluation of A for atomic A. ■

Corollary 13.3.3 *For real numbers time or for integer time, for any monadic $\beta(t, Q_1, \dots, Q_n)$ there exists a formula B of US such that for any t, $x_1, \dots, x_{m-1}$, $\|B\|^h_{x_1,\dots,x_{m-1},t} = 1$ iff $\beta(t, h(q_1), \dots, h(q_n))$ is valid.*

Proof. This is clear, since the above logic is one-dimensionally functionally complete over the reals and the integers. ■

Theorem 13.3.4 *US^m is functionally complete (in m dimensions) for real time or for integer time.*

Proof. Let $\alpha(x_1, \dots, x_{m-1}, t, Q_1, \dots, Q_k)$ be any monadic formula. Rather as in lemma 13.2.3, define

$$\beta(t, Q_1, \dots, Q_k, R_1, \dots, R_{m-1}) = \exists x_1, \dots, x_{m-1}(\bigwedge_{i<m} R_i(x_i) \wedge \alpha),$$

where the R_i are new monadic predicates. By the corollary there exists a formula $B(q_1, \ldots, q_k, r_1, \ldots, r_{m-1})$ of US such that in integer or real time, for any $x_1, \ldots, x_{m-1}, t$ and all h,

$$\|B\|^h_{x_1,\ldots,x_{m-1},t} = 1 \text{ iff } \beta(t, h(q_j), h(r_i)) \text{ is valid.}$$

In particular this holds for all h satisfying $h(r_i) = \{x_i\}$ for each $i < m$. Now observe that if h has this property, then for all $x_1, \ldots, x_{m-1}, t$, and all $i < m$,

$$\|r_i\|^h_{x_1,\ldots,x_{m-1},t} = \|E_i\|^h_{x_1,\ldots,x_{m-1},t}.$$

As the reference points $x_1, \ldots, x_{m-1}$ are not touched in the evaluation of B, if we let A be the formula obtained by replacing each atomic subformula of B of the form r_i by the formula E_i, we have

$$\|B\|^h_{x_1,\ldots,x_{m-1},t} = \|A\|^h_{x_1,\ldots,x_{m-1},t}$$

for all h with the property described. Hence for such an h,

$$\alpha(x_1, \ldots, x_{m-1}, t, h(q_j)) \text{ is valid iff } \beta(t, h(q_j), h(r_i)) \text{ is valid,}$$

$$\text{iff } \|B\|^h_{x_1,\ldots,x_{m-1},t} = 1, \text{ iff } \|A\|^h_{x_1,\ldots,x_{m-1},t} = 1.$$

But the first and the last formulae in this equivalence do not involve the atoms r_i or monadic predicates R_i. Hence for all h,

$$\alpha(x_1, \ldots, x_{m-1}, t, h(q_j)) \text{ is valid iff } \|A\|^h_{x_1,\ldots,x_{m-1},t} = 1,$$

so that A is equivalent to α over integer or real time. This completes the proof. ■

Example 13.3.5 If $m = 2$ and α is $Q(x_1) \wedge x_1 < t$, then β is $\exists x_1(R_1(x_1) \wedge Q(x_1) \wedge x_1 < t)$, and we can clearly take B to be $P(q \wedge r_1)$ (or more strictly, $S(q \wedge r_1, \top)$). We then obtain $A = P(q_1 \wedge E_1)$. For all x_1, t, if $h(r_1) = \{x_1\}$ we have

$$Q(x_1) \wedge x_1 < t \text{ is valid iff } \exists u(u \in h(q_1) \cap h(r_1) \wedge u < t) \text{ is valid,}$$

$$\text{iff } \|P(q_1 \wedge r_1)\|^h_{x_1,t} = 1, \text{ iff } \|P(q_1 \wedge E_1)\|^h_{x_1,t} = 1.$$

The method of proof shows the following theorem to be true.

Theorem 13.3.6 *Let L be an $m+1$-dimensional logic that is $m+1$-functionally complete in one dimension. Let $n > 0$ be a natural number. Then L can be transformed into an $m+n$-dimensional logic L' which is functionally complete in n dimensions. So for a class of flows of time, FC_1^{m+1} implies FC_n^{m+n} for all $n > 0$.*

Proof. The idea is that the extra $n-1$ reference points will not be touched during evaluation, so they can be dealt with in the same way as $x_1, \ldots, x_{m-1}$ above. ∎

13.4 H-dimension and interpretations

13.4.1 Introduction

Here we will give a method of obtaining a structure from another in a way that preserves finite H-dimension and the existence of a finite set of expressively complete connectives. We will need it in sections 13.5 and 13.7, where, having proved that certain structures have these properties, we will be able to deduce that others do too.

Given a structure A, we may define a new structure B from A as follows. The domain of B will be a definable subset of A, and we will require that all atomic formulae on B are definable in A. So given any formula θ of B, by relativizing quantifiers and replacing atomic subformulae with appropriate A-formulae we can find a formula $\theta\uparrow$ of A that is equivalent to θ on B.

Suppose that A has finite H-dimension. Then certainly, $\theta\uparrow$ can be replaced modulo A by a formula with a bounded number of bound variables. However, we require a formula of B equivalent to θ in B. $\theta\uparrow$ is a formula of A, so may involve quantification over elements of A. So we will assume further that the elements of A have *representatives* in B which mimic their properties in a definable way. (As $B \subseteq A$, this is slightly stronger than A's being one-dimensionally interpretable in B in the usual first-order sense.) We then say that B *fits in* A. In this case, quantification over A can be reduced to quantification over B, and it will follow that B has finite H-dimension. Similarly, if there is a finite set of expressively complete one-dimensional connectives for A, then we can construct from it a finite set of expressively complete connectives for B. Converse results can also be established.

We will carry through this process over entire classes $\mathcal{K}$ of structures, if all formulae involved work uniformly. We will show that if $\mathcal{K}$ has finite H-dimension and/or a finite expressively complete set of temporal connectives, then so does any class of structures obtained from structures in $\mathcal{K}$ in the way described, and conversely.

13.4.2 Defining a structure in another

We now fix first-order signatures L, L', and a class $\mathcal{K}$ of L-structures. Note: as is usual, we write $\theta(x_1, \ldots, x_n)$ to indicate that the free variables of the formula θ lie in $\{x_1, \ldots, x_n\}$; not all x_i need actually occur free in θ. Firstly, we show how to define a new structure in any $A \in \mathcal{K}$.

Definition 13.4.1 An *(L'-)definition over* $\mathcal{K}$ is a pair $\delta = \langle \varphi, \alpha \rangle$, where:

(i) $\varphi(v)$ is a non-monadic L-formula with one free variable, v,

(ii) α is a map providing a non-monadic L-formula $\alpha_{R'}(x_1, \ldots, x_n)$ for each non-monadic atomic formula $R'(x_1, \ldots, x_n)$ of L',

such that for every $A \in \mathcal{K}$,

(iii) $\{a \in A \mid A \vDash \varphi(a)\}$ is the (non-empty) domain of an L'-structure B, such that for each non-monadic atomic formula $R'(x_1, \ldots, x_n)$ of L' and all $b_1, \ldots, b_n \in B$, $B \vDash R'(b_1, \ldots, b_n)$ iff $A \vDash \alpha_{R'}(b_1, \ldots, b_n)$.

If $A \in \mathcal{K}$ we will write $\delta(A)$ for the L'-structure B of (iii); it is uniquely defined, given $A \in \mathcal{K}$.

Note that it follows from (iii) that $\alpha_{x=y}$ is equivalent to $x = y$.

Lemma 13.4.2 *Let $\delta = \langle \varphi, \alpha \rangle$ be a definition over $\mathcal{K}$, and let $\psi(x_1, \ldots, x_n, P_1, \ldots, P_m)$ be a monadic L'-formula. Then there is a monadic L-formula $\psi\!\uparrow\!(x_1, \ldots, x_n, P_1, \ldots, P_m)$ such that for any $A \in \mathcal{K}$, if $B = \delta(A)$, $\bar{b} = b_1, \ldots, b_n \in B$ and h assigns subsets of B to the monadic predicates $P_1, \ldots, P_m$, we have*

$$B \vDash \psi(\bar{b}, h(P_1), \ldots, h(P_m)) \Leftrightarrow A \vDash \psi\!\uparrow\!(\bar{b}, h(P_1), \ldots, h(P_m)).$$

Proof. Notice that as $B \subseteq A$, h is also an assignment into A, so the result makes sense. Suppose that ψ is given. First we replace each monadic subformula $P(t(y_1, \ldots, y_m))$ of ψ, where P is a monadic predicate symbol and $t(y_1, \ldots, y_m)$ is an L'-term, by the formula $\exists z(z = t(y_1, \ldots, y_m) \wedge P(z))$, where z is a new variable. Now all monadic subformulae have the form $P(z)$ (z a variable). Then we simultaneously do the following. We relativize all quantifiers of ψ to $\varphi(v)$. At the same time we replace each non-monadic atomic subformula $R'(x_1, \ldots, x_n)$ of ψ by $\alpha_{R'}(x_1, \ldots, x_n)$. The result is a new monadic L-formula $\psi\!\uparrow$.

More formally, $\psi\!\uparrow$ is defined by induction on the structure of ψ. If ψ is the monadic formula $P(t(y_1, \ldots, y_m))$, then $\psi\!\uparrow$ is defined to be

$$\exists z[\alpha_{z=t(y_1,\ldots,y_m)}(z, y_1, \ldots, y_m) \wedge P(z) \wedge \exists v(v = z \wedge \varphi(v))],$$

where z is any variable distinct from $y_1, \ldots, y_m$. If ψ is a non-monadic atomic L'-formula $R'(x_1, \ldots, x_n)$ then $\psi\!\uparrow$ is defined to be $\alpha_{R'}(x_1, \ldots, x_n)$. We then define $(\psi_1 \wedge \psi_2)\!\uparrow = \psi_1\!\uparrow \wedge \psi_2\!\uparrow$, $(\sim\!\psi)\!\uparrow = \sim\!(\psi\!\uparrow)$, as usual. For

$\exists$ we split into two cases. We let $(\exists v\psi)\!\uparrow = \exists v(\varphi(v) \wedge \psi\!\uparrow)$, and if $x \neq v$, $(\exists x\psi)\!\uparrow = \exists x[\exists v(v = x \wedge \varphi(v)) \wedge \psi\!\uparrow(x)]$.

The lemma now follows easily from the definitions by induction on ψ. ■

Remark 13.4.3 It is easy to see how the bound variables of $\psi\!\uparrow$ arise. Suppose that V is a set of variables containing all the variables of φ and the bound variables of ψ and of the $\alpha_{R'(y_1,\ldots,y_m)}$ for each atomic subformula $R'(y_1,\ldots,y_m)$ of θ; assume also that for each atomic subformula of ψ of the form $P(t(y_1,\ldots,y_m))$ where P is a monadic predicate and t a term, at least one variable $z \in V$ is distinct from $v, y_1, \ldots, y_m$, and V contains the bound variables of $\alpha_{z=t(y_1,\ldots,y_m)}$. Then $\psi\!\uparrow$ can evidently be written in such a way that all its bound variables lie in V.

13.4.3 Constructions

Secondly, we must arrange that elements of each $A \in \mathcal{K}$ have representatives in $B = \delta(A)$.

Definition 13.4.4 Let $r \geq 1$.

1. If X is a set of variables, an *X, r-sort* is a map $\nu : X \to \{1, \ldots, r\}$. If $Y \supseteq X$ is another set of variables, a Y, r-sort ν' is said to *extend* ν if $\nu' \restriction X = \nu$.

2. An *r-ary (L'-)construction over $\mathcal{K}$* is a triple

$$\xi = \langle \delta, (\chi_i \mid i \leq r), \beta \rangle,$$

where

(a) $\delta = \langle \varphi, \alpha \rangle$ is an L'-definition over $\mathcal{K}$.

(b) Each $\chi_i(x, y)$ $(1 \leq i \leq r)$ is an L-formula. We require that for each $A \in \mathcal{K}$:

(i) $A \models \forall x x' y(\chi_i(x, y) \wedge \chi_i(x', y) \to x = x' \wedge \varphi(y))$ for all $i \leq r$.

(ii) $A \models \forall y(\varphi(y) \to \exists x \chi_i(x, y))$ for each $i \leq r$.

(iii) $A \models \forall x \exists y(\varphi(y) \wedge \bigvee_{i \leq r} \chi_i(x, y))$.

(iv) $A \models \forall x(\varphi(x) \leftrightarrow \chi_1(x, x))$.

(v) If $1 < i \leq r$ then $A \models \forall xy(\chi_i(x, y) \to {\sim}\varphi(x))$.

By (i) and (ii), each χ_i defines a function from $B = \delta(A)$ into A. For notational simplicity, if $b \in B$ we will write $\chi_i(b)$ for the unique $a \in A$ satisfying $\chi_i(a, b)$ in A.

(c) β is a map providing for each non-monadic atomic L-formula $R(x_1, \ldots, x_n)$ and each $\{x_1, \ldots, x_n\}, r$-sort ν, a non-monadic formula $\beta(R, \nu)(x_1, \ldots, x_n)$ of L'. We will write $\beta(R, \nu)$ simply as $R^\nu(x_1, \ldots, x_n)$ where β is understood. We require that for all $A \in \mathcal{K}$, if $B = \delta(A), b_i \in B$ and $\chi_{\nu(x_i)}(b_i) = a_i \in A$ for each $i \leq n$, then

$$A \models R(a_1, \ldots, a_n) \text{ iff } B \models R^\nu(b_1, \ldots, b_n).$$

3. Let $\xi = \langle \delta, (\chi_i \mid i \leq r), \beta \rangle$ be an r-ary L'-construction over $\mathcal{K}$. If $A \in \mathcal{K}$ we write $\xi(A)$ for the L'-structure $\delta(A)$. If $a \in A \in \mathcal{K}$, then any $b \in \delta(A)$ satisfying $A \models \bigvee_{i \leq r} \chi_i(a, b)$ is called a *representative* of a. An element a may have more than one representative, but by (b(iii)) it has at least one.

4. Let ξ be an L'-construction over $\mathcal{K}$. We will say that a class $\mathcal{K}'$ of L'-structures *fits in* $\mathcal{K}$ *via* ξ if $\mathcal{K}' = \{\xi(A) \mid A \in \mathcal{K}\}$ up to isomorphism. So for each $B \in \mathcal{K}'$ there is $A \in \mathcal{K}$ with $\xi(A) \cong B$ — $\xi(A)$ and B are isomorphic as L'-structures — and for each $A \in \mathcal{K}$, $\xi(A)$ is isomorphic to some structure in $\mathcal{K}'$.

5. We say that $\mathcal{K}'$ *fits in* $\mathcal{K}$ if it fits in $\mathcal{K}$ via ξ for some L'-construction ξ over $\mathcal{K}$.

6. Let A, B be structures. We say that B *fits in* A if the class $\{B\}$ fits in the class $\{A\}$.

In figure 13.2, B fits in A. We have written $R^{(3,2)}(x, y)$ for $R^\nu(x, y)$, where $\nu(x) = 3$ and $\nu(y) = 2$. We will see more concrete examples later. Let $\xi = \langle \langle \varphi, \alpha \rangle, (\chi_i \mid i \leq r), \beta \rangle$ be an r-ary L'-construction over $\mathcal{K}$. We will prove a converse to lemma 13.4.2, namely that all properties of elements of any $A \in \mathcal{K}$ reduce to properties of their representatives in $B = \xi(A)$.

Definition 13.4.5

1. For each monadic predicate letter P, we introduce r new monadic predicate letters $P^1, \ldots, P^r$. The predicate P^j will describe the interpretation of P on the part $\chi_j(B)$ of A. We will arrange that for all $b \in B$ and $j \leq r$, $B \models P^j(b)$ iff $A \models P(\chi_j(b))$. Formally, if $A \in \mathcal{K}$, $B = \xi(A)$ and h assigns subsets of A to the 'old' monadic predicates P, we define an assignment h' of the 'new' predicates P^j by:

$$h'(P^j) = \{b \in B \mid \chi_j(b) \in h(P)\} = \chi_j^{-1}(h(P))$$

for each old P and each $j \leq r$.

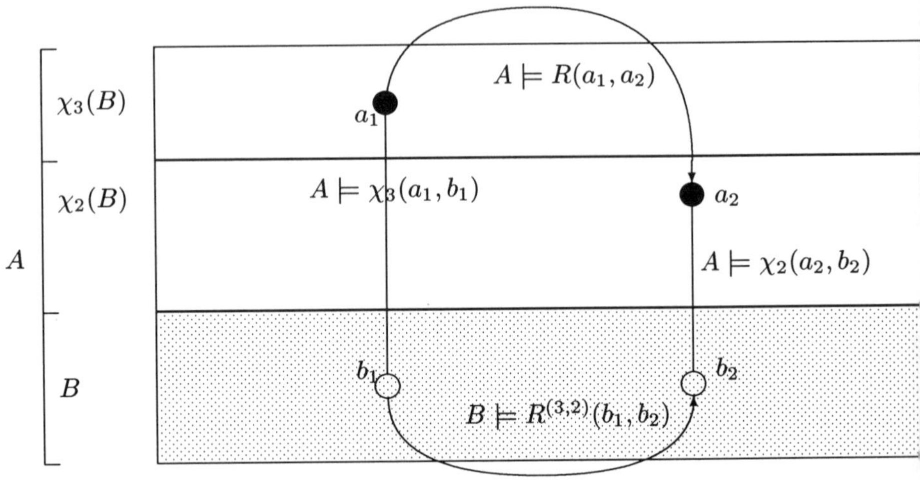

Figure 13.2

2. For each monadic L-formula $\theta(x_1, \ldots, x_n)$ and $\{x_1, \ldots, x_n\}$, r-sort ν, we will define a monadic L'-formula $\theta^\nu(x_1, \ldots, x_n)$. We may replace any monadic subformula of θ of the form $P(t(y_1, \ldots, y_m))$ by a formula of the form $\exists z(z = t(y_1, \ldots, y_m) \wedge P(z))$, for some variable $z \notin \{y_1, \ldots, y_m\}$. So we can assume that all monadic subformulae of θ are of the form $P(z)$. We then define θ^ν by induction:

   - If θ is atomic (non-monadic) then θ^ν is defined to be $\beta(\theta, \nu)$ (cf. definition 13.4.4(2c)).
   - $[P(x_i)]^\nu = P^{\nu(x_i)}(x_i)$
   - $(\sim\theta)^\nu = \sim\theta^\nu$
   - $(\theta \wedge \theta')^\nu = \theta^\nu \wedge \theta'^{\,\nu}$
   - $[\exists y \theta(x_1, \ldots, x_n, y)]^\nu = \bigvee_{\nu'} \exists y[\theta^{\nu'}(x_1, \ldots, x_n, y)]$, where the (finite) disjunction is taken over all $\{x_1, \ldots, x_n, y\}$, r-sorts ν' extending ν.

Remark 13.4.6 We can bound the number of bound variables required in θ^ν as follows. Suppose that all monadic subformulae of $\theta(x_1, \ldots, x_n)$ are of the form $P(z)$ for some variable z, and θ uses k bound variables. Assume also that for each atomic subformula $R(y_1, \ldots, y_m)$ of θ, the formula $\beta(R,\nu)(y_1,\ldots,y_m)$ uses at most k^*-k bound variables. Then θ^ν can be written with at most k^* bound variables. For if $v_{k+1},\ldots,v_{k^*}$ are distinct variables not occurring in θ, we can assume that the $\beta(R,\nu)$ use these as their bound variables. If the bound variables of θ are $v_1,\ldots,v_k$, then clearly θ^ν uses $v_1,\ldots,v_{k^*}$ as bound variables.

Lemma 13.4.7 *Let $\theta(x_1,\ldots,x_n,P_1,\ldots,P_m)$ be a monadic L-formula and ν be an $\{x_1,\ldots,x_n\},r$-sort. Let $A\in\mathcal{K}$ and $B=\xi(A)$. Let h be an assignment of subsets of A to $P_1,\ldots,P_m$. Then for all elements $b_1,\ldots,b_n\in B$, if $\chi_{\nu(x_i)}(b_i)=a_i\in A$ $(i\le n)$, then*

$$A\models\theta(a_1,\ldots,a_n,P_1,\ldots,P_m)\iff B\models\theta^\nu(b_1,\ldots,b_n,h'(P_1^1),\ldots,h'(P_m^r)),$$

where h' is as in definition 13.4.5(1).

Proof. By induction on θ. If θ is atomic (non-monadic) this is given by definition 13.4.4(2c). If θ is $P(x_j)$ then $A\models P(a_j)$ iff $a_j\in h(P)$, iff $b_j\in h'(P^{\nu(x_j)})$ iff $B\models P^{\nu(x_j)}(b_j)$. The boolean connectives are easy to deal with.

Finally we assume that θ is the formula $\exists y\theta'(x_1,\ldots,x_n,y)$. Then $A\models\theta(a_1,\ldots,a_n)$ iff for some $a'\in A$, $A\models\theta'(a_1,\ldots,a_n,a')$. As any a' has a representative in B, this holds iff for some $\{x_1,\ldots,x_n,y\},r$-sort ν' extending ν and some $b'\in B$, $A\models\theta'(a_1,\ldots,a_n,\chi_{\nu'(y)}(b'))$. By the inductive hypothesis, this last holds iff there are $b'\in B$ and an $\{x_1,\ldots,x_n,y\},r$-sort ν' extending ν such that $B\models\theta'^{\nu'}(b_1,\ldots,b_n,b')$, i.e.
iff $B\models\bigvee_{\nu'\supseteq\nu}\exists y[\theta'^{\nu'}(b_1,\ldots,b_n,y)]$,
iff $B\models\theta^\nu(b_1,\ldots,b_n)$, as required. ■

13.4.4 Constructions and H-dimension

From now until the end of section 13.4, we assume that $\mathcal{K},\mathcal{K}'$ are classes of L-, L'-structures respectively. Here we will show the significance of constructions for H-dimension.

Theorem 13.4.8 *Assume that $\mathcal{K}'$ fits in $\mathcal{K}$ via the r-ary L'-construction $\xi=\langle\langle\varphi,\alpha\rangle,(\chi_i\mid i\le r),\beta\rangle$ over $\mathcal{K}$.*[1] *Assume further that $\mathcal{K}$ has finite H-dimension d, and that $d^*\ge d$ is a whole number such that for any atomic formula $R(x_1,\ldots,x_n)$ of L and any $\{x_1,\ldots,x_n\}$-sort ν, the formula $\beta(R,\nu)$ uses fewer than d^*-d bound variables.*

Then $\mathcal{K}'$ also has finite H-dimension, which is at most d^*.

[1] In fact, we need only assume that $\mathcal{K}'\subseteq\{\xi(A)\mid A\in\mathcal{K}\}$ up to isomorphism.

Proof. We will show that for every monadic L'-formula ψ there is a monadic L'-formula ψ_* using at most d^* bound variables and $\mathcal{K}'$-equivalent to ψ.

Suppose that $\psi(x_1, \ldots, x_n, P_1, \ldots, P_m)$ is given. As $\mathcal{K}$ has H-dimension d, there is a monadic L-formula $\psi\!\uparrow^{(d)}(x_1, \ldots, x_n, P_1, \ldots, P_m)$, using at most d bound variables and $\mathcal{K}$-equivalent to $\psi\!\uparrow(x_1, \ldots, x_n, P_1, \ldots, P_m)$. Here, $\psi\!\uparrow$ is as in lemma 13.4.2. Let ν be the $\{x_1, \ldots, x_n\}, r$-sort given by $\nu(x_i) = 1$ for all $i \leq n$. Obtain $[\psi\!\uparrow^{(d)}]^\nu(x_1, \ldots, x_n, P_1^1, \ldots, P_m^r)$ as in definition 13.4.5(2). Let $\psi_*(x_1, \ldots, x_n, P_1, \ldots, P_m)$ be the formula obtained from $\psi\!\uparrow^{(d)\nu}$ by replacing P_j^1 by P_j and P_j^l by **false** if $l > 1$, for each $1 \leq j \leq m$.
Claim: ψ_* is $\mathcal{K}'$-equivalent to ψ.
Proof of claim: Let $B \in \mathcal{K}'$. As $\mathcal{K}'$ fits in $\mathcal{K}$ there is $A \in \mathcal{K}$ such that $B \cong \xi(A)$, so we can assume that $B = \xi(A)$. Let $b_1, \ldots, b_n \in B$ and let h assign subsets of B to the P_j. Then

$$
\begin{array}{lll}
 & B \models \psi(b_1, \ldots, b_n, h(P_1), \ldots, h(P_m)) & \\
\text{iff} & A \models \psi\!\uparrow(b_1, \ldots, b_n, h(P_1), \ldots, h(P_m)) & \text{by lemma 13.4.2,} \\
\text{iff} & A \models \psi\!\uparrow^{(d)}(b_1, \ldots, b_n, h(P_1), \ldots, h(P_m)), & \text{as the formulae } \psi\!\uparrow, \psi\!\uparrow^{(d)} \\
 & & \text{are } \mathcal{K}\text{-equivalent.}
\end{array}
$$

Let h' be obtained from h as in definition 13.4.5(1). By part 2b(iv) of definition 13.4.4, $b_i = \chi_1(b_i)$ for all $i \leq n$. So by lemma 13.4.7, the last formula holds iff $B \models \psi\!\uparrow^{(d)\nu}(b_1, \ldots, b_n, h'(P_1^1), \ldots, h'(P_m^r))$. As h is an assignment into B, we have $h(P_j) \subseteq B$ for all j. Hence by definition 13.4.4(2b(iv,v)), for each $j \leq m$,

$$
\begin{array}{l}
h'(P_j^1) = h(P_j) \\
h'(P_j^l) = \varnothing \text{ for all } 1 < l \leq r.
\end{array}
$$

So the last expression holds iff $B \models \psi_*(b_1, \ldots, b_n, h(P_1), \ldots, h(P_m))$. Hence ψ and ψ_* are $\mathcal{K}'$-equivalent, proving the claim.

By remark 13.4.6, ψ_* can be taken to use at most d^* bound variables. Hence $\mathcal{K}'$ has H-dimension at most d^*, as required. ∎

There is a converse to this result.

Theorem 13.4.9 *Suppose that $\xi = \langle\langle\varphi, \alpha\rangle, (\chi_i \mid i \leq r), \beta\rangle$ is a r-ary construction over L such that $\mathcal{K}'$ fits in $\mathcal{K}$ via ξ.*[2] *Assume that $\mathcal{K}'$ has finite H-dimension, and that for some finite d, each $\alpha_{R'}$ (for each atomic R' of L') can be written with at most d bound variables. Then $\mathcal{K}$ also has finite H-dimension.*

[2] In fact, $\{\xi(A) \mid A \in \mathcal{K}\} \subseteq \mathcal{K}'$ is enough.

Proof. By increasing d if need be, we can suppose that $\mathcal{K}'$ has H-dimension at most d. Let $\sigma(P_1, \ldots, P_m)$ be a monadic L-sentence. The unique $\varnothing, r$-sort is $\varnothing$. As $\mathcal{K}'$ has finite H-dimension, there is a monadic L'-sentence $\sigma^{\varnothing(d)}(P_1^1, \ldots, P_m^r)$ using at most d bound variables and $\mathcal{K}'$-equivalent to $\sigma^{\varnothing}(P_1^1, \ldots, P_m^r)$. Recall that each χ_j has free variables x, y. For each $j \leq r$ and monadic predicate P let $\pi_j(y, P)$ be the monadic L-formula $\exists x(\chi_j(x, y) \wedge P(x))$. We now let

$$\sigma^*(P_1, \ldots, P_m) = \sigma^{\varnothing(d)}\!\uparrow\!(P_i^j / \pi_j(-, P_i)),$$

where we have substituted $\pi_j(y, P_i)$ for each monadic subformula of $\sigma^{\varnothing(d)}\!\uparrow$ of the form $P_i^j(y)$, and $\exists y(y = x \wedge \pi_j(y, P_i))$ for each subformula of $\sigma^{\varnothing(d)}\!\uparrow$ of the form $P_i^j(x)$ for $x \neq y$ (for each $i \leq m$ and $j \leq r$).

Claim: σ and σ^* are $\mathcal{K}$-equivalent.

Proof of claim: Let $A \in \mathcal{K}$ and let h be an assignment into A. Let $B = \xi(A)$; by assumption, $B \in \mathcal{K}'$. Define the assignment h' into B as in definition 13.4.5(1). By lemma 13.4.7 we have $A \models \sigma(h(P_1), \ldots, h(P_m))$ iff $B \models \sigma^{\varnothing}(h'(P_i^j))$. As $B \in \mathcal{K}'$, this is iff $B \models \sigma^{\varnothing(d)}(h'(P_i^j))$, which by lemma 13.4.2 is iff $A \models \sigma^{\varnothing(d)}\!\uparrow\!(h'(P_i^j))$. Now by definition of h', for all $b \in A$ we have $b \in h'(P_i^j)$ iff $A \models \pi_j(b, h(P_i))$. Hence the above holds iff $A \models \sigma^*(h(P_1), \ldots, h(P_m))$. The claim is proved.

Observe that σ^* can be written with a bounded number of bound variables. This is evident from its construction, using remark 13.4.3 and an argument similar to that of the preceding theorem. The theorem now follows from lemma 13.2.3. ∎

13.4.5 Constructions and expressive completeness

Finally, suppose that $\xi = \langle\langle\varphi, \alpha\rangle, (\chi_i \mid i \leq r), \beta\rangle$ is an r-ary construction over $\mathcal{K}$. If $\mathcal{K}'$ fits in $\mathcal{K}$ via ξ, we will show how to derive a finite expressively complete set $C^{\mathcal{K}'}$ of one-dimensional connectives for $\mathcal{K}'$, given such a set for $\mathcal{K}$. We will also prove a converse, corresponding to theorem 13.4.9.

Definition 13.4.10 Let $C^{\mathcal{K}}$ be a finite set of one-dimensional connectives for $\mathcal{K}$.

1. Let $\sharp \in C^{\mathcal{K}}$ be a connective with table $\tau_\sharp(x, P_1, \ldots, P_m)$. We identify an $\{x\}, r$-sort ν with the number $i = \nu(x) \leq r$. For each $\{x\}, r$-sort i we define a new (mr)-ary connective $\sharp^i$ with table $(\tau_\sharp)^i(y, P_1^1, \ldots, P_m^r)$ (a monadic formula of L' as in definition 13.4.5(2)).

2. We define $C^{\mathcal{K}'} = \{\sharp^i \mid \sharp \in C^{\mathcal{K}}, 1 \leq i \leq r\}$.

3. As in definition 13.4.5(1), for each atom p we introduce new atoms p^i $(1 \leq i \leq r)$; and given an assignment h of subsets of $A \in \mathcal{K}$ to the atoms p, we define an assignment h' of subsets of $\xi(A)$ to the new atoms p^i, by:

$$h'(p^i) = \{b \in \xi(A) \mid \chi_i(b) \in h(p)\}.$$

4. If D is a tense logical formula of $C^{\mathcal{K}}$, and $1 \leq i \leq r$, define D^i, a formula of $C^{\mathcal{K}'}$, by induction on D:

   - if p is an atom, then p^i is given by (3) above
   - $(\sim D)^i = \sim D^i$
   - $(D \wedge D')^i = D^i \wedge D'^{\,i}$
   - $\sharp(D_1, \ldots, D_m)^i = \sharp^i(D_1^1, \ldots, D_m^r)$.

Lemma 13.4.11 *Let D be any $C^{\mathcal{K}}$-formula. Let $A \in \mathcal{K}$ and write B for $\xi(A)$. Suppose that $b \in B$, $i \leq r$, $\chi_i(b) = a \in A$, and h is an assignment into A. Then $\|D\|_a^h = \|D^i\|_b^{h'}$. (Here, formulae of $C^{\mathcal{K}}$ are evaluated in A, and formulae of $C^{\mathcal{K}'}$ in B.)*

Proof. By induction on D. If D is atomic the result holds by definition of h'. The boolean connectives are easy to handle. Consider the case $D = \sharp(D_1, \ldots, D_m)$. If $j \leq m$, $l \leq r$ write S_j^l for $\{b \in B \mid \|D_j\|_{\chi_l(b)}^h = 1\}$. By the inductive hypothesis, for all $j \leq m$ and $k \leq r$ we have $S_j^k = h'(D_j^k)$ $(=_{\text{def.}} \{b \in B \mid \|D_j^k\|_b^{h'} = 1\})$. So by lemma 13.4.7,

$$\|D\|_a^h = 1 \text{ iff } A \models \tau_\sharp(a, h(D_1), \ldots, h(D_n)),$$

$$\text{iff } B \models \tau_\sharp^i(b, S_1^1, \ldots, S_m^r), \text{ iff } B \models \tau_\sharp^i(b, h'(D_1^1), \ldots, h'(D_m^r)),$$

$$\text{iff } \|\sharp^i(D_1^1, \ldots, D_m^r)\|_b^{h'} = 1 \text{ iff } \|D^i\|_b^{h'} = 1,$$

as required. ■

Theorem 13.4.12 *Suppose that $\mathcal{K}'$ fits in $\mathcal{K}$ via ξ.*[3] *Assume that $C^{\mathcal{K}}$ is expressively complete over $\mathcal{K}$. Then $C^{\mathcal{K}'}$ is expressively complete over $\mathcal{K}'$.*

---

[3]As usual, $\mathcal{K}' \subseteq \{\xi(A) \mid A \in \mathcal{K}\}$ suffices.

Proof. Let $\psi(x, P_1, \ldots, P_m)$ be a monadic L'-formula. By expressive completeness of $C^{\mathcal{K}}$ for $\mathcal{K}$, there is a formula $D(p_1, \ldots, p_m)$ of $C^{\mathcal{K}}$ whose table is $\mathcal{K}$-equivalent to the L-formula $\psi\uparrow(x, P_1, \ldots, P_m)$. Define $D_*(p_1, \ldots, p_m)$ to be the result of replacing p_i^1 by p_i and p_i^j $(j > 1)$ by **false** in $D^1(p_1^1, \ldots, p_m^r)$, for each $i \leq m$.

We claim that D_* has table $\mathcal{K}'$-equivalent to ψ. For let $B \in \mathcal{K}'$. As $\mathcal{K}'$ fits in $\mathcal{K}$, there is $A \in \mathcal{K}$ such that $B \cong \xi(A)$. Let h be any assignment of subsets of B to the atoms $p_1, \ldots, p_m$, and let $b \in B$. Then by lemma 13.4.2, $B \models \psi(b, h(P_1), \ldots, h(P_m))$ iff $A \models \psi\uparrow(b, h(P_1), \ldots, h(P_m))$, iff $\|D\|_b^h = 1$ in A. But $\chi_1(b) = b$, so by lemma 13.4.11, $\|D\|_b^h = 1$ in A iff $\|D^1\|_b^{h'} = 1$ in B. Also, as in the claim of theorem 13.4.8, $h'(p_j^1) = h(p_j)$, and if $l > 1$, $h'(p_j^l) = \varnothing$. Hence $\|D^1(p_1^1, \ldots, p_m^r)\|_b^{h'} = \|D_*\|_b^h$. So $B \models \psi(b, h(P_1), \ldots, h(P_m))$ iff $\|D_*(p_1, \ldots, p_m)\|_b^h = 1$, as required. ∎

Now for the converse.

Theorem 13.4.13 *Assume that $\mathcal{K}'$ fits in $\mathcal{K}$ via the construction $\xi = \langle\langle\varphi, \alpha\rangle, (\chi_i \mid i \leq r), \beta\rangle$ over L.*[4] *Suppose we are given a finite set $C^{\mathcal{K}'}$ of expressively complete one-dimensional connectives for $\mathcal{K}'$. Then we can effectively construct a finite set $C^{\mathcal{K}}$ of expressively complete connectives for $\mathcal{K}$.*

Proof. $C^{\mathcal{K}}$ consists of:

- an m-ary connective $\sharp\uparrow$, with table $\tau\uparrow$, for each m-ary connective $\sharp \in C^{\mathcal{K}'}$ with table τ;
- a unary connective U_i ('up') with table $\upsilon_i(y, P) = \exists x(\chi_i(x, y) \wedge P(x))$, for each $i \leq r$;
- a unary connective D_i ('down') with table $\delta_i(x, Q) = \exists y(\chi_i(x, y) \wedge Q(y))$, for each $i \leq r$.

Let $\theta(x, P_1, \ldots, P_m)$ be a monadic L-formula. By expressive completeness, for each $i \leq r$ there is a tense logical formula $E_i(p_1^1, ..., p_m^r)$ of $C^{\mathcal{K}'}$ that is $\mathcal{K}'$-equivalent to $\theta^i(x, P_1^1, ..., P_m^r)$. Let $E_i\uparrow(p_1^1, ..., p_m^r)$ be the $C^{\mathcal{K}}$-formula obtained by replacing each connective $\sharp$ in E_i by $\sharp\uparrow$. An easy induction shows that if the table of E_i is τ then the table of $E_i\uparrow$ is $\tau\uparrow$. Let $E_i^*(p_1, \ldots, p_m)$ be obtained from $E_i\uparrow$ by replacing each atom p_j^l by $U_l(p_j)$. Finally, let E be the formula $\bigvee_{i \leq r} D_i(E_i^*)$. The theorem will be established if we show:

Claim: θ and E are $\mathcal{K}$-equivalent.

[4] In fact, $\{\xi(A) \mid A \in \mathcal{K}\} \subseteq \mathcal{K}'$ suffices.

Proof of claim: Let $A \in \mathcal{K}$ and write B for $\xi(A) \in \mathcal{K}'$. Let h be an assignment into A and let h' be obtained as in definition 13.4.10(3). Let $a \in A$. Then by lemma 13.4.7, $A \models \theta(a, h(P_1), ..., h(P_m))$ iff for some $i \leq r$ and $b \in B$ such that $\chi_i(b) = a$, $B \models \theta^i(b, h'(P_1^1), ..., h'(P_m^r))$, iff for some such i and b, $\|E_i(p_1^1, ..., p_m^r)\|_b^{h'} = 1$ in B. By the above, this is true iff for some i, b, $\|E_i\uparrow(p_1^1, ..., p_m^r)\|_b^{h'} = 1$ in A (of course, h' can be regarded as an assignment into A).

Now observe that for all $c \in A$, $j \leq m$, and $l \leq r$, we have $c \in h'(p_j^l)$ iff $\|U_l(p_j)\|_c^h = 1$ in A. Hence $A \models \theta(a, h(P_1), ..., h(P_m))$ iff $\|E_i^*\|_b^h = 1$ in A for some $i \leq r$ and $b \in B$ with $\chi_i(b) = a$. This clearly holds iff $\| \bigvee_{i \leq r} D_i(E_i^*)\|_a^h = 1$, proving the claim. ■

13.5 Covering theorems

We want to investigate how we can piece together flows of time in such a way that (a) if the pieces have finite H-dimension then the resulting flow also has finite H-dimension, and (b) if the pieces have a finite expressively complete set of one-dimensional temporal connectives then so does the resulting flow. We give some sufficient conditions for this. This information is important. We saw in section 13.2 that for a flow of time, the notions of finite H-dimension and expressive completeness are equivalent. Results in this section generalize those of [Amir and Gabbay, 1987].

We begin with some examples of 'piecing-together' operations.

Definition 13.5.1 Let $(R, <_R)$ be a flow of time. We say that $(R, <_R)$ can be *covered* by the family $\langle (B_t, <_t) \mid t \in T \rangle$ iff the following conditions hold:

1. $(T, <_T)$ and $(B_t, <_t)$ $(t \in T)$ are flows of time, and $<_T$ is irreflexive.
2. $R = \bigcup_{t \in T} B_t$; and $t \neq s \Rightarrow B_t \cap B_s = \varnothing$.
3. For $x, y \in R$ let $t(x)$, $t(y)$ be the unique elements of T such that $x \in B_{t(x)}$ and $y \in B_{t(y)}$. Then $x <_R y$ iff $t(x) <_T t(y) \vee [t(x) = t(y) \wedge x <_{t(x)} y]$.
4. The relation '$\equiv$' on $(R, <_R)$ given by $x \equiv y$ iff $t(x) = t(y)$ is definable by some first-order formula $\varepsilon(x, y)$ of $(R, <, =)$, perhaps using quantifiers over elements of R.

Notice that we are ordering $R = \bigcup_{t \in T} B_t$ *lexicographically*. We will prove that if $(T, <_T)$ is taken from a class $\mathcal{T}$ that has finite H-dimension, and the $(B_t, <_t)$ $(t \in T)$ are all taken from a class $\mathcal{B}$ also having finite H-dimension, then $(R, <_R)$ also has finite H-dimension. We will also construct from given finite sets of expressively complete connectives on $\mathcal{T}$ and on $\mathcal{B}$, a finite set of expressively complete connectives for the class of all $(R, <_R)$ obtained like this. See theorem 13.5.34 below.

Examples 13.5.2 Since the natural numbers $\mathbb{N}$ and the integers $\mathbb{Z}$ have $\{U, S\}$ as a finite expressively complete set of connectives (see chapter 10), we can conclude that the following flows of time, $(R, <_R)$, obtained as in definition 13.5.1, are also expressively complete. We will always find that $\equiv$ is definable. But in most cases there will even be a (non-monadic) first-order formula $\xi(t)$ of $(R, <_R)$ which defines the integers as a subset of R. ξ will choose a canonical representative of each $\equiv$-class. In the terminology of the preceding theorem, $\xi(t)$ defines $(T, <_T)$ and we can always find formulae $\beta(t, x)$ defining '$x \in B_t$'. All the $(B_t, <_t)$ are isomorphic.

1. The flow $(R, <)$ of figure 13.3 where $R = \mathbb{N} \times \mathbb{Z}$ and $<$ is the lexicographic ordering. Explicitly, take

$$x \equiv y \text{ iff } \forall z(\zeta(z) \to (z > x \leftrightarrow z > y))$$

where $\zeta(z) =_{\text{def.}} \forall x(x < z \to \exists t(x < t < z))$ defines the 'zero point' of each copy of $\mathbb{N}$.

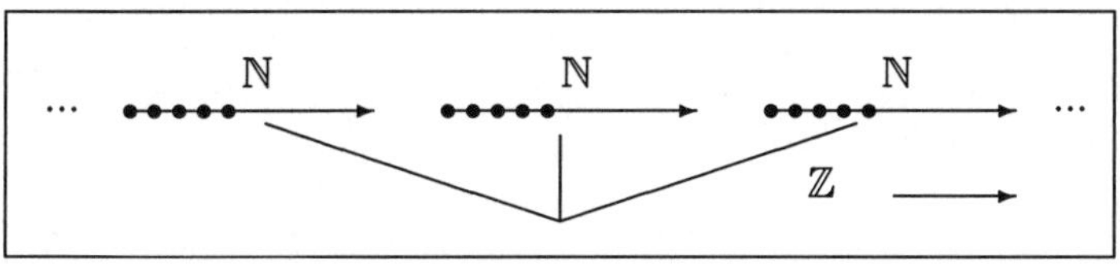

Figure 13.3. $\mathbb{N} \times \mathbb{Z}$: each $z \in \mathbb{Z}$ replaced by a copy of $\mathbb{N}$.

We will show how to construct expressively complete connectives for this flow. Note, however, that in this case U and S are already expressively complete. For by the results of chapter 11, U, S, U' and S' are certainly expressively complete for $(R, <)$; but as it has no gaps, U' and S' are equivalent to $\bot$. Cf. proposition 6.9.9, where an analogous result for $\mathbb{R}$ is proved.

2. Take the flow of time $(T, <)$ to be as in figure 13.4.

This is finite, so has finite H-dimension; and we can easily find expressively complete connectives for it. We can use it to build a

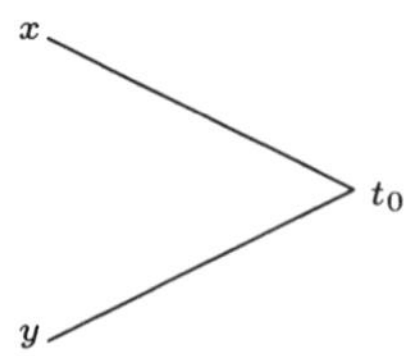

Figure 13.4. $(T, <)$.

new flow, as follows. Let $B_x = B_y = \{0, -1, -2, \ldots\}$, the negative integers; let $B_{t_0} = \{t_0\}$. Clearly, Until and Since are expressively complete for $\{B_t \mid t \in T\}$. Then we obtain the flow in figure 13.5.

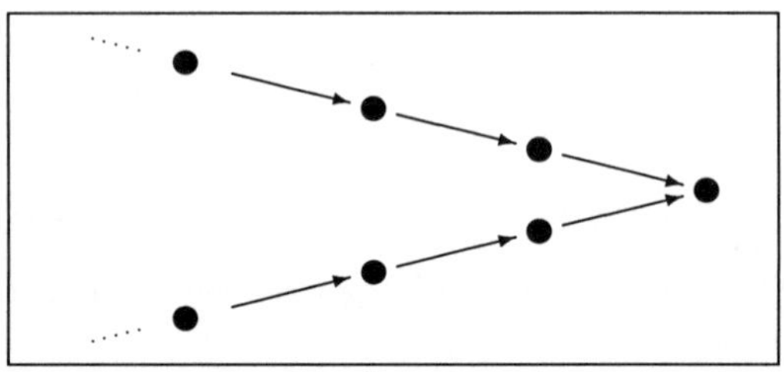

Figure 13.5. Replace x, y in T by {negative integers}.

$\equiv$ is clearly easy to define. So this flow has finite H-dimension and a finite expressively complete set of one-dimensional connectives.

3. Figures 13.6, 13.7 and 13.8 picture some more examples of flows of time that arise from the same construction, and so are expressively complete.

4. If B_t is the flow pictured in figure 13.9, and T is $\mathbb{Z}$, we obtain figure 13.10.

There are other flows that are clearly similar to these examples, in that they arise by replacing each point of a 'spine' T by a new flow B_t. For example, the 'comb' of figure 13.11.

In this flow, each point t of $\mathbb{Z}$ has been replaced by a copy $\mathbb{N}_t$ of $\mathbb{N}$, as in example 13.5.2(1). But now, the ordering on the result is defined differently. Within each copy of $\mathbb{N}$ the ordering is as before, and is the ordering induced from $\mathbb{N}$. But between copies of $\mathbb{N}$ the ordering is defined

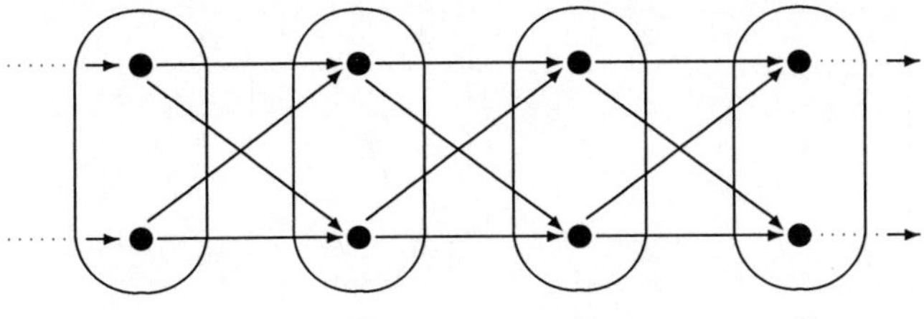

Figure 13.6. $T = \mathbb{Z}, B_t$ is a 2-element set.

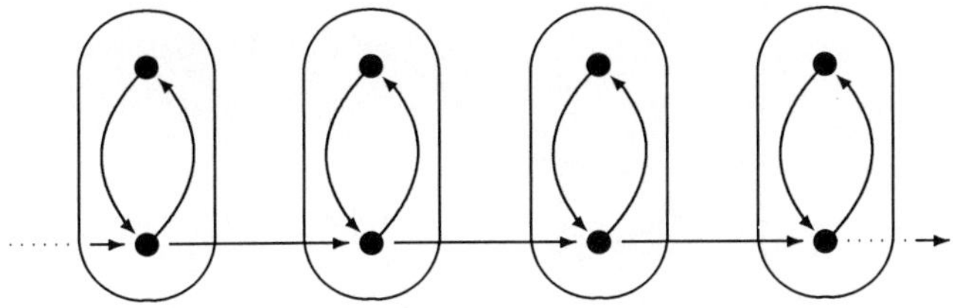

Figure 13.7. $T = \mathbb{Z}, B_t$ is $\{0,1\}$ with $0 < 1 < 0$.

as follows. If $a \in \mathbb{N}_t, b \in \mathbb{N}_u$ for $t, u \in \mathbb{Z}$, then $a < b$ iff $t < u$ in $\mathbb{Z}$ and $a = 0$ in $\mathbb{N}_t$. Thus the copies of $\mathbb{N}$ do not 'lie flush' in $\mathbb{Z}$ but 'stick out' at the side. In effect, we attach a copy of $\mathbb{N}$ to each element of $\mathbb{Z}$.

One can iterate this idea by attaching an entire comb to each element of $\mathbb{Z}$, and so on.

As for the examples above, combs are based closely on $\mathbb{N}$ and $\mathbb{Z}$, for which U and S are expressively complete. So we would like to find an expressively complete set of connectives for them also.

13.5.1 Canonical structures

In order to handle these examples, we will prove a fairly general result. We begin by defining the class $\mathcal{T}[\mathcal{B}]$ of 'canonical' structures made of the disjoint union of flows of time $T \in \mathcal{T}$ and $B_t \in \mathcal{B}$ (all $t \in T$). Our aim is to prove:

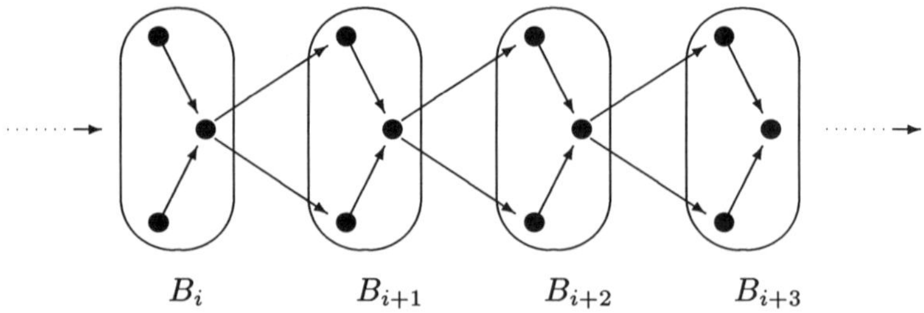

Figure 13.8. $T = \mathbb{Z}$, B_t is the flow of figure 13.5.2.

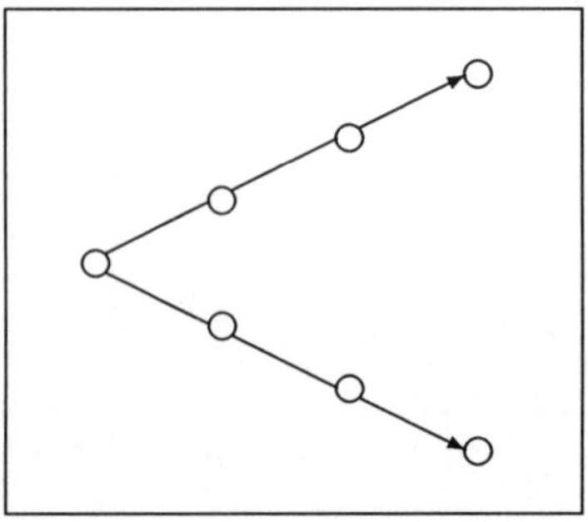

Figure 13.9 A finite flow of time.

Theorem 13.5.3 *Let $\mathcal{B}, \mathcal{T}$ be classes of flows of time. If $\mathcal{K}$ is a class of flows of time arising from some $T \in \mathcal{T}$ and $B_t \in \mathcal{B}$ (for each $t \in T$) along the lines of the examples (formally, it fits in $\mathcal{T}[\mathcal{B}]$), then*

- *if $\mathcal{B}, \mathcal{T}$ have finite H-dimension then so does $\mathcal{K}$;*
- *if $\mathcal{B}, \mathcal{T}$ have finite expressively complete sets of one-dimensional temporal connectives, then so does $\mathcal{K}$.*

We will show that the class $\mathcal{T}[\mathcal{B}]$ has finite H-dimension, and a finite expressively complete set of connectives, if $\mathcal{T}$ and $\mathcal{B}$ have. The theorem will then follow from the results of section 13.4, because any class $\mathcal{K}$ of structures that fits in the class $\mathcal{T}[\mathcal{B}]$ will then also have finite H-dimension and a finite set of expressively complete connectives.

The structures in $\mathcal{T}[\mathcal{B}]$ will not be simply flows of time. But, as before, our proofs apply to more general structures than flows of time. Hence we

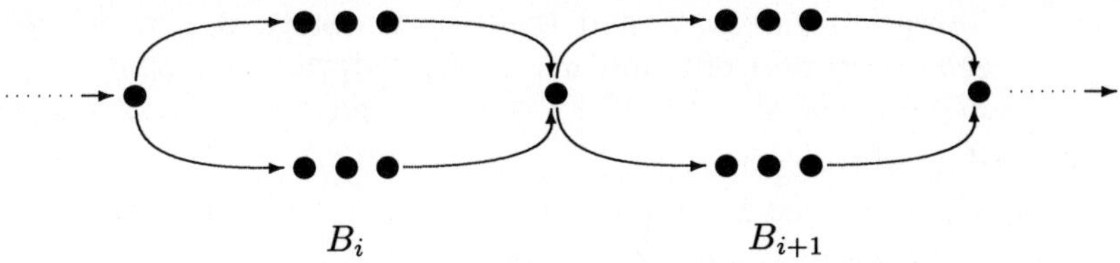

Figure 13.10. The flow obtained from $\mathbb{Z}$ and the preceding flow.

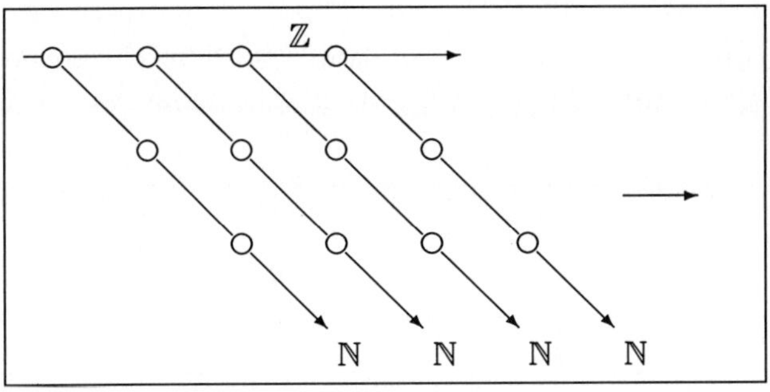

Figure 13.11. The comb.

now fix until further notice a first-order signature L without any function or constant symbols,[5] and assume that $\mathcal{T}, \mathcal{B}$ are classes of L-structures. For example, if $\mathcal{T}, \mathcal{B}$ are classes of flows of time then L is just $\{<\}$. (We could even allow $\mathcal{T}, \mathcal{B}$ to have different signatures.)

Definition 13.5.4

1. The *canonical signature* L_Ξ is $L \cup \{\pi\}$, where π is a new unary function symbol (assumed not to occur in L).

2. Let $\mathcal{T}, \mathcal{B}$ be classes of L-structures. A structure A is said to be a *canonical* $\mathcal{TB}$*-structure* if:

   - A is an L_Ξ-structure.
   - there are $T \in \mathcal{T}$ and $B_t \in \mathcal{B}$ for each $t \in T$, such that the domain of A is the *disjoint* union of T and the B_t.

[5]This restriction ensures that the disjoint union of L-structures is a well-defined L-structure. In fact we could allow L to have unary function symbols.

- the relations of L are interpreted in A as the union of their interpretations in T and the B_t. Formally, if R is an n-ary relation symbol of L, and $a_1, \ldots, a_n \in A$, then $A \models R(a_1, \ldots, a_n)$ iff for some $X \in \{T, B_t \mid t \in T\}$, we have $a_1, \ldots, a_n \in X$ and $X \models R(a_1, \ldots, a_n)$.
- π relates each B_t to its corresponding index t. Formally,
  - if $t \in T$ and $b \in B_t$ then $A \models \pi(b) = t$;
  - $A \models \pi(t) = t$ for all $t \in T$.

3. We write $\mathcal{T}[\mathcal{B}]$ for the class of canonical $\mathcal{TB}$-structures.

In a picture, the situation is as in figure 13.12, for the purpose of which we have assumed that T and the B_t are flows of time (so $L = \{<\}$).

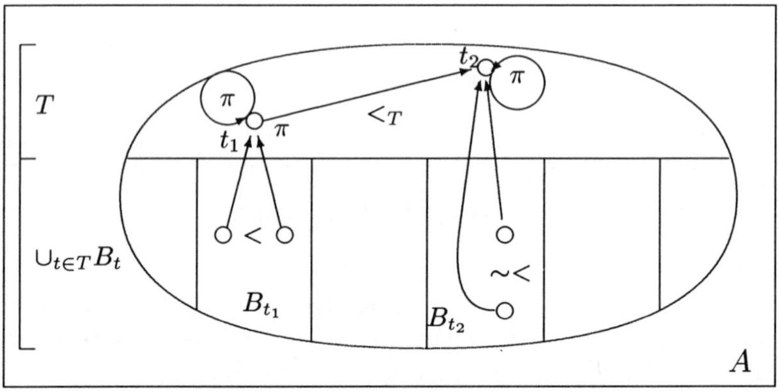

Figure 13.12. A canonical structure.

We are going to rewrite any monadic formula of $\mathcal{T}[\mathcal{B}]$ in terms of monadic formulae of $\mathcal{B}$ and $\mathcal{T}$. These can then be replaced by $\mathcal{B}$- and $\mathcal{T}$-equivalent formulae with a bounded number of bound variables, so $\mathcal{T}[\mathcal{B}]$ has finite H-dimension if $\mathcal{T}$ and $\mathcal{B}$ do. It will also follow that $\mathcal{T}[\mathcal{B}]$ has a finite set of expressively complete connectives if $\mathcal{T}$ and $\mathcal{B}$ do. We will then be able to apply the results of section 13.4 to complete the argument.

We will need some definitions for canonical structures.

Definition 13.5.5

1. We write $x \equiv y$ for the L_{Ξ}-formula:

$$(x \neq \pi(x) = \pi(y) \neq y) \vee (x = \pi(x) \wedge y = \pi(y)).$$

In any L_Ξ-structure A, $\equiv$ defines an equivalence relation.

2. The *basic theory* is the set Ξ of L_Ξ-sentences:

   - $\forall x(\pi^2(x) = \pi(x))$ (i.e. π is *idempotent*);
   - $\forall x(\pi(x) = x \to \exists y(x = \pi(y) \wedge x \neq y))$;
   - $\forall x_1, \ldots, x_n[R(x_1, \ldots, x_n) \to \bigwedge_{i<n} x_i \equiv x_{i+1}]$ for all n and all n-ary relation symbols $R \in L$.

   Ξ expresses that a structure has the form of figure 13.5.10. It is valid in $\mathcal{T}[\mathcal{B}]$ for any classes $\mathcal{T}, \mathcal{B}$.

3. Let $A \models \Xi$. We define:

   - $T(A) = \{t \in A \mid A \models \pi(t) = t\}$;
   - $B_t = \{b \in A \mid A \models \pi(b) = t \neq b\}$ for each $t \in T(A)$;
   - $B(A) = \bigcup_{t \in T} B_t$.

   The equivalence classes of $\equiv$ in A are $T(A)$ and the B_t $(t \in T(A))$.

4. Because L has no function or constant symbols, each of T, B, and the B_t is an L-substructure of A, so an L-structure in its own right. Formally, let $X \in \{T(A), B(A), B_t \mid t \in T(A)\}$. If $R \in L$ is an n-ary relation symbol and $a_1, \ldots, a_n \in X$, then we define $X \models R(a_1, \ldots, a_n)$ iff $A \models R(a_1, \ldots, a_n)$. This makes X into an L-structure.

5. We will write simply T or B, if A is understood.

Definition 13.5.6 (relativization) Let $\theta(x_1, \ldots, x_n, P_1, \ldots, P_m)$ be a monadic L-formula.

1. Let $\varphi(y, \bar{v})$ be an L_Ξ-formula whose free variables are taken from $y, \bar{v}$. Assume that no variable in $\bar{v}$ occurs bound in θ. We define a monadic formula θ^φ of L_Ξ, the *relativization of θ to φ*, by induction on θ as follows:

   - if θ is atomic (e.g. a monadic predicate letter) then θ^φ is just θ;
   - $(\sim\theta)^\varphi$ is $\sim(\theta^\varphi)$;
   - $(\theta_1 \wedge \theta_2)^\varphi$ is $\theta_1^\varphi \wedge \theta_2^\varphi$;
   - $[\exists y \theta]^\varphi$ is $\exists y[\varphi(y) \wedge \theta^\varphi]$;
   - If $x \neq y$, $[\exists x \theta]^\varphi$ is $\exists x[\exists y(y = x \wedge \varphi(y)) \wedge \theta^\varphi]$.

2. We write $\theta^T(x_1, \ldots, x_n, P_1, \ldots, P_m)$ for the relativization $\theta^{y=\pi(y)}$.

3. Let $z(v)$ be an L_Ξ-term, necessarily of the form $\pi(\pi(\ldots\pi(v))\ldots)$ for some (possibly zero) number of πs, whose variable v does not occur bound in θ. We write

$$\theta^{B[z(v)]}(x_1, \ldots, x_n, v, P_1, \ldots, P_m)$$

for the relativization $\theta^{\pi(y)=z(v)\wedge y\neq\pi(y)}$. If the term z is just a variable and is clear from the context, we will simply write θ^B.

Remark 13.5.7 Note that v occurs free in $\theta^{B[z(v)]}$. We are relativizing all quantifiers of θ to the set $\{y \mid \pi(y) = z(v) \wedge y \neq \pi(y)\}$, i.e. to some B_t, where $t = z(v)$. The value of v will determine what t is. See lemma 13.5.8(ii) below.

In general, the free variables of θ^φ are those of θ, plus perhaps any of φ in addition to y. The bound variables of θ^φ are at most those of θ plus those of φ, plus y.

Clearly, $[\exists x\theta]^{\varphi(y)}$ is equivalent to $\exists x(\varphi(y/x) \wedge \theta^\varphi)$, where $\varphi(y/x)$ is the result of free substitution of x for y in $\varphi(y)$.

We give a standard lemma on relativization, tailored to our situation.

Lemma 13.5.8 *Let $A \models \Xi$. We write B, T for $B(A)$ and $T(A)$ respectively, following definition 13.5.5(5). Let $\theta(x_1, \ldots, x_n, P_1, \ldots, P_m)$ be any monadic formula of L.*

(i) *Let $t_1, \ldots, t_n \in T$ and $S_1, \ldots, S_m \subseteq A$. Then:*

$$A \models \theta^T(t_1, \ldots, t_n, S_1, \ldots, S_m) \text{ iff } T \models \theta(t_1, \ldots, t_n, T \cap S_1, \ldots, T \cap S_m).$$

(ii) *Let $t \in T$, $\bar{b} = b_1, \ldots, b_n \in B_t$, and $\bar{S} = S_1, \ldots, S_m \subseteq A$. Let $z(v)$ be a term of L_Ξ and suppose that $a \in A$ is such that $A \models z(a) = t$. Then:*

$$A \models \theta^{B[z(v)]}(\bar{b}, v/a, \bar{S}) \iff B_t \models \theta(\bar{b}, B_t \cap S_1, \ldots, B_t \cap S_m),$$

where $\theta^{B[z(v)]}(x_1, \ldots, x_n, v, P_1, \ldots, P_m)$ is as in definition 13.5.6(3), and we use v/a to denote substitution of the element $a \in A$ for the variable v.

Proof. The proofs are by induction on θ.

(i) If θ is a non-monadic atomic formula, the result follows from definition 13.5.5(4), and if θ is a monadic predicate letter the result is clear. The boolean connectives present no difficulties. So consider the case $\exists x\theta$. We have:

$$T \models \exists x\theta(x, t_1, \ldots, t_n, T \cap S_1, \ldots, T \cap S_m)$$

iff for some $a \in T, T \models \theta(a, t_1, \ldots, t_n, T \cap S_1, \ldots, T \cap S_m)$
iff for some $a \in T, A \models \theta^T(a, t_1, \ldots, t_n, S_1, \ldots, S_m)$ [by inductive hypothesis]
iff $A \models \exists x(\pi(x) = x \wedge \theta^T(x, t_1, \ldots, t_n, S_1, \ldots, S_m))$
iff $A \models (\exists x\theta)^T(t_1, \ldots, t_n, S_1, \ldots, S_m)$, as required.

(ii) The atomic and boolean cases are as above. Consider the case $\exists x\theta$. We have:

$$B_t \models \exists x\theta(x, \bar{b}, B_t \cap S_1, \ldots, B_t \cap S_m)$$

iff for some $a \in B_t$, $B_t \models \theta(a, \bar{b}, B_t \cap S_1, \ldots, B_t \cap S_m)$
iff $A \models \exists x(\pi(x) = t \wedge x \neq t \wedge \theta^{B[z(v)]}(x, \bar{b}, v/a, S_1, \ldots, S_m))$, by the inductive hypothesis. Because v does not occur bound in $\exists x\theta$, this holds iff $A \models (\exists x\theta)^{B[z(v)]}(\bar{b}, v/a, \bar{S})$, as required.

■

Definition 13.5.9 Let X be a finite set of variables. A *partition formula* is a maximal Ξ-consistent conjunction $N(X)$ of L_Ξ-formulae of the form $x \equiv y, \pi(x) = x$ and their negations (for $x, y \in X$).

If X is empty, any partition formula $N(X)$ is equivalent to $\top$. Otherwise, $N(X)$ specifies a unique partition of X into $\equiv$-classes $X_1, \ldots, X_k, Y$, where:

- $k \geq 0$;
- $X_i \neq \varnothing$ for each $i \leq k$;
- for each i and $x \in X_i, N(X) \vdash \pi(x) \neq x$;
- for each $y \in Y, N(X) \vdash \pi(y) = y$;
- if $x_1, x_2 \in X, N(X) \vdash x_1 \equiv x_2$ iff $x_1, x_2 \in X_i$ for some i, or $x_1, x_2 \in Y$.

We call the X_i and Y the *components of* N.

In what follows, when we say $N(X)$ is a partition formula with components $X_1, \ldots, X_k, Y$ for $k \geq 0$, it will be implicit that Y is the T-component, i.e. for all $x \in X$, $N \vdash \pi(x) = x$ iff $x \in Y$. So if $A \models \Xi$

and we have given to us an assignment of elements of A to the variables of X such that $N(X)$ holds in A, then $Y \subseteq T$ and there are distinct elements $t_i \in T$ such that $X_i \subseteq B_{t_i}$ for each i.

Definition 13.5.10 We define a *mixed formula* to be a formula of the form

$$\mu(X) \quad = \quad N(X) \wedge \bigwedge_{1 \leq i \leq k} \theta_i^{B[t_i]}(X_i)$$

$$\wedge\, \tau^T(Y, z_1/t_1, \ldots, z_k/t_k, P_1, \ldots, P_m, Q_1/\sigma_1^B, \ldots, Q_n/\sigma_n^B),$$

where:

1. $N(X)$ is a partition formula with components $X_1, \ldots, X_k, Y$.
2. $\tau(Y, z_1, \ldots, z_k, P_1, \ldots, P_m, Q_1, \ldots, Q_n)$ is an L-formula with monadic predicates $P_1, \ldots, P_m, Q_1, \ldots, Q_n$, such that *no variable in X occurs bound in τ*. (We can always rename the variables to meet this condition.)
3. For each j $(1 \leq j \leq n)$, σ_j is a monadic L-sentence.

and for each i with $1 \leq i \leq k$:

4. $\theta_i(X_i)$ is a monadic L-formula such that *no $x \in X$ occurs bound in θ_i*.
5. For each $i \leq k$, z_i is a variable not occurring in θ_i.
6. t_i is an L_Ξ-term of the form $\pi(x_i)$ for arbitrary $x_i \in X_i$. See remark 13.5.11 below.

We explain our notation further.

7. z_i/t_i denotes substitution of t_i for all free occurrences of z_i in $\theta_i^{B[z_i]}$ and in τ^T. We can do this as by remark 13.5.7, no $x \in X$ occurs bound in θ_i^B or in τ^T.
8. Q_j/σ_j^B denotes simultaneous substitution of σ_j^B for Q_j in τ^T. We explain how this is done.

   First observe that all occurrences of Q_j in τ^T are of the form $Q_j(v)$ for some variable v, since the only terms of L are variables. Let $Q_j(v)$ be such an occurrence. If necesssary rewrite σ_j so that it does not use the variable v. Hence $\sigma_j^{B[v]}$ can be taken to have v as its sole free variable (cf. definition 13.5.6(3)). Now replace the atomic subformula $Q_j(v)$ of τ^T by $\sigma_j^{B[v]}$. We write $\tau^T(Y, z_1, \ldots, z_k, P_1, \ldots, P_m, Q_1/\sigma_1^B, \ldots, Q_n/\sigma_n^B)$ for the result of doing this simultaneously for all occurrences of all Q_j in τ^T.

Notice that substitution is performed on τ^T, not on τ. That is, first we obtain τ^T from τ, and then we substitute, not vice versa. (Otherwise any quantifiers in the σ_i would become relativized to T, which is not what we want.)

Remark 13.5.11 Clearly, when choosing $t_i = \pi(x_i)$, the choice of x_i is arbitrary. Replacing x_i in t_i by $x_i' \in X_i$ gives an equivalent formula $\mu'(X)$, since $N \vdash \pi(x_i) = \pi(x_i')$.

Note also that not all the θ_i or τ need to be present, as $\top^B = \top^T = \top$.

Definition 13.5.12 Let X be a finite set of variables. Clearly, there are only finitely many Ξ-inequivalent partition formulae $N(X)$. Let $\Delta(X)$ be a set of representatives of partition formulae $N(X)$, so that any partition formula $N(X)$ is Ξ-equivalent to exactly one formula in $\Delta(X)$.

We will now show how to rewrite any monadic L_Ξ-formula as a Ξ-equivalent boolean combination (disjunction) of mixed formulae.

Lemma 13.5.13 *Let $\psi(X, P_1, \ldots, P_r)$ be a monadic formula of L_Ξ. Then ψ is Ξ-equivalent to a disjunction of mixed formulae, the disjunction having the same free variables as ψ.*

Proof. Our method is to show that any atomic monadic formula ψ of L_Ξ is Ξ-equivalent to a disjunction of mixed formulae, and then to prove that modulo Ξ, the class of mixed formulae is closed under the boolean operations and quantification.

(a) First let $\psi(X)$ be an atomic (possibly monadic) L_Ξ-formula, where X is the set of variables in ψ. Clearly, ψ is Ξ-equivalent to $\bigvee_{N \in \Delta(X)}(N(X) \wedge \psi(X))$. So it is enough to show that for any partition formula $N(X)$, $N(X) \wedge \psi(X)$ is Ξ-equivalent to $\bot$ or a mixed formula μ. (Falsity will not contribute to the disjunction; and if all disjuncts are false, ψ is Ξ-equivalent to the empty disjunction of mixed formulae.)

Let $X_1, \ldots, X_k, Y$ be the components of $N(X)$. As $\Xi \vdash$ 'π is idempotent', for all $r \geq 1$ we may replace any term $\pi^r(x) = \pi\pi\pi\ldots\pi(x)$ (with r πs) in ψ by $\pi(x)$. Moreover, if $y \in Y$ we can replace $\pi(y)$ by y. Thus we can assume that ψ is an atomic formula of the form

$$\psi(\pi^{i_1}(x_1), \ldots, \pi^{i_n}(x_n), y_1, \ldots, y_m)$$

where $X - Y = \{x_1, \ldots, x_n\}, Y = \{y_1, \ldots, y_m\}$ and $i_j \in \{0,1\}$ for each j, where $\pi^0(x) =_{\text{def.}} x$.

Case I: $n = 0$ or all i_j are 1. Then $N \wedge \psi$ is talking about T. Define $\mu(y_1, \ldots, y_m, z_1, \ldots, z_k) = N \wedge \psi^T(z_{k_1}, \ldots, z_{k_n}, y_1, \ldots, y_m)$, where $x_i \in X_{k_i}$ for each $i \leq n$. Note that $\psi^T = \psi$ since ψ is atomic.

Case II: $n > 0, k = 1$, all i_j are 0, and $m = 0$. In this case, $N \wedge \psi$ is talking about some B_t. Define $\mu = N \wedge \psi^B(x_1, \ldots, x_n)$. Note that $\psi^B = \psi$ here.

Case III: Otherwise, not all the terms $\pi^{i_1}(x_1), \ldots, \pi^{i_n}(x_n), y_1, \ldots, y_m$ are $\equiv$-equivalent according to N. So by definition 13.5.5(2), $N(X) \wedge \psi(X)$ is Ξ-equivalent to $\perp$.

In cases I and II we obtain a mixed formula Ξ-equivalent to $N \wedge \psi$.

(b) Clearly, the class of disjunctions of mixed formulae is closed under $\vee$.

(c) We have to check negation now. First we need the following observation.

Remark 13.5.14 Let $N(X)$ be a partition formula and let $X' \supseteq X$ contain finitely many extra variables. Then N is logically equivalent to a finite disjunction of partition formulae $\bigvee_i N_i(X')$. This disjunction expresses all the possible ways that the extra variables of $X' - X$ can be arranged around those of X via $\equiv$. The $\equiv$-type of the variables in X is already fixed by N.

Hence any mixed formula $N(X) \wedge \delta$ is Ξ-equivalent to $\bigvee_i (N_i(X') \wedge \delta)$, which is a disjunction of mixed formulae but using the extra variables.

To proceed, let $M = \bigvee_i (N_i \wedge \bigwedge_j \theta_{ij}^B \wedge \tau_i^T)$ be a disjunction of mixed formulae. Let X be the set of all free variables of M. By the remark, we can assume that each N_i is a partition formula of the form $N_i(X)$. For each $N \in \Delta(X)$ let

$$\iota(N) = \{i \mid N_i \text{ is } \Xi\text{-equivalent to } N\}$$

$$\psi_N = \bigvee_{i \in \iota(N)} [\bigwedge_j \theta_{ij}^B \wedge \tau_i^T].$$

So ψ_N is false if there is no i such that N_i and N are Ξ-equivalent.

Claim:

(i) M is Ξ-equivalent to $\bigvee_{N \in \Delta} N \wedge \psi_N$.

(ii) $\sim M$ is Ξ-equivalent to $\bigvee_{N \in \Delta} N \wedge \sim\psi_N$.

Proof of claim: (i) follows from the definition of M and ψ_N. (ii) follows from (i), given that modulo Ξ, the $N \in \Delta$ are pairwise inconsistent and $\bigvee_{N \in \Delta} N = \top$. This proves the claim.

Hence we need only check that for each $N \in \Delta$, $N \wedge \sim\psi_N$ is equivalent to a disjunction of mixed formulae.

If $\iota(N) = \varnothing$ then $N \wedge {\sim}\psi_N$ is equivalent to N, which is mixed. Otherwise, $N \wedge {\sim}\psi_N = N \wedge {\sim}\bigvee_{i\in\iota(N)}[\bigwedge_j \theta_{ij}^B \wedge \tau_i^T]$, which is equivalent to

$$\bigwedge_{i\in\iota(N)} \bigvee_j (N \wedge {\sim}\theta_{ij}^B) \vee (N \wedge {\sim}\tau_i^T).$$

Treating each subformula $(N \wedge {\sim}\theta_{ij}^B)$ and $(N \wedge {\sim}\tau_i^T)$ as atomic, rewrite this last expression in disjunctive normal form. By the clauses for $\wedge$ and $\sim$ in definition 13.5.6, any conjunction of these 'atoms' is a mixed formula. Hence $N \wedge {\sim}\psi_N$ is Ξ-equivalent to a disjunction of mixed formulae, as required.

(d) We will now show that the class of finite disjunctions of mixed formulae is closed under existential quantification. Since $\exists$ distributes across $\bigvee$, it is enough to show that if $\mu(X) =$

$$\begin{gathered} N(X) \wedge \textstyle\bigwedge_{1\le i\le k} \theta_i^{B[t_i]}(X_i) \\ \wedge\tau^T(Y, z_1/t_1, \ldots, z_k/t_k, P_1, \ldots, P_m, Q_1/\sigma_1^B, \ldots, Q_n/\sigma_n^B) \end{gathered}$$

is a single mixed formula, and $x \in X$, then $\exists x\mu(X)$ is equivalent to a mixed formula $\mu'(X - \{x\})$.

Let $X_1, \ldots, X_k, Y$ be the components of N. Let $N'(X - \{x\}) \in \Delta(X - \{x\})$ be the unique partition formula that is Ξ-consistent with N. N' is obtained from N by deleting all conjuncts involving x.

Case I: $x \in Y$.
Then $\exists x\mu$ is Ξ-equivalent to $N' \wedge \bigwedge_{1\le i\le k} \theta_i^{B[t_i]}(X_i) \wedge \exists x[\pi(x) = x \wedge \tau^T(Y, z_1/t_1, \ldots, z_k/t_k, P_1, \ldots, P_m, Q_1/\sigma_1^B, \ldots, Q_n/\sigma_n^B)]$. This is Ξ-equivalent to the mixed formula

$$\begin{gathered} N'(X - \{x\}) \wedge \textstyle\bigwedge_{1\le i\le k} \theta_i^{B[t_i]}(X_i) \\ \wedge\ (\exists x\tau)^T(Y - \{x\}, z_1/t_1, \ldots, z_k/t_k, P_1, \ldots, P_m, Q_1/\sigma_1^B, \ldots, Q_n/\sigma_n^B), \end{gathered}$$

since by definition 13.5.6, $[\exists x\tau]^T$ is equivalent to $\exists x(\pi(x) = x \wedge \tau^T)$.

Case II: $x \in X_i$ for some i, and $X_i - \{x\} \neq \varnothing$.
We can assume that $i = k$. By remark 13.5.11 we can also assume that t_k is $\pi(x')$ for some $x' \in X_k - \{x\}$. In this case, $\exists x\mu$ is Ξ-equivalent to

$$\begin{gathered} N'(X - \{x\}) \wedge \textstyle\bigwedge_{1\le i<k} \theta_i^{B[t_i]}(X_i) \wedge \exists x\left[\pi(x) = t_k \neq x \wedge \theta_k^{B[t_k]}(X_k)\right] \\ \wedge\ \tau^T(Y, z_1/t_1, \ldots, z_k/t_k, P_1, \ldots, P_m, Q_1/\sigma_1^B, \ldots, Q_n/\sigma_n^B). \end{gathered}$$

As the central conjunct $\exists x\Big[\pi(x) = t_k \neq x \wedge \theta_k^{B[t_k]}(X_k)\Big]$ is Ξ-equivalent to $(\exists x\theta_k)^{B[t_k]}(X_k - \{x\})$, we obtain a mixed formula.

Case III: for some $i, X_i = \{x\}$.
Again we can assume that $i = k$, so that θ_k is just $\theta_k(x)$ and t_k is $\pi(x)$. Here, $\exists x\mu$ is Ξ-equivalent to

$$\begin{gathered} N'(X - \{x\}) \wedge \textstyle\bigwedge_{1\le i<k} \theta_i^{B[t_i]}(X_i) \\ \wedge\exists x\Big[x \neq \pi(x) \wedge (\textstyle\bigwedge_{i<k} \pi(x) \neq t_i) \wedge \theta_k^{B[\pi(x)]}(x) \\ \wedge\ \tau^T(Y, z_1/t_1, \ldots, z_{k-1}/t_{k-1}, z_k/\pi(x), P_j, Q_1/\sigma_1^B, \ldots, Q_n/\sigma_n^B)\Big]. \end{gathered}$$

The last, '$\exists x$'-conjunct here is Ξ-equivalent to $\exists z_k[\bigwedge_{i<k} z_k \neq t_i \wedge \exists x(\pi(x) = z_k \neq x \wedge \theta_k^{B[z_k]}(x)) \wedge \tau^T(Y, z_1/t_1, \ldots, z_{k-1}/t_{k-1}, z_k, P_1, \ldots, P_m, Q_1/\sigma_1^B, \ldots, Q_n/\sigma_n^B)]$, i.e. to

$$\begin{gathered} \exists z_k\Big[\pi(z_k) = z_k \wedge \textstyle\bigwedge_{i<k} z_k \neq t_i \wedge (\exists x\theta_k)^{B[z_k]}(z_k) \\ \wedge\tau^T(Y, z_1/t_1, \ldots, z_{k-1}/t_{k-1}, z_k, P_j, Q_1/\sigma_1^B, \ldots, Q_n/\sigma_n^B)\Big]. \end{gathered}$$

But this is equivalent to
$\tau'^T(Y, z_1/t_1, \ldots, z_{k-1}/t_{k-1}, P_j, Q_1/\sigma_1^B, \ldots, Q_n/\sigma_n^B, Q_{n+1}/[\exists x\theta_k]^B)$,
where Q_{n+1} is a new monadic predicate symbol, and
$\tau'(Y, z_1, \ldots, z_{k-1}, P_j, Q_1, \ldots, Q_{n+1})$ is
$\exists z_k[\bigwedge_{i<k} z_k \neq z_i \wedge Q_{n+1}(z_k) \wedge\tau(Y, z_1, \ldots, z_k, P_1, \ldots, P_m, Q_1, \ldots, Q_m)]$.
This completes the case of $\exists x$, and the lemma is proved. ■

13.5.2 Canonical structures and H-dimension

Now we will prove that if $\mathcal{T}$ and $\mathcal{B}$ have finite H-dimension then so does $\mathcal{T}[\mathcal{B}]$. We can assume that $\mathcal{T}$ and $\mathcal{B}$ are non-empty, for otherwise $\mathcal{T}[\mathcal{B}] = \varnothing$ and the result is trivial.

Lemma 13.5.15 *Let $\mu_\ell(X) =$*

$$\begin{gathered} N(X) \wedge \textstyle\bigwedge_{1\le i\le k} \theta_{i,\ell}^{B[t_i]}(X_i)\wedge \\ \tau_\ell^T(Y, z_1/t_1, \ldots, z_k/t_k, P_1, \ldots, P_m, Q_1/\sigma_{1,\ell}^B, \ldots, Q_n/\sigma_{n,\ell}^B) \end{gathered}$$

for $\ell = 1, 2$, be mixed formulae with the same partition formula N. Assume that:

(*a*) *τ_1 and τ_2 are $\mathcal{T}$-equivalent,*

(b) *for all* $i \leq k, \theta_{i,1}$ *and* $\theta_{i,2}$ *are* $\mathcal{B}$*-equivalent,*

(c) *for all* $j \leq n, \sigma_{j,1}$ *and* $\sigma_{j,2}$ *are* $\mathcal{B}$*-equivalent.*

Then μ_1 *and* μ_2 *are* $\mathcal{T}[\mathcal{B}]$*-equivalent.*

Proof. This follows from lemma 13.5.8. ■

Corollary 13.5.16 *Assume that:*

(a) *the H-dimension of* $\mathcal{T}$ *is a finite number* $h(\mathcal{T})$*;*

(b) *the H-dimension of* $\mathcal{B}$ *is a finite number* $h(\mathcal{B})$*.*

Then the H-dimension of the class $\mathcal{T}[\mathcal{B}]$ *of canonical* $\mathcal{TB}$*-structures is finite, and is at most* $d = \max(1 + h(\mathcal{B}), h(\mathcal{T}))$.

Proof. By lemma 13.5.13 any monadic formula of A is equivalent over $\mathcal{T}[\mathcal{B}]$ to a disjunction of mixed formulae μ. It therefore suffices to show that for any mixed formula μ there is a $\mathcal{T}[\mathcal{B}]$-equivalent formula μ' using at most $h = \max(1 + h(\mathcal{B}), h(\mathcal{T}))$ bound variables.

Let $\mu(X) =$

$$N(X) \wedge \textstyle\bigwedge_{1\leq i\leq k} \theta_i^{B[t_i]}(X_i)$$
$$\wedge \tau^T(Y, z_1/t_1, \ldots, z_k/t_k, P_1, \ldots, P_m, Q_1/\sigma_1^B, \ldots, Q_n/\sigma_n^B)$$

be a mixed formula. Let $V = \{v_1, \ldots, v_h\}$ be a set of h distinct new variables. We will rewrite μ to a $\mathcal{T}[\mathcal{B}]$-equivalent formula using bound variables only from V, so proving the corollary.

Now $\mathcal{T}$ and $\mathcal{B}$ have H-dimension at most h. So as bound variables can always be renamed, by lemma 13.5.15 we can assume that:

- the bound variables of the $\theta_i(X_i)$, $\tau(Y, z_1, \ldots, z_k)$ and the σ_j all come from V,
- no $v \in V$ occurs free in these formulae.

By remark 13.5.7, the same holds for the relativizations $\theta_i^{B[t_i]}$, σ_j^B, and τ^T in $\mu(X)$.

We check that the substitutions of σ_i^B in τ^T do not affect this. Let $Q_j(v)$ be an atomic subformula of τ^T. Since $V - \{v\}$ has at least $h(\mathcal{B})$ elements, we can rewrite σ_j using only variables from $V - \{v\}$. As v does not occur in σ_j, we can obtain $\sigma_j^{B[v]}$ as in definition 13.5.6(3) and substitute it for $Q_j(v)$ in τ^T. Thus $\tau^T(Y, \bar{z}, P_1, \ldots, P_m, Q_1/\sigma_1^B, \ldots, Q_n/\sigma_n^B)$ involves bound variables from V only. This proves the corollary. ■

13.5.3 Connectives for canonical structures

We can also use lemma 13.5.13 to obtain constructively a set of expressively complete connectives for $\mathcal{T}[\mathcal{B}]$, based on expressively complete connectives for $\mathcal{T}$ and for $\mathcal{B}$. All connectives that we consider are *one-dimensional.*

Assume that we are given finite sets of one-dimensional connectives

$$C^{\mathcal{B}} = \{\sharp_i^{\mathcal{B}} \mid i = 1, \ldots, n^{\mathcal{B}}\}, \qquad C^{\mathcal{T}} = \{\sharp_j^{\mathcal{T}} \mid j = 1, \ldots, n^{\mathcal{T}}\}$$

that are expressively complete for $\mathcal{B}, \mathcal{T}$ respectively. Let $\beta_i(x)$ be the table of $\sharp_i^{\mathcal{B}}$, and $\tau_j(y)$ the table for $\sharp_j^{\mathcal{T}}$ for each i, j. We will derive some connectives for the class $\mathcal{T}[\mathcal{B}]$, with tables in the language L_{Ξ} of definition 13.5.4(1), and show them to be expressively complete for $\mathcal{T}[\mathcal{B}]$. First we deal with the connectives from $C^{\mathcal{T}}$.

Definition 13.5.17

1. For each connective $\sharp_j^{\mathcal{T}}$ of $C^{\mathcal{T}}$ we define a new connective $(\sharp_j^{\mathcal{T}})^T$, with table $\tau_j^T(y)$.

2. Let E be a tense logical formula of $C^{\mathcal{T}}$. Write E^T for the formula of $C^{\mathcal{T}[\mathcal{B}]}$ obtained by replacing every occurrence of $\sharp_j^{\mathcal{T}}$ in E by the corresponding connective $(\sharp_j^{\mathcal{T}})^T$.

Lemma 13.5.18 *Let E be a tense logical formula of $C^{\mathcal{T}}$. Suppose that E has table $\varepsilon(y, P_1, \ldots, P_m)$. Then E^T has table $\varepsilon^T(y, P_1, \ldots, P_m)$.*

Proof. By induction on the structure of E. An atom p has table $P(y)$, and as $p^T = p$ and $P(y)^T = P(y)$ the result is clear. Boolean combinations are easy to handle.

Let E be $\sharp_j^{\mathcal{T}}(E_1, \ldots, E_n)$. This has table $\varepsilon(y) = \tau_j(y, \varepsilon_1, \ldots, \varepsilon_n)$, where E_i has table $\varepsilon_i(y)$ (each $i \leq n$). Clearly, E^T is $(\sharp_j^{\mathcal{T}})^T(E_1^T, \ldots, E_n^T)$. By definition, $(\sharp_j^{\mathcal{T}})^T$ has table $\tau_j^T(y, P_1, \ldots, P_n)$, and by the inductive hypothesis, E_i^T has table $\varepsilon_i^T(y)$ (all $i \leq n$). So E^T has table $\tau_j^T(y, \varepsilon_1^T, \ldots, \varepsilon_n^T)$. But it is easily seen by induction on the structure of τ_j that this last formula is equivalent to $[\tau_j(y, \varepsilon_1, \ldots, \varepsilon_n)]^T$, i.e. to $\varepsilon(y)^T$. This finishes the proof. ■

We want to prove a similar lemma for $C^{\mathcal{B}}$. We will work semantically this time. By renaming bound variables we can assume that for each $i \leq n^{\mathcal{B}}$, the variable x does not occur bound in the table $\beta_i(x)$ of $\sharp_i^{\mathcal{B}}$. Thus, $\beta_i^{B[\pi(x)]}(x)$ is defined and is the relativization of β_i to the set $\{y \mid \pi(y) = \pi(x) \neq y\}$, i.e. to $B_{\pi(x)}$. For brevity we will write β_i^x for $\beta_i^{B[\pi(x)]}$.

Definition 13.5.19

1. For each connective $\sharp_i^{\mathcal{B}}$ of $C^{\mathcal{B}}$ we define a new connective $(\sharp_i^{\mathcal{B}})^B$, with table $\beta_i^x(x)$.

2. Let D be a tense logical formula of $C^{\mathcal{B}}$. Write D^B for the tense logical formula of $C^{\mathcal{T}[\mathcal{B}]}$ obtained from D by replacing every occurrence of a connective $\sharp_i^{\mathcal{B}}$ of $C^{\mathcal{B}}$ by the corresponding connective $(\sharp_i^{\mathcal{B}})^B$.

Lemma 13.5.20 *Let $A \in \mathcal{T}[\mathcal{B}]$ be a canonical structure, and let $t \in T(A)$. Let h be an assignment into A, and write h_t for the assignment into B_t given by $h_t(q) = B_t \cap h(q)$ for all atoms q. Then for any temporal formula D of $C^{\mathcal{B}}$ and any $a \in B_t$, we have*

$$\|D\|_a^{h_t} = \|D^B\|_a^h,$$

the left-hand side being evaluated in B_t and the right-hand side in A.

Proof. By induction on D. The cases of atoms and boolean combinations are clear. Let D be $\sharp_i^{\mathcal{B}}(D_1, \ldots, D_n)$, and assume the result for the D_i. For each $i \le n$ let $S_i = \{a \in B_t \mid \|D_i\|_a^{h_t} = 1\}$, and $S_i' = \{a \in A \mid \|D_i^B\|_a^h = 1\}$. By the inductive hypothesis, we have $S_i = B_t \cap S_i'$ for each $i \le n$. Then $\|D\|_a^{h_t} = 1$ (in B_t) iff $B_t \models \beta_i(a, S_1, \ldots, S_n)$. By lemma 13.5.8(ii), this holds iff $A \models \beta_i^x(a, S_1', \ldots, S_n')$. As $(\sharp_i^{\mathcal{B}})^B$ has table β_i^x, this is equivalent to $\|(\sharp_i^{\mathcal{B}})^B(D_1^B, \ldots, D_n^B)\|_a^h = 1$, i.e. to $\|D^B\|_a^h = 1$, as required. ■

It follows from this and from lemma 13.5.8(ii) that the table of D^B is Ξ-equivalent to $\delta^x(x)$, where $\delta(x)$ is the table of D after renaming bound variables so that x does not occur bound.

Definition 13.5.21

1. Let R, S be the unary one-dimensional connectives for $\mathcal{T}[\mathcal{B}]$ with tables τ_R, τ_S as follows:
   - $\tau_R(x, P)$ is $\exists x(P(x) \wedge \pi(x) = y \wedge y \neq x)$;
   - $\tau_S(y, P)$ is $\pi(x) \neq x \wedge P(\pi(x))$.

2. We define the set $C^{\mathcal{T}[\mathcal{B}]}$ of connectives for $\mathcal{T}[\mathcal{B}]$ as follows:

$$C^{\mathcal{T}[\mathcal{B}]} = \{R, S, (\sharp_i^{\mathcal{B}})^B, (\sharp_j^{\mathcal{T}})^T\}.$$

Theorem 13.5.22 *Assume that the sets $C^{\mathcal{T}}$, $C^{\mathcal{B}}$ of one-dimensional connectives are expressively complete for $\mathcal{T}, \mathcal{B}$ respectively. Then $C^{\mathcal{T}[\mathcal{B}]}$ is expressively complete for $\mathcal{T}[\mathcal{B}]$.*

Proof. Let $\theta(x)$ be any monadic L_{Ξ}-formula with one free variable x. By lemma 13.5.13, $\theta(x)$ can be equivalently written as a disjunction of mixed formulae $\mu(x)$ of the form

(i) $\pi(x) = x \wedge \tau^T(x, P_1, \ldots, P_n, Q_j/\sigma_j^B)$, or

(ii) $\pi(x) \neq x \wedge \theta^x(x, P_1, \ldots, P_n) \wedge \tau^T(z/\pi(x), P_1, \ldots, P_n, Q_j/\sigma_j^B)$,

where θ and the σ_j are in the monadic language of $\mathcal{B}$, and $\tau(x, P_i, Q_j)$ is in the monadic language of $\mathcal{T}$. Notice that since the only free variable of each mixed formula μ is x, its partition formula N has only to say whether $x \in T$ or not. We can assume that the monadic predicates occurring in the σ_j and in θ are also taken from $P_1, \ldots, P_n$. Let $p_1, \ldots, p_n$ be atoms corresponding to $P_1, \ldots, P_n$. Clearly, it is enough to find a tense logical formula of $C^{\mathcal{T}[\mathcal{B}]}$ whose table is Ξ-equivalent to $\mu(x)$.

Case I: μ is $\pi(x) = x \wedge \tau^T(x, P_1, \ldots, P_n, Q_j/\sigma_j^B)$.
Using expressive completeness of $C^{\mathcal{B}}$, for each j let $D_j(p_1, \ldots, p_n)$ be a formula of $C^{\mathcal{B}}$ with table $\mathcal{B}$-equivalent to $\sigma_j \wedge (x = x)$. Similarly, using the expressive completeness of $C^{\mathcal{T}}$ let $E_\tau(p_i, q_j)$ be a formula of $C^{\mathcal{T}}$ whose table is $\mathcal{T}$-equivalent to $\tau(x, P_i, Q_j)$. Then by lemmas 13.5.18 and 13.5.20, the table of $E_\tau^T(p_1, \ldots, p_n, q_j/RD_j^B)$ is Ξ-equivalent to $\tau^T(x, P_1, \ldots, P_n, Q_j/\sigma_j^B)$. So

$$RT \wedge E_\tau^T(p_1, \ldots, p_n, q_j/RD_j^B)$$

is Ξ-equivalent to μ.

Case II: μ is $\pi(x) \neq x \wedge \theta^x(x) \wedge \tau^T(z/\pi(x), P_1, \ldots, P_n, Q_j/\sigma_j^B)$.
Let D_j and E_τ be as in Case I, and let D_θ be a formula of $C^{\mathcal{B}}$ with table $\mathcal{B}$-equivalent to $\theta(x, P_1, \ldots, P_n)$. Then, as before,

$$ST \wedge D_\theta^B \wedge SE_\tau^T(q_j/RD_j^B)$$

has table Ξ-equivalent to μ. This completes the proof. ∎

Proof of theorem 13.5.3. This now follows from theorems 13.4.8 and 13.4.12, corollary 13.5.16 and theorem 13.5.22. ∎

13.5.4 The covering theorem

Here we prove the covering theorem:

Theorem 13.5.23 *Let $\mathcal{T}, \mathcal{B}$ be classes of flows of time ($\mathcal{T}$ being a class of irreflexive flows). Let $\mathcal{R}$ be the class of flows of time of the form $(R, <)$, where $R = \bigcup_{t \in T} B_t$, $T \in \mathcal{T}$, $B_t \in \mathcal{B}$ for each $t \in T$, and the ordering $<$ on R is defined as in definition 13.5.1 (it is the lexicographic ordering). Assume that there is a fixed formula of the non-monadic language with $<, =$ that defines the relation $\equiv$ in every $R \in \mathcal{R}$. Then:*

(*i*) *If $\mathcal{T}$ and $\mathcal{B}$ both have finite H-dimension, then so does $\mathcal{R}$.*

(*ii*) *From finite expressively complete one-dimensional sets of connectives for $\mathcal{T}$ and for $\mathcal{B}$, we can effectively obtain a finite expressively complete set of one-dimensional connectives for $\mathcal{R}$.*

Proof. By theorem 13.5.3 it suffices to show that $\mathcal{R}$ fits in $\mathcal{T}[\mathcal{B}]$.

The language $L_{\mathcal{R}}$ of $\mathcal{R}$ has as its only non-logical symbol the binary relation symbol $<$. The language L_{Ξ} of $\mathcal{T}[\mathcal{B}]$ is $\{<, \pi\}$. We define a construction $\xi = \langle\langle\varphi, \alpha\rangle, (\chi_i \mid i \leq r), \beta\rangle$ over $\mathcal{T}[\mathcal{B}]$, that fits $\mathcal{R}$ in $\mathcal{T}[\mathcal{B}]$, as follows.

- $\varphi(x)$ is $\pi(x) \neq x$;
- $\alpha_<(x_1, x_2)$ is $x_1 < x_2 \vee \pi(x_1) < \pi(x_2)$;
- $\chi_1(x, y)$ is $x = y$ as usual;
- $\chi_2(x, y)$ is $x = \pi(y)$ (so '$x = \chi_i(y)$' is $x = \pi^{i-1}(y)$).

We must now define $\beta(R, \nu)$ for each atomic formula R of L_{Ξ} and 2-sort ν defined on at least the free variables of R. The possible R are $\pi^i(x) = \pi^j(y)$, and $\pi^i(x) < \pi^j(y)$, for $i, j \geq 0$. We will define $\beta(R, \nu)$ so that it only depends on the restriction of ν to the free variables in R. Thus we only need specify $\beta(R, \nu)$ for ν with domain $\{x, y\}$. If $p, q \in \{1, 2\}$ write (p, q) for the $\{x, y\}$ 2-sort ν such that $\nu(x) = p$ and $\nu(y) = q$.

First consider $\beta(\pi^i(x) = \pi^j(y), (p, q))$. By definition of χ_1 and χ_2, this is asking if $\pi^i(\pi^{p-1}(x)) = \pi^j(\pi^{q-1}(y))$. We must therefore find a formula of $L_{\mathcal{R}}$ that holds for $x, y \in R$ iff $\pi^{i+p-1}(x) = \pi^{j+q-1}(y)$. So we define $\beta(\pi^i(x) = \pi^j(y), (p, q))$ to be:

- $x = y$ if $i + p = j + q = 1$;
- $\epsilon(x, y)$ if $i + p = j + q = 2$;
- $\perp$ otherwise.

Similarly, using the irreflexivity of $(T, <)$, we define $\beta(\pi^i(x) < \pi^j(y), (p, q))$ to be

- $x < y \wedge \epsilon(x, y)$ if $i + p = j + q = 1$;
- $x < y \wedge \sim\epsilon(x, y)$ if $i + p = j + q = 2$;
- $\perp$ otherwise.

Hence $\mathcal{R}$ fits in $\mathcal{T}[\mathcal{B}]$, as required. Notice that the irreflexivity of $(T, <)$ and definability of $\equiv$ in $L_{\mathcal{R}}$ are used in recovering the ordering on T from that on R. ∎

13.5.5 The comb

Theorem 13.5.23 above covers most of the cases in examples 13.5.2. Earlier in this section we discussed the comb. It is not obtainable from $\mathbb{Z}$ and $\mathbb{N}$ by the 'covering' construction of definition 13.5.1, and so the theorem does not apply. None the less, we will now show that the comb M on $\mathbb{Z}$ and $\mathbb{N}$ does fit in the canonical $\{\mathbb{Z}\}\{\mathbb{N}\}$-structure A (pictured in figure 13.13). Hence by theorem 13.5.3 it has finite H-dimension and a finite expressively complete set of temporal connectives.

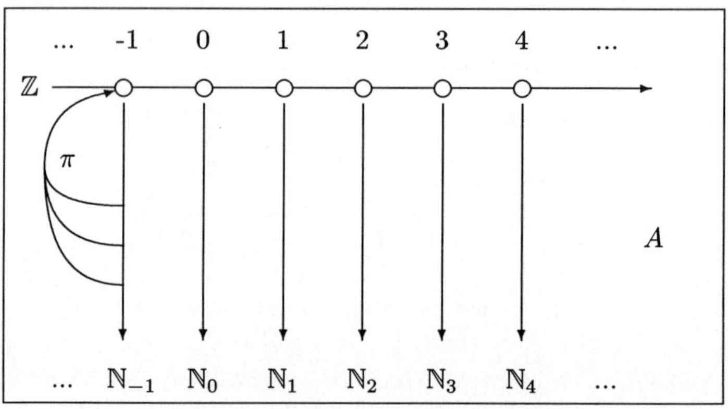

Figure 13.13. The comb's canonical structure.

Recall that the comb M has domain $\bigcup_{t \in \mathbb{Z}} \mathbb{N}_t$, where $\mathbb{N}_t$ is the tth copy of $\mathbb{N}$, and if $a \in \mathbb{N}_t, b \in \mathbb{N}_u$ then $a < b$ in M iff (i) $t = u$ and $a < b$ in $\mathbb{N}_t$, or (ii) $t < u$ in $\mathbb{Z}$ and $a = 0$ in $\mathbb{N}_t$.

First note that $x \equiv y$, meaning that x and y lie in the same copy $\mathbb{N}_t$ of $\mathbb{N}$, is definable in M. If $x, y \in M$, say that x and y are *comparable* if $x = y$, $x < y$ or $x > y$. Let us write $x \approx y$ if for all $z \in M$, z is comparable to x iff z is comparable to y. Clearly, $\approx$ is an equivalence relation on M, with equivalence classes of the form (a) $\mathbb{N}_t - \{0\}$ for each t, and (b) $\{0\} \subseteq \mathbb{N}_t$ for each $t \in \mathbb{Z}$. It is now easy to define $x \equiv y$. $x \equiv y$ iff either $x \approx y$, or exactly one of x, y (x, say) lies in a one-element $\approx$-class (so it is the zero-element of its $\mathbb{N}_t$), and every z with $x < z \leq y$ is $\approx$-equivalent to y.

We show that M fits in A. We can find a 2-ary construction $\xi = \langle\langle\varphi, \alpha\rangle, (\chi_i \mid i \leq r), \beta\rangle$ over $\{A\}$ very similar to the one for $\mathcal{R}$ above. The formula $\varphi(x)$ is as before. We define $\alpha_<(x_1, x_2)$ to be $x_1 < x_2 \vee (\pi(x_1) < \pi(x_2) \wedge {\sim}\exists z(z < x_1))$. The definitions of χ_1, χ_2 and $\beta(\pi^i(x) < \pi^j(y), (p, q))$ are as before. Let $\zeta(v, z)$ be the formula $v \equiv z \wedge \forall w(z \approx w \rightarrow z = w)$. If $b \in \mathbb{N}_t$ and $c \in M$, $M \models \zeta(b, c)$ iff c is the zero-element of $\mathbb{N}_t$. We define $\beta(\pi^i(x) < \pi^j(y), (p, q))$ to be

- $x < y \wedge x \equiv y$ if $i + p = j + q = 1$;
- $\forall z(\zeta(x, z) \rightarrow [z \leq y \wedge {\sim}(z \equiv y)])$ if $i + p = j + q = 2$;
- $\perp$ otherwise.

This clearly works: the comb M fits in A. Hence M has finite H-dimension and a finite expressively complete set of one-dimensional connectives.

13.5.6 Counterexamples

We give an example to show that there are flows of time $(R, <)$ covered by flows B_t ($t \in T$), such that T and each B_t have finite H-dimension, but R does not have finite H-dimension.

The reason for this failure is that the B_t may not have *uniformly* finite H-dimension, so that while each B_t individually has finite H-dimension, the class $\{B_t \mid t \in T\}$ does not.

We will take $T = \mathbb{N}$, the natural numbers $\{0, 1, 2, \ldots\}$. There are two ways in which the $B_t (t \in \mathbb{N})$ can fail to have uniformly finite H-dimension. The first, trivial way is that the H-dimension of B_t increases without bound with t. Hence the flow in figure 13.14 does not have finite H-dimension. Here, $x \equiv y$ is definable by ${\sim}(x < y) \wedge {\sim}(y < x)$.

The second possibility is that although the B_t all have H-dimension $\leq k$ for some fixed k, so that any monadic formula $\varphi(x)$ is equivalent in B_t to some formula $\psi_t(x)$ with only k bound variables, none the less the ψ_t become more and more complex as t increases. Maybe the depth of quantifier nesting in ψ_t increases without bound with t.

An example of this is when we take B_t to be a set of $t+3$ points ordered in a circle. So $B_t = \{b_0, b_1, \ldots, b_{t+2}\}$, with $b_i < b_j$ iff $j = i + 1 \pmod{t + 3}$. Note that this 'ordering' is not transitive.

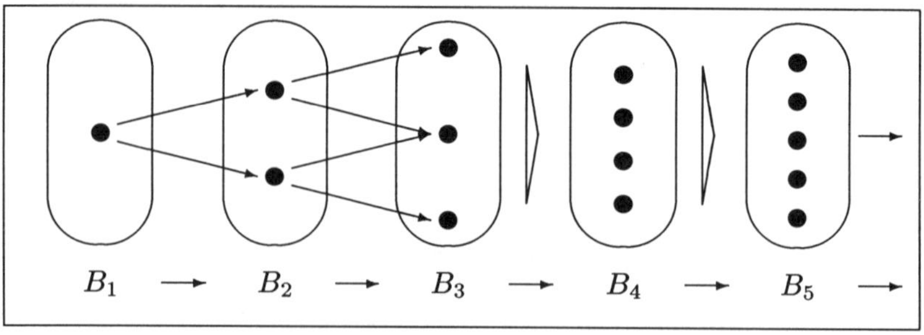

Figure 13.14. A flow with infinite H-dimension.

Now each B_t has H-dimension at most 3. Moreover, since $|B_t| \geq 3$ for all t, it is not hard to see that $\equiv$ is definable in $\bigcup_{t \in \mathbb{N}} B_t$ by:

$$x \equiv y \text{ iff } {\sim}\alpha(x, y) \wedge {\sim}\alpha(y, x),$$

where $\alpha(x, y)$ is $x < y \wedge \exists z(x < z < y)$. However, a compactness argument shows that $\bigcup_t B_t$ does not have finite H-dimension. We leave the details an an exercise for the reader.

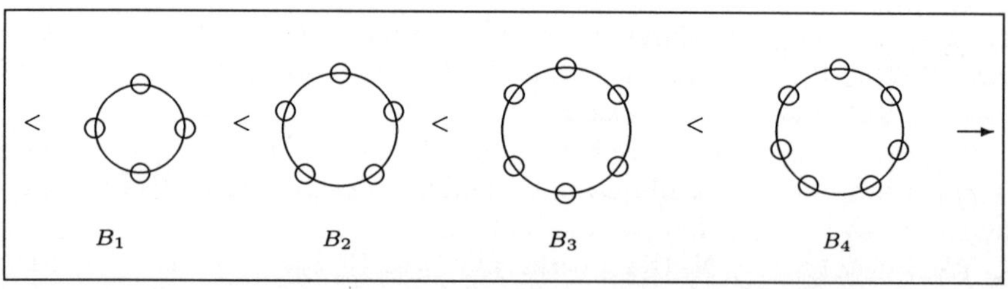

Figure 13.15. Another flow with infinite H-dimension.

13.6 The k-variable property

This is another property similar to that of finite H-dimension discussed already. It was introduced in [Immerman and Kozen, 1987]. It also concerns the monadic theory of classes of flows of time and is related to expressive completeness. We will discuss its relationship with H-dimension. We also establish model-theoretic and game-theoretic characterizations of the k-

variable property; arguably, it is more natural to do this for the k-variable property than for H-dimension.

Throughout this section there will be available a fixed countable set $\{x_1, x_2, \ldots\}$ of distinct variables. Until further notice, L will be a first-order signature with no n-place function symbols for any $n > 1$. Recall that the *arity* of L, if it exists, is the largest n such that L has an n-place relation symbol. Here we take the equality relation $=$ to be part of L, so that L has arity at least 2. Because of the restriction on function symbols, we could define the arity of L to be the least n (if one exists) such that no atomic formula of L has more than n free variables.

Definition 13.6.1 Let k be a positive number.

1. We write L_k for the *k-variable (monadic) fragment of* L: the set of all monadic first-order L-formulae that are written using only variables from the set $\{x_1, \ldots, x_k\}$.

   More generally, if $X \subseteq \{x_1, x_2, \ldots\}$, we write L_X for the set of all monadic L-formulae written with variables only from X.

2. Define L_ω to be $\bigcup_{k<\omega} L_k$.

3. Let $\mathcal{K}$ be a class of L-structures. $\mathcal{K}$ is said to have the *k-variable property* if for every monadic formula $\varphi(x_1, \ldots, x_k, P_1, \ldots, P_n)$ with free variables among $\bar{x} = x_1, \ldots, x_k$, there is a monadic formula $\psi(x_1, \ldots, x_k, P_1, \ldots, P_n)$ of L_k that is equivalent to φ over $\mathcal{K}$: i.e. for all $M \in \mathcal{K}$ and all subsets $S_1, \ldots, S_n$ of the domain of M, we have

$$M \models \forall x_1 \ldots \forall x_k(\varphi(\bar{x}, S_1, \ldots, S_n) \leftrightarrow \psi(\bar{x}, S_1, \ldots, S_n)).$$

Of course, any given monadic L-formula φ is in L_k for all sufficiently large k. That is, L_ω is the set of all monadic L-formulae. Clearly, if $\mathcal{K}$ has the k-variable property then so does any subclass of $\mathcal{K}$.

13.6.1 The k-variable property and H-dimension

We can immediately see that the k-variable property is at least as strong as having H-dimension k. Essentially, 'H-dimension at most k' is 'the k-variable property for formulae without free variables (sentences)'.

Theorem 13.6.2 *Let $\mathcal{K}$ be a class of L-structures with the k-variable property for some k. Then $\mathcal{K}$ has H-dimension at most k.*

Proof. Assume that $\mathcal{K}$ has the k-variable property. Then any monadic L-sentence $\sigma(P_1, \ldots, P_n)$ is $\mathcal{K}$-equivalent to a sentence $\sigma^*(P_1, \ldots, P_n)$ of L_k. So by lemma 13.2.3, $\mathcal{K}$ has H-dimension at most k. ∎

Remark 13.6.3 In [Hodkinson and Simon, 1992] an example is given of a class $\mathcal{K}$ of flows of time with the following properties:

- $\mathcal{K}$ has H-dimension 3.
- $\mathcal{K}$ admits a finite expressively complete set of one-dimensional temporal connectives.
- $\mathcal{K}$ does not have the k-variable property for any finite k.

Hence the converse to the theorem does not hold, and the k-variable property is strictly stronger than H-dimension k. Also see [Flum, 1991].

13.6.2 Syntactic aspects of the k-variable property

We examine some simple consequences of the k-variable property. Among other things, we will see that the k-variable property has implications for formulae with more than k free variables.

Proposition 13.6.4 *Let $\mathcal{K}$ be a class of L-structures with the k-variable property. Then $\mathcal{K}$ has the $k+1$-variable property.*

Proof. Let $\varphi(x_1, \ldots, x_{k+1}, P_1, \ldots, P_n)$ be a monadic L-formula with at most $k+1$ free variables. We must find $\varphi^* \in L_{k+1}$ that is $\mathcal{K}$-equivalent to φ.

Let Q be a new monadic predicate letter not occurring in φ, and let $\psi(x_1, \ldots, x_k, P_1, \ldots, P_n, Q)$ be the formula $\exists x_{k+1}(Q(x_{k+1}) \wedge \varphi)$. As $\mathcal{K}$ has the k-variable property there is $\psi^*(x_1, \ldots, x_k, P_1, \ldots, P_n, Q) \in L_k$ that is $\mathcal{K}$-equivalent to ψ. Replace each atomic subformula of ψ^* of the form $Q(t)$ (t a term of L) by the formula $t = x_{k+1}$, so obtaining a formula $\varphi^*(x_1, \ldots, x_{k+1}, P_1, \ldots, P_n)$ of L_{k+1}.

Claim: φ^* is $\mathcal{K}$-equivalent to φ.

Proof of claim: Let $M \in \mathcal{K}$, let $S_1, \ldots, S_n$ be subsets of the domain of M and let $a_1, \ldots, a_{k+1} \in M$. Then $M \models \varphi(a_1, \ldots, a_{k+1}, S_1, \ldots, S_n)$
iff $M \models \exists x_{k+1}(x_{k+1} = a_{k+1} \wedge \varphi(a_1, \ldots, a_k, x_{k+1}, S_1, \ldots, S_n))$
iff $M \models \psi(a_1, \ldots, a_k, S_1, \ldots, S_n, \{a_{k+1}\})$ by definition of ψ,
iff $M \models \psi^*(a_1, \ldots, a_k, S_1, \ldots, S_n, \{a_{k+1}\})$, as ψ and ψ^* are $\mathcal{K}$-equivalent,
iff $M \models \varphi^*(a_1, \ldots, a_{k+1}, S_1, \ldots, S_n)$ by definition of φ^* (the argument here is as in lemma 13.2.3; a full proof uses induction on the structure of ψ^*).

The claim is proved, and with it the proposition. ■

Lemma 13.6.5 *Let $X \subseteq \{x_1, x_2, \ldots\}$ have cardinality k, where k is greater than the arity of L. Then every formula $\varphi \in L_X$ is a boolean combination of formulae ψ with at most $k-1$ free variables.*

Proof. The proof is by induction on φ. We assume for simplicity that $X = \{x_1, \ldots, x_k\}$. If φ is atomic, we are given this by the condition on k. The boolean connectives clearly preserve the given property. Finally, let $\psi(x_1, \ldots, x_k)$ be a monadic formula of L_k. Then if $1 \leq i \leq k$, $\exists x_i \psi$ clearly has at most $k-1$ free variables. ∎

Note that this lemma fails if L has function symbols of arity 2 or more. For example, in L_4, the formula $f(f(x_1, x_2), x_3) = x_4$ is not a boolean combination of formulae with at most two free variables.

We can now easily obtain:

Theorem 13.6.6 *Let $\mathcal{K}$ be a class of L-structures and let k be a positive number greater than the arity of L. Then the following are equivalent:*

1. *$\mathcal{K}$ has the k-variable property.*
2. *$\mathcal{K}$ has the n-variable property for all $n \geq k$.*
3. *For all monadic L-formulae $\varphi(X)$ with free variables from some set $X \subseteq \{x_1, x_2, \ldots\}$ of cardinality $\geq k$, φ is $\mathcal{K}$-equivalent to a boolean combination of formulae in $\bigcup\{L_Y \mid Y \subseteq X,\ Y \text{ has cardinality } k\}$.*

Proof. (3) $\Rightarrow$ (1) follows by taking $X = \{x_1, \ldots, x_k\}$ in (3), and (1) $\Rightarrow$ (2) follows from proposition 13.6.4 by induction. Assume (2); we prove (3) by induction on the cardinality $|X|$ of X (we can assume that X is finite).

If $|X| = k$ then the statement of (3) is just the k-variable property. Let $|X| = n+1 > k$ and assume (3) for all $Y \subseteq \{x_1, x_2, \ldots\}$ of cardinality n. Let $\varphi(X)$ be given. As $\mathcal{K}$ has the $n+1$-variable property, there is a formula of L_X that is $\mathcal{K}$-equivalent to φ. Hence we can assume that $\varphi \in L_X$ already. By lemma 13.6.5, φ is a boolean combination of monadic formulae ψ with at most n free variables. Hence it is enough to deal with each ψ. But we can do this immediately, using the inductive hypothesis. This completes the proof. Observe that by lemma 13.6.5, each formula of the boolean combination in (3) has at most $k-1$ free variables. ∎

13.6.3 Expressive completeness

The k-variable property is also connected to expressive completeness of temporal logics. The analogy of theorem 13.2.4 holds for the k-variable property. We prove this next.

Theorem 13.6.7 *Let $\mathcal{K}$ be a class of flows of time that has the k-variable property for some k (in the monadic first-order language over the signature $\{=, <\}$). Then:*

1. *There is a finite set of k-dimensional temporal connectives of arity at most 1 that is expressively complete over $\mathcal{K}$ in k dimensions (FC^k_k).*

2. *If $k \geq 3$, there is a finite set of $(k-1)$-dimensional temporal connectives of arity at most $k-1$ that is expressively complete over $\mathcal{K}$ in k dimensions (FC^{k-1}_{k-1}).*

Proof. 1. As $\mathcal{K}$ has the k-variable property, it is enough if we prove that for every monadic formula $\varphi(x_1, \ldots, x_k, Q_1, \ldots, Q_n) \in \{=, <\}_k$ whose free variables lie in $\{x_1, \ldots, x_k\}$, there is a temporal formula $A = A_\varphi(q_1, \ldots, q_n)$ using the connectives below, whose monadic table $\tau(x_1, \ldots, x_k, Q_1, \ldots, Q_n)$ is $\mathcal{K}$-equivalent to φ.

In addition to the boolean connectives, we use the following:

- M_σ for all maps $\sigma : \{1, 2, \ldots, k\} \to \{1, 2, \ldots, k\}$ (1-ary);
- E (nullary);
- L (nullary);
- $\Box_i$ for all $1 \leq i \leq k$ (1-ary).

Each connective $\sharp$ above has table $\tau[\sharp]$ as follows:

- $\tau[M_\sigma](x_1, \ldots, x_k, Q)$ is $\exists y_1 \ldots \exists y_k [Q(y_1, \ldots, y_k) \wedge \bigwedge_i y_i = x_{\sigma(i)}]$;
- $\tau[E](x_1, \ldots, x_k)$ is $x_1 = x_2$;
- $\tau[L](x_1, \ldots, x_k)$ is $x_1 < x_2$;
- $\tau[\Box_i](x_1, \ldots, x_k, Q)$ is $\forall x_i Q(x_1, \ldots, x_k)$.

As this logic is k-dimensional, each temporal atom q appears in the tables as a k-place predicate letter Q. However, as in section 13.1, the value of q at a k-tuple of points $a_1, \ldots, a_k$ in a flow of time will depend only on a_1. Thus, the temporal atom q corresponds to a k-adic relation symbol $Q(x_1, \ldots, x_k)$ whose assignment is restricted so that it depends only on x_1. Since the quantified variables in $\tau[M_\sigma]$ are introduced so that in $Q(y_1, \ldots, y_k)$, the arguments $y_1, \ldots, y_k$ are all distinct (cf. definition 13.1.2(1)), any temporal formula has a *monadic* first-order table — cf. theorem 13.1.6.

The proof now goes by induction on the structure of $\varphi \in L_k$. If φ is a monadic predicate letter $Q(x_i)$ then we set $A = M_\sigma q$, where σ is an arbitrary map such that $\sigma(1) = i$. The table of A is $\exists y_1, \ldots, y_k(Q(y_1, \ldots, y_k) \wedge \bigwedge_j y_j = x_{\sigma(j)})$, which is $\mathcal{K}$-equivalent to $Q(x_i)$ given our restriction on assignments to atoms.[6] If φ is $x_i = x_j$ we set $A = M_\sigma E$ where σ is

[6] We abuse notation by letting Q denote both the monadic predicate and its k-ary version.

arbitrary so long as $\sigma(1) = i$ and $\sigma(2) = j$. Then A has table $\exists y_1 \ldots y_k(y_1 = y_2 \wedge \bigwedge_i y_i = x_{\sigma(i)})$, which is equivalent to $x_i = x_j$. The formula for $x_i < x_j$ is $M_\sigma L$ where σ is as before.

Inductively if A and B have tables $\mathcal{K}$-equivalent to φ and ψ, then $A \wedge B$ and $\sim A$ have tables $\mathcal{K}$-equivalent to $\varphi \wedge \psi$ and $\sim\varphi$ respectively. Finally, if the table of A is $\mathcal{K}$-equivalent to $\varphi(x_1, \ldots, x_k)$, then clearly $\forall x_i \varphi(x_1, \ldots, x_k)$ is $\mathcal{K}$-equivalent to the table of $\Box_i A$. This ends the proof.

2. We introduce the following $(k-1)$-dimensional connectives:

- M_σ for all maps $\sigma : \{1, 2, \ldots, k-1\} \to \{1, 2, \ldots, k-1\}$ (1-ary);
- E (nullary);
- L (nullary);
- C ($(k-1)$-ary).

The tables of M, E, and L are as before. The table

$$\tau[C](x_1, \ldots, x_{k-1}, Q_1, \ldots, Q_{k-1})$$

is $\exists x_k(\bigwedge_{i<k} Q_i(x_1, \ldots, x_{i-1}, x_{i+1}, \ldots, x_k))$. We prove expressive completeness by induction on $\varphi(x_1, \ldots, x_{k-1}) \in \{=, <\}_k$, as before. The only interesting case is $\exists x_k \varphi(x_1, \ldots, x_k)$. By lemma 13.6.5, φ is a boolean combination of formulae with at most $k-1$ free variables. We may write it in disjunctive normal form: $\bigvee_i \bigwedge_j \varphi_{ij}$, where each φ_{ij} has at most $k-1$ free variables. Thus, $\exists x_k \varphi$ is equivalent to $\bigvee_i \exists x_k \bigwedge_j \varphi_{ij}$; and so it suffices to find a temporal formula equivalent to $\exists x_k \bigwedge_j \varphi_{ij}$. Clearly, this formula is equivalent to one of the form

$$\psi \quad = \quad \exists x_k \Big(\bigwedge_{1 \le i < k} \psi_i(x_1, \ldots, x_{i-1}, x_{i+1}, \ldots, x_k)\Big) \wedge \psi_k(x_1, \ldots, x_{k-1}).$$

By the inductive hypothesis, there are temporal formulae $A_1, \ldots, A_k$ $\mathcal{K}$-equivalent to $\psi_1(x_1, \ldots, x_{k-1}), \ldots, \psi_k(x_1, \ldots, x_{k-1})$, respectively, where we have permuted the variables in these formulae so that their free variables are $x_1, \ldots, x_{k-1}$ in each case. Then ψ is equivalent to the temporal formula $C(A_1, \ldots, A_{k-1}) \wedge A_k$. ■

In the case $k = 2$, the logic in (2) of the theorem is the logic $CC\lambda$ of Venema, and the argument in (2) is essentially due to Venema. The dimension $k-1$ in (2) cannot be reduced in general: in section 13.7 we give an example of a class of flows of time ('circles') that has the three-variable property but is not FC_1^1. The converse of the theorem is false in general. For example, the class $\mathcal{K}$ of all linear flows of time is FC_1^1, because Until, Since and the Stavi connectives are expressively complete for $\mathcal{K}$ in one dimension (see chapter 11). But $\mathcal{K}$ does not have the two-variable property (by theorem 13.6.21 below, it has the k-variable property iff $k \geq 3$).

13.6.4 First-order definable classes

We now consider the special case where $\mathcal{K}$ is a *first-order definable class.* In this case we can give model-theoretic characterizations of the k-variable property.

Definition 13.6.8

1. A *theory* is a set of sentences. An L-*theory* is a set of L-sentences.
2. If Σ is an L-theory, we write $mod_L(\Sigma)$ for the class of all L-structures that are models of Σ. We write simply $\text{mod}(\Sigma)$ if L is clear from the context.
3. We say that a class $\mathcal{K}$ of L-structures is *first-order definable* if there is an L-theory Σ such that $\mathcal{K} = \text{mod}(\Sigma)$.

Examples of first-order definable classes $\mathcal{K}$ occur frequently, and include the class of all flows of time, the class of all irreflexive linear flows of time, the class of all dense linear flows of time, and certain classes of trees.

Notation: We will fix a countably infinite sequence $P_1, P_2, \ldots,$ of monadic predicate letters.

Definition 13.6.9

1. Let M be an L-structure. An *assignment into* M is a map h providing for each monadic predicate letter P_i ($i \geq 1$) a subset $h(P_i)$ of the domain of M.
2. If S is a set, a *tuple in* S is a finite sequence $\bar{a}$ of (not necessarily distinct) elements of S. Formally, a tuple is a map from $\{1, \ldots, m\}$ into S, for some $m < \omega$. The *length* of the tuple is then m, and the tuple is said to be an m-*tuple.* We let $\langle\rangle$ denote the unique empty tuple (of length 0). We write $\bar{a} \in S$ if $\bar{a}$ is a tuple in S. If M is an L-structure we write $\bar{a} \in M$ to mean that $\bar{a}$ is a tuple of the domain of M. An *element* of $\bar{a}$ is an element of the range $rng(\bar{a})$ of $\bar{a}$; if $j \leq m$, the jth element of $\bar{a}$ is just $\bar{a}(j)$. If $\bar{a}, \bar{b}$ are tuples of length m, n, respectively, we write $\bar{a}\bar{b}$ for their concatenation: a tuple $\bar{c}$ of

length $n + m$ given by $\bar{c}(i) = \bar{a}(i)$ for $i \leq m$, and $\bar{c}(m + i) = \bar{b}(i)$ for $i \leq n$. We will often consider *elements* of a structure as tuples of length 1.

Definition 13.6.10 Let M, N be L-structures, and let $\kappa \leq \omega$ be a cardinal. Let h, h' be assignments into M, N, respectively, and let $\bar{a} \in M$, $\bar{b} \in N$ be tuples of equal (finite) length $m \leq \kappa$. We write

$$(M, h, \bar{a}) \equiv_\kappa (N, h', \bar{b})$$

if for all formulae $\varphi(x_1, \ldots, x_m, P_1, \ldots, P_n)$ of L_κ,[7] we have

$$M \models \varphi(\bar{a}, h(P_1), \ldots, h(P_n)) \text{ iff } N \models \varphi(\bar{b}, h'(P_1), \ldots, h'(P_n)).$$

We often write $(M, h, \bar{a}) \equiv_\omega (N, h', \bar{b})$ more simply as $(M, h, \bar{a}) \equiv (N, h', \bar{b})$. This is the monadic version of 'elementary equivalence'. If $k < \omega$, $\equiv_k$ is (roughly speaking) 'monadic elementary equivalence in the k-variable fragment'. Note, however, that $(M, h, \bar{a}) \equiv_k (N, h', \bar{b})$ is formally weaker than another notion suggested by the notation: namely that if we add new constant symbols $\bar{c}$ to L and interpret them in M as $\bar{a}$ and in N as $\bar{b}$, obtaining structures $(M, \bar{a})$, $(N, \bar{b})$, respectively, then $((M, \bar{a}), h) \equiv_k ((N, \bar{b}), h')$. This is because variables in L_k-formulae are reassigned by the quantifiers during evaluation, so that the values of $\bar{a}, \bar{b}$ may be lost. This 'forgetfulness' is typical of finite-variable logics, and suggests a difference between 'H-dimension at most d' and 'the d-variable property'.

Fix a class $\mathcal{K} = \text{mod}_L(\Sigma)$ for some first-order L-theory Σ. We can now give our first model-theoretic characterization of the k-variable property. It is due to [Immerman and Kozen, 1987]. The idea is very simple. Suppose $M \in \mathcal{K}$, h is an assignment into M, and $\bar{a} \in M$ is a tuple. If $\mathcal{K}$ has the k-variable property, the L_k-formulae that are true of $\bar{a}$ clearly determine which L_ω- formulae are true of $\bar{a}$. But because $\mathcal{K}$ is first-order definable, the converse also holds; the proof of this uses compactness for first-order logic.

Theorem 13.6.11 *Let k be a positive whole number. The following are equivalent.*

(i) $\mathcal{K}$ has the k-variable property.

(ii) For all $M, N \in \mathcal{K}$, all assignments h, h' into M, N, respectively, and all k-tuples $\bar{a} \in M$ and $\bar{b} \in N$,

$$(M, h, \bar{a}) \equiv_k (N, h', \bar{b}) \quad \Rightarrow \quad (M, h, \bar{a}) \equiv (N, h', \bar{b}).$$

[7] Recall that L_ω is $\bigcup_{k<\omega} L_k$.

Proof. (i) $\Rightarrow$ (ii): Assume (i), and let M, N, h, h', $\bar{a}$, and $\bar{b}$ as in (ii) be given. Assume that $(M, h, \bar{a}) \equiv_k (N, h', \bar{b})$. Let $\varphi(\bar{x}, P_1, \ldots, P_n)$ be any monadic formula of L, where $\bar{x}$ has length k. We must prove that $M \models \varphi(\bar{a}, h(P_1), \ldots, h(P_n))$ iff $N \models \varphi(\bar{b}, h'(P_1), \ldots, h'(P_n))$.

As $\mathcal{K}$ has the k-variable property, by theorem 13.6.6 φ is equivalent in (M, h) and in (N, h') to a monadic formula $\psi(\bar{x}, P_1, \ldots, P_n)$ of L_k. So as $(M, h, \bar{a}) \equiv_k (N, h', \bar{b})$, we have $M \models \psi(\bar{a}, h(P_1), \ldots, h(P_n))$ iff $N \models \psi(\bar{b}, h'(P_1), \ldots, h'(P_n))$. Thus, we have $M \models \varphi(\bar{a}, h(P_1), \ldots, h(P_n))$ iff $N \models \varphi(\bar{b}, h'(P_1), \ldots, h'(P_n))$, as required.

(ii) $\Rightarrow$ (i): Assume (ii), and let $\chi(x_1, \ldots, x_k, P_1, \ldots, P_n)$ be any monadic L-formula with free variables among $x_1, \ldots, x_k$. We will show that χ is $\mathcal{K}$-equivalent to a formula of L_k.

We let

$$\begin{aligned}\Phi = \; & \{\varphi(x_1, \ldots, x_k, P_1, \ldots, P_n) \in L_k \\ & \mid \Sigma \models \forall x_1, \ldots, x_k(\chi(x_1, \ldots, x_k) \to \varphi(x_1, \ldots, x_k))\}.\end{aligned}$$

Φ is the set of k-variable monadic consequences of χ (modulo Σ) that involve only at most the monadic letters $P_1, \ldots, P_n$. It suffices to prove that $\Sigma \cup \Phi \models \chi$. For given this, by compactness for first-order logic there are $\varphi_1, \ldots, \varphi_r \in \Phi$ such that $\Sigma \cup \{\varphi_1, \ldots, \varphi_r\} \models \chi$, so that $\Sigma \models \forall x_1, \ldots,$ x_k $([\bigwedge_{1 \le i \le r} \varphi_i(x_1, \ldots, x_k)] \to \chi(x_1, \ldots, x_k))$. Since $\varphi_1, \ldots, \varphi_r \in \Phi$, we have $\Sigma \models \forall x_1, \ldots, x_k(\chi(x_1, \ldots, x_k) \to \bigwedge_{1 \le i \le r} \varphi_i(x_1, \ldots, x_k))$, so that the L_k-formula $\bigwedge_{1 \le i \le r} \varphi_i(x_1, \ldots, x_k, P_1, \ldots, P_n)$ is $\mathcal{K}$-equivalent to χ.

To prove that $\Sigma \cup \Phi \models \chi$, let $M \in \mathcal{K}$, let h be an assignment into M, let $\bar{a} = (a_1, \ldots, a_k)$ be a k-tuple of elements of M, and suppose that $M \models \varphi(\bar{a}, h(P_1), \ldots, h(P_n))$ for all $\varphi \in \Phi$. We show that $M \models \chi(\bar{a}, h(P_1),$ $\ldots, h(P_n))$.

Let $\Psi = \{\psi(x_1, \ldots, x_k, P_1, \ldots, P_n) \mid \psi \in L_k, M \models \psi(a_1, \ldots, a_k, h(P_1),$ $\ldots, h(P_n))\}$. Let $\bar{c} = (c_1, \ldots, c_k)$ be a k-tuple of distinct constant symbols not occurring in L. Write $\Psi(\bar{c})$ for the set $\{\psi(c_1, \ldots, c_k, P_1, \ldots, P_n) \mid \psi \in \Psi\}$.

We claim that the theory

$$\Sigma \cup \Psi(\bar{c}) \cup \{\chi(\bar{c})\}$$

is consistent. For if not, by compactness again there are $\psi_1, \ldots, \psi_s \in \Psi$ such that writing ψ for $\psi_1 \wedge \ldots \wedge \psi_s$, we have $\Sigma \models \chi(\bar{c}) \to {\sim}\psi(\bar{c})$. Hence $\Sigma \models$ $\forall \bar{x}\ (\chi(\bar{x}) \to {\sim}\psi(\bar{x}))$, so that ${\sim}\psi(\bar{x}) \in \Phi$ by definition of Φ. By choice of M, $\Phi \subseteq \Psi$, so that $M \models {\sim}\psi(\bar{a})$. But by definition of Ψ, $M \models \psi_1(\bar{a}) \wedge \ldots \wedge \psi_s(\bar{a})$, i.e. $M \models \psi(\bar{a})$. This contradiction proves the claim.

So we can take $N \in \mathcal{K}$ and an assignment h' into N such that for some k-tuple $\bar{b} \in N$, $N \models \psi(\bar{b}, h'(P_1), \ldots, h'(P_n))$ for every $\psi \in \Psi$, and also

$N \models \chi(\bar{b}, h'(P_1), \ldots, h'(P_n))$. By definition of $\Psi, (M, h, \bar{a}) \equiv_k (N, h', \bar{b})$. By assumption (ii), $(M, h, \bar{a}) \equiv (N, h', \bar{b})$, so that $M \models \chi(a_1, \ldots, a_k, h(P_1), \ldots, h(P_n))$, as required. ∎

Of course, the same proof shows that a first-order definable class $\mathcal{K}$ has H-dimension at most d if and only if for all $M, N \in \mathcal{K}$ and all assignments h, h' into M, N, respectively,

$$(M, h) \equiv_d (N, h') \quad \Rightarrow \quad (M, h) \equiv (N, h').$$

13.6.5 ω-saturated structures

There is another model-theoretic characterization of the k-variable property, due to van Benthem [van Benthem, 1992]. Rather than referring to formulae, it uses partial isomorphisms between structures. This can make it easier to use in practice, when establishing that a first-order definable class $\mathcal{K}$ has the k-variable property. However, the notion of isomorphism is in general stronger than the notion of elementary equivalence, so we must work with special 'homogeneous' structures in $\mathcal{K}$, on which the two notions agree. We will use the so-called ω-saturated structures for this purpose.

Definition 13.6.12 Let M be an L-structure, and let h be an assignment into M.

1. Let N be another L-structure, and h' an assignment into N. (N, h') is said to be an *elementary extension* of (M, h), and (M, h) an *elementary substructure* of (N, h'), if $dom(M) \subseteq dom(N)$ and for all monadic L-formulae $\varphi(\bar{x}, P_1, \ldots, P_n)$ and all tuples $\bar{a} \in M$ of the same length as $\bar{x}$, $M \models \varphi(\bar{a}, h(P_1), \ldots, h(P_n))$ iff $N \models \varphi(\bar{a}, h'(P_1), \ldots, h'(P_n))$. We write $(M, h) \preceq (N, h')$ in this case.

2. Let A be a subset of the domain of M. A *type over* A (*in* (M, h)) is a set p of monadic formulae of the form $\varphi(x, \bar{a}, P_1, \ldots, P_n)$ where $\bar{a}$ is a tuple of elements of A, such that for any finite number r and formulae $\varphi_i(x, \bar{a}_i, P_1, \ldots, P_{n_i}) \in p$ for $1 \leq i \leq r$,

$$M \models \exists x \bigwedge_{1 \leq i \leq r} \varphi_i(x, \bar{a}_i, h(P_1), \ldots, h(P_{n_i})).$$

A type p in (M, h) is said to be *realized in* (M, h) if there is $b \in M$ such that $M \models \varphi(b, \bar{a}, h(P_1), \ldots, h(P_n))$ for all $\varphi(x, \bar{a}, P_1, \ldots, P_n) \in p$. Such a b is said to *realize* p.

3. (M, h) is said to be *ω-saturated* if whenever A is a *finite* subset of the domain of M, any type over A in (M, h) is realized in (M, h).

Examples 13.6.13 We review some simple facts concerning these notions. See [Chang and Keisler, 1990] for more details.

1. $\preceq$ is a reflexive and transitive relation on the class of pairs (M, h). If M is finite and $(M, h) \preceq (N, h')$ then $(M, h) = (N, h')$. If $(M, h) \preceq (N, h')$ then M is a substructure of N (i.e. the interpretations of the symbols of L in M and N agree where meaningful), and for all $i, h'(P_i) \cap dom(M) = h(P_i)$.

2. It follows from the definitions that if $(M, h) \preceq (N, h')$, and $\bar{a} \in M$, then for any κ, $(M, h, \bar{a}) \equiv_k (N, h', \bar{a})$. If $\mathcal{K}$ is a first-order definable class of L-structures, $M \in \mathcal{K}$ and $(M, h) \preceq (N, h')$ then also $N \in \mathcal{K}$.

3. *Elementary chains.* If δ is a limit ordinal, M_i are L-structures, h_i assignments into M_i for each $i < \delta$, and $(M_i, h_i) \preceq (M_j, h_j)$ for each $i < j < \delta$, then we can define an L-structure M_δ with domain $\bigcup_{i<\delta} dom(M_i)$, by defining for each atomic formula $\varphi(\bar{x})$ and $\bar{a} \in M_\delta$, $M_\delta \models \varphi(\bar{a})$ iff $M_i \models \varphi(\bar{a})$ for all $i < \delta$ such that $\bar{a} \in M_i$. An assignment h_δ into M_δ can be similarly defined. Then $(M_i, h_i) \preceq (M_\delta, h_\delta)$ for all $i < \delta$.

4. If $(M, h) \preceq (N, h')$ and p is a type in (M, h), then p is also a type in (N, h'), and if p is realized in (M, h) then it is also realized in (N, h').

5. If M is a finite structure (i.e. its domain is finite), then (M, h) is ω-saturated for any h. For let p be a type. If no element of M realizes p, then for each $a \in M$ there is $\varphi_a(x, P_1, \ldots, P_{n_a}) \in p$ such that $M \models {\sim}\varphi_a(a, h(P_1), \ldots, h(P_{n_a}))$. Thus $M \models {\sim}\exists x \bigwedge_{a \in M} \varphi_a(x, h(P_1), \ldots, h(P_{n_a}))$, a contradiction.

6. The structure $(\mathbb{N}, <, h)$ is not ω-saturated for any h. For if p is the type $\{\varphi_n(x) \mid n < \omega\}$ over $\varnothing$, where $\varphi_n(x)$ is $\exists x_0 \ldots x_n (x_0 < \ldots < x_n < x)$, then p is not realized in $(\mathbb{N}, <, h)$.

The existence of many ω-saturated structures is given by the following standard results of elementary model theory.

Lemma 13.6.14 *Let M be an L-structure and h an assignment into M. Let A be a subset of the domain of M, and p a type over A. Then there exists an elementary extension (N, h') of (M, h) in which p is realized.*

Proof. We regard the elements of the domain of M as constant symbols naming themselves; we assume that no such element is in L. Let $\Delta(M, h)$

be the set of all sentences $\psi(\bar{a}, P_1, \ldots, P_n)$ (where $\bar{a} \in M$) that are true in (M, h). Let c be a new constant symbol and let

$$p(c) = \{\varphi(c, \bar{a}, P_1, \ldots, P_n) \mid \varphi(x, \bar{a}, P_1, \ldots, P_n) \in p\}.$$

Then $\Delta(M, h) \cup p(c)$ is consistent, for since p is consistent, (M, h) can be expanded to a model of any finite subset of it. Let $(N, h') \models \Delta(M, h) \cup p(c)$. Each element $a \in M$, being a constant symbol occurring in $\Delta(M, h)$, has an interpretation $v(a)$ in N. We identify $a \in M$ with $v(a)$; thus $dom(M) \subseteq dom(N)$. Since $(N, h') \models \Delta(M, h)$, we have $(M, h) \preceq (N, h')$. And the interpretation in N of c realizes p. ■

Corollary 13.6.15 *Let M be an L-structure and h an assignment into M. Then there exists an ω-saturated elementary extension (N, h') of (M, h).*

Proof (sketch) If M is finite it is already ω-saturated. Assume that M is infinite. We suppress the assignments h throughout.

List the types over finite subsets of the domain of M as p_i $(i < \kappa)$ where κ is the number of such types. Define $M_i \succeq M$ $(i < \kappa)$ by induction: $M_0 = M$, $M_\delta = \bigcup_{i<\delta} M_i$ for limit $\delta < \kappa$, as in example 13.6.13(3), and if M_i is defined, choose $M_{i+1} \succeq M_i$ realizing p_i (using example 13.6.13(4) and lemma 13.6.14). Let $M^* = \bigcup_{i<\kappa} M_i$. Then $M^* \succeq M$, and M^* realizes all types over finite subsets of M.

Now define a new chain by induction: $M_0 = M$, $M_{i+1} = (M_i)^*$. Then $\bigcup_{i<\omega} M_i$ is an ω-saturated elementary extension of M. ■

13.6.6 Back-and-forth systems

We can now apply ω-saturation to the k-variable property.

Definition 13.6.16 Let M, N be L-structures and let h, h' be assignments into M, N respectively. Let $\kappa \leq \omega$ be a cardinal.

1. A *partial isomorphism from (M, h) to (N, h')* is a partial map $\theta : M \to N$ such that for all atomic (or equivalently, quantifier-free) L-formulae $\varphi(\bar{x}, P_i)$ and all tuples $\bar{a} \in dom(\theta)$ of the same length as $\bar{x}$, we have

   $$M \models \varphi(\bar{a}, h(P_i)) \iff N \models \varphi(\theta(\bar{a}), h'(P_i)),$$

   where $\theta(\bar{a})$ is the tuple in N obtained by applying θ to each element of $\bar{a}$ in order.

   Example: the empty map $\varnothing$ is a partial isomorphism from (M, h) to (N, h') iff $M \models \sigma(h(P_i)) \iff N \models \sigma(h'(P_i))$ for all monadic atomic sentences $\sigma(P_i)$ of L.

2. A *κ-back-and-forth system* from (M, h) to (N, h') is a non-empty set $\mathcal{C}$ of partial isomorphisms from (M, h) to (N, h') having the following properties:

   (a) if $\theta \in \mathcal{C}$ then $|dom(\theta)| \leq \kappa$;

   (b) if $\theta \in \mathcal{C}$ then any restriction of θ to a subset of its domain also lies in $\mathcal{C}$;

   (c) if $\theta \in \mathcal{C}$, $a \in M - dom(\theta)$ and $|dom(\theta)| < \kappa$ then there is $\theta^+ \in \mathcal{C}$ extending θ and defined on a;

   (d) if $\theta \in \mathcal{C}$, $b \in N - rng(\theta)$ and $|dom(\theta)| < \kappa$ then there is $\theta^+ \in \mathcal{C}$ extending θ and whose range includes b.

   Items (d) and (c) are the 'back' and 'forth' parts respectively.

3. Let $\bar{a} \in M, \bar{b} \in N$ be tuples of the same length $m \leq \kappa$. The structures $(M, h, \bar{a})$ and $(N, h', \bar{b})$ are said to be *κ-back-and-forth equivalent* if there exists a κ-back-and-forth system $\mathcal{C}$ from (M, h) to (N, h') containing a map θ such that $\theta(\bar{a}) = \bar{b}$.

Still we assume that $\mathcal{K}=$ mod(Σ) for some first-order L-theory Σ.

Lemma 13.6.17 *Let $\kappa \leq \omega$, let $M, N \in \mathcal{K}$, let h, h' be assignments into M, N, respectively, and let $\bar{a} \in M$, $\bar{b} \in N$ be tuples of equal (finite) length of at most κ. Assume that (M, h) and (N, h') are ω-saturated. Then the following are equivalent.*

(i) $(M, h, \bar{a})$ and $(N, h', \bar{b})$ are κ-back-and-forth equivalent.

(ii) $(M, h, \bar{a}) \equiv_\kappa (N, h', \bar{b})$.

Proof. (i) ⇒ (ii) *holds without the assumption of ω-saturation.* Suppose that $\mathcal{C}$ is a κ-back-and-forth system from (M, h) to (N, h'). By definitions 13.6.16(3) and 13.6.10, it suffices to show that whenever $k \leq \kappa$ is finite, $\varphi(x_1, \ldots, x_k, P_1, \ldots, P_n) \in L_\kappa$ and $\theta \in \mathcal{C}$, then for any k-tuple $\bar{a} \in dom(\theta)$,

$$M \models \varphi(\bar{a}, h(P_1), \ldots, h(P_n)) \iff N \models \varphi(\theta(\bar{a}), h'(P_1), \ldots, h'(P_n)).$$

We prove this by induction on φ. If φ is atomic the result holds because any $\theta \in \mathcal{C}$ is a partial isomorphism from (M, h) into (N, h'). The boolean connectives give no trouble.

Now consider the case $\varphi(x_1, \ldots, x_k, P_1, \ldots, P_n) = \exists x_i \psi \in L_\kappa$, where $i \leq \kappa$ is finite. Assume the result for ψ. Clearly, if $i \leq k$ then the value of $\bar{a}(i)$ is irrelevant to the truth of $\varphi(\bar{a}, P_1, \ldots, P_n)$, since x_i is bound. So by deleting the ith entry from $\bar{a}$ if necessary, and permuting variables, we can assume that $i = k + 1$. Hence $k < \kappa$.

Suppose that $\bar{a} \in dom(\theta) \in \mathcal{C}$. By (b) of definition 13.6.16(2) we can assume that θ is defined only on $\bar{a}$, so that $|dom(\theta)| < \kappa$. Assume that $M \models \varphi(\bar{a}, h(P_1), \ldots, h(P_n))$. Choose $b \in M$ such that $M \models \psi(\bar{a}, b, h(P_1), \ldots, h(P_n))$. By definition 13.6.16(2c) there is θ^+ extending θ and defined on b. By the inductive hypothesis, $N \models \psi(\theta^+(\bar{a}), \theta^+(b), h'(P_1), \ldots, h'(P_n))$. Hence $N \models \varphi(\theta^+(\bar{a}), y, h'(P_1), \ldots, h'(P_n))$. As θ^+ extends θ, we have $N \models \varphi(\theta(\bar{a}), y, h'(P_1), \ldots, h'(P_n))$, as required.

The case where $N \models \varphi(\theta(\bar{a}), h'(P_1), \ldots, h'(P_n))$ is similar and uses definition 13.6.16(2d).

(ii) $\Rightarrow$ (i). First a definition.

Definition 13.6.18 If $\bar{a} \in M$ and $\bar{b} \in N$ are tuples of length m, then they yield a relation $\{(\bar{a}(i), \bar{b}(i)) : i \leq m\} \subseteq M \times N$, which we write as $\langle \bar{a}, \bar{b} \rangle$.

Let $\mathcal{C}$ be the set of all relations $\langle \bar{a}, \bar{b} \rangle$ such that:

- $\bar{a} \in M$, $\bar{b} \in N$ are tuples of equal finite length $\leq \kappa$;
- $(M, h, \bar{a}) \equiv_\kappa (N, h', \bar{b})$.

Assuming (ii), we have $\langle \bar{a}, \bar{b} \rangle \in \mathcal{C}$ already. So to finish the proof it suffices to show that $\mathcal{C}$ is a κ-back-and-forth system from (M, h) to (N, h').

Clearly, $\mathcal{C}$ is a non-empty set of partial isomorphisms from (M, h) to (N, h'), and is closed under restrictions. To prove that the 'forth' extension property (c) of definition 13.6.16(2) holds, we show that if $\bar{a} \in M$, $\bar{b} \in N$ have finite length $< \kappa$, $(M, h, \bar{a}) \equiv_\kappa (N, h', \bar{b})$, and $c \in M$, then there is $d \in N$ such that $(M, h, \bar{a}, c) \equiv_\kappa (N, h', \bar{b}, d)$. (This makes sense as $\bar{a}c$ has length at most κ.)

Let p be the set

$$\{\varphi(x, \bar{b}, P_1, \ldots, P_n) \in L_\kappa \mid M \models \varphi(c, \bar{a}, h(P_1), \ldots, h(P_n))\}.$$

Claim: p is a type over $\bar{b}$ in (N, h').
Proof of claim: Let $r < \omega$ and $\varphi_i(x, \bar{b}, P_1, \ldots, P_{n_i}) \in p$ for $i < r$. Then $M \models \varphi_i(c, \bar{a}, h(P_1), \ldots, h(P_{n_i}))$ for all $i < r$, so that $M \models \exists x \bigwedge_{i<r} \varphi_i(x, \bar{a}, h(P_1), \ldots, h(P_{n_i}))$. Since $\exists x \bigwedge_{i<r} \varphi_i(x, \bar{y}, P_1, \ldots, P_{n_i}) \in L_\kappa$ and $(M, h, \bar{a}) \equiv_\kappa (N, h', \bar{b})$, we have $N \models \exists x \bigwedge_{i<r} \varphi_i(x, \bar{b}, h'(P_1), \ldots, h'(P_{n_i}))$. Hence p is a type in (N, h'), as claimed.

So as $\bar{b}$ is finite and N is ω-saturated there is $d \in N$ realizing p. Then by definition of p, $(M, h, \bar{a}, c) \equiv_\kappa (N, h', \bar{b}, d)$, as required.

The property (d) follows by symmetry. This completes the proof. Note that only saturation for L_κ-formulae was required. ■

Example 13.6.19 The implication (ii) $\Rightarrow$ (i) of lemma 13.6.17 can fail if ω-saturation is not assumed. For example, let (N, h') be the natural numbers $\mathbb{N}$ with the ordering $<$ and an arbitrary assignment h', and let (M, h) be an elementary extension of (N, h') in which there is an element m realizing the type p of example 13.6.13(6). (M, h) exists by lemma 13.6.14. We have $M \models m > n$ for all $n \in \mathbb{N}$. Then $(N, h') \models \sigma$ iff $(M, h) \models \sigma$ for all sentences σ of L_ω. But (M, h) and (N, h') are not ω-back-and-forth equivalent. For if $\mathcal{C}$ were an ω-back-and-forth system from (M, h) to (N, h'), let $\theta \in \mathcal{C}$ have finite domain including m. Let $\theta(m) = n$. By repeated application of definition 13.6.16(2,iii), there is $\theta' \in \mathcal{C}$ extending θ whose domain includes $0, 1, \ldots, n$. As θ' is a partial isomorphism we have $\theta'(0) < \theta'(1) < \ldots < \theta'(n) < \theta'(m) = n$, an impossibility. Here the assignments h and h' play no role.

A variant of this argument shows that M and N are not even two-back-and-forth equivalent.

Theorem 13.6.20 *Let $k < \omega$. The following are equivalent.*

(*i*) *$\mathcal{K}$ has the k-variable property.*

(*ii*) *Let $M, N \in \mathcal{K}$, let h, h' be assignments into M, N respectively, and suppose that (M, h) and (N, h') are ω-saturated. If $\bar{a} \in M$ and $\bar{b} \in N$ are k-tuples such that $(M, h, \bar{a})$ and $(N, h', \bar{b})$ are k-back-and-forth equivalent, then they are ω-back-and-forth equivalent.*

Proof. (i) $\Rightarrow$ (ii). Assume (i); let M, N, h, h' be as in (ii), and let $\bar{a} \in M$ and $\bar{b} \in N$ be k-tuples such that $(M, h, \bar{a})$ and $(N, h', \bar{b})$ are k-back-and-forth equivalent. By the lemma, $(M, h, \bar{a}) \equiv_k (N, h', \bar{b})$. As $\mathcal{K}$ has the k-variable property, by theorem 13.6.11 we have $(M, h, \bar{a}) \equiv (N, h', \bar{b})$. By the lemma again (using ω-saturation), $(M, h, \bar{a})$ and $(N, h', \bar{b})$ are ω-back-and-forth equivalent, as required.

(ii) $\Rightarrow$ (i): Assume (ii). Let $M, N \in \mathcal{K}$, let h, h' be assignments into M, N, respectively, let $\bar{a} \in M$, $\bar{b} \in N$ be k-tuples, and suppose that $(M, h, \bar{a}) \equiv_k (N, h', \bar{b})$. We will show that $(M, h, \bar{a}) \equiv (N, h', \bar{b})$; the result will then follow by theorem 13.6.11.

By corollary 13.6.15 we can take ω-saturated elementary extensions $(M^*, h^*) \succeq (M, h)$ and $(N^*, h'^*) \succeq (N, h')$. As the extensions are elementary, by example 13.6.13(2) we have M^*, $N^* \in \mathcal{K}$ and $(M^*, h^*, \bar{a}) \equiv_k (N^*, h'^*, \bar{b})$. It now follows from lemma 13.6.17 (using ω-saturation) that $(M^*, h^*, \bar{a})$ and $(N^*, h'^*, \bar{b})$ are k-back-and-forth equivalent. But we assumed that (ii) holds, so they are ω-back-and-forth equivalent. Hence $(M^*, h^*, \bar{a}) \equiv (N^*, h'^*, \bar{b})$ by the lemma again, and using example 13.6.13(2) once more we obtain $(M, h, \bar{a}) \equiv (N, h', \bar{b})$ as required. This completes the proof. ∎

We can use the same proof to show that $\mathcal{K}$ has H-dimension at most d if and only if for all $M, N \in \mathcal{K}$ and all assignments h, h' into M, N such that (M, h) and (N, h') are ω-saturated, if (M, h) and (N, h') are k-back-and-forth equivalent, then they are ω-back-and-forth equivalent.

13.6.7 Linear orders have the three-variable property

We give an application of theorem 13.6.20. The proof is also due to van Benthem [van Benthem, 1992]. We let $\mathcal{K}$ be the class of all irreflexive linear flows of time.

Proposition 13.6.21 *$\mathcal{K}$ has the three-variable property. (So any subclass of $\mathcal{K}$ also has the three-variable property.)*

Proof. In this case L has '=' plus the single binary relation '<', the earlier–later relation. Let $T, T' \in \mathcal{K}$ and let h, h' be assignments into T, T' respectively. Suppose that $\mathcal{C}$ is a three-back-and-forth system from (T, h) to (T', h'). We prove that $\mathcal{C}$ extends to an ω-back-and-forth system $\mathcal{C}^*$ from (T, h) to (T', h'). By theorem 13.6.20, this gives the result. Notice that we can even find $\mathcal{C}^*$ extending $\mathcal{C}$, and that (T, h) and (T', h') need not be ω-saturated.

First note that it suffices to find $\mathcal{C}^*$ meeting only conditions (a), (c), and (d) of definition 13.6.16(2), and such that $dom(\theta^*)$ is finite for all $\theta^* \in \mathcal{C}^*$. For given such a $\mathcal{C}^*$, let $\mathcal{C}' \supseteq \mathcal{C}^*$ be the set of restrictions of partial isomorphisms in $\mathcal{C}^*$. We claim that $\mathcal{C}'$ meets all four conditions of the definition and is an ω-back-and-forth system. Conditions (a) and (b) are evident. We check the 'forth' extension property (c) for $\mathcal{C}'$. If θ' is a restriction of $\theta^* \in \mathcal{C}^*$, and $a \in T - dom(\theta')$, then by (c) for $\mathcal{C}^*$ there is $\theta^{*+} \in \mathcal{C}^*$ extending θ^* and defined on a, and thus θ' has an extension in $\mathcal{C}'$ defined on a, namely θ^{*+} itself. Property (d) is similarly proved. Thus condition (b) is only of significance for a κ-back-and-forth system when κ is finite.

So let $\mathcal{C}^*$ be the set of all partial isomorphisms $\theta^* : (T, h) \to (T', h')$ with finite domain $\{a_1, \ldots, a_n\}$ with $a_1 < \ldots < a_n$, such that:

- for all $1 \leq i \leq n$, $\theta^* \lceil \{a_i\} \in \mathcal{C}$;
- for all $1 \leq i < n$, $\theta^* \lceil \{a_i, a_{i+1}\} \in \mathcal{C}$.

(We need both clauses in case $n = 1$.) Clearly, $\mathcal{C}^* \supseteq \mathcal{C}$, and because the structures in $\mathcal{K}$ are linear orders the elements of $\mathcal{C}^*$ are all partial isomorphisms. Property (a) of definition 13.6.16(2) trivially holds for $\mathcal{C}^*$. For (c), let $\theta^* \in \mathcal{C}^*$ have domain $\{a_1, \ldots, a_n\}$ with $a_1 < \ldots < a_n$, and let $a \in T - \{a_1, \ldots, a_n\}$. Suppose, for example, that $a_i < a < a_{i+1}$. Let $\theta = \theta^* \lceil \{a_i, a_{i+1}\}$. As $\theta \in \mathcal{C}$, by property (c) for $\mathcal{C}$ there is $\theta^+ \in \mathcal{C}$ extending θ and defined on a. Define θ^{*+} by: $dom(\theta^{*+}) = \{a_1, \ldots, a_n, a\}$, $\theta^{*+}\lceil\{a_1, \ldots, a_n\} = \theta^*$, and $\theta^{*+}(a) = \theta^+(a)$. As T and T' are linearly ordered, θ^{*+} is a partial isomorphism from (T, h) into (T', h'); it extends θ^* and is defined on a. Moreover, by property (b) for $\mathcal{C}$ applied to θ^+, $\theta^{*+}\lceil\{a_i, a\}$, $\theta^{*+}\lceil\{a\}$ and $\theta^{*+}\lceil\{a, a_{i+1}\}$ are in $\mathcal{C}$. Hence $\theta^{*+} \in \mathcal{C}^*$. The cases when $a < a_1$ and $a > a_n$ are proved similarly. Thus property (c) for $\mathcal{C}^*$ is established; and property (d) holds by symmetry. This proves the proposition. ■

Theorem 13.6.22 *$\mathcal{K}$ has the k-variable property iff $k \geq 3$; and $\mathcal{K}$ has H-dimension 3.*

Proof. The class of all order-preserving maps from $\mathbb{Q}$ to $\mathbb{Z}$ with domain of size at most 2 is a two-back-and-forth system from $\mathbb{Q}$ to $\mathbb{Z}$. By lemma 13.6.17 ((i) $\Rightarrow$ (ii)), $\mathbb{Z} \equiv_2 \mathbb{Q}$. So if the sentence $\forall xy(x < y \to \exists z(x < z < y))$ were equivalent in linear orders to one with two (bound) variables then its truth value would be the same in $\mathbb{Q}$ and $\mathbb{Z}$ — which it clearly is not. Hence the class of linear orders does not have H-dimension 2, and so by theorem 13.6.2, neither does it have the two-variable property.

By theorem 13.6.21, $\mathcal{K}$ has the three-variable property, and so by theorem 13.6.2, H-dimension at most 3. Hence it has H-dimension exactly 3. ■

13.6.8 Games and the *k*-variable property

It is known (see e.g. [Karp, 1965]) that back-and-forth systems are closely related to games. The Ehrenfeucht–Fraïssé game provides an alternative characterization of the equivalence $\equiv_\kappa$, and so using theorem 13.6.11 we can establish a connection between games and the k-variable property. Again this work is due to [Immerman and Kozen, 1987].

For the k-variable property, games are a 'finitary' alternative to ω-saturation, and as such, require that there are only finitely many inequivalent atomic formulae, even including monadic predicates. So until the end of the section we will assume that:

- L is a *finite relational signature:* one with no function or constant symbols, and finitely many relation symbols;
- we admit an arbitrary but finite number — $P_1, \ldots, P_r$, say — of the monadic predicate letters $P_1, P_2, \ldots$.

Later, we will indicate how these restrictions can be avoided. Often L is already relational — as in temporal logic, where it will be $\{=, <\}$ plus finitely many monadic predicate symbols.

Lemma 13.6.23 *For each $m, n < \omega$ there are up to logical equivalence only finitely many formulae of L with free variables taken from $x_1, \ldots, x_m$ and of quantifier depth at most n.*

Proof. By induction on n. Because L is now assumed finite and relational, for any m there are only finitely many — ℓ, say — monadic atomic L-formulae using at most the variables $x_1, \ldots, x_m$ and predicates $P_1, \ldots, P_r$. Hence by considering disjunctive normal forms we see that there are at most 2^{2^ℓ} pairwise inequivalent quantifier-free monadic L-formulae, which finishes the case $n = 0$. If inductively $\psi_i(x_1, \ldots, x_{m+1}, P_1, \ldots, P_r)$ $(i \leq \ell')$ are representatives (up to logical equivalence) of the monadic L-formulae of quantifier depth n with free variables taken from $x_1, \ldots, x_{m+1}$, then the finitely many boolean combinations of formulae from $\{\exists x_j \psi_i \mid i \leq \ell', j \leq m+1\}$ are representatives of the L_κ-formulae $\varphi(x_1, \ldots, x_m, P_1, \ldots, P_r)$ of quantifier depth $n + 1$ (because by renaming variables we can assume that the bound variable giving φ quantifier depth $n + 1$ is x_i for some $i \leq m + 1$). This completes the proof. ■

Now we introduce the games.

Definition 13.6.24 Let $\kappa \leq \omega$ be a cardinal, $n \geq 0$, M, N be L-structures, h, h' be assignments into M, N respectively, and let $\bar{a} \in M$, $\bar{b} \in N$, respectively, be m-tuples for some finite $m \leq \kappa$.

1. We define the *κ-pebble game* $G(\bar{a}, \bar{b}, \kappa, n, r)$ of length n. This is a restricted version of the well-known Ehrenfeucht–Fraïssé game. Strictly, we should write $G(M, h, N, h', \bar{a}, \bar{b}, \kappa, n, r)$, but we take M, h, N, and h' to be implicitly understood.

   $G(\bar{a}, \bar{b}, \kappa, n, r)$ is a game for two players, $\forall$ (male) and $\exists$ (female). There are available 2κ coloured *pebbles*. If $\kappa < \omega$ the pebbles are coloured $1, 2, \ldots, \kappa$, while if $\kappa = \omega$ the pebbles are of colours $1, 2, \ldots$. There are two pebbles of each colour.

   The structures M, N (or rather their domains, which are assumed disjoint) form the *board* on which the game is played, and pebbles will be placed on elements of M and N. At the start of the game, for each $i \leq m$ one pebble of colour i is on $\bar{a}(i) \in M$, and the other is on $\bar{b}(i) \in N$. Any remaining pebbles of colour $> m$ are spare, and are kept at the side.

   The game has n rounds. In each round, player $\forall$ can either:

- move a pebble already in play;
- bring into play a spare pebble of the first available colour (if any remain).

He puts his chosen pebble on an element of M or of N, at his choice.

Player $\exists$ responds by moving or bringing into play the other pebble of the same colour. She must place it on the other structure — so if $\forall$ put his pebble on an element of M, $\exists$ must place hers on an element of N, and vice versa. Her move completes the round. After n rounds the game ends.

2. We can describe a *play* of $G(\bar{a}, \bar{b}, \kappa, n, r)$ by a sequence $((\bar{a}_i, \bar{b}_i) \mid 0 \leq i \leq n)$ of pairs of tuples. $(\bar{a}_0, \bar{b}_0)$ is $(\bar{a}, \bar{b})$, representing the initial position. For each $1 \leq i \leq n$, $\bar{a}_i$ and $\bar{b}_i$ are tuples of length equal to the number of pebbles in play immediately after round i, and $\bar{a}_i(j)$, $\bar{b}_i(j)$ are the elements of M, N, respectively, that were then covered by a pebble of colour j.

3. $\exists$ wins the play $((\bar{a}_i, \bar{b}_i) \mid 0 \leq i \leq n)$ if for each $i \leq n$, $\langle \bar{a}_i, \bar{b}_i \rangle$ is a partial isomorphism from (M, h) to (N, h') with respect to the relation symbols from L and the monadic predicates $P_1, \ldots, P_r$ (see definitions 13.6.16 and 13.6.18). Otherwise, $\forall$ wins.

4. A *strategy* for a player in a game $G(\bar{a}, \bar{b}, \kappa, n, r)$ is a collection of rules telling him/her what to do in any position. It can be formalized as a family of functions. A strategy is said to be *winning* if following the strategy always results in a win for the player concerned.

Definition 13.6.25 Let M, N be L-structures, $\kappa \leq \omega$ and $n < \omega$. Let h, h' be assignments into M, N respectively. Let $\bar{a} \in M$, $\bar{b} \in N$ be tuples of equal length $m \leq \kappa$. We write

$$(M, h, \bar{a}) \equiv_{\kappa,n,r} (M, h', \bar{b})$$

if for all L_κ-formulae $\psi(x_1, \ldots, x_m, P_1, \ldots, P_r)$ of quantifier depth at most n, we have $M \models \psi(\bar{a}, h(P_1), \ldots, h(P_r))$ iff $N \models \psi(\bar{b}, h'(P_1), \ldots, h'(P_r))$.

The key fact on games that we need is as follows.

Proposition 13.6.26 *Let $\kappa \leq \omega$ and $n < \omega$. Let M, N be L-structures, let h, h' be assignments into M, N, respectively, and let $\bar{a} = a_1, \ldots, a_m \in M$, $\bar{b} = b_1, \ldots, b_m \in N$ be m-tuples, where $m \leq \kappa$. Then the following are equivalent.*

(*i*) *∃ has a winning strategy in the game* $G(\bar{a}, \bar{b}, \kappa, n, r)$.

(*ii*) $(M, h, \bar{a}) \equiv_{\kappa,n,r} (N, h', \bar{b})$.

Proof. By induction on n. If $n = 0$ then $G(\bar{a}, \bar{b}, \kappa, n, r)$ has no moves, so ∃ has a winning strategy iff $\langle \bar{a}, \bar{b} \rangle$ is already a partial isomorphism from M to N. As $m \leq \kappa$, this is equivalent to the condition $(M, h, \bar{a}) \equiv_{\kappa,0,r} (N, h', \bar{b})$.

Assume the equivalence of (i) and (ii) for n; we prove it for $n+1$. Suppose first that ∃ has a winning strategy in $G(\bar{a}, \bar{b}, \kappa, n+1, r)$. We must establish that $(M, h, \bar{a}) \equiv_{\kappa,n+1,r} (N, h', \bar{b})$. By taking disjunctive normal forms we see that it is enough to prove that if $\varphi(x_1, \ldots, x_m, P_1, \ldots, P_r) \in L_\kappa$ has quantifier depth $\leq n$, then for any finite $i \leq \kappa$,

$$(M, h) \models \exists x_i \varphi(a_1, \ldots, a_m) \iff (N, h') \models \exists x_i \varphi(b_1, \ldots, b_m).$$

(We write $(M, h) \models \exists x_i \varphi(a_1, \ldots, a_m)$ rather than the more cumbersome $M \models \exists x_i \varphi(a_1, \ldots, a_m, h(P_1), \ldots, h(P_r))$.)

Suppose then that $(M, h) \models \exists x_i \varphi(a_1, \ldots, a_m)$. There are two cases, according to whether $i \leq m$ or $i > m$. Assume that $i \leq m$. Let $c \in M$ be such that $(M, h) \models \varphi(a_1, \ldots, a_{i-1}, c, , a_{i+1}, \ldots, a_m)$. Suppose that ∀ decides to move a pebble of colour i on to c (if $i > m$, he will bring into play a new pebble). Let ∃ respond using her winning strategy, moving the other pebble of that colour on to $d \in N$, say. Let $\bar{a}_1$, $\bar{b}_1$ be the tuples describing the new state of play: so, for example, $\bar{a}_1$ is obtained from $\bar{a}$ by replacing the ith entry of $\bar{a}$ by c. Clearly, ∃ has a winning strategy for $G(\bar{a}_1, \bar{b}_1, \kappa, n, r)$: namely, *continue with the current winning strategy in* $G(\bar{a}, \bar{b}, \kappa, n+1, r)$. Hence by the inductive hypothesis, $(M, h, \bar{a}_1) \equiv_{\kappa,n,r} (N, h', \bar{b}_1)$. So $(N, h') \models \varphi(\bar{b}_1)$, and thus $(N, h') \models \exists x_i \varphi(b_1, \ldots, b_m)$, as required. The case where $(N, h') \models \exists y \varphi(\bar{b}, y)$ is similar, but now ∀ moves his pebble on to an appropriate element of N.

Conversely, assume that $(M, h, \bar{a}) \equiv_{\kappa,n+1,r} (N, h', \bar{b})$. We must find a winning strategy for ∃ in $G(\bar{a}, \bar{b}, \kappa, n+1, r)$. Suppose that ∀ moves a pebble of colour i on to an element $c \in M$ (if he should choose an element of N, the argument is similar).

Again, there are two cases: $i \leq m$ or $i = m+1$. Assume the latter this time. By lemma 13.6.23 there are up to equivalence only finitely many L_κ-formulae $\varphi(x_1, \ldots, x_{m+1}, P_1, \ldots, P_r)$ of quantifier depth at most n, such that $(M, h) \models \varphi(\bar{a}, c)$. Let $\psi \in L_\kappa$ be their conjunction; note that ψ also has quantifier depth n. Then $(M, h) \models \psi(a_1, \ldots, a_m, c)$, so

$$(M, h) \models \exists x_i \psi(a_1, \ldots, a_m, x_i).$$

By assumption,

$$(N, h') \models \exists x_i \psi(b_1, \ldots, b_m, x_i),$$

so there is $d \in N$ such that $(N, h') \models \psi(b_1, \ldots, b_m, d)$. By definition of ψ we have $(M, h, \bar{a}c) \equiv_{\kappa,n,r} (N, h', \bar{b}d)$, so by inductive hypothesis $\exists$ has a winning strategy σ in $G(\bar{a}c, \bar{b}d, \kappa, n, r)$. Hence the strategy *choose d in response to c, then use σ* is winning for $\exists$ in $G(\bar{a}, \bar{b}, \kappa, n+1, r)$. This completes the proof. ■

Example 13.6.27 The implication (ii) ⇒ (i) of the proposition can fail for infinite or non-relational signatures — cf. example 13.6.19. For example, suppose that $L = \{<, P_1, P_2, \ldots\}$, and let $(\mathbb{N}, h)$ be such that $h(P_i) = \{i, i+1, \ldots\}$ for all $i \in \mathbb{N}$. Then $p = \{P_i(x) \mid i \in \mathbb{N}\}$ is a type in $(\mathbb{N}, h)$; let $(M, h') \succeq (\mathbb{N}, h)$ have an element c realizing it (lemma 13.6.14). Then (M,h) and $(\mathbb{N}, h')$ satisfy the same monadic L-sentences of quantifier depth ≤ 1. But if $\forall$ chooses c in $G(\langle\rangle, \langle\rangle, \omega, 1, \omega)$ then $\exists$ has no winning response. The same problem can occur if the signature has function symbols. For example, if $s(x)$ is interpreted as the successor function in $\mathbb{N}$, then we can consider the type $\{x \neq s^n(0) \mid n \in \mathbb{N}\}$.

We can now derive our last characterization of the k-variable property.

Theorem 13.6.28 *Let $\mathcal{K} = Mod_L(\Sigma)$ for some L-theory Σ. The following are equivalent.*

(i) $\mathcal{K}$ has the k-variable property.

(ii) Let $M, N \in \mathcal{K}$, h, h' be assignments into M, N, respectively (so they are defined on the same finite set of atoms), and $\bar{a} \in M$, $\bar{b} \in N$ be of equal length $\leq k$. Suppose that $\exists$ has a winning strategy in $G(\bar{a}, \bar{b}, k, n, r)$ for all $n, r < \omega$. Then she has a winning strategy in $G(\bar{a}, \bar{b}, m, n, r)$ for all $m, n, r < \omega$.

Proof. By proposition 13.6.26, for any $\kappa \leq \omega$, $\exists$ has a winning strategy in $G(\bar{a}, \bar{b}, \kappa, n, r)$ for all $n, r < \omega$ iff $(M, h, \bar{a}) \equiv_\kappa (N, h', \bar{b})$ in the sense of definition 13.6.10. Hence (ii) is equivalent to: $(M, h, \bar{a}) \equiv_k (N, h', \bar{b}) \Rightarrow (M, h, \bar{a}) \equiv (N, h', \bar{b})$. The theorem now follows immediately from theorem 13.6.11. ■

How can we use games if L is infinite or not relational? If L has function symbols, create a new signature L^* with a relation symbol $R_\alpha(\bar{x})$ for every atomic L-formula $\alpha(\bar{x})$. We may translate Σ into L^*, and add axioms to say that functions behave as functions: for example, if f is a binary function symbol of L, we would add an axiom $\forall xy \exists_{=1} z (R_{f(x,y)=z}(x, y, z))$. (Here, $\exists_{=1} z \varphi(z)$ abbreviates the formula $\exists z \forall v (\varphi(v) \leftrightarrow v = z)$, saying that φ is satisfied by a unique z.) Thus, any first-order definable class $\mathcal{K} = \mathrm{Mod}_L(\Sigma)$ can be transformed into a first-order definable class $\mathcal{K}^*$ of L^*-structures.

Since R_α has the same variables as α, $\mathcal{K}^*$ has the k-variable property if and only if $\mathcal{K}$ does. Now, since any given formula can use only finitely many relation symbols, it is not difficult to see that $\mathcal{K}^*$ has the k-variable property if and only if: for any *finite* subsignature $L' \subseteq L^*$ and any $r < \omega$, for any $M, N \in \mathcal{K}^*$, assignments h, h' of r atoms into M, N, respectively, and k-tuples $\bar{a} \in M$, $\bar{b} \in N$, if $\exists$ has a winning strategy in $G(\bar{a}, \bar{b}, k, n, r)$ *in the sense of* L' for all $n, r < \omega$, then she has a winning strategy in $G(\bar{a}, \bar{b}, m, n, r)$ in the sense of L' for all $m, n, r < \omega$.

13.7 H-dimension and expressive completeness

We have seen that if a class of flows of time admits a finite set of expressively one-dimensional complete connectives then it has finite H-dimension. In this section we will prove that the converse of this fails. Thus the situation is not the same as for many-dimensional connectives (see theorem 13.2.4). Our proof is in two stages. First we show that the class of finite 'circles' has H-dimension 3 but has no finite expressively complete set of temporal connectives. Circles are very natural 'flows of time', of interest in their own right. However, the ordering relation on a circle is not transitive, so we then convert the class into a new one in which the ordering is transitive (and irreflexive), preserving the two properties cited. This latter class will establish the result.

Definition 13.7.1 A *circle* is a structure $(C, <)$, where $<$ is a binary relation on C satisfying the following (first-order expressible) axioms:

1. Trichotomy: For all $x, y \in C$, exactly one of $x < y, x = y, x > y$ holds.

2. Local linearity: For all $x \in C$, $\{y \in C \mid y > x\}$ and $\{y \in C \mid y < x\}$ (i.e. the future and past of x) are linearly ordered by $<$.

3. Circularity: $\forall xy(x < y \to \exists z(y < z \wedge z < x))$.

Any linear flow of time satisfies the first two axioms, but not the third. A structure $(C, <)$ satisfying the first two axioms is sometimes called a *local order;* $(C, <)$ is a linear order iff $<$ satisfies trichotomy and is transitive. If C is a set of cities of differing longitudes modulo π, containing Paris (longitude 2° E), Los Angeles (118° W) and Beijing (114° E), and we let $x < y$ ($x > y$) if y is east (respectively west) of x, then $(C, <)$ is a non-linear local order, as the three cities p, la, b cited form a *three-cycle:* we have $p < b < la < p$.

Non-trivial circles (i.e. with at least three elements) are local orders in which any two distinct points lie in a three-cycle. As an example of a finite circle, if C is the set of days of the week, and we let $x \leq y$ iff y is at most three days ahead of x, and $y < x$ otherwise, then $(C, <)$ is a circle of size seven. We can form an infinite circle $(D, <)$ by distributing countably many points densely around the unit circle in the complex plane (for any open interval $I \subseteq [0, 2\pi]$, a point p is added with $arg(p) \in I$), in such a way that no two points are opposite. Then define $p \leq q$ iff $0 \leq arg(q/p) < \pi$, and $p > q$ otherwise (so $p \leq q$ iff the shortest route around the circle from p to q is anticlockwise). D is known as the *countable homogeneous local order*. For those familiar with Fraïssé's amalgamation constructions, it can be obtained as the limit of the amalgamation class consisting of all finite local orders.

13.7.1 Circles

First we show how to localize a circle at a point, to give a linear structure.

Definition 13.7.2 Let $(C, <)$ be a circle, and let $e \in C$. We define the structure $C_e = (C, \prec, H)$, where $\prec$ is a binary relation symbol and H is a unary relation symbol, as follows. H is interpreted as $\{c \in C \mid c < e\}$, the past of c. We let

$$x \prec y \text{ iff } x \neq y \wedge [(H(x) \leftrightarrow H(y)) \leftrightarrow x < y].$$

Lemma 13.7.3 *Let $(C, <)$ be a circle, and let $e \in C$. Then $\prec$ linearly orders C_e.*

Proof. As $\prec$ clearly satisfies the trichotomy axiom, it is enough if we show it to be transitive. Let $a \prec b \prec c$ in C_e. We require $a \prec c$. By symmetry we can assume $H(a)$ holds, i.e. $a < e$.

There are now four cases. Firstly, if $b, c < e$ then the result is immediate, as the past of e is linearly ordered by $<$. Secondly, if $b < e \leq c$, then we have $a < b > c$ and must show $a > c$. But $a \leq c$ implies that b, c, e are all in a's linearly ordered future, contradicting $b < e \leq c < b$. Thirdly, if $c < e \leq b$, then we have $c < b < a$ and want $a < c$. But if $a \geq c$ then a, b, e are all in c's future, contradicting $a < e \leq b < a$. Finally, if $e \leq b, c$ then we have $a > b < c$ and want $a > c$. But $a \leq c$ would yield $a, b, e \leq c$, contradicting $a < e \leq b < a$.

■

Note that e is the $\prec$-least element of C_e.

Lemma 13.7.4 *Let $(C, <), (C', <)$ be circles, let $e \in C$ and let $\theta : C \to C'$ be a partial map defined on e. Then θ is a partial isomorphism from $(C, <)$ to $(C', <)$ iff it is a partial isomorphism from $(C_e, \prec, H)$ to $(C_{\theta(e)}, \prec, H)$.*

Proof. We must check that θ preserves $<$ iff it preserves $\prec$ and H. If $a \in C$ write a' for $\theta(a) \in C'$. Clearly, if θ preserves $<$ then for any $a \in C$, $C_e \models H(a)$ iff $a < e$, iff $a' < e'$, iff $C_{e'} \models H(a')$, so θ preserves H. But if θ preserves $<$ and H then it evidently preserves $\prec$, since this is defined in terms of them. Conversely, assume that θ preserves $\prec$ and H. Then if $a, b \in C$ are distinct, $a < b$ iff $C_e \models (H(a) \leftrightarrow H(b)) \leftrightarrow a \prec b$, iff $C_{e'} \models (H(a') \leftrightarrow H(b')) \leftrightarrow a' \prec b'$, iff $a' < b'$, as required. ■

13.7.2 Circles and H-dimension

Linear orders have H-dimension 3. We can use localizations to establish the same result for the class of all circles. Assume that C, C' are circles, h, h' are assignments into C, C', respectively, and that Ξ is a three-back-and-forth system from (C, h) to (C', h') (see section 13.6). We will show that there exists an ω-back-and-forth system from (C, h) to (C', h'). Theorem 13.6.20 will then yield the result.

Choose non-empty $\xi \in \Xi$ and $e \in dom(\xi)$, and write e' for $\xi(e)$. We could observe that if Ξ were a four-back-and-forth system from (C, h) to (C', h'), the set of restrictions to a domain of size ≤ 3 of elements of Ξ defined on e and taking it to e' would by lemma 13.7.4 be a three-back-and-forth system from (C_e, h) to $(C'_{e'}, h')$. By the proof of proposition 13.6.21, this would extend to an ω-back-and-forth system from (C_e, h) to $(C'_{e'}, h')$, which (by the lemma again) would be an ω-back-and-forth system from $(C, <, h)$ to $(C', <, h')$. But this only shows that the class of all circles has H-dimension at most 4. We can do better by constructing an ω-back-and-forth system directly from Ξ.

Definition 13.7.5

1. If $p : C \to C'$ is a partial map defined on e, we say that $a, b \in dom(p)$ are *p-adjacent* if either (i) $a = e$ and b is the $\prec$-maximal element of $dom(p)$ in C_e, or (ii) $a \prec b$ in C_e and there is no $d \in dom(p)$ with $a \prec d \prec b$.

2. Let Ξ^* be the set of all partial isomorphisms $p : (C, h) \to (C', h')$ such that:

- p has finite domain;
- $e \in dom(p)$ and $p(e) = e'$;
- for all $a, b \in dom(p)$, if a, b are p-adjacent then $p\lceil\{a, b\} \in \Xi$.

Lemma 13.7.6 *Ξ^* is an ω-back-and-forth system from (C, h) to (C', h').*

Proof. Let $p \in \Xi^*$ have domain D, and suppose that $c \in C \setminus D$. We wish to find $p^+ \in \Xi^*$ extending p and defined on c. (The case where $c \in C' \setminus rng(p)$ is similar.) If D has size at most 2 then this is clear, as Ξ is a three-back-and-forth system. Assume not. There are unique adjacent $a \prec b$ in D such that either $a = e, b = max_{\prec}(D)$ and $b \prec c$, or else $a \prec c \prec b$. Choose $q \in \Xi$ extending $p\lceil\{a, b\}$ and defined on c, and let $p^+ = p \cup q$ (regarding p, q as sets of ordered pairs, as indeed they are). Clearly, a, c (or vice versa) and c, b (or vice versa) are p^+-adjacent, and the other adjacent pairs of p^+ are the same as for p. So to show that $p^+ \in \Xi^*$, it remains to check that p^+ is a partial isomorphism from (C, h) to (C', h').

Since every restriction of p^+ to singletons is in Ξ, for every atom r and each $d \in dom(p^+)$ we have $d \in h(r) \iff p^+(d) \in h'(r)$. So by lemma 13.7.4, we need only show that p^+ is a partial isomorphism from $C_e = (C, \prec, H)$ to $C_{e'} = (C', \prec, H)$. If $x \in dom(p^+)$ write x' for the image $p^+(x)$.

Case I: $a = e$ and $b = max_{\prec}(D)$. Then, clearly, q preserves H; and as it preserves $<$ also, it preserves $\prec$. Hence we have $b' \prec c'$. As $b = max_{\prec}(D)$ and p preserved $\prec$, we have $c \succ d$ and $c' \succ d'$ for all $d \in D$, and are done.

Case II: Otherwise. If $H(c)$ is not preserved (by q), then as $<$ is preserved, and p preserves H, neither of $a \prec c$, $c \prec b$ would be preserved. So we have $a' \succ c' \succ b'$. By transitivity of $\succ$, $a' \succ b'$, which is a contradiction. So $H(c)$ is preserved by q, and so also $a' \prec c' \prec b'$. But (a, b) is a $\prec$-interval of C_e disjoint from D, and p preserves $\prec$. It follows as in the linear case (proposition 13.6.21) that p^+ preserves $\prec$, as required. ■

We have shown:

Theorem 13.7.7 *The class of all circles has H-dimension 3 (and indeed the 3-variable property).* ■

13.7.3 Circles and expressive completeness

We now show that no finite set of connectives is expressively complete for circles.

Definition 13.7.8 If $n \geq 3$ is odd, we let C_n be the circle consisting of the nth roots of unity, ordered as before. (C_n consists of n points spaced equally around a ring, with $x < y$ iff the shortest route from x to y is anticlockwise.)

We let $\mathcal{K} = \{C_n \mid n \geq 3 \text{ is odd}\}$.

Theorem 13.7.9 *The class $\mathcal{K}$ does not admit a finite expressively complete set of temporal connectives; and nor does any larger class of structures in the same signature $(<)$, such as the class $\mathcal{C}$ of all circles.*

Proof. Expressively complete connectives for a class are trivially expressively complete for any smaller class (see remark 13.1.12), so it suffices to prove the result for $\mathcal{K}$. Fix a finite set Φ of connectives. Let A be a formula using these connectives, without atoms. (We can assume that $\top, \bot$ are among the connectives of Φ, so that such formulae exist.) If $(C, <) \in \mathcal{K}$, then for any assignments h, h' into C and $c, c' \in C$ we have $\|A\|_c^h = \|A\|_{c'}^{h'}$. (The truth value of A is clearly independent of h, and its independence of c is because of the symmetry of these circles.) Hence for $C \in \mathcal{K}$ we can write $C \models A$ if $\|A\|_c^h = 1$ for some (any) h, c.

There are finitely many formulae of the form $B = \sharp(B_1, ..., B_n)$ for $\sharp \in \Phi$, $B_1, ..., B_n \in \{\top, \bot\}$. As $\mathcal{K}$ is infinite, we can choose distinct $C, C' \in \mathcal{K}$ such that $C \models B$ iff $C' \models B$ for all such B.

Claim: For all formulae A without atoms, $C \models A$ iff $C' \models A$.

Proof of claim: By induction on A. If $A \in \{\top, \bot\}$ this is clear, and the case of the boolean connectives is simple. Suppose $A = \sharp(A_1, ..., A_n)$ for $\sharp \in \Phi$ and A_i without atoms. Inductively, C and C' agree on the A_i. Let $B = \sharp(B_1, ..., B_n)$, where $B_i = \top$ if $C \models A_i$, and $\bot$ otherwise. Then, clearly, $C \models A \leftrightarrow B$ and similarly for C'. But by choice of C, C', they agree on B, and it follows that they agree on A. This proves the claim.

Now assume for contradiction that Φ is expressively complete over $\mathcal{K}$. Suppose that C has size n. Let A be a formula of Φ equivalent in $\mathcal{K}$ to $\nu(x) = x = x \wedge \exists_{=n} y(y = y)$, saying that the domain has exactly n elements. We can assume A has no atoms, by replacing them by $\top$; the semantics of A are unaffected as ν is not monadic. Then $C \models A$, so by the claim, $C' \models A$, and C' has size n. But by definition, there is a unique element of $\mathcal{K}$ of size n. Hence $C' = C$, a contradiction. ■

13.7.4 Circle-like flows of time

The second stage is to construct a class of transitive partial orders (flows of time) from circles, and to show the same properties for it.

Definition 13.7.10 Let $(C, <)$ be a circle. We define a poset $(C^*, \sqsubset)$ as follows:

- $C^* = C \times \{0, 1, 2\}$.
- We write c_i for $(c, i) \in C^*$. We identify c with c_0 $(c \in C)$.
- $c_i \sqsubset c_j$ iff $i < j$, for all $c \in C$.

- $c_0 \sqsubset c'_2$ iff $c < c'$ (all $c, c' \in C$).
- no other elements of C^* are ordered by $\sqsubset$.

Example 13.7.11 Figure 13.16 shows what becomes of the three-element circle C when transformed into C^*.

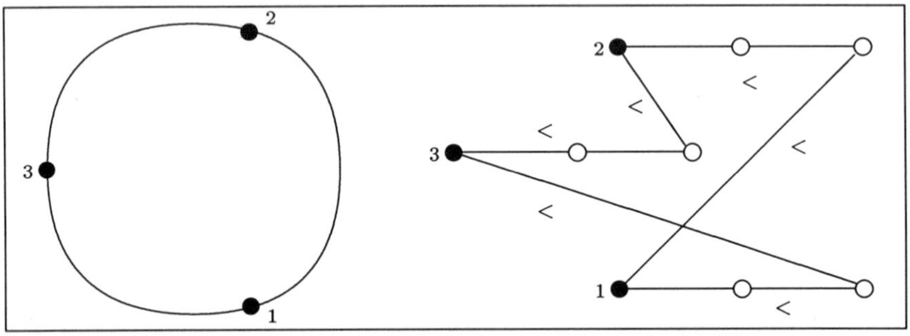

Figure 13.16. A circle and its corresponding flow of time.

Clearly, $(C^*, \sqsubset)$ is an irreflexive transitive poset. We will show that the class $\mathcal{C}^* = \{(C^*, \sqsubset) \mid (C, <) \text{ a circle}\}$ has H-dimension 3 and yet admits no finite expressively complete set of connectives.

The key observation is that we can *construct* $\mathcal{C}$ from $\mathcal{C}^*$, the class of all circles, as in section 13.4. For let $(C^*, \sqsubset)$ be as above. We check the provisions of definition 13.4.4. Firstly, the formula $\varphi(x) = {\sim}\exists y(y \sqsubset x)$ defines the domain of C in $(C^*, \sqsubset)$, and the relation $x < y$ on C is defined in $(C^*, \sqsubset)$ by $\alpha(x,y) = x \neq y \wedge \exists z(x \sqsubset z \wedge \exists x(y \sqsubset x \sqsubset z))$. Thus (φ, α) defines $\mathcal{K}$ over $\mathcal{K}^*$.

It is best to reindex the formulae χ of section 13.4, so that the base formula χ_1 becomes χ_0, χ_2 becomes χ_1, etc. In this numbering scheme we may let $\chi_0(x,y)$ be $\varphi(y) \wedge x = y$, $\chi_1(x,y)$ be $y \sqsubset x \wedge \exists z(x \sqsubset z)$ and $\chi_2(x,y)$ be $\exists z(y \sqsubset z \sqsubset x)$. Thus the χ_i define functions from C in to C^* that cover C^*. We have $(C^*, \sqsubset) \models \chi_i(c,d)$ iff $d = c_i$ (for all $c \in C, d \in C^*$ and $i = 0,1,2$).

Finally, we must define the formulae $\sqsubset^{ij}(x,y)$ for all possible $i,j < 2$ so that $(C,<) \models \sqsubset^{ij}(a,b)$ iff $(C^*, \sqsubset) \models a_i \sqsubset b_j$. We define $\sqsubset^{ij}(x,y)$ to be $\perp$ if $i \geq j$, $\sqsubset^{01}(x,y)$ and $\sqsubset^{12}(x,y)$ to be $x = y$, and $\sqsubset^{02}(x,y)$ to be $x \leq y$.

It is evident that $\mathcal{C}$ fits in $\mathcal{C}^*$ under this definition.

Theorem 13.7.12 *There is no finite expressively complete set of temporal connectives for $\mathcal{C}^*$.*

Proof. By theorem 13.4.12, any finite expressively complete set of connectives for $\mathcal{C}^*$ would yield a finite expressively complete set of connectives for $\mathcal{C}$, which by theorem 13.7.9 is impossible. ■

Theorem 13.7.13 *$\mathcal{C}^*$ has H-dimension at most* 3.

Proof. By theorems 13.4.9 and 13.7.7. ■

We remark that although the argument of theorem 13.4.9 does not prove that $\mathcal{C}^*$ has the three-variable property, it is not hard to see that it does have this property, and indeed that theorem 13.4.9 can be generalised accordingly. We leave the proof as an exercise. From this and remark 13.6.3, we deduce that, even for classes of flows of time, the k-variable property is independent of the existence of a finite expressively complete set of one-dimensional temporal connectives.

Chapter 14

Adding a temporal dimension to a logic system

M. Finger[1] and D. Gabbay

14.1 Introduction

We introduce a methodology whereby an arbitrary logic system **L** can be enriched with temporal features to create a new system **T**(**L**). The new system is constructed by amalgamating **L** with a pure propositional temporal logic **T** (such as linear temporal logic with 'Since' and 'Until') in a special way. We refer to this method as 'adding a temporal dimension to **L**' or just 'temporalizing **L**'.

We show that the logic system **T**(**L**) preserves several properties of the original temporal logic such as soundness, completeness, separation, and decidability.

When the process is applied to classical predicate logic, we obtain the traditional predicate temporal logic. When the process is applied to a predicate temporal logic, we obtain a two-dimensional predicate temporal logic.

We are interested in describing the way that a system $\mathcal{S}$, specified in a logic **L**, changes over time. There are two main methods for doing so.

---

[1] M. Finger's work was supported by CAPES-Brazil grant 1481/89 and the ESPRIT Basic Research Action 3096 (SPEC).

In the *external* method, snapshots of $\mathcal{S}$ are taken at different moments of time as describing the state of $\mathcal{S}$ at those times. We can write $\mathcal{S}_t$ for the way $\mathcal{S}$ is at time t, and use $\mathbf{L}$ to describe $\mathcal{S}_t$. We then externally add a temporal system that allows us to relate different $\mathcal{S}_t$ at different times t.

In the *internal* method, instead of considering $\mathcal{S}$ as a whole, we observe how $\mathcal{S}$ is built up from internal components and we transform these components into time-dependent building blocks. The internal temporal description of each component will give us the temporal description of the whole system $\mathcal{S}$. We can assume that $\mathcal{S}$ can be completely described through its components and that the way the components are put together to make $\mathcal{S}$ into a whole is also a (possibly time varying) component.

Both the external and the internal methods have their counterpart in logic as well. A temporal logical systems with temporal connectives such as 'Since' and 'Until' is the result of externally turning classical logic into a temporal (time varying) system. The use of a two-sorted predicate logic with one time variable in which atoms are of the form $A(t, x)$, with t time and x an element of a domain, is an internal way of making classical logic into a temporal system.

The purpose of this chapter is to investigate the external way of temporalizing a logic system. In the external approach, we do not need to have detailed knowledge about the components of the system $\mathcal{S}$ or about the logical components of its description in $\mathbf{L}$. An arbitrary logic system $\mathbf{L}$ can be enriched with temporal features to create a new system $\mathbf{T(L)}$. The new system is constructed by combining $\mathbf{L}$ with a pure propositional temporal logic $\mathbf{T}$ (e.g. linear-time temporal logic with 'Since' and 'Until') in a special way. We refer to this method as 'adding a temporal dimension to $\mathbf{L}$' or just 'temporalizing $\mathbf{L}$'. The method we use is not confined to temporal features only, but is a methodology of combining two logics by substituting one in another. Thus in the general case we can combine any two logic systems $\mathbf{L_1}$ and $\mathbf{L_2}$ to form $\mathbf{L_1(L_2)}$.

In classical propositional temporal logic we add to the language of classical propositional logic the connectives P and F and we are able to express statements like '*in the future* a certain proposition a will hold' by constructing sentences of the form Fa. The idea we develop here is to apply temporal operators not only to propositions but also to sentences from an arbitrary logic system $\mathbf{L}$.

Our aim can be viewed as describing both the 'statics' and the 'dynamics' of a logic system, while still remaining in a logical framework. The 'statics' is given by the properties of the underlying logic system $\mathbf{L}$; in propositional temporal logic $\mathbf{T}$, we already have the ability to describe the 'dynamics', i.e. changes in time of a set of atomic propositions. This point of view leads us to combine the upper-level temporal $\mathbf{T}$ system with an underlying logic system **L** so as to describe the evolution in time of a theory in **L** and its models.

Another more general point of view comes from the work in [Gabbay, 1991b] about networks of logic databases. A database is considered to be a model of a theory in some logic system $\mathbf{L}_2$ and the interaction between databases is modelled by another logic system $\mathbf{L}_1$; therefore, two basic logic levels can be identified, namely the local logic $\mathbf{L}_2$ and the global logic $\mathbf{L}_1$. The two systems are illustrated in figure 14.1 with a temporal upper-level system **T** in place of $\mathbf{L}_1$ and an arbitrary underlying logic system **L** in place of $\mathbf{L}_2$.

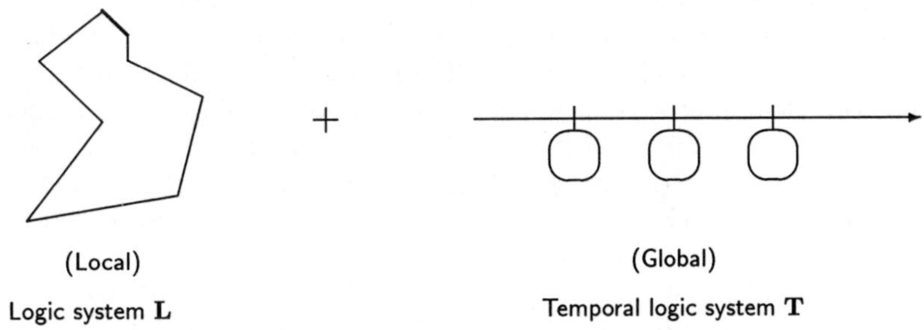

Figure 14.1. Two logic levels in a database network.

We consider a network of databases distributed in time, as an extension of the more usual idea of a network of databases distributed in space. The underlying logic system **L** characterizes the local behaviour of a database, i.e. the way queries are answered by a single element of the network. The upper-level logic system describes how one local system (at some moment in time) relates to another local system (at some other moment in time). We combine those two logic systems to be able to reason about the 'temporal network' as a whole, creating a logic system **T(L)**. The result of this combination is the addition of a temporal dimension to system **L**, as illustrated in figure 14.2.

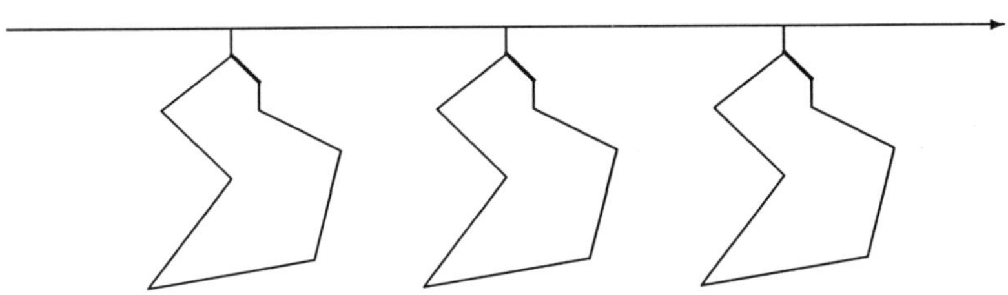

Figure 14.2. The logic system **T(L)**.

The above point of view is not yet the most general setting for our operations. One may ask a general question: given two logics $\mathbf{L}_1$ and $\mathbf{L}_2$, can we combine them into one logic? Suppose we take a disjoint union of the two systems, for example a modal logic system **K**, with modality $\Box_1$, and a modal logic system **S4**, with modality $\Box_2$. Here $\mathbf{L}_1 = \mathbf{K}$ and $\mathbf{L}_2 = \mathbf{S4}$. Form a language with $\{\Box_1, \Box_2\}$ and the separate axioms on $\Box_1$ (**K** axioms) and on $\Box_2$ (**S4** axioms). What do we know about the union? What is the semantics? These questions have been recently investigated by Fine and Schurz [Fine and Schurz, 1992] and by Kracht and Wolter [Kracht and Wolter, 1991], in a framework in which several independently axiomatizable monomodal systems were syntactically combined. The temporal case, however, differs from those since temporal logic is a bimodal system where the two modalities, one for the past and one for the future, always interact. The methods in [Kracht and Wolter, 1991] do not immediately apply. The work in this chapter differs from that in the above papers in two respects. Firstly, we are dealing with binary connectives *Since* (S) and *Until* (U). Secondly, and most importantly, we are not combining two logics arbitrarily but, rather, embedding one logic inside the other. If we were to embed one modality within another in the framework above we would syntactically combine them ruling out the formulae containing $\Box_1$ within the scope of $\Box_2$. This yields what we call $\mathbf{L_1(L_2)}$ ($\Box_1$ is *externally* applied to $\mathbf{L_2}$). The special case where $\mathbf{L}_1$ is a temporal logic **T** and $\mathbf{L}_2$ is an arbitrary logic **L**, gives us $\mathbf{T(L)}$, which we study in this chapter.

General combinations of logics have been addressed in the literature in various forms. Combinations of tense and modality were discussed in [Thomason, 1984], without explicitly providing a general methodology for doing so. A methodology for constructing logics of belief based on existing deductive systems was proposed by Konolige [Konolige, 1986]; in this case, the language of the original system was the base for the construction of a new modal language, and the modal logic system thus generated had its semantics defined in terms of the inferences of the original system. The model theory used by Konolige, called a *deductive model*, was the connection between the original system and the modal one. Here we present a quite different methodology, in which the language, inference system, and semantics of $\mathbf{T(L)}$ are based on, respectively, the language, the inference system, and the semantics of **T** and **L**. Recently we have developed a general methodology for combining any two logics through fibring their semantics [Gabbay, 1992a; Gabbay, 1992b]; the assumptions on the semantics of the candidate logics are very general and yield many known results.

Extensions of temporal logic are also found in the literature. A family of formal languages was generated, in [Casanova and Furtado, 1982], by means of certain mechanisms to define temporal modalities; the approach there was based on grammars and the resulting family of languages was claimed to be useful in expressing transition constraints for databases. Gabbay [Gabbay, 1991a] mixes two predicate languages **G** and **L**, generating the language $\mathbf{L}_k^*(\mathbf{G})$, a two-sorted predicate language in which one sort comes from terms originated in **G** and the other sort comes from terms originated in **L**; in the case where the original language **G** is supposed to describe an order relation $<$, the resulting system $\mathbf{L}_k^*(\mathbf{G})$ can be seen as a predicate-logic-like approach to temporal logic. Such a construction corresponds to an *internal* way of adding a temporal dimension to a logic system. In this work we propose a different approach, in which temporal modalities are applied to an existing logic system and thence a temporal dimension is added. Eventually, we are going to compare informally the internal and external approaches in section 14.7.

The rest of the chapter is organized as follows. In section 14.2 we formalize the idea of temporalizing a logic system **L** in terms of the S, U-temporal logic and we show the soundness and completeness of the resulting system **T(L)** over linear time. Section 14.3 shows that **T(L)** preserves the decidability property of system **L** over linear time, and the complexity of the decision procedure is estimated. section 14.4 shows that **T(L)** is a conservative extension of **L**. Section 14.5 shows that **T(L)** has the separation property, which is useful to specify how the past states of a database influence its future states. In section 14.6 we discuss the temporalization of first-order logic as a particularly interesting application; two different temporalizations of first-order logic are shown, yielding two expressively different logics. Finally, in section 14.7 we show how the added temporal dimension can be internalized in first-order logic and we compare the temporalized approach with the internalized first-order one.

14.2 Temporalizing an existing logic

This section will construct **T(L)** out of **T** and **L**. Our **T** is one of the temporal systems with 'Since' and 'Until', described below. Our **L** is in general any logic and in particular it can be classical predicate logic. We construct **T(L)** by allowing substitution of formulae of **L** for the atoms of formulae of **T**. We are *not* allowing the substitution of formulae of **T** or even formulae of **T(L)** for atoms of **L**. Thus the temporal connectives of **T** are never within the scope of connectives of **L**.

Next we first define **T**, both syntactically and semantically. Then we define **T(L)** syntactically and semantically and we prove soundness and completeness for **T(L)**.

14.2.1 The logic T

We allow several possibilities for the logic **T**. They are all propositional temporal logics of 'Since' and 'Until' but vary in the nature of the flow of time they describe. The language is the usual one built up with U and S from a set of propositional letters $\mathcal{P}$. Call the set of propositional formulae $\mathcal{L}_{\mathbf{S},\mathbf{U}}$.

The semantics of the logic is defined recursively as usual but for notational convenience in this chapter we will write $\mathcal{M}, t \models A$ if the formula A holds over model $\mathcal{M}$ at time point t.

We use a minimal axiomatic system for the S, U-temporal logic $\vdash_{\mathbf{S},\mathbf{U}}$ containing the following axioms:

A0	all classical tautologies;
A1a	$G(p \to q) \to (U(p,r) \to U(q,r))$;
A1b	$H(p \to q) \to (S(p,r) \to S(q,r))$;
A2a	$G(p \to q) \to (U(r,p) \to U(r,q))$;
A2b	$H(p \to q) \to (S(r,p) \to S(r,q))$;
A3a	$(p \wedge U(q,r)) \to U(q \wedge S(p,r),r)$;
A3b	$(p \wedge S(q,r)) \to S(q \wedge U(p,r),r)$.

Note that the axioms above come in pairs, represented by a and b, such that one is the mirror image of the other. The inference rules are:

Subst Uniform Substitution, i.e. let $A(q)$ be an axiom containing the propositional letter q and let B be any formula, then from $\vdash A(q)$ infer $\vdash A(q \backslash B)$ by substituting all appearances of q in A by B.

MP Modus ponens: from $\vdash A$ and $\vdash A \to B$ infer $\vdash B$.

TG Temporal Generalization: from $\vdash A$ infer $\vdash HA$ and $\vdash GA$.

The axioms of $\vdash_{\mathbf{S},\mathbf{U}}$ can be extended by a set of axioms Σ so as to impose restrictions on the flow of time, therefore generating the inference system $\vdash_{\mathbf{S},\mathbf{U}(\boldsymbol{\Sigma})}$. When Σ is the empty set we have $\vdash_{\mathbf{S},\mathbf{U}} = \vdash_{\mathbf{S},\mathbf{U}(\varnothing)}$. A set of formulae is consistent if we cannot deduce falsity ($\bot$) from it.

We say that an inference system is *sound* and *complete* with respect to a class $\mathcal{K}$ of flows of time if

$$\mathcal{K} \models A \text{ iff } \vdash A,$$

or, equivalently,

$$A \text{ is consistent iff } A \text{ has a model over } \mathcal{K},$$

soundness corresponding to the *if* part and completeness [2] to the *only if* part. We write **S,U**$/\mathcal{K}$ to indicate that fact.

If we consider $\mathcal{K}_0$, the class of all flows of time, we have the following well-known result.

Theorem 14.2.1 (soundness and completeness of S,U/$\mathcal{K}_0$) *The inference system* $\vdash_{\mathbf{S,U}}$ *is sound and complete with respect to the class* $\mathcal{K}_0$.

An elegant proof of the above is given by Xu [Xu, 1988]. A proof of completeness for the class of transitive linear flows of time, $\mathcal{K}_{lin}$, is given by Burgess [Burgess, 1982a], adding the following set Σ of axioms together with their mirror images (b axioms).

A4a $U(p,q) \rightarrow U(p, q \wedge U(p,q))$;

A5a $U(q \wedge U(p,q), q) \rightarrow U(p,q)$;

A6a $(U(p,q) \wedge U(r,s))$
$\rightarrow (U(p \wedge r, q \wedge s) \vee U(p \wedge s, q \wedge s) \vee U(q \wedge r, q \wedge s))$.

Burgess actually used an extra axiom, but Xu [Xu, 1988] proved the same result omitting it and axiom (A5b). Axioms (A4ab) and (A5a) are responsible for restricting the class of flows of time to a transitive one. The pair of axioms (A6ab) are responsible for restricting the class of flows of time to a linear one. Adding the axiom

A7a $(p \wedge H p) \rightarrow F H p$

and its mirror image restricts the flow of time to a discrete one. Extending original proofs of completeness to include new axioms over a more restricted flow of time is discussed by Burgess [Burgess, 1984a]. With axioms (A0)–(A7) we have soundness and completeness results for a class of linear, discrete and transitive flows of time. There are also complete axiomatizations **S,U/R** over the reals [Gabbay and Hodkinson, 1990; Reynolds, 1992c] and **S,U/Z** over the integers [Reynolds, 1993a].

14.2.2 Logic systems and their temporalized form

Having defined a family of S,U-temporal logics, we now externally apply such logic systems to any other logic system $\mathbf{L}$, i.e. we 'temporalize' $\mathbf{L}$.

A logical system is a pair $\mathbf{L} = \langle \mathcal{L}_{\mathbf{L}}, \vdash_{\mathbf{L}} \rangle$, where $\mathcal{L}_{\mathbf{L}}$ is its language and $\vdash_{\mathbf{L}}$ is its inference system; the set $\mathcal{L}_{\mathbf{L}}$ must be countable. A model for the

[2]This is sometimes called *weak completeness*; strong completeness says that for any (possibly infinite) set of formula Γ, if Γ is consistent then Γ has a model. Strong completeness implies weak completeness but the converse is not true.

logic system **L** is a structure $\mathcal{M}_{\mathbf{L}}$ and we denote $\mathcal{M}_{\mathbf{L}} \models \alpha$ when a formula $\alpha \in \mathcal{L}_{\mathbf{L}}$ is true under the model $\mathcal{M}_{\mathbf{L}}$. The class of all models of **L** is denoted by $\mathcal{CM}_{\mathbf{L}}$ and a formula α is said to be valid if $\mathcal{M}_{\mathbf{L}} \models \alpha$ for all $\mathcal{M}_{\mathbf{L}} \in \mathcal{CM}_{\mathbf{L}}$.

A logical system **L** is said to be sound if, whenever $\vdash_{\mathbf{L}} \alpha$, we have $\mathcal{M}_{\mathbf{L}} \models \alpha$ for all $\mathcal{M}_{\mathbf{L}} \in \mathcal{CM}_{\mathbf{L}}$. The logical system **L** is said to be complete if, whenever $\mathcal{M}_{\mathbf{L}} \models \alpha$ for all $\mathcal{M}_{\mathbf{L}} \in \mathcal{CM}_{\mathbf{L}}$, we have that $\vdash_{\mathbf{L}} \alpha$.

We constrain the logic system **L** to be an extension of classical logic, i.e. all propositional tautologies must be valid in it. This constraint is due to the fact that all S, U-temporal logics presented above are extensions of classical logic and any of them can be taken as the logic **T** in which we base the temporalization. We discuss later in this section what should be the case if **L** were not an extension of classical logic.

Definition 14.2.2 (boolean combinations and monolithic formulae) The set $\mathcal{L}_{\mathbf{L}}$ is partitioned in two sets, $BC_{\mathbf{L}}$ and $ML_{\mathbf{L}}$. A formula $A \in \mathcal{L}_{\mathbf{L}}$ belongs to the set of *boolean combinations*, $BC_{\mathbf{L}}$, iff it is built up from other formulae by the use of one of the boolean connectives $\sim$ or $\wedge$ or any other connective defined only in terms of those; it belongs to the set of *monolithic formula* $ML_{\mathbf{L}}$ otherwise.

We can proceed then to the definition of the temporalized language. In the following we will use α, β, γ, ... , to range over formulae of $\mathbf{T(L)}$.

The result of temporalizing the logic system **L** is the logic system $\mathbf{T(L)} = \langle \mathcal{L}_{\mathbf{T(L)}}, \vdash_{\mathbf{T(L)}} \rangle$ and its models by $\mathcal{M}_{\mathbf{T(L)}}$. The alphabet of the temporalized language uses the alphabet of **L** plus the two-place operators S and U, if they are not part of the alphabet of **L**; otherwise, we use S_2 and U_2 or any other proper renaming.

Definition 14.2.3 (temporalized formulae) The set $\mathcal{L}_{\mathbf{T(L)}}$ of formulae of the logic system **L** is the smallest set such that:

1. If $\alpha \in ML_{\mathbf{L}}$, then $\alpha \in \mathcal{L}_{\mathbf{T(L)}}$.
2. If $\alpha, \beta \in \mathcal{L}_{\mathbf{T(L)}}$ then $\sim\alpha \in \mathcal{L}_{\mathbf{T(L)}}$ and $(\alpha \wedge \beta) \in \mathcal{L}_{\mathbf{T(L)}}$.
3. If $\alpha, \beta \in \mathcal{L}_{\mathbf{T(L)}}$ then $S(\alpha, \beta) \in \mathcal{L}_{\mathbf{T(L)}}$ and $U(\alpha, \beta) \in \mathcal{L}_{\mathbf{T(L)}}$.

The set of *maximal monolithic subformulae of* α, $Mon(\alpha)$, is the set of all monolithic subformulae of α that are used to build α up by the rules above.

It is obvious from the definition above that the set $\mathcal{L}_{\mathbf{T(L)}}$ is denumerably infinite. Note that from items 1 and 2 of the definition above, it follows that $\mathcal{L}_{\mathbf{L}} \subset \mathcal{L}_{\mathbf{T(L)}}$. The reason to define the base case in item 1 in terms of monolithic formulae of **L** instead of simply defining it in terms of any formula in $\mathcal{L}_\mathbf{L}$ is that we would have a double parsing problem. In fact, suppose an item $1'$ that would state that:

$1'$. If $\alpha \in \mathcal{L}_\mathbf{L}$, then $\alpha \in \mathcal{L}_{\mathbf{T}(\mathbf{L})}$.

Suppose we want to define a function over the set of formulae, e.g. the depth of the parsing tree of a formula. Consider the formula $(\alpha \wedge \beta) \in \mathcal{L}_\mathbf{L}$; it would belong to $\mathcal{L}_{\mathbf{T}(\mathbf{L})}$ both by items $1'$ and 2. If we parse it by $1'$, then its depth will be 0, but if we parse it by 2, its depth will be 1, i.e. depth is not a well-defined function. To avoid such problems we introduce the restriction to monolithic formulae in item 1. We also note that, for instance, if $\Box$ is an operator of the alphabet of **L** and α and β are two formulae in $\mathcal{L}_\mathbf{L}$, the formula $\Box U(\alpha, \beta)$ is *not* in $\mathcal{L}_{\mathbf{T}(\mathbf{L})}$.

There is nothing to prevent us from defining the temporalization in terms of some F, P-temporal language, but since the language with S and U is more expressive it received our preference.

If **L** is an extension of classical logic, we must pay attention to some details before being able to describe the semantics of $\mathbf{T}(\mathbf{L})$. First, if $\mathcal{M}_\mathbf{L}$ is a model in the class of models of **L**, for every formula $\alpha \in \mathcal{L}_\mathbf{L}$ we must have either $\mathcal{M}_\mathbf{L} \models \alpha$ or $\mathcal{M}_\mathbf{L} \models {\sim}\alpha$. For example, if **L** is a modal logic system, e.g. **S4**, we must consider a 'current world' o as part of its model to achieve that condition. Second, we must be careful about the semantics of boolean connectives in the temporalized system. The construction of temporalized formulae based on monolithic formulae of $\mathcal{L}_\mathbf{L}$ guarantees that the semantics of the boolean connectives is the same in both the upper-level temporal logic system **T** and in the temporalized system $\mathbf{T}(\mathbf{L})$.

The language of $\mathbf{T}(\mathbf{L})$ is independent of the underlying flow of time, but not its semantics and inference system, so we must fix a class $\mathcal{K}$ of flows of time over which the temporalization is defined; this is equivalent to fixing one logic **T** among the family of temporal logics presented above. We are then in a position to define the semantics of the temporalized logic system $\mathbf{T}(\mathbf{L})$.

Definition 14.2.4 (semantics of the temporalized logic) Consider a flow of time $(T, <) \in \mathcal{K}$ and a function $g : T \to \mathcal{CM}_\mathbf{L}$, mapping every time point in T to a model in the class of models of **L**. A model of $\mathbf{T}(\mathbf{L})$ is a triple $\mathcal{M}_{\mathbf{T}(\mathbf{L})} = (T, <, g)$ and the fact that α is true in the model $\mathcal{M}_{\mathbf{T}(\mathbf{L})}$ at time point t is represented by $\mathcal{M}_{\mathbf{T}(\mathbf{L})}, t \models \alpha$. The semantics of $\mathbf{T}(\mathbf{L})$ is given by:

$\mathcal{M}_{\mathbf{T(L)}}, t \models \alpha$, $\alpha \in ML_{\mathbf{L}}$	iff $g(t) = \mathcal{M}_{\mathbf{L}}$ and $\mathcal{M}_{\mathbf{L}} \models \alpha$.
$\mathcal{M}_{\mathbf{T(L)}}, t \models \sim\alpha$	iff it is not the case that $\mathcal{M}_{\mathbf{T(L)}}, t \models \alpha$.
$\mathcal{M}_{\mathbf{T(L)}}, t \models (\alpha \wedge \beta)$	iff $\mathcal{M}_{\mathbf{T(L)}}, t \models \alpha$ and $\mathcal{M}_{\mathbf{T(L)}}, t \models \beta$.
$\mathcal{M}_{\mathbf{T(L)}}, t \models S(\alpha, \beta)$	iff there exists $s \in T$ such that $s < t$ and $\mathcal{M}_{\mathbf{T(L)}}, s \models \alpha$ and for every $u \in T$, if $s < u < t$ then $\mathcal{M}_{\mathbf{T(L)}}, u \models \beta$.
$\mathcal{M}_{\mathbf{T(L)}}, t \models U(\alpha, \beta)$	iff there exists $s \in T$ such that $t < s$ and $\mathcal{M}_{\mathbf{T(L)}}, s \models \alpha$ and for every $u \in T$, if $t < u < s$ then $\mathcal{M}_{\mathbf{T(L)}}, u \models \beta$.

We write $\mathbf{T(L)} \models \alpha$ if, for every model $\mathcal{M}_{\mathbf{T(L)}}$ whose underlying flow of time $(T, <) \in \mathcal{K}$ and for every time point $t \in T$, it is the case that $\mathcal{M}_{\mathbf{T(L)}}, t \models \alpha$.

The inference system of $\mathbf{T(L)}/\mathcal{K}$ is given by the following:

Definition 14.2.5 (axiomatization for T(L))

- The axioms of $\mathbf{T}/\mathcal{K}$;
- The inference rules of $\mathbf{T}/\mathcal{K}$;
- For every formula α in $\mathcal{L}_{\mathbf{L}}$, if $\vdash_{\mathbf{L}} \alpha$ then $\vdash_{\mathbf{T(L)}} \alpha$.

The third item above constitutes a new inference rule needed to preserve the theoremhood of formulae of the logic system **L**. Therefore we call it **Preserve**. The only inference rules we are considering here are **Subst**, **MP** and **TG**, but other rules such as the irreflexivity rule **IRR** [Gabbay and Hodkinson, 1990], can also be added.

The first concern about the axiomatization is its soundness, i.e. if whenever $\vdash_{\mathbf{T(L)}} \alpha$ we have $\mathbf{T(L)} \models \alpha$.

Theorem 14.2.6 (soundness of T(L)) *If the logic system* **L** *is sound and* **S,U**$/\mathcal{K}$ *is sound over the class of flows of time* $\mathcal{K}$, *then so is the logic system* $\mathbf{T(L)}/\mathcal{K}$.

Proof. Soundness of **S,U**$/\mathcal{K}$ gives us the validity of the axioms over $\mathcal{K}$. As for the inference rules, soundness of **L** guarantees that all formulae generated by **Preserve** are valid; soundness of **S,U**$/\mathcal{K}$ guarantees that the other inference rules, when applied to valid formulae, always generate valid formulae. ■

Completeness is discussed in section 14.2.4. Let us first present a few examples of the temporalization of an existing logic system.

Example 14.2.7 (temporalizing modal logic of belief) Suppose we have a propositional modal logic of belief $\mathbf{B} = \langle \mathcal{L}_{\mathbf{B}}, \vdash_{\mathbf{B}} \rangle$ with the modal operator B, in which Bp is intended to mean that p is a proposition that is believed by an agent. The axiomatization, $\vdash_{\mathbf{B}}$, is given by the basic modal logic system **K** plus the transitivity axiom **4** as one of the introspective properties of belief systems in [Hintikka, 1962]:

$$K \begin{cases} \text{All propositional tautologies} \\ B(p \to q) \to (Bp \to Bq) \\ \text{Rules: } \textbf{Subst}, \textbf{MP}, \textbf{Generalization} \end{cases} \qquad + \; Bp \to BBp$$

The transitivity axiom means that, if some fact is believed, it is believed to be believed, which represents a positive introspection of the believing agent; for a discussion on modal logics of belief, see [Halpern and Moses, 1985]. This system is provided with a standard Kripke semantics for modal logics [Hughes and Cresswell, 1968], with a set of possible worlds W, an accessibility relation R, and a valuation function V, so that $\mathcal{M}_{\mathbf{B}} = (W, R, V)$ is a model structure in which the accessibility relation R is transitive. Actually, we are considering $\mathcal{M}_{\mathbf{B}} = (W, R, V, o)$, where o is a 'current world' from which the observations are made, so that we may have both validity and satisfiability in the model theory of **B**.

Consider the temporalized logic system $\mathbf{T}(\mathbf{B})$ over the class $\mathcal{K}_0$ of all flows of time. Its inference system $\vdash_{\mathbf{T}(\mathbf{B})}$, for example, gives us as theorems

$$\begin{array}{l} B(p \to q) \to (Bp \to Bq) \\ \sim(Bp \wedge \sim Bp) \\ GB\sim(Bp \wedge \sim Bp) \\ G(Bp \to q) \to (U(Bp, Bq) \to U(q, Bq)). \end{array}$$

If we have a theory $\Gamma = \{GBp, Bp \to Fp, U(q, Bp)\}$. We construct one possible model $\mathcal{M}_{\mathbf{T}(\mathbf{B})}$ by choosing a flow of time with $T = \{a, b, c, d\}$ and the partial order $< = \{(a, b), (b, c), (a, c), (a, d)\}$. We construct the assignment g such that:

$$\begin{array}{lcl} g(a) = \mathcal{M}^a_{\mathbf{B}} \models p, & & g(b) = \mathcal{M}^b_{\mathbf{B}} \models Bp \wedge p, \\ g(c) = \mathcal{M}^c_{\mathbf{B}} \models Bp \wedge q & \text{and} & g(d) = \mathcal{M}^d_{\mathbf{B}} \models Bp. \end{array}$$

In the resulting model $\mathcal{M}_{\mathbf{T}(\mathbf{B})}$ we have $\mathcal{M}_{\mathbf{T}(\mathbf{B})}, a \models \Gamma$ as illustrated in figure 14.3.

Example 14.2.8 (temporalizing propositional logic) Consider classical propositional logic $\mathbf{PL} = \langle \mathcal{L}_{\mathbf{PL}}, \vdash_{\mathbf{PL}} \rangle$. Its temporalization generates the logic system $\mathbf{T}(\mathbf{PL}) = \langle \mathcal{L}_{\mathbf{T}(\mathbf{PL})}, \vdash_{\mathbf{T}(\mathbf{PL})} \rangle$.

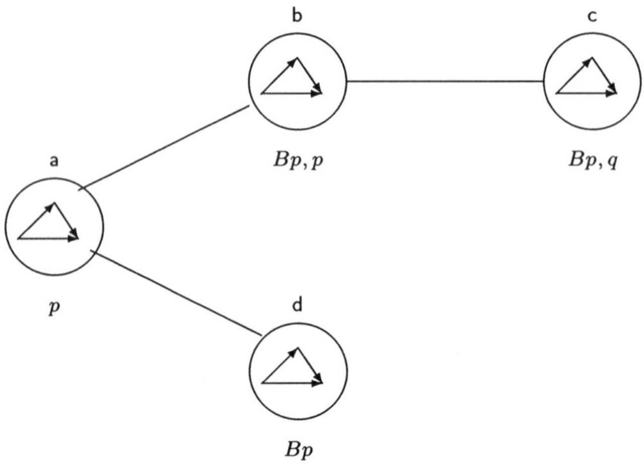

Figure 14.3.

It is not difficult to see that $\mathcal{L}_{\mathbf{T(PL)}} = \mathcal{L}_{\mathbf{S,U}}$ and $\vdash_{\mathbf{T(PL)}} = \vdash_{\mathbf{S,U}}$, i.e. the temporalized version of **PL** over any $\mathcal{K}$ is actually the temporal logic $\mathbf{T} = \mathbf{S,U}/\ \mathcal{K}$. With respect to $\mathcal{M}_{\mathbf{T(L)}}$, the function h actually assigns, for every time point, a **PL** model.

Example 14.2.9 (temporalizing S, U-temporal logic) If we temporalize over $\mathcal{K}$ the one-dimensional logic system $\mathbf{S,U}/\mathcal{K}$ we obtain the logic system $\mathbf{T(S,U)} = \langle \mathcal{L}_{\mathbf{T(S,U)}}, \vdash_{\mathbf{T(S,U)}} \rangle = \mathbf{T^2(PL)}/\mathcal{K}$. In this case we have to rename the two-place operators S and U of the temporalized alphabet to, say, S_2 and U_2.

In order to obtain a model for $\mathbf{T(S,U)}$, we must fix a 'current time', o, in $\mathcal{M}_{\mathbf{S,U}} = (T_1, <_1, g_1)$, so that we can construct the model $\mathcal{M}_{\mathbf{T(S,U)}} = (T_2, <_2, g_2)$ as previously described. Note that, in this case, the flows of time $(T_1, <_1)$ and $(T_2, <_2)$ need not be the same. $(T_2, <_2)$ is the flow of time of the upper-level temporal system, whereas $(T_1, <_1)$ is the flow of time of the underlying logic which, in this case, happens to be a temporal logic.

The logic system we obtain by temporalizing S, U-temporal logic is the two-dimensional temporal logic described in [Finger, 1992].

Example 14.2.10 (n-dimensional temporal logic) If we repeat the process started in the last two examples, we can construct an n-dimensional temporal logic $\mathbf{T^n(PL)/K}$ (its alphabet including S_n and U_n) by temporalizing an $(n-1)$-dimensional temporal logic.

Every time we add a temporal dimension, we are able to describe changes in the underlying system. Temporalizing the system **L** once, we are creating a way of describing the history of **L**; temporalizing for the second time, we are describing how the history of **L** is viewed in different moments of time. We can go on indefinitely, although it is not clear what is the purpose of doing so.

The assumption that the underlying logic system **L** is an extension of classical logic allows us to make a clear distinction between boolean and monolithic formulae, avoiding double parsing and reconstructing the boolean formulae and its semantics in the temporalized system $\mathbf{T(L)}$. If we were to temporalize a logic system that is not an extension of classical logic, or any system in which we do not have the notion of satisfiability, only validity, we could consider all its formulae as being monolithic. The problem would then be the different semantics of the boolean connectives in the underlying system and in the upper-level (classical) temporal system, if those symbols were identical in both systems. The solution would be renaming the boolean connectives, say, in the underlying system. The applications of such a hybrid logic system are not clear so, to avoid extra difficulties in the results we are going to prove, we will stick to the constraint on **L** being an extension of classical logic.

14.2.3 The correspondence mapping

We are now going to relate the temporalized logic system $\mathbf{T(L)}$ with the original S, U-temporal logic used as a base for the temporalization process. Consider $\mathcal{P}$, a denumerably infinite set of propositional letters, and let **S,U** be the propositional temporal logic system induced by $\mathcal{P}$. The following defines a relationship between a temporalized language $\mathcal{L}_{\mathbf{T(L)}}$ and a propositional temporal language $\mathcal{L}_{\mathbf{S,U}}$.

Definition 14.2.11 (the correspondence mapping) Consider an enumeration $p_1, p_2, \ldots,$ of elements of $\mathcal{P}$ and consider an enumeration α_1, $\alpha_2, \ldots,$ of formulae in $ML_{\mathbf{L}}$. The *correspondence mapping* $\sigma : \mathcal{L}_{\mathbf{T(L)}} \to \mathcal{L}_{\mathbf{S,U}}$ is given by:

$$\begin{array}{rcl} \sigma(\alpha_i) & = & p_i \text{ for every } \alpha_i \in ML_{\mathbf{L}}, i = 1, 2\ldots \\ \sigma({\sim}\alpha) & = & {\sim}\sigma(\alpha) \\ \sigma(\alpha \wedge \beta) & = & \sigma(\alpha) \wedge \sigma(\beta) \\ \sigma(S(\alpha, \beta)) & = & S(\sigma(\alpha), \sigma(\beta)) \\ \sigma(U(\alpha, \beta)) & = & U(\sigma(\alpha), \sigma(\beta)). \end{array}$$

The following is the *correspondence lemma.*

Lemma 14.2.12 *The correspondence mapping is a bijection.*

Proof. By two straightforward structural inductions we can prove that σ is both injective and surjective. Details are omitted. ■

As a consequence, we can always refer to an element Q of $\mathcal{L}_{\mathbf{S,U}}$ as $\sigma(\alpha)$, because there is guaranteed to be a unique $\alpha \in \mathcal{L}_{\mathbf{T(L)}}$ such that α is mapped into Q by σ. We can then establish a connection between consistent formulae in $\mathbf{T(L)}/\mathcal{K}$ and in $\mathbf{S,U}/\mathcal{K}$.

Lemma 14.2.13 *If α is $\mathbf{T(L)}$-consistent then $\sigma(\alpha)$ is $\mathbf{S,U}$-consistent.*

Proof. Suppose $\sigma(\alpha)$ is inconsistent. Since all axioms and inference rules in $\mathbf{S,U}/\mathcal{K}$ are also in $\mathbf{T(L)}$ $/\mathcal{K}$, the derivation of $\vdash_{\mathbf{S,U}} \sigma(\alpha) \to \bot$ can be imitated to derive $\vdash_{\mathbf{T(L)}} \alpha \to \bot$, which contradicts α being $\mathbf{T(L)}$-consistent. ■

The results above are very useful for the proof of completeness and decidability of $\mathbf{T(L)}$.

14.2.4 Completeness of T(L)

We are going to show here that whenever there exists complete axiomatization for $\mathbf{S,U}/\mathcal{K}$ and for $\mathbf{L}$, where $\mathcal{K} \subseteq \mathcal{K}_{lin}$ is any linear class of flows of time, then the temporalized logic system $\mathbf{T(L)}/\mathcal{K}$ is also complete.

The strategy of the completeness proof is illustrated in figure 14.4. We prove the completeness of $\mathbf{T(L)}/\mathcal{K}$ indirectly by transforming a consistent formula of $\mathbf{T(L)}$ and then mapping it into a consistent formula of $\mathbf{S,U}$. Completeness of $\mathbf{S,U}/\mathcal{K}$ is used to find a model for the mapped formula that is used to construct a model for the original $\mathbf{T(L)}$ formula.

The transformation function ε is introduced to deal with the differences between deductions in $\mathbf{S,U}$ and $\mathbf{T(L)}$ due to the presence of the inference rule **Preserve** in $\mathbf{T(L)}$. This inference rule states that theorems in $\mathbf{L}$ are also theorems in $\mathbf{T(L)}$. The model theoretic counterpart of this property is that valid formulae in $\mathbf{L}$ are also valid in $\mathbf{T(L)}$. The idea behind the transformation ε is to extract 'valid and contradictory content' that formulae of $\mathbf{T(L)}$ may have due to the validity or unsatisfiability of some set of its subformulae in $\mathbf{L}$.

Definition 14.2.14 (the transformations η and ε) Given a formula $\alpha \in \mathcal{L}_{\mathbf{T(L)}}$, consider the following sets:

$$\begin{aligned} Lit(\alpha) &= Mon(\alpha) \cup \{{\sim}\beta \mid \beta \in Mon(\alpha)\}. \\ Inc(\alpha) &= \{\textstyle\bigwedge \Gamma \mid \Gamma \subseteq Lit(\alpha) \text{ and } \Gamma \vdash_{\mathbf{L}} \bot\}. \end{aligned}$$

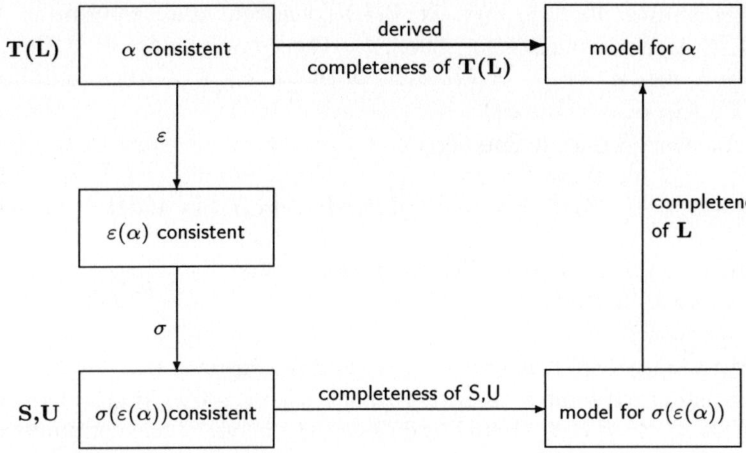

Figure 14.4. Strategy for the proof of completeness.

where $Mon(\alpha)$ is the set of maximal monolithic subformulae of α. We define the operator $\Box$ (always) and the formulae $\eta(\alpha)$ and $\varepsilon(\alpha)$:

$$\begin{aligned} \Box\beta &= \beta \wedge G\beta \wedge H\beta. \\ \eta(\alpha) &= \bigwedge_{\beta\in Inc(\alpha)} \Box{\sim}\beta. \\ \varepsilon(\alpha) &= \alpha \wedge \eta(\alpha). \end{aligned}$$

Since $\eta(\alpha)$ is a theorem of $\mathbf{T(L)}$, we have the following lemma.

Lemma 14.2.15 $\vdash_{\mathbf{T(L)}} \varepsilon(\alpha) \leftrightarrow \alpha$

If $\mathcal{K}$ is a subclass of linear flows of time, we also have the following property:

Lemma 14.2.16 *Let $\mathcal{M}_{\mathbf{S,U}}$ be a temporal model over $\mathcal{K} \subseteq \mathcal{K}_{lin}$ such that for some $o \in T$, $\mathcal{M}_{\mathbf{S,U}}, o \models \sigma(\Box\alpha)$. Then, for every $t \in T$, $\mathcal{M}_{\mathbf{S,U}}, t \models \sigma(\Box\alpha)$.*

Therefore, if some subset of $Lit(\alpha)$ is inconsistent, the transformed formula $\varepsilon(\alpha)$ puts that fact in evidence so that, when σ maps it into $\mathbf{S,U}$, inconsistent subformulae will be mapped into falsity.

To prove the completeness of $\mathbf{T(L)}/\mathcal{K}$ given the completeness of $\mathbf{S,U}/\mathcal{K}$, we fix an α and assume it is a $\mathbf{T(L)}$-consistent formula. We have then to construct a model for α over $\mathcal{K}$.

By lemma 14.2.15, $\varepsilon(\alpha)$ is $\mathbf{T(L)}$-consistent and, by lemma 14.2.13, $\sigma(\varepsilon(\alpha))$ is $\mathbf{S,U}$-consistent. Then, by the completeness of $\mathbf{S,U}/\mathcal{K}$, there exists a model $\mathcal{M}_{\mathbf{S,U}} = (T, <, h)$ with $(T, <) \in \mathcal{K}$ such that for some $o \in T$, $\mathcal{M}_{\mathbf{S,U}}, o \models \sigma(\varepsilon(\alpha))$.

For every $t \in T$, define $G_\alpha(t)$:

$$G_\alpha(t) \quad = \quad \{\beta \in Lit(\alpha) \mid \mathcal{M}_{\mathbf{S,U}}, t \models \sigma(\beta)\}.$$

Lemma 14.2.17 *If α is $\mathbf{T(L)}$-consistent, then for every $t \in T$, $G_\alpha(t)$ is finite and $\mathbf{L}$-consistent.*

Proof. Since $Lit(\alpha)$ is finite, $G_\alpha(t)$ is finite for every t. Suppose $G_\alpha(t)$ is inconsistent for some t, then there exist $\{\beta_1, \ldots, \beta_n\} \subseteq G_\alpha(t)$ such that $\vdash_{\mathbf{L}} \bigwedge \beta_i \to \bot$. So $\bigwedge \beta_i \in Inc(\alpha)$ and $\Box{\sim}(\bigwedge \beta_i)$ is one of the conjuncts of $\varepsilon(\alpha)$. Applying lemma 14.2.16 to $\mathcal{M}_{\mathbf{S,U}}, o \models \sigma(\varepsilon(\alpha))$ we obtain that for every $t \in T$, $\mathcal{M}_{\mathbf{S,U}}, t \models {\sim}(\bigwedge \sigma(\beta_i))$ but by, the definition of G_α, $\mathcal{M}_{\mathbf{S,U}}, t \models \bigwedge \sigma(\beta_i)$, which is a contradiction. ∎

We are finally ready to prove the completeness of $\mathbf{T(L)}/\mathcal{K}$.

Theorem 14.2.18 (completeness for T(L)) *If the logical system $\mathbf{L}$ is complete and $\mathbf{S,U}/\mathcal{K}$ is complete over a subclass of linear flows of time $\mathcal{K} \subseteq \mathcal{K}_{lin}$, then the logical system $\mathbf{T(L)}/\mathcal{K}$ is complete over $\mathcal{K}$.*

Proof. Assume that α is $\mathbf{T(L)}$-consistent. By lemma 14.2.17, we have $(T, <) \in \mathcal{K}$ and associated to every time point in T we have a finite and $\mathbf{L}$-consistent set $G_\alpha(t)$. By (weak) completeness of $\mathbf{L}$, every $G_\alpha(t)$ has a model, so we define the temporalized valuation function g:

$$g(t) \quad = \quad \{\mathcal{M}^t_{\mathbf{L}} \mid \mathcal{M}^t_{\mathbf{L}} \text{ is a model of } G_\alpha(t)\}.$$

Consider the model $\mathcal{M}_{\mathbf{T(L)}} = (T, <, g)$ over $\mathcal{K}$. By structural induction over β, we show that for every β that is a subformula of α and for every time point t,

$$\mathcal{M}_{\mathbf{S,U}}, t \models \sigma(\beta) \text{ iff } \mathcal{M}_{\mathbf{T(L)}}, t \models \beta.$$

We show only the basic case, $\beta \in Mon(\alpha)$. Suppose $\mathcal{M}_{\mathbf{S,U}}, t \models \sigma(\beta)$; then $\beta \in G_\alpha(t)$ and $\mathcal{M}^t_{\mathbf{L}} \models \beta$, and hence $\mathcal{M}_{\mathbf{T(L)}}, t \models \beta$. Suppose $\mathcal{M}_{\mathbf{T(L)}}, t \models \beta$ and assume $\mathcal{M}_{\mathbf{S,U}}, t \models {\sim}\sigma(\beta)$; then ${\sim}\beta \in G_\alpha(t)$ and $\mathcal{M}^t_{\mathbf{L}} \models {\sim}\beta$, which contradicts $\mathcal{M}_{\mathbf{T(L)}}, t \models \beta$; hence $\mathcal{M}_{\mathbf{S,U}}, t \models \sigma(\beta)$. The inductive cases are straightforward and details are omitted.

So, $\mathcal{M}_{\mathbf{T(L)}}$ is a model for α over $\mathcal{K}$ and the proof is finished. ∎

Theorem 14.2.18 gives us sound and complete axiomatizations for **T(L)** over many interesting classes of flows of time, such as the class of all linear flows of time, $\mathcal{K}_{lin}$, the integers, $\mathbb{Z}$, and the reals, $\mathbb{R}$. These classes are, in their **S,U** versions, decidable and the corresponding decidability of **T(L)** is dealt with in section 14.3. Integer and real flows of time also have the separation property, which is discussed in section 14.5.

14.3 The decidability of T(L) and its complexity

The main goal of this section is to show that, if the logic system **L** is decidable and the logic system **S,U** is decidable over $\mathcal{K} \subseteq \mathcal{K}_{lin}$, then the logic system **T(L)** is also decidable over $\mathcal{K}$. We assume throughout this section that **S,U**$/\mathcal{K}$ is complete.

Definition 14.3.1 (decidability of a logic system) A logic system **L** is said to be *decidable* if there exists an algorithm (a decision procedure) that, for every formula $\alpha \in \mathcal{L}_{\mathbf{L}}$, outputs 'yes' if α is a theorem in the logic system **L** and 'no' otherwise.

There are several results of decidability of **S,U** over several linear classes of flows of time, among which are the class $\mathcal{K}_{lin}$ of all linear flows of time [Burgess, 1984a], the integer, and the real flows of time [Burgess and Gurevich, 1985].

As in the proof of completeness, we are going to prove the decidability result using the correspondence mapping σ and the transformation η. Recall definition 14.2.4, in which the sets $Mon(\alpha)$, $Lit(\alpha)$ and $Inc(\alpha)$ were all finite, so that we have the following result about $\eta(\alpha)$.

Lemma 14.3.2 *For any $\alpha \in \mathcal{L}_{\mathbf{T(L)}}$, if the logic system* **L** *is decidable then there exists an algorithm for constructing $\eta(\alpha)$.*

The relationship between **T(L)** and **S,U** that we need to prove the decidability of **T(L)** is the following:

Lemma 14.3.3 *Over a linear flow of time, for every $\alpha \in \mathcal{L}_{\mathbf{T(L)}}$,*

$$\vdash_{\mathbf{T(L)}} \alpha \text{ iff } \vdash_{\mathbf{S,U}} \sigma(\eta(\alpha) \to \alpha).$$

Proof. The *if* case comes trivially from the definition of $\vdash_{\mathbf{T(L)}}$. For the *only if* part, suppose $\vdash_{\mathbf{T(L)}} \alpha$. We prove by induction on the deduction of α that $\vdash_{\mathbf{S,U}} \sigma(\eta(\alpha) \to \alpha)$.

Basic cases:

1. α is obtained using the inference rule **Preserve**. Then $\eta(\alpha) = \sim\sim\alpha$ and $\vdash_{\mathbf{S,U}} \sigma(\sim\sim\alpha \to \alpha)$.

2. α is obtained using the inference rule **Subst**. Suppose α was obtained by substituting p_i by β_i in some axiom **A**. Then $\vdash_{\mathbf{S,U}} \alpha$ can be obtained by substituting $\sigma(p_i)$ by $\sigma(\beta_i)$ in axiom **A**.

Inductive cases:

1. $\alpha = G\beta$ is obtained using the inference rule **TG**. Note that $\eta(\alpha) = \eta(\beta)$. Then

$\vdash_{\mathbf{S,U}} \sigma(\eta(\alpha)) \to \sigma(\beta)$	by induction hypothesis
$\vdash_{\mathbf{S,U}} G(\sigma(\eta(\alpha)) \to \sigma(\beta))$	by **T(G)**
$\vdash_{\mathbf{S,U}} G(\sigma(\eta(\alpha))) \to \sigma(\alpha)$	by temporal logic and $\alpha = G\beta$
$\vdash_{\mathbf{S,U}} \sigma(\eta(\alpha)) \to G(\sigma(\eta(\alpha)))$	by the definition of η and $\mathcal{K}$ linear
$\vdash_{\mathbf{S,U}} \sigma(\eta(\alpha) \to \alpha)$	from the two preceding lines

Similarly for $\alpha = H\beta$.

2. α is obtained from β and $\beta \to \alpha$ by **MP**. Then

$\vdash_{\mathbf{S,U}} \sigma(\eta(\beta)) \to \sigma(\beta)$	by induction hypothesis
$\vdash_{\mathbf{S,U}} \sigma(\eta(\beta \to \alpha)) \to \sigma(\beta \to \alpha)$	by induction hypothesis
$\vdash_{\mathbf{S,U}} \sigma(\eta(\beta \to \alpha)) \to \sigma(\eta(\beta))$	by the definition of η
$\vdash_{\mathbf{S,U}} \sigma(\eta(\beta \to \alpha)) \to \sigma(\beta)$	from the 3rd and 1st lines
$\vdash_{\mathbf{S,U}} \sigma(\eta(\beta \to \alpha)) \to \sigma(\alpha)$	from the 4th and 2nd lines

Let p be a proposition that occurs in $\sigma(\beta)$ but not in $\sigma(\alpha)$. If we eliminate from $\sigma(\eta(\alpha \to \beta))$ all the conjuncts in which p occurs, obtaining $\sigma(\gamma)$, using the completeness of $\mathbf{S,U}/\mathcal{K}$ we can obtain $\vdash_{\mathbf{S,U}} \sigma(\gamma) \to \sigma(\alpha)$. If we do that for all such propositions, we end up with $\vdash_{\mathbf{S,U}} \sigma(\eta(\alpha) \to \alpha)$.

■

Theorem 14.3.4 (decidability of T(L)) *If* **L** *is a decidable logic system, and* **S,U** *is decidable over* $\mathcal{K} \subseteq \mathcal{K}_{lin}$, *then the logic system* **T(L)** *is also decidable over* $\mathcal{K}$.

Proof. Consider $\alpha \in \mathcal{L}_{\mathbf{T(L)}}$. Since **L** is decidable, by lemma 14.3.2 there is an algorithmic procedure to build $\eta(\alpha)$. Since σ is a recursive function, we have an algorithm to construct $\sigma(\eta(\alpha) \to \alpha)$, and, due to the decidability of **S,U** over $\mathcal{K}$, we have an effective procedure to decide if it is a theorem or not. Since $\mathcal{K}$ is linear, by lemma 14.3.3 this is also a procedure for deciding whether α is a theorem or not. ■

Once we have a decidability result, the next natural question is about the complexity of the decision procedure. We briefly discuss here an upper bound for the complexity analysis. Let N be the number of (boolean and modal) connectives in a formula, let the complexity of the decision procedure in **L** be $\mathcal{O}(f_{\mathbf{L}}(N))$ and in **S,U** be $\mathcal{O}(f_{\mathbf{S,U}}(N))$. The decision procedure for **T(L)** as given by the proof above basically consists of two steps:

1. constructing $\eta(\alpha)$;
2. deciding whether $\sigma(\eta(\alpha) \to \alpha)$ is a theorem or not.

The construction of $\eta(\alpha)$ involves generating all subsets of $Lit(\alpha)$ and applying the decision procedure for each subset; therefore its complexity is $\mathcal{O}(2^N \times f_{\mathbf{L}}(N))$. The second step is dominated by the decision procedure of **S,U** since the application of σ can be done in polynomial time; in the worst case, when all tests in **L** succeed, the size of $\eta(\alpha)$ is $\mathcal{O}(2^N)$, and therefore the decision procedure is $\mathcal{O}(f_{\mathbf{S,U}}(2^N))$. So an upper bound for the decision procedure for **T(L)** is given by the dominating term of $\mathcal{O}(2^N \times f_{\mathbf{L}}(N))$ and $\mathcal{O}(f_{\mathbf{S,U}}(2^N))$. As for a lower bound for the decision procedure of **T(L)**, it cannot be any lower than the highest of the lower bounds for **S,U** and **L**.

14.4 Conservativeness of T(L)

Conservativeness can easily be derived from the soundness of **S,U** and the completeness of **L**, without any assumptions on the flow of time.

Definition 14.4.1 (conservative extension) A logic system $\mathbf{L_1}$ is an *extension* of a logic system $\mathbf{L_2}$ if $\mathcal{L}_{\mathbf{L_2}} \subseteq \mathcal{L}_{\mathbf{L_1}}$ and if $\vdash_{\mathbf{L_2}} \alpha$ then $\vdash_{\mathbf{L_1}} \alpha$. A logic $\mathbf{L_1}$ is a *conservative extension* of $\mathbf{L_2}$ if it is an extension of $\mathbf{L_2}$ such that if $\alpha \in \mathcal{L}_{\mathbf{L_2}}$, then $\vdash_{\mathbf{L_1}} \alpha$ only if $\vdash_{\mathbf{L_2}} \alpha$.

We know that all complete **S,U** are conservative extensions of predicate logic **PL**. Clearly, **T(L)** is an extension of **L**. We prove that it is also conservative.

Theorem 14.4.2 (conservativeness of T(L)) *Let* **L** *be a complete logic system and* **S,U** *be sound over* $\mathcal{K}$. *The logic system* **T(L)** *is a conservative extension of* **L**.

Proof. Let $\alpha \in \mathcal{L}_{\mathbf{L}}$ such that $\vdash_{\mathbf{T(L)}} \alpha$. Suppose by contradiction that $\not\vdash_{\mathbf{L}} \alpha$, so by completeness of **L**, there exists a model $\mathcal{M}_{\mathbf{L}}$ such that $\mathcal{M}_{\mathbf{L}} \models {\sim}\alpha$. We construct a temporalized model $\mathcal{M}_{\mathbf{T(L)}} = (T, <, g)$ by making $g(t) = \mathcal{M}_{\mathbf{L}}$ for all $t \in T$. $\mathcal{M}_{\mathbf{T(L)}}$ clearly contradicts the soundness of **T(L)** and therefore that of **S,U**, so $\vdash_{\mathbf{L}} \alpha$. ■

14.5 Separation over the added dimension

The separation property of the S,U-temporal logic allows us to rewrite any temporal formula into a conjunction of formulae of the form

past formula and present formula $\rightarrow$ future formula.

Once a formula is in the format above, it can be interpreted imperatively against a partial temporal model according to [Gabbay, 1989a], so that if the antecedent holds in the past and present in the model, then we must execute the consequent in the future so as to make the formula true in the model. The imperative interpretation of a formula (also called the execution of a temporal specification) is based on an asymmetric view of the flow of time; in a symmetric view of time, whenever the antecedent is true in the past and present, we could either make the consequent true in the future or we could try to falsify the antecedent itself, in both cases maintaining the validity of the temporal specification. In this asymmetric view of time, we discard the latter possibility and remain with the former as the only possibility for the execution of a temporal specification.

In this section we want to extend this imperative interpretation of a temporal formula over a logic system **L** so that, after temporalizing **L** over a flow of time that is like the integers or reals, we can execute temporal specifications in **T(L)**. The concept of a separated formula is based on the notion of a pure formula, so we present the definitions of pure formula and separated formula for the **S,U** logic.

Definition 14.5.1 (pure formulae in S,U)

1. A *pure present formula* is a boolean combination of propositional letters.
2. A *pure past formula* is a boolean combination of formulae of the form $S(\alpha, \beta)$ where α and β are either pure present or pure past formulae.
3. A *pure future formula* is a boolean combination of formulae of the form $U(\alpha, \beta)$ where α and β are either pure present or pure future formulae.

A *separated formula* is a formula that is a boolean combination of pure formulae only.

Once we have a separated formula, it can be brought to a conjunctive normal form, i.e. a conjunction of disjuncts, so that each conjunct can be finally brought to the form:

pure-present and pure-past $\rightarrow$ pure-future.

The following is the basic result about separation over the integers.

Theorem 14.5.2 (separation theorem) *For any formula $A \in \mathcal{L}_{\mathbf{S},\mathbf{U}}$ there exists a separated formula $B \in \mathcal{L}_{\mathbf{S},\mathbf{U}}$ such that A is equivalent to B over an integer-like flow of time.*

A proof of the separation theorem can be found in [Gabbay, 1989a] or chapter 9. It also holds for the reals.

The generalization of pure formula for a temporalized logic system $\mathbf{T}(\mathbf{L})$ is given below.

Definition 14.5.3 (pure temporalized formulae)

1. every formula $\alpha \in \mathcal{L}_{\mathbf{L}}$ is a *pure present temporalized formula.*
2. A *pure past temporalized formula* is a boolean combination of formulae of the form $S(\alpha, \beta)$ where α and β are either pure present or pure past temporalized formulae.
3. A *pure future temporalized formula* is a boolean combination of formulae of the form $U(\alpha, \beta)$ where α and β are either pure present or pure future temporalized formulae.

A *separated temporalized formula* is a boolean combination of pure formulae of $\mathbf{T}(\mathbf{L})$.

Example 14.5.4 (temporalizing a modal logic of belief) Suppose $\mathbf{L}$ is the modal logic system of belief, with the modal operator B. Here are some examples of pure temporalized formulae in $\mathbf{T}(\mathbf{L})$:

1. Pure present: $Bp \to p$, $\sim(p \wedge \sim p)$, and any other formula of the logic $\mathbf{L}$.
2. Pure past: $P(Bp) \to S(Bp, \sim p)$.
3. Pure future: $F(Bp) \to \sim Fp \vee G(Bp \to \sim p)$.

In order to prove the separation theorem for the temporalized logic $\mathbf{T}(\mathbf{L})$ we will use the correspondence mapping. The basic strategy of the proof is illustrated in figure 14.5.

The following is a helpful result that will lead us to the proof of separation for the temporalized logic $\mathbf{T}(\mathbf{L})$.

Lemma 14.5.5 *Let σ be a correspondence mapping between $\mathcal{L}_{\mathbf{T}(\mathbf{L})}$ and $\mathcal{L}_{\mathbf{S},\mathbf{U}}$. $\sigma(\alpha)$ is a separated formula in the logic* **S,U** *iff α is a separated formula in $\mathbf{T}(\mathbf{L})$.*

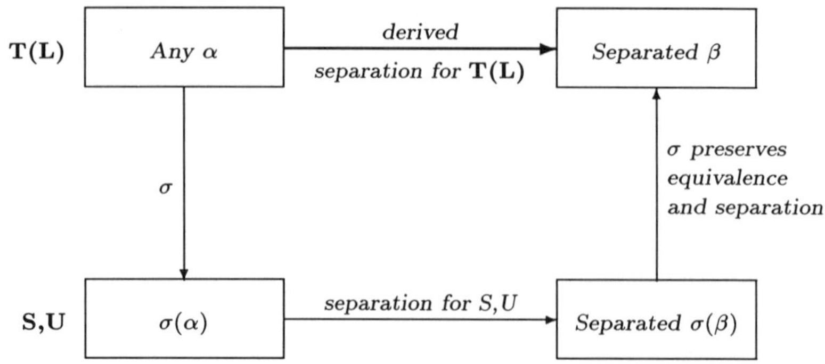

Figure 14.5. Separation of **T(L)**-formulae via separation of **S,U**-formulae.

Proof. From the definition of the correspondence mapping it follows that if α is a boolean combination of $\alpha_1, \ldots, \alpha_n \in \mathcal{L}_{\mathbf{T(L)}}$ then $\sigma(\alpha)$ is a boolean combination of $\sigma(\alpha_1), \ldots, \sigma(\alpha) \in \mathcal{L}_{\mathbf{S,U}}$. The converse is also true since σ is a bijection.

Therefore, to show that α is separated in **T(L)** iff $\sigma(\alpha)$ is separated in **S,U**, all we have to do is to prove that $\sigma(\alpha)$ is a pure formula iff α is a pure formula. We show the proof for the *only if* case; the *if* part is completely analogous.

Suppose $\sigma(\alpha)$ is a pure present, then it is a boolean combination of propositional letters. Therefore α is a boolean combination of monolithic formulae of **L**, therefore α is a formula of **L**, and pure present in **T(L)**.

Suppose $\sigma(\alpha)$ is pure past, then it is a boolean combination of formulae in $\mathcal{L}_{\mathbf{S,U}}$ of the form $S(\sigma(\beta), \sigma(\gamma))$ where $\sigma(\beta)$ and $\sigma(\gamma)$ are pure present or pure past. Therefore, α must be a boolean combination of formulae in $\mathcal{L}_{\mathbf{T(L)}}$ of the form $S(\gamma, \delta)$, where γ and δ are, by induction hypothesis, either pure present or pure past. Therefore, α is a pure past formula in $\mathcal{L}_{\mathbf{T(L)}}$.

Suppose $\sigma(\alpha)$ is pure future, then by an argument analogous to the preceding case, α is a pure future formula. Therefore, we have proved that if $\sigma(\alpha)$ is a pure formula in $\mathcal{L}_{\mathbf{S,U}}$, α is a pure formula in $\mathcal{L}_{\mathbf{T(L)}}$. ■

Theorem 14.5.6 (separation theorem for T(L)) *If α is any formula in $\mathcal{L}_{\mathbf{T(L)}}$, then there exists a separated formula $\beta \in \mathcal{L}_{\mathbf{T(L)}}$ such that β is equivalent to α over an integer-like flow of time.*

Proof. All we have to do is to prove that if α and β are formulae of **T(L)** and $\vdash_{\mathbf{S,U}} \sigma(\alpha) \leftrightarrow \sigma(\beta)$, then $\vdash_{\mathbf{T(L)}} \alpha \leftrightarrow \beta$. In fact, since all axioms and inference rules of **S,U** also belong to **T(L)**, the deduction of $\vdash_{\mathbf{S,U}} \sigma(\alpha) \leftrightarrow \sigma(\beta)$ also leads to $\vdash_{\mathbf{T(L)}} \alpha \leftrightarrow \beta$.

Thus let α be any formula of **T(L)**. From the separation theorem of **S,U**, we know that there exists a separated β, such that $\vdash_{\mathbf{S,U}} \sigma(\alpha) \leftrightarrow \sigma(\beta)$ and $\sigma(\beta)$ is separated. So by lemma 14.5.5, β is also a separated formula equivalent to α. ■

Once we have the separation property for the temporalized system **T(L)**, we can rewrite any temporalized formula into a separated equivalent one of the form

pure temporalized past and present $\to$ pure temporalized future.

The imperative interpretation of such a formula is the following. If the antecedent holds in past and present models of the logic system **L**, then we execute the temporalized formula by constructing a future model (or a series of future models) of **L** so as to make the consequent true.

Since the separation property also holds for a real flow of time, the proof above can be trivially adapted to a real flow of time. Note that the separation property for the temporalized system was obtained without any assumptions on the underlying logic system **L**, as opposed to the results of soundness, completeness, and decidability, all of which depend on whether the property holds for the underlying logic system **L**.

14.6 Temporalizing first-order logic

In this section we examine in more detail the addition of a temporal dimension to a first-order language. We will consider a first-order language with the quantifier $\forall$, an equality symbol $=$, a countable set of variables $X = \{x_1, x_2, \ldots\}$, a countable set of predicate symbols $P = \{p_1, p_2, \ldots\}$ such that every predicate symbol has an associated natural number $n > 0$, called its arity, a set C of constant symbols, and a set F of functional symbols; C and F are possibly empty. The quantifier $\exists$ can be defined in the normal way as $\exists = {\sim}\forall{\sim}$. A term is either a variable, a constant symbol, or an n-ary function symbol applied to n terms. The notion of the set of free variables of a formula is the usual one. A sentence is a formula with no free variables.

A first-order domain $\mathcal{D}$ is a non-empty set. An interpretation $\mathcal{I}$ is a mapping that associates, for every constant in the language an element in the domain, and for every n-place predicate symbol an n-ary relation over $\mathcal{D}^n$. An assignment function $\mathcal{A}$ is a mapping that associates every variable with an element of the domain. A first-order model is a pair $\mathcal{M} = (\mathcal{D}, \mathcal{I})$.

If t is a term, $[\![t]\!]^{\mathcal{I},\mathcal{A}} \in \mathcal{D}$ represents its extension over the domain $\mathcal{D}$ under interpretation $\mathcal{I}$ and assignment $\mathcal{A}$. The semantics of a first-order language is then defined in the usual way, where $\mathcal{M}, \mathcal{A} \models_{\text{FOL}} \alpha$ reads '$\mathcal{M}$ is a model of the formula α under assignment $\mathcal{A}$':

$$
\begin{array}{lll}
\mathcal{M}, \mathcal{A} \models_{\text{FOL}} p_i(t_1, \ldots, t_n) & \text{iff} & \langle [\![t_1]\!]^{\mathcal{I},\mathcal{A}}, \ldots, [\![t_n]\!]^{\mathcal{I},\mathcal{A}} \rangle \in \mathcal{I}(p_i), \text{ for all } n\text{-ary predicate symbols } p_i \in P. \\
\mathcal{M}, \mathcal{A} \models_{\text{FOL}} \sim\alpha & \text{iff} & \mathcal{M}, \mathcal{A} \not\models_{\text{FOL}} \alpha. \\
\mathcal{M}, \mathcal{A} \models_{\text{FOL}} \alpha \wedge \beta & \text{iff} & \mathcal{M}, \mathcal{A} \models_{\text{FOL}} \alpha \text{ and } \mathcal{M}, \mathcal{A} \models_{\text{FOL}} \beta. \\
\mathcal{M}, \mathcal{A} \models_{\text{FOL}} t_1 = t_2 & \text{iff} & [\![t_1]\!]^{\mathcal{I},\mathcal{A}} = [\![t_2]\!]^{\mathcal{I},\mathcal{A}}. \\
\mathcal{M}, \mathcal{A} \models_{\text{FOL}} \forall x\alpha & \text{iff} & \text{for any assignment } \mathcal{A}' \text{ which agrees with } \mathcal{A}, \text{ except possibly on variable } x,\ \mathcal{M}, \mathcal{A}' \models_{\text{FOL}} \alpha.
\end{array}
$$

We say that α has a model $\mathcal{M}$, and write $\mathcal{M} \models_{\text{FOL}} \alpha$, if $\mathcal{M}, \mathcal{A} \models_{\text{FOL}} \alpha$ for all assignments $\mathcal{A}$ (this is always the case when α is a sentence).

The derivability relation, $\vdash_{\text{FOL}}$, can be any of the existing ones for first-order logic. It can be an axiomatic system, but it need not be.

Since in first-order logic we have a basic distinction between sentences and ordinary formulae, we have to consider both cases of adding a temporal dimension to monolithic sentences and to monolithic formulae in general.

14.6.1 Temporalizing first-order sentences

If we temporalize first-order sentences, we have no problems in following the methodology we have developed so far. We first identify the monolithic sentences as those that are not in the format $\alpha \wedge \beta$ or $\sim\alpha$. For instance, $\forall x p(x)$ and $\forall x {\sim}(q(x) \wedge {\sim}q(x))$ are monolithic sentences, whereas $\exists x p(x)$ (implicit negation) and $\forall x p(x) \wedge \forall y {\sim} q(y)$ are boolean combinations. We then follow the procedure described in section 14.2, obtaining the logic system $\mathbf{T(FOs)}$. Note that in $\mathbf{T(FOs)}$ a temporal operator never occurs inside the scope of a quantifier.

The structure of the first-order models that compose the temporalized model deserves some special attention, since one model may differ form another in several different ways, as if we had various 'degrees of freedom' in generating a temporalized version of first-order models. Those degrees of freedom are illustrated in Table 14.6.1.

If all first-order models that compose a temporalized model $\mathcal{M}_{\mathbf{T(FOs)}}$ refer to the same domain, a constant domain assumption is satisfied; otherwise, we have varying domains. We may have rigid constant and rigid functional symbols, i.e. they have the same interpretation in every model of the temporalized structure; they are called non-rigid or flexible otherwise.

Table 14.1. Degrees of freedom in temporalizing first-order models

Element	*Fixed*	*Variable*
Domain	constant domains	variable domains
Constant and functional symbols	rigid	non-rigid or flexible
Predicate symbols	rigid	non-rigid or flexible
Assignment	global	local

A rigid predicate symbol has the same interpretation at all times; otherwise it is a flexible predicate symbol. And, finally, the assignment function may be global, i.e. all variables are assigned the same domain element in all models of the temporalized structure (global assignments make sense only under a constant domain assumption); otherwise, it is a local assignment.

In fact, constant domains or rigid terms or predicates are not a consequence of the temporalization; they are, actually, further assumptions on the temporalized first-order model made so as to impose some external intended meaning of adding a temporal dimension to a logic system. All the previously established results of soundness, completeness, and separation are valid for unconstrained **T(FOs)**; decidability is obviously not applicable.

Nevertheless, there is no quantification over the temporal operators in **T(FOs)**, which means that the expressivity of this logic is clearly limited. In the following, we examine one step further in increasing this expressivity, while still keeping the original idea of adding a temporal dimension to a logic system.

14.6.2 Temporalizing first-order formulae

We now take general monolithic first-order formulae as a basis for the addition of a temporal dimension, i.e. all first-order formulae that are not of the form $\sim\alpha$ or $\alpha \wedge \beta$. We thus generate the logic system **T(FOf)**. Note that the language of **T(FOs)** is contained in the language of **T(FOf)**.

The particular feature that distinguishes this system from all the previously considered systems is that, since we are considering first-order formulae that may contain free variables, monolithic formulae with free variables only have a defined semantics over a first-order model $\mathcal{M}_{\mathbf{FOL}}$ if a variable assignment function is provided, and the free variables of a first-order formula used to build a temporalized formula α remain free in α.

Therefore, while constructing a model for the system $\mathbf{T(FOf)}$, we must consider the existence of a *global assignment function*, $\mathcal{A}_g$, to cope with the free variables. A global assignment function makes sense only in a constant domain context, so we must have this assumption as well; we further assume that all terms are rigid. The effect of the global assignment $\mathcal{A}_g$ is to ground all the free variables of a temporalized formula α. Only the interpretation of predicate symbols changes among the models of $\mathbf{L}$ in the temporalized model structure. We write

$$\mathcal{M}_{\mathbf{T(FOf)}} \models \alpha \text{ iff } \mathcal{M}_{\mathbf{T(FOf)}}, \mathcal{A}_g \models \alpha \text{ for any } \mathcal{A}_g.$$

Since the construction of its temporalized model and inference system does not follow exactly the way other temporal systems were constructed, the results previously established of soundness, completeness, and separation cannot be applied directly.

We know that the more expressive full first-order temporal logic has no possible finite axiomatization over several useful classes of linear flows of time such as $\{\mathbb{R}\}$, $\{\mathbb{Z}\}$ and $\{\mathbb{N}\}$, e.g. see [Garson, 1984], but we do have a finite axiomatization for $\mathbf{T(FOs)}$. The logic system $\mathbf{T(FOf)}$ has an intermediary expressive power and it can be shown that $\mathbf{T(FOf)}$ cannot be finitely axiomatized over linear flows of time that contain the natural numbers, although we will not do it here. Perhaps more interesting is that separation can be achieved for this logic through model theory.

Since the concept of separated formula is purely syntactic and does not depend on the model or the inference system, the definition of a separated temporalized formulae given by definition 14.5.3 is also valid for $\mathbf{T(FOf)}$. For the same reasons, the definition of a correspondence mapping σ and the correspondence lemma 14.2.12 stating that σ is a bijection are also valid in $\mathbf{T(FOf)}$.

Definition 14.6.1 (corresponding model) Let $\mathcal{M}_{\mathbf{T(FOf)}} = (T, <, g)$ be a model of $\mathbf{T(FOf)}$, and let $\mathcal{A}$ be a global assignment. We construct the valuation function g_σ such that, for every time point $t \in T$ and for every propositional letter $p = \sigma(\alpha) \in \mathcal{P}$ we have

$$\sigma(\alpha) \in g_\sigma(t) \text{ iff } \mathcal{M}_{\mathbf{T(FOf)}}, \mathcal{A}, t \models \alpha.$$

A model of the temporal logic system **S,U**, $\mathcal{M}^{\sigma}_{\mathbf{S,U}} = (T, <, g_\sigma)$, is then called the *corresponding model* of $\mathcal{M}_{\mathbf{T(L)}}$ under the corresponding mapping σ and assignment $\mathcal{A}$.

Lemma 14.6.2 *If $\mathcal{M}^{\sigma}_{\mathbf{S,U}}$ is the corresponding model of $\mathcal{M}_{\mathbf{T(FOf)}}$ under σ and $\mathcal{A}$ then*

$$\mathcal{M}^{\sigma}_{\mathbf{S,U}}, t \models \sigma(\alpha) \textit{ iff } \mathcal{M}_{\mathbf{T(FOf)}}, \mathcal{A}, t \models \alpha$$

for every $\alpha \in \mathcal{L}_{\mathbf{T(L)}}$ and for every $t \in T$.

Proof. Straightforward by structural induction on α. ■

Theorem 14.6.3 (separation for T(FOf)) *For every $\alpha \in \mathcal{L}_{\mathbf{T(FOf)}}$ there exists a separated formula $\beta \in \mathcal{L}_{\mathbf{T(FOf)}}$ such that β is equivalent to α over an integer-like flow of time.*

Proof. Let σ be a correspondence mapping and $\mathcal{A}$ an arbitrary global assignment. Consider a temporalized model $\mathcal{M}_{\mathbf{T(FOf)}} = (T, <, h)$, $(T, <) \in \mathbb{Z}$, and let $\mathcal{M}^{\sigma}_{\mathbf{S,U}} = (T, <, g_\sigma)$ be its correspondent model under σ and A. By lemma 14.6.2, we have

$$\mathcal{M}^{\sigma}_{\mathbf{S,U}}, t \models \sigma(\alpha) \text{ iff } \mathcal{M}_{\mathbf{T(FOf)}}, \mathcal{A}, t \models \alpha \tag{14.1}$$

for every $\alpha \in \mathcal{L}_{\mathbf{T(FOf)}}$ and for every $t \in T$.

By the separation theorem for **S,U** we obtain that, for every formula $\sigma(\alpha) \in \mathcal{L}_{\mathbf{S,U}}$ there exists a separated formula $\sigma(\beta) \in \mathcal{L}_{\mathbf{S,U}}$ such that

$$\mathcal{M}^{\sigma}_{\mathbf{S,U}}, t \models \sigma(\alpha) \text{ iff } \mathcal{M}^{\sigma}_{\mathbf{S,U}}, t \models \sigma(\beta) \tag{14.2}$$

for all time points $t \in T$.

By lemma 14.5.5, we have that the corresponding mapping preserves separation, i.e. β is a separated formula iff $\sigma(\beta)$ is a separated formula and, by application of (14.1):

$$\mathcal{M}^{\sigma}_{\mathbf{S,U}}, t \models \sigma(\beta) \text{ iff } \mathcal{M}_{\mathbf{T(FOf)}}, \mathcal{A}, t \models \beta \tag{14.3}$$

for all time points $t \in T$.

Combining (14.1), (14.2), and (14.3) we obtain that, for every $\alpha \in \mathcal{L}_{\mathbf{T(FOf)}}$ there exists a separated $\beta \in \mathcal{L}_{\mathbf{T(FOf)}}$ such that, for all $t \in T$,

$$\mathcal{M}_{\mathbf{T(FOf)}}, \mathcal{A}, t \models \alpha \text{ iff } \mathcal{M}_{\mathbf{T(FOf)}}, \mathcal{A}, t \models \beta. \tag{14.4}$$

Since the assignment $\mathcal{A}$ was arbitrarily chosen and the separated β does not depend on the particular choice of $\mathcal{A}$, expression (14.4) holds for any global assignment $\mathcal{A}$, and separation for **T(FOf)** remains proved. ■

We note that if we fix a current time, o, and a global assignment $\mathcal{A}_g$, we can apply the temporalization process to the logic system $\mathbf{T(FOf)}$, obtaining a two-dimensional temporal predicate system, $\mathbf{T^2(FOf)}$, as a predicate version of the two-dimensional propositional system described in example 14.2.2.

14.7 Internalizing the temporal dimension

There are three basic approaches to adding a temporal dimension to a logic system, namely:

1. The temporal operators approach.
2. The first-order internalization of the temporal dimension.
3. A mixed approach combining the two approaches above.

Those three different approaches are discussed in detail in [Gabbay, 1989b] in the context of propositional temporal logic. The first approach is the one we have been following so far. Here we briefly present the others in the context of temporalized formulae.

Consider the temporalized first-order formula in $\mathbf{T(FOf)}$

$$believed(x) \to F\ happens(x),$$

expressing that whatever is believed now will become true in the future. This statement could actually be completely coded in the original first-order language by adding a temporal argument to the predicates *believed* and *happens*. The resulting formulation would be

$$believed^*(t,x) \to \exists s(t < s \land happen^*(s,x)).$$

This process of getting rid of the temporal operators by adding a new temporal argument to the predicates plus some extra conditions on those arguments can be done systematically by an *internalization function* $*$ defined inductively over the structure of a formula of $\mathbf{T(FOf)}$ and also taking as argument a reference time point, generating a two-sorted predicate formula, one sort over time and the other sort over domain elements. We call this process the *internalization* of the temporal dimension. The internalization of the temporal dimension is basically obtained by the standard translation of temporal logic into predicate logic, e.g. [van Bentham, 1983], with an extra argument to incorporate the temporal reference; this extra argument can be interpreted as the result of Quine's 'eternalization' of first-order sentences [Quine, 1960].

In the internalized version it is necessary to incorporate a theory expressing the properties of the flow of time $\mathcal{K} = (T, <)$ to restore the deductive capability of temporal formulae. However, there are several flows of time over which there are complete temporal axiomatizations that are not definable in first-order logic, e.g. the integers and the reals.

Another way of getting to a first-order predicate logic approach to temporal logic, as proposed by Gabbay [Gabbay, 1991a], is by mixing two predicate logic languages in the following way. Let **G** (for global) and **L** (for local) be two first-order languages. The two-sorted predicate language $\mathbf{L}_k^*(\mathbf{G})$ is the result of mixing the **G** and **L** (in our present notation it would be $\mathbf{G}(\mathbf{L}_k^*)$). If we consider the language $\mathbf{L}_1^*(\mathbf{G})$, then a formula of the form $P^*(t, x_1, \ldots, x_n)$ means that $P(x_1, \ldots, x_n)$ holds at time t. This language is the same language as that of the internalized temporal dimension system. But this approach gives us a way of creating an internalized logic system in a very similar way to that in which a temporalized system was created, i.e. as a result of putting two languages together. In fact, the original languages **G** and **L** can be seen as two linked languages 'sharing variables' in the language $\mathbf{L}_1^*(\mathbf{G})$. One of the original languages, **G**, has the exclusivity of dealing with temporal facts, as the upper-level S, U-temporal logic system, whereas the language **L** is responsible only for the local behaviour at each point in the flow of time.

The temporal operators' approach to a temporalized formula can be seen as treating time points implicitly, always referring to a current time. The first-order internalization refers explicitly to the points in the flow of time. A hybrid form of internalization of the temporal dimension can be obtained by combining temporal operators with first-order internalized formulae, mixing the explicit reference with the implicit reference of time.

In the combined approach, every temporalized formula α is associated with a first-order atomic formula $holds(t, \alpha)$, where α is now treated as a first-order term, and the free variables of α are considered free in $holds(t, \alpha)$. A set of axioms is added to combine the $holds(t, \alpha)$ formulae with the first-order internalized formulae, for example:

$$\begin{array}{lcl} holds(t, \alpha) & \leftrightarrow & (\alpha)^*[t], \text{ for all monolithic } \alpha \in \mathcal{L}_{\mathbf{L}} \\ holds(t, \alpha \wedge \beta) & \leftrightarrow & holds(t, \alpha) \wedge holds(t, \beta) \\ holds(t, \alpha \wedge \beta) & \leftrightarrow & \exists s[s < t \wedge holds(s, \alpha) \\ & & \wedge \forall u(s < u < t \to holds(u, \beta))], \end{array}$$

etc.

As in the internalized approach, in the combined approach we still have to provide axioms for the flow of time.

14.8 Generalizations

We have shown a way of composing an upper-level temporal logic system with a generic underlying logic system **L** and the resulting logic system **T**(**L**) was called the temporalization of system **L**. We used the correspondence mapping method to prove soundness, completeness, decidability, conservativeness, and separation for the temporalized logic system over linear flows of time. All those properties were initially properties of the temporal logic system. Many other properties remain to be analysed, such as compactness, finite model property, and interpolation, among others; the properties discussed here over classes of linear flows of time remain to be expanded for all classes of flows of time.

We need by no means restrict the upper-level logic system to temporal logic. In fact, the temporalization presented here can be generalized to any propositional modal logic system **M** in the role of the upper-level logic system, so as to create a modalized logic system **M**(**L**). Its language and inference system can be obtained following the method we used to derive those of **T**(**L**), based on the monolithic formulae of **L**. If the logic **L** has a possible world semantics, each possible world may be substituted by a model of **L**, so as to construct a model for the system **M**(**L**) in the same way that a model was built for **T**(**L**). The correspondence mapping method may then be used to study how the properties of the modal logic system **M** are preserved in the modalized logic system **M**(**L**).

Chapter 15

Decidability in temporal logic

15.1 Introduction

This chapter is not intended as a survey of the huge field of decidability. There are many fine surveys, such as [Rabin, 1977] and [Gurevich, 1985], in the literature already. Rather, it is a discussion of the decidability of the temporal systems used in the rest of the book. In section 15.7 we will list these systems, and consider the decidability of each one. Most will turn out to be decidable, the main exceptions being the predicate systems of chapter 4 and the second-order systems of chapter 8. Some indications of the complexity of decision procedures will be given, but for a fuller treatment the reader is referred to the extensive literature on complexity, and to the second volume of this work.

In the early part of this chapter we will amass some techniques for determining decidability. Specifically, we will consider the decidability, with respect to various languages, of the following classes of flows of time:

- the class of all flows of time;
- the class of all linear flows;
- the class of all finite linear flows;
- the class of all Dedekind complete flows;
- the class of all continuous flows (Dedekind complete, dense, and without endpoints);

- the class of all discrete flows (for every time point with a future, there is a next point (a 'tomorrow'), and similarly in the past);
- the class of all well-ordered flows (there are no infinite sequences leading into the past);
- the class of all circles (see section 13.7);
- the classes $\{\mathbb{N}\}, \{\mathbb{Z}\}, \{\mathbb{Q}\}$, and $\{\mathbb{R}\}$;
- classes of flows of time obtained from other classes by the covering theorem (sections 13.4–13.5).

While we do not intend to be comprehensive, we will none the less briefly mention the techniques that have been used in these decision problems, and will incidentally note the decidability of some other structures and classes along the way.

To prove that a given temporal system is decidable is important. By establishing the *effectiveness* (that is, the algorithmic nature) of deciding truth in the system, it allows the effectiveness of other things to be established. For an application to expressive completeness see chapter 12, where the decidability of the monadic first-order logic over linear time was used to prove that the translation of a monadic first-order formula into a linear-time-equivalent temporal formula using U, S, and the Stavi connectives is effective. Moreover, decidability provides an algorithm that can in principle be used to validate a temporal specification. However, even the decision problem for classical propositional logic is currently intractable — it is NP-complete to decide whether a formula of propositional logic has a model — so one should not expect too much in practice.

15.2 Monadic second-order logic

We will need to use *full monadic second-order logic.* The formation rules of full monadic second-order logic are:

1. the usual first-order formation rules;
2. for any monadic predicate P, $P(x)$ is an atomic formula;
3. if ψ is a formula and P a monadic predicate then $\forall P\psi$ is a formula.

Of course, which formulae we actually get depends on the first-order signature. For example, for the signature $\{=, <\}$ of flows of time, the formulae obtained by (1) alone would be all first-order formulae written with $=, <$ only.

The semantics of this logic are defined by induction in the obvious way. If M is a structure in the appropriate signature, we have the following:

1. The first-order semantics are defined as usual.
2. If $\psi(x, P)$ is $P(x)$, S is a subset of, and a is an element of, the domain of M (we write $S \subseteq M$, $a \in M$ respectively), then $M \models \psi(a, S)$ if and only if $a \in S$.
3. If $\psi(x_1, \ldots, x_n, P, P_1, \ldots, P_m)$ is a formula, $a_1, \ldots, a_n \in M$ and S_1, $\ldots, S_m \subseteq M$, we have $M \models \forall P\psi(a_1, \ldots, a_n, P, S_1, \ldots, S_m)$ if and only if $M \models \psi(a_1, \ldots, a_n, S, S_1, \ldots, S_m)$ for all subsets $S \subseteq M$.

It will often be convenient to use an assignment h of subsets of the domain of M to the monadic predicate variables, so that $h(P) \subseteq M$ for each monadic predicate variable P. In this notation the last two clauses become:

2. $(M, h) \models P(a)$ if and only if $a \in h(P)$;
3. $(M, h) \models \forall P\psi(a_1, \ldots, a_n, P, P_1, \ldots, P_m)$ if and only if $(M, h') \models \psi(a_1, \ldots, a_n, P, P_1, \ldots, P_m)$ for all h' agreeing with h except perhaps on P.

A *universal monadic second-order* formula is a formula of the form $\forall P_1 \ldots \forall P_n\psi$, where $\psi(x_1, \ldots, x_k, P_1, \ldots, P_{n'})$ is a monadic first-order formula with free monadic predicate variables $P_1, \ldots, P_{n'}$ and free first-order element variables $x_1, \ldots, x_k$. Clearly, universal monadic second-order logic is a fragment of full monadic second-order logic. In particular, if M is a structure, $a_1, \ldots, a_k \in M$ and $S_{n+1}, \ldots, S_{n'} \subseteq M$, we have $M \models \forall P_1 \ldots P_n\psi(a_1, \ldots, a_k, P_1, \ldots, P_n, S_{n+1}, \ldots, S_{n'})$ if and only if $M \models \psi(a_1, \ldots, a_k, S_1, \ldots, S_{n'})$ for all subsets $S_1, \ldots, S_n \subseteq M$.

A *sentence* of (universal) monadic second-order logic is a formula with no free element variables and no free monadic predicate variables.

We will find the following notation useful.

Notation 15.2.1

1. Let P, Q be monadic predicates. We write $P \subseteq Q$ for the formula $\forall x(P(x) \to Q(x))$.
2. Let $\mu(x), \alpha$ be formulae of monadic second-order logic. We define α^μ, the *relativization of α to μ*, by induction on α:
   - if α is quantifier-free then $\alpha^\mu = \alpha$;
   - $(\alpha \wedge \beta)^\mu = \alpha^\mu \wedge \beta^\mu$;
   - $({\sim}\alpha)^\mu = {\sim}(\alpha^\mu)$;

- $(\forall x\alpha)^\mu = \forall x(\mu(x) \to \alpha^\mu)$;
- $(\forall P\alpha)^\mu = \forall P(\forall x(P(x) \to \mu(x)) \to \alpha^\mu)$.

For any structure M in a relational signature, let $S = \{a \in M \mid M \models \mu(a)\}$, and write $M\lceil S$ for the structure with domain S and relations defined by their restrictions to S in M. A simple induction shows that if S is non-empty, for any formula α of monadic second-order logic and any h assigning elements and subsets of S to the variables and monadic predicates of α we have

$$(M, h) \models \alpha^\mu \iff (M\lceil S, h) \models \alpha.$$

15.3 Decision problem for temporal logic

Here we explain what the decision problem is, and describe the results that will concern us, and their methods of proof.

15.3.1 Definitions

Given a class $\mathcal{K}$ of flows of time and a set of temporal connectives used for expressing properties of $\mathcal{K}$, the *decision problem* asks whether there is an algorithm that, given as input any temporal formula A, decides whether or not A is valid in $\mathcal{K}$. That is, the algorithm answers 'yes' if we have $\|A\|_t^h = 1$ for every flow of time $T \in \mathcal{K}$, every assignment h into T and every $t \in T$, and 'no' otherwise (i.e. if $\sim A$ has a *model* with flow of time in $\mathcal{K}$). We say that the temporal logic in question is *decidable* over $\mathcal{K}$ if there is such an algorithm, and *undecidable* otherwise. The notion of 'algorithm' can be formalized by, for example, the Turing machine. Note that the algorithm must terminate either way; algorithms that answer 'yes' if A is valid, and 'loop' if not, are not allowed.

$\mathcal{K}$ has a decision problem in other logics, too. If $\mathcal{L}$ is either first-order logic, universal monadic second-order logic, or full monadic second-order logic, the $\mathcal{L}$*-theory* of $\mathcal{K}$ is the set of all $\mathcal{L}$-sentences of the signature $\{=, <\}$ that are *valid in* $\mathcal{K}$, i.e. are true in every structure in $\mathcal{K}$. We say that the $\mathcal{L}$-theory of $\mathcal{K}$ is *decidable* if there is an algorithm that, given as input any $\mathcal{L}$-sentence σ, decides whether or not σ is in the $\mathcal{L}$-theory of $\mathcal{K}$.

Most of the temporal languages that we have seen employ (i) monadic assignments to atoms, and (ii) connectives with first-order tables. Recall, from theorem 13.1.6, for example, that for any temporal formula A of k-dimensional logic, there is a first-order formula $\delta A(x_1, \ldots, x_k, P_1, \ldots, P_n)$, the *table* of A, such that for any flow of time T, assignment h into T, and tuple $\bar{a} = a_1, \ldots, a_k \in T$, we have $\|A\|^h_{\bar{a}} = 1$ if and only if $(T, h) \models \delta A(\bar{a})$. For example, the formula $A = U(p, S(q, r))$ has table

$$\begin{aligned}\delta A(x) = \quad & \exists y > x(P(y) \wedge \forall z[x < z < y \\ & \rightarrow \exists x < z(Q(x) \wedge \forall y[x < y < z \rightarrow R(y)])]).\end{aligned}$$

Of course, the formula δA is always obtainable effectively from A — in fact very efficiently, by induction on the structure of A. Clearly, to ask whether A is valid in the class $\mathcal{K}$ of flows of time, is the same as to ask whether $\forall P_1, \ldots, P_n \forall x_1, \ldots, x_k \; \delta A(x_1, \ldots, x_k, P_1, \ldots, P_n)$ is valid in $\mathcal{K}$. We have:

Theorem 15.3.1 *Let $\mathcal{K}$ be a class of flows of time. If the universal monadic second-order theory of $\mathcal{K}$ is decidable, then any propositional temporal logic (even a many-dimensional logic) in which atoms are evaluated at time points, and whose connectives have first-order tables, is decidable over $\mathcal{K}$.*

Thus our main interest will be in the decision problem for universal monadic second-order sentences over the classes of flows of time cited.

This will not be our only concern, however. Firstly, the universal monadic logic over the class of all flows of time is undecidable; yet the Prior logic with F and P only is decidable over this class. In this case we have to use other techniques to prove decidability; see section 15.7 below. Secondly, in chapter 8 we saw propositional quantification and the use of fixed points in temporal logic. These are second-order operations; to establish their decidability we will use *full monadic second-order logic,* and not just the universal fragment. An example in $\mathbb{N}$ is the logic $S1S$, described in chapter 8: recall that section 8.4 showed the expressive equivalence of *USF* and $S1S$, so that decidability of *USF* is equivalent to decidability of $S1S$. Besides, very powerful decidability results for full monadic second-order logic are known, and we can derive theorems about the universal fragment directly from these.

15.3.2 Results

So we will examine the decidability of our classes of flows of time with respect to:

- universal monadic second-order logic;
- full monadic second-order logic.

Table 15.1 summarizes the known results of interest to us. 'Yes' means decidable, 'no' undecidable. Attributions of the key results in the table are

Table 15.1

A decidable class?	*Universal monadic second-order logic*	*Full monadic second-order logic*
All flows of time	no	no
Reals	yes (Burgess, Gurevich)	no (Shelah, Gurevich)
Dedekind complete flows	yes (Gurevich)	no
Continuous flows	yes (Gurevich)	no
Circles	yes (Reynolds)	no
Dense flows	yes (Gurevich)	no
Discrete flows	yes (Gurevich)	no
All linear flows	yes (Gurevich)	no (Shelah, Gurevich)
Rationals	yes	yes (Rabin)
Finite linear flows	yes	yes (Büchi)
Natural numbers, integers	yes	yes (Büchi)
Well-ordered flows	yes (Rabin)	refer to text

to whoever proved them originally. (Shelah proved the undecidability of the full monadic theory of linear flows in [Shelah, 1975], assuming the continuum hypothesis; this assumption was removed in [Gurevich and Shelah, 1982], so we have attributed the result to both Shelah and Gurevich.)

It can be seen that all these classes except one are decidable in universal monadic second-order logic, whereas all but three are undecidable in full monadic second-order logic. This uniformity is not surprising, for a result for one class will often immediately yield results for others. The case of the class of all flows of time will follow from the undecidability of first-order predicate logic; and the treatment of the full monadic second-order theory of well-ordered flows of time is necessarily rather special, as we shall see in section 15.5.4. But all the other results in table 15.1 follow from three deep theorems. For decidability, a clutch of results of Burgess and Gurevich deals with Dedekind complete, continuous and real flows of time, and the famous theorem of Rabin on $S2S$ (fact 15.4.1 below) handles all the rest. All the undecidability results follow from the impressive theorem of Shelah, later improved with Gurevich, on the undecidability of the full monadic second-order theory of $(\mathbb{R}, <)$. The order of entries in the table reflects these methods of proof, but the original proofs were sometimes different.

In section 15.4, working down the first column (universal monadic second-order logic), we discuss the Burgess/Gurevich theorems, and then, working up from the bottom of the first column, the Rabin-style proofs (except that the well-ordered flows are dealt with rather later). In section 15.5 we will proceed down the second column (full monadic second-order logic).

15.3.3 Techniques

What techniques can we use to obtain new decidability (undecidability) results from old? Clearly, any class whose full monadic second-order theory is decidable also has decidable universal monadic second-order theory. Thus, any 'yes' in the second column forces a 'yes' in the first column, and vice versa for 'no'.

The other important technique here is the following observation. We fix a logic $\mathcal{L}$, which could be either first-order logic, universal monadic second-order logic, or full monadic second-order logic.

Definition 15.3.2 Let $\mathcal{K}, \mathcal{K}'$ be classes of structures. We say that $\mathcal{K}'$ *reduces* to $\mathcal{K}$ if:

- there is $\mathcal{D} \subseteq \mathcal{K}$ and a map $\nu : \mathcal{D} \to \mathcal{K}'$ such that for all $N \in \mathcal{K}'$ there is $M \in \mathcal{D}$ such that $\nu(M)$ is isomorphic to N (we say that ν is *surjective up to isomorphism*);
- there is an $\mathcal{L}$-sentence σ such that for all $M \in \mathcal{K}$, $M \in \mathcal{D} \iff \mathcal{M} \models {\sim}\sigma$;
- there is an effective map $* : \{\text{sentences of } \mathcal{L} \text{ in the signature of } \mathcal{K}'\} \to \{\text{sentences of } \mathcal{L} \text{ in the signature of } \mathcal{K}\}$, such that for all $N \in \mathcal{K}'$ and all sentences α of $\mathcal{L}$ in the signature of $\mathcal{K}'$,

$$N \models \alpha \iff \text{for all } M \in \mathcal{D} \text{ such that } \nu(M) \cong N, M \models \alpha^*.$$

Proposition 15.3.3 *Let $\mathcal{K}$, $\mathcal{K}'$ be classes of structures, and suppose that $\mathcal{K}'$ reduces to $\mathcal{K}$. Then if $\mathcal{K}$ is decidable (with respect to $\mathcal{L}$-sentences), so is $\mathcal{K}'$.*

Proof. Let $\mathcal{A}$ be an algorithm deciding $\mathcal{L}$-sentences for $\mathcal{K}$. Given a sentence α of $\mathcal{L}$, α is valid in $\mathcal{K}'$ if and only if $\sigma \vee \alpha^*$ is valid in $\mathcal{K}$. Note that $\sigma \vee \alpha^*$ is obtainable effectively from α, and for each of the three possibilities for $\mathcal{L}$ is effectively equivalent to an $\mathcal{L}$-sentence. Applying $\mathcal{A}$ to this sentence decides α on $\mathcal{K}'$. ■

Corollary 15.3.4 *Let $\mathcal{K}' \subseteq \mathcal{K}$ be classes of structures, and suppose that there is some $\mathcal{L}$-sentence σ such that $\mathcal{K}' = \{M \in \mathcal{K} \mid M \models \sim\sigma\}$. Then if $\mathcal{K}$ is decidable, so is $\mathcal{K}'$.*

Proof. Take $\nu = id_{\mathcal{K}'}$. ■

In the next two sections we will look at the key results in table 15.1, and use these methods to obtain the rest.

15.4 Decidability results

Now we examine the arguments for the 'yes' entries in table 15.1.

15.4.1 The real numbers

John Burgess and Yuri Gurevich [Burgess and Gurevich, 1985] offered two different proofs of the decidability of the temporal logic with U and S over the real numbers. (Also see [Burgess, 1979], [Bull, 1968a].) As they point out, this is actually equivalent to the decidability of the universal monadic second-order theory of the reals, because U and S are expressively complete over the reals (see chapter 10). We will not discuss these proofs in detail. One proof shows that any US-formula is satisfiable in $(\mathbb{R}, <)$ if and only if it is satisfiable in $(\mathbb{Q}, <)$ under a 'nice' assignment, and that the set of formulae satisfiable in $(\mathbb{Q}, <)$ under nice assignments is decidable. The second proof uses a variant of the finite model property, together with a Feferman–Vaught-style theorem showing that the truth value of universal monadic second-order sentences in a lexicographic sum of orders is determined by their truth values in the components of the sum. The argument of section 9 of chapter 6 is taken from this. Cf. section 15.6 below.

Using the second method it was also shown that the following classes have decidable universal monadic second-order theories:

- the class of all Dedekind complete linear flows;
- the class of all continuous flows (linear, Dedekind complete, dense, and without endpoints);
- the class of all Dedekind complete linear orders that are *separable* (they have a countable dense subset).

15.4.2 Rabin's theorem

In [Rabin, 1969] Michael Rabin proved a very strong theorem, from which all our remaining decidability results follow. [Gabbay, 1975a] uses the theorem to prove many more still.

The *binary tree* is the structure (T, l, r) where T is the set of all finite sequences of zeros and ones, and l, r are unary function symbols interpreted in T by:

$$l(a) = a^\wedge 0, \quad r(a) = a^\wedge 1 \quad \text{for all } a \in T,$$

where $a^\wedge 0$, $a^\wedge 1$ are the sequences obtained by adding 0 (respectively, 1) on to the end of a. The set of all sentences of full monadic second-order logic that are true in this structure is called $S2S$ (the Second-order theory of 2 Successor functions).

Fact 15.4.1 [Rabin, 1969] $S2S$ is decidable: there is an algorithm to decide whether a monadic second-order sentence is true or false in the binary tree.

We will not attempt to prove this theorem. Rabin's original proof is very difficult; simpler versions can be found in [Gurevich and Harrington, 1982], [Muchnik, 1984], [Yakhnis and Yakhnis, 1990] and [Zeitman, to appear], for example. A detailed discussion can be found in [Gurevich, 1985], while [Rabin, 1977] surveys applications. One can even show that $S\omega S$, the second-order theory of ω successor functions, is decidable; [Gabbay, 1975a] uses this theorem to prove decidability of many different modal logics.

We will need some simple definitions to do with the binary tree. We will write $\varnothing$ for the empty sequence of zeros and ones. We let $x \sqsubseteq y$ be the formula

$$\forall P(\forall zt[P(z) \wedge (z = l(t) \vee z = r(t)) \to P(t)] \wedge P(y) \to P(x)).$$

Then if $a, b \in T$ are sequences of zeros and ones, we have $(T, l, r) \models a \sqsubseteq b$ if and only if a is an initial segment of b. Of course we can also define 'a is a proper initial segment of b', written $a \sqsubset b$, by $(a \sqsubseteq b \wedge a \neq b)$.

15.4.3 The integers and natural numbers

Büchi [Büchi, 1962] proved that the full monadic second-order theory of the classes $\{\mathbb{N}\}$ and $\{\mathbb{Z}\}$ are decidable. (For $\mathbb{N}$, the logic is equivalent to the system $S1S$ that we saw in section 7 of chapter 8.) His proof used automata. We discussed a similar proof in corollary 8.7.9 above; see [Gurevich, 1985] for more details. Büchi's theorem was also reproved by Shelah in [Shelah, 1975] using a quite different argument, involving Feferman–Vaught-style reasoning as in the case of $(\mathbb{R}, <)$. See remark 15.6.3.

Nowadays we can prove the theorem of Büchi more easily using Rabin's theorem, from which it follows trivially (though of course, some proofs of Rabin's theorem use Büchi's).

Theorem 15.4.2 *The full monadic second-order theory of the class* $\mathcal{N} = \{(\mathbb{N}, <)\}$ *is decidable.*

Proof. We observe that we can define the natural numbers in the binary tree (T, l, r), and so reduce $\{(\mathbb{N}, <)\}$ to $\{(T, l, r)\}$. Let $\varphi(Q)$ be the formula $\forall x[\forall y(y \sqsupseteq x) \to Q(x)] \land \forall x(Q(x) \to Q(r(x)))$. Then let

$$\psi(P) \quad = \quad \varphi(P) \land \forall Q(\varphi(Q) \to P \subseteq Q)$$

(see notation 15.2.1 for the meaning of $P \subseteq Q$). Clearly, if $S \subseteq T$ then $(T, l, r) \models \psi(S)$ if and only if $S = \{\varnothing, r(\varnothing), rr(\varnothing), \ldots\}$, the sequences without zeros. If $(T, l, r) \models \psi(S)$, we can identify $(S, \sqsubset)$ with $(\mathbb{N}, <)$.

Now given any monadic second-order sentence σ of the signature $\{=, <\}$, we can obtain the formula $\sigma'(P)$ by replacing each atomic subformula of σ^P of the form $x < y$ by $x \sqsubset y$. (See notation 15.2.1 for the definition of σ^P.) Then

$$(\mathbb{N}, <) \models \sigma \quad \iff \quad (T, l, r) \models \forall P(\psi(P) \to \sigma'(P)).$$

Hence by applying the algorithm deciding $S2S$ to $\sigma^* = \forall P(\psi \to \sigma')$, which is effectively constructable from σ, we can decide σ in $(\mathbb{N}, <)$. ■

We can decide the full monadic second-order theory of $(\mathbb{Z}, <)$ in a similar way, by identifying it with the set of sequences of T consisting only of zeros or only of ones; or we can simply interpret it in $(\mathbb{N}, <)$. We leave the details to the reader.

15.4.4 Finite linear orders

The decidability of this class follows easily from Büchi's theorem.

Theorem 15.4.3 *The full monadic second-order theory of the class of all finite linear orders is decidable.*

Proof. Given a sentence σ of the signature $\{=, <\}$ of order, let x be a new element variable and let $\sigma^*(x)$ be the result of relativizing all element quantifiers of σ to the set $\{y \mid y < x\}$. That is, we replace $\exists y \ldots$ by $\exists y(y < x \land \ldots)$ and $\forall y \ldots$ by $\forall y(y < x \to \ldots)$. Clearly, σ is true in all finite linear flows of time if and only if $\forall x(\exists y(y < x) \to \sigma^*(x))$ is true in $(\mathbb{N}, <)$. The theorem now follows from theorem 15.4.2. ■

15.4.5 The rationals

Theorem 15.4.4 *The full monadic second-order theory of* $(\mathbb{Q}, <)$ *is decidable.*

Proof. This was proved by Rabin [Rabin, 1969]. We view the binary tree T as in Figure 15.1.

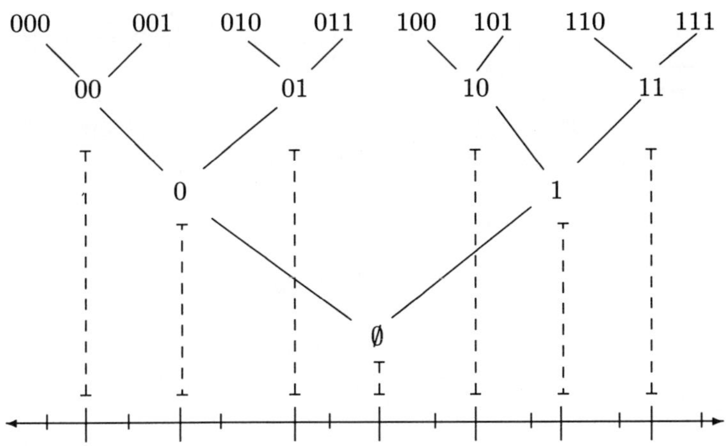

Figure 15.1

The bottom node is the empty sequence $\varnothing$; the two next highest nodes are the sequences 0, 1, etc. The nodes in T can be 'projected' down on to a horizontal axis, as shown in the figure. Clearly, the linear left–right order on this axis induces on T a countably infinite dense linear order without endpoints, i.e. a linear order isomorphic to $(\mathbb{Q}, <)$. We can define this order in the language of $S2S$ by: $x \prec y$ if and only if

$$l(y) \sqsubseteq x \quad \vee \quad r(x) \sqsubseteq y \quad \vee \quad \exists z(l(z) \sqsubseteq x \wedge r(z) \sqsubseteq y).$$

Now, given a monadic second-order sentence σ of the signature $\{=, <\}$ of order, we obtain σ^* by replacing each atomic subformula of σ of the form $x < y$ by the formula $x \prec y$ as defined above. Clearly, $(\mathbb{Q}, <) \models \sigma \iff (T, l, r) \models \sigma^*$, so by fact 15.4.1 we can decide whether $(\mathbb{Q}, <) \models \sigma$. ■

15.4.6 Other classes of linear orders

The fact that the universal monadic theory of the class $\mathcal{L}$ of all linear flows of time is decidable was first proved by Gurevich in [Gurevich, 1964]. We will derive it from the case of $\mathbb{Q}$. We will need the downward Löwenheim–Skolem theorem:

Fact 15.4.5 If M is any model-theoretic structure in a countable signature, then there is a countable substructure N of M satisfying the same first-order sentences as M.

Proof. This is proved in [Chang and Keisler, 1990], for example. ■

Theorem 15.4.6 *The universal monadic second-order theory of the class $\mathcal{L}$ of all linear flows of time is decidable.*

Proof. Let $\sigma = \forall P_1 \ldots \forall P_n \psi$ be a universal monadic second-order sentence, where $\psi(P_1, \ldots, P_n)$ is first-order in the signature $\{=, <, P_1, \ldots, P_n\}$. We claim that σ is valid in $\mathcal{L}$ if and only if it is valid in the class of all countable linear flows. Left to right is obvious. For the other direction, assume $(T, <)$ is a linear flow of time in which σ is not valid. Thus there are $S_1, \ldots, S_n \subseteq T$ such that $(T, <) \models \sim\psi(S_1, \ldots, S_n)$. By fact 15.4.5 there is countable $T_0 \subseteq T$ such that $(T_0, <) \models \sim\psi(T_0 \cap S_1, \ldots, T_0 \cap S_n)$. (Here and below, we write $(T_0, <)$ for the flow of time with domain T_0 and earlier–later ordering $<$ induced from that of $(T, <)$.) Hence σ is not valid in the countable linear flow $(T_0, <)$. This proves our claim.

Now it is well-known that any countable linear order is isomorphic to a suborder of the rational order $(\mathbb{Q}, <)$. Hence σ is valid in $\mathcal{L}$ if and only if $(\mathbb{Q}, <) \models \sigma^*$, where

$$\sigma^* \quad = \quad \forall Q(\exists x Q(x) \rightarrow (\forall P_1 \ldots P_n \psi)^Q),$$

where Q is a new predicate variable and $(\forall P_1 \ldots P_n \psi)^Q$ indicates that the quantifiers of $\forall P_1 \ldots P_n \psi$ have been relativized to Q. Hence we can decide whether σ is $\mathcal{L}$-valid by deciding whether $(\mathbb{Q}, <) \models \sigma^*$. By theorem 15.4.4 there is an algorithm to do this. ■

Corollary 15.4.7 *Any class of the form $\{M \in \mathcal{L} \mid M \models \sigma\}$, where σ is an existential monadic second-order sentence (i.e. a sentence of the form $\exists P_1 \ldots \exists P_n \psi$, where ψ is first-order), has decidable universal monadic theory. In particular, the universal monadic second-order theories of the following classes of linear flows of time are decidable:*

- *flows with (or without) minimum (or maximum) elements;*

- *dense flows with (or without) minimum (or maximum) elements;*
- *discrete flows: those satisfying*

$$\forall x(\exists y(y < x) \to \exists y[y < x \land \forall z(z > y \to z \geq x)]),$$

and its mirror image (swap $<$ and $>$);

- *the class of infinite linear flows of time;*
- *the class of linear, non-well-ordered flows of time.*

Proof. Immediate from theorem 15.4.6 and corollary 15.3.4. ■

15.4.7 The class of all well-ordered flows

A flow of time $(T, <)$ is said to be *well-ordered* if it is linear and any non-empty subset of T has a $<$-least element. That is, $(T, <) \vDash \forall P(\exists x P(x) \to \exists x(P(x) \land \forall y(y < x \to \sim P(y))))$. Equivalently, $(T, <)$ is linear and there are no infinite sequences of time points leading strictly into the past. Note that any subflow of a well-ordered flow is also well-ordered. The proof of theorem 15.4.6 can now be easily adapted to show that the universal monadic second-order theory of the class of all well-ordered flows is decidable.

15.4.8 Circles

In [Reynolds, to appear] it was shown that the class of all circles has decidable universal monadic second-order theory. The proof is similar to that for the linear flows (theorem 15.4.6).

15.4.9 Trees

Finally, we mention a result on trees, useful for logics of historical necessity (section 7.7). A *tree* is a flow of time $(T, <)$ such that the past $\{u \in T \mid u < t\}$ of any point $t \in T$ is linear (when ordered by $<$). A *branch* of T is a maximal linearly ordered subset of T. We consider the full monadic second-order language in the signature with $=$ and $<$. The semantics are as usual except that the set quantifiers range over branches of trees. That is, if X is a set variable free in $\varphi(X)$, then $(T, <) \vDash \forall X \varphi(X)$ if and only if $(T, <) \vDash \varphi(\beta)$ for all branches β of T.

In [Gurevich and Shelah, 1985] it was shown that the theory of the class of all trees in this logic is decidable. Moreover, this is so even for sentences with free unary predicates, which may be interpreted as arbitrary sets of time points, not necessarily branches. That is, there is an algorithm to decide the set

$$\{\sigma(P_1, \ldots, P_n) \mid \forall \text{ trees } (T, <), \forall S_1, \ldots, S_n \subseteq T, (T, <) \vDash \sigma(S_1, \ldots, S_n)\},$$

where the notation $\sigma(P_1, \ldots, P_n)$ indicates that σ has no free element variables, its free set variables lying among $P_1, \ldots, P_n$. The proof is technically interesting and combines the use of automata with the 'composition method' of [Shelah, 1975].

15.5 Undecidability results

Now we examine the negative results in table 15.1.

15.5.1 The class of all flows of time

It follows from the undecidability of first-order predicate logic (proved by Church) that the class $\mathcal{T}$ of all flows of time has undecidable first-order theory; the second-order results for $\mathcal{T}$ in table 15.1 follow from this.

Proposition 15.5.1 *The first-order theory of $\mathcal{T}$ is undecidable.*

Proof. It is well known that any first-order structure can be interpreted in a *graph* — a non-empty set G of 'nodes', equipped with a single irreflexive symmetric binary relation R. (This uses a trick similar to that of definition 13.7.10.) Hence the first-order theory of the class $\mathcal{G}$ of all graphs is undecidable.

Any flow $(T, <) \in \mathcal{T}$ with maximal elements can be regarded as a graph $(G, R) = \nu(T, <)$ as follows. Let $\mu(x)$ be the formula $\sim\exists y(y > x)$. We let $G = \{t \in T \mid (T, <) \vDash \mu(t)\}$, and let $R(t, u)$ hold if and only if $t \neq u \wedge \exists v(v < t \wedge v < u)$ holds in $(T, <)$.

Now let $(G, R) \in \mathcal{G}$ be a graph. As R is irreflexive and symmetric, we can identify it with a subset of the pair-sets (sets of size 2) of elements of G. That is, we let $\{a, b\} \in R$ if and only if $R(a, b)$ holds. We construct a flow of time $(T, <)$ as follows. We let T be the disjoint union of G and R, and define $t > u$ if and only if $t \in G, u \in R$, and $t \in u$. Clearly, $\nu(T, <) \cong (G, R)$. Hence the map $\nu : \mathcal{T} \to \mathcal{G}$ is surjective up to isomorphism. Its domain $\mathcal{D}$ is defined by the first-order sentence $\exists x \mu(x)$. If for any first-order sentence α of the signature $\{=, R\}$ of graphs, we let α^* be the sentence obtained by replacing any atomic subformula of α^μ of the form $R(x, y)$ by $\exists z(z < x \wedge z < y)$, we clearly have $(T, <) \vDash \alpha^* \iff \nu(T, <) \vDash \alpha$ for any $(T, <) \in \mathcal{T}$. (See notation 15.2.1 for the definition of α^μ.) We can obtain α^* effectively from α. Thus $\mathcal{G}$ reduces to $\mathcal{T}$ in first-order logic. Since the first-order theory of $\mathcal{G}$ is undecidable, it follows from proposition 15.3.3 that the first-order theory of $\mathcal{T}$ is also undecidable.

We could avoid the use of graphs by directly interpreting first-order structures in flows of time, using the technique given in section 13.7. ■

However, some temporal logics over $\mathcal{T}$ are weaker than first-order logic, and are in fact decidable. An example is the basic Prior logic K_t with connectives F and P only, which is sound and complete for the class of all flows of time. We proved that K_t is decidable in section 3 of chapter 3.

15.5.2 The real numbers

The crucial undecidability result from which all the rest follow is:

Fact 15.5.2 The full monadic second-order theory of $(\mathbb{R}, <)$ is undecidable.

This deep theorem was proved by Saharon Shelah in [Shelah, 1975], under the assumption of the *continuum hypothesis*.[1] The proof is beyond the scope of this book. Gurevich and Shelah [Gurevich and Shelah, 1982] later provided a new proof that did not use the continuum hypothesis. It involves a sophisticated reduction of arithmetic to $(\mathbb{R}, <)$, and is discussed in the survey [Gurevich, 1985].

15.5.3 Other classes

We can now clean up with the following lemma. The argument for the class of all linear flows of time is in [Shelah, 1975].

Lemma 15.5.3 *Let $\mathcal{K}$ be a class of flows of time containing a flow of time that embeds $(\mathbb{R}, <)$ (i.e. one having a sub-order isomorphic to $\mathbb{R}$). Then the full monadic second-order theory of $\mathcal{K}$ is undecidable.*

Proof. Recall that a flow of time $(T, <)$ is *continuous* if it is a linear order and is dense, Dedekind complete, and without endpoints. We claim that:

1. Any continuous flow of time has a suborder isomorphic to $(\mathbb{R}, <)$.

2. Any continuous suborder of $(\mathbb{R}, <)$ is isomorphic to $(\mathbb{R}, <)$.

---

[1] That is, he assumed that $2^\omega = \omega_1$ — a statement that cannot be proved or refuted from the usual axioms of set theory, ZFC. See [Cohen, 1966].

For (1), we note that any continuous flow embeds $(\mathbb{Q}, <)$ and therefore its Dedekind completion, $(\mathbb{R}, <)$. For (2), let $(S, <)$ be a continuous suborder of $(\mathbb{R}, <)$. List the open intervals of $\mathbb{R}$ with rational endpoints as $\{I_n \mid n < \omega\}$. For each $n < \omega$ define an element $d_n \in S$ as follows. If $S \cap I_n$ is non-empty then let d_n be any element of $S \cap I_n$; otherwise let d_n be any element of S. Then $\{d_n \mid n < \omega\}$ is a countable dense subset of S — so S is *separable*. But any continuous separable linear order is isomorphic to $(\mathbb{R}, <)$ — cf. [Rosenstein, 1982]. This proves the claim.

For a predicate variable P, let $\chi(P)$ be the formula expressing that the elements satisfying P, ordered by $<$, form a continuous flow (it is easy to write such a formula in monadic second-order logic). For each monadic second-order sentence σ of the language of order we define a new sentence

$$\sigma^* \quad = \quad \forall P(\chi(P) \to \exists Q(Q \subseteq P \wedge \chi(Q) \wedge \sigma^Q)).$$

It follows from the claim that for any such σ and any flow of time $(T, <)$, $(\mathbb{R}, <) \models \sigma \Rightarrow (T, <) \models \sigma^*$, and that if $(T, <)$ embeds $(\mathbb{R}, <)$ then the converse holds. So for any σ, $(\mathbb{R}, <) \models \sigma$ if and only if σ^* is valid in $\mathcal{K}$. Hence $\{(\mathbb{R}, <)\}$ reduces to $\mathcal{K}$, so by proposition 15.3.3 and fact 15.5.2, the full monadic second-order theory of $\mathcal{K}$ is undecidable. ■

Corollary 15.5.4 *The following classes of flows of time have undecidable full monadic second-order theory.*

1. *any first-order definable class (i.e. the class of all flows of time satisfying Σ, where Σ is some given first-order theory in the signature $\{=, <\}$) that contains flows of time with infinite linearly ordered subsets;*
2. *$\mathcal{L}$, the class of all linear flows;*
3. *the class of all dense flows;*
4. *the class of all discrete flows;*
5. *the class of all circles;*
6. *the class of all Dedekind complete flows;*
7. *the class of all continuous flows (Dedekind complete, dense, and without endpoints).*

Proof. For (1) we can use first-order compactness. If Σ has a model T with an infinite linearly ordered subset, and $\{c_r \mid r \in \mathbb{R}\}$ is a set of distinct constant symbols, the theory $\Xi = \{c_r < c_s \mid (\mathbb{R}, <) \models r < s\}$ is consistent with Σ, as T can be expanded to a model of any finite subset of $\Sigma \cup \Xi$. Hence there is $T' \models \Sigma \cup \Xi$; and clearly T' has a subordering isomorphic to $(\mathbb{R}, <)$. Items (2)–(5) follow from this and lemma 15.5.3. The last two items follow immediately from the lemma. ■

15.5.4 The class of all well-ordered flows

The position for the decision problem for the class $\mathcal{W}$ of well-ordered flows of time is complicated. Roughly, the statement 'the full monadic second-order theory of the third infinite cardinal, ω_2, is decidable' is independent of (i.e. neither provable or refutable from) the usual set-theoretic axioms ZFC. This has consequences for $\mathcal{W}$. We have:

Fact 15.5.5

1. ZFC does not determine the monadic second-order theory of ω_2. That is, there is a sentence σ of full monadic second-order logic that is true (in ω_2) in some models of ZFC but false in others.

2. ZFC cannot prove that the full monadic second-order theory of ω_2 is undecidable.

3. ZFC cannot prove that the full monadic second-order theory of ω_2 is decidable.

Remarks 15.5.6

1. We will certainly not need to know the form of σ in fact 15.5.5(1), but as an aside we will mention that one such σ says 'whenever $X \subseteq \omega_2$ is stationary[2] and each $\alpha \in X$ has countable cofinality,[3] then there is $\alpha < \omega_2$ of uncountable cofinality such that $\alpha \cap X$ is a stationary subset of α'. This is expressible in monadic second-order logic. The independence was proved by Shelah [Shelah, 1975] using results of Baumgartner and Jensen.

In a way, this means that we cannot even make up our own minds what sentences should be valid in ω_2, let alone find an algorithm to judge them. Any putative decision algorithm $\mathcal{A}$ for the monadic second-order theory of ω_2 will answer either 'yes' or 'no' when fed σ. Let us assume it says 'yes' (the situation is similar if it says 'no'). Now $\mathcal{A}$, σ, and the computation of $\mathcal{A}$ on σ leading to this answer are all finite and can be formalized in

[2]If α is a limit ordinal, a *club* in α is an unbounded subset $C \subseteq \alpha$ such that for all limit $\delta < \alpha$, if $C \cap \delta$ is unbounded in δ then $\delta \in C$. A subset $S \subseteq \alpha$ is said to be *stationary* if it has non-empty intersection with every club in α.

[3]The *cofinality* of an ordinal α, $cf(\alpha)$, is the least ordinal β such that α has an unbounded (cofinal) subset of order type β. $cf(\alpha)$ is necessarily a cardinal.

ZFC. Hence ZFC $\vdash$ '$\mathcal{A}$ says "yes" to σ', and so $\mathcal{A}$ will answer 'yes' to σ in any model of ZFC. But the truth value of '$\omega_2 \models \sigma$' varies among models of ZFC! Thus, in some models of ZFC $\mathcal{A}$ will be wrong and will not decide the theory of ω_2, whereas in other models it will be right, at least about σ.

2. In spite of this confusion, one might still hope to prove directly that the monadic theory of ω_2 is undecidable, perhaps by showing that for any algorithm $\mathcal{A}$ there is some sentence (other than σ!) that $\mathcal{A}$ decides wrongly in any model of ZFC. In fact, this is not possible. Several different models of ZFC in which the monadic second-order theory of ω_2 is *decidable* are now known. See [Shelah, 1978], for example, where it is shown that a certain axiom, additional to (and consistent with) those of ZFC, forces the theory of ω_2 to be decidable. Magidor proved a similar result, but assumed the existence of a weakly compact cardinal;[4] the theory of ω_2 here is different! Thus we cannot prove from ZFC that ω_2 has undecidable monadic second-order theory.

3. In [Gurevich *et al.*, 1983], again assuming that there exists a weakly compact cardinal, it is proved that there are continuum-many (i.e. 2^ω) possible theories of ω_2 as the model of ZFC varies. (This is as many as possible, since there are only this many sets of monadic second-order sentences!) Indeed, there is an algorithm that, given a number n, constructs a monadic second-order sentence σ_n, the sentences σ_n having the following property: for every $S \subseteq \omega$ there is a 'forcing extension of the ground model of set theory' (see [Cohen, 1966]) in which $\{n < \omega \mid \omega_2 \models \sigma_n\} = S$. A similar result was proved in [Lifsches and Shelah, 1992], without assuming that there is a weakly compact cardinal. Thus, for example, by choosing a non-recursive set S we can show that there are models of ZFC in which ω_2 has undecidable monadic second-order theory. So we cannot prove from ZFC that the monadic second-order theory of ω_2 is decidable, either.

Proposition 15.5.7 *The class $\{\omega_2\}$ reduces to $\mathcal{W}$ in the sense of definition 13.3.1.*

Proof. Let P be a predicate variable. We claim that for all $n < \omega$ we can write a monadic second-order formula $\sigma_n(P)$ such that for all $(T, <) \in \mathcal{W}$ and non-empty $S \subseteq T$, we have $(T, <) \models \sigma_n(S) \iff (S, <) \cong \omega_n$. First define $\chi(P, Q)$ by $Q \subseteq P \wedge \forall x(P(x) \to \exists y\ (Q(y) \wedge y \geq x))$, saying that Q is

[4] A cardinal κ is *weakly compact* if it is uncountable, and for any linear flow of time $(T, <)$ where T is of size κ, there is either a strictly increasing or a strictly decreasing sequence of time points $t_i \in T$ $(i < \kappa)$. For example, 2^ω is not weakly compact (consider $(\mathbb{R}, <)$). Weakly compact cardinals must be rather large; it is not known whether 'there is a weakly compact cardinal' is consistent with ZFC. Any text on set theory, such as [Jech, 1978], will give more details.

cofinal in P. Let $\sigma_{-1}(P)$ be the formula $\exists x \forall y(P(y) \leftrightarrow y = x)$, saying that P has a single element. Let $n < \omega$ and assume that $\sigma_m(P)$ are defined for all $-1 \leq m < n$. Let $\psi_n(P)$ be the formula

$$\exists Q[\chi(P, Q) \wedge \bigvee_{-1 \leq m < n} \sigma_m(Q)],$$

saying that P has a cofinal subset that is either a singleton or of order-type ω_m for some $m < n$. We now define

$$\sigma_n(P) \quad = \quad \forall Q \subseteq P[\psi_n(Q) \leftrightarrow {\sim}\chi(x = x, Q)].$$

This is easily seen to work, and proves our claim.

Write σ for the sentence $\sigma_2(x = x)$. Then $\{T \in \mathcal{W} \mid T \vDash \sigma\} = \{T \in \mathcal{W} \mid T \cong \omega_2\}$. So any monadic second-order sentence α is true in $(\omega_2, <)$ if and only if $\sigma \to \alpha$ is valid in $\mathcal{W}$. ■

Thus the monadic second-order theory of $\mathcal{W}$ is also not determined by ZFC. There are models of ZFC in which $\mathcal{W}$ is undecidable (e.g. any model in which the theory of ω_2 is undecidable) and so we cannot prove in ZFC that $\mathcal{W}$ is decidable. At the time of writing, whether it is consistent with ZFC that $\mathcal{W}$ is decidable is an open question. Cf. [Shelah, 1975].

15.6 Covering theorem and decidability

In chapter 14 it was shown that **T(L)** is decidable if the logic **L** is. Here we will prove another preservation result, to do with the covering theorem of section 13.5. Recall that from classes $\mathcal{B}, \mathcal{T}$ of flows of time (or arbitrary L-structures) we constructed the class $\mathcal{T}[\mathcal{B}]$ of 'canonical' L_{Ξ}-structures made of the disjoint union of flows of time $T \in \mathcal{T}$ and $B_t \in \mathcal{B}$ (all $t \in T$).

Proposition 15.6.1 *Assume that $\mathcal{B}, \mathcal{T}$ have decidable universal monadic second-order theories. Then so does $\mathcal{T}[\mathcal{B}]$.*

Proof. Let $\chi = \forall P_1 \ldots \forall P_m \psi(P_1, \ldots, P_m)$ be a universal monadic sentence, where $\psi(P_1, \ldots, P_m)$ is a monadic first-order sentence. Lemma 13.5.13 proved that ψ is $\mathcal{T}[\mathcal{B}]$-equivalent to a disjunction of mixed sentences which is constructable effectively from ψ. Thus we can assume that ψ has the form

$$\tau^T(P_1, \ldots, P_m, Q_1/\sigma_1^B, \ldots, Q_n/\sigma_n^B),$$

where $\tau(P_1, \ldots, P_m, Q_1, \ldots, Q_n)$ is a monadic sentence of $\mathcal{T}$ and the σ_i are monadic sentences of $\mathcal{B}$, all constructable effectively from χ. The σ_i

may also involve $P_1, \ldots, P_m$. The notations τ^T, σ^B indicate relativization (see definition 13.5.10); we can combine the disjuncts into one formula because relativization commutes with disjunction. Hence if $M \in \mathcal{T}[\mathcal{B}]$ is the canonical structure arising from $T \in \mathcal{T}$, $B_t \in \mathcal{B}$ $(t \in T)$, and h^* is an assignment into M, the truth value of ψ in (M, h^*) is determined by

- the sets $h^*(P_i) \cap T$ for $i = 1, \ldots, m$;
- the sets $S_j = \{t \in T \mid (B_t, h^* \lceil B_t) \models \sigma_j\}$ for $j = 1, \ldots, n$,

where $(h^* \lceil B_t)(P_i) = h^*(P_i) \cap B_t$ for all $i \leq m$. We have

$$(*) \qquad (M, h^*) \models \psi \iff T \models \tau(h^*(P_1) \cap T, \ldots, h^*(P_m) \cap T, S_1, \ldots, S_n).$$

For details see section 13.5.

Now, for each $X \subseteq \{1, \ldots, n\}$ let $\xi_X = \bigwedge_{i \in X} \sigma_i \wedge \bigwedge_{j \notin X} \sim\sigma_j$. Let $\mathcal{C} = \{X \subseteq \{1, \ldots, n\} \mid \forall P_1 \ldots \forall P_m \sim\xi_X$ is *not* valid in $\mathcal{B}\}$. We can construct $\mathcal{C}$ effectively from χ, as the construction of the sentences σ_i is effective and the universal monadic second-order theory of $\mathcal{B}$ is assumed decidable. $\mathcal{C}$ is the set of all subsets X of $\{1, \ldots, n\}$ such that ξ_X has a model with flow of time in $\mathcal{B}$. Define

$$\zeta(x, Q_1, \ldots, Q_n) \quad = \quad \bigvee_{X \in \mathcal{C}} \left[\bigwedge_{i \in X} Q_i(x) \wedge \bigwedge_{j \notin X} \sim Q_j(x) \right].$$

Notice that we can also construct ζ effectively. Now let χ^* be the universal monadic second-order sentence

$$\forall P_1 \ldots P_m Q_1 \ldots Q_n (\forall x \zeta(x, Q_1, \ldots, Q_n) \to \tau(P_1, \ldots, P_m, Q_1, \ldots, Q_n)).$$

Of course, χ^* is effectively constructable from χ.

Claim: χ is valid in $\mathcal{T}[\mathcal{B}]$ if and only if χ^* is valid in $\mathcal{T}$.

Proof of claim: First suppose that χ is valid in $\mathcal{T}[\mathcal{B}]$. Let $T \in \mathcal{T}$; we evaluate χ^* in T. Let h assign subsets of T to $P_1, \ldots, P_m$, $Q_1, \ldots, Q_n$. We will prove that $(T, h) \models \forall x \zeta(x, Q_1, \ldots, Q_n) \to \tau(P_1, \ldots, P_m, Q_1, \ldots, Q_n)$. So assume that $(T, h) \models \forall x \zeta(x, Q_1, \ldots, Q_n)$ holds; we will show that $(T, h) \models \tau(P_1, \ldots, P_m, Q_1, \ldots, Q_n)$.

By assumption, for all $t \in T$ there is $X_t \in \mathcal{C}$ such that for all $j \leq n$, $t \in h(Q_j) \iff j \in X_t$. For each t choose a model (B_t, h_t) of ξ_{X_t} with $B_t \in \mathcal{B}$. Thus, for all $j \leq n$, $(B_t, h_t) \models \sigma_j \iff j \in X_t$. Now form the canonical structure M from T and the B_t $(t \in T)$, and define an assignment h^* of subsets of M to $P_1, \ldots, P_m$ by $h^*(P_i) = h(P_i) \cup \bigcup_{t \in T} h_t(P_i)$ for each $i \leq m$. Then for all $j \leq n$ we have $h(Q_j) = \{t \in T \mid (B_t, h_t) \models \sigma_j\} = S_j$ in the notation above. So by $(*)$ above, $(M, h^*) \models \psi$ if and only if $(T, h) \models \tau$.

As χ is assumed valid in $\mathcal{T}[\mathcal{B}]$, we do have $(M, h^*) \models \psi$. Hence $(T, h) \models \tau$, as required.

Conversely, assume that χ^* is valid in $\mathcal{T}$. Let $M \in \mathcal{T}[\mathcal{B}]$ and let h^* assign subsets of M to $P_1, \ldots, P_m$. If M is the canonical structure made from T, B_t $(t \in T)$, let $h_t = h^* \lceil B_t$ for all t, and write h for the assignment into T given by

- $h(P_i) = h^*(P_i) \cap T$ for all $i \leq m$;
- $h(Q_j) = \{t \in T \mid (B_t, h_t) \models \sigma_j\}$ for all $j \leq n$.

For each $t \in T$ write X_t for $\{j \leq n \mid t \in h(Q_j)\}$. Clearly, $B_t \models \exists P_1 \ldots P_m$ ξ_{X_t} for all $t \in T$. Hence $X_t \in \mathcal{C}$, and so $(T, h) \models \zeta(t, Q_1, \ldots, Q_n)$, for all $t \in T$. But χ^* is assumed $\mathcal{T}$-valid. Therefore, $(T, h) \models \tau(P_1, \ldots, P_m, Q_1, \ldots, Q_n)$. By $(*)$ we obtain $(M, h^*) \models \psi$. As h^* was arbitrary, $M \models \chi$. As M was arbitrary, we have shown χ to be valid in $\mathcal{T}[\mathcal{B}]$. This proves the claim.

Thus we can apply the algorithm for $\mathcal{T}$ to χ^* in order to decide χ in $\mathcal{T}[\mathcal{B}]$. ■

Proposition 15.6.2 *Assume $\mathcal{K}, \mathcal{K}'$ are classes of structures, $\mathcal{K}$ is decidable in universal monadic second-order logic, and $\delta = \langle \varphi, \alpha \rangle$ is a definition over $\mathcal{K}$ in the sense of definition 13.4.1, such that $\mathcal{K}' = \{\delta(A) \mid A \in \mathcal{K}\}$. Then $\mathcal{K}'$ is also decidable in universal monadic second-order logic.*

Proof. Let $\sigma = \forall P_1 \ldots P_m \psi(P_1, \ldots, P_m)$ be a universal monadic second-order sentence of the signature of $\mathcal{K}'$, where ψ is a first-order monadic sentence. Then, clearly, σ is $\mathcal{K}'$-valid if and only if $\forall P_1 \ldots P_m \psi\!\uparrow\!(P_1, \ldots, P_m)$ is valid in $\mathcal{K}$, where $\psi\!\uparrow$ is as in lemma 13.4.2. Hence $\mathcal{K}'$ reduces to $\mathcal{K}$, so we can use proposition 15.3.3. ■

Hence any class $\mathcal{K}'$ of flows of time obtained by the methods of section 13.5 has decidable universal monadic second-order theory, if the original classes do. All the examples of the early part of section 13.5 (before 13.5.1) are thus seen to have decidable universal monadic second-order theory. Using similar techniques, we can show that the temporal logics on $\mathcal{K}'$ obtained by the methods of section 13.5.3 are decidable, if the original temporal logics are decidable.

Remark 15.6.3 As we mentioned, the technique of deducing decidability of classes of 'lexicographic sums' of structures from the decidability of the components was also at the heart of Gurevich's proof of decidability of the universal monadic second-order theory of $\mathbb{R}$. The reader may be wondering if the method can be extended to full monadic second-order logic. The answer is 'yes': it was done by Shelah in [Shelah, 1975]. See [Gurevich, 1985] for a description. Shelah's technique is very powerful and allows uniform proofs of most second-order decidability results.

15.7 Decidability of earlier systems

Here we will run through the systems mentioned in earlier chapters and discuss their decidability.

1. *The system K_t of section 3.2.* An axiomatic presentation of this system was given in section 3.2, and its soundness and completeness for the class of all flows of time established. K_t was proved to be decidable in section 3.3, by showing that it is also sound and complete for the class of all finite flows of time (the *finite model property*). (One can also use Rabin's theorem — see [Gabbay, 1975a].) The reader may have noticed that the proof of the finite model property (theorem 3.3.1) actually shows more: namely, that if A has a model then it has a model with flow of time of cardinality at most 2^n, where A, regarded as a text string, has length n. This is because the size of the flow of time of the model built in the theorem is at most 2^k, where k is the number of subformulae of A. Clearly, k is just the number of atoms in A plus the number of connectives in A, and this sum is, clearly, at most n. We can use this information in two ways. Firstly, to find out whether a formula A is valid in the class of all flows of time, we merely evaluate A in all structures of size 2^n, where n is the length of A. If it is true in all of them, it is valid; otherwise not. (We do not need to consider structures smaller than 2^n because any such structure can be extended to one of size 2^n by adding new points, and if we ensure that these extra points are not related by '$<$' to the old ones, the truth of A will not be affected.) Secondly, we can obtain a rough upper bound of $2^{2^{2n}}$ on the time that it will take to evaluate A in all models of size 2^n and so to decide the validity of A. In [Pratt, 1979], [Pratt, 1980], Pratt improved this bound to about 2^n. Later we will examine complexity in a little more detail.

2. *Löb's system Y1 of section 3.4.* This was an axiom system that was proved sound and complete for the class of all finite linear flows of time. By theorems 15.3.1 and 15.4.3, it is decidable.

3. *The system Y2 of section 3.5.* This is decidable. (Hence we obtain a recursive axiomatization of it, namely the set of valid formulae. It is an open question to find a finite axiomatization.)

It will be recalled that H is the only temporal connective of $Y2$. A model was essentially of the form $(\mathbb{N}, <, h, F)$ where h is a propositional assignment and F a non-principal ultrafilter on $\mathbb{N}$. The semantics of H were given by $\|HA\|_t^h = 1$ if and only if either (i) $t > 0$ and $\|A\|_u^h = 1$ for all $0 < u < t$, or (ii) $t = 0$ and $\{u \mid \|A\|_u^h = 1\} \in F$. A formula A is valid if it is true at 0 in all models.

We can decide this system as follows. Given a formula A, rewrite it in conjunctive normal form, treating each maximal subformula of the form HB as atomic. We obtain $\bigwedge_i \bigvee_j B_{ij}$, where each B_{ij} is atomic, negated atomic, or an 'H-formula' of the form HC or $\sim HC$. Clearly, A is valid if and only if each conjunct $\bigvee_j B_{ij}$ is valid. Now for any formulae B, C and any model $(\mathbb{N}, <, h, F)$, we have $\|\sim HB\|_0^h = \|H\sim B\|_0^h$ and $\|HB \vee HC\|_0^h = \|H(B \vee C)\|_0^h$. (This is because F is an ultrafilter.) Hence we can rewrite the H-formulae in $\bigvee_j B_{ij}$ to a single formula HC, and so it suffices to decide the validity of disjunctions of the form $\bigvee_i \pm p_i \vee HC$, where the p_i are propositional atoms, the '$\pm$' indicating that they may be negated, and C is any formula.

Now $\| \bigvee_i \pm p_i \|_0^h$ depends only on the values $\|p_i\|_0^h$, while $\|HC\|_0^h$ depends only on the values $\|C\|_t^h$ for $t > 0$. Hence $\bigvee_i \pm p_i \vee HC$ is valid if and only if $\bigvee_i \pm p_i$ is valid or HC is valid. It is trivial to decide the former: it is valid if and only if some atom involved occurs both positively and negatively. For the latter, we observe that HC is valid if and only if $\|HC\|_0^h = 1$ for all h and all non-principal ultrafilters F on $\mathbb{N}$, if and only if for all h there is a cofinite set of $t \in \mathbb{N}$ such that $\|C\|_{t+1}^h = 1$. (This is because a set $S \subseteq \mathbb{N}$ is cofinite if and only if $S \in F$ for all non-principal ultrafilters F on $\mathbb{N}$.) This latter condition can be decided quite easily, using (say) the decision procedure for $S1S$ (theorem 15.4.2), or by deciding the validity of the formula FGC in its conventional interpretation in natural numbers time.

4. *The predicate system Q_1K_t of section 4.2.* This is undecidable: first-order predicate logic is undecidable, and so no extension of that system is decidable. More formally, any sentence σ of predicate logic is already a sentence of Q_1K_t; and examination of the semantics in definition 4.2.2 shows that σ is classically valid if and only if it is temporally valid. So any decision procedure for Q_1K_t would be able to decide all valid first-order sentences as a special case; and Church proved that there is no algorithm to do this.

5. *The predicate system Q_2K_t of section 4.2.* This is undecidable for the same reason.

6. *The systems $QFP/\mathcal{K}$, QFP/L and QFP/Q of sections 4.3–4.5.* The system $QFP/\mathcal{K}$ was introduced at the beginning of section 4.3. It is sound and complete for the class $\mathcal{K}$ of all flows of time (theorem 4.3.1). QFP/L is an extension of $QFP/\mathcal{K}$ by two more axioms, that make it sound and complete for linear time (theorem 4.4.1); and QFP/Q has four more axioms to make it sound and complete for rational time $(\mathbb{Q}, <)$ (see section 4.5). Again, all these systems extend classical predicate logic and so are undecidable.

7. *Predicate systems over real, integer, or natural number time (section 4.6).* These are not recursively axiomatizable (corollary 4.6.5) so are certainly not decidable (otherwise, the set of valid sentences would constitute a recursive axiomatization).

8. *The extensions of K_t defined in definition 6.3.1.* These propositional systems involved adding new axioms to the basic temporal system K_t. Sound and complete axiomatic systems were thus obtained for various classes of flows of time, namely:

   (a) linear time without endpoints;

   (b) dense linear time;

   (c) Dedekind complete real time;

   (d) discrete linear time.

   The connectives are still F and P, and propositional atoms are still evaluated at points of time (i.e. monadically). So by theorem 15.3.1 and the results of table 15.1, all these systems are decidable.

9. *The system US_t of definition 6.3.4, and its extensions as in (8) above.* By theorem 6.3.5, each of the systems defined is sound and complete for the class of flows of time that we had in mind, these being as in (8) above. Atoms are evaluated at time points. As U, S are connectives with first-order tables, again these systems are all decidable.

10. *The Stavi connectives over linear time (definition 6.3.6(A)).* The Stavi connectives have first-order tables, and atoms are evaluated monadically. Hence this system is decidable, for the same reason as before.

11. *The system for the connective γ^- (definition 6.3.6(B)).* The axioms for this connective are sound and complete for linear time. Since γ^- has a first-order table, the system is decidable.

12. *The integer time system of definition 6.3.7.* The system is sound and complete for integer time, by theorem 6.3.8. Theorem 15.4.2 shows that the monadic second-order theory of the integers is decidable. The decidability of the system of definition 6.3.7 now follows from theorem 15.3.1, as usual.

13. *The system of section 7.1.* This is a subsystem of the Vlach–Åqvist system below, so is decidable over linear time.

14. *Various interval systems mentioned in examples 7.1.5.* If the atoms are evaluated dyadically, giving true interval logic, the systems tend to be (very) undecidable. See [Halpern and Shoham, 1986].

15. *The system* s *of Vlach and Åqvist (section 7.2).* This is a two-dimensional system, but the atoms are still evaluated monadically (in terms of section 7.1 the validity is *weak validity*). Moreover, the connectives all have first-order tables. The system presented is shown in theorem 7.2.4 to be sound and complete over irreflexive linear time. Hence by theorems 15.3.1 and 15.4.6, it is decidable.

16. *The system* **FP** *of section 7.3.* Over linear time, **FP** is inter-translatable with the system US of (9) above, and the translation is effective. Hence it is decidable over linear time. Alternatively, we can apply theorem 15.3.1, using the fact that atoms are evaluated monadically and the connectives all have first-order tables.

17. *The system* η(**FP**) *of definition 7.3.13.* This is sound and complete over rational time. The semantics are as before. Hence by theorems 15.3.1 and 15.4.4, it is decidable.

18. *The various systems of section 7.4.* The same considerations apply.

19. *The system* **b** *of definition 7.5.1.* This is a three-dimensional system, but still atoms are evaluated monadically and the connectives F^o, P^o have first-order tables. Thus every formula has a first-order table. (Sometimes the reference points are taken to be $\pm\infty$, but by adding $\infty, -\infty$ as new endpoints we can ignore this quirk.) Hence the system is decidable over linear time.

20. *The system of section 7.6.* The situation is similar — the two-dimensional system with a binary connective E is decidable over linear time.

21. *The historical necessity system of section 7.7.* Decidability of 'strong Ockhamist validity' was proved in [Burgess, 1979]. This is the logic

HN2 of section 7.7, but with condition 4 of definition 7.7.1 dropped. That is, it is a system in which histories are *maximal* linear sets of time points, and in which the value of an atom may depend on the history in which it is evaluated. The proof uses Rabin's theorem; also see [Burgess, 1980]. Clearly, a formula A is valid in **HN2** iff $[\Box \bigwedge(q \to Lq)] \to A$ is valid in the strong Ockhamist sense, where the $\bigwedge$ is taken over all atoms occurring in A, and $\Box X$ abbreviates $LX \wedge LGX \wedge H(LX \wedge LGX)$, expressing that X is true at all times in all histories. Hence **HN2** is itself decidable.

The decidability of the system **HN3** of Remark 7.7.14 follows from the result of Gurevich and Shelah noted in section 15.4.9. In this system, the histories are precisely the maximal linear sets of time points, and condition 4 of definition 7.7.1 is reinstated.

We thank Alberto Zanardo for helpful communications on this subject.

22. *The propositional quantification systems of sections 8.1–8.3.* It was proved in corollary 8.3.6 that the logic *PQ(atoms, named)*, where $\exists x$ means that there is an *atom* x, and $\exists x(x \wedge H{\sim}x)$ is valid, is undecidable. The technique was much the same as that of proposition 15.5.1. We remarked in section 8.3 that even the logic *PQ(atoms, linear, named)*, with irreflexive linear time semantics, is undecidable.

We now consider the logic *PQ(full)*, in which we interpret $\exists x$ to mean that there exists a *subset* x, so that in the notation of section 8.1, $\mathbf{S} = 2^T$. We will show that this logic is equivalent to full monadic second-order logic in its expressive power, at least with respect to sentences. First, we show that any temporal formula $A(x_1, \ldots, x_n, q_1, \ldots, q_m)$ has a monadic second-order table $A^*(X_1, \ldots, X_n, Q_1, \ldots, Q_m, t)$. Here, $X_1, \ldots, X_n, Q_1, \ldots, Q_m$ are just monadic predicate variables. We construct A^* by induction on A.

- For an atom q, we let $q^* = Q(t)$.
- For a variable x, we let $x^* = X(t)$.
- The clauses for the boolean connectives are as expected.
- $(FA)^* = \exists u > t(A^*(u))$.
- $(\exists x A)^* = \exists X A^*$.

Clearly, a temporal sentence $A(q_1, \ldots, q_n)$ is valid over a class $\mathcal{K}$ of flows of time iff $\forall t \forall Q_1, \ldots, Q_n A^*$ is valid in $\mathcal{K}$.

A kind of converse also holds, so long as the temporal language can express $\Diamond A$, with semantics $\|\Diamond A\|_t^h = 1$ iff $\|A\|_u^h = 1$ for some time point u. (For example, $\Diamond A$ is expressible in linear time by $A \vee FA \vee PA$.) First let $O(x)$ abbreviate the temporal formula

$$\Diamond x \wedge \forall y(\Box(y \to x) \to (\Box(x \to y) \vee \Box{\sim}y)).$$

In any flow of time $(T, <)$, and any $t \in T$, we have $\|O(q)\|_t^h = 1$ iff there is a unique $u \in T$ with $\|q\|_u^h = 1$. We will now translate a monadic second-order formula $\varphi(x_1, \ldots, x_n, X_1, \ldots, X_m)$, with free point variables $x_1, \ldots, x_n$ and free monadic predicate variables $X_1, \ldots, X_m$, into a temporal formula $\varphi^{\#}(x_1, \ldots, x_n, X_1, \ldots, X_m)$, where in $\varphi^{\#}$ we regard $x_1, \ldots, x_n$ and $X_1, \ldots, X_m$ as propositional variables. We define $\varphi^{\#}$ as follows:

- $(x = y)^{\#}$ is $\Box(x \leftrightarrow y)$.
- $(x < y)^{\#}$ is $\Box(x \to Fy)$.
- $(X(x))^{\#}$ is $\Box(x \to X)$.
- The boolean connectives are treated as usual.
- $(\exists x\varphi)^{\#}$ is $\exists x(O(x) \wedge \varphi^{\#})$.
- $(\exists X\varphi)^{\#}$ is $\exists X(\varphi^{\#})$.

It is clear that for any monadic second-order sentence σ, $\sigma^{\#}$ is a sentence of propositional quantifier logic, and given any flow of time $(T, <)$, we have $(T, <) \models \sigma$ iff $(T, <) \models \sigma^{\#}$. Hence σ is valid in a class $\mathcal{K}$ of flows of time iff $\sigma^{\#}$ is valid over $\mathcal{K}$.

Obviously the maps $A \mapsto A^*$ and $\varphi \mapsto \varphi^{\#}$ are effective. Hence — assuming that $\Diamond$ is expressible — the system *PQ*(*full*) is decidable over a class $\mathcal{K}$ of flows of time iff $\mathcal{K}$ has decidable full monadic second-order theory. This latter condition can be checked by referring to table 15.1. Thus for example, *PQ*(*full,linear*) is undecidable, while on the other hand, the system *PQ*(*full,rational*) with semantics $\mathbb{Q}$ is decidable.

23. *The systems UYF and USF of section 8.4.* We showed at the end of section 8.4 that *USF* is as expressive as *UYF*, so it suffices to consider only the latter logic. It was shown in proposition 8.7.3 that every formula of *UYF* has a monadic second-order table. Hence by Büchi's result on the decidability of full monadic second-order logic over the natural numbers (theorem 15.4.2 above), *UYF* (and also *USF*) is decidable over its intended semantics, $\mathbb{N}$. This was proved in corollary 8.7.9. To decide the satisfiability of a formula A of *USF*, we could instead apply a decision procedure for U, S over $\mathbb{N}$ to the US-formula $\hat{A} \wedge \Box A\dagger$ (see lemma 8.8.1), where $\Box A$ abbreviates $A \wedge GA \wedge HA$.

15.8 Complexity

We briefly discuss some known results on the complexity of decision procedures. We gratefully acknowledge the assistance of Edith Spaan in preparing this section.

Ladner [Ladner, 1977] proved that the validity problem for any modal logic between K and the extension (S4) of K by the reflexivity and transitivity axioms ($\Box A \to A$ and $\Box A \to \Box\Box A$) is *PSPACE*-complete with respect to log-space reduction. Spaan [Spaan, 1993] proves that the same holds for the corresponding temporal logics. Indeed, she gives algorithms that work in space $O(n^2)$ for K_t and its extension by $GA \to A$, and $O(n^4)$ for $K_t + GA \to GGA$ and its extension by $GA \to A$, where the formula to be checked has length n. The approach in both cases is tableau-based.

In [Ono and Nakamura, 1980] it is shown that the complexity of the satisfiability problems in the language with F and P with respect to the following classes of flows of time are all NP-complete:

- linear transitive flows of time;
- linear transitive flows of time without endpoints;
- dense linear transitive flows of time without endpoints.

The satisfiability problem for the flow of time $(\mathbb{N}, \leq)$ in the language with just F (in effect a modal language) was shown to be NP-complete in [Sistla and Clarke, 1985]. The satisfiability problem is also NP-complete for formulae using F and the connective T, where TA ('tomorrow, A') holds at n iff A holds at $n + 1$, so long as all negations are next to atoms. (So, for example, it is impossible to express GA.)

Also, [Sistla and Clarke, 1985] showed that the satisfiability problem for $(\mathbb{N}, <)$ is *PSPACE*-complete for each of the following four systems:

- F, T;
- U (until);
- U, S, T (until, since, tomorrow);
- the logic ETL of Wolper (see [Wolper, 1983], where another proof of the complexity is given).

Note, however, that the tables of the connectives F, U and S used in [Sistla and Clarke, 1985] are not quite the same as ours. For example, for an assignment h into $\mathbb{N}$, and $n \in \mathbb{N}$, they have $\|U(p,q)\|^h_n = 1$ iff there is $m \geq n$ with $m \in h(p)$ and $k \in h(q)$ for all k with $n \leq k < m$.

The expressive power of the four logics cited above varies widely, right up to monadic second-order logic in the case of ETL. None the less, the decision procedures all have the same complexity. It is important that while monadic second-order logic over $\mathbb{N}$ is decidable (by theorem 15.4.2), any decision procedure is non-elementary — cf. [Robertson, 1974]. Hence it is substantially easier to decide the temporal logics than the naïve application of theorem 15.4.2 to the tables of formulae would suggest.

For further information on future temporal logics, see [Emerson, 1990].

15.8.1 PSPACE-completeness

We conclude with the proof of [Sistla and Clarke, 1985] of the *PSPACE*-completeness of the four logics above, but for the tables of the connectives used in this book. That is, the flow of time will always be $(\mathbb{N}, <)$, and for an assignment h into $\mathbb{N}$, and $n \in \mathbb{N}$:

- $\|Tq\|_n^h = 1$ iff $\|q\|_{n+1}^h = 1$;
- $\|Fq\|_n^h = 1$ iff there is $m > n$ with $m \in h(q)$;
- $\|U(p,q)\|_n^h = 1$ iff there is $m > n$ with $m \in h(p)$ and $k \in h(q)$ for all k with $n < k < m$;
- $\|S(p,q)\|_n^h = 1$ iff there is $m < n$ with $m \in h(p)$ and $k \in h(q)$ for all k with $m < k < n$.

We will prove:

Theorem 15.8.1

1. *The satisfiability problem in* $(\mathbb{N}, <)$ *for the language with* F, T *is PSPACE-hard with respect to polynomial time reduction.*
2. *The satisfiability problem in* $(\mathbb{N}, <)$ *for the language with* U, S *is in PSPACE.*

Recall the logic *USF* of section 8.4, with U, S and a fixed point operator φ. We showed its decidability in section 15.7.23 above.

Corollary 15.8.2

1. *The satisfiability problem for any sublanguage of USF that is able to express* F *and* T *is PSPACE-complete (with respect to polynomial time reduction).*

2. *The satisfiability problems over* $(\mathbb{N}, <)$ *for USF and for the four languages with (respectively) connectives* $\{U, S\}$, U, $\{F, T, S\}$ *and* $\{F, T\}$ *are all PSPACE-complete.*

Proof. (1). Clearly the satisfiability problem for any such language $\mathcal{L}$ is *PSPACE*-hard since it can express F, T which have *PSPACE*-hard satisfiability problem. But given a formula A of *USF*, we can convert it in polynomial time into the US-formula $\hat{A} \wedge \Box(A\dagger)$ as in lemma 8.8.1, which is satisfiable iff A is. Here, $\Box X$ abbreviates $X \wedge GX \wedge HX$. Thus the satisfiability problem for $\mathcal{L}$ reduces in p-time to the satisfiability problem for US, and so is in *PSPACE*. Hence it is *PSPACE*-complete. (2) is immediate from (1), as FA is equivalent to $U(A, \top)$ and TA to $U(A, \bot)$ in $(\mathbb{N}, <)$. ∎

Proof of theorem 15.8.1(1) Let Σ be an alphabet, and write Σ^* for the set of words (finite sequences of elements) of Σ. Let $L \subseteq \Sigma^*$ be a language in *PSPACE*. We seek a function $f : \Sigma^* \to \{F, T\text{-formulae}\}$, such that:

- for all words $w \in \Sigma^*$, $w \in L \iff f(w)$ is satisfiable in $(\mathbb{N}, <)$;
- we can compute $f(w)$ in time polynomial in the length of w.

We know that there is a one-tape Turing machine $M = (Q, \Sigma, \delta, q_0, q_A, q_R)$ and a polynomial $p(n)$ such that, given a word w of Σ of length n, M runs in space at most $p(n)$, reaching state q_A if $w \in L$ and state q_R otherwise. (We say that M accepts w in the former case, and that M rejects w in the latter.) Q is the (finite) set of states of M, and $q_0, q_A, q_R \in Q$, q_0 being the initial state. $\delta : Q \times \Sigma \to Q \times \Sigma \times \{0, \pm 1\}$ is the transition table — $\delta(q, a) = (q', a', d)$ means that if M is in state q and its head reads an a in the current tape square, then M writes an a' in that square, moves its head one square in direction d (0 = stay put, 1 = right, -1 = left) and goes to state q'. We may make some harmless simplifying assumptions. First, we can assume that $p(n) \geq n$. M can be assumed to have a single one-way infinite tape, with squares indexed by $\mathbb{N}$, and we can suppose that M never tries to move its head left from square 0 or right from square $p(n) - 1$ on an input of length n. We may also add a 'blank' character to Σ: a tape square having this symbol indicates that it is empty. Finally, we can suppose that $\delta(q_A, a) = (q_A, a, 0)$ for all $a \in \Sigma$, and similarly for q_R. So, once in a final state, M stays there, the tape remaining the same forever, and so its run is always infinite.

So let w be a word of input of length n', and let $n = p(n')$. We will construct in polynomial time a formula of F, T that is satisfiable iff M accepts w.

If we regard the elements of $\Sigma \cup (Q \times \Sigma)$ as propositional atoms, we may code a run of M on w by their values in $\mathbb{N}$. We only need to describe the contents of the first n squares of the tape of M at each stage of the run, since n bounds the space that M uses. So we let the first n numbers of $\mathbb{N}$ code the initial configuration of M (i.e. its state, head position, and tape contents initially), the next n numbers code the configuration of M after one instruction, the next n the configuration of M after two, and so on. The coding works like this. Consider the configuration of M after m instructions ($m \geq 0$). Suppose the tape of M contains symbols $a_0, \ldots, a_{n-1}$ in squares $0, \ldots, n-1$ respectively. Suppose also that M is now in state q and its head over square k. Then the mth block in $\mathbb{N}$ (the atoms true at $mn, mn+1, \ldots, mn+n-1$) looks like this:

$$a_0, a_1, \ldots, a_{k-1}, (q, a_k), a_{k+1}, \ldots, a_{n-1}.$$

Notice how we code the head position by the pair (q, a_k). No other atoms are true except those indicated.

The trick now is to write a formula of F, T expressing that a structure $(\mathbb{N}, <, h)$ codes a successful run of M on the input word w. To do this, we first write a formula stating explicitly what is true at the first n numbers — a conjunction of the form $\bigwedge_{i<n} T^i p_i$, where p_i is the atom we want true at i. (Here, $T^0 A = A$ and $T^{i+1} A = TT^i A$, as usual.) This formula states in effect that the initial state is q_0, the initial head position is 0, and the initial tape contents are the characters of w followed by blanks up to square $n-1$. Given w, it can be written in time quadratic in n.

Then, we write a formula expressing the relation between each block of n numbers and the next. So, for example, if $\delta(q, a) = (q', a', -1)$ then we would include a conjunct of the form $G[b \wedge T(q, a) \to T^n((q', b) \wedge Ta')]$ for each $b \in \Sigma$. This will enforce that if b and (q, a) are true at locations $i-1$ and i in a block, respectively, then $(q', b), a'$ are true at the same locations in the next block. It reflects the fact that if M is in state q and its head is in square i reading an a, then M will write a' in square i, go to state q', and move left to square $i-1$. The number of such formulae required is independent of w, and each can be written in time linear in n.

Finally, we say that the run is accepting: $F \bigvee_{a \in \Sigma} (q_A, a)$. We also have to say that exactly one atom is true at each time: $G^*(x \to \bigwedge_{y \neq x} {\sim} y)$ where x and y are atoms and $G^* A$ is $A \wedge GA$. The conjunction $f(w)$ of all these formulae can be written in time polynomial in n, and so in n'. Moreover, $f(w)$ is satisfiable iff it is true at 0 in some structure (as there is no P in the language), iff some structure codes an accepting run of M on w, iff M accepts w, iff $w \in L$. Thus we have shown that any L in *PSPACE* reduces to the satisfiability problem for F, T in polynomial time, so that the latter problem is *PSPACE*-hard with respect to polynomial time reduction. ■

Proof of theorem 15.8.1(2) To show that the satisfiability problem for U, S over $\mathbb{N}$ is in *PSPACE*, we conduct a Feferman–Vaught-style analysis of truth of formulae in a structure (cf. section 6.9.4). The position is simpler than before because the logic U, S is simpler than first-order logic, yet more complex because we require effective bounds on the size of structures.

We want a *PSPACE* algorithm to decide the satisfiability of an arbitrary formula A of U, S. We may replace A by $A \vee FA$, whose length is polynomial in the length of A. Clearly, A is satisfiable in $\mathbb{N}$ iff $A \vee FA$ is satisfiable at 0, i.e. there is an assignment h such that $\|A \vee FA\|_0^h = 1$. So we need only show that 'satisfiability at 0' is in *PSPACE*. The proof now is in two stages.

1. First we show that if A (of size n) is satisfiable at 0 then in fact it is satisfiable at 0 in a structure of the form $T + (T' * \mathbb{N})$, where T, T' are linear temporal structures, T has size at most 2^n and T' has size at most $n.2^n$. (For the notation see section 6.9.4; it means that $(\mathbb{N}, <, h)$ is, up to isomorphism, a copy of T followed by a copy of T', followed by another copy of T', and so on, ω times.)

2. Then we give a non-deterministic *PSPACE* algorithm that guesses such a structure and checks that A is true at 0 in it.

Hence 'satisfiability at 0' is in *NPSPACE*. But by a theorem of Savitch [Savitch, 1970], *NPSPACE* = *PSPACE*, so we are done.

So fix a formula A of length n.

Definition 15.8.3

1. Given $x < y$ in a temporal structure M, we write $M(x, y]$ for the structure obtained from M by deleting the half-open interval $(x, y] = \{z \mid x < z \leq y\}$ of M. That is,

$$M(x,y] \quad = \quad M\lceil\{t \mid t \leq x \vee t > y\}.$$

2. Let $\mathcal{S}$ be the set of subformulae of A.

3. If $(T, <, h)$ is any linear temporal structure, and $t \in T$, we define the *colour* of t in $(T, <, h)$ to be the set $\{B \in \mathcal{S} \mid \|B\|_t^h = 1\}$.

Clearly, $\mathcal{S}$ has size at most n. There are at most 2^n possible colours — the number of subsets of $\mathcal{S}$ is at most this number. The colour of a point depends on the structure in which we evaluate.

Lemma 15.8.4 (snipping lemma) *Suppose that $M = (T, <, h)$ is a linear temporal structure, and $x < y$ in T have the same colour in M. Then each $x \notin (x, y]$ has the same colour in the substructure $M(x, y]$ as it does in M.*

Proof. A simple induction on the structure of a subformula of A. ■

The lemma says that we can 'snip out' a part $(x, y]$ of a structure and preserve the colours of all remaining points, so long as x and y have the same colour originally.

Now assume that A is satisfiable at 0. Let $M = (\mathbb{N}, <, h)$ be such that $||A||_0^h = 1$. Clearly, we can write M as $T + T' + T''$, where

- T and T' are finite temporal structures and T'' has flow of time $(\mathbb{N}, <)$;
- the last point of T and the last point of T' have the same colour in M;
- if $U(B, C) \in \mathcal{S}$ is true in M at the last point t of T, then there is $u \in T'$ such that B is true at u in M.

For example, we can choose points $t_0 < t_1 < \ldots$ in $\mathbb{N}$ all of the same colour in M, and let $T = M\lceil[0, t_0]$, $T' = M\lceil(t_0, t_\ell]$ and $T'' = M\lceil\{u \mid u > t_\ell\}$ for large enough ℓ. Choose $M = T + T' + T''$ so that T is as small as possible, of size k, say. Then fix $M = T + T' + T''$ so that T has size k and T' is as small as possible — of size k', say — subject to this constraint.

Lemma 15.8.5 *$k \leq 2^n$ and $k' \leq n.2^n$.*

Proof. If $k > 2^n$ then there are distinct points $x < y$ in T of the same colour in M. Use the preceding lemma to snip out $(x, y]$ from M; as colours are preserved, we obtain a structure with smaller k, a contradiction.

If $k' > n.2^n$ then T' contains $n + 1$ points $a_0 < \ldots < a_n$ of the same colour in M. These define n pairwise disjoint intervals $(a_0, a_1]$, $(a_1, a_2]$, $\ldots$, $(a_{n-1}, a_n]$ in T'. Now, clearly, there are fewer than n formulae in $\mathcal{S}$ of the form $U(B, C)$ and true at the last point of T. For each such formula $U(B, C)$ choose a point in T' where B is true in M. So one of the n intervals does not contain any of the points chosen. If we snip out this interval from M we obtain a structure with the same k and smaller k', a contradiction. ■

Proposition 15.8.6 *A is true at 0 in the periodic structure $M^* = T + (T' * \mathbb{N})$.*

Proof. Define a map $\theta : M^* \to M$ as follows. There is an obvious surjection from M^* on to $T + T'$: it is the identity on T and maps all copies of T' in M^* on to the first copy. But $T + T'$ embeds into M; and we take θ to be the composition of these two maps.

We now show that θ preserves colours. We prove by induction on the complexity of $B \in \mathcal{S}$ that for all $t \in M^*$, the truth value of B at t in M^* is the same as its value at $\theta(t)$ in M. This will complete the proof, as $A \in \mathcal{S}$ and A is true in M at $\theta(0)$.

The atomic and boolean cases are easy. Consider the case $U(B, C) \in \mathcal{S}$, and assume the result for B and C (which are also in $\mathcal{S}$). There are three subcases.

- Suppose that $t \in M^*$ is in T. If $U(B, C)$ is true in M^* at t, then in M^*, there is $u > t$ at which B is true, C being true between t and u. Note that $\theta(u) > \theta(t)$. By the inductive hypothesis, B is true in M at $\theta(u)$ and C is true between $\theta(t)$ and $\theta(u)$. Hence $U(B, C)$ holds at $\theta(t)$ in M. Conversely, if $U(B, C)$ holds at $\theta(t)$ in M then there is $u \in T + T'$ where B holds, C holding between t and u in M. For otherwise, $U(B, C)$ holds at the last point of T in M but B does not hold at any point in T' in M, contradicting the choice of T'. It now follows from the inductive hypothesis that $U(B, C)$ holds at t in M^*.

- Suppose that $t \in M^*$ is the last (i.e. maximal) element of a copy of T'. By the inductive hypothesis, all copies of T' in M^* agree with respect to the truth of B and C, since their images under θ are the same. As U is pure future, they all agree with respect to $U(B, C)$, too. It follows that $U(B, C)$ holds at t in M^* iff it holds at the last point of T in M^*, iff it holds at the last point s of T in M, by the first case. But s and the last point $\theta(t)$ of T' in M have the same colour. Hence $U(B, C)$ holds at t in M^* iff it holds at $\theta(t)$ in M.

- Suppose that $t \in M^*$ is in a copy of T', but not its last element. Let $t' > t$ be the last element of this copy. Assume that $U(B, C)$ holds in M^* at t. If there is u with $t < u \leq t'$ at which B holds (in M^*), C holding between t and u, then by the inductive hypothesis, B holds in M at $\theta(u)$, C holding between $\theta(t)$ and $\theta(u)$. So $U(B, C)$ holds in M at $\theta(t)$. Otherwise, C holds in M^* between t and t', and $C \wedge U(B, C)$ holds at t'. By the inductive hypothesis and the preceding case, this is iff C holds in M between $\theta(t)$ and $\theta(t')$, and $C \wedge U(B, C)$ holds at $\theta(t')$, so that $U(B, C)$ holds at $\theta(t)$ in M. The converse is similar.

Now consider the case $S(B, C)$. It is clear (using the inductive hypothesis) that the value of $S(B, C)$ is the same at t in M^* as at $\theta(t)$ in M, whenever $t \in T + T'$. Hence it suffices to show that if t, u are in copies of T' in M^* and $\theta(t) = \theta(u)$ then t and u agree on the value of $S(B, C)$. Inductively, all copies of T' in M^* agree with respect to B, C.

Assume first that t, u are the last elements of their respective copies, and that $S(B, C)$ is true at t in M^*.

1. If there is $s < t$ in the same copy of T', such that B holds at s and C holds between s and t, then the same pattern can be found at u, so that $S(B, C)$ is true at u in M^*.

2. If C holds at every $s < t$ in the same copy of T', and B holds at the element x of M^* just before this copy, then B holds at the last element of every copy of T', and also at the last element of T. (Recall that the last element of T has the same colour in M as the last element of T'.) Hence this pattern can be found at u, too, so that $U(B, C)$ holds at u.

3. If C holds at every $s < t$ in the same copy of T' as t, and $C \wedge S(B, C)$ holds at the element x of M^* just before this copy, then in fact C holds throughout every copy of T', and $C \wedge S(B, C)$ holds at the last point of T. Hence $S(B, C)$ holds everywhere in every copy of T'.

Finally, if t, u are not the last elements of their copies of T', either (1) above holds (in which case the same argument works), or else C holds at every $s < t$ in the same copy of T' as t, and $C \wedge S(B, C)$ holds at the element x of M^* just before this copy. By the above, and the inductive hypothesis, $C \wedge S(B, C)$ will hold in M^* at the last element of T and every copy of T'. So again, the same pattern can be found at u, too, so that $U(B, C)$ holds at u. ■

We have shown:

Theorem 15.8.7 *If a US-formula A of length n is satisfiable at 0 in a structure with flow of time $\mathbb{N}$, then it is satisfiable in a structure of the form $M = T + (T' * \mathbb{N})$ where:*

1. *T is linear and has k points, for some $k \leq 2^n$.*

2. *T' is linear and has k' points, for some $k' \leq n.2^n$.*

3. *The same subformulae of A are true (in M) at any point in a copy of T' as are true at the corresponding point in any other copy of T'.*

4. *The same subformulae of A are true (in M) at the last point of T as are true at the last point of T'.*

5. *If a subformula $U(B,C)$ of A is true in M at the last point of T, then there is $u \in T'$ such that B is true at u in M.*

We remark that the same can be shown by a different argument, using Ramsey's theorem [Ramsey, 1930].

We now describe a non-deterministic Turing machine N that, given A of length n, runs in space polynomial in n, guesses such a structure, and checks that A is true at 0 in it. Recall that $\mathcal{S}$ is the set of all subformulae of A.

First, N guesses k and k'. Writing them in binary, it uses polynomial space in n. It also initializes a counter to 0. It then guesses the set S_0 of formulae in $\mathcal{S}$ that are true at 0. (It takes space $O(n^2)$ to write down S_0.) It checks the propositional consistency of S_0: i.e. that if $\sim B \in \mathcal{S}$ then $\sim B \in S_0$ iff $B \notin S_0$, and that if $B \wedge C \in \mathcal{S}$ then $B \wedge C \in S_0$ iff $B, C \in S_0$. It also checks that $A \in S_0$, and also checks that S_0 contains no formulae of the form $S(B,C)$ (which are false at 0). Then at each stage i with $1 \leq i < k+k'$, it guesses $S_i \subseteq \mathcal{S}$ (the formulae it hopes are true at i) and checks its propositional consistency and two relationships with S_{i-1}:

- if $U(B,C) \in \mathcal{S}$ then $U(B,C) \in S_{i-1}$ iff $B \in S_i$ or $(C, U(B,C) \in S_i)$;
- if $S(B,C) \in \mathcal{S}$ then $S(B,C) \in S_i$ iff $B \in S_{i-1}$ or $(C, S(B,C) \in S_{i-1})$.

If these conditions are met, it discards S_{i-1} and increments its counter. However, at the stage when it guesses S_{k-1}, which it knows by comparing its counter with k, it saves the result (S_{k-1}) for later, and re-initializes its counter. At this stage it also makes up a list of all formulae in S_{k-1} of the form $U(B,C)$. From then on, as soon as S_i has been guessed (for each $k \leq i < k+k'$), it deletes from this list any $U(B,C)$ with $B \in S_i$.

When it has finally guessed $S_{k+k'-1}$, which it knows by comparing its counter with k', and made the checks above, it tests to see if the list is empty and if $S_{k+k'-1}$ is equal to the saved copy of S_{k-1}. If so, it accepts; otherwise it rejects.

If A is satisfiable at 0 then by theorem 15.8.7 it is satisfiable in a structure $T + (T' * \mathbb{N})$ with k, k' as above. So N will accept A if it makes the right choices (as suggested by $T + (T' * \mathbb{N})$). Conversely, if N accepts A then we can easily construct a structure in which A is true at 0. We define

$$
\begin{aligned}
T &= (\{0,1,\ldots,k-1\},<,h) && \text{where } h(q) = \{i < k \mid q \in S_i\},\\
T' &= (\{0,1,\ldots,k'-1\},<,h') && \text{where } h'(q) = \{i < k' \mid q \in S_{k+i}\}.
\end{aligned}
$$

We then let $M^* = T + (T' * \mathbb{N})$. M^* has flow of time isomorphic to $\mathbb{N}$. Define a map $\theta : \mathbb{N} \to \mathbb{N}$ by : θ is the identity on T and takes all copies of

T' to the first copy. Then it is simple to show by induction on $B \in \mathcal{S}$ that B is true at t in M^* iff $B \in S_{\theta(t)}$. Hence A is true at 0 in M^*.

It is clear that N runs in space linear in n. Hence 'satisfiability at 0' is in *NPSPACE*, as required. ■

Bibliography

[Abadi, 1987a] M. Abadi. The power of temporal proofs. In *Proc. Symp. on Logic in Computer Science*, pages 176–186, Ithaca, June 1987.

[Abadi, 1987b] M. Abadi. *Temporal-logic theorem proving.* PhD thesis, Department of Computing, Stanford University, March 1987. STAN-CS-87-1151.

[Abadi, 1989] M. Abadi. The power of temporal proofs. *Theoretical Computer Science*, 64:35–84, 1989.

[Abadi, 1990] M. Abadi. Errata for 'the power of temporal proofs'. *Theoretical Computer Science*, 70:275, 1990.

[Abadi and Manna, 1985] M. Abadi and Z. Manna. Non-clausal temporal deduction. *Lecture Notes in Computer Science*, 193:1–15, June 1985.

[Abadi and Manna, 1986a] M. Abadi and Z. Manna. Modal theorem proving. In *Proc. 8th CADE.* Springer, 1986. LNCS 230.

[Abadi and Manna, 1986b] M. Abadi and Z. Manna. A timely resolution. In *Proc. Symp. on Logic in Computer Science*, pages 176–186, Boston, June 1986.

[Abadi and Manna, 1987] M. Abadi and Z. Manna. Temporal logic programming. In *Proc. Int. Symp. on Logic Programming*, IEEE, 1987.

[Abadi and Manna, 1989] M. Abadi and Z. Manna. Temporal logic programming. *J. Symbolic Computation*, 8, 1989.

[Abadi and Manna, 1990] M. Abadi and Z. Manna. Non-clausal deduction in first-order temporal logic. *J. ACM*, 37, 1990.

[Abadi *et al.*, 1989] M. Abadi, L. Lamport, and P. Wolper. Realizable and unrealizable specifications of reactive systems. In *Proc. 16th Colloquium on Automata, Languages and Programming*, 1989. Springer LNCS 372.

[Allen, 1981a] J. F. Allen. A general model of action and time. Technical report, University of Rochester, 1981. TR97.

[Allen, 1981b] J. F. Allen. An interval based representation of temporal knowledge. In *Proc. 7th IJCAI*, 1981.

[Allen, 1983] J. F. Allen. Maintaining knowledge about temporal intervals. *Communications of the ACM*, 26(11):832–843, November 1983.

[Allen, 1984] J. F. Allen. Towards a general theory of action and time. *Artificial Intelligence*, 23(2):123–154, 1984.

[Allen and Hayes, 1985a] J. F. Allen and P. Hayes. A commonsense theory of time: the longer paper. Technical report, University of Rochester, 1985.

[Allen and Hayes, 1985b] J. F. Allen and P. J. Hayes. A commonsense theory of time. In *Proceedings IJCAI*, pages 528–531, 1985.

[Allen and Hayes, 1985c] J. F. Allen and P. J. Hayes. Moments and points in an interval-based temporal logic. Technical Report TR180, Department of Computer Science, University of Rochester, 1985.

[Allen and Hayes, 1987] J. F. Allen and P. Hayes. Short time periods. In *Proc. 10th IJCAI*, 1987.

[Allen and Kautz, 1983] J. F. Allen and H. Kautz. A model of naïve temporal reasoning. In R. Moore J. Hobbs, editor, *Contributions to Artificial Intelligence, Volume 1.* Ablex, Norwood, N.J., 1983.

[Allen and Koomen, 1983] J. F. Allen and J. Koomen. Planning using a temporal world model. *IJCAI*, 8:711–714, 1983.

[Almeida, 1983] J. Almeida. Reasoning about the temporal structure of narratives. Technical report, University at Buffalo, 1983. 87-10.

[Almeida and Shapiro, 1983] J. Almeida and S. Shapiro. Reasoning about the temporal structure of narrative texts. In *Proc. 5th Meeting, Cognitive Sci. Soc.*, 1983.

[Alur and Henzinger, 1989a] R. Alur and T. A. Henzinger. Real-time logics: Complexity and expressiveness. *Submitted to LICS*, 1989.

[Alur and Henzinger, 1989b] R. Alur and T. A. Henzinger. A really temporal logic. In *Proc. 30th IEEE. Symp. Found. of Comp. Sci.*, pages 164–169, 1989.

[Amir, 1982a] A. Amir. Expressive completeness failure in branching time structures. *Journal of Computer and System Sciences*, 34(1), 1982.

[Amir, 1982b] A. Amir. *Functional completeness in tense Logic*. PhD thesis, Bar-Ilan University, Ramat-Gan, Israel, 1982.

[Amir, 1985] A. Amir. Separation in nonlinear time models. *Information and Control*, 66:177 – 203, 1985.

[Amir and Gabbay, 1987] A. Amir and D. M. Gabbay. Preservation of expressive completeness in temporal models. *Information and Computation*, 72:66–83, 1987.

[Anderson, 1938] J. Anderson. The problem of causality. *Austr. J. Phil.*, 2, 1938.

[Anderson *et al.*, 1982] T. L. Anderson, A. Bolour, L. J. Dekeyser, and H. K. T. Wong. The role of time in information processing: A survey. *ACM SIGMOD. Review*, 12(3), April 1982.

[Andréka and Németi, 1977] H. Andréka and I. Németi. Completeness of Floyd method w.r.t. nonstandard time models. *Seminar Notes, Math. Inst. Hungar. Acad. Sci. — SZKI*, 1977. in Hungarian.

[Andréka *et al.*, 1978] H. Andréka, I. Németi, and I. Sain. On the completeness of Floyd–Hoare logic. *Preprint Math. Inst. Hungar. Acad. Sci. Budapest*, 1978.

[Andréka *et al.*, 1979] H. Andréka, I. Németi, and I. Sain. Completeness problems in verification of programs and program schemes. In J. Bečvár, editor, *Mathematical Foundations of Computer Science '79 (Proc. Conf. Olomouc Czechoslovakia, 1979)*, volume 74 of *Lecture Notes in Computer Science*. Springer-Verlag, Berlin, 1979.

[Andréka *et al.*, 1982a] H. Andréka, I. Németi, and I. Sain. A complete logic for reasoning about programs via nonstandard model theory, part I. *Theoretical Computer Science*, 17:193–212, 1982.

[Andréka *et al.*, 1982b] H. Andréka, I. Németi, and I. Sain. A complete logic for reasoning about programs via nonstandard model theory, part II. *Theoretical Computer Science*, 17:259–278, 1982.

[Andréka *et al.*, 1984] H. Andréka, I. Németi, and I. Sain. Abstract model theoretic approach to algebraic logic, 1984. Unpublished manuscript.

[Andréka *et al.*, 1991] H. Andréka, I. Németi, and I. Sain. On the strength of temporal proofs. *Theoretical Computer Science*, 80:125–151, 1991.

[Andréka *et al.*, 1992] H. Andréka, V. Goranko, S. Mikulas, I. Németi, and I. Sain. Effective temporal logics of programs, 1992.

[Anscombe, 1964] G. Anscombe. Before and after. *Phil. Review*, 73, 1964.

[Åqvist, 1966] L. Åqvist. 'Next' and 'Ought': alternative foundations for von Wright's tense logic with an application to deontic logic. *Logique et Analyse*, 9, 1966.

[Åqvist, 1976] L. Åqvist. Formal semantics for verb tenses as analyzed by Reichenbach. In T. van Dijk, editor, *Pragmatics of Language and Literature*, pages 229–236. North Holland, Amsterdam, 1976.

[Åqvist, 1979] L. Åqvist. A conjectured axiomatization of two-dimensional Reichenbachian tense logic. *J. of Philosophical Logic*, 8:1–45, 1979.

[Åqvist and Guenthner, 1977] L. Åqvist and F. Guenthner, editors. *Tense Logic*. Nauwelaerts, Louvain, 1977.

[Åqvist and Guenthner, 1978] L. Åqvist and F. Guenthner. Fundamentals of a theory of verb aspect and events within the setting of an improved tense logic. In F. Guenthner and C. Rohrer, editors, *Studies in Formal Semantics*, pages 167–200. North Holland, Amsterdam, 1978.

[Åqvist and Hoepelman, 1981] L. Åqvist and J. Hoepelman. Some theorems about a 'tree' system of deontic tense logic. In R. Hilpinen, editor, *New Studies in Deontic Logic*, pages 187–221. Reidel, Dordrecht, 1981.

[Babb, 1990] E. Babb. An incremental pure logic language with constraints and classical negation. In T. Dodd, R. Owens, and S. Torrance, editors, *Logic Programming: Expanding the Horizons*, pages 14–63. Blackwells, Oxford, 1990.

[Babb *et al.*, 1991] E. Babb, H. Barringer, and D. M. Gabbay. HF-PLL – a metalevel rewrite language for parallel planning. Technical report, Imperial College and ICL. Bracknell, 1991.

[Bach, 1981] E. Bach. On time, tense and aspect: An essay in English metaphysics. In Cole, editor, *Radical pragmatics*, pages 63–81. Academic Press, New York, 1981.

[Balzer, 1978] W. Balzer. *Empirische Geometrie und Raum-Zeit- Theorie in megentheoretischer Darstellung*. Scriptor Verlag, Kronberg, 1978.

[Bannieqbal and Barringer, 1986] B. Bannieqbal and H. Barringer. A study of an extended temporal language and a temporal fixed point calculus. Technical Report UMCS-86-10-2, Department of Computer Science, University of Manchester, 1986.

[Barringer, 1987] H. Barringer. The use of temporal logic in the compositional specification of concurrent systems. In A Galton, editor, *Temporal Logics and Their Applications.* Academic Press, 1987.

[Barringer and Gabbay, 1990] H. Barringer and D. M. Gabbay. The imperative future: Past success → future actions. Technical report, University of Manchester and Imperial College, 1990.

[Barringer *et al.*, 1990] H. Barringer, M. Fisher, D. Gabbay, G. Gough, and R. Owens. Metatem: A framework for programming in temporal logic. In *Proc. of REX Workshop: Stepwise Refinement of Distributed Systems – Models, Formalisms and Correctness*, pages 94–129. LNCS Vol 430, Springer-Verlag, 1990.

[Baudinet, 1988] M. Baudinet. On the semantics of temporal logic programming. Technical report, Stanford University, 1988. STAN-CS-88-1203.

[Baudinet, 1989] M. Baudinet. Temporal logic programming is complete and expressive. In *Proc. of ACM Symposium on Principles of Programming Languages.* Austin, Texas, January 1989.

[Bäuerle, 1979a] R. Bäuerle. *Temporale Deixis, Temporale Frage.* Gunter Narr Verlag, Tübingen, 1979.

[Bäuerle, 1979b] R. Bäuerle. Tense logics and natural language. *Synthese*, 40:226–230, 1979.

[Bäuerle and von Stechow, 1980] R. Bäuerle and A. von Stechow. Finite and non-finite temporal constructions in german. In C. Rohrer, editor, *Proc. of the Stuttgart Conference on the Logic of Tense and Quantification.* Max Niemayer Verlag, Tübingen, 1980.

[Bayart, 1970] A. Bayart. On truth-tables for M, B, S4 and S5. *Logique et Analyse*, 13, 1970.

[Ben-Ari *et al.*, 1981a] M. Ben-Ari, Z. Manna, and A. Pnueli. The logic of 'nexttime'. In *8th ACM Symposium on Principles of Programming Languages*, 1981.

[Ben-Ari *et al.*, 1981b] M. Ben-Ari, Z. Manna, and A. Pnueli. The temporal logic of branching time. In *Eighth ACM Symposium on Principles of Programming Languages*, pages 164–176. Williamburg, Va, 1981.

[Ben-David, 1987] S. Ben-David. The global time assumption and semantics for concurrent systems. Technical report, Department of Computer Science, Technion, Haifa, 1987.

[Bennett, 1977] M. Bennett. A guide to the logic of tense and aspect in English. *Logique et Analyse*, 20:137–163, 1977.

[Bennett, 1981] M. Bennett. Of tense and aspect: One analysis. in [Tedeschi and Zaenen, 1981], 1981. pages 13–30.

[Bennett, 1982] J. Bennett. Counterfacturals and temporal direction. Technical report, Xerox, Syracuse University, 1982.

[Bennett and Partee, 1972] M. Bennett and B. Partee. Toward the logic of tense and aspect in English. Technical report, Systems Development Corporation Santa Monica, California, reprinted by Indiana University Linguistics Club, Bloomington, 1972.

[Bernstein and Harter, 1981] A. Bernstein and P. K. Harter. Proving real-time properties of programs with temporal logic. In *8th ACM Symposium on Operating System Principles*, 1981.

[Bethke and Richards, 1987] I. Bethke and B. Richards. The temporal logic IQ. In J. Oberlander, editor, *Temporal Reference and Quantification: An IQ Perspective*, pages 114–132. Centre for Cognitive Science, University of Edinburgh, 1987.

[Biró and Sain, to appear] B. Biró and I. Sain. Peano arithmetic for the time scale of nonstandard models for logics of programs. *Annals of Pure and Applied Logic*, to appear.

[Blackburn, 1989] P. Blackburn. *Reasoning in and about Time.* PhD thesis, Centre for Cognitive Science, University of Edinburgh, 1989.

[Blok, 1980] W. Blok. The lattice of modal logics: an algebraic investigation. *J. Symbolic Logic*, 45, 1980.

[Boolos and Jeffrey, 1974] G. Boolos and R. Jeffrey. *Computability and logic.* Cambridge University Press, London, 1974.

[Boolos and Sambin, 1985] G. Boolos and G. Sambin. An incomplete system of modal logic. *J. Philosophical Logic*, 14, 1985.

[Bowen, 1979] K. Bowen. *Model theory for modal logic.* Reidel, 1979.

[Bowen and de Jongh, 1986] K. Bowen and D. de Jongh. Some complete logics for branched time. part 1. Technical Report 86-05, Institute for Language, Logic and Information, 1986.

[Bressan, 1987] A. Bressan. *A general interpreted modal calculus.* Yale University Press, 1987.

[Brown *et al.*, 1987] M. Brown, E. Clarke, and O. Grumberg. Characterising Kripke structures in temporal logic. In *Coll. on trees in algebra and programming, Pisa*, 1987.

[Brzoka, 1989] C. Brzoka. Temporal logic programming: a survey. Technical report, Universität Karlsruhe, 1989.

[Büchi, 1962] J.R. Büchi. On a decision method in restricted second order arithmetic. In *Logic, Methodology, and Philosophy of Science: Proc. 1960 Intern. Congress*, pages 1–11. Stanford University Press, 1962.

[Büchi and Landweber, 1969] J.R. Büchi and L.H. Landweber. Solving sequential conditions by finite state operators. *Trans. AMS*, 138:295–311, 1969.

[Bull, 1964a] R. A. Bull. Axiomatisations of the modal calculus Q. *Notre Dame J. Formal Logic*, 5, 1964.

[Bull, 1964b] R. A. Bull. A note on the modal calculi S4.2 and S4.3. *Zeitschrift Math. Logic und Grundlagen der Mathematik*, 10, 1964.

[Bull, 1965a] R. A. Bull. An algebraic study of Diodorean modal systems. *J. Symbolic Logic*, 30, 1965.

[Bull, 1965b] R. A. Bull. A class of extensions of the modal system S4 with the finite model property. *Zeitschrift Math. Logic und Grundlagen der Mathematik*, 11, 1965.

[Bull, 1966] R. A. Bull. That all normal extensions of S4 have the finite model property. *Zeitschrift Math. Logic und Grundlagen der Mathematik*, 12, 1966.

[Bull, 1967] R. A. Bull. On the extension of S4 with CLMpMLp. *Notre Dame J. Formal Logic*, 8, 1967.

[Bull, 1968a] R. A. Bull. An algebraic study of tense logic with linear time. *J. Symbolic Logic*, 33:27–38, 1968.

[Bull, 1968b] W. Bull. Time, tense and the verb. Technical report, University of California Press, Berkeley, 1968.

[Bull, 1968c] W. Bull. *Time, tense and the verb.* University of California Press, Berkeley, 1968.

[Bull, 1969] R. A. Bull. Notes on a paper on tense logic. *J. Symbolic Logic*, 34, 1969.

[Bull, 1978] R. A. Bull. An approach to tense logic. *Theoria*, 36:282–300, 1978.

[Bull and Segerberg, 1984] R. A. Bull and K. Segerberg. Basic modal logic. In D. Gabbay and F. Günthner, editors, *Handbook of Philosophical Logic II.* Reidel, 1984.

[Bunge, 1959] M. Bunge. *Causality.* Harvard University Press, 1959.

[Burgess, 1978] J. Burgess. The unreal future. *Theoria*, 44, 1978.

[Burgess, 1979] J. P. Burgess. Logic and time. *J. Symbolic Logic*, 44:566–582, 1979.

[Burgess, 1980] J. P. Burgess. Decidability and branching time. *Studia Logica*, 39:203–218, 1980.

[Burgess, 1982a] J. P. Burgess. Axioms for tense logic I: 'since' and 'until'. *Notre Dame J. Formal Logic*, 23(2):367–374, 1982.

[Burgess, 1982b] J. P. Burgess. Axioms for tense logic II: Time periods. *Notre Dame J. Formal Logic*, 23:375–383, 1982.

[Burgess, 1984a] J. P. Burgess. Basic tense logic. In D. M. Gabbay and F. Guenthner, editors, *Handbook of Philosophical Logic, vol II: Extensions of Classical Logic*, pages 89–113. Reidel, Dordrecht, 1984.

[Burgess, 1984b] J. P. Burgess. Beyond tense logic. *J. of Philosophical Logic*, 13:235–248, 1984.

[Burgess, to appear] J. P. Burgess. The unreal future. *Theoria*, to appear.

[Burgess and Gurevich, 1985] J. P. Burgess and Y. Gurevich. The decision problem for linear temporal logic. *Notre Dame J. Formal Logic*, 26(2):115–128, 1985.

[Carnap, 1946] R. Carnap. Modalities and quantification. *J. Symbolic Logic*, 11, 1946.

[Carnap, 1947] R. Carnap. *Meaning and necessity.* University of Chicago Press, 1947.

[Casanova and Furtado, 1982] M. A. Casanova and A. L. Furtado. A family of temporal languages for the description of transition constraints. In *3rd Workshop on Logical Bases for Databases*, Toulouse, France, December 1982.

[Casley, 1986] R. Casley. A proof editor for propositional temporal logic. Technical report, Stanford University, 1986. STAN-CS-86-1109.

[Cavalli and del Cerro, 1984] A. Cavalli and L. Farinas del Cerro. *A decision method for linear temporal logics.* Springer, 1984. LNCS 170.

[Chang and Keisler, 1990] C.C. Chang and H.J. Keisler. *Model Theory.* North-Holland, Amsterdam, 3rd edition, 1990.

[Chapman, 1978] T. Chapman. A modal logic with temporal variables. *Notre Dame J. Formal Logic*, 19, 1978.

[Cheeseman, 1983] P. Cheeseman. A representation of time for planning. Technical report, Stanford University AI Center, 1983. Technical note 278.

[Cheeseman, 1984] P. Cheeseman. A representation of time for automatic planning. In *Proc. IEEE. Conference on Robotics*, 1984.

[Chellas, 1969] B. Chellas. *The Logical Form of Imperatives.* Perry Lane Press, Stanford, 1969.

[Chellas, 1971] B. Chellas. Imperatives. *Theoria*, 37:114–129, 1971.

[Chellas, 1980] B. Chellas. *Modal logic: an introduction.* Cambridge University Press, 1980.

[Clark and McCabe, 1984] K. L. Clark and F. G. McCabe. *Micro-PROLOG, Programming in Logic.* Prentice-Hall International, Englewood Cliffs, 1984.

[Clarke and Emerson, 1981] E. Clarke and E. Emerson. Synthesis of synchronization skeletons for branching time temporal logic. In *Proc. Workshop on Logic of Programs, Yorktown Heights, NY*, pages 52–71. Springer, Berlin, 1981.

[Clarke *et al.*, 1985] E. M. Clarke, M. C. Browne, E. A. Emerson, and A. P. Sistla. Using temporal logic for formal verification of finite state systems. In *Logics and models of concurrent systems*, 1985. NATO. ASI. series, volume F13.

[Clarke *et al.*, 1986] E. Clarke, E. Emerson, and A. Sistla. Automatic verification of finite-state concurrent systems using temporal logic specifications. *ACM Toplas*, 8:244–263, 1986.

[Clifford, 1966] J. E. Clifford. Tense logic and the logic of change. *Logique et analyse*, 9, 1966.

[Clifford and Warren, 1983] J. Clifford and D. Warren. Formal semantics for time in databases. *ACM TODS*, 8, 1983.

[Cocchiarella, 1965] N. Cocchiarella. *Tense and modal logic: a study of the topology of temporal reference.* PhD thesis, University of California, Los Angeles, 1965.

[Cocchiarella, 1966a] N. Cocchiarella. Completeness theorems for tense logics. *J. Symbolic Logic*, 31:688–691, 1966.

[Cocchiarella, 1966b] N. Cocchiarella. Modality within tense logic. *J. Symbolic Logic*, 31, 1966.

[Cocchiarella, 1966c] N. Cocchiarella. *Tense logics.* PhD thesis, University of California, Los Angeles, 1966.

[Cohen, 1966] P.J. Cohen. *Set Theory and the Continuum Hypothesis.* Benjamin, New York, 1966.

[Comrie, 1976] B. Comrie. *Aspect.* Cambridge University Press, 1976.

[Cresswell, 1973] M. Cresswell. *Logics and languages.* Methuen, 1973.

[Cresswell, 1975] M. Cresswell. Frames and models in modal logic. In J. N. Crossley, editor, *Algebra and Logic.* Springer, 1975. LNM. 450.

[Cresswell, 1977a] M. Cresswell. *Adverbial modification, interval semantics and its rivals.* Reidel, 1977.

[Cresswell, 1977b] M. J. Cresswell. Interval semantics and logical words. In C. Rohrer, editor, *On the Logical Analysis of Tense and Aspect*, pages 7–29. Gunter Narr Verlag, Tübingen, 1977.

[Cresswell, 1984] M. Cresswell. An incomplete decidable modal logic. *J. Symbolic Logic*, 49, 1984.

[Csirmaz, 1985] L. Csirmaz. A completeness theorem for dynamic logic. *Notre Dame J. Formal Logic*, 26:51–60, 1985.

[Csirmaz, 1990] L. Csirmaz. Induction and Peano models. Technical report, DIMACS. Technical Report 90–28, Rutgers University, Dept of Comp. Sci., 1990.

[Csirmaz and Paris, 1984] L. Csirmaz and J. Paris. A property of 2–sorted Peano models and program verification. *Zeitschrift für Math. Logic u. Grundlagen der Math.*, 30:324–334, 1984.

[Curry, 1963] H. B. Curry. *Foundations of Mathematical Logic.* McGraw Hill, New York, 1963.

[Davidson, 1967] D. Davidson. The logical form of action sentences. In *The Logic of Decision and Action.* Pittsburgh University Press, 1967.

[Davies, 1974] P. C. W. Davies. *The physics of time asymmetry.* University of California Press, 1974.

[Davis, 1979] L. Davis. *Theory of action.* Prentice Hall, 1979.

[Davis, 1987] E. Davis. Constraint propagation with interval labels. *Artificial Intelligence*, 32, 1987.

[de Jongh and Bowen, 1986] D. de Jongh and K. Bowen. Some complete logics for branched time. Technical report, Institute for Logic Language and Information, University of Amsterdam, 1986. Report 86-05.

[de Jongh *et al.*, 1988] D. de Jongh, F. Veltman, and R. Verbrugge. The logics of dense and discrete time. Technical report, Institute for Language, Logic and Information, University of Amsterdam, 1988.

[Dean, 1983] T. Dean. Time map maintenance. Technical report, Yale University, 1983.

[Dean, 1984a] T. Dean. Managing time maps. In *Proc. Canad. Soc. for Computat. St. of Intell.*, 1984.

[Dean, 1984b] T. Dean. Planning and temporal reasoning under uncertainty. In *Proc. IEEE. Wkshp. on Princ. of Knowledge- Based Systems*, 1984.

[Dean, 1985] T. Dean. Temporal reasoning involving counterfactuals and disjunctions. In *Proc. 9th IJCAI*, 1985.

[Dean, 1986a] T. Dean. Intractability and time dependent planning. In *Proc. AAAI Workshop on Planning and Reasoning about Action*, 1986.

[Dean, 1986b] T. Dean. *Temporal imagery: an approach to reasoning about time for planning and problem solving.* PhD thesis, Yale University, 1986.

[Dean and McDermott, 1987] T. Dean and D. McDermott. Temporal database management. *AI*, 32:1–55, 1987.

[Decker, 1988] R. Decker. Modelling the temporal behaviour of technical systems. In *Proc. GWAI*, 1988.

[del Cerro, 1982] L. Farinas del Cerro. A simple deduction method for modal logic. *Information Processing Letters*, 14, 1982.

[del Cerro, 1983] L. Farinas del Cerro. Temporal reasoning and termination of programs. In *Proc. IJCAI*, 1983.

[del Cerro, 1985] L. Farinas del Cerro. Resolution modal logics. In K. Apt, editor, *Logics and Models of Concurrent Systems.* NATO ASI Series, 1985. Vol. 13.

[Doets, 1989] K. Doets. Monadic Π^1_1-theories of Π^1_1-properties. *Notre Dame J. Formal Logic*, 30:224–240, 1989.

[Downing, 1959] P. Downing. Subjunctive conditionals, time order and causation. In *Proc. Aristotelian Society, 59*, pages 125–140, 1959.

[Dowty, 1977a] D. Dowty. Toward a semantic analysis of verb aspect and the English imperfect. *Linguistics and Philosophy*, 1:45–77, 1977.

[Dowty, 1977b] D. R. Dowty. Toward a semantic analysis of verb aspect and the English 'imperfect' progressive. *Linguistics and Philosophy*, 1, 1977.

[Dry, 1979] H. Dry. Four temporal effects of narrative 'now'. *Texas Linguistic Forum*, 15:60–69, 1979.

[Dummet, 1977] M. Dummet. *Elements of intuitionism.* Clarendon Press, 1977.

[Dummet and Lemmon, 1959] M. Dummet and E. Lemmon. Modal logic between S4 and S5. *Zeitschrift für Math. Logik und Grundlagen der Math.*, 5, 1959.

[Earman *et al.*, 1977] J. Earman, C. Glymour, and J. Stachel, editors. *Foundations of Space–Time Theories.* University of Minnesota Press, Minneapolis, 1977.

[Ehrenfeucht, 1961] A. Ehrenfeucht. An application of games to the completeness problem for formalized theories. *Fund. Math.*, 49:128–141, 1961.

[Emerson, 1983] E. Emerson. Alternative semantics for temporal logics. *TCS*, 26, 1983.

[Emerson, 1988] E. Emerson. Branching time temporal logics. In G. Rosenberg, editor, *REX Workshop on Temporal Logic and Concurrency.* Springer, Berlin, 1988.

[Emerson, 1990] E.A. Emerson. Temporal and modal logic. In J. van Leeuwen, editor, *Handbook of Theoretical Computer Science*, volume B. Elsevier, Amsterdam, 1990.

[Emerson and Clarke, 1982] E. Emerson and E. C. Clarke. Using branching time temporal logic to synthesise synchronisation skeletons. *Sci. of Computer Programming*, 2, 1982.

[Emerson and Halpern, 1982] E. Emerson and J. Halpern. Decision procedures and expressiveness in the temporal logic of branching time. In *Proc. 14th ACM Symp. on Theory of Computing*, 1982.

[Emerson and Halpern, 1986] E. Emerson and J. Halpern. 'sometimes' and 'not never' revisited: on branching versus linear time. *J. ACM*, 33, 1986.

[Emerson and Sistla, 1983] E. Emerson and A. Sistla. *Deciding branching time logic: a triple exponential decision procedure for CTL^*.* Springer, 1983. LNCS 164.

[Emerson and Sistla, 1984a] E. Emerson and A. Sistla. Deciding branching time logic. In *Proc. 16th ACM Symposium on Theory of Computing*, 1984.

[Emerson and Sistla, 1984b] E. Emerson and A. Sistla. Deciding full branching time logic. *Information and Control*, 61:175 – 201, 1984.

[Emerson and Srinivasan, 1988] E. Emerson and J. Srinivasan. *Branching time temporal logic.* Springer, 1988. LNCS 354.

[Emerson *et al.*, 1989a] E. Emerson, T. Sadler, and J. Srinivasan. Efficient temporal reasoning. In *Proc. 16th ACM Symposium on Principles of Programming Languages*, 1989.

[Emerson *et al.*, 1989b] E. A. Emerson, A. K. Mok, A. P. Sistla, and J. Strinivasan. Quantitative temporal reasoning. In *Workshop on Automatic Verification of Finite-State Systems.* Grenoble, France, 1989.

[Enc, 1986] M. Enc. Towards a referential analysis of temporal expressions. *Linguistics and Philosophy*, 9:405–427, 1986.

[Fagin and Vardi, 1985] R. Fagin and M. Vardi. An internal semantics for modal logics. Technical report, Stanford University, 1985. Report 85-25.

[Fenstad *et al.*, 1987] J-E. Fenstad, P-K. Halvorsen, T. Langholm, and J. van Benthem. *Situations, language and logic.* Reidel, 1987.

[Filshie, 1987] D. S. C. Filshie. The logic USF. Master's thesis, Department of Computing, Imperial College, 1987.

[Findler and Chen, 1985] N. Findler and D. Chen. On the problem of temporal retrieval, temporal relations, causality and coexistence. In *Proc. IJCAI*, 1985.

[Findley, 1941] J. Findley. Time: a treatment of some puzzles. *Austr. J. Psychol. and Philos.*, 19, 1941.

[Fine, 1971] K. Fine. The logics containing S4.3. *Zeit. Math. Log. Grund.*, 17, 1971.

[Fine, 1974a] K. Fine. An ascending chain of S4 logics. *Theoria*, 40, 1974.

[Fine, 1974b] K. Fine. An incomplete logic containing S4. *Theoria*, 40, 1974.

[Fine, 1974c] K. Fine. Logics extending K4. *J. Symbolic Logic*, 39, 1974.

[Fine, 1975] K. Fine. Normal forms in modal logic. *Notre Dame J. Formal Logic*, 16, 1975.

[Fine, 1977] K. Fine. Model theory for modal logic. *J. Philosophical Logic*, 7, 1977.

[Fine, 1979] K. Fine. Failure of the interpolation lemma in quantified modal logic. *J. Symbolic Logic*, 44:201–206, 1979.

[Fine, 1981a] K. Fine. First-order modal theories. *Noûs*, 15, 1981.

[Fine, 1981b] K. Fine. Model theory for modal logic. *J. Philosophical Logic*, 10, 1981.

[Fine, 1985] K. Fine. Modal logics containing K4. *J. Symbolic Logic*, 50, 1985.

[Fine and Schurz, 1992] Kit Fine and Gerhard Schurz. Transfer theorems for stratified multimodal logics, 1992. Unpublished manuscript.

[Finger, 1992] Marcelo Finger. Handling database updates in two-dimensional temporal logic. *J. of Applied Non-Classical Logic*, pages 201–224, 1992.

[Finger and Gabbay, 1991] M. Finger and D. M. Gabbay. Adding a temporal dimension to a logic. Technical report, Imperial College, 1991.

[Fisher, 1987] M. Fisher. *Temporal logics for abstract semantics.* PhD thesis, University of Manchester, 1987.

[Fisher, 1988] M. D. Fisher. Implementing a prototype metatem interpreter. Technical report, Temple Group, Department of Computer Science, University of Manchester, November 1988.

[Fitting, 1973] M. Fitting. Model existence theorem for modal and intuitionistic logics. *J. Symbolic Logic*, 38:613–627, 1973.

[Fitting, 1983] M. Fitting. *Proof methods for modal and intuitionistic logics.* Reidel, 1983.

[Fitting, 1984] M. Fitting. Linear reasoning in modal logic. *J. Symbolic Logic*, 49:1363–1378, 1984.

[Fitting, 1990] Melvin Fitting. *First-Order Logic and Automated Theorem Proving.* Texts and Monographs in Computer Science. Springer-Verlag, 1990.

[Flum, 1991] J. Flum. On bounded theories. In E. Boerger, G. Jaeger, H. Kleine Buening, and M.M.Richter, editors, *Computer Science Logic '91, Berne*, Lecture Notes in Computer Science 626, pages 111–118. Springer-Verlag, 1991.

[Fujita *et al.*, 1986] M. Fujita, S. Kono, H. Tanaka, and T. Moto-Oka. Logic programming based on temporal logic and its compilation to Prolog. In *Proc. 3rd Conf. on Logic Programming.* Springer, 1986. LNCS 225.

[Gabbay, 1971a] D. M. Gabbay. On decidable finitely axiomatisable modal and tense logics without the finite model property, part II. *Israel Journal of Mathematics*, 10:496–503, 1971.

[Gabbay, 1971b] D. M. Gabbay. On decidable finitely axiomatisable modal and tense loigcs without the finite model property, part I. *Israel Journal of Mathematics*, 10:478–495, 1971.

[Gabbay, 1972a] D. Gabbay. Craig's interpolation theorem for modal logics. In *Conference in Math. Logic, London*, pages 111–127. Springer, 1972. LNM. 255.

[Gabbay, 1972b] D. M. Gabbay. Tense systems with discrete moments of time. *J. Philosophical Logic*, 1:35–44, 1972.

[Gabbay, 1973] D. M. Gabbay. A survey of decidability results for modal, tense and intermediate logics. In P. Suppes *et al.*, editors, *Proceedings of the Fourth International Congress on Logic, Methodology and Philosophy of Science, Bucarest*, pages 29–43. North Holland Publishing Co, 1973.

[Gabbay, 1974] D. M. Gabbay. Tense logics and the tenses of English. In Moravcsik, editor, *Readings in Logic.* Mouton, The Hague, 1974. reprinted as Chapter 12 in [Gabbay, 1976a].

[Gabbay, 1975a] D. M. Gabbay. Decidability results in non-classical logics, part I. *Ann. Math. Logic*, 8:237–295, 1975.

[Gabbay, 1975b] D. M. Gabbay. Model theory for tense logics. *Ann. Math. Logic*, 8:185–235, 1975.

[Gabbay, 1976a] D. M. Gabbay. *Investigations in Modal and Tense Logics with Applications to Problems in Philosophy and Linguistics.* Synthese. vol 92. D. Reidel, Dordrecht, 1976.

[Gabbay, 1976b] D. M. Gabbay. Two dimensional propositional tense logics. In A. Kasher, editor, *Language in focus: Foundation, Methods and Systems—Essays in Memory of Yehoshua Bar-Hillel*, pages 569–583. D. Reidel, Dordrecht, 1976. reprinted as Chapter 10 in [Gabbay, 1976a].

[Gabbay, 1978] D. M. Gabbay. A tense logic with split truth tables. *Logique et analyse*, 21:5–39, 1978.

[Gabbay, 1979] D. M. Gabbay. The separation property of tense logics. Unpublished manuscript, September 1979.

[Gabbay, 1981a] D. M. Gabbay. Expressive functional completeness in tense logic (preliminary report). In U. Monnich, editor, *Aspects of Philosophical Logic*, pages 91–117. Reidel, Dordrecht, 1981.

[Gabbay, 1981b] D. M. Gabbay. An irreflexivity lemma with applications to axiomatizations of conditions on tense frames. In U. Monnich, editor, *Aspects of Philosophical Logic*, pages 67–89. Reidel, Dordrecht, 1981.

[Gabbay, 1981c] D. M. Gabbay. Temporal logic: Mathematical foundations and computational aspects, 1981. draft manuscript.

[Gabbay, 1987a] D. M. Gabbay. Executable temporal logic for interactive systems —2nd draft. Technical report, Imperial College, 1987.

[Gabbay, 1987b] D. M. Gabbay. Modal and temporal logic programming 1. In A. Galton, editor, *Temporal Logics and their Applications*, pages 197–237. Academic Press, 1987.

[Gabbay, 1989a] D. M. Gabbay. Declarative past and imperative future: Executable temporal logic for interactive systems. In B. Banieqbal, H. Barringer, and A. Pnueli, editors, *Proceedings of Colloquium on Temporal Logic in Specification, Altrincham, 1987*, pages 67–89. Springer-Verlag, 1989. Springer Lecture Notes in Computer Science 398.

[Gabbay, 1989b] D. M. Gabbay. Temporal logics: Tense or non-tense? In R. Spencer-Smith and S. Torrance, editors, *Machinations, Computational Studies of Logic, Language and Cognition.* Ablex Publishing Co, 1989. Inaugural Lecture at Imperial College, 17 May 1988.

[Gabbay, 1990] D. M. Gabbay. Modal and temporal logic programming II: A temporal logic programming machine. In T. Dodd, R. P. Owens, and S. Torrance, editors, *Logic Programming—Expanding the Horizons.* Blackwells, Oxford, 1990.

[Gabbay, 1991a] Dov M. Gabbay. Modal and Temporal Logic Programming III — metalevel features in the object level. In L. F. del Cerro and M. Penttonen, editors, *Non-Classical Logic Programming.* Oxford University Press, 1991.

[Gabbay, 1991b] Dov M. Gabbay. Theoretical Foundations for Non-monotonic Reasoning Part 2: Structured Non-monotonic Theories. In *SCAI 91—Third Scandinavian Conference on AI*, pages 19–40, May 1991.

[Gabbay, 1992a] D. M. Gabbay. *Labelled Deductive Systems.* Oxford University Press, 1992.

[Gabbay, 1992b] Dov M. Gabbay. Fibred semantics and the weaving of logics. Unpublished manuscript, Imperial College, 1992.

[Gabbay, forthcoming] D. Gabbay. Classical versus non-classical logic. In D. Gabbay, C. Hogger, and J. Robinson, editors, *Handbook of Logic in Artificial Intelligence*, volume 2. Oxford University Press, forthcoming.

[Gabbay, to appear] D. Gabbay. What is a logical system. In D. Gabbay and F. Guenthner, editors, *What is a logical system.* Oxford University Press, to appear.

[Gabbay and Guenthner, 1982] D. M. Gabbay and F. Guenthner. A note on many-dimensional tense logics. In T. Pauli, editor, *Philosophical Essays Dedicated to Lennart Åqvist on his Fiftieth Birthday*, pages 63–70. University of Uppsala, 1982.

[Gabbay and Hodkinson, 1989] D. M. Gabbay and I. M. Hodkinson. Nonmonotonic coding of the declarative past. Technical report, Imperial College, 1989. Draft report.

[Gabbay and Hodkinson, 1990] D. M. Gabbay and I. M. Hodkinson. An axiomatisation of the temporal logic with until and since over the real numbers. *Journal of Logic and Computation*, 1(2):229 – 260, 1990.

[Gabbay and Hodkinson, to appear] D. M. Gabbay and I. M. Hodkinson. Temporal logic in context. In J. Copeland, editor, *Arthur Prior Memorial Volume*. Oxford University Press, to appear.

[Gabbay and Moravcsik, 1980] D. M. Gabbay and J. Moravcsik. Verbs, events and the flow of time. In C. Rohrer, editor, *Time Tense and Quantifiers: Proceedings of the Stuttgart Conference on the Logic of Tense and Quantification*. Max Niemayer Verlag, Tübingen, 1980.

[Gabbay and Rohrer, 1978] D. M. Gabbay and C. Rohrer. Relative tenses: The interpretation of tense forms which occur in the scope of temporal or in embedded sentences. In C. Rohrer, editor, *Papers on tense, aspect and verb clarification*, pages pp 99–111. TBL. Verlag, G. Narr, Tübingen, 1978.

[Gabbay and Rohrer, 1979] D. M. Gabbay and C. Rohrer. *Semantics from Different Points of View*, volume 6 of *Language and Communications*, chapter Do we really need tenses other than future and past?, pages 15–20. Springer-Verlag, Berlin, 1979.

[Gabbay *et al.*, 1980] D. M. Gabbay, A. Pnueli, S. Shelah, and J. Stavi. On the temporal analysis of fairness. In *7th ACM Symposium on Principles of Programming Languages, Las Vegas*, pages 163–173, 1980.

[Gabbay *et al.*, 1993] D. M. Gabbay, I. M. Hodkinson, and M. A. Reynolds. Temporal expressive completeness in the presence of gaps. In *Proceedings ASL European Meeting 1990, Helsinki*, Lecture Notes in Logic. Springer-Verlag, 1993.

[Gallagher, 1989] J. K. Gallagher. Separating US and real time. Master's thesis, Imperial College, 1989.

[Gallin, 1975] D. Gallin. *Intensional and higher order modal logics.* North Holland, 1975.

[Galton, 1983] A. Galton. *The logic of aspect: an inquiry into the semantic structure of ordinary temporal discourse.* PhD thesis, University of Leeds, 1983.

[Galton, 1984] A. Galton. *The Logic of Aspect.* Clarendon Press, Oxford, 1984.

[Galton, 1987a] A. Galton. The logic of occurrence. In A. Galton, editor, *Temporal logics and their applications.* Academic Press, 1987.

[Galton, 1987b] A. Galton. Temporal logic and computer science: an overview. In A. Galton, editor, *Temporal logics and their applications.* Academic Press, 1987.

[Galton, 1990] A. Galton. A critical examination of Allen's theory of action and time. *AI*, 42:159–188, 1990.

[Garson, 1984] J. W. Garson. Quantification in modal logic. In D. Gabbay and F. Guenthner, editors, *Handbook of Philosophical Logic.* D. Reidel, Dordrecht, 1984.

[Geach, 1965] P. Geach. Some problems about time. *Proc. British Academy*, 51, 1965.

[Geach, 1972] P. Geach. *Logic matters.* Blackwell, 1972.

[Ginsberg and Smith, 1988] M. Ginsberg and D. Smith. Reasoning about action I: a possible world approach. In *AI 35*, page 165, 1988.

[Goldblatt, 1975] R. Goldblatt. First order definability in modal logic. *J. Symbolic Logic*, 40, 1975.

[Goldblatt, 1976] R. Goldblatt. Metamathematics of modal logic. *Reports on Math. Logic*, 6:4–77, 1976.

[Goldblatt, 1980] R. Goldblatt. Diodorean modality in Minkowski spacetime. *Studia Logica*, 39:219–236, 1980.

[Goldblatt, 1987a] R. Goldblatt. *Logics of Time and Computation*, volume 7 of *CSLI. Lecture Notes.* The Chicago University Press, Chicago, 1987.

[Goldblatt, 1987b] R. Goldblatt. *Orthogonality and Space–Time Geometry.* Springer, New York, 1987.

[Goldblatt and Thomason, 1975] R. Goldblatt and S. Thomason. *Axiomatic classes in propositional modal logic.* Springer, 1975. LNM. 480.

[Goranko, 1991] V. Goranko. Applications of quasi-structural rules to axiomatizations in modal logic (abstract). In *Logic Colloquium '91, Abstracts of the 9th International Congress of Logic, Methodology and Philosophy of Sciences, Uppsala, 1991*, page 119, 1991. Volume I: Logic.

[Gough, 1984a] G. Gough. *Decision procedures for temporal logics.* PhD thesis, University of Manchester, 1984. M.Sc. dissertation.

[Gough, 1984b] G. D. Gough. Decision procedures for temporal logic. Master's thesis, Department of Computer Science, University of Manchester, 1984.

[Greenspan, 1975] P. Greenspan. Conditional oughts and hypothetical imperatives. *J. Philosophy*, 72:259–276, 1975.

[Grimbront and Wolper, 1990] E. Grimbront and P. Wolper. An introduction to temporal logic. In A. Thayse, editor, *From Natural Language Processing to Logic for Expert Systems.* John Wiley and Sons, Chichester, 1990.

[Grünbaum, 1967] A. Grünbaum. *Modern science and Zeno's paradoxes.* Wesleyan University Press, 1967.

[Grünbaum, 1971] A. Grünbaum. The meaning of time. In E. Freman and W. Sellers, editors, *Basic issues in the philosophy of time.* Open Court, 1971.

[Guckenbiehl, 1990] T. Guckenbiehl. Reasoning with persistent facts. Technical report, Fraunhofer-Institut, 1990.

[Guenthner, 1978] F. Guenthner. Tense logic, time schemes and the tenses of English. In F. Guenthner and S. Schmidt, editors, *Formal Semantics and Pragmatics for Natural Languages.* Reidel, Dordrecht, 1978.

[Guenthner, 1980] F. Guenthner. Remarks on the present perfect in English. In C. Rohrer, editor, *Time, Tense and Quantifiers: Proceedings of the Stuttgart Conference on the Logic of Tense and Quantification.* Max Niemayer Verlag, Tübingen, 1980.

[Gupta and Thomason, 1981] A. Gupta and R. Thomason. A theory of conditionals in the context of branching time. In W. Harper, G. Pearce, and R. Stalnaker, editors, *Ifs: Conditionals, Beliefs, Decision, Chance and Time*, pages 299–322. Reidel, Dordrecht, 1981.

[Gurevich, 1964] Y. Gurevich. Elementary properties of ordered abelian groups. *Algebra and Logic*, 3:5–39, 1964. (Russian; an English version is in *Trans. Amer. Math. Soc.* 46 (1965), 165–192).

[Gurevich, 1977] Y. Gurevich. Expanded theory of ordered abelian groups. *Annals Math. Logic*, 12:193–228, 1977.

[Gurevich, 1985] Y. Gurevich. Monadic second-order theories. In J. Barwise and S. Feferman, editors, *Model-Theoretic Logics*, pages 479–507. Springer-Verlag, New York, 1985.

[Gurevich and Harrington, 1982] Y. Gurevich and L.A. Harrington. Trees, automata, and games. In *Theory of Computing* (*Proc. 14th annual ACM Symposium, San Francisco*), *Association for Computing Machinery*, pages 60–65, 1982.

[Gurevich and Shelah, 1982] Y. Gurevich and S. Shelah. Monadic theory of order and topology in ZFC. *Annals of Math. Logic*, 23:179–198, 1982.

[Gurevich and Shelah, 1985] Y. Gurevich and S. Shelah. The decision problem for branching time logic. *J. Symbolic Logic*, 50:668–681, 1985.

[Gurevich *et al.*, 1983] Y. Gurevich, M. Magidor, and S. Shelah. The monadic theory of ω_2. *J. Symbolic Logic*, 48:387–398, 1983.

[Hájek, 1986] P. Hájek. Some conservativeness results for nonstandard dynamic logic. In J.Demetrovics, G. Katona, and A. Salomaa, editors, *Algebra, combinatorics, and logic in computer science* (*Proc. Conf. Győr Hungary 1983*), *Colloq. Math. Soc.*, volume 42, pages 443–449. J. Bolyai (North–Holland), 1986.

[Hale, 1987] R. Hale. Temporal logic programming. In A. Galton, editor, *Temporal Logics and their Applications*, chapter 3, pages 91–119. Academic Press, London, December 1987.

[Halpern and Moses, 1985] Joseph Y. Halpern and Yoram Moses. A Guide to the Modal Logics of Knowledge and Belief. In *Proceedings of the 9th International Joint Conference on Artificial Intelligence* (*IJCAI-85*), pages 480–490, 1985.

[Halpern and Shoham, 1986] J. Halpern and Y. Shoham. A propositional modal logic of time intervals. In *Proceedings Symposium on Logic in Computer Science*. IEEE, Boston, 1986.

[Halpern *et al.*, 1983a] J. Halpern, Z. Manna, and B. Moszkowski. *A hardware semantics based on temporal intervals.* Springer, 1983. LNCS 154.

[Halpern *et al.*, 1983b] J.Y. Halpern, Z. Manna, and B. Moszkowski. A high-level semantics based on interval logic. In *Proc. 10th ICALP*, pages 278–291, 1983.

[Hamblin, 1971] C. Hamblin. Starting and stopping. In E. Freman and W. Sellers, editors, *Basic issues in the philosophy of time.* Open Court, 1971.

[Hamblin, 1972] C. L. Hamblin. Instants and intervals. In J. T. Fraser, F. C. Haber, and G. H. Muller, editors, *The study of Time.* Springer-Verlag, New York, 1972.

[Hanks and McDermott, 1986] S. Hanks and D. McDermott. *Temporal reasoning and default logic.* AAAI, 1986.

[Harel, 1979] D. Harel. First–order dynamic logic. *Lecture Notes in Computer Science*, 68, 1979.

[Harel, 1988] E. Harel. Temporal analysis of real-time systems. Master's thesis, Weizmann Institute, 1988.

[Harel and Pnueli, 1988] E. Harel and A. Pnueli. Applications of temporal logic to the specification of real time systems. In M. Joseph, editor, *Formal Techniques in Real-time and Fault-tolerant Systems*, pages 84–98. Springer, Berlin, 1988. Lec. Notes in Comp. Sci. 331.

[Harel *et al.*, 1990] E. Harel, O. Lichtenstein, and A. Pnueli. Explicit clock temporal logic. Technical report, Weizmann Institute of Science, Rehovot 76100, Israel, 1990. in LICS. 1990.

[Harper *et al.*, 1981] W. Harper, R. Stalnaker, and G. Pearce, editors. *Ifs: Conditionals, Belief, Decision, Chance and Time.* Reidel, Dordrecht, 1981.

[Harrison, 1969] C. Harrison. Kripke's semantics for modal logic. In J. Davis et. al., editor, *Philosophical logic.* Reidel, 1969.

[Hart and Sharir, 1984] S. Hart and M. Sharir. Probabilistic temporal logics for finite and bounded models. In *Proc. 16th ACM Symposium on Theory of Computing*, 1984.

[Hattori *et al.*, 1986] T. Hattori, R. Nakajima, N. Niide, and K Takenaka. RACCO: A modal-logic programming language for writing models of real time process-control systems. Technical report, Research Institute for Mathematical Sciences, Kyoto University, 1986.

[Haugh, 1987] B. Haugh. Simple causal minimizations for temporal persistence and projection. In *Proc. 6th AAAI*, 1987.

[Heny, 1982] F. Heny. Tense, aspect and time adverbials II. *Linguistics and Philosophy*, 5:109–154, 1982.

[Hinman, 1987] P. G. Hinman, editor. *Recursion Theory*, volume 4 of *[Ω]-bibliography of Mathematical Logic,* edited by Gert H. Muller in collob-oration with Wolfgang Lenski. Springer-Verlag, Berlin, 1987.

[Hinrichs, 1988] E. Hinrichs. Tense, quantifiers and context. *Computat. Linguistics*, 14, 1988.

[Hintikka, 1962] J. Hintikka. *Knowledge and Belief.* Cornell University Press, 1962.

[Hintikka, 1963] J. Hintikka. The modes of modality. *Acta Philos. Fennica*, 16, 1963.

[Hintikka, 1964] J. Hintikka. Aristotle and the master argument of Diodorus. *Amer. Philos. Quarterly*, 1, 1964.

[Hintikka, 1969] J. Hintikka. *Models for modalities.* Reidel, 1969.

[Hintikka, 1973] J. Hintikka. *Time and necessity.* Oxford University Press, 1973.

[Hirsch, 1990] R. Hirsch. Temporal representation in intervals and planning. Technical report, Department of Computing, Imperial College, 1990.

[Hirschmann, 1981] L. Hirschmann. Representing implicit and explicit time relations in narrative. In *Proc. 7th IJCAI*, 1981.

[Hodkinson, 1989a] I. Hodkinson. Decidability and elimination of fixed point operators in the temporal logic USF. Technical report, Imperial College, 1989.

[Hodkinson, 1989b] I. Hodkinson. Elimination of fixpoint operators in the temporal logic YF. Technical report, Department of Computing, Imperial College, 1989.

[Hodkinson, 1992a] I. Hodkinson. On Gabbay's temporal fixed point operator. 1992. To appear in *J. Theoretical Computer Science.*

[Hodkinson, 1992b] I. M. Hodkinson. Finite H-dimension does not imply expressive completeness. 1992. To appear in *J. of Philosophical Logic* .

[Hodkinson and Simon, 1992] I. Hodkinson and A. Simon. The k-variable property is stronger than H-dimension k. Preprint, Budapest, 1992.

[Hoepelman, 1974] J. Hoepelman. Tense logics and the semantics of the Russian aspects. *Theoretical Linguistics*, 1:158–182, 1974.

[Hortalá-González and Rodriguez-Artalejo, 1989] T. Hortalá-González and M. Rodriguez-Artalejo. A nonstandard completeness theorem. *Theoretical Computer Science*, 68:277–302, 1989.

[Hrycej, 1986] T. Hrycej. An efficient algorithm for reasoning about time intervals. Technical report, PCS, München, 1986. TEX-B-9.

[Hrycej, 1988] T. Hrycej. Extended temporal Prolog. Technical report, PCS, München, 1988. TEX-B-48.

[Huddlestone, 1969] K. Huddlestone. Some observations on tense and deixes in English. *Language*, 45, 1969.

[Hughes and Cresswell, 1968] G. Hughes and M. Cresswell. *An Introduction to Modal Logic.* Methuen, London, 1968.

[Hughes and Cresswell, 1984] G.E. Hughes and M.J. Cresswell. *A Companion to Modal Logic.* Methuen, London, 1984.

[Humberstone, 1979] I. L. Humberstone. Interval semantics for tense logics. *J. Philosophical Logic*, 8:171–196, 1979.

[Immerman and Kozen, 1987] N. Immerman and D. Kozen. Definability with bounded number of bound variables. In *LICS87, Proceedings of the Symposium on Logic in Computer Science, Ithaca, New York*, pages 236–244, Washington, 1987. Computer Society Press.

[Jackson and Michon, 1985] J. Jackson and J. Michon, editors. *Time, Mind and Behavior.* Springer, Berlin, 1985.

[Janssen, 1983] T.M.V. Janssen. *Foundations and applications of Montague grammar.* PhD thesis, Mathematisch Centrum, Amsterdam, 1983.

[Jech, 1978] T. Jech. *Set Theory.* Academic Press, New York, 1978.

[Kahn and Gorry, 1977] K. Kahn and G. Gorry. Mechanizing temporal knowledge. *AI*, 9, 1977.

[Kamp, 1968a] H. Kamp. Seminar notes on tense logics. *J. Symbolic Logic*, 1968.

[Kamp, 1968b] H. Kamp. *Tense logic and the theory of linear order.* PhD thesis, University of California, Los Angeles, 1968.

[Kamp, 1971] H. Kamp. Formal properties of 'now'. *Theoria*, 37:237–273, 1971.

[Kamp, 1978] H. Kamp. Completeness results in temporal logic, 1978. Unpublished manuscript.

[Kamp, 1979] H. Kamp. Instants, events and temporal discourse. In R. Bäuerle, C. Schwarze, and A. von Stechow, editors, *Semantics from Different Points of View*, pages 376–417. Springer, Berlin, 1979.

[Kamp, 1980a] H. Kamp. A theory of truth and semantics representation. In Groenendijk *et al.*, editors, *Formal Methods in the Study of Language, Part I*, pages 227–231. Mathematisch Centrum, Amsterdam, 1980.

[Kamp, 1980b] J. A. W. Kamp. Some remark on the logic of change. In Guenthner and Rohrer, editors, *Time, tense and quantifiers*, 1980. Proc. Stuttgart Conference on the logic of tense and quantification.

[Kamp, 1983] H. Kamp. Discourse representation and temporal reference, 1983. manuscript.

[Kamp and Rohrer, 1983] J. A. W. Kamp and C. Rohrer. Tense in texts. In Bäuerle et. al., editor, *Meaning, use and interpretation of language.* de Gruyter, 1983.

[Kamp and Rohrer, 1988] H. Kamp and C. Rohrer. Tense in text. Technical report, Institut fuer Computerlinguistik, Universitaet Stuttgart, 1988.

[Kandrashina, 1983] E. Kandrashina. Representation of temporal knowledge. In *Proc. 8th IJCAI*, 1983.

[Kanger, 1957] S. Kanger. A note on quantification and modalities. *Theoria*, 23, 1957.

[Karp, 1965] Carol Karp. Finite-quantifier equivalence. In J.W. Addison *et al.*, editors, *The Theory of Models, Proc. 1963 Internat. Symposium at Berkeley*, pages 407–412, Amsterdam, 1965. North-Holland.

[Katz, 1988] S. Katz. Exploiting interleaving set temporal logic to simplify correctness proofs. In G. Rosenberg, editor, *REX Workshop on Temporal Logic and Concurrency.* Springer, Berlin, 1988.

[Kautz and Vilain, 1986] H. Kautz and M. Vilain. Constraint propagation algorithms for temporal reasoning. *AAAI*, 86:377–382, 1986.

[Konolige, 1986] K. Konolige. *A Deductive Model of Belief.* Research notes in Artificial Intelligence. Morgan Kaufmann, 1986.

[Koomen, 1988] J. Koomen. The TIMELOGIC temporal reasoning system. Technical report, University of Rochester, 1988. TR. 231.

[Kowalski, 1986] R. Kowalski. Database updates in the event calculus. Technical report, Imperial College, London, 1986. DOC. 86/12.

[Kowalski and Sergot, 1986] R. Kowalski and M. Sergot. A logic-based calculus of events. *New Generation Computing* (*Japan*), 4, 1986.

[Koyman, 1986] R. L. C. Koyman. *Specifying message passing and time-critical systems with temporal logic.* PhD thesis, Technical University Eindhoven, 1986.

[Krablin, 1979] L. Krablin. A temporal analysis of fairness. Master's thesis, University of Pennsylvania, 1979.

[Kracht and Wolter, 1991] Marcus Kracht and Frank Wolter. Properties of independently axiomatizable bimodal logics. *J. Symbolic Logic*, 56(4):1469–1485, 1991.

[Krauss and Lehmann, 1983] S. Krauss and D. Lehmann. Reasoning with time and chance. *Information and Control*, 53, 1983.

[Kripke, 1959] S. Kripke. A completeness theorem in modal logic. *J. Symbolic Logic*, 24, 1959.

[Kripke, 1963] S. Kripke. Semantical analysis of modal logic: normal propositional calculi. *Zeitschrift Math. Logic und Grundlagen der Mathematik*, 9, 1963.

[Kripke, 1971] S. Kripke. Semantical considerations on modal logics. In L. Linsky, editor, *Reference and modality.* Oxford University Press, 1971.

[Kripke, 1980] S. Kripke. *Naming and necessity.* Harvard University Press, 1980.

[Kröger, 1984] F. Kröger. A generalized nexttime operator in temporal logic. *J. CSS*, 29, 1984.

[Kröger, 1988] F. Kröger. *Temporal logic of programs.* EATCS Monographs on Theoretical Computer Science. 1988.

[Kuhn, 1979] S. Kuhn. The pragmatics of tense. *Synthese*, 40:237–263, 1979.

[Kuhn, 1983] S. Kuhn. Where does tense belong. Unpublished manuscript, Georgetown University, Washington DC, 1983.

[Kuhn, 1989] S. T. Kuhn. Time and tense. In D. Gabbay and F. Guenthner, editors, *Handbook of Philosophical Logic IV*. Reidel, 1989.

[Kvart, 1980] I. Kvart. Formal semantics for temporal logic and counterfactuals. *Logique et Analyse*, 23:35–62, 1980.

[Ladkin, 1985] P. Ladkin. Comments on the representation of time. In *Proc. Distr. AI Workshop*, 1985.

[Ladkin, 1986a] P. Ladkin. Primitives and units for time specification. In *Proc. 5th AAAI*, 1986.

[Ladkin, 1986b] P. Ladkin. Time representation: a taxonomy of interval relations. In *Proc. AAAI*, 1986.

[Ladkin, 1987a] P. Ladkin. The completeness of a natural system for reasoning with time intervals. In *Proc. 10th IJCAI*, 1987.

[Ladkin, 1987b] P. Ladkin. The logic of time representation. Technical report, Department of Mathematics, University of California at Berkeley, 1987. Dissertation.

[Ladkin, 1987c] P. Ladkin. Models of axioms for time intervals. Technical report, The Kestrel Institute, Palo Alto, 1987.

[Ladkin and Maddux, 1987] P. Ladkin and R. Maddux. The algebra of convex time intervals. Technical report, The Kestrel Institute, Palo Alto, Department of Mathematics, Iowa State University at Ames, 1987.

[Ladner, 1977] R. Ladner. The computational complexity of provability in systems of modal propositional logic. *SIAM. J. Comput.*, 6:467–480, 1977.

[Lamport, 1983] L. Lamport. What good is temporal logic? In R. Mason, editor, *Information processing 83 IFIP*. North-Holland, 1983.

[Lascarides, 1987] A. Lascarides. Temporal reference and the internal structure of intervals. Technical report, Centre for Cognitive Science, Edinburgh, 1987. EUCCS/RP-10.

[Lascarides, 1989] A. Lascarides. The semantics of aspectual classes with reference to intervals. Technical report, Centre for Cognitive Science, Edinburgh, 1989. EUCCS/RP-22.

[Läuchli and Leonard, 1966] H. Läuchli and J. Leonard. On the elementary theory of linear order. *Fundamenta Mathematicae*, 59:109–116, 1966.

[Lehman and Shelah, 1983] D. Lehman and S. Shelah. Reasoning with time and chance. In *Lecture Notes in Computer Science 154*. Springer Verlag, 1983.

[Lehrer and Taylor, 1965] K. Lehrer and R. Taylor. Time, truth and modalities. *Mind*, 74:390–398, 1965.

[Leivant, 1985] D. Leivant. Logical and mathematical reasoning about imperative programs. In *Proc. 12th Ann. ACM Symp. on Principles of Programming Languages, New Orleans*, pages 132–140, 1985.

[Lemmon, 1966] E. J. Lemmon. Algebraic semantics for modal logics I,II. *J. Symbolic Logic*, 31, 1966.

[Lemmon, 1967] E. J. Lemmon. Comments on Davidson's 'The logical form of action sentences'. In *The logic of decision and action.* Pittsburgh, 1967.

[Lemmon, 1977] E. J. Lemmon. *An introduction to modal logic.* Blackwell, 1977. Amer. Philos. Quarterly Monograph Series 11.

[Lewis, 1968] D. Lewis. Counterpart theory for quantified modal logic. *J. Philos.*, 65, 1968.

[Lewis, 1979] D. Lewis. Counterfactual dependence and time's arrow. *Nous*, 13, 1979.

[Lichtenstein *et al.*, 1985] O. Lichtenstein, A. Pnueli, and L. Zuck. The glory of the past. In R. Parikh, editor, *Logics of Programs* (*Proc. Conf. Brooklyn USA 1985*), volume 193 of *Lecture Notes in Computer Science*, pages 196–218. Springer-Verlag, Berlin, 1985.

[Lifsches and Shelah, 1992] Shmuel Lifsches and Saharon Shelah. The monadic theory of ω_2 may be complicated. *Arch. Math. Logic*, 31:207–213, 1992.

[Linsky, 1971] L. Linsky. *Reference and modality.* Oxford University Press, 1971.

[Lloyd, 1988] J. W. Lloyd. Directions for metaprogramming. Technical Report CS-88-10, University of Bristol, 1988.

[Lodaya and Thiagarajan, 1987] K. Lodaya and P. Thiagarajan. A modal logic for a subclass of event structures. In *Proc. 14th ICALP*, 1987. LNCS 267.

[Long and Russ, 1983] W. Long and T. Russ. A control structure for time dependent reasoning. In *Proc. 8th IJCAI*, 1983.

[Loux, 1968] J. Loux. *The possible and the actual.* Cornell University Press, 1968.

[Luce, 1966] D. Luce. A calculus of 'before'. *Theoria*, 32, 1966.

[Makinson, 1966a] D. Makinson. Completeness theorem for modal logics. *Zeitschrift Math. Logic und Grundlagen der Mathematik*, 12, 1966.

[Makinson, 1966b] D. Makinson. There are infinitely many Diodorean modal logics. *J. Symbolic Logic*, 31, 1966.

[Makinson, 1969] D. Makinson. A normal modal calculus between T and S4 without the finite model property. *J. Symbolic Logic*, 34, 1969.

[Makowsky and Sain, 1989] J.A. Makowsky and I. Sain. Weak second order characterizations of various program verification systems. *Theoretical Computer Science*, 66:299–321, 1989.

[Malik and Binfold, 1983] J. Malik and T. Binfold. Reasoning in time and space. In *Proc. 8th IJCAI*, 1983.

[Manna, 1981] Z. Manna. Verification of sequential programs: Temporal axiomatisation. In F. L. Bauer, E. W. Dijkstra, and C. A. R. Hoare, editors, *Theoretical Foundations of Programming Methodology.* Reidel, Holland, 1981. NATO Scientific Series.

[Manna and Moszkowski, 1984] Z. Manna and B. Moszkowski. Reasoning in interval temporal logic. In *AMC/NSF/ONR. Workshop on Logics of Programs*, pages 371–383. Springer-Verlag, Berlin, 1984. Volume 164, LNCS.

[Manna and Pnueli, 1979] Z. Manna and A. Pnueli. The modal logic of programs. In *International Colloquium on Automata, Languages and Programming'79* (*Proc. Conf. Graz 1979*), volume 71 of *Lecture Notes in Computer Science*, pages 385–409. Springer Verlag, Berlin, 1979.

[Manna and Pnueli, 1981] Z. Manna and A. Pnueli. Verification of concurrent programs: The temporal framework. In R. S. Boyer and J. S. Moore, editors, *The Correctness Problem in Computer Science*, pages 215–273. Academic Press, London, 1981. International Lecture Series in Computer Science.

[Manna and Pnueli, 1983] Z. Manna and A. Pnueli. Verification of concurrent programs: a temporal proof system. Technical report, Mathematische Zentrum, Amsterdam, 1983.

[Manna and Pnueli, 1987] Z. Manna and A. Pnueli. A hierarchy of temporal properties. Technical report, STAN–CS–87– 1186, Dept of Comp. Sci., Stanford Univ., 1987.

[Manna and Pnueli, 1988] Z. Manna and A. Pnueli. The anchored version of the temporal framework. In *REX Workshop, Noordwijkerh.*, 1988. LNCS 354.

[Manna and Pnueli, 1992] Z. Manna and A. Pnueli. *The Temporal Logic of Reactive and Concurrent Systems: Specification.* Springer-Verlag, New York, 1992.

[Manna and Waldinger, 1978] Z. Manna and R. Waldinger. Is 'sometime' sometimes better than 'always'? *Comm. ACM*, 21:159–172, 1978.

[Manna and Wolper, 1984] Z. Manna and P. Wolper. Synthesis of communicating processes from temporal logic specifications. *ACM Transactions on Programming Languages and Systems*, 6(1):68–93, January 1984.

[Massey, 1969] G. Massey. Tense logic! why bother? *Nous*, 3:17–32, 1969.

[Masui *et al.*, 1983] S. Masui, D. McDermott, and A. Sobel. Decision-making in time-critical situations. In *Proc. 8th IJCAI*, 1983.

[McArthur, 1976] R. P. McArthur. *Tense Logic.* Reidel, Dordrecht, 1976.

[McBrien, 1990] P. McBrien. Implementing logic languages by graph rewriting. In T. Dodd, R. P. Owens, and S. Torrance, editors, *Logic Programming – Expanding the Horizons*, pages 164–189. Blackwells, Oxford, 1990.

[McCall, 1968] S. McCall. On what it means to be future. *J. Symbolic Logic*, 33, 1968.

[McCawley, 1971] J. McCawley. Tense and time reference in English. In Fillmore and Langendoen, editors, *Studies in Lingusitic Semantics*, pages 96–113. Holt Rinehart and Winston, New York, 1971.

[McDermott, 1982] D. McDermott. A temporal logic for reasoning about process and plans. *Cognitive Science*, 6(2):101–155, 1982.

[McKinney, 1975] A. McKinney. *Conditional obligation and temporally dependent necessity: a study in conditional deontic logic.* PhD thesis, University of Pennsylvania, 1975.

[McNaughton, 1966] R. McNaughton. Testing and generating infinite sequences by finite automata. *Information and Control*, 9:521–530, 1966.

[McTaggart, 1908] J. McTaggart. The unreality of time. *Mind*, 18:457–474, 1908.

[McTaggart, 1927] J. McTaggart. *The nature of existence.* Cambridge University Press, 1927.

[Meyer, 1975] A.R. Meyer. Weak monadic second order theory of successor is not elementary-recursive. In *Proceedings, Logic Colloquium, Springer Lecture Notes in Math. 453*, pages 132–154, 1975.

[Moens and Steedman, 1988] M. Moens and M. Steedman. Temporal ontology and temporal reference. *Journal of Computational Linguistics*, 14, 1988.

[Moore, 1980] R. Moore. Reasoning about knowledge and action. Technical report, SRI. International, 1980. Technical Note 191.

[Moszkowski, 1983] B. Moszkowski. *Reasoning about digital circuits.* PhD thesis, Stanford University, July 1983.

[Moszkowski, 1986] B. Moszkowski. *Executing Temporal Logic Programs.* Cambridge University Press, 1986.

[Mourelatos, 1978] A. P. D. Mourelatos. Events, processes and states. *Linguistics and Philosophy*, 2:415–434, 1978.

[Muchnik, 1984] A.A. Muchnik. Games on infinite trees and automata with dead-ends: a new proof of the decidability of the monadic theory of two successors. *Semiotics and Information*, 24:17–40, 1984. (In Russian).

[Myers and Pnueli, 1979] T. J. Myers and A. Pnueli. The temporal situation until now. Technical report, University of Pennsylvania, 1979.

[Needham, 1975] P. Needham. *Temporal Perspective: A Logical Analysis of Temporal Reference in English*, volume 25 of *Philosophical Studies*. University of Uppsala, 1975.

[Németi, 1982] I. Németi. Nonstandard dynamic logic. In D. Kozen, editor, *Logics of Programs* (*Proc. Conf. New York 1981*), volume 131 of *Lecture Notes in Computer Science*, pages 311–348. Springer Verlag,Berlin, 1982.

[Newton-Smith, 1980] W. Newton-Smith. *The structure of time*. Routledge and Kegan Paul, 1980.

[Nishimura, 1979a] H. Nishimura. Is the semantics of branching structures adequate for chronological modal logics? *J. of Philosophical Logic*, 8:469–475, 1979.

[Nishimura, 1979b] H. Nishimura. Is the semantics of branching structures adequate for non-metric Ockhamist tense logics? *J. of Philosophical Logic*, 8:477–478, 1979.

[Nishimura, 1980] H. Nishimura. Interval logics with applications to study of tense and aspect in English. *Publications of the Research Institute for Mathematical Sciences, Kyoto University*, 16:417–459, 1980.

[Nökel, 1988a] K. Nökel. Convex relations between time intervals. Technical report, University of Kaiserslautern, 1988. SEKI. Report SR-88-17.

[Nökel, 1988b] K. Nökel. Temporeal matching: recognising dynamic situations from discrete measurements. Technical report, University of Kaiserslautern, 1988. SEKI. Report SR-88-20.

[Normore, 1982] C. Normore. Future contingents. In N. Kretzman *et al.*, editors, *Cambridge History of Later Medieval Philosophy*, pages 358–381. Cambridge University Press, 1982.

[Oberlander, 1987a] J. Oberlander. *The semantics of temporal indexicals*. PhD thesis, Centre for Cognitive Science, University of Edinburgh, 1987.

[Oberlander, 1987b] J. Oberlander. Temporal reference and quantification: An IQ perspective. Technical report, Centre for Cognitive Science, University of Edinburgh, 1987.

[Ono and Nakamura, 1980] H. Ono and A. Nakamura. On the size of refutation Kripke models for some linear modal and tense logics. *Studia Logica*, 39:325–333, 1980.

[Ostroff, 1989] J. S. Ostroff. *Temporal Logic for Real-Time Systems*. Advanced Software Development Series. John Wiley and Sons, 1989.

[Parikh, 1978] R. Parikh. A decidability result for second order process logic. In *IEEE. Symposium on Foundation of Computer Science*, pages 177–183, 1978.

[Parry, 1949] W. Parry. Modalities in the survey system of strict implication. *J. Symbolic Logic*, 4, 1949.

[Passoneau, 1987] R. Passoneau. Situations and intervals. In *Proc. 25th meeting of Assoc. for Computat. Linguistics, Stanford*, 1987.

[Pasztor, 1986] A. Pasztor. Non–standard algorithmic and dynamic logic. *J. Symbolic Computation*, 2:59–81, 1986.

[Pasztor, 1991] A. Pasztor. An infinite hierarchy of program verification methods. In J.Tucker and K.Meinke, editors, *Many Sorted Logic and its Applications* (*Proc. Conf. Leeds, Great Britain 1988*). John Wiley and Sons, 1991.

[Pasztor and Sain, 1990] A. Pasztor and I. Sain. A streamlined temporal completeness theorem. In E. Börger, H. Kleine-Bühning, and M.M. Richter, editors, *Computer Science Logic 1989* (*Proc. Conf. Kaiserslautern Germany 1989*), volume 440 of *Lecture Notes in Computer Science*. Springer-Verlag, 1990.

[Penberthy, 1987] J. Penberthy. Temporal unification and the temporal partial order. In *Proc. 5th IEEE. — Conf. on AI applications*, 1987.

[Penczek, 1987a] W. Penczek. A temporal logic for event structures. *Fundamenta Informaticae*, 11, 1987.

[Penczek, 1987b] W. Penczek. The temporal logic of event structures. Technical report, Inst. of Comp. Sci. Polish Acad. of Sc., 1987. Warszawa.

[Perlis, 1985] D. Perlis. Languages with self reference I: Foundations. *Artificial Intelligence*, 25:301–322, 1985.

[Perlis, 1988] D. Perlis. Languages with self reference II: Knowledge, belief and modality. *Artificial Intelligence*, 34:179–212, 1988.

[Petrovic, 1988] S. Petrovic. The temporal logic of AI. *Automatika*, 29, 1988.

[Pinter and Wolper, 1984] J. Pinter and P. Wolper. A temporal logic for reasoning about partially ordered computations. In *Proc. 3rd ACM Symp. on Princ. of Distrib. Comp.*, 1984.

[Plaisted, 1983a] D. Plaisted. A decision procedure for combinations of propositional temporal logic and other specialized theories. Technical report, SRI. International, 1983.

[Plaisted, 1983b] D. Plaisted. A low-level language for obtaining decision procedures for classes of temporal logics. Springer, 1983. LNCS 164.

[Pnueli, 1977] A. Pnueli. The temporal logic of programs. In *Proceedings of the Eighteenth Symposium on Foundations of Computer Science*, pages 46–57, 1977. Providence, RI.

[Pnueli, 1979] A. Pnueli. The temporal semantics of concurrent programs. In *Int. Symp. Semantics of Concurrent computation*, July 1979. Evian.

[Pnueli, 1981a] A. Pnueli. The temporal logic of programs. Technical report, Preprint Weizman Institute of Science, Dept of Applied Math, 1981.

[Pnueli, 1981b] A. Pnueli. The temporal semantics of concurrent programs. *Theoretical Computer Science*, 13:45–60, 1981.

[Pnueli, 1986] A. Pnueli. Specification and development of reactive systems. In H.-J. Kugler, editor, *Information Processing (IFIP'86)*, volume 86, pages 845–858. North–Holland, 1986.

[Pnueli, 1988] A. Pnueli. Linear time temporal logic. In G. Rosenberg, editor, *REX Workshop on temporal logic and concurrency*. Springer, Berlin, 1988.

[Pratt, 1979] V. R. Pratt. Models of program logics. In *Proc. 20th IEEE. Symposium on Foundations of Computer Science, San Juan*, pages 115–122, 1979.

[Pratt, 1980] V.R. Pratt. A near optimal method for reasoning about action. *J. Computer and System Sciences*, 20:231–254, 1980.

[Pratt, 1988] W. Pratt. Dynamic algebras as a well-behaved fragment of relation algebras. In C.H.Bergman, R.D. Maddux, and D. L. Pigozzi, editors, *Algebraic Logic and Universal Algebra in Computer Science*, volume 425 of *LNCS*, pages 77–110. Springer, 1988.

[Prior, 1953] A. Prior. Three valued logic and future contingents. *Philos. Quart.*, 3, 1953.

[Prior, 1955] A. Prior. Diodorean modalities. *Philos. Quart.*, 5, 1955.

[Prior, 1956] A. Prior. Modality and quantification in S5. *J. Symbolic Logic*, 21, 1956.

[Prior, 1957] A. Prior. *Time and Modality.* Oxford University Press, 1957.

[Prior, 1958a] A. Prior. Diodorus and modal logics: a correction. *Philos. Quart.*, 8, 1958.

[Prior, 1958b] A. Prior. The syntax of time distinctions. *Franciscan Studies*, 18, 1958.

[Prior, 1959] A. Prior. Notes on a group of new modal systems. *Logique et Analyse*, 2, 1959.

[Prior, 1960] A. Prior. Identifiable individuals. *Review of Metaphysics*, 13, 1960.

[Prior, 1962a] A. Prior. Possible worlds. *Philos. Quart.*, 12, 1962.

[Prior, 1962b] A. Prior. Tense logic and the continuity of time. *Studia Logica*, 13, 1962.

[Prior, 1964a] A. Prior. Axiomatizations of the modal calculus Q. *Notre Dame J. Formal Logic*, 5, 1964.

[Prior, 1964b] A. Prior. K1, K2 and related modal systems. *Notre Dame J. Formal Logic*, 5, 1964.

[Prior, 1966] A. Prior. Postulates for tense logic. *Amer. Philos. Quart.*, 3, 1966.

[Prior, 1967a] A. Prior. *Past, Present and Future.* Oxford University Press, 1967.

[Prior, 1967b] A. Prior. Stratified metric tense logic. *Theoria*, 33, 1967.

[Prior, 1968a] A. Prior. Now. *Noûs*, 2, 1968.

[Prior, 1968b] A. Prior. *Papers on Time and Tense.* Oxford University Press, 1968.

[Prior, 1968c] A. Prior. Time and change. *Ratio*, 10, 1968.

[Prior, 1969] A. Prior. Tensed propositions as predicates. *Amer. Philos. Quart.*, 6, 1969.

[Putnam, 1967] H. Putnam. Time and physical geometry. *J. Philos.*, 64, 1967.

[Quine, 1960] Willard Van Orman Quine. *Word and Object.* MIT Press, 1960.

[Rabin, 1969] M. O. Rabin. Decidability of second order theories and automata on infinite trees. *American Mathematical Society Transactions*, 141:1–35, 1969.

[Rabin, 1977] M.O. Rabin. Decidable theories. In J. Barwise, editor, *Handbook of Mathematical Logic*, pages 595–629. North-Holland, 1977.

[Ramsey, 1930] F. P. Ramsey. On a problem of formal logic. *Proc. London Math. Soc.*, 30:264–286, 1930.

[Rayner, 1985] M. Rayner. On the applicability of non-monotonic logic to formal reasoning in continuous time. Technical report, Swedish Institute of Computer Science, 1985.

[Reichenbach, 1957] H. Reichenbach. *The Philosophy of Space and Time.* Dover, New York, 1957.

[Reichgelt, 1987] H. Reichgelt. Semantics for reified temporal logic. In J. Hallam and C. Mellish, editors, *Advances in artificial intelligence.* Wiley and Sons, 1987.

[Rescher, 1966] N. Rescher. On the logic of chronological propositions. *Mind*, 75, 1966.

[Rescher, 1967] N. Rescher. *The logic of decision and action.* Pittsburgh University Press, 1967.

[Rescher and Garson, 1967] N. Rescher and J. Garson. A note on chronological logic. *Theoria*, 33, 1967.

[Rescher and Garson, 1968] N. Rescher and J. Garson. Topological logic. *J. Symbolic Logic*, 33, 1968.

[Rescher and Urquhart, 1971] N. Rescher and A. Urquhart. *Temporal Logic.* Springer, Berlin, 1971.

[Reynolds, 1992a] M. Reynolds. Axiomatising first-order temporal logic: Until and since over linear time, 1992. submitted to Notre Dame J. Formal Logic.

[Reynolds, 1992b] M. Reynolds. Axiomatising predicate temporal logic, 1992. technical report, Imperial College.

[Reynolds, 1992c] M. Reynolds. An axiomatization for Until and Since over the reals without the IRR rule. *Studia Logica*, 51:165–194, 1992.

[Reynolds, 1993a] M. Reynolds. Axiomatizing U and S over integer time. Technical report, Imperial College, 1993.

[Reynolds, 1993b] M. Reynolds. Metatem in intensive care, 1993. To appear in *J. of Applied Non-classical Logics*, currently being revised.

[Reynolds, to appear] M. Reynolds. Axiomatisation and decidability of F and P in cyclical time. *Journal of Philosophical Logic*, to appear.

[Richards, 1974] E. B. Richards. A point of reference. *Synthese*, 28:361–454, 1974.

[Richards, 1982] B. Richards. Tense, aspect and time adverbials I. *Linguistics and Philosophy*, 5:59–107, 1982.

[Richards, 1986] B. Richards. Tenses, temporal quantifiers and semantic innocence. In E. Lepore, editor, *The Philosophy of Donald Davidson: A. perspective on inquiries into truth and interpretation.* Oxford: Basil Blackwell, 1986.

[Richards, 1989] B. Richards. *Temporal Representation and Inference.* Academic Press, London, 1989.

[Richter and Szabo, 1986] M.M. Richter and M.E. Szabo. Nonstandard computation theory. In J.Demetrovics, G.Katona, and A.Salomaa, editors, *Algebra, combinatorics, and logic in computer science* (*Proc. Conf. Győr Hungary 1983*), pages 667–693. Colloq. Math. Soc. J. Bolyai (North–Holland), 1986.

[Rit, 1986] J. Rit. Propagating temporal constraints for scheduling. In *Proc. AAAI*, 1986.

[Robb, 1914] A. A. Robb. *A theory of time and space.* Cambridge University Press, 1914.

[Robertson, 1974] E.L. Robertson. Structure of complexity in weak monadic second order theories of the natural numbers. In *Proc. 6th Symp. on Theory of Computing*, pages 161–171, 1974.

[Rohrer, 1977a] C. Rohrer. How to define temporal conjuctions. *Linguistische Berichte*, pages 1–12, 1977.

[Rohrer, 1977b] C. Rohrer, editor. *On the Logical Analysis of Tense and Aspect.* Gunter Narr Verlag, Tübingen, 1977.

[Rohrer, 1977c] C. Rohrer, editor. *Time, Tense and Quantifiers: Proceedings of the Stuttgart Conference on the Logic of Tense and Quantification.* Max Niemayer Verlag, Tübingen, 1977.

[Ron, 1984] D. Ron. Temporal verification of communication protocols. Master's thesis, Weizmann Institute, 1984.

[Röper, 1980] P. Röper. Intervals and tenses. *J. of Philosophical Logic*, 9:451–469, 1980.

[Rosenberg, 1988] G. Rosenberg, editor. *REX Workshop on Temporal Logic and Concurrency.* Springer, Berlin, 1988.

[Rosenstein, 1982] J. G. Rosenstein. *Linear Orderings.* Academic Press, New York, 1982.

[Saarinen, 1979] E. Saarinen. Backwards-looking operators in tense logic and in natural language. In *Proceedings of the Fourth Scandinavian Logic Symposium*, 1979. The Hague.

[Sadri, 1987] F. Sadri. Three recent approaches to temporal reasoning. In A. Galton, editor, *Temporal Logics and Their Applications.* Academic Press, 1987.

[Sahlqvist, 1975] H. Sahlqvist. Completeness and correspondence in the first and second order semantics for modal logic. In S. Kanger, editor, *Proceedings of the Third Scandinavian Logic Symposium, Uppsala, 1973*, pages 110–143, Amsterdam, 1975. North Holland.

[Sain, 1983a] I. Sain. Successor axioms for time increase the program verifying power of full computational induction. Technical report, Preprint Math. Inst. Hungar. Acad. Sci. Budapest, No. 23/1983, 1983.

[Sain, 1983b] I. Sain. Total correctness in nonstandard dynamic logic. *Bull. Section of Logic* (*Wrocław-Lódz*), 12:64–70, 1983.

[Sain, 1984a] I. Sain. On Burstall's method, parts I and II. Technical report, Preprint Math. Inst. Hungar. Acad. Sci. Budapest, No. 57/1984, 1984.

[Sain, 1984b] I. Sain. Structured nonstandard dynamic logic. *Zeitschrift für Math. Logic u. Grundlagen der Math.*, 3:481–497, 1984.

[Sain, 1985a] I. Sain. The reasoning powers of Burstall's (modal logic) and Pnueli's (temporal logic) program verification methods. In R. Parikh, editor, *Logics of Programs* (*Proc. Conf. Brooklyn USA. 1985*), volume

193 of *Lecture Notes in Computer Science*, pages 302–319. Springer-Verlag, Berlin, 1985.

[Sain, 1985b] I. Sain. Relative program verifying powers of the various temporal logics. Technical report, Preprint No 40/1985, Math. Inst. Hungar. Acad. Sci., 1985. An extended abstract of this is [Sain, 1985a].

[Sain, 1986] I. Sain. *Nonstandard logics of programs.* PhD thesis, Dissertation, Hungarian Academy of Sciences, Budapest, 1986. in Hungarian.

[Sain, 1987] I. Sain. Total correctness in nonstandard logics of programs. *Theoretical Computer Science*, 50:285–321, 1987.

[Sain, 1988] I. Sain. Is 'some–other–time' sometimes better than 'sometime' in proving partial correctness of programs? In M.M. Richter and M.E. Szabo, editors, *Nonstandard methods in Mathematics — Special Volume of Studia Logica*, volume 47, pages 279–301. 1988.

[Sain, 1990] I. Sain. Past proves more invariance properties but not pca's. In J. Dassaw and J. Kelemen, editors, *Aspects and Prospects of Theoretical Computer Science* (*Proc. 6th International Meeting of Young Computer Scientists, Smolenice, Czechoslovakia, 1990*), volume 464 of *Lecture Notes in Computer Science*, pages 80–92. Springer-Verlag, Berlin, 1990.

[Sain, 1991a] I. Sain. Comparing and characterizing the power of established program verification methods. In J. Tucker and K. Meinke, editors, *Many Sorted Logic and its Applications* (*Proc. Conf. Leeds, Great Britain 1988*). John Wiley and Co, 1991.

[Sain, 1991b] I. Sain. Temporal logics need their clocks. *Theoretical Computer Science*, 97, 1991.

[Sakuragawa, 1986] T. Sakuragawa. Temporal prolog. Technical report, Research Institute for Mathematical Sciences, Kyoto University, 1986. To appear in Computer Software.

[Sambin and Vaccaro, 1989] G. Sambin and V. Vaccaro. A new proof of Sahlqvist's theorem on modal definability and completeness. *J. Symbolic Logic*, 54:992–999, 1989.

[Sandewall, 1988] E. Sandewall. Non-monotonic entailment for reasoning about time and action, Part I–III. Technical report, Linköping University, 1988. LiTH-IDA-R-88-27 (28,29).

[Sato, 1976] M. Sato. A study of Kripke-type models for some modal logics by Gentzen's sequential method. Technical report, Research Institute for Mathematical Sciences, Kyoto University, 1976.

[Savitch, 1970] W. J. Savitch. Relationships between non-deterministic and deterministic tape complexities. *J. Comput. Syst. Sci.*, 4:177–192, 1970.

[Schindler, 1970] P. Schindler. Tense logic for discrete future time. *J. Symbolic Logic*, pages 105–118, 1970.

[Schlingloff, 1992] B-H. Schlingloff. Expressive completeness of temporal logic over trees. *Journal of Applied Non-classical Logics*, 2:157 – 180, 1992.

[Schmiedel, 1988a] A. Schmiedel. A temporal constraint handler for the BACK system. Technical report, TU. Berlin, KIT. group, 1988.

[Schmiedel, 1988b] A. Schmiedel. Temporal constraint networks. Technical report, TU. Berlin, KIT. group, 1988.

[Schulz, 1987] K. Schulz. *Event and Interval Structures: A Mathematical Comparison.* Universitaet Tübingen, 1987. Seminar Fuer Natuerlich-Sprachliche Systeme.

[Schwartz *et al.*, 1983a] R. Schwartz, P. Melliar-Smith, and F. Vogt. An interval-based temporal logic. Springer, 1983. LNCS 164.

[Schwartz *et al.*, 1983b] R.L. Schwartz, P.M. Melliar-Smith, and F.H. Vogt. *An Interval Logic for Higher-level Temporal Reasoning.* SRI. International, Computer Science Laboratory, 1983.

[Scott, 1965] D. Scott. The logic of tenses. Technical report, Princeton University, 1965. Unpublished.

[Scott, 1976] D. Scott. Advice on modal logic. In K. Lambert, editor, *Philosophical Problems in Logic.* Reidel, 1976.

[Seeskin, 1971] K. Seeskin. Many-valued logic and future contingencies. *Logique et analyse*, 14:759–773, 1971.

[Segerberg, 1967] K. Segerberg. On the logic of tomorrow. *Theoria*, 33, 1967.

[Segerberg, 1968] K. Segerberg. Decidability of S4.1. *Theoria*, 34:7–20, 1968.

[Segerberg, 1970] K. Segerberg. Modal logics with linear alternative relations. *Theoria*, 36:301–322, 1970.

[Segerberg, 1971] K. Segerberg. An essay in classical modal logic. *Filosofiska Studier, Uppsala*, 13:301–322, 1971.

[Segerberg, 1973] K. Segerberg. Two-dimensional modal logic. *J. Philosophical Logic*, 2:77–96, 1973.

[Segerberg, 1980] K. Segerberg. Applying modal logic. *Studia Logica*, 39, 1980.

[Shelah and Stavi, 1980] S. Shelah and J. Stavi. Expressive completeness for temporal logic, 1980. Unpublished.

[Shelah, 1975] S. Shelah. The monadic theory of order. *Annals of Mathematics*, 102:379–419, 1975.

[Shelah, 1978] S. Shelah. A weak generalisation of MA to higher cardinals. *Israel J. Math.*, 30:297–306, 1978.

[Shimura, 1979] M. Shimura. Resolution in a new modal logic. In *Proc. IJCAI*, 1979.

[Shoham and McDermott, 1988] Y. Shoham and D. McDermott. Problems in formal temporal reasoning. *Artificial Intelligence*, 36, 1988.

[Shoham, 1985] Y. Shoham. Ten requirements for a theory of change. *New Generation Computing*, 3-4, 1985.

[Shoham, 1986a] Y. Shoham. Chronological ignorance: time, nonmonotonicity, necessity and causal theory. In *Proc. AAAI*, 1986.

[Shoham, 1986b] Y. Shoham. Reified temporal logics: semantical and ontological considerations. In *Proc. 7th ECAI*, 1986.

[Shoham, 1987a] Y. Shoham. Nonmonotonic logics: meaning and utility. In *Proc. IJCAI*, 1987.

[Shoham, 1987b] Y. Shoham. Temporal logics in AI: semantical and ontological considerations. *Artificial Intelligence*, 33, 1987.

[Shoham, 1988] Y. Shoham. Chronological ignorance: experiments in nonmonotonic temporal reasoning. *Artificial Intelligence*, 36, 1988.

[Sistla, 1987] A. Sistla. On the eventuality operator in temporal logic. In *Proc. Symp. Logic in Comp. Sci.*, 1987.

[Sistla and Clarke, 1985] A. Sistla and E. Clarke. Complexity of propositional linear temporal logics. *J. ACM*, 32:733–749, 1985.

[Sistla *et al.*, 1987] A. Sistla, M. Vardi, and P. Wolper. The complementation problem for Buchi automata with applications to temporal logic. *Theoretical Computer Science*, 49:217–237, 1987.

[Sklar, 1974] L. Sklar. *Space, time and spacetime.* University of California Press, Berkeley & Los Angeles, 1974.

[Slote, 1978] M. Slote. Time and counterfactuals. *Philosophical Review*, 87:3–27, 1978.

[Smart, 1964] J. Smart, editor. *Problems of Space and Time.* McMillan, New York, 1964.

[Smart, 1969] J. Smart. Causal theories of time. *Monist*, 53, 1969.

[Smets, 1988] P. Smets. *Non-standard logics for automated reasoning.* Academic Press, 1988.

[Smirnov, 1982] V. Smirnov. The definition of modal operators by means of tense operators. *Acta Philosophica Fennica*, 35:50–69, 1982.

[Snyder, 1971] D. Snyder. *Modal logic and its applications.* Van Nostrand, New York, 1971.

[Sorabji, 1976] R. Sorabji. Aristotle and the instant of change. *Proc. Aristotelian Soc.*, 50, 1976.

[Spaan, 1993] E. Spaan. The complexity of propositional tense logics. In M. de Rijke, editor, *Diamonds and Defaults, Studies in Logic, Language and Information.* Kluwer, Dordrecht, 1993.

[Stavi, 1979] J. Stavi. Functional completeness over the rationals. Unpublished, Bar-Ilan University, Ramat-Gan, Israel, 1979.

[Steedman, 1977] M. Steedman. Verb, time and modality. *Cognitive Science*, 1, 1977.

[Steedman, 1982] M. J. Steedman. Reference to past time. In R. Jarvella and W. Klein, editors, *Speech, Place and Action*, pages 125–157. John Wiley and Sons, New York, 1982.

[Stirling, 1988] C. Stirling. Expressibility and definability in branching and linear time temporal logics. Technical report, Department of Computer Science, University of Edinburgh, 1988.

[Strulo, 1990] B. Strulo. Probabalistic temporal logic. Master's thesis, Imperial College, 1990.

[Suppes, 1973] P. Suppes. Some open problems in the philosophy of space and time. In P. Suppes, editor, *Space, time and geometry.* Reidel, 1973.

[Szalas, 1986] A. Szalas. Concerning the semantic consequence relation in first-order temporal logic. *Theoretical Computer Science*, 47:329–334, 1986.

[Szalas, 1987a] A. Szalas. Arithmetical axiomatization of first-order temporal logic. *Information Processing Letters*, 26, 1987.

[Szalas, 1987b] A. Szalas. A complete axiomatic characterization of first-order temporal logic of linear time. *Theoretical Computer Science*, 54:199–214, 1987.

[Szalas and Holenderski, 1988] A. Szalas and L. Holenderski. Incompleteness of first-order temporal logic with until. *Theoretical Computer Science*, 57:317 -325, 1988.

[Taylor, 1977] B. Taylor. Tense and continuity. *Linguistics and Philosophy*, 1:199–220, 1977.

[Tedeschi and Zaenen, 1981] Tedeschi and Zaenen, editors. *Tense and Aspect.* Syntax and Semantics 14. Academic Press, New York, 1981.

[Thayse, 1988] A. Thayse. *From standard logic to logic programming, introducing a logic-based approach to AI.* Wiley, 1988.

[Thayse, 1989] A. Thayse. *From modal logic to deductive databases.* Wiley, 1989.

[Thomason, 1970] R. Thomason. Indeterminist time and truth value gaps. *Theoria*, 36:264–281, 1970.

[Thomason, 1972] S. K. Thomason. Semantic analysis of tense logic. *J. Symbolic Logic*, 37:150–158, 1972.

[Thomason, 1974a] R. H. Thomason, editor. *Formal Philosophy: Selected Papers of Richard Montague.* Yale University Press, 1974.

[Thomason, 1974b] S. Thomason. An incompleteness theorem in modal logic. *Theoria*, 40, 1974.

[Thomason, 1974c] S. Thomason. Noncompactness in propositional modal logic. *J. Symbolic Logic*, 39, 1974.

[Thomason, 1974d] S. K. Thomason. Reduction of tense logic to modal logic. *J. Symbolic Logic*, 39:549–551, 1974.

[Thomason, 1975a] S. Thomason. The logical consequence relation of propositional tense logic. *Zeitschrift Math. Logic und Grundlagen der Mathematik*, 21, 1975.

[Thomason, 1975b] S. Thomason. Reduction of second order logic to modal logic. *Zeitschrift Math. Logic und Grundlagen der Mathematik*, 21, 1975.

[Thomason, 1976] S. Thomason. Some completeness results for modal predicate calculi. In K. Lambert, editor, *Philosophical Problems in Logic*. Reidel, 1976.

[Thomason, 1979] S. K. Thomason. Possible worlds, time and tenure. Technical report, Department of Mathematics, Simon Fraser University, Burnaby, BC, 1979.

[Thomason, 1981] R. Thomason. Deontic logic as founded on tense logic. In R. Hilpinen, editor, *New Studies in Deontic Logic*, pages 165–176. Reidel, Dordrecht, 1981.

[Thomason, 1982a] R. Thomason. Counterfactuals and temporal direction. Technical report, University of Pittsburgh, 1982.

[Thomason, 1982b] S. Thomason. Undecidability of the completeness problem of modal logic. Technical report, Banach Center Publications 9, Warszawa, 1982.

[Thomason, 1984] R. Thomason. Combinations of tense and modality. In D. Gabbay and F. Guenthner, editors, *Handbook of Philosophical Logic, Vol II: Extensions of Classical Logic*, pages 135–165. Reidel, Dordrecht, 1984.

[Thomason, 1987] S. K. Thomason. Free construction of time from events. Technical report, Department of Mathematics, Simon Fraser University, Burnaby, BC, 1987.

[Thomason and Gupta, 1981] S. Thomason and A. Gupta. A theory of conditionals in the context of branching time. In R. Stalnaker W. Harper and GH. Pearce, editors, *Ifs: conditionals, beliefs, decision, chance and time*. Reidel, 1981.

[Tichy, 1980] P. Tichy. The logic of temporal discourse. *Linguistics and Philosophy*, 3:343–369, 1980.

[Tichy, 1985] P. Tichy. Do we need interval semantics? *Linguistics and Philosophy*, 8:263–282, 1985.

[Tsang, 1986] E. Tsang. The interval structure of Allen's tense logic. Technical report, University of Essex Cognitive Studies Centre, 1986. CSCM-24.

[Turner, 1984] R. Turner. *Logics for artificial intelligence.* Ellis Horwood, 1984.

[Valdéz-Peréz, 1986] R. Valdéz-Peréz. Spatio-temporal reasoning and linear inequalities. Technical report, MIT, 1986. AI-Memo 875.

[Valdéz-Peréz, 1987] R. Valdéz-Peréz. The satisfiability of temporal constraint networks. In *Proc. AAAI*, 1987.

[van Bentham, 1977a] J. van Bentham. Tense logic and standard logic. In L. Åqvist and F. Guenthner, editors, *Tense Logic.* Nauwelaerts, Louvain, 1977.

[van Bentham, 1977b] J. van Bentham. Tense logic and standard logic. *Logique et'Analyse*, 80:47–83, 1977.

[van Bentham, 1981] J. van Bentham. Tense logic, second order logic and natural language. In U. Monnich, editor, *Aspects of Philosophical Logic*, pages 1–20. Reidel, Dordrecht, 1981.

[van Bentham, 1983] J. van Bentham. *The Logic of Time.* Reidel, Dordrecht, 1983.

[van Bentham, 1984a] J. van Bentham. Correspondence theory. In D. M. Gabbay and F. Guenthner, editors, *Handbook of Philosophical Logic*, pages 167–247. D. Reidel, 1984.

[van Bentham, 1984b] J. van Bentham. Tense logic and time. *Notre Dame J. Formal Logic*, 25:1–16, 1984.

[van Bentham, 1986] J. van Bentham. Tenses in real time. *Zeitschrift für mathematische Logik und Grundlagen der Mathematik*, 32:61–72, 1986.

[van Benthem, 1977] J. van Benthem. Tense logic as a system of logic. In *Proc. Tübingen Conference on Formal Semantics*, 1977.

[van Benthem, 1978] J. van Benthem. Two simple incomplete modal logics. *Theoria*, 44, 1978.

[van Benthem, 1980] J. van Benthem. Points and periods. In C. Rohrer, editor, *Time, Tense and Quantifiers.* Niemayer, 1980.

[van Benthem, 1981] J. van Benthem. *Modal logic as second order logic.* Indices, Napoli, 1981.

[van Benthem, 1982] J. van Benthem. Later than late: on the logical origin of the temporal order. *Pacific Philosophical Quarterly*, 1982.

[van Benthem, 1984a] J. van Benthem. Minimal conditions for interval models. Technical report, Centre for the Study of Languages and Information, Stanford University, 1984.

[van Benthem, 1984b] J. van Benthem. Possible world semantics: a research program that cannot fail? *Studia Logica*, 43, 1984.

[van Benthem, 1985] J. F. A. K. van Benthem. *Modal Logic and Classical Logic.* Bibliopolis, Naples, 1985.

[van Benthem, 1986a] J. van Benthem. Notes on modal definability. Technical report, Math. Institute, University of Amsterdam, 1986. Report 86-11.

[van Benthem, 1986b] J. van Benthem. Tense and modality. In J. van Benthem, editor, *Essays in Logical Semantics.* Reidel, 1986.

[van Benthem, 1988a] J. van Benthem. *A manual of intensional logic.* Centre for the Study of Languages and Information, 1988. Lecture Notes 1.

[van Benthem, 1988b] J. van Benthem. *Time, logic and computation.* Springer, 1988. LNCS 354.

[van Benthem, 1989] J. van Benthem, 1989. private communication.

[van Benthem, 1992] J. van Benthem. Temporal logic. In D. Gabbay, editor, *Handbook of Logic in Artificial Intelligence and Logic Programming.* Oxford University Press, 1992.

[van Benthem, to appear] J.F.A.K. van Benthem. General dynamics. *Theoretical Linguistics*, to appear.

[van Benthem and Blok, 1978] J. van Benthem and W. Blok. Transitivity follows from dummett's axiom. *Theoria*, 44, 1978.

[van Eck, 1977] J. van Eck. *A System of Temporally relative modal and denotic predicate logic and its philosophical applications.* PhD thesis, Department of Philosophy, University of Groningen, 1977. Also appeared in *J. Logique et Analyse*, 1983.

[van Fraassen, 1981] B. van Fraassen. A temporal framework for conditionals and chance. In W. Harper, G. Pearce, and R. Stalnaker, editors, *Ifs: Conditionals, Belief, Decision, Chance and Time.* Reidel, Dordrecht, 1981.

[Vardi, 1988] M. Vardi. A temporal fixpoint calculus. In *Proc. 13th ACM Symp. on Principles of Programming languages*, pages 250–259, 1988.

[Vendler, 1967] Z. Vendler. *Verbs and Times*, chapter 4, pages 97–121. Cornell University Press, Ithach, NY, 1967.

[Venema, 1988] Y. Venema. Expressiveness and completeness of an interval tense logic. Technical Report 88-02, Institute for language, logic and Information, University of Amsterdam, 1988.

[Venema, 1990a] Y. Venema. Completeness by completeness for a two-dimensional temporal logic, 1990. See also section 3.4 of [Venema, 1992].

[Venema, 1990b] Y. Venema. Expressiveness and completeness of an interval tense logic. *Notre Dame J. Formal Logic*, 31:529–547, 1990.

[Venema, 1991a] Y. Venema. Completeness via completeness. In M. de Rijke, editor, *Colloquium on Modal Logic, 1991.* ITLI-Network Publication, Instit. for Lang., Logic and Information, University of Amsterdam, 1991.

[Venema, 1991b] Y. Venema. A logic with the chop operator. *J. Logic Computat.*, 1:453–476, 1991.

[Venema, 1992] Y. Venema. *Many-dimensional modal logic.* PhD thesis, University of Amsterdam, 1992.

[Vere, 1981] S. Vere. Planning in time: windows and duration for activities and goals. Technical report, Jet Propulsion Lab, Pasadena, CA, November 1981.

[Verkuyl, 1972] H. J. Verkuyl. *On the Compositional Nature of Aspect.* D. Reidel, 1972.

[Verkuyl, 1976] H. J. Verkuyl. Interpretive rules and the description of the aspects. *Foundations of Language*, 14:471–503, 1976.

[Vilain, 1982] M. Vilain. A system for reasoning about time. In *Proc. AAAI*, 1982.

[Vilain and Kautz, 1986] M. Vilain and H. Kautz. Constraint propagation algorithms for temporal reasoning. In *Proc. AAAI*, 1986.

[Vlach, 1973] F. Vlach. *Now and then: A formal study in the logic of tense anaphora.* PhD thesis, UCLA, 1973.

[Vlach, 1979] F. Vlach. The semantics of tense and aspect. Technical report, University of New South Wales, 1979.

[Vlach, 1980] F. Vlach. The semantics of tense and aspect in English. Technical report, University of New South Wales, 1980.

[Vlach, 1981] F. Vlach. The semantics of the progressive. In Tedeschi and Zaenen, editors, *Tense and Aspect*, pages 271–292. Academic Press, 1981.

[von Kutschera, 1976] F. von Kutschera. *Einführung in die intensionale Semantik.* de Gruyter, 1976.

[von Wright, 1965] G. von Wright. And next. *Acta Philos. Fennica*, 18, 1965.

[von Wright, 1966] G. von Wright. And then. *Commentationes Physico-Mathematicae of the Finnish Soc. of Sciences*, 32, 1966.

[von Wright, 1968a] G. von Wright. Always. *Theoria*, 34, 1968.

[von Wright, 1968b] G. von Wright. Time, change and contradiction. *A S. Eddington Memorial Lecture*, 22, 1968.

[von Wright, 1969] G. von Wright. An essay in deontic logic and the general theory of action. *Acta Philos. Fennica*, 21, 1969.

[Voß, 1987] H. Voß. Representing and analyzing causal, temporal and hierarchical relations of devices. Technical report, University of Kaiserslautern, 1987. SEKI. Report SR-87-17.

[Werner, 1974] B. Werner. *Foundations of temporal modal logic.* PhD thesis, University of Wisconsin at Madison, 1974.

[White, 1981] M. J. White. Modal tense logic incompleteness and the master argument. Technical report, University of Arizona, 1981.

[White, 1985] M. White. The necessity of the past and modal tense logic incompleteness. *Notre Dame J. Formal Logic*, 25, 1985.

[Whitrow, 1980] G. Whitrow. *The Natural Philosophy of Time.* Clarendon Press, Oxford, 1980.

[Whorf, 1956] B. L. Whorf. *Language, thought and reality.* MIT Press, 1956.

[Williams, 1986] B. Williams. Doing time: putting qualitative reasoning on firmer ground. In *Proc. AAAI National conference on AI*, 1986.

[Winnie, 1977] J. A. Winnie. The causal theory of space-time. In J. S. Earman *et al.*, editors, *Foundations of Space-time Theories*, pages 134–205. University of Minnesota Press, 1977.

[Winskel, 1988] G. Winskel. Event structures. In G. Rozenberg, editor, *REX Workshop on Temporal Logic and Concurrency.* Springer, Berlin, 1988.

[Wolper, 1982a] P. Wolper. Specification and synthesis of communicating processes using an extended temporal logic. In *Ninth Symp. on Principles of Programming Languages*, pages 20–33, Albuquerque, New Mexico, 1982.

[Wolper, 1982b] P. Wolper. *Synthesis of communicating processes from temporal logic specifications.* PhD thesis, Stanford University, 1982.

[Wolper, 1983] P. Wolper. Temporal logic can be more expressive. *Information and computation*, 56(1–2):72–99, 1983.

[Wolper, 1985] P. Wolper. The tableau method for temporal logic: an overview. *Logique et Analyse*, June–Sept 1985.

[Woolhouse, 1973] R. Woolhouse. Tensed modalities. *J. Philosophical Logic*, 2:393–415, 1973.

[Xu, 1988] Ming Xu. On some U, S-tense logics. *J. of Philosophical Logic*, 17:181–202, 1988.

[Yakhnis and Yakhnis, 1990] A. Yakhnis and V. Yakhnis. Extension of Gurevich–Harrington's restricted memory determinacy theorem: a criterion for the winning player and an explicit class of winning strategies. *Annals of Pure and Applied Logic*, 48:277–297, 1990.

[Zanardo, 1985] A. Zanardo. A finite axiomatization of the set of strongly valid Ockamist formulas. *J. of Philosophical Logic*, 14:447–468, 1985.

[Zanardo, 1992] A. Zanardo. Branching-time logic with quantification over branches: the point of view of modal logic, 1992. In preparation.

[Zeitman, to appear] S. Zeitman. Unforgettable forgetful determinacy. *J. Logic Computat.*, to appear.

[Zeman, 1973] J. Zeman. *Modal logic: the Lewis modal systems.* Oxford University Press, 1973.

[Zuck, 1986] L. Zuck. *Past temporal logic.* PhD thesis, Weizmann Institute, 1986.

Notation index

Index of systems

Index